PRELIMINARY EDITION

Exploring Differential Equations via Graphics and Data

PRELIMINARY EDITION

Exploring Differential Equations via Graphics and Data

David Lomen
University of Arizona

David Lovelock
University of Arizona

WILEY

John Wiley & Sons, Inc.
New York • Chichester • Brisbane • Toronto • Singapore

ACQUISITION EDITOR	Barbara Holland
MARKETING MANAGER	Debra Riegert
PRODUCTION EDITOR	Ken Santor
TEXT DESIGNER	Nancy Field
COVER DESIGNER	Levavi/Levavi
MANUFACTURING COORDINATOR	Dorothy Sinclair

This book was set in 10/12 Times Roman by Eigentype Compositors,
and printed and bound by Courier-Stoughton.
The cover was printed by NEBC.

Recognizing the importance of preserving what has been written, it is a
policy of John Wiley & Sons, Inc. to have books of enduring value published
in the United States printed on acid-free paper, and we exert our best
efforts to that end.

The paper on this book was manufactured by a mill whose forest management programs include sustained
yield harvesting of its timberland. Sustained yield harvesting principles ensure that the number of trees
cut each year does not exceed the amount of new growth.

ISBN: 0-471-07649-X

Printed in the United States of America

10 9 8 7 6 5 4 3 2 1

Preface

To the Student from the Authors

We have written this book to enable you to be an active participant in your own learning. Our objective is to transform the introductory ordinary differential equations course from the traditional one where you simply learn formal methods of solution, to one where you think, experiment, and comprehend. We want you to not only understand ordinary differential equations — their origins, qualitative behavior and interpretation of solutions — but also to see them as a natural tool for investigating many aspects of science and engineering. We encourage you to explore differential equations by employing technology as a tool to discover and interpret the behavior of solutions.

Visualization and the Rule of Four

We emphasize visual exploration of solutions via slope fields, direction fields, phase plane solutions, and other graphical interpretations. We cover and use graphical and numerical techniques early in the book because they are essential whenever analytical solutions cannot be found in terms of familiar functions. Wherever appropriate we also implement a "rule of four," treating topics from numerical, graphical, analytical, and descriptive viewpoints. Each of these viewpoints brings its unique contribution and perspective to the topic at hand. We develop results in a manner that demonstrates that differential equations are a logical extension of calculus.

Real Data

Most topics in this book are problem and data driven, and we give careful attention to mathematical modeling, solution techniques, and interpretation of results. We use data sets to develop differential equations, to obtain values of parameters in differential equations, and to check the accuracy of mathematical models. We encourage you to generate your own data from simple experiments involving common items, or from library sources. Examples of data sets from experiments include the oscillations of a pendulum or spring (like a "slinky"), heating a probe to human body temperature, draining a container of water through a small hole in the bottom, and the rise of moisture on a paper towel dipped in water. Examples of data sets from library sources include the change in world population, the growth of bacteria, the spread of diseases, and the change in the maximum speed of an aircraft. We encourage you to work projects that allow you to explore applications in your major field of study. By developing differential equations from data sets, then comparing the solutions with the data sets, you are better equipped to understand the origin and interpretation of the solution as thoroughly as the techniques of the solution. Nevertheless, the emphasis of the book is on differential equations, not modeling.

Technology Integration

We feel that the appropriate use of technology is vital to using differential equations as a tool. Thus, we expect you to have access to mathematical software or to a calculator that draws slope fields or direction fields, generates numerical solutions, and simultaneously displays data and graphs of functions. A computer algebra system is not necessary, nor is familiarity with any specific mathematical software.

We have developed computer programs appropriate for this differential equations course. This software is free of charge, runs on IBM compatible personal computers, and may be downloaded using ftp from SOFTWARE@MATH.ARIZONA.EDU. One program, "Twiddle", allows you to analyze data easily and to construct and test mathematical models. Other programs plot slope fields ("Slopes") and the solutions of systems of up to six first order differential equations ("Systems"). Another program, called "Are You Ready for Differential Equations?", tests your proficiency with the background material necessary for the book and identifies potential trouble spots.

A Lively Approach

Our intent is to provide lively reading, with compelling applications, projects, and experiments that supply you with opportunities to explore the relationship between the parameters in a differential equation and the process it models.

We do not use a "cookbook" presentation nor do we use a sequence of tricks. This book is not organized as "Theorem: Proof: Example: Application". Our organization is best illustrated by describing some of the items that appear in each chapter.

- Each chapter begins with an overview entitled, "Where are We Going?" and ends with a summary entitled, "What have We Learned?". This allows you to see the big picture of the methods you are learning.

- Clear summaries of important techniques are given in "How to..." boxes for handy reference and review.

- "Comments about .." guide you to important points that you might otherwise overlook, and "Words of Caution" prevent possible dilemmas.

- Exercises vary from those that help you hone your skills with analytical techniques to those that emphasize graphical or numerical techniques. Some exercises utilize technology and there are many exploratory problems.

Prerequisites

The prerequisites for this book are a course in single variable calculus, from either a traditional or a reform approach. Linear algebra and multivariable calculus are not required. This book is appropriate for any major, although many of the applications are from engineering and the sciences (physical, biological, and social).

To the Student from Other Students

by David Brokl, Ian Scott, and Michael VanZeeland, students at the University of Arizona

We found a review of the main ideas from our previous courses in calculus to be very valuable for learning from this book. Appendix A.1 featured some of these ideas. Also, the time we spent using the "Are You Ready for Differential Equations?" computer program paid big dividends (it has great HELP screens).

Each chapter in this book started with a section called "Where Are We Going?". We read this section carefully to get a sense of direction for the entire chapter. Then we read through the rest of the chapter as if reading a story: We looked for the main ideas, but did not dwell on them. At the end of the chapter there was a section "What Have We Learned?" which emphasized the important concepts and results of the chapter. As we read this we asked ourselves "Were did this come from?" and "How does that work?". After finishing this, we went back and carefully worked through the chapter, making sure to understand all the details. In many places there were questions asking why something was true. We found it very worthwhile to figure out the answer. The words "Comments about ... " appeared at various places in the text. These were not just comments, but either emphasized an important point or made useful remarks about a prior result. We paid close attention to these!

We found the book often developed new mathematical ideas while discussing a specific example. Thus, while working through each example, we looked for a basic approach or principle that applied to other situations. Many times we discovered these situations in our other science and engineering classes.

We took advantage of technology and used computer generated graphs and slope fields to help us understand newly presented ideas. We questioned what we saw on the screen, and then tried to investigate it by numerical or analytical methods. Computers can be misleading so we questioned things which didn't make sense. We found that answering these questions led to a much better understanding of the material. It was a great feeling to discover that we could explain why the computer was misleading.

We read the book thoroughly and found it combined the best parts of traditional math books and reformed calculus books. Thus, there was a clear explanation, with examples, of how to use specific techniques of solution, yet the material was presented in a manner that emphasized derivations and understanding of the concepts.

We found it very useful to study with other students who were using this book. Explaining things to each other was a very effective way of learning the material. Having several points of view available was also a big help to us in solving some of the problems. We really enjoyed using this book. We hope you do also.

To the Instructor from the Authors

The philosophy, prerequisites, and use of technology that pervades this book is described under "To the Student from the Authors". We will not repeat it here.

This book may be covered completely in a two quarter course, but contains more material than may be covered in a one semester course. The following are suggestions for a one semester course:

- Chapters 1 through 7, 9.1 through 9.4, 11, and 12 or 13.

- Chapters 1 through 7, and 12 through 14.

- Chapters 1 through 11, if you are very selective over omitting sections on applications.

As far as selecting material from Chapters 1 through 7 is concerned, you should note the following:

- Sections 1.4, 3.4, 4.4, 4.6, 4.7, 4.8, 5.3, 5.4, and 6.4 are not extensively used in later sections, and so may be selectively omitted.

- The material in Section 4.5 is used to motivate a technique of solution in Section 5.1.

- The general solutions of $y'' \pm a^2 y = 0$ are derived in Section 5.5 and then used in Section 7.2.

- The method of undetermined coefficients for first order linear differential equations is introduced in Section 6.3. This method is then used in Section 7.2. However, it is not necessary that Section 6.3 be covered first.

- Second order differential equations are introduced in Section 7.1 as a means of solving a system of first order equations.

- All the cases for solving homogeneous second order differential equations with constant coefficients are given in Section 7.2. It assumes an elementary knowledge of complex numbers. The Appendix contains a brief introduction to complex numbers for students lacking this background.

Some comments on the dependence of Chapters 8 through 14 follow, assuming that you have covered Chapters 1 through 7 in the manner previously described.

- Chapter 8. All sections are essentially independent. Some of the examples in Sections 8.1 through 8.3 are used in later chapters for motivational purposes and as the basis for applications. However, none of these is critical for later sections. Section 8.4 is not used later.

- Chapter 9. This chapter uses some examples from Chapter 8 for motivational purposes only. You could choose other examples in their place. Sections 9.5 and 9.6 are used only in Chapter 10.

- Chapter 10. This chapter uses Sections 9.5 and 9.6. It is not used later.

- Chapter 11. This chapter is based on Sections 9.1 through 9.4. It is not used later.

- Chapter 12. This chapter follows from Chapter 7. It is not needed later, although Section 12.3 is useful for Chapter 13.

- Chapter 13. This chapter follows from Chapter 7, although a passing knowledge of Section 12.3 is helpful. This chapter uses elementary properties of matrices. The Appendix contains a brief introduction to matrices for students lacking this background.

- Chapter 14. This chapter follows from Chapter 7.

Preliminary versions of this material have been class tested by various instructors at five institutions over the past two years. All instructors have noted that most examples in this book are very rich. Therefore, it is easy to spend too much time gleaning every possible idea from every example. Doing so will definitely limit the amount of material you can cover in one semester. However, one of our major goals has been to write a book that students can read. Thus, you could cover the major points of an example in class, and then assign the other points of the example for out-of-class study.

Because material is presented in a way in which results are obtained in a logical manner using ideas from calculus, it is easy for students to be lulled into thinking they understand the material without doing much work. Results are very believable in this book, but understanding follows only after working exercises. To obtain this understanding we suggest assigning a considerable number of exercises. This is especially true at the beginning of the course where we suggest you assign several exercises from Sections 1.1 and 1.2 on the first day of class. We also recommend having an examination over prerequisite material (such as is contained in the "Are You Ready for Differential Equations?" computer program) about a week after the start of classes. The Appendix also contains a summary of background material, including techniques of differentiation, techniques of integration, partial fraction decomposition, and properties of power series

The following supplements will be available.

- A resource manual containing detailed suggestions together with additional exercises, experiments, data sets, and projects.

- A solution manual for students.

- Free DOS-based computer programs developed for this differential equations course which may be downloaded using ftp from SOFTWARE@MATH.ARIZONA.EDU.

Acknowledgments

We would like to thank the following for their constructive ideas and valuable assistance.

Alicia Acevedo, Gabriel Aldaz, David Anthony, Frank Avenoso, Bruce Bayly, Mark Biedrzycki, David Brokl, Erik Brown, Jessie Cameron, Ken Cardell, Danny Carillo, Elizabeth Cheney, Haiqian Cheng, Robert Cole, Michael Colvin, Steve Connor, Paul Deneke, Scott Downs, Mason Dykstra, Wade Ellis Jr., William Emerson, Melisa Enrico, Melissa Erickson, Jason Figueroa, David Gay, Rick Greenberg, Larry Grove, Tom Hallam, Lisa Hanley, Brink Harrison, Steve Hiller, Barbara Holland, Shih-Chieh Huang, Ralph Huarto, Simone Jacobsen, Stephen Koester, Ira Lackow, Kathy Lackow, Alaric Lebaron, Clay Lines, Connie Lomen, Cynthia Lovelock, Denise Lovelock, Fiona Lovelock, Danielle Manuszak, Jim Mark, Elizabeth McBride, Sean McHaney, Terri McSweeney, Lang Moore, Bill Mueller, Charlie Nafzigger, Thanh Nguyen, Michael Oddy, Lesley Perg, Thai Phan, Anthony Phillips, Yuri Pinelis, Zwi Reznick, Cynthia Rhoads, Paul Richards, Arturo Ruiz, Joe Scionti, Ian Scott, Steve Sheldon, Chris Sinclair, Cathy Steffens, Michelle Switala, Hal Tharp, Michael Van Zeeland, Tracy Wood, Jue Wu, David Yabi, Chris Yetman, Weijun Zhu, Richard Zimmer.

Contents

CHAPTER 1

Basic Concepts

Where Are We Going?

We will develop ways of analyzing ordinary differential equations that take full advantage of the power of calculus and technology. We do this by treating topics from graphical, numerical, and analytical points of view.

In this first chapter, we introduce these points of view by considering differential equations of the form

$$\frac{dy}{dx} = g(x),$$

which means we are simply looking at antidifferentiation problems. We start with a brief introduction, definitions and examples, which make use of prior knowledge of antiderivatives. Then we spend the next two sections illustrating graphical techniques, including slope fields and isoclines. These techniques often let us discover many properties of the solution of a specific differential equation by simply analyzing the differential equation from a graphical point of view. We end this chapter with a discussion of the use of Taylor series when functions are given by integrals that have no simple antiderivatives.

The purpose of this chapter is to illustrate graphical, numerical, and analytical approaches in the familiar setting of antiderivatives, so it will be easy to understand this approach with the new situations that occur in later chapters.

1.1   Simple Differential Equations and Explicit Solutions

The simple definition of a differential equation is an equation that involves a derivative. Thus many differential equations are solved in beginning calculus courses, perhaps without anyone stating it. We start with a very familiar example, one which we will frequently return to in this chapter.

EXAMPLE 1.1

Consider the problem of finding an antiderivative $y(x)$ of the function $1/x$. This could be cast as finding $y(x)$ as a solution of the differential equation

$$\frac{dy}{dx} = \frac{1}{x}. \tag{1.1}$$

1

Because any antiderivative of $1/x$ may be written as

$$y(x) = \ln|x| + C, \tag{1.2}$$

or

$$y(x) = \begin{cases} \ln x + C & \text{if } x > 0, \\ \ln(-x) + C & \text{if } x < 0, \end{cases} \tag{1.3}$$

where C is an arbitrary constant, it seems reasonable to call (1.2) or (1.3) solutions of our differential equation (1.1). Because of the arbitrary constant, we have an infinite number of solutions, a different one for each choice of C (collectively called a family of solutions). Some of these solutions are graphed in Figure 1.1, where we notice that the role of the arbitrary constant is to determine the vertical position. If $x > 0$ all solutions of this differential equation have the same general shape, and any two solutions will differ from each other by a vertical translation. The same is true for $x < 0$. □

This is one of many examples of differential equations covered in calculus. All problems where we found the indefinite integral (or antiderivative) of a function, $g(x)$, could have been formulated as finding y as the solution of

$$\frac{dy}{dx} = g(x). \tag{1.4}$$

The solutions of (1.4) all have the form

$$y(x) = \int g(x)\, dx + C, \tag{1.5}$$

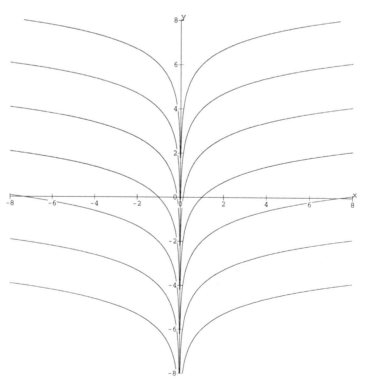

FIGURE 1.1   Some solutions of $dy/dx = 1/x$

where $\int g(x)\,dx$ is any specific antiderivative of $g(x)$.[1] The arbitrary constant C indicates that we have an infinite number of solutions, related to each other by a vertical translation.

Before developing any methods for finding solutions of differential equations, we give some formal definitions. These are mostly common-sense definitions, but they will be helpful later when we consider more complicated situations.

Differential equations such as (1.1) and (1.4) are called first order differential equations, because the first derivative is the highest one that occurs in the equation. Thus we have

Definition 1.1: **A FIRST ORDER ORDINARY DIFFERENTIAL EQUATION is an equation that involves at most the first derivative of an unknown function. If y, the unknown function, is a function of x, then we write the first order differential equation as**

$$\frac{dy}{dx} = g(x,y) \tag{1.6}$$

where $g(x,y)$ is a given function of the two variables x and y.

Comments about first order differential equations

- The right-hand side of (1.6) may contain x and y explicitly, for example, $x^2 + y^2$. However, in this chapter we will consider the case where $g(x, y)$ is a function of x alone.

We previously noted that (1.2) and (1.5) are solutions of (1.1) and (1.4) respectively. We know this because if we differentiate these functions and substitute the result into the proper differential equation, we obtain an identity. These solutions are called explicit because they have the dependent variable, y, solely given in terms of the independent variable, x. This prompts our second definition.

Definition 1.2: **An EXPLICIT SOLUTION of the first order ordinary differential equation**

$$\frac{dy}{dx} = g(x,y) \tag{1.7}$$

is any function $y = y(x)$ (with a derivative in some interval $a < x < b$) that identically satisfies the differential equation (1.7).

Comments about explicit solutions

- Because an explicit solution has a derivative in the interval $a < x < b$, it must be continuous in that interval. (Why?)

- An explicit solution may contain an arbitrary constant. If it does, we have an infinite number of solutions, called a FAMILY OF EXPLICIT SOLUTIONS. If it does not contain an arbitrary constant, we have a PARTICULAR EXPLICIT SOLUTION. In either case we may write an equation giving the dependent variable y explicitly in terms of the independent variable x. Often particular explicit solutions are just called particular solutions.

- The graph of a particular solution is called a SOLUTION CURVE of the differential equation. This is also called an integral curve.

---

[1] In calculus it is customary to have the symbol $\int g(x)\,dx$ include the arbitrary constant C. Here we add the constant explicitly to emphasize geometrical ideas.

The explicit solutions of the two differential equations mentioned so far both contain an arbitrary constant, so they are families of explicit solutions. This constant may be determined if we know the value of the solution at some specific value of x.

Thus, on the one hand, if we specify that the solution of the differential equation (1.1) must pass through the point $(-1, 0)$, the value of the constant C in the solution (1.2) will be 0. This appears to give the particular solution $\ln|x|$. However, $\ln|x|$ consists of two disconnected branches (one with $x < 0$, and the other with $x > 0$) and so is not continuous, whereas our solution must be continuous. Because our initial point is given on the left branch of $\ln|x|$, the particular solution that passes through $(-1, 0)$ is $\ln|x|$ on the interval $-\infty < x < 0$; that is, $\ln(-x)$. On the other hand, however, if the solution of (1.1) is to pass through the point $(e, 6)$, the value of C would be 5. With similar reasoning to that above, the particular solution passing through $(e, 6)$ is $\ln|x| + 5$ on the interval $0 < x < \infty$; that is, $\ln x + 5$. These two particular solutions are shown in Figure 1.2. If we look at Figure 1.1 through these eyes we see that it represents 14 particular solutions, not 7, as a cursory glance might indicate.

The problem of finding a solution of a differential equation that must pass through a given point — such as $(-1, 0)$ or $(e, 6)$ in the case of (1.1) — is called an INITIAL VALUE PROBLEM, and the point is called an INITIAL VALUE, INITIAL CONDITION, or INITIAL POINT.

There is another way of expressing the solution of (1.1) when an initial point is specified as (x_0, y_0). If we use the fact that $\int_{x_0}^{x} g(t)\,dt$ is an antiderivative of $g(x)$, our solution in (1.2) may also be expressed as

$$y(x) = \int_{x_0}^{x} \frac{1}{t}\,dt + C, \qquad (1.8)$$

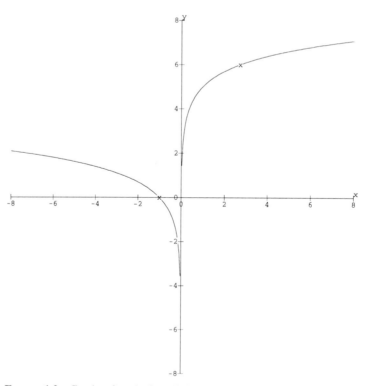

FIGURE 1.2   Graphs of particular solutions of $dy/dx = 1/x$ through $(-1, 0)$ and $(e, 6)$

if x_0 and x have the same sign. If we substitute $x = x_0$ into (1.8) and note the initial condition, $y(x_0) = y_0$, and the fact that $\int_{x_0}^{x_0} 1/t \, dt = 0$, (if $x_0 \neq 0$) the value of C can be determined as $C = y_0$. Thus (1.8) can be written as

$$y(x) = \int_{x_0}^{x} \frac{1}{t} \, dt + y_0.$$

From this example we see that the explicit solution of the differential equation

$$\frac{dy}{dx} = g(x)$$

subject to the condition that

$$y(x_0) = y_0,$$

can be written in the form

$$y(x) = \int_{x_0}^{x} g(t) \, dt + y(x_0) = \int_{x_0}^{x} g(t) \, dt + y_0, \tag{1.9}$$

if $g(t)$ is continuous for t between x_0 and x. This form of the solution is particularly useful when we are unable to evaluate the integral in terms of familiar functions. We demonstrate this in the following example.

EXAMPLE 1.2: The Error Function

An important function, used extensively in applications in probability and diffusion, is the solution of the differential equation

$$\frac{dy}{dx} = \frac{2}{\sqrt{\pi}} e^{-x^2} \tag{1.10}$$

subject to the initial condition that

$$y(0) = 0. \tag{1.11}$$

This example will recur throughout this chapter.
    The explicit solution of (1.10) can be written as

$$y(x) = \frac{2}{\sqrt{\pi}} \int e^{-x^2} \, dx + C. \tag{1.12}$$

The usual way to evaluate the constant C so (1.11) is satisfied is to substitute $x = 0$ and $y = 0$ into the solution (1.12) and solve for C. However, the integral in (1.12) cannot be expressed in terms of familiar functions, so this usual way to evaluate C does not work. To bypass this problem we change the form of our solution to the one given in (1.9) and use the fact that $x_0 = 0$ and $y_0 = 0$ to obtain

$$y(x) = \frac{2}{\sqrt{\pi}} \int_{0}^{x} e^{-t^2} \, dt.$$

The integral form of this solution is called the Error Function, and it is usually denoted by $erf(x)$, that is

$$erf(x) = \frac{2}{\sqrt{\pi}} \int_0^x e^{-t^2} \, dt. \tag{1.13}$$

We might ask how we can determine the graph of this function from its form in (1.13). One way would be to construct a table of values of $(x, erf(x))$ by using a numerical method of approximating the integral for specific choices of x (see Exercise 2). However, numerical techniques require considerable computation to plot enough points to be confident of the shape of the graph (see Exercises 3, 4, and 5). For that reason, in the next sections we develop methods for obtaining the graph of the solution of our differential equation directly from the differential equation.   □

EXERCISES

1. Solve each of the following first order differential equations. Sketch the explicit solution for three different values of the arbitrary constant C. Then find the specific value of C and the formula for $y(x)$ giving the particular explicit solution that passes through the point P.

   (a) $dy/dx = x^3$. P has coordinates $(1, 1)$.

   (b) $dy/dx = x^4$. P has coordinates $(1, 1)$.

   (c) $dy/dx = \cos x$. P has coordinates $(0, 0)$.

   (d) $dy/dx = \sin x$. P has coordinates $(\pi, 2)$.

   (e) $dy/dx = e^{-x}$. P has coordinates $(0, 1)$.

   (f) $dy/dx = 1/x^2$. P has coordinates $(1, 1)$.

   (g) $dy/dx = 1/x$. P has coordinates $(-1, 1)$.

   (h) $dy/dx = 1/[x(x-1)]$. P has coordinates $(2, 1)$.

   (i) $dy/dx = 1/(1+x^2)$. P has coordinates $(1, \pi/4)$.

   (j) $dy/dx = \ln x$. P has coordinates $(1, 1)$.

   (k) $dy/dx = x^2 e^{-x}$. P has coordinates $(0, 1)$.

2. The purpose of this exercise is to graph the Error Function defined by (1.13), namely

$$erf(x) = \frac{2}{\sqrt{\pi}} \int_0^x e^{-t^2} \, dt,$$

   by constructing a table of values of $(x, erf(x))$.

   (a) What is the value of $erf(0)$?

   (b) What is the relationship between $erf(x)$ and $erf(-x)$?

   (c) Use a computer/calculator package that performs numerical integration to obtain approximate values (say to 3 decimal places) for $erf(x)$ at $x = 2, 4, 6, 8,$ and 10. Use this information to plot $erf(x)$ in the interval $[-10, 10]$. How confident are you that the graph you have is fairly accurate?

   (d) Now repeat part (c) for $x = 1, 3, 5, 7,$ and 9. Did this change the accuracy of your previous graph for $erf(x)$?

   (e) Now repeat part (c) for $x = 0.5, 1.5, 2.5,$ and 3.5. Did this change the accuracy of your previous graph for $erf(x)$?

   (f) What do you think happens to $erf(x)$ as $x \to \infty$?

3. Using (1.9), write down an integral that represents the solution of

$$\frac{dy}{dx} = \sqrt{\frac{2}{\pi}} \sin x^2$$

subject to the condition that

$$y(0) = 0.$$

This solution, known as the Fresnel Sine Integral, cannot be expressed in terms of familiar functions. Use the ideas from Exercise 2 to graph the solution of this differential equation. What do you think happens to $y(x)$ as $x \to \infty$?

4. Using (1.9), write down an integral that represents the solution of

$$\frac{dy}{dx} = \sqrt{\frac{2}{\pi}} \cos x^2$$

subject to the condition that

$$y(0) = 0.$$

This solution, known as the Fresnel Cosine Integral, cannot be expressed in terms of familiar functions. Use the ideas from Exercise 2 to graph the solution of this differential equation. What do you think happens to $y(x)$ as $x \to \infty$?

5. Using (1.9), write down an integral that represents the solution of

$$\frac{dy}{dx} = g(x)$$

subject to the condition that

$$y(0) = 0,$$

where

$$g(x) = \begin{cases} \sin x / x & \text{if } x \neq 0 \\ 1 & \text{if } x = 0 \end{cases}.$$

This solution cannot be expressed in terms of familiar functions. Use the ideas from Exercise 2 to graph the solution of this differential equation. What do you think happens to $y(x)$ as $x \to \infty$?

6. Find the family of solutions for each of the following two differential equations

$$\frac{dy}{dx} = e^x$$

and

$$\frac{dy}{dx} = -e^{-x}.$$

Graph these two families of solutions on one plot, using the same scale for the x- and y-axes. What do you notice about the angle of intersection[2] between these two families of solutions? Could you have seen that directly from the differential equations, without solving them?

---

[2] The angle of intersection between two curves at a point is the angle between the tangent lines to the curves at that point.

7. Write down some odd[3] functions and find their antiderivatives. What property do these antiderivatives share? Make a conjecture that starts: *The antiderivative of an odd function is always* $\cdots$. Prove your conjecture.

8. Show that the family of antiderivatives of an even[4] function is symmetric with respect to the origin. Under what conditions will an antiderivative of an even function be an odd function? Give some examples.

1.2   Graphical Solutions Using Calculus

In the previous section we used antiderivatives to determine the behavior of solutions of $dy/dx = g(x)$. In this section we discover that there is a wealth of information available about the behavior of such solutions by considering the differential equation itself.

EXAMPLE 1.3

We return to the first example

$$y' = \frac{dy}{dx} = \frac{1}{x}, \tag{1.14}$$

where $'$ indicates differentiation with respect to x. Look at the 14 particular solutions in Figure 1.1. They were sketched directly from the functions $\ln x + C$ for $x > 0$, and $\ln(-x) + C$ for $x < 0$, for different values of C. Based on Figure 1.1, answer the following questions.

1. *Monotonicity*.[5] Where are the solutions increasing and where are they decreasing?

2. *Concavity*. Where are the solutions concave up and where are they concave down?

3. *Symmetry*. Are there any symmetries?

4. *Singularities*.[6] Is it possible to start on a solution curve where $x < 0$ and proceed along this curve and eventually arrive at positive values of x?

5. *Uniqueness*. Do any solutions intersect?

   Based on the graphs in Figure 1.1, the conjectures seem to be:

1. *Monotonicity*. Decreasing for $x < 0$ and increasing for $x > 0$.

2. *Concavity*. Concave down for $x \neq 0$.

3. *Symmetry*. Yes — about the y-axis. However, no particular solution has this symmetry. It is the family of solutions that has this symmetry.

4. *Singularities*. Not if the y-axis is a vertical asymptote, as it appears to be.

5. *Uniqueness*. On this graph, the answer appears to be yes — near the y-axis.

   Now imagine that we are unable to integrate (1.14) explicitly, and so we are unable to draw the particular solutions in Figure 1.1. How much of this information can we obtain directly from the differential equation (1.14) using our knowledge of calculus?

---

[3]Odd functions have the property that $f(-x) = -f(x)$.

[4]Even functions have the property that $f(-x) = f(x)$.

[5]A function is monotonic on an interval if it is either increasing or decreasing on the entire interval.

[6]A function $f(x)$ is singular at $x = a$ if $\lim_{x \to a} f(x) = \pm\infty$.

From calculus we know that $y' > 0$ implies that y is increasing and $y' < 0$ implies that y is decreasing. We also know that $y'' > 0$ implies the function is concave up, whereas $y'' < 0$ means the function is concave down. In Figure 1.3 we show the general solution curves for the four cases: $y' < 0$, $y'' > 0$; $y' > 0$, $y'' > 0$; $y' > 0$, $y'' < 0$; and $y' < 0$, $y'' < 0$.

With this information, let's return to our original questions.

1. *Monotonicity.* From (1.14) we see that the derivative of $y(x)$ is positive for $x > 0$, and so y increases in this region. Similar reasoning shows that y decreases when $x < 0$.

2. *Concavity.* If we differentiate (1.14) with respect to x, we have

$$y'' = \frac{d^2 y}{dx^2} = -\frac{1}{x^2},$$

which is negative for $x \neq 0$. Thus, the solutions must be concave down for all $x \neq 0$.

3. *Symmetry.* To have a symmetry about the y-axis means that we have no change in the family of solutions if x is replaced by $-x$ on both sides of (1.14). If in (1.14) we replace x by $-x$ we have

$$\frac{dy}{-dx} = \frac{1}{-x},$$

or $y' = 1/x$, which is exactly (1.14). Thus, the family of solutions that satisfies (1.14) is unchanged by the interchange of x with $-x$, and so must be symmetric about the y-axis.

4. *Singularities.* Because (1.14) is undefined at $x = 0$, we anticipate problems at $x = 0$.

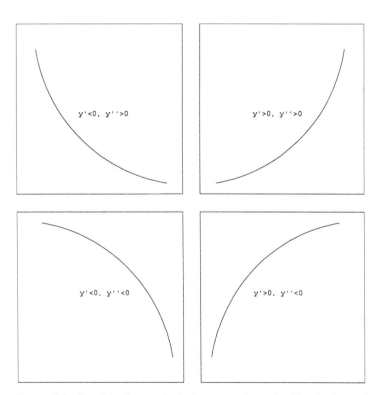

FIGURE 1.3 Possible shapes of solution curves determined by the first and second derivatives

5. *Uniqueness.* The statement that two solutions touch means that there is a common point (x_0, y_0) through which two distinct particular solutions of (1.14), say $y_1(x)$ and $y_2(x)$, pass. Because both y_1 and y_2 are solutions of (1.14), we must have $y_1' = 1/x$ and $y_2' = 1/x$, so that $y_1' = y_2'$, or $\left(y_1 - y_2\right)' = 0$. From this we have

$$y_1(x) = y_2(x) + C.$$

The fact that $y_0 = y_1(x_0)$ and $y_0 = y_2(x_0)$ implies that $C = 0$, so that $y_1(x) = y_2(x)$. In other words the two curves $y_1(x)$ and $y_2(x)$ are one and the same. This means that only one solution of (1.14) can pass through any point (x_0, y_0). Consequently, contrary to our conjecture based on Figure 1.1, solutions do not touch. In fact this argument can be used on any differential equation of the form $y' = g(x)$ to show that solutions cannot intersect. ☐

Words of Caution: Care must be exercised when drawing conclusions about curves touching from a graph alone.

From the preceding analysis we see that just by using calculus we can obtain a lot of information about solution curves without knowing the explicit solution. Let's see how we can use this to sketch the Error Function, *erf(x)*.

EXAMPLE 1.4: The Error Function

Using the techniques of calculus, sketch the family of solutions of the differential equation

$$\frac{dy}{dx} = \frac{2}{\sqrt{\pi}} e^{-x^2}. \tag{1.15}$$

- *Monotonicity.* The derivative of y is always positive, so all solutions are increasing.

- *Concavity.* If we differentiate (1.15) with respect to x we find

$$\frac{d^2y}{dx^2} = -\frac{4}{\sqrt{\pi}} x e^{-x^2}.$$

  From this we see that $y'' > 0$ when $x < 0$, and $y'' < 0$ when $x > 0$. Thus all solutions are concave up when $x < 0$ and concave down when $x > 0$.

- *Symmetry.* If we replace x with $-x$ on both sides of (1.15), the right-hand side is unchanged but the left-hand side changes sign. So the family of solutions is not symmetric about the y-axis. However, if we simultaneously replace x with $-x$ and replace y with $-y$, then we obtain (1.15) back again. So the family of solutions of (1.15) is unchanged under simultaneous interchange of x with $-x$ and y with $-y$. This means that the family of solutions is symmetric about the origin.[7]

- *Singularities.* There are no obvious points where the derivative fails to exist.

- *Uniqueness.* From our arguments at the end of Example 1.3, we see that solutions cannot intersect.

  Based on this information, we can sketch by hand[8] the family of solutions of (1.15), which is shown in Figure 1.4. This family of solutions contains the particular solution curve that passes through the point P with coordinates $(0, 0)$ — namely, $erf(x)$. ☐

---

[7] A graph is symmetric about the origin if it is unchanged when rotated 180° about the origin.

[8] Throughout the text we make reference to hand-drawn solutions. Of course, they were drawn by machine.

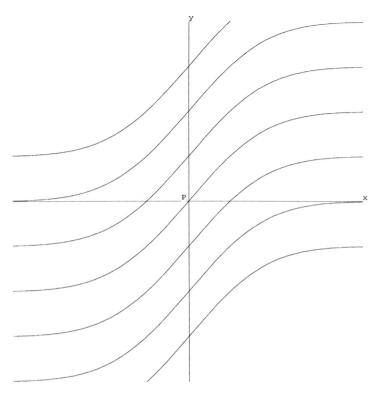

FIGURE 1.4 Some hand-drawn solution curves of $dy/dx = 2/\pi^{1/2} e^{-x^2}$

EXERCISES

1. Use monotonicity, concavity, symmetry, singularities, and uniqueness to sketch various solution curves for each of the following first order differential equations. Then draw the particular solution curve that passes through the point P. When you have finished, compare your answers with those you found for Exercise 1, Section 1.1.

   (a) $dy/dx = x^3$. P has coordinates $(1, 1)$.

   (b) $dy/dx = x^4$. P has coordinates $(1, 1)$.

   (c) $dy/dx = \cos x$. P has coordinates $(0, 0)$.

   (d) $dy/dx = \sin x$. P has coordinates $(\pi, 2)$.

   (e) $dy/dx = e^{-x}$. P has coordinates $(0, 1)$.

   (f) $dy/dx = 1/x^2$. P has coordinates $(1, 1)$.

   (g) $dy/dx = 1/x$. P has coordinates $(-1, 1)$.

   (h) $dy/dx = 1/[x(x - 1)]$. P has coordinates $(2, 1)$.

   (i) $dy/dx = 1/(1 + x^2)$. P has coordinates $(1, \pi/4)$.

   (j) $dy/dx = \ln x$. P has coordinates $(1, 1)$.

   (k) $dy/dx = x^2 e^{-x}$. P has coordinates $(0, 1)$.

2. Use monotonicity, concavity, symmetry, singularities, and uniqueness to sketch various solution curves for the differential equation

$$\frac{dy}{dx} = \sqrt{\frac{2}{\pi}} \sin x^2.$$

   Then draw the graph of the particular solution that satisfies $y(0) = 0$. What do you think happens to $y(x)$ as $x \to \infty$? Compare your answer with the one you found for Exercise 3, Section 1.1.

3. Use monotonicity, concavity, symmetry, singularities, and uniqueness to sketch various solution curves for the differential equation

$$\frac{dy}{dx} = \sqrt{\frac{2}{\pi}} \cos x^2.$$

Then draw the graph of the particular solution that satisfies $y(0) = 0$. What do you think happens to $y(x)$ as $x \to \infty$? Compare your answer with the one you found for Exercise 4, Section 1.1.

4. Use monotonicity, concavity, symmetry, singularities, and uniqueness to sketch various solution curves for the differential equation

$$\frac{dy}{dx} = g(x),$$

where

$$g(x) = \begin{cases} \sin x/x & \text{if } x \neq 0 \\ 1 & \text{if } x = 0 \end{cases}.$$

Then draw the graph of the particular solution that satisfies $y(0) = 0$. What do you think happens to $y(x)$ as $x \to \infty$? Compare your answer with the one you found for Exercise 5, Section 1.1.

1.3   Slope Fields and Isoclines

Using the techniques of calculus, we can sketch solutions of $y' = g(x)$ in very broad strokes using the signs of the first and second derivatives. However, there is still more information contained in the differential equation, because it also gives us the magnitude of the slope at each point on a solution curve. In this section we exploit this information.

EXAMPLE 1.5

We start with a very simple differential equation that describes the function whose rate of change is always 1, namely

$$\frac{dy}{dx} = 1. \tag{1.16}$$

Let's use our knowledge of calculus to sketch the solution curves of (1.16). Because the right-hand side of (1.16) is positive (namely, 1), we know that the solutions of (1.16) are increasing everywhere. Equation (1.16) also means that for all values of x and y, the solution of this differential equation has a tangent line whose slope is 1. To transfer this information to a graph we can select various coordinates (x, y) and draw short line segments with slope 1, as shown in Figure 1.5.

There is another way of stating that a differentiable function may be approximated near a point on the curve by its tangent line at that point: Each tangent segment gives the slope of the solution at that point. Such a collection of short line segments is known as a SLOPE FIELD of the differential equation, as it gives a short segment of the tangent line to the solution curve at each selected point. Slope fields are sometimes called direction fields.

We now construct a solution curve such that the tangent lines to this curve are consistent with the slope field. If we try to draw a curve whose tangent line has the slope 1 everywhere, we will end up drawing a straight line with slope 1. In fact the solution curves of (1.16) are the family of straight lines $y = x + C$.   □

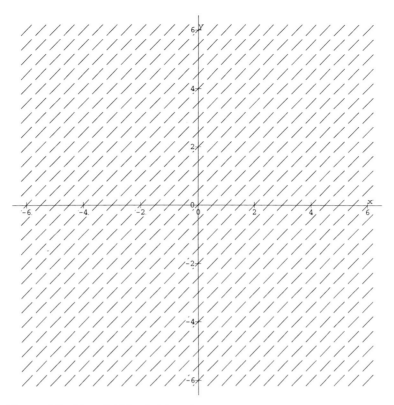

FIGURE 1.5   Slope field for $dy/dx = 1$

How to Sketch the Slope Field for $dy/dx = g(x, y)$

Purpose: To sketch the slope field for

$$\frac{dy}{dx} = g(x, y).$$

Technique:

1. Select the rectangular window in the xy plane in which to view the slope field.

2. Subdivide the rectangular region into a grid of equally spaced points (x, y). The number of points in the x and y directions may be different.

3. At each of these points (x, y), find the numerical value of $g(x, y)$ and draw short line segments at (x, y) with slope $g(x, y)$.

Comments about slope fields

- All the slope fields in this book were computer generated.

- **From now on we assume you either have access to a computer/calculator program that displays slope fields or are willing to construct slope fields by hand.**

EXAMPLE 1.6

Now consider the differential equation that models the situation where the rate of change of the unknown function is equal to the value of the independent variable,

$$\frac{dy}{dx} = x. \tag{1.17}$$

Again using our knowledge of calculus we see that (1.17) tells us that the solution $y = y(x)$ increases if $x > 0$ and decreases if $x < 0$. Moreover, because $d^2y/dx^2 = 1$, the second derivative of y is always positive, so y is concave up everywhere. Finally if we replace x by $-x$ in (1.17), the differential equation remains unchanged, so the family of solutions is symmetric about the y-axis.

If we construct the slope field as shown in Figure 1.6, we can obtain more information; we have a zero slope when $x = 0$, and the slopes of the short line segments of the slope field increase as x increases. Notice that the slope field appears to be symmetric about the y-axis.

Because the slope field for a differential equation gives the inclination of the tangent line to solutions at many points, the graph of a solution of this differential equation must be consistent with these tangent lines. Notice that the solution curves for (1.17) have horizontal tangents for $x = 0$, positive slopes for $x > 0$, and negative slopes for $x < 0$. Also note that these slopes become larger as x increases.

To manually draw a solution curve on the graph of the slope field, we start at some point, say where we have already drawn a short tangent line. Because the solution will be a differentiable function, its graph is well approximated by its tangent line near every point. Thus, we may proceed along this tangent line for a short distance and then see what the slope field looks like near the end of this distance. We then should adjust our curve so it changes in a manner consistent with the slope field.

To show this for a specific case in Figure 1.6, consider the solution curve that passes through the point $(0, 1)$. As we move to the right from this point, the curve changes from being horizontal in such a manner that the slope is continually increasing. This gives the curve labeled A shown in Figure 1.7. (Note that this results in a curve that is concave up.) Figure 1.7 also shows some other hand-drawn solution curves (all of which have the shape of parabolas), each having a different y-intercept. □

How to Manually Sketch Solution Curves from the Slope Field for $dy/dx = g(x, y)$

Purpose: To sketch, by hand, solution curves from the slope field for

$$\frac{dy}{dx} = g(x, y). \qquad \textbf{(1.18)}$$

Technique:

1. Sketch the slope field for (1.18).

2. Start at some initial point and put a dot there. Estimate the direction of the slope field at this point.

3. Proceed in this direction for a short distance. Place a dot at the point where you finished.

4. Adjust your direction to the direction of the slope field at the point where you finished.

5. Repeat steps 3 and 4, as often as needed.

6. Join the dots with a curve.

7. Start with a new initial point, and return to step 2.

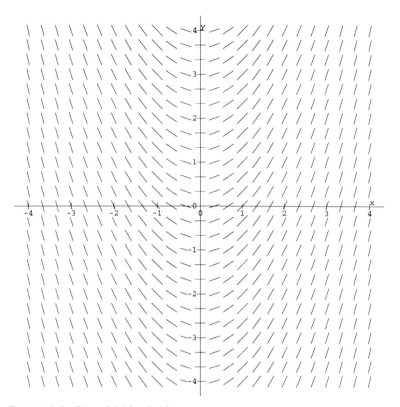

FIGURE 1.6 Slope field for $dy/dx = x$

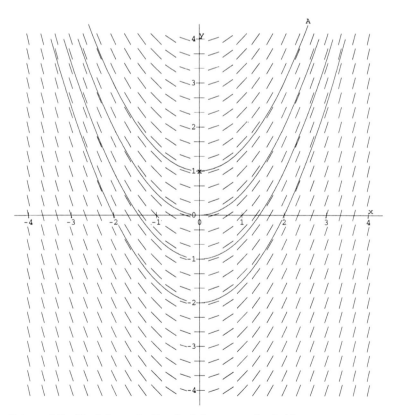

FIGURE 1.7 Hand-drawn family of solution curves for $dy/dx = x$

EXAMPLE 1.7

Once more we return to our first differential equation

$$\frac{dy}{dx} = \frac{1}{x}.$$
(1.19)

Based on our previous analysis we expect a concave down, decreasing shape for $x < 0$, and a concave down, increasing shape for $x > 0$. We also expect the family of solutions to be symmetric about the y-axis. We now draw the slope field for (1.19) (see Figure 1.8) and from it confirm the major properties of its solution.

We see from Figure 1.8 that the solution curves consistent with this slope field are increasing for $x > 0$ and decreasing for $x < 0$ and that all the slopes above a specific x location are equal. Also note that the solution curves are concave down everywhere, and that the slope field appears to be symmetric about the y-axis.

Figure 1.9 shows a few solution curves drawn on this slope field, and they appear to be vertical translations of each other. You should measure the vertical distances between the curves to verify this conjecture. □

Now that we have some solution curves on the slope field in Figure 1.9, it is apparent that **we can think of a slope field as what remains after plotting many solution curves and then erasing parts of them, leaving only some short segments here and there.** In this sense the challenge is to fill in the gaps between the short line segments of the slope field. This is the major use of slope fields — namely, to determine the graph of a solution of a differential equation whether or not an explicit solution is readily obtainable in terms of familiar functions.

EXAMPLE 1.8

We already have an example of this need in the case of the differential equation that generates the Error Function,

$$\frac{dy}{dx} = \frac{2}{\sqrt{\pi}} e^{-x^2}.$$
(1.20)

Because we want the solution of this equation, which starts at the point $(0, 0)$, we construct the slope field that includes this point, as shown in Figure 1.10.

As expected, the slope field indicates that the solution curve that passes through $(0, 0)$ is increasing, concave up when $x < 0$ and concave down when $x > 0$. Also notice that the slope field appears to be symmetric about the origin. Figure 1.11 shows a hand-drawn solution curve of (1.20) that passes through the origin, the graph of $y = erf(x)$.

We could also use a numerical integration technique to obtain values for $erf(x)$ at different values of x from its definition, namely

$$erf(x) = \frac{2}{\sqrt{\pi}} \int_0^x e^{-t^2} \, dt.$$

For example, we used Simpson's rule[9] to create Table 1.1. (Here we set the number of subintervals to 16 and rounded the answers to three decimal places.) Figure 1.12 shows the slope field, these numerical values, and a hand-drawn solution curve. Notice the agreement between this solution curve and these numerical values. □

---

[9]See a calculus text to remind yourself of Simpson's rule.

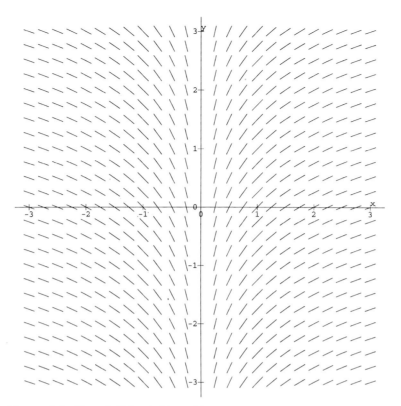

FIGURE 1.8 Slope field for $dy/dx = 1/x$

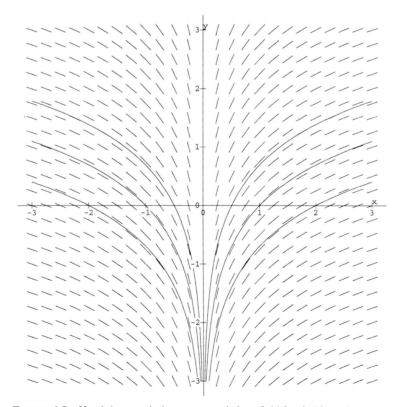

FIGURE 1.9 Hand-drawn solution curves and slope field for $dy/dx = 1/x$

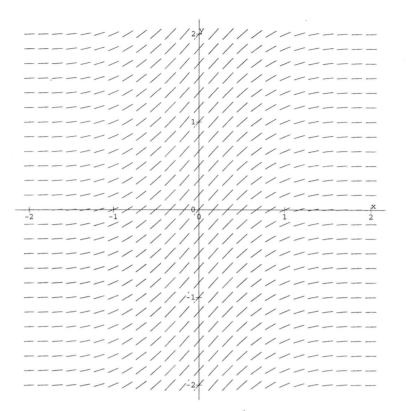

FIGURE 1.10   Slope field for $dy/dx = 2/\pi^{1/2} e^{-x^2}$

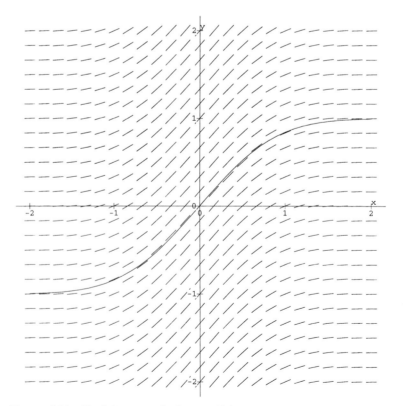

FIGURE 1.11   Hand-drawn graph of $y = erf(x)$

TABLE 1.1
Simpson's rule for *erf*(*x*)

x	*y*(*x*)
0.0	0.000
0.5	0.520
1.0	0.843
1.5	0.966
2.0	0.995

Words of Caution: We must make sure that any conclusions drawn from slope fields are confirmed by other means. One way is to make use of the analytical and graphical techniques learned in calculus, where the first and second derivatives of a function give us information about the function itself.

We now exploit the slope field concept in a different way, to obtain additional properties of solution curves directly from the differential equation.

EXAMPLE 1.9

We start by reconsidering our first differential equation

$$\frac{dy}{dx} = \frac{1}{x}, \tag{1.21}$$

which has the slope field shown in Figure 1.8. Our objective is to show that this slope field is reasonable.

In calculus, a first step in plotting the graph of a function was to compute its derivative, but (1.21) supplies the derivative of all the solution curves without any further work. In calculus we then set the derivative equal to 0 to find horizontal tangents. Because the derivative in this case $(1/x)$ is never equal to 0, there are no horizontal tangents to the solution curve.

Even though there are no points on this solution curve that have a horizontal tangent, we can set the derivative equal to another constant and see for what value (or values) of x the solution curve would have that constant slope. For example, because $1/x = 1$ for $x = 1$, the slope field will have a slope of 1 at all points on the graph where x equals 1. In general we can say that the solution curve will have a slope equal to m whenever

$$\frac{1}{x} = m,$$

that is

$$x = \frac{1}{m},$$

which is a vertical line through the point $(1/m, 0)$.

This vertical line is called an ISOCLINE (equal inclination) of the differential equation (1.21). All solution curves of (1.21) will have the same slope, m, as they cross the isocline at this value of x. For example, the solution curves of (1.21) will have a slope of 1 when $x = 1$, a slope of $1/2$ when $x = 2$, a slope of 2 when $x = 1/2$, a slope of -1 when $x = -1$, a slope of $-1/2$ when $x = -2$, and a slope of -2 when $x = -1/2$. We can see that Figure 1.13 is consistent with this information, which shows isoclines for $m = \pm 1/2$ and $m = \pm 1$. To make sure you understand isoclines, add the isoclines for $m = 2$ and $m = -2$ to this figure. Is there an isocline for $m = 0$?

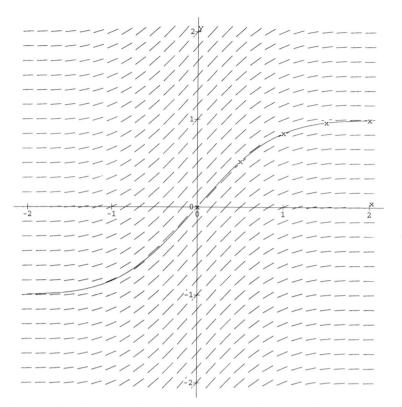

FIGURE 1.12   Numerical values and hand-drawn graph of $y = erf(x)$

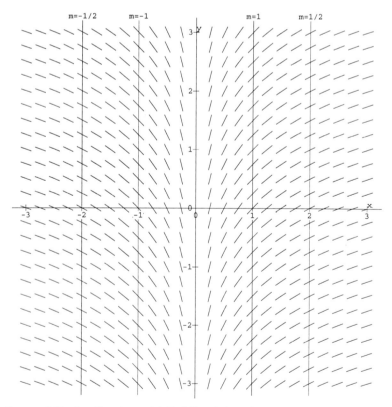

FIGURE 1.13   Isoclines ($m = \pm 1/2, \pm 1$) and slope field for $dy/dx = 1/x$

With this additional information we hope you feel very confident in drawing solution curves consistent with the slopes, the isoclines, the monotonicity, and the concavity (Figure 1.14). □

Definition 1.3: **An ISOCLINE CORRESPONDING TO SLOPE *m* of the differential equation**

$$\frac{dy}{dx} = g(x,y)$$

is the curve characterized by the equation

$$g(x,y) = m.$$

Comments about isoclines

- For any particular *m*, the isocline corresponding to slope *m* may consist of more than one curve.

- An isocline corresponding to slope *m* is also called an isocline for slope *m*.

- In the general case where $g(x, y)$ depends on both x and y, isoclines may not be lines. For example, if $g(x, y) = x^2 + y^2$ then the isoclines $x^2 + y^2 = m$ are circles with radius $\sqrt{m}$.

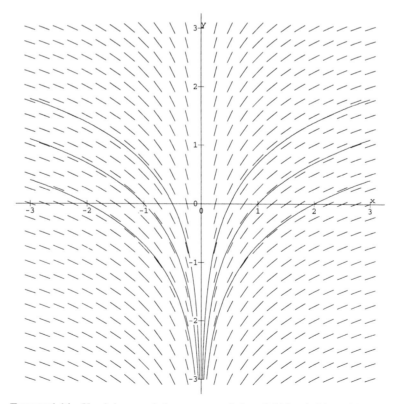

FIGURE 1.14   Hand-drawn solution curves and slope field for $dy/dx = 1/x$

How to Sketch Isoclines for $dy/dx = g(x, y)$

Purpose: To sketch isoclines for

$$\frac{dy}{dx} = g(x, y).$$

Technique:

1. Set

$$g(x, y) = m, \qquad\qquad\qquad (1.22)$$

   where m is constant.

2. Pick several values for m. For each m try to solve (1.22) for y in terms of x and m, or for x in terms of y and m. If this cannot be done, try to identify the curves defined implicitly by (1.22). Plot these curves. This is the isocline corresponding to slope m.

3. For different values of m, plot the isocline corresponding to slope m.

4. If you are constructing slope fields by hand, draw short line segments with slope m crossing the isoclines.

EXAMPLE 1.10

Now we return to the differential equation giving rise to the Error Function ,

$$\frac{dy}{dx} = \frac{2}{\sqrt{\pi}} e^{-x^2}.$$

The isocline corresponding to slope m is given by

$$\frac{2}{\sqrt{\pi}} e^{-x^2} = m. \qquad\qquad\qquad (1.23)$$

Notice that this implies that there are no isoclines for slope $m \leq 0$ or for slope $m > 2/\sqrt{\pi} \approx$ 1.128. (Why?) If we solve (1.23) for x we obtain $x = \pm \left(\ln \left(2/ \left(m\sqrt{\pi} \right) \right) \right)^{1/2}$ as the equation of the isocline corresponding to slope m. The isoclines for slope $0.1, 0.3$, and 0.7 are shown in Figure 1.15. Notice that in this case each isocline consists of two vertical lines.   □

Finally, we gather together some general observations about differential equations of the special form

$$\frac{dy}{dx} = g(x).$$

Comments about $y' = g(x)$

- All solutions are explicit.

- Once we have found one member of the family of solutions, other members of the family can be generated from this member by vertical translations.

- If $dy/dx = g(x)$ remains unchanged after the interchange of x with $-x$, then the family of solutions is symmetric about the y-axis.

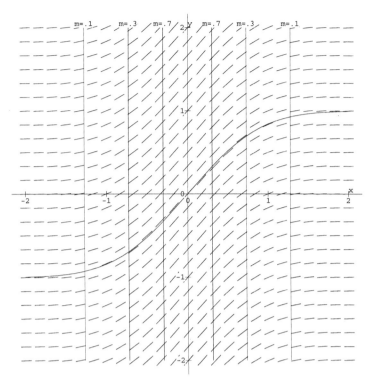

FIGURE 1.15 Isoclines ($m = 0.1, 0.3, 0.7$) and slope field for $dy/dx = 2/\pi^{1/2}e^{-x^2}$

- If $dy/dx = g(x)$ remains unchanged after the simultaneous interchange of y with $-y$ and x with $-x$, then the family of solutions is symmetric about the origin.

- For the case $y' = g(x)$ all isoclines are vertical — that is, parallel to the y-axis.

EXERCISES

1. Sketch the slope field for each of the following first order differential equations. In each case draw some isoclines to confirm your sketch. Use your sketch to draw various solution curves. Then draw the solution curve that passes through the point P. When you have finished, compare your answers with those you found for Exercise 1, Section 1.1, and Exercise 1, Section 1.2.

    (a) $dy/dx = x^3$. P has coordinates $(1, 1)$.

    (b) $dy/dx = x^4$. P has coordinates $(1, 1)$.

    (c) $dy/dx = \cos x$. P has coordinates $(0, 0)$.

    (d) $dy/dx = \sin x$. P has coordinates $(\pi, 2)$.

    (e) $dy/dx = e^{-x}$. P has coordinates $(0, 1)$.

    (f) $dy/dx = 1/x^2$. P has coordinates $(1, 1)$.

    (g) $dy/dx = 1/x$. P has coordinates $(-1, 1)$.

    (h) $dy/dx = 1/[x(x - 1)]$. P has coordinates $(2, 1)$.

    (i) $dy/dx = 1/(1 + x^2)$. P has coordinates $(1, \pi/4)$.

    (j) $dy/dx = \ln x$. P has coordinates $(1, 1)$.

    (k) $dy/dx = x^2 e^{-x}$. P has coordinates $(0, 1)$.

2. Use slope fields and isoclines for the differential equation

$$\frac{dy}{dx} = \sqrt{\frac{2}{\pi}} \sin x^2$$

to draw various solution curves. Then draw the solution curve that satisfies $y(0) = 0$. What do you think happens to $y(x)$ as $x \to \infty$? Compare your answer with those you found for Exercise 3, Section 1.1, and Exercise 2, Section 1.2.

3. Use slope fields and isoclines for the differential equation

$$\frac{dy}{dx} = \sqrt{\frac{2}{\pi}} \cos x^2$$

to draw various solution curves. Then draw the solution curve that satisfies $y(0) = 0$. What do you think happens to $y(x)$ as $x \to \infty$? Compare your answer with those you found for Exercise 4, Section 1.1, and Exercise 3, Section 1.2.

4. Consider the differential equation

$$\frac{dy}{dx} = g(x),$$

where

$$g(x) = \begin{cases} \sin x / x & \text{if } x \neq 0 \\ 1 & \text{if } x = 0 \end{cases}.$$

Use slope fields and isoclines to draw various solution curves. Then draw the solution curve that satisfies $y(0) = 0$. What do you think happens to $y(x)$ as $x \to \infty$? Compare your answer with those you found for Exercise 5, Section 1.1, and Exercise 4, Section 1.2.

5. Figure 1.16 shows one member of a family of solutions of the differential equation

$$\frac{dy}{dx} = g(x),$$

where $g(x)$ is a given function.

(a) Use this information to plot other members of the family of solutions. Do not attempt to find $y(x)$ or $g(x)$.

(b) Can every solution of the differential equation be obtained by the technique used in part (a)?

6. Consider the following four differential equations

   i. $\dfrac{dy}{dx} = x + 1$.

   ii. $\dfrac{dy}{dx} = x - 1$.

   iii. $\dfrac{dy}{dx} = \ln |x|$.

   iv. $\dfrac{dy}{dx} = x^2 - 1$.

(a) The slope field of one of the preceding equations is given in Figure 1.17. Match the correct equation with the figure, carefully stating your reasons. Do not plot the slope fields for i through iv to answer this question.

(b) Briefly outline a general strategy for this matching process.

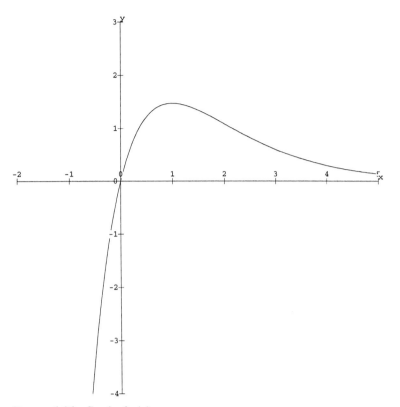

FIGURE 1.16   Graph of $y(x)$

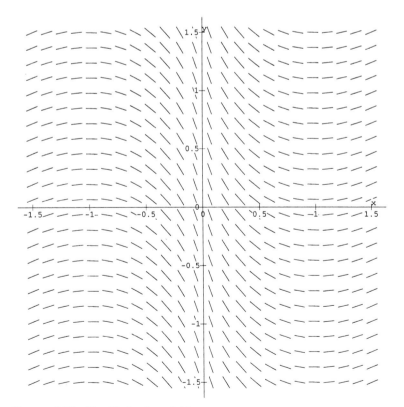

FIGURE 1.17   Identify the slope field

7. Figures 1.18 and 1.19 are mystery slope fields, believed to be the slope fields for two of the following differential equations

$$\frac{dy}{dx} = \frac{x^2 + 1}{x^2 - 1}, \quad \frac{dy}{dx} = \frac{x^2 - 1}{x^2 + 1}, \quad \frac{dy}{dx} = -\frac{x^2 + 1}{x^2 - 1}, \quad \frac{dy}{dx} = -\frac{x^2 - 1}{x^2 + 1}.$$

(a) Identify to which differential equation each of the mystery slope fields belongs. Confirm all the information using calculus and isoclines.

(b) Now superimpose the two mystery slope fields, perhaps by placing one on top of the other, or by plotting the slope fields for

$$\frac{dy}{dx} = a\frac{x^2 + 1}{x^2 - 1} + b\frac{x^2 - 1}{x^2 + 1}$$

with $a = \pm 1$ and $b = 0$, and then with $a = 0$ and $b = \pm 1$. What do you notice? Would this have made your previous analysis in part (a) easier?

8. If an object falls out of an airplane, its downward velocity after t seconds is often approximated by

$$\frac{dy}{dt} = \frac{g}{k}\left(1 - e^{-kt}\right),$$

where $g = 9.8$ m/sec^2 and $k = 0.2$ sec^{-1}. If it falls from 5000 meters above the ground, estimate how many seconds it falls before it hits the ground, by

(a) using slope fields, isoclines, and concavity,

(b) using numerical approximations, and

(c) finding the exact solution.

1.4   Functions and Power Series Expansions

So far in this chapter we have developed methods for sketching solution curves of $y' = g(x)$ that pass through specific points. For the case when we were unable to discover an antiderivative in terms of familiar functions, we still could write down the explicit solution

$$y(x) = \int_{x_0}^{x} g(t)\,dt + y_0$$

that satisfied the initial condition $y(x_0) = y_0$. To evaluate $y(x)$ at specific values of x from this expression requires the use of a numerical approximation for the definite integral.

However, there is an alternative expression for this explicit solution that takes the form of an infinite series. To obtain such an expression, we use Taylor series[10] to expand the integrand $g(t)$, and then integrate the resulting powers of t term by term.

To illustrate this procedure, we consider the Error Function defined by

$$erf(x) = \frac{2}{\sqrt{\pi}} \int_0^x e^{-t^2}\,dt. \tag{1.24}$$

---

[10]A summary of Taylor series results is in the appendix.

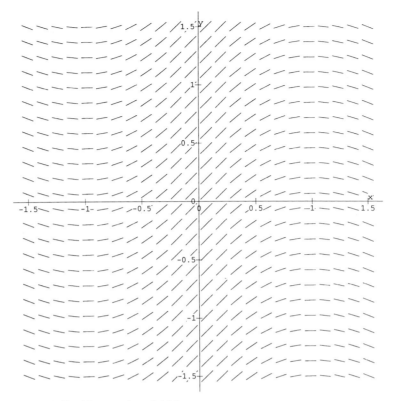

FIGURE 1.18   Mystery slope field 1

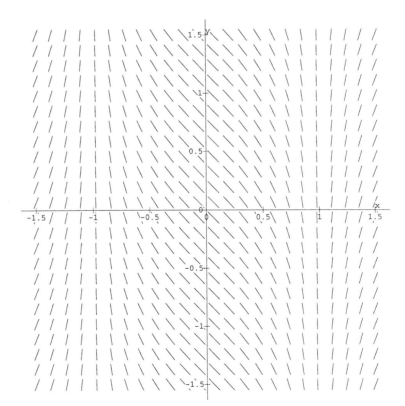

FIGURE 1.19   Mystery slope field 2

First recall that the function e^x has the Taylor series expansion [11]

$$e^x = \sum_{k=0}^{\infty} \frac{x^k}{k!} = 1 + x + \frac{x^2}{2!} + \frac{x^3}{3!} + \frac{x^4}{4!} + \frac{x^5}{5!} + \cdots, \qquad (1.25)$$

valid for all x. If we replace x by $-t^2$ in (1.25), we find

$$e^{-t^2} = \sum_{k=0}^{\infty} (-1)^k \frac{t^{2k}}{k!} = 1 - t^2 + \frac{t^4}{2!} - \frac{t^6}{3!} + \frac{t^8}{4!} - \frac{t^{10}}{5!} + \cdots,$$

which, when integrated from 0 to x, gives

$$\int_0^x e^{-t^2}\, dt = \sum_{k=0}^{\infty} (-1)^k \int_0^x \frac{t^{2k}}{k!}\, dt = \sum_{k=0}^{\infty} (-1)^k \frac{x^{2k+1}}{(2k+1)k!}. \qquad (1.26)$$

(Recall that the integral of a convergent power series is also convergent.) Combining (1.24) and (1.26), we see that the Error Function can be written in the form

$$erf(x) = \frac{2}{\sqrt{\pi}} \sum_{k=0}^{\infty} (-1)^k \frac{x^{2k+1}}{(2k+1)k!} = \frac{2}{\sqrt{\pi}} \left(x - \frac{x^3}{3} + \frac{x^5}{5 \cdot 2!} - \frac{x^7}{7 \cdot 3!} + \frac{x^9}{9 \cdot 4!} - \frac{x^{11}}{11 \cdot 5!} + \cdots \right),$$

$$\qquad (1.27)$$

valid for all x. This is an alternative representation for the Error Function.

The first term on the right-hand side of (1.27), denoted by

$$P_1(x) = \frac{2}{\sqrt{\pi}} x,$$

is the Taylor polynomial of degree one for $erf(x)$. The first two terms on the right-hand side of (1.27), denoted by

$$P_3(x) = \frac{2}{\sqrt{\pi}} \left(x - \frac{x^3}{3} \right),$$

are the Taylor polynomial of degree three for $erf(x)$, and so on. Figure 1.20 shows the graph of the Error Function together with the Taylor polynomials of degrees one, three, five, and so on to eleven — that is $P_1(x)$, $P_3(x)$, $P_5(x)$, $\cdots$, $P_{11}(x)$. Notice how the graph of the Error Function gradually emerges as the curve common to these graphs. It appears that $erf(x)$ is trapped by the other curves. Also notice that near the origin $erf(x)$ behaves like the first term in the Taylor series (1.27), namely the straight line $y = 2x/\sqrt{\pi}$.

EXERCISES

1. Calculate the power series expansion for the Fresnel Sine Integral $S(x)$, defined as

$$S(x) = \sqrt{\frac{2}{\pi}} \int_0^x \sin t^2\, dt.$$

Confirm that it is consistent with the answers you found for Exercise 3, Section 1.1; Exercise 2, Section 1.2; and Exercise 2, Section 1.3.

---

[11] Remember $0! = 1$, $n! = n(n-1)(n-2)\cdots 3 \cdot 2 \cdot 1$.

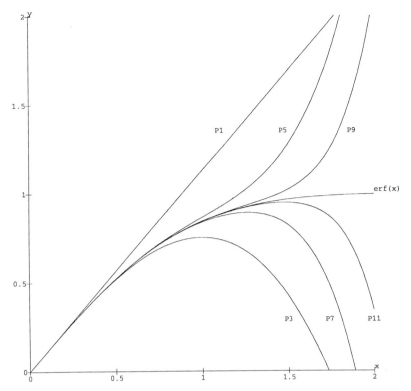

FIGURE 1.20 The Error Function trapped by its Taylor polynomials $P_1(x)$, $P_3(x)$, $P_5(x), \ldots, P_{11}(x)$

2. Calculate the power series expansion for the Fresnel Cosine Integral $C(x)$, defined as

$$C(x) = \sqrt{\frac{2}{\pi}} \int_0^x \cos t^2 \, dt.$$

Confirm that it is consistent with the answers you found for Exercise 4, Section 1.1; Exercise 3, Section 1.2; and Exercise 3, Section 1.3.

3. Consider the differential equation

$$\frac{dy}{dx} = g(x)$$

where

$$g(x) = \begin{cases} \sin x / x & \text{if } x \neq 0 \\ 1 & \text{if } x = 0 \end{cases}.$$

Explain why the solution of this differential equation can be written as

$$y(x) = \int_0^x \frac{\sin t}{t} \, dt,$$

paying particular attention to the lower limit of integration. Calculate the power series expansion for this function $y(x)$. Confirm that it is consistent with the answers you found for Exercise 5, Section 1.1; Exercise 4, Section 1.2; and Exercise 4, Section 1.3.

What Have We Learned?

MAIN IDEAS

- A FIRST ORDER ORDINARY DIFFERENTIAL EQUATION is an equation that involves at most the first derivative of an unknown function. If y, the unknown function, is a function of x, then we write the first order equation as

$$\frac{dy}{dx} = g(x, y)$$

  where $g(x, y)$ is a given function of x and y.

- An EXPLICIT SOLUTION of

$$\frac{dy}{dx} = g(x, y) \tag{1.28}$$

  is any function $y = y(x)$ (differentiable in some interval $a < x < b$) that identically satisfies the differential equation (1.28).

- If an explicit solution contains an arbitrary constant, the infinite number of solutions it generates is called a FAMILY OF EXPLICIT SOLUTIONS. If an explicit solution contains no arbitrary constant, it is called a PARTICULAR EXPLICIT SOLUTION.

- The graph of a particular solution is called a SOLUTION CURVE of the differential equation.

- Looking for a solution of a differential equation that must pass through a given point is called an INITIAL VALUE PROBLEM, and the point is called an INITIAL VALUE, INITIAL CONDITION, or INITIAL POINT.

- To sketch slope fields, see *How to Sketch the Slope Field for $dy/dx = g(x, y)$* on page 13.

- To hand-draw solution curves from slope fields, see *How to Manually Sketch Solution Curves from the Slope Field for $dy/dx = g(x, y)$* on page 14.

- An ISOCLINE CORRESPONDING TO SLOPE m of the differential equation

$$\frac{dy}{dx} = g(x, y)$$

  is the curve characterized by the equation

$$g(x, y) = m.$$

- To sketch isoclines, see *How to Sketch Isoclines for $dy/dx = g(x, y)$* on page 22.

- When the explicit solution is left in the form of a definite integral it is frequently possible to put it in an alternative form. This is done by expanding the integrand using Taylor series, and then integrating the result term by term. See page 26.

WORDS OF CAUTION

- Exercise care when drawing conclusions about curves touching from a graph alone.

- Make sure that any conclusions drawn from slope fields are confirmed by other means.

CHAPTER 2

Autonomous Differential Equations

Where Are We Going?

In Chapter 1 we considered the simple differential equation

$$\frac{dy}{dx} = g(x), \tag{2.1}$$

for which we could always write our solution as

$$y(x) = \int g(x)\,dx + C. \tag{2.2}$$

The arbitrary constant in (2.2) indicates that we have an infinite number of solutions. Even though we cannot always evaluate the integral in (2.2) in terms of familiar functions, we showed that we may still determine the constant C so the solution passes through the point (x_0, y_0). This gave the form

$$y(x) = \int_{x_0}^{x} g(t)\,dt + y_0, \tag{2.3}$$

if $g(t)$ is continuous for t between x_0 and x. Besides this analytical approach, we developed graphical ways to discover properties of the solution directly from (2.1).

In this chapter we consider first order differential equations of the form

$$\frac{dy}{dx} = g(y), \tag{2.4}$$

where the right-hand side of (2.4) depends only on the dependent variable y. Such differential equations are called autonomous. One of the objectives of this chapter is to show how the techniques developed in Chapter 1 can be applied to help us fully understand the behavior of solutions of these autonomous differential equations.

Although we always have a unique solution of (2.1) passing through a specified point (x_0, y_0), given by (2.3), such is not the case for differential equations of the form (2.4). Thus, we need an existence and uniqueness theorem for initial value problems associated with (2.4). This is given, and its usefulness is very apparent when we hand-draw solution curves using information given by a slope field. Equilibrium solutions and phase line analysis are two new topics introduced in this chapter to help determine properties of solution curves of $dy/dx = g(y)$.

2.1   Autonomous Equations

In this section we apply the techniques developed in Chapter 1 — graphical analysis and explicit solutions — to differential equations of the form (2.4) where the right-hand side depends only on y. We first consider a simple example, and then make some general observations about what we have learned. The important concept of an equilibrium solution is introduced.

We start with a definition that applies to all the differential equations considered in this chapter.

Definition 2.1:   **A first order differential equation of the form**

$$\frac{dy}{dx} = g(y), \tag{2.5}$$

where $g(y)$ is a given function of y alone, is called AUTONOMOUS.

EXAMPLE 2.1

As our first example of an autonomous differential equation, we look at the counterpart of our very first example in Chapter 1 — $dy/dx = 1/x$ — which in this case is

$$\frac{dy}{dx} = \frac{1}{y}. \tag{2.6}$$

Let's see what (2.6) tells us about its solution curves if we apply the ideas of calculus. If $y > 0$, then $y' > 0$ and so the solution $y(x)$ is increasing. If $y < 0$, then $y' < 0$ and the solution $y(x)$ is decreasing. To consider the concavity of the solution curves, we need d^2y/dx^2. If we differentiate (2.6) with respect to x, we find

$$\frac{d^2y}{dx^2} = -\frac{1}{y^2}\frac{dy}{dx}. \tag{2.7}$$

This differs from the situation in Chapter 1, where all second derivatives were explicit functions of x. Here we find the second derivative in terms of the first. However, we can substitute (2.6) into (2.7) to find

$$\frac{d^2y}{dx^2} = -\frac{1}{y^3}. \tag{2.8}$$

Even though the right-hand side of (2.8) is not an explicit function of x, it does tells us that the sign of the second derivative is determined by the sign of y. So if $y > 0$, the solution curves are concave down, and if $y < 0$ they are concave up.

Now let's consider the construction of the slope field for (2.6). As we calculate the slopes of solution curves at equally spaced points (x, y), we see that along the line $y = 0$ (the x-axis), the slopes are all infinite. Along the horizontal line $y = 1$ the slopes are all 1. Along the line $y = 2$ the slopes are all $1/2$. Along the line $y = -1$ the slopes are all -1. Along the line $y = -2$ the slopes are all $-1/2$. Therefore, we can still construct a slope field, which is shown in Figure 2.1.

We now consider the isocline question — namely, where is the slope of the solution curve equal to m? This will be when $y = 1/m$. These are horizontal lines, and Figure 2.1 is consistent with this information. To make sure you understand this, add isoclines for $m = 1, m = -1, m = 2,$ and $m = -2$ to Figure 2.1.

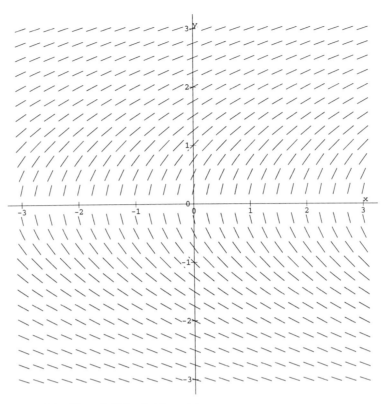

FIGURE 2.1   Slope field for $dy/dx = 1/y$

Notice that the slope field in Figure 2.1 appears symmetric about the x-axis. How can we tell whether this is actually correct? Symmetry about the x-axis means that interchanging y with $-y$ gives the same picture. If in (2.6) we interchange y with $-y$, on both sides of the equation, we find exactly (2.6) again, so the family of solutions corresponding to it must be symmetric about the x-axis.

Thus, without solving (2.6) explicitly, we have been able to use graphical techniques to obtain the general shape of the solution curves, which are shown in Figure 2.2. Explain why this figure contains 12 particular solutions, and not 6.

We now look at the analytical approach. The counterpart of (2.2) for (2.6), namely

$$y(x) = \int \frac{1}{y}\, dx,$$

will not give our solution for $y(x)$ directly. To evaluate the integral we would have to know y explicitly as a function of x. But, if we knew y explicitly as a function of x we would already have the solution. This doesn't mean that (2.6) has no explicit solutions. We are able to solve

$$\frac{dy}{dx} = \frac{1}{y}$$

if we first multiply by y, giving

$$y\frac{dy}{dx} = 1.$$

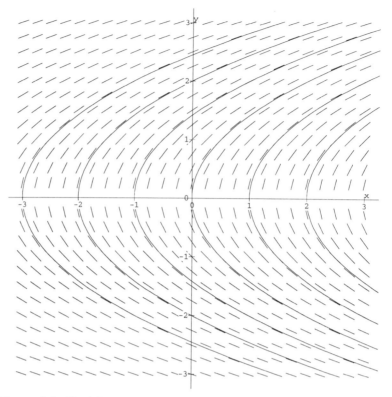

FIGURE 2.2   Hand-drawn solutions and slope field for $dy/dx = 1/y$

In this form we now integrate with respect to x and find

$$\int y\frac{dy}{dx}\,dx = \int\,dx,$$

which gives

$$\frac{1}{2}y^2 = x + C, \tag{2.9}$$

where C is an arbitrary constant. Notice that this can be expressed as an explicitly defined function for $x(y)$. We can also solve (2.9) for y, and we find the family of solutions

$$y(x) = \pm\sqrt{2\,(x + C)}. \tag{2.10}$$

It should be apparent that contrary to the situation in (2.2), the arbitrary constant in (2.10) does not represent a vertical translation between any two solutions, but represents a horizontal translation. Several particular solutions of this family — with initial values $(0, 2)$, $(0, -1)$, $(2, 2)$, $(2, -1)$ — are shown in Figure 2.3. (Why do these particular solutions remain in either the upper or lower half-plane?) The agreement between our hand-drawn solutions (Figure 2.2) and graphs of our explicit solutions (Figure 2.3) is obvious.   □

From the results of the preceding example we can make some general observations about autonomous differential equations

$$\frac{dy}{dx} = g(y).$$

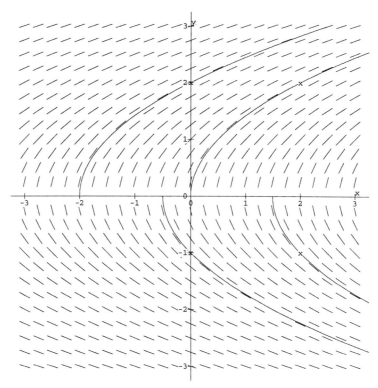

FIGURE 2.3   Analytical solution curves and slope field for $dy/dx = 1/y$

Comments about autonomous differential equations

- Usually solutions of autonomous differential equation are explicitly defined functions for $x(y)$ but are not necessarily explicitly defined functions for $y(x)$. Sometimes we may be able to solve $x = x(y)$ for $y = y(x)$.

- Isoclines of autonomous differential equations are always horizontal lines.

- If $dy/dx = g(y)$ remains unchanged after the interchange of y with $-y$ on both sides of the differential equation, then the family of solutions is symmetric about the x-axis.

- If $dy/dx = g(y)$ remains unchanged after the simultaneous interchange of y with $-y$ and x with $-x$ on both sides of the differential equation, then the family of solutions is symmetric about the origin.

- We use the expression ANALYTICAL SOLUTION CURVE to describe the graph of an explicit solution, to distinguish it from the hand-drawn solution curves obtained directly from the differential equation.

EXAMPLE 2.2

Often beginning calculus courses contain examples of differential equations that concern the growth or decay of some substance. These applications model situations where the rate of change of a substance is proportional to the amount of substance present, that is,

$$\frac{dy}{dx} = ay, \tag{2.11}$$

where y is the amount of substance, x is time, and a is a given constant, $a \neq 0$. If the application concerns growth of bacteria, then $a > 0$ (the number of bacteria is increasing), whereas for radioactive decay $a < 0$. The understanding in (2.11) is that we are given a and want to find $y = y(x)$. For the purposes of this example, we will consider all values of y, although for applications dealing with growth or decay we would consider only nonnegative values of y, because y gives the amount of substance present at time x.

First let's see what the ideas of calculus tell us about the solution curves of (2.11). If $a > 0$, then y is increasing when $y > 0$ and decreasing when $y < 0$. The reverse is true if $a < 0$. To consider the concavity of the solution curves, we need d^2y/dx^2. If we differentiate (2.11) with respect to x, we find

$$\frac{d^2y}{dx^2} = a\frac{dy}{dx},$$

which substituting from (2.11) becomes

$$\frac{d^2y}{dx^2} = a^2y. \tag{2.12}$$

Because $a^2 > 0$, (2.12) tells us that the sign of the second derivative is determined by the sign of y. So if $y > 0$, the solution curves are concave up, and if $y < 0$, they are concave down.

We now consider the construction of the slope field for (2.11). As we calculate the slopes of solution curves at equally spaced points (x, y), we see that along the line $y = 0$ (the x-axis), the slopes are all 0. Along the horizontal line $y = 1$ the slopes are all a. Along the line $y = 2$ the slopes are all $2a$. Along the horizontal line $y = -1$ the slopes are all $-a$. Therefore, we can still construct slope fields, which for $a = 1$ and $a = -1$ are shown in Figures 2.4 and 2.5, respectively. Notice also that both slope fields in these figures appear symmetric about the x-axis. How would you prove that this observation is correct?

We now consider the isocline question — namely, where is the slope of the solution curve equal to m? As before, this will be when $y = m/a$. These are horizontal lines, and Figures 2.4 and 2.5 are consistent with this information.

Thus, without solving (2.11) explicitly, we have been able to use graphical techniques to obtain the general shape of the solution curves, which are shown in Figure 2.6 for $a = 1$ and in Figure 2.7 for $a = -1$.

We now look at the analytical approach. If we divide

$$\frac{dy}{dx} = ay \tag{2.13}$$

by y, we have

$$\frac{1}{y}\frac{dy}{dx} = a. \tag{2.14}$$

In this form we now integrate with respect to x, and find

$$\ln|y(x)| = ax + c, \tag{2.15}$$

where c is an arbitrary constant. We apply the exponential function to both sides of this equation to get

$$|y(x)| = e^{ax+c} = e^c e^{ax},$$

or

$$y(x) = \pm e^c e^{ax}.$$

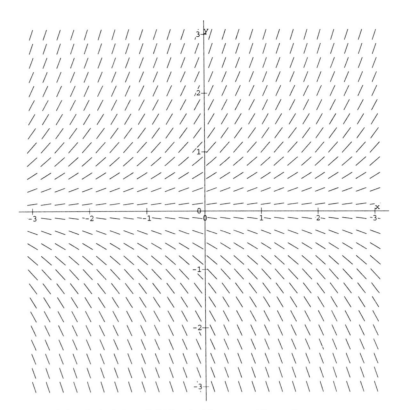

FIGURE 2.4   Typical slope field for $dy/dx = ay$ with $a = 1$

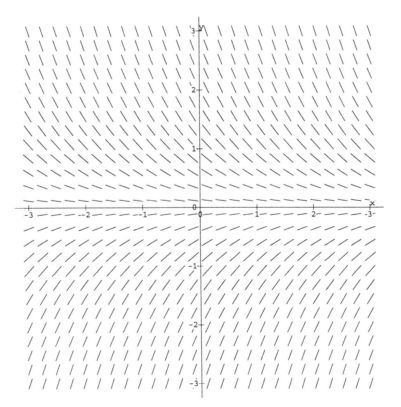

FIGURE 2.5   Typical slope field for $dy/dx = ay$ with $a = -1$

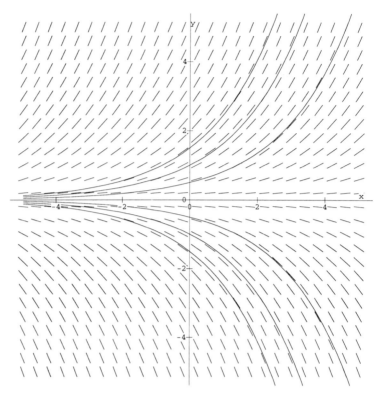

FIGURE 2.6   Hand-drawn solution curves and slope field for $dy/dx = ay$ with $a = 1$

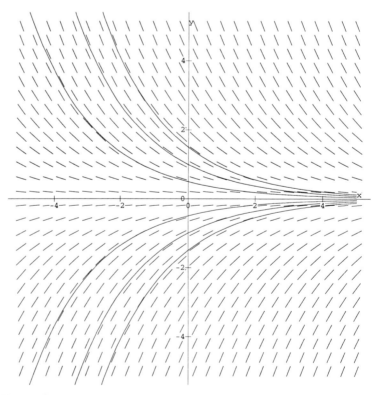

FIGURE 2.7   Hand-drawn solution curves and slope field for $dy/dx = ay$ with $a = -1$

If we define $C = \pm e^c$ (which is never 0), we have the family of solutions

$$y(x) = Ce^{ax}, \qquad C \neq 0, \tag{2.16}$$

which is in good agreement with Figures 2.6 and 2.7 .

At first sight this may seem acceptable. However, this analysis requires closer scrutiny. The problem is that in going from (2.13) to (2.14) we implicitly assumed that $y(x) \neq 0$. But (2.13) is clearly satisfied by

$$y(x) = 0, \tag{2.17}$$

which we lost in going from (2.13) to (2.14). This is a solution that is not contained in the solution given by (2.16), although it could be if we allow $C = 0$. Thus, we can combine the two distinct solutions (2.16) and (2.17) into the single equation

$$y(x) = Ce^{ax}, \tag{2.18}$$

where C can now take on any value, including 0.

Note that the value of C is the y-intercept because $y(0) = C$. Several members of this family, with $y(0) = 0.0, 0.5, -1$, and 1.5, are shown in Figure 2.8 for $a = 1$, and in Figure 2.9 for $a = -1$. The agreement between our hand-drawn solutions (Figures 2.6 and 2.7) and our analytical solution curves (Figures 2.8 and 2.9) is obvious.

In the application of this solution to population growth (where $a > 0$), we see that if $y(0) > 0$, the population grows without bound as $x \to \infty$. In the case of radioactive decay (where $a < 0$ and $y(0) > 0$), we see that the amount of radioactive substance goes to 0 as $x \to \infty$. In both cases, if there is no substance present at the beginning — that is, $y(0) = 0$ — then there is never any substance present, $y(x) = 0$ for all x. □

In the preceding example we found that the constant function $y(x) = 0$ was a particular explicit solution of (2.11). This solution is an example of an equilibrium solution, as defined below.

Definition 2.2: **If the differential equation**

$$\frac{dy}{dx} = g(x,y)$$

has a solution of the form $y(x) = $ *constant*, then this solution is called an EQUILIBRIUM SOLUTION.

Words of Caution: It is a common oversight to mislay the equilibrium solutions.

EXERCISES

1. This exercise deals with the differential equation

$$\frac{dy}{dx} = \frac{1}{y^2}.$$

   (a) Use graphical methods (calculus, slope field, isoclines) to sketch some solution curves.

   (b) Use analytical methods to solve the differential equation for $y(x)$.

   (c) Confirm that parts (a) and (b) are consistent.

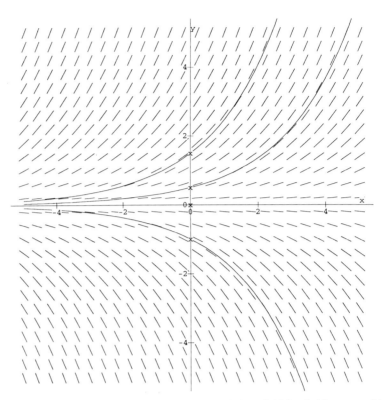

FIGURE 2.8 Four analytical solution curves and slope field for $dy/dx = ay$ with $a = 1$

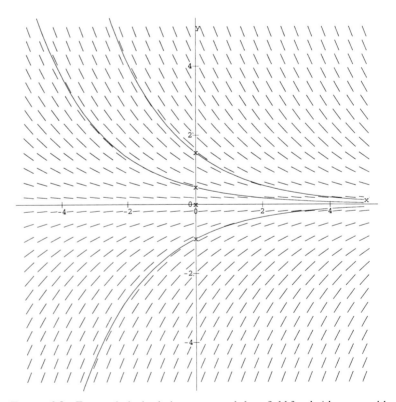

FIGURE 2.9 Four analytical solution curves and slope field for $dy/dx = ay$ with $a = -1$

2. This exercise deals with the differential equation

$$\frac{dy}{dx} = 1 + y^2.$$

   (a) Use graphical methods (calculus, slope field, isoclines) to sketch some solution curves.
   (b) Use analytical methods to solve the differential equation for $y(x)$.
   (c) Confirm that parts (a) and (b) are consistent.

3. This exercise deals with the differential equation

$$\frac{dy}{dx} = y^2.$$

   (a) Use graphical methods (calculus, slope field, isoclines) to sketch some solution curves.
   (b) Use analytical methods to solve the differential equation for $y(x)$.
   (c) Confirm that parts (a) and (b) are consistent.

4. This exercise deals with the differential equation

$$\frac{dy}{dx} = y^{5/2}.$$

   (a) Use graphical methods (calculus, slope field, isoclines) to sketch some solution curves.
   (b) Use analytical methods to solve the differential equation for $y(x)$.
   (c) Confirm that parts (a) and (b) are consistent.

5. This exercise deals with the differential equation

$$\frac{dy}{dx} = y^{3/2}.$$

   (a) Use graphical methods (calculus, slope field, isoclines) to sketch some solution curves.
   (b) Use analytical methods to solve the differential equation for $y(x)$.
   (c) Confirm that parts (a) and (b) are consistent.

6. This exercise deals with the differential equation

$$\frac{dy}{dx} = \sqrt{1 - y^2}.$$

   (a) Use graphical methods (calculus, slope field, isoclines) to sketch some solution curves.
   (b) Use analytical methods to solve the differential equation for $y(x)$.
   (c) Confirm that parts (a) and (b) are consistent.

7. Find the family of explicit solutions for each of the following two differential equations

$$\frac{dy}{dx} = e^y$$

and

$$\frac{dy}{dx} = -e^{-y}.$$

Graph these two families of solutions on one plot, using the same scale for the x- and y-axes. What do you notice about the angle of intersection between these two families of solutions? Could you have seen that directly from the differential equations, without solving them?

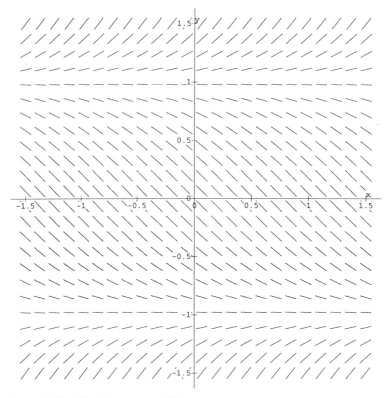

FIGURE 2.10   Identify the slope field

8. Solve

$$\frac{dy}{dx} = 1 - y,$$

(a) subject to the initial condition $y(0) = 0$.

(b) subject to the initial condition $y(0) = 1$.

9. Consider the following four differential equations.

  i. $\dfrac{dy}{dx} = y + 1$.

 ii. $\dfrac{dy}{dx} = y - 1$.

iii. $\dfrac{dy}{dx} = \ln|y|$.

 iv. $\dfrac{dy}{dx} = y^2 - 1$.

(a) The slope field of one of these equations is given in Figure 2.10. Match the correct equation with the figure, carefully stating your reasons. Do not plot the slope fields for i through iv to answer this question.

(b) Briefly outline a general strategy for this matching process.

10. Figures 2.11 and 2.12 are mystery slope fields, believed to be the slope fields for two of the following differential equations

$$\frac{dy}{dx} = \frac{y^2 + 1}{y^2 - 1}, \quad \frac{dy}{dx} = \frac{y^2 - 1}{y^2 + 1}, \quad \frac{dy}{dx} = -\frac{y^2 + 1}{y^2 - 1}, \quad \frac{dy}{dx} = -\frac{y^2 - 1}{y^2 + 1}.$$

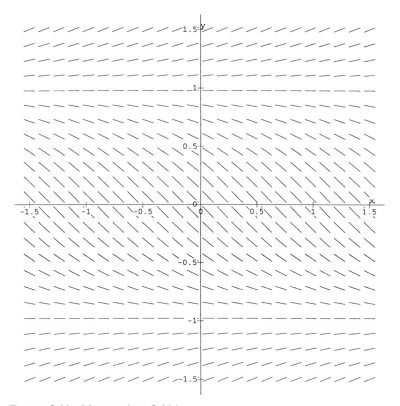

FIGURE 2.11 Mystery slope field 1

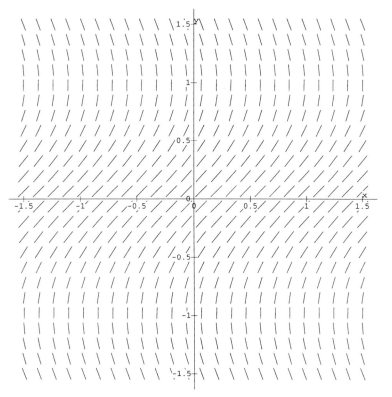

FIGURE 2.12 Mystery slope field 2

(a) Identify to which differential equation each of the mystery slope fields belongs. Confirm all the information using calculus and isoclines.

(b) Now superimpose the two mystery slope fields, perhaps by placing one on top of the other, or by plotting the slope fields for

$$\frac{dy}{dx} = a\frac{y^2 + 1}{y^2 - 1} + b\frac{y^2 - 1}{y^2 + 1}$$

with $a = \pm 1$ and $b = 0$, and then with $a = 0$ and $b = \pm 1$. What do you notice? Would this have made your previous analysis in part (a) easier?

11. Can the slope field of the differential equations discussed in Chapter 1 — namely, $dy/dx = g(x)$ — be symmetric about the x-axis? If your answer is yes, prove it. If your answer is no, give an example where it is not symmetric.

12. Can the slope field of the autonomous differential equation $dy/dx = g(y)$, be symmetric about the y-axis? If your answer is yes, prove it. If your answer is no, give an example where it is not symmetric.

2.2   Simple Applications

In this section we consider two examples that illustrate how the differential equation $dy/dx = ay$ is applied; to human populations and to the strength of a drug in the blood stream. We will also discover how to use real data sets to estimate numerical values for parameters in a differential equation.

EXAMPLE 2.3: The Population of Botswana (1975–1990)

The population of Botswana[1] from 1975 to 1990 is shown in Table 2.1 and Figure 2.13. It is claimed that this data set is consistent with the model where the population growth per unit population is equal to a constant. Let y be the population in millions at time x in years. Because the population growth per unit population is the quantity $(1/y)dy/dx$, we have the differential equation (2.11) studied before, namely,

$$\frac{1}{y}\frac{dy}{dx} = k,$$

where k is a positive constant (the growth rate). We know the solution of this differential equation [see (2.18)] is

$$y(x) = Ce^{kx}, \tag{2.19}$$

where C is a constant and $x = 0$ corresponds to the year 1975.

How do we estimate values for C and k from the data set? If the data set is approximated by a line then we could fit the data set by eye and estimate the slope and intercept. Unfortunately these data are not linear. However, if we take the logarithm of (2.19) we find

$$\ln y(x) = \ln C + kx.$$

---

[1]"World Population Growth and Aging" by N. Keyfitz, University of Chicago Press, 1990, page 118.

TABLE 2.1
Population of Botswana

Year	Population (millions)
1975	0.755
1980	0.901
1985	1.078
1990	1.285

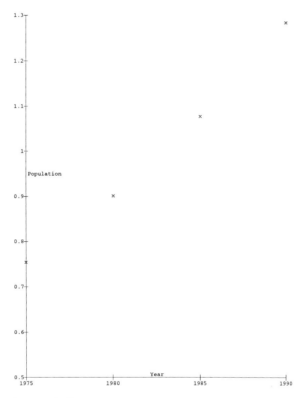

FIGURE 2.13   Population of Botswana

[Note, this is just (2.15).] Consequently, if we plot $\ln y(x)$ versus x, we would have a straight line[2] with slope k and y-intercept $\ln C$, from which we can estimate both C and k. This is done in Figure 2.14, from which we estimate $\ln C \approx -0.281$ (so $C \approx 0.755$) and $k \approx 0.0355$.

Figure 2.15 shows the population data as well as the function

$$y(x) = 0.755e^{0.0355x} \tag{2.20}$$

(adjusted so $x = 0$ corresponds to the year 1975), which seem to be in good agreement.

Let's see where this model leads. It predicts that the population of Botswana is doubling about every 20 years (see Exercise 1 on page 49), compared with the world population's doubling about every 34 years. According to this model, the population of Botswana will increase to about 25 million by 2075. If we continue in this way, we find

---

[2]Throughout the text we will be fitting straight lines through data sets. This can be done by eye, or by using the least squares method described in the appendix. We will consistently use the latter without further reference to it.

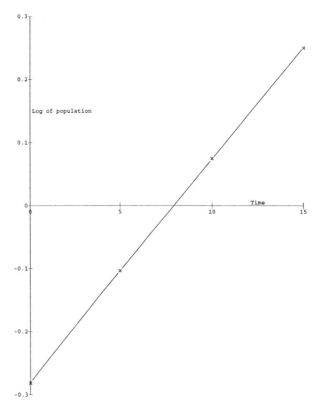

FIGURE 2.14   Log of population of Botswana versus time

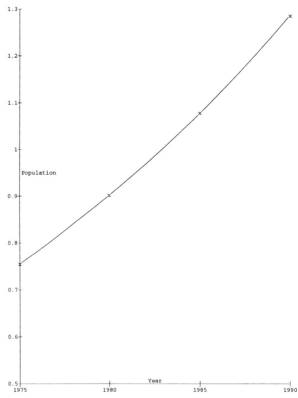

FIGURE 2.15   Actual and theoretical population of Botswana

that, about 160 years from 1975, the population of Botswana will be about 255 million, which exceeds the 1990 population of the entire United States. However, the area of Botswana is about that of the state of Texas. Although this model seems reasonable for 1975–1990, it is unlikely to be valid forever. □

In the next section we will develop an alternative model for population growth that will predict bounded populations for all time.

EXAMPLE 2.4: Administering Drugs[3]

Table 2.2 shows the concentration of theophylline, a common asthma drug, in the blood stream as a function of time after injection of a 300 mg initial dose given to a subject weighing 50 kg. Figure 2.16 shows these data plotted against time.

TABLE 2.2   Concentration of theophylline in the blood

Time (hours)	Concentration (mg/l)
0	12.0
1	10.0
3	7.0
5	5.0
7	3.5
9	2.5
11	2.0
13	1.5
15	1.0
17	0.7
19	0.5

Figure 2.17 shows the logarithm of the concentration plotted against time with a straight line fit. This suggests that the concentration $y(x)$ at time x might decay exponentially and so is governed by the differential equation

$$\frac{dy}{dx} = -ky, \tag{2.21}$$

with solution

$$y(x) = Ce^{-kx}.$$

Figure 2.17 allows us to estimate $\ln C$ because it is the y-intercept. Thus we have $\ln C \approx 2.485$, so $C \approx 12$. Similarly $-k$ is the slope so $k \approx 0.167$. This gives

$$y(x) = 12e^{-0.167x}. \tag{2.22}$$

Figure 2.18 shows these data plotted against time with the exponential function (2.22). □

---

[3]Based on "Applying Mathematics" by D.N. Burghes, I. Huntley, and J. McDonald, Ellis Horwood 1982, page 125.

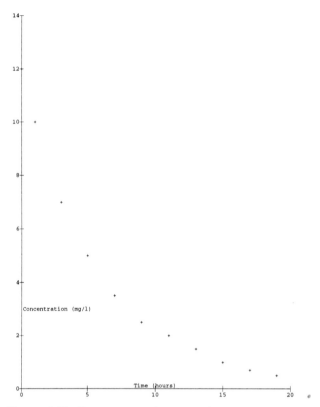

FIGURE 2.16   Drug concentration versus time

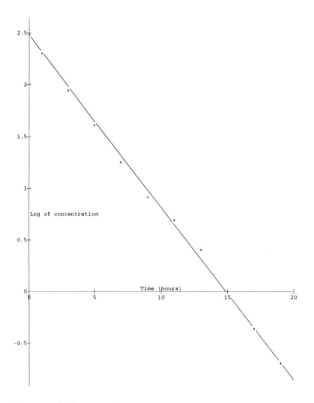

FIGURE 2.17   Log of drug concentration versus time with straight line fit

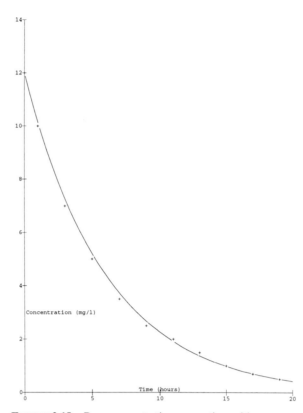

FIGURE 2.18 Drug concentration versus time with exponential fit

EXERCISES

1. A measure of exponential growth, $y(x) = Ce^{kx}$ where $k > 0$, is the DOUBLING TIME T_d — the time it takes for the value of y at any particular time x_0, that is, $y(x_0)$, to double to $2y(x_0)$. Show that $T_d = \ln 2/k$. Notice that it is independent of $y(x_0)$.

2. There are various measures[4] of exponential decay, $y(x) = Ce^{kx}$ where $k < 0$.

   (a) The HALF-LIFE, T_h — the time it takes for the value of y at any particular time x_0, that is, $y(x_0)$, to halve to $y(x_0)/2$. Show that $T_h = -\ln 2/k$. Notice that it is independent of $y(x_0)$.

   (b) The TIME CONSTANT, T_c — the time where the tangent through the point $(0, y(0))$ crosses the x-axis. Show that $T_c = -1/k$. Confirm that the half-life is about 70% of the time constant. In some disciplines, the value of y at five time constants is regarded as zero. What is the value of y at five time constants?

   (c) The SETTLING TIME, T_s — the time after which the value of y never exceeds 1% of the maximum value of y. Show that, for exponential decay, $T_s = -2\ln 10/k$. Confirm that the settling time lies between four and five time constants.

   (d) Figure 2.19 shows the graph of an exponentially decaying function $y = y(x)$. From the preceding definitions identify the half-life, the time constant, and the settling time for $y = y(x)$. [Do this geometrically. Do not attempt to identify the function.]

   (e) Example 2.4 deals with administering the drug theophylline. Using (2.22), namely, $y(x) = 12e^{-0.167x}$, evaluate the half-life, the time constant, and the settling time for theophylline.

   (f) What is the settling time for $y(x) = xe^{-x}$?

---

[4]The half-life and the time constant apply only to exponential decay. The settling time is more general and applies to any quantity that has the property that $y \to 0$ as $x \to \infty$.

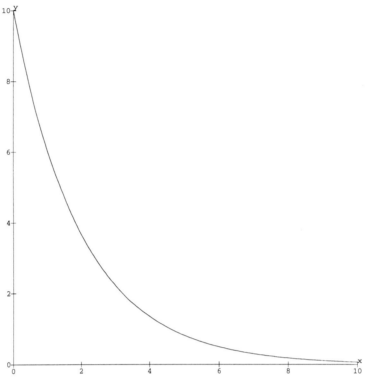

FIGURE 2.19   Exponentially decaying function

3. In 1800 the world's population was approximately 1 billion, while in 1900 it was 1.7 billion. If the population P at time t obeys the differential equation

$$\frac{dP}{dt} = kP,$$

estimate the world's population in the year 2000. Do you think this is a good estimate? What is the doubling time?

4. A culture of bacteria doubles its size every 6 hours. How many hours will it take to be 5 times its original size?

5. At time $t = 0$ a bacteria culture has N_0 bacteria. One hour later the population has grown by 25%. If the population P at time t obeys the differential equation

$$\frac{dP}{dt} = kP,$$

how long will it take the population to double?

6. After administration of a dose, the concentration of a drug in one's body decreases by 50% in 10 hours. How long will it take for the concentration of the drug to reach 10% of its original value, if the concentration C at time t obeys the differential equation

$$\frac{dC}{dt} = kC?$$

7. The land area of Botswana is 220,000 square miles (about the size of Texas). Show that based on the model characterized by (2.20), the time when the population density of Botswana will be one person per square foot will occur in about 450 years from 1975.

8. Global Warming.[5] Table 2.3 shows the rise in temperature (in ^{o}C) of the earth (from the 1860 figure) as a function of time since 1900. Figure 2.20 shows these data plotted against time. How well does the exponential model fit this data set?

TABLE 2.3
Temperature rise of earth
over 1860 figure

Year	Temperature ^{o}C
1900	0.03
1910	0.04
1920	0.06
1930	0.08
1940	0.10
1950	0.13
1960	0.18
1970	0.24

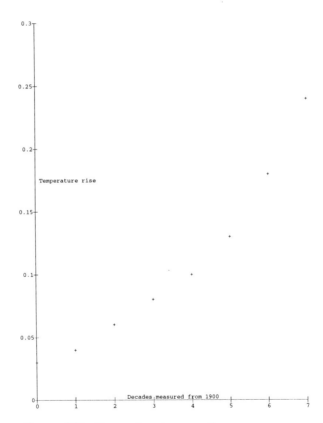

FIGURE 2.20   Temperature rise versus time

---

[5]Based on "Applying Mathematics" by D.N. Burghes, I. Huntley, and J. McDonald, Ellis Horwood 1982, page 175.

9. A bullet passes through a piece of wood 10 cm thick. It enters the wood at 200 m/sec, and it leaves at 80 m/sec. It is assumed that the velocity v of the bullet while in the wood obeys the differential equation

$$\frac{dv}{dt} = -kv^2$$

where k is a given positive constant. How long did it take the bullet to pass through the wood? How thick should the wood be to bring the bullet to rest? Does this differential equation produce a realistic model?

10. Torricelli's law.[6] Consider the following experiment. An empty 2 liter clear plastic soda bottle is rinsed, and a small hole (about 1/8 inch in diameter with no burrs) is made in the side, near the bottom of the bottle. A tape measure is attached vertically to the bottle (with rubber bands), with 0 cm set at the position of the hole. The bottle is filled with water, and the hole is covered by a finger. The finger is removed, and the time that the water level passes each centimeter mark on the measuring tape is recorded.

(a) Table 2.4 shows the results of such an experiment, which are also shown in Figure 2.21. It is claimed that $h(t)$, the height of the water at time t, is governed by TORRICELLI'S LAW,

$$\frac{dh}{dt} = a\sqrt{h},$$

where a is a constant. Should a be positive or negative? Solve this differential equation for $h(t)$, and see how well the data set in Table 2.4 fits Torricelli's law.

(b) Conduct your own experiment, and see what sort of agreement you get with Torricelli's law.

TABLE 2.4 Height of water as a function of time

Height (cm)	Time (sec)
10	17.3
9	29.0
8	41.3
7	53.7
6	67.7
5	83.5
4	101.0
3	120.7
2	146.5
1	179.7

11. It is known from experiments that theophylline has hardly any therapeutic effect if its concentration is below 5 mg/l and that concentrations above 20 mg/l are likely to be toxic. The problem is to administer the drug to the individual in Example 2.4 in such a way that the concentration remains between 5 and 20 mg/l. We want to do this by administrating equal doses of theophylline at equal intervals of time. What advice should we give the doctor, assuming that the concentration is governed by the differential equation (2.21) with $k = 0.167$?

---

[6]Based on "Modelling with Differential Equations" by D.N. Burghes and M.S. Borrie, Ellis Horwood 1981, page 59.

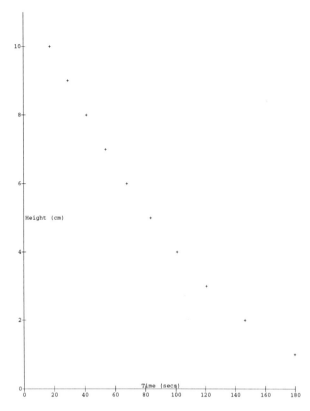

FIGURE 2.21   Water height versus time

12. From your own experience, or a recent newspaper article or magazine, develop or find a data set, of at least five data points, that exhibits exponential growth. Some possibilities include the out-of-state tuition at your local college, the weight of your new puppy, the cost of mailing a first-class letter. Comment on how long this exponential growth could continue. Now repeat this for a data set that exhibits exponential decay.

2.3   The Logistics Equation

Using $y' = ay$ to model the growth of any substance y always gives an exponential solution which is unbounded as $x \to \infty$. Thus, this model does not yield realistic results over long periods of time for substances whose growth is bounded. In our next example we consider the growth of sunflowers, which is clearly limited.

EXAMPLE 2.5: Growth of Sunflower Plants[7]

Table 2.5 shows the height of sunflower plants, y, as a function of time, x, and we have plotted this data set in Figure 2.22. How can we explain this?

From this figure, we see that while for small values of time the height increases rapidly, for larger values of time the increases diminish. In fact, the last two data points show very little change in height.

---

[7]"Growth of Sunflower Seeds" by H.S. Reed and R.H. Holland, Proc. Nat. Acad. Sci., **5**, 1919, page 140.

TABLE 2.5   Height of sunflower plants

Time (days)	Height (cm)
7	17.93
14	36.36
21	67.76
28	98.10
35	131.00
42	169.50
49	205.50
56	228.30
63	247.10
70	250.50
77	253.80
84	254.50

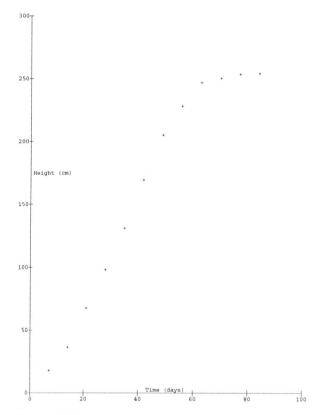

FIGURE 2.22   Height of sunflower versus time

One way to account for this growth behavior is to add a term to the right-hand side of $y' = ay$ that would decrease the growth rate for large values of y. This should be done in such a way that the rate of change of height per unit height is decreasing as the height increases. The equation

$$\frac{1}{y}\frac{dy}{dx} = a(b - y),$$

where a and b are given positive constants, describes such a situation. We can rewrite this differential equation in the form

$$\frac{dy}{dx} = ay(b - y). \tag{2.23}$$

Equation (2.23) is called the LOGISTICS DIFFERENTIAL EQUATION, and it is used commonly for population models where the population's growth rate decreases with increasing population due to factors such as limited food supply, overcrowding, and disease.

Let's consider the construction of the slope field for (2.23). Because we are dealing with heights of sunflower plants, we consider only nonnegative values of y. As we use (2.23) to evaluate the slopes of solution curves at equally spaced points (x, y), we see that along the horizontal lines $y = 0$ (the x-axis) and $y = b$, the slopes are all 0. Along the horizontal line $y = 1$ the slopes are all $a(b - 1)$. Along the horizontal line $y = 2$ the slopes are all $2a(b - 2)$, and so on. The slope field for (2.23), with $a = 1$ and $b = 2$, is shown in Figure 2.23. Because we are concerned only with $x \geq 0$ and $y \geq 0$, we don't look for symmetries.

The isoclines for (2.23) are given by

$$ay(b - y) = m. \tag{2.24}$$

With $m = 0$ we see that the only places where the solution curves have horizontal tangents are along the horizontal lines

$$y = 0, \text{ and } y = b.$$

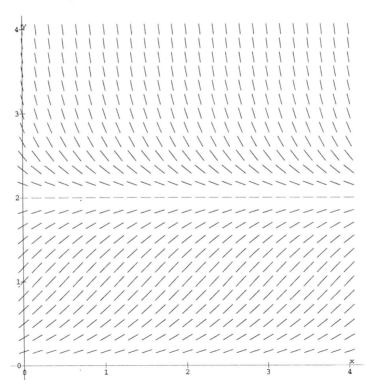

FIGURE 2.23 Slope field for $dy/dx = ay(b - y)$ with $a = 1, b = 2$

To consider other values of m we rewrite (2.24) as a quadratic equation in y,

$$y^2 - by + \frac{m}{a} = 0,$$

and use the quadratic formula to solve for our isoclines as

$$y = \frac{1}{2}\left(b \pm \sqrt{\frac{ab^2 - 4m}{a}}\right), \qquad \textbf{(2.25)}$$

where m is the specified slope. This tells us that when $ab^2 - 4m < 0$ — that is, $m > ab^2/4$ — there are no solution curves with slope m. In other words the only solution curves of (2.23) with slope m occur if m is in the interval $-\infty < m \leq ab^2/4$. Equation (2.25) also tells us that the maximum slope, $m = ab^2/4$, will occur at $y = b/2$, halfway (in the y direction) between the horizontal lines $y = 0$ and $y = b$. The slope field shown in Figure 2.23 is consistent with this information.

The slope field also suggests that all solutions of (2.23) are either increasing or decreasing depending on the initial value of y. This may be verified from (2.23), namely,

$$\frac{dy}{dx} = ay(b - y),$$

in which the derivative is positive if $0 < y < b$ and negative if $y > b$. Thus, if we start with an initial value of y_0 at $x = 0$, where $y_0 > b$, the solution will decrease to the limiting value b. Likewise, if we start with an initial value of y_0, where $0 < y_0 < b$, the solution will increase toward its limiting value b. In contrast to the exponential growth model for populations, all solutions of the logistics differential equation for $y_0 > 0$ will approach the value of b for large values of time. This limiting value of b is often called the CARRYING CAPACITY of a specific population for situations where y denotes a population.

There are two specific initial conditions for which the preceding analysis breaks down — namely, $y_0 = 0$ and $y_0 = b$. If $y_0 = b$, then from the differential equation we see that its rate of change is always 0 (recall that $y = b$ was an isocline for slope 0). Because $y(x) = b$ is a solution of (2.23) for all values of time, it is an equilibrium solution. Note that $y(x) = 0$ is also an equilibrium solution, and it makes sense that if we have a process in which the rate of change is proportional to the number present, and we start with 0 present, we will stay at 0. Likewise, as seen from the slope field, for any $y_0 > 0$ all the solutions tend toward b as time increases, so if we start at this limiting value of b, there will be no change in the population.

Figure 2.24 shows an enlargement of the slope field between the two equilibrium solutions, where we see that there are inflection points. To discover the exact location of the inflection points, we differentiate the original differential equation (2.23) to obtain

$$\frac{d^2y}{dx^2} = a\left[\frac{dy}{dx}(b - y) + y\left(-\frac{dy}{dx}\right)\right] = a(b - 2y)\frac{dy}{dx} = a^2y(b - 2y)(b - y).$$

There are three places where this second derivative is zero, two coinciding with the equilibrium solutions and the third when $b = 2y$. Therefore, the inflection point must occur when

$$y = \frac{b}{2},$$

that is, at one-half the carrying capacity b.

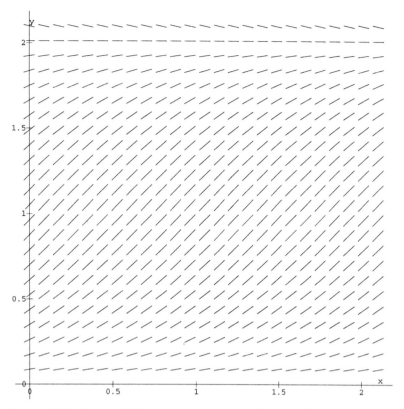

FIGURE 2.24 Slope field for $dy/dx = ay(b - y)$ with $a = 1, b = 2, 0 < y < 2$

Because y is monotonically increasing in $0 < y < b$, we have just discovered that solutions of the logistics equation for $y > 0$ will have an inflection point only when

$$0 < y_0 < \frac{b}{2}.$$

For all other situations the solution curves are either increasing and concave down (for $b/2 < y_0 < b$) or decreasing and concave up (for $y_0 > b$). Some hand-drawn solution curves for the logistics equation are shown in Figure 2.25. Note that we have obtained many essential features of solutions of the logistics equation without obtaining an explicit solution.

However, we can find explicit solutions of the logistics differential equation

$$\frac{dy}{dx} = ay(b - y). \tag{2.26}$$

First we observe that there are two equilibrium solutions, namely,

$$y(x) = 0 \text{ and } y(x) = b.$$

If we concentrate on the nonequilibrium solutions, from (2.26) we find

$$\int \frac{dy}{y(b - y)} = \int a\, dx,$$

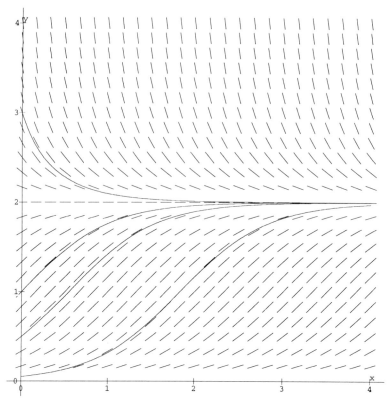

FIGURE 2.25   Hand-drawn solution curves and slope field for $dy/dx = ay(b - y)$ with $a = 1, b = 2, 0 < y < 4$

which, by partial fractions,[8] leads to the equation

$$\int \frac{1}{b} \left(\frac{1}{y} + \frac{1}{b - y} \right) dy = \int a \, dx.$$

After integration this becomes

$$\ln |y| - \ln |b - y| = abx + c,$$

where c is the constant of integration. Using the properties of logarithmic functions, we can rewrite this as

$$\ln \left| \frac{y}{b - y} \right| = abx + c. \tag{2.27}$$

This is an implicit solution of (2.26). In this case it is possible to solve (2.27) for y to obtain an explicit solution. Taking the exponential of each side of (2.27) we have

$$\frac{y}{b - y} = Ce^{abx}, \tag{2.28}$$

---

[8] A summary of techniques of integration is in the appendix.

where $C = \pm e^c \neq 0$. Now we multiply (2.28) by $b - y$ and solve for y to find

$$y(x) = \frac{bCe^{abx}}{1 + Ce^{abx}}.$$

This family of solutions can be rewritten in the alternative form

$$y(x) = \frac{bC}{e^{-abx} + C}. \tag{2.29}$$

Notice that one equilibrium solution, $y(x) = 0$, can be absorbed into (2.29) if we let $C = 0$, but that the other one, $y(x) = b$, is not contained in (2.29) for any finite choice of C.

If we are given the initial condition $y(0) = y_0$, then we can read off the value of the constant C from (2.28), namely,

$$C = \frac{y_0}{b - y_0},$$

which can be substituted into (2.29) to give

$$y(x) = \frac{by_0}{(b - y_0)e^{-abx} + y_0}. \tag{2.30}$$

This is the explicit solution of (2.26).

Let's look at this solution (2.30) a little more carefully for the physically meaningful case when $0 < y_0 < b$. We see that $y(x) \to b$ as $x \to \infty$, in complete agreement with our previous graphical analysis, which suggested b was the carrying capacity of the population.

By dividing (2.30) by b, we find

$$y(x) = \frac{y_0}{(1 - y_0/b)e^{-abx} + y_0/b}.$$

From this equation we can see that if the carrying capacity b is large compared with the initial population y_0, so that y_0/b can be neglected when compared to 1, then $y \approx y_0 e^{abx}$ for small x. In other words, in this case, the initial growth of the logistics model is approximately exponential. Consequently, at the beginning of the growth of a plant or a population, it is difficult to distinguish between exponential and logistical growth.

Some analytical solution curves for the logistics equation are shown in Figure 2.26. There is good agreement with our hand-drawn solutions in Figure 2.25. $\qquad \square$

The two equilibrium solutions in this last example — namely, $y(x) = b$ and $y(x) = 0$ — have different characteristics. If we start near the equilibrium solution $y(x) = b$, we move closer to $y = b$ as x increases. Such an equilibrium solution is called a **stable equilibrium solution**. If we start near the equilibrium solution $y(x) = 0$, we move further from $y = 0$ as x increases. Such an equilibrium solution is called an **unstable equilibrium solution**.

This leads to the following definition.

Definition 2.3: **Let the differential equation**

$$\frac{dy}{dx} = g(x, y)$$

have an equilibrium solution, that is, a solution of the form $y(x) = constant$. An equilibrium solution is called STABLE if all solutions of the differential equation that start near this equilibrium solution remain near this equilibrium solution as $x \to \infty$. An equilibrium

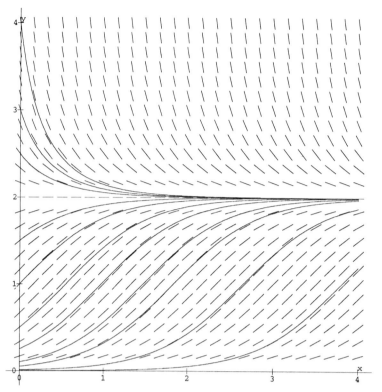

FIGURE 2.26 Analytical solution curves and slope field for $dy/dx = ay(b - y)$ with $a = 1, b = 2$

solution is called UNSTABLE **if all other solutions of the differential equation that start near this equilibrium solution move away from this equilibrium solution as $x \to \infty$. If an equilibrium solution is neither stable nor unstable, it is called** SEMISTABLE.

Let us now return to the sunflower data set and see if the explicit solution (2.29), namely,

$$y(x) = \frac{bC}{e^{-abx} + C},\qquad(2.31)$$

where C is an arbitrary constant, gives a good fit. Notice that there are three unknowns in the solution — namely, a, b, and C. Unlike the last two examples involving exponential growth and decay, there is no way to manipulate (2.31) to construct a linear function involving three arbitrary constants. (Why?) However, from an intermediate step of the solution, (2.27), that is,

$$\ln\left[\frac{y(x)}{b - y(x)}\right] = abx + c,$$

where $e^c = C$, we see that if we can estimate b, the carrying capacity, we could plot $\ln\left[y/(b - y)\right]$ against x to see whether the data are approximately linear. If they are, we can then estimate ab and c.

There are two ways we can estimate b: directly, by estimating the value of y where the logistics curve levels off, in which case b equals this value; or indirectly, by estimating the value of y where the curve has an inflection point, in which case b equals twice this value. (Remember, for the logistics equation, the point of inflection occurs at $b/2$.) Looking at

Figure 2.22 on page 54, we can try to estimate b based on these two estimates. It appears that $b \approx 260$ is reasonable from both points of view. Figure 2.27 shows a plot of $\ln[y/(260-y)]$ against x, together with a straight line fit. From this we estimate $ab \approx 0.087$, because ab is the y-intercept. Similarly, c is the slope, so $c \approx -3$ and thus, $C = e^c \approx 0.0498$. In this case, the height of the sunflower over time is given by

$$y(x) = \frac{(260)(0.0498)}{e^{-0.087x} + 0.0498} = \frac{12.94}{e^{-0.087x} + 0.0498}. \tag{2.32}$$

Figure 2.28 shows the data set plotted against time with the function (2.32).

EXERCISES

1. **Change of Scale.** Show that the logistics differential equation

$$\frac{dy}{dx} = ay(b-y),$$

where a and b are positive constants, can be transformed into

$$\frac{dY}{dX} = Y(1-Y)$$

by the substitutions $x = X/(ab)$, $y = bY$. Solve this differential equation for $Y = Y(X)$. Now substitute for X and Y in terms of the original variables x and y. Compare your analysis to the derivation of (2.29) from (2.26). Comment on this process.

2. In each of the following cases, imagine that you are given the solution of the first differential equation involving X and Y. Explain how you could immediately write down the solution of the second differential equation involving x and y. (Constants a and b are positive.)

   (a) $\dfrac{dY}{dX} = Y$ $\qquad$ $\dfrac{dy}{dx} = ay$

   (b) $\dfrac{dY}{dX} = X$ $\qquad$ $\dfrac{dy}{dx} = ax$

   (c) $\dfrac{dY}{dX} = -Y \ln Y$ $\qquad$ $\dfrac{dy}{dx} = -by \ln \dfrac{y}{a}$

   (d) $\dfrac{d^2Y}{dX^2} \pm Y = 0$ $\qquad$ $\dfrac{d^2y}{dx^2} \pm a^2 y = 0$

3. A differential equation that arises in the study of a chemical of concentration y (as a function of time x) that is undergoing both first and second order reactions is

$$\frac{dy}{dx} = ay - by^2,$$

   subject to the initial condition $y(0) = y_0$. (The quantities a and b are constants.)

   (a) Find the solution of this differential equation assuming $a \neq 0$ and $b \neq 0$.

   (b) Find the solution of this differential equation assuming $a \neq 0$ and $b = 0$. Is this solution contained in part (a)?

   (c) Find the solution of this differential equation assuming $a = 0$ and $b \neq 0$. Is this solution contained in part (a)?

   (d) Find the equilibrium solutions of this differential equation. Could they be included in the results of parts (a), (b), and (c)?

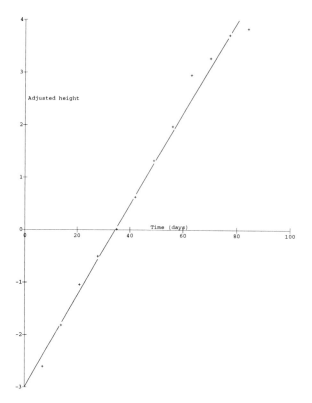

FIGURE 2.27   Adjusted height of sunflower versus time

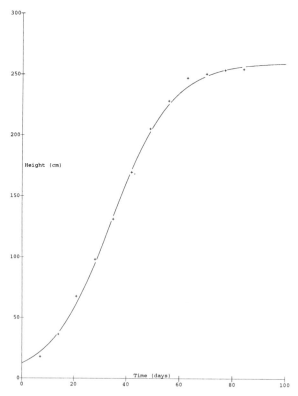

FIGURE 2.28   Height of sunflower versus time with logistics fit

4. Of 10,000 companies, 100 have adopted a new development at time $t = 0$. If the number, $y(t)$, of companies that have adopted the innovation at time t (units of years) satisfies the differential equation

$$y' = 0.005y(10,000 - y),$$

find the number of companies that can be expected to adopt the innovation after 10 years.

5. The University of Arizona has 35,000 students. On the first day of the semester, a class of 35 students thought they heard their mathematics professor say that everyone would receive the grade of A for the course. The next day a carefully conducted survey of the entire student population showed that by now 700 students had heard this rumor. If the rumor continues to spread according to the logistics equation

$$\frac{dy}{dt} = ay(b - y),$$

at what time will 90% of the students be aware of this rumor?

6. Suppose there is a homogeneous group of N individuals which at time t (measured in days) consists of x individuals who are susceptible to infection — say by some virus — and y individuals who are already infective. Also assume that there is no removal from circulation by recovery, isolation, or death. (This would be the situation for the early stages of an upper respiratory infection.) This means we have $x + y = N$. If we assume that the rate of change of infected individuals is proportional to both the number of infected individuals and the number of susceptible individuals, we have the differential equation

$$\frac{dy}{dt} = kxy = k(N - y)y.$$

If we assume that at time $t = 0$, one person becomes ill and the rest are not infected, our initial condition is $y(0) = 1$.

(a) By analyzing the differential equation directly, find the number of people who are infected at the time the infection has the most rapid rate of change.

(b) Solve the initial value problem for y.

(c) Public records for epidemics record the number of new cases appearing each day. The quantity is given by $-dx/dt$ and is called the epidemic curve. Compute this quantity and graph the result. What is the maximum value of this function? Show how you could have found this maximum value without solving the differential equation.

7. Observations on the growth of animal tumors indicate that the size y of the tumor obeys the differential equation

$$\frac{dy}{dt} = -ky \ln \frac{y}{b},$$

where k and b are positive constants. This differential equation is sometimes called the Gompertz growth law.

(a) Look at the slope field for this equation to obtain some idea on how solution curves behave. What sort of behavior does this remind you of?

(b) Determine regions in which the solutions of the differential equation are increasing, decreasing, concave up, or concave down.

(c) Solve this differential equation subject to the initial condition $y(0) = y_0$, $y_0 > 0$.

(d) Find a relation between y_0 and b such that the graph of y versus t will have no inflection point.

(e) Does this differential equation have any equilibrium solutions? If so, what are they? If not, explain why not.

(f) The solution of the logistics equation, $dy/dt = ay(b - y)$, was given in (2.30) as $by_0 / \left[(b - y_0) e^{-abt} + y_0 \right]$, where $y(0) = y_0$. Graph this solution and the solution of the Gompertz equation for the same initial condition, the same carrying capacity, and the same time for the inflection point. Describe the similarities and differences between the two graphs. Give some general guidelines as to when one model would be more appropriate to use than the other.

8. Experiments with water bugs suggest a model of population growth of a species that is limited by the food supply. If $P(t)$ represents the population of water bugs at time t, the basic equation says the rate of change of $P(t)$ is proportional to the difference between the available food, $A(t)$, and the food necessary for subsistence, $N(t)$, as well as being proportional to the current population — that is,

$$\frac{dP}{dt} = k \left(A(t) - N(t) \right) P(t).$$

It is assumed that the available food is a constant, $A(t) = a$, and that the food necessary for subsistence, $N(t)$, is equal to the sum of two terms. The first term gives the need for food as proportional to the existing population (bP), whereas the second term accounts for the need of more food during periods of rapid growth ($r\, dP/dt$). Under these assumptions the differential equation becomes

$$\frac{dP}{dt} = k \left(a - bP - r \frac{dP}{dt} \right) P,$$

where k, a, b, and r are positive constants. (If $r = 0$ we have the standard logistics equation.) We may rearrange this equation as

$$\frac{dP}{dt} = k \frac{aP - bP^2}{1 + krP}.$$

What can you tell about the behavior of the solution subject to an initial condition $P(0) = P_0$? (The slope field might be of some help here.) Does this behavior depend on the value of P_0? Please explain fully.

9. Using the values for C and k we obtained for the exponential model of population growth of Botswana, show that the logistics model could also approximate the population of Botswana with many different carrying capacities.

10. **Bacteria.**[9] Table 2.6 shows the size of a bacterial colony as a function of time. Figure 2.29 shows these data plotted against time. How well does the logistics model fit this data set?

TABLE 2.6   Size of
bacteria colony

Time (days)	Area (sq cm)
0	0.24
1	2.78
2	13.53
3	36.30
4	47.50
5	49.40

[9]"Growth of Bacterial Colony" by H.G. Thornton, Ann. Appl. Biol., 1922, page 265.

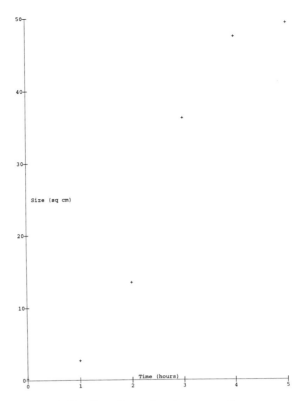

FIGURE 2.29   Size of bacteria colony versus time

11. **Yeast.**[10] Table 2.7 shows the amount of yeast as a function of time. Figure 2.30 shows these data plotted against time. How well does the logistics model fit this data set?

12. **Collared doves.**[11] During the first half of the 20th century, the collared dove spread across Europe from east to west. This bird was very rare in Britain before 1955, so rare that it is not even included in the *1952 Checklist of the Birds of Great Britain and Ireland*. Thus, the spread of the collared dove in Britain from 1955 was of great interest to bird watchers. Consequently, there are very good records of its spread throughout Britain from 1955 to 1964, when the dove was so plentiful that little attention was then paid to it. The population of the collared dove is believed to be directly proportional to the number of locations where it was resident. Table 2.8 shows the number of locations where it was resident each year. Figure 2.31 shows these data plotted against time. How well does the logistics model fit this data set?

13. Parents frequently record the height of a child as a function of time. This is usually not regarded as a scientific experiment, so very accurate measurements are seldom taken. Cynthia's parents are no different from anyone else. Cynthia was born in September 1972, and Table 2.9 shows Cynthia's approximate height (in inches) from age 1 to age 12, when she thought she was too grown-up for such childishness. Figure 2.32 shows these data plotted against time. How well does the logistics model fit this data set? What would you estimate Cynthia's height to be today?

    (a) A rule of thumb, used by pediatricians to estimate the ultimate height of a growing child, is "Take the height at age 2 and double it." Assuming that height as a function of time obeys the logistics model, discuss what this tells you about the inflection point.

[10]"Growth of Yeast" by T. Carlson, Biochemische Zeitschrift, **57**, 1913, page 313.

[11]"The spread of the Collared Dove in Britain and Ireland" by R. Hudson, British Birds, **58** No 4, April 1965, pages 105–139.

TABLE 2.7   Amount of yeast population

Time (hours)	Amount
0	9.6
1	18.3
2	29.0
3	47.2
4	71.1
5	119.1
6	174.6
7	257.3
8	350.7
9	441.0
10	513.3
11	559.7
12	594.8
13	629.4
14	640.8
15	651.1
16	655.9
17	659.6
18	661.8

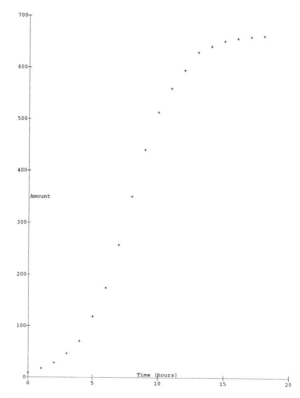

FIGURE 2.30   Amount of yeast versus time

(b) Try to obtain your own data for someone's height against time (preferably your own). How well does the logistics model fit it?

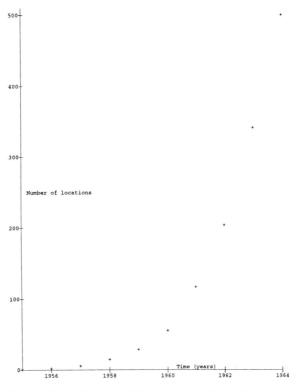

FIGURE 2.31 Number of locations of collared doves by year

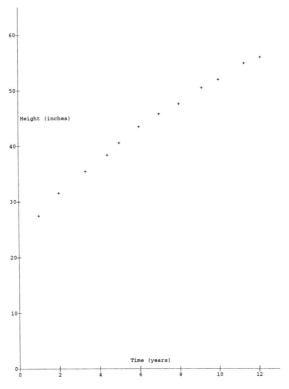

FIGURE 2.32 Cynthia's height by year

TABLE 2.8 Number of locations of collared doves

Year	Total
1955	1
1956	2
1957	6
1958	15
1959	29
1960	58
1961	117
1962	204
1963	342
1964	501

TABLE 2.9 Cynthia's height

Date	Height (in.)
September 1973	29.5
September 1974	31.6
January 1976	35.3
February 1977	38.4
September 1977	40.6
September 1978	43.5
September 1979	45.8
September 1980	47.6
November 1981	50.5
September 1982	52.0
January 1984	54.9
October 1984	56.0

TABLE 2.10 Beans' height

Time (days)	Height (in.)
10	0.125
11	0.250
12	0.750
13	1.125
14	1.500
15	2.500
16	4.000
17	5.500
18	7.000
19	9.000
20	10.000
21	10.500
22	11.000
23	11.250
24	11.500
40	12.000

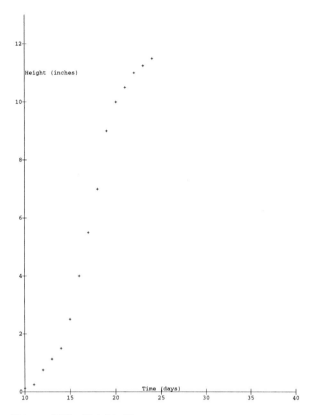

FIGURE 2.33   Height of beans

14. A student[12] placed beans in a styrofoam cup, covered them lightly with soil, and watered them regularly. On the 10th day sprouts emerged, and from then he carefully measured the height (in inches) as a function of time (in days). Table 2.10 shows the approximate height of the beans. Figure 2.33 shows these data plotted against time. How well does the logistics model fit this data set? Conduct your own experiment, by growing your own beans, and recording the height as a function of time. Compare your data set to the logistics model.

15. Construct a simple experiment that will result in a graph that resembles that of a solution of the logistics equation. See if you can give rational arguments as to why the logistics equation is a reasonable model for your experiment. [Hint: Two examples of such experiments are (a) the number of kernels of popcorn popped in a popper as a function of time, and (b) the rise of moisture in a paper towel suspended vertically with an edge in water.]

2.4   Existence and Uniqueness of Solutions and More Words of Caution

In this section we consider examples where things can go wrong. We first look at an example which illustrates that the graphical analysis may not show everything about the solutions. This leads to a discussion of initial conditions and the uniqueness of solutions for an autonomous equation. Uniqueness guarantees that solutions do not touch. The last

---

[12] Ian Scott, University of Arizona.

three examples expose different problems that can occur because we fail to use all available information.

EXAMPLE 2.6

Let us look at what happens if the right-hand side of $y' = g(y)$ is a quadratic function of y, namely,

$$\frac{dy}{dx} = y^2. \tag{2.33}$$

The slope field for (2.33) is shown in Figure 2.34. Notice it is symmetric about the origin.

Equation (2.33) implies solution curves are increasing for $y \neq 0$. Also, by differentiating (2.33), we find

$$\frac{d^2y}{dx^2} = 2y\frac{dy}{dx} = 2y^3,$$

so that solution curves are concave up if $y > 0$ and concave down if $y < 0$. The slope field in Figure 2.34 is consistent with this information. Figure 2.35 shows some hand-drawn solution curves.

Now let's find the explicit solution. We can see that $y(x) = 0$ is an equilibrium solution. (Why?) If $y(x) \neq 0$, then (2.33) can be written as

$$y^{-2}\frac{dy}{dx} = 1,$$

which, after integration, yields

$$-\frac{1}{y} = x + C. \tag{2.34}$$

If we solve (2.34) for y, we find the family of solutions

$$y(x) = \frac{-1}{x + C}. \tag{2.35}$$

The family of solutions (2.35) does not contain the equilibrium solution $y(x) = 0$ for any finite C.

Notice that the denominator of (2.35) is 0 when $x + C = 0$, so that the function defined by (2.35) has a vertical asymptote at

$$x = -C.$$

For example, the solution that passes through the point $y(1) = 1$ is

$$y(x) = \frac{1}{2 - x},$$

which goes to $+\infty$ as x goes from 1 to 2. Don't forget that a solution cannot continue past its vertical asymptote. (Why?)

Also notice that as $x \to \infty$, the function given by (2.35) goes to 0. For example, the solution that passes through the point $y(-1) = -1$ is

$$y(x) = -\frac{1}{x + 2},$$

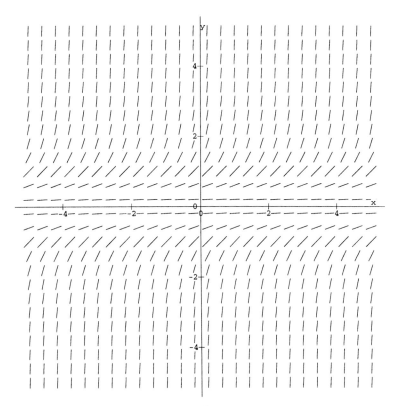

FIGURE 2.34   Slope field for $dy/dx = y^2$

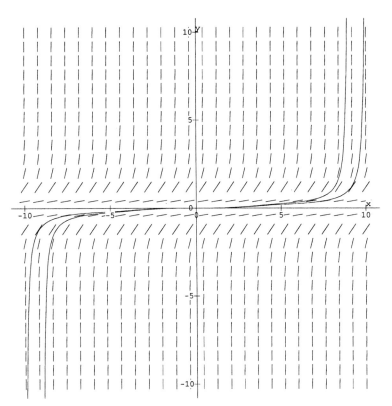

FIGURE 2.35   Hand-drawn solution curves and slope field for $dy/dx = y^2$

which goes to 0 (through negative values) as x goes from -1 to ∞.

We can evaluate the arbitrary constant in (2.35) for any initial condition. If our initial condition is $y(0) = y_0 \neq 0$, we have $C = -1/y_0$ and so

$$y(x) = \frac{-1}{x - 1/y_0}.$$

Thus, we have two cases: (a) if $y_0 > 0$, then $y(x) \to \infty$ as x approaches the finite value of $1/y_0$; (b) if $y_0 < 0$, then $y(x) \to 0$ (from below) as $x \to \infty$. However, if $y_0 = 0$, then we have only the equilibrium solution $y(x) = 0$.

Notice that when we drew Figure 2.35 we missed the vertical and horizontal asymptotes. The moral is that the **graphical analysis may not show everything**. Solutions need not go on forever.

Figure 2.36 shows the original slope field and some particular analytical solution curves passing through the initial points $(1, 1)$, $(-1, -1)$, $(3, 1/3)$, $(-3, -1/3)$, $(5, 1/5)$, and $(-5, -1/5)$, which demonstrate these properties.

Wait! It's not over yet. Carefully compare Figure 2.36 with Figure 2.35, which shows solution curves drawn by hand from the slope fields. However, something else is different, but what? Do you notice that in Figure 2.35 our hand-drawn curves cross the solution $y(x) = 0$, whereas the analytical solution curves in Figure 2.36 do not? Had we known before drawing our solutions by hand that these solutions do not cross, we would not have drawn the curves we did.   □

Words of Caution: Graphical analysis will not show whether a solution has a vertical asymptote at a finite value of x. Solutions need not go on forever.

In the last example we observed that the knowledge that solution curves cannot cross each other would have been very valuable when sketching solution curves. We now state an important theorem that guarantees that solution curves cannot touch each other.

Theorem 2.1: *If $g(x, y)$ and $\partial g / \partial y$ are defined* [13] *and continuous in a finite rectangular region containing the point (x_0, y_0) in its interior, then the differential equation*

$$\frac{dy}{dx} = g(x, y)$$

has a unique solution passing through the point

$$y(x_0) = y_0.$$

This solution is valid for all x for which the solution remains inside the rectangle.

Comments about Theorem 2.1

- As we mentioned in Chapter 1, solving a first order differential equation

$$\frac{dy}{dx} = g(x, y)$$

subject to an initial condition

$$y(x_0) = y_0$$

---

[13] $\partial g / \partial y$ is the partial derivative of $g(x, y)$ with respect to y and is obtained by differentiating $g(x, y)$ with respect to y, treating x as a constant.

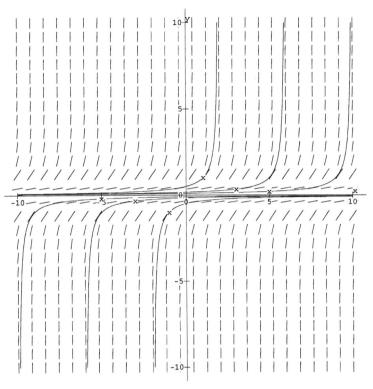

FIGURE 2.36   Analytical solution curves and slope field for $dy/dx = y^2$

is called an INITIAL VALUE PROBLEM.

- This theorem is an existence and uniqueness theorem because it guarantees that a solution exists and that there is only one solution. Theorem 2.1 says that if we want to solve the initial value problem

$$\frac{dy}{dx} = g(x, y)$$

subject to

$$y(x_0) = y_0,$$

then we are guaranteed that there is a solution curve that passes through this point (x_0, y_0) and no other solution curve can pass through this point, provided that $g(x, y)$ and $\partial g/\partial y$ are continuous in the vicinity of (x_0, y_0).

- This theorem gives no hints on how to find a solution. However, it guarantees that there is one to look for.

- The uniqueness part of this theorem implies that no matter how we find a solution, whether it be by diligence, luck, intuition, or skullduggery, it is THE solution.

- This theorem does **not** say that if $g(x, y)$ and $\partial g/\partial y$ are continuous for all x and y then there is a unique solution valid for all x and y. There is a unique solution, but it may not be valid for all x and y.

How to See Whether Unique Solutions of $dy/dx = g(x, y)$ *Exist*

Purpose: To see whether the differential equation

$$\frac{dy}{dx} = g(x, y)$$

has a unique solution subject to

$$y(x_0) = y_0.$$

Technique:

1. Confirm that $g(x, y)$ is continuous in the vicinity of (x_0, y_0).

2. Compute $\partial g/\partial y$, which may be a function of both x and y. Confirm that this function is continuous in the vicinity of (x_0, y_0).

3. If steps 1 and 2 are confirmed, then there is a unique solution through (x_0, y_0), valid in the vicinity of (x_0, y_0). If either step 1 or 2 fails, a unique solution may or may not exist.

EXAMPLE 2.7

Let's see how Theorem 2.1 applies to the differential equation

$$\frac{dy}{dx} = y^2. \tag{2.36}$$

Here $g(x, y) = y^2$, so $\partial g/\partial y = 2y$. Both $g(x, y)$ and $\partial g/\partial y$ are continuous for all x and y. The theorem thus states that there is a unique solution satisfying the initial condition $y(x_0) = y_0$, or, to put it another way, there is a unique curve that passes through the point $y(x_0) = y_0$. This means that solution curves cannot cross each other, because otherwise there would be two distinct solutions with the same initial condition at the point where they cross. Now the function $y(x) = 0$ is a solution of (2.36), so no other solutions can cross it. If we return to the slope field for (2.36), namely, Figure 2.34, armed with this information, and draw in the solution $y(x) = 0$, the fact that we must never cross it shows us that Figure 2.35 is incorrect. Thus, when we draw solution curves by hand, we must check to see that this theorem is satisfied and, if so, make sure that any solution curve we draw does not cross another.   □

Words of Caution: If this existence and uniqueness theorem is satisfied, then solution curves cannot touch or cross.

The existence and uniqueness theorem allows us to settle a subtle point that we have avoided mentioning so far. To illustrate the point, let's return to (2.11), namely,

$$\frac{dy}{dx} = ay, \tag{2.37}$$

which satisfies the conditions of the existence and uniqueness theorem, so solutions do not cross each other.

When we analyzed (2.37), we first noted that $y(x) = 0$ was an equilibrium solution. Then we searched for nonequilibrium solutions by dividing (2.37) by y and integrating. However, how do we know that there isn't a particular x, say x_0, for which $y(x_0) = 0$? Our division by $y(x)$ would not be valid for this x_0, and we might miss a particular solution because of this. The answer is that the equilibrium solution $y(x) = 0$ passes through the point $(x_0, 0)$, so no other solution can. Thus, we have not lost any solutions in this way.

We now look at three different examples, all of which expose different problems that can occur. The first demonstrates that without the existence and uniqueness theorem, graphical analysis may not show everything. The second shows that without the graphical analysis, the solution we obtain analytically may be wrong. The third shows that not only can we get our analytical calculations wrong, but so can most of the computer algebra system software packages that give solutions analytically for differential equations.

EXAMPLE 2.8

We consider the autonomous differential equation

$$\frac{dy}{dx} = (y - 1)^{2/3} \tag{2.38}$$

subject to the initial condition

$$y(1) = 1. \tag{2.39}$$

Figure 2.37 shows the slope field[14] for (2.38), and in Figure 2.38 we have hand-drawn a solution curve through the point $(1, 1)$.

Now let's look at the analytical approach. Although the function $(y - 1)^{2/3}$ is continuous everywhere, its derivative with respect to y — namely, $(2/3)(y - 1)^{-1/3}$ — is discontinuous whenever $y = 1$, which includes our initial condition. Thus, in spite of Figure 2.38, we are not guaranteed that there is only one solution through the point $(1, 1)$.

We can write (2.38) in the form

$$\frac{1}{(y - 1)^{2/3}} \frac{dy}{dx} = 1,$$

which can be integrated to yield

$$3(y - 1)^{1/3} = x + C,$$

or

$$y(x) = \frac{1}{27}(x + C)^3 + 1. \tag{2.40}$$

If we apply the initial condition (2.39) to (2.40), we find $C = -1$, so that (2.40) reduces to

$$y(x) = \frac{1}{27}(x - 1)^3 + 1. \tag{2.41}$$

If we stop here, we would conclude that there is only one solution through $(1, 1)$ and that Figure 2.38 is correct. We would be wrong, because (2.38) has an additional solution

---

[14]Some computer programs may draw the slope field only for $y > 1$. If this happens, try using $\left[(y - 1)^2\right]^{1/3}$ in place of $(y - 1)^{2/3}$.

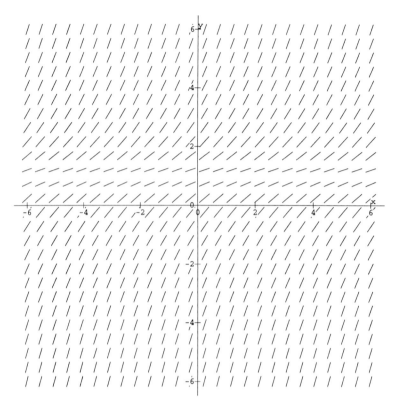

FIGURE 2.37   Slope field for $dy/dx = (y - 1)^{2/3}$

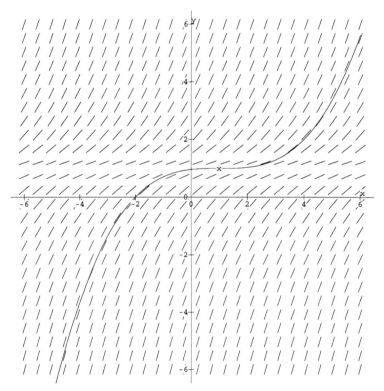

FIGURE 2.38   Hand-drawn solution curve through $(1, 1)$ and slope field for $dy/dx = (y - 1)^{2/3}$

that we have overlooked, the equilibrium solution $y(x) = 1$. Furthermore, it also satisfies the initial condition (2.39). Thus,

$$y(x) = 1 \qquad \text{(2.42)}$$

is another solution through $(1, 1)$.

Figure 2.39 shows the slope field for (2.38), together with the two solutions (2.41) and (2.42). Notice that we missed the equilibrium solution $y(x) = 1$ when we drew Figure 2.38.

If we look at Figure 2.39, we can actually see four solutions. The two obvious ones are (2.41) and (2.42). However, taking $y(x) = 1$ for $x \leq 1$ and otherwise taking the $y(x) = (x - 1)^3 / 27 + 1$ branch is a third. Finally, taking $y(x) = (x - 1)^3 / 27 + 1$ for $x \leq 1$ and otherwise taking $y(x) = 1$ is a the fourth. In fact these last two are not only continuous at $x = 1$, but also differentiable there. (See Exercise 7, page 84.) □

Words of Caution: It is possible to miss nonunique solutions using graphical analysis.

EXAMPLE 2.9

Now we consider the autonomous differential equation

$$\frac{dy}{dx} = y^{3/2}. \qquad \text{(2.43)}$$

Both $y^{3/2}$ and its derivative with respect to y, $(3/2)\, y^{1/2}$, are continuous throughout their domain, so the existence and uniqueness theorem applies. Equation (2.43) has the equilibrium solution $y(x) = 0$. In the nonequilibrium solution case, if $y > 0$ the solution is increasing, and if $y < 0$ the differential equation is undefined. (Why?) Because

$$\frac{d^2 y}{dx^2} = \frac{3}{2} y^{1/2} \frac{dy}{dx} = \frac{3}{2} y^{1/2} y^{3/2} = \frac{3}{2} y^2,$$

the solution is concave up for $y \neq 0$. The slope field for (2.43) is shown in Figure 2.40. Notice nothing is drawn for $y < 0$. (Why?)

Now let's look at the explicit solution for (2.43) in the nonequilibrium case when $y(x) \neq 0$. We divide (2.43) by $y^{3/2}$ to find

$$y^{-3/2} \frac{dy}{dx} = 1,$$

which when integrated yields

$$-2y^{-1/2} = x + C,$$

or

$$\sqrt{y} = \frac{-2}{x + C}. \qquad \text{(2.44)}$$

Squaring (2.44) gives

$$y(x) = \frac{4}{(x + C)^2}. \qquad \text{(2.45)}$$

Notice that there is no finite value of C that allows the equilibrium solution $y(x) = 0$, to be absorbed into (2.45).

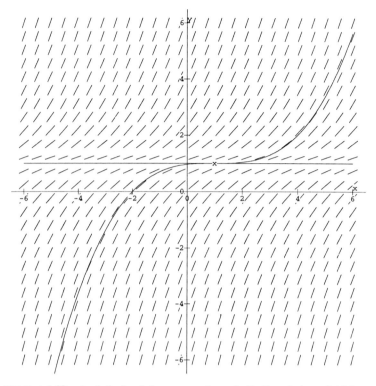

FIGURE 2.39 Analytical solution curves through $(1, 1)$ and slope field for $dy/dx = (y - 1)^{2/3}$

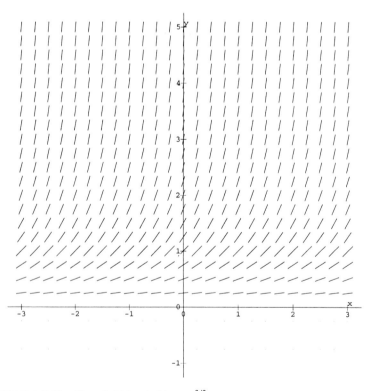

FIGURE 2.40 Slope field for $dy/dx = y^{3/2}$

If we consider the particular case of (2.45) when $C = 0$ — namely, $y(x) = 4/x^2$ — we see that it passes through the points $(\pm 2, 1)$, for example. The slope field for (2.43), along with the function (2.45) with $C = 0$ — that is, $y(x) = 4/x^2$ — and the points $(\pm 2, 1)$, are shown in Figure 2.41.

There is something wrong here! The right-hand part of $y(x) = 4/x^2$ and the slope field disagree. Where is our mistake? Well, we have made a very common oversight. In fact, had we merely obtained the explicit solution, without doing the graphical analysis, we wouldn't even realize that anything was wrong. Our mistake is going from (2.44) to (2.45). The right-hand side of (2.44) must be positive, which means that (2.44) is valid only for $x < -C$. Consequently, the correct way of expressing the family of solutions using (2.45) is

$$y(x) = \frac{4}{(x+C)^2}, \qquad x < -C.$$

Thus, $y = 4/x^2$ is a solution if $x < 0$ and is plotted in Figure 2.42.

To illustrate what happens for another initial point, we note the correct solution that passes through the point $(2, 1)$ is $4/(x - 4)^2$, valid for $x < 4$. It is shown in Figure 2.42.   □

Words of Caution: Be very careful when solving for $y(x)$. It is a common mistake to introduce extra functions that are not solutions.

This last example has something in common with the solution (2.29) of the logistics equation

$$\frac{dy}{dx} = ay(b - y),$$

namely,

$$y(x) = \frac{bC}{e^{-abx} + C}.$$

In both cases there is a particular solution (in each case an equilibrium solution) that is not obtained from the family of solutions by selecting a finite value for C. Such solutions are called singular.

Definition 2.4:   **A particular solution of**

$$\frac{dy}{dx} = g(x, y)$$

that cannot be obtained from the family of solutions (containing the arbitrary constant C) by selecting a finite value for C is called a SINGULAR SOLUTION.

EXAMPLE 2.10

Consider the autonomous differential equation

$$\frac{dy}{dx} = \sqrt{1 - y^2} \tag{2.46}$$

subject to the initial condition

$$y(0) = 0. \tag{2.47}$$

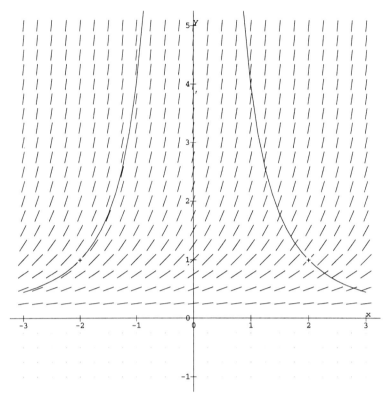

FIGURE 2.41   The function $4/x^2$ and slope field for $dy/dx = y^{3/2}$

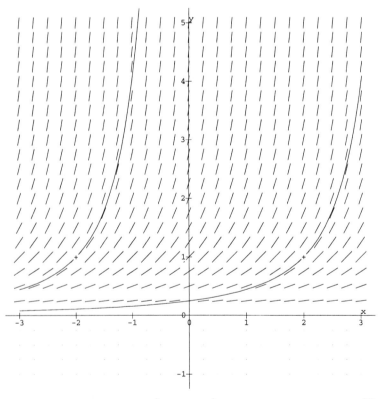

FIGURE 2.42   The functions $4/x^2$, $4/(x-4)^2$, and slope field for $dy/dx = y^{3/2}$

Both $\sqrt{1-y^2}$ and its derivative, $-y/\sqrt{1-y^2}$, are continuous in the vicinity of $(0, 0)$, so we don't anticipate any nonuniqueness problems.

Equation (2.46) has the equilibrium solutions

$$y(x) = \pm 1. \tag{2.48}$$

If we divide (2.46) by $\sqrt{1-y^2}$ and integrate, we have

$$\int \frac{dy}{\sqrt{1-y^2}} = \int dx,$$

which yields

$$\operatorname{arcsin} y = x + C. \tag{2.49}$$

If we use the initial condition (2.47) in (2.49) we find $C = 0$, so (2.49) reduces to

$$\operatorname{arcsin} y = x, \tag{2.50}$$

from which we get

$$y = \sin x. \tag{2.51}$$

But wait a minute! According to (2.46) all solutions should be increasing (or have horizontal tangent lines when $y = \pm 1$), whereas the solution we have found, (2.51), oscillates. Something is wrong. This can also be seen from Figure 2.43 where the slope field for (2.46) and the solution (2.51) are sketched together.

We have made a similar mistake in going from (2.50) to (2.51) as we did in Example 2.9. The range of the arcsin function is $[-\pi/2, \pi/2]$, so that (2.50) is valid only for $-\pi/2 \le x \le \pi/2$. Thus, (2.51) is valid only for $-\pi/2 \le x \le \pi/2$. The correct solution of (2.46) subject to (2.47) is

$$y = \sin x \quad \text{where} \quad -\frac{\pi}{2} \le x \le \frac{\pi}{2}. \tag{2.52}$$

This mistake is easy to make. In fact, most software packages that solve differential equations give (2.51) as the solution, which it is not.

In Figure 2.44 we have sketched the slope field for (2.46), the solution (2.52), and the equilibrium solutions (2.48). (Why is the slope field drawn only in the region $-1 \le y \le 1$?) At the points $(\pi/2, 1)$ and $(-\pi/2, -1)$, there appear to be nonunique solutions. How is this consistent with the existence and uniqueness theorem?

In fact, it is possible to piece together parts of the equilibrium solutions and the solution (2.52) to create a solution valid for $-\infty < x < \infty$. The function

$$y(x) = \begin{cases} -1 & \text{if } -\infty < x < -\pi/2 \\ \sin x & \text{if } -\pi/2 \le x \le \pi/2 \\ 1 & \text{if } \pi/2 < x < \infty \end{cases} \tag{2.53}$$

is a differentiable function that satisfies (2.46) subject to (2.47) for $-\infty < x < \infty$. See Figure 2.45. You should check that (2.53) is differentiable at $x = \pm \pi/2$. $\square$

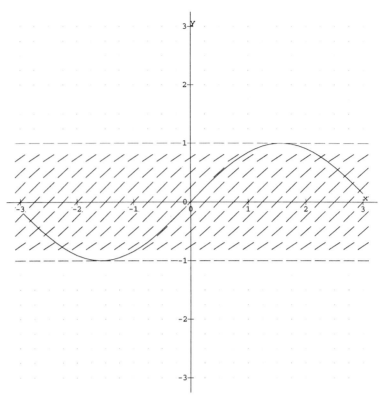

FIGURE 2.43   The function $\sin x$ and slope field for $dy/dx = (1 - y^2)^{1/2}$

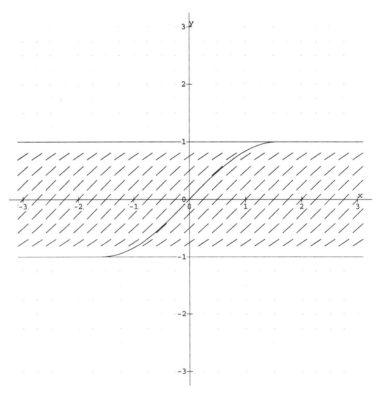

FIGURE 2.44   Part of $\sin x$, the equilibrium solutions $y = \pm 1$, and slope field for $dy/dx = (1 - y^2)^{1/2}$

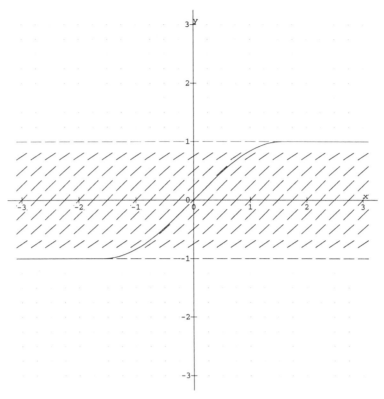

FIGURE 2.45 A solution of $dy/dx = (1 - y^2)^{1/2}$ through $(0, 0)$

Words of Caution: Software packages that find solutions of differential equations analytically are not infallible. Many of the popular ones give $y(x) = \sin x$ as the solution of $y' = \sqrt{1 - y^2}$, subject to $y(0) = 0$. Slope fields are useful!

EXERCISES

1. Does

$$\frac{dy}{dx} = 3y - 2$$

have any constant solutions? For what initial condition(s) would this constant solution be the unique solution of an initial value problem?

2. Solve the differential equation

$$\frac{dy}{dx} = y^{1/3}.$$

Someone says that there are at least three different solutions passing through the point $(0, 0)$. Is that person right? Comment.

3. This exercise deals with the logistics equation

$$\frac{dy}{dx} = ay(b - y).$$

(a) If $y(0) > b$, show there is an earlier time $\tilde{x}$ (< 0) for which $y(x) \to +\infty$ as $x \to \tilde{x}$ from the right (values greater than $\tilde{x}$). [Hint: Use (2.30).]

(b) If $y(0) < 0$, show there is a later time $\tilde{x}$ (> 0) for which $y(x) \to -\infty$ as $x \to \tilde{x}$ from the left (values less than $\tilde{x}$).

4. Consider the equilibrium solution

$$y = b$$

of the logistics equation

$$\frac{dy}{dx} = ay(b - y).$$

Is it possible for a nearby solution to reach this solution for a finite value of x? [Hint: Use the uniqueness theorem.]

5. Is it possible to write down a continuous function $g(y)$, where

$$\frac{dy}{dx} = g(y),$$

that satisfies all the following conditions? If it is, write down $g(y)$ and comment on its uniqueness. If it is not, explain your reasoning.

(a) There are exactly three equilibrium solutions — namely, $y(x) = 1$, $y(x) = 2$, and $y(x) = 3$.

(b) The solutions $y(x) = 1$ and $y(x) = 3$ are stable.

(c) The solution $y(x) = 2$ is unstable.

6. Is it possible to write down a continuous function $g(y)$, where

$$\frac{dy}{dx} = g(y),$$

that satisfies all the following conditions? If it is, write down $g(y)$ and comment on its uniqueness. If it is not, explain your reasoning.

(a) There are exactly three equilibrium solutions — namely, $y(x) = 1$, $y(x) = 2$, and $y(x) = 3$.

(b) The solutions $y(x) = 1$ and $y(x) = 2$ are stable.

(c) The solution $y(x) = 3$ is unstable.

7. Show that the function defined by

$$f(x) = \begin{cases} 0 & \text{if } x \leq 1 \\ \frac{1}{27}(x - 1)^3 + 1 & \text{if } x > 1 \end{cases}$$

is differentiable everywhere.

8. Figure 2.46 shows one member of a family of solutions of the differential equation

$$\frac{dy}{dx} = g(y),$$

where $g(y)$ is a given continuous function. Use this information to plot the family of solutions. Do not attempt to find $y(x)$ or $g(y)$. How do you know that there are an infinite number of points where dg/dy is discontinuous?

9. When savings are compounded continuously at rate r, the principal, P, changes with time, t, according to the differential equation $dP/dt = rP$. In order to attract new business, a local bank offers to do better than this by compounding the principal of a new investor according to

$$\frac{dP}{dt} = rP^n,$$

where n is a constant, and $n > 1$. Discuss the pros and cons of opening a new account at this bank.

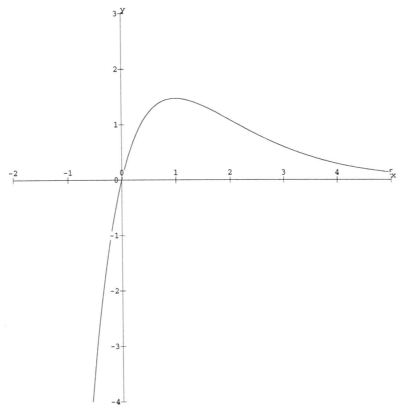

FIGURE 2.46   Graph of $y(x)$

2.5   Phase Line Analysis for Autonomous Equations

There is another way of analyzing autonomous differential equations, using what is called a PHASE LINE ANALYSIS. This method does not produce an explicit solution, but it can give the qualitative behavior of the solutions with very little calculation. The key to this technique is the uniqueness theorem, which guarantees that solutions cannot cross. In this section we will demonstrate this phase analysis by concentrating on a special case of the logistics equation.

EXAMPLE 2.11: Phase Line Analysis of

$$\frac{dy}{dx} = y(2 - y) \tag{2.54}$$

As usual we first find the equilibrium solutions of (2.54), which are $y(x) = 0$ and $y(x) = 2$. From the uniqueness theorem we know that no other solutions can cross these two solutions. Thus, we will concentrate on the three regions created by these equilibrium solutions — namely, $y < 0$, then $0 < y < 2$, and finally $y > 2$.

1. $y < 0$. If $y < 0$ then by (2.54) $y' < 0$, so in this region y is a decreasing function of x. Thus, if $y < 0$ at any time — that is, if $y(x) < 0$ for any value of x — then, as x increases y moves away from the equilibrium solution $y(x) = 0$.

2. $0 < y < 2$. In this case we see that (2.54) implies that $y' > 0$, so if at any value of x we have $0 < y < 2$, then the solution $y(x)$ will be an increasing function of x. It cannot cross $y(x) = 2$.

3. $y > 2$. In this case $y' < 0$, so if at any value of x we have $y > 2$, then the solution $y(x)$ will be a decreasing function of x. It cannot cross $y(x) = 2$.

We can represent this graphically. We draw a horizontal straight line to represent the y values and identify the equilibrium solutions by the points 0 and 2 — called EQUILIBRIUM POINTS. Then in the three regions $y < 0$, $0 < y < 2$, and $y > 2$, we place arrows to indicate whether a solution $y(x)$ is moving away from or toward the equilibrium solution at its boundary, as x increases. For example, in case 1, if $y < 0$, then y moves away from the equilibrium solution $y = 0$; we indicate this with an arrow pointing left. We do this for all three regions, and the result is Figure 2.47, called the PHASE LINE of (2.54).

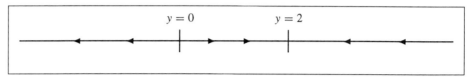

FIGURE 2.47   Phase line analysis for $y' = y(2 - y)$

That $y = 0$ is an unstable equilibrium solution is characterized by arrows on either side of the equilibrium point pointing away from it. Arrows pointing toward an equilibrium point indicate a stable equilibrium solution.

If initially $y(x_0) > 2$, we expect the solution $y(x)$, which passes through the point $(x_0, y(x_0))$, to decrease toward 2 as x increases. However, this solution $y(x)$ cannot reach 2 for any value of x, because $y(x) = 2$ is another solution, and the uniqueness theorem guarantees that two solutions cannot cross.

Figure 2.48 shows typical solutions consistent with the phase line analysis. In Exercise 3 of the last section we saw that solutions of the logistics equation had vertical asymptotes. If initially $y(0) > b$, then there is a negative value of x at which y came in from $+\infty$. If $y(0) < 0$, then there is a positive value of x at which $y \to -\infty$. Thus, if $y(0) > b$, the solution does not go back all the way to $x = -\infty$, and if $y(0) < 0$ the solution does not go forward all the way to $x = +\infty$. However, Exercise 4 shows that if $0 < y(0) < b$, then the solution extends from $-\infty$ to $+\infty$. The phase line analysis, like the graphical analysis, does not show these properties.   □

Comments about phase lines

- A simple way to construct the phase line for $dy/dx = g(y)$ is to graph the function $g(y) = y(2 - y)$ as a function of y (see Figure 2.49), making y the horizontal axis.

  (a) The equilibrium points are where the graph of $g(y)$ crosses the y-axis, so identify those points on the y-axis.

  (b) The values of y for which $g(y) > 0$ are where the function y is increasing. Identify those regions on the y-axis with arrows pointing to the right.

  (c) The values of y for which $g(y) < 0$ are where the function y is decreasing. Identify those regions on the y-axis with arrows pointing to the left.

  (d) The y-axis is now the phase line.

- Another way of thinking of the phase line is to imagine the solution curves projected onto the y-axis. In Figure 2.48, if we imagine looking down the x-axis from $x = +\infty$, and then project the directions of all solution curves onto the y-axis, we could characterize the properties of these solutions by the phase line in Figure 2.47.

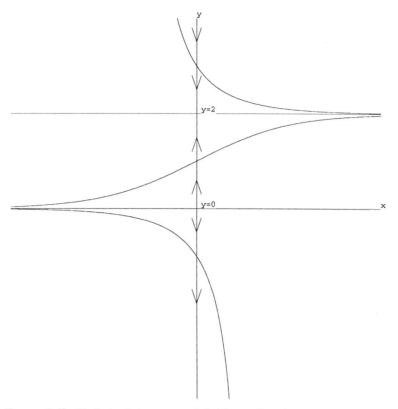

FIGURE 2.48 Typical solution curves of $dy/dx = y(2 - y)$

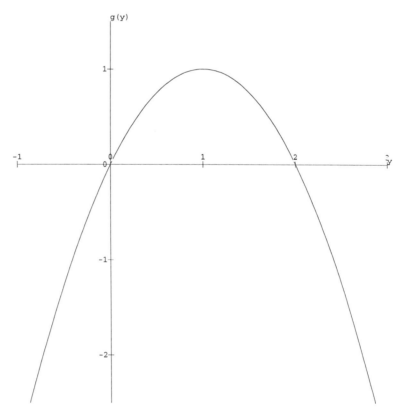

FIGURE 2.49 The function $g(y) = y(2 - y)$ versus y

EXAMPLE 2.12: Phase Line Analysis of

$$\frac{dy}{dx} = y \tag{2.55}$$

We now look at the phase line for (2.55). There is only one equilibrium point— namely, $y = 0$—and the phase line in Figure 2.50 is constructed, by realizing that the straight line $g(y) = y$ is negative for $y < 0$ and positive for $y > 0$. Now let us analyze this phase line. From the direction of the arrows we see that $y = 0$ is an unstable equilibrium. If a solution curve starts out with $y < 0$, it will decrease as x increases and move away from the equilibrium solution $y(x) = 0$. In the same way, a solution that starts with $y > 0$ increases with x. If we compare this observation with Figure 2.8 (on page 40), we see that Figure 2.50 can be obtained by projecting Figure 2.8 onto the y-axis and rotating 90° clockwise.   □

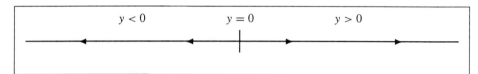

FIGURE 2.50   Phase line analysis for $y' = y$

EXAMPLE 2.13: Phase Line Analysis of

$$\frac{dy}{dx} = y^2 \tag{2.56}$$

There is one equilibrium solution — namely, $y(x) = 0$. Because the curve $g(y) = y^2$ is a parabola opening up that touches the y-axis at $y = 0$, the phase line for (2.56) is given in Figure 2.51. Solutions of (2.56) that start with $y < 0$ increase toward the equilibrium solution $y = 0$ as x increases, whereas solutions that start with $y > 0$ also increase and move away from the equilibrium solution $y = 0$ as x increases. We see that the equilibrium solution $y(x) = 0$ is neither a stable nor an unstable equilibrium. It is semistable.

We should not think that the arrows in the phase line in Figure 2.51 imply that as x increases, a solution that starts with $y < 0$ passes through $y = 0$ and continues increasing. The equation $y(x) = 0$ is a solution and, because of the uniqueness theorem, no other solution can pass through it. So Figure 2.51 tells us that if a solution starts with $y < 0$, it stays with $y < 0$. If we compare this observation with Figure 2.36 (on page 73), we see that Figure 2.51 can be obtained by projecting Figure 2.36 onto the y-axis and rotating 90° clockwise.   □

Words of Caution: When analyzing the phase line, realize that if the uniqueness theorem is satisfied, no solutions cross equilibrium solutions.

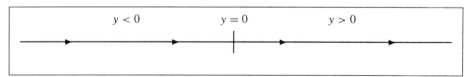

FIGURE 2.51   Phase line analysis for $y' = y^2$

How to Perform a Phase Line Analysis on $dy/dx = g(y)$

Purpose: To construct and analyze the phase line of the autonomous differential equation

$$\frac{dy}{dx} = g(y). \tag{2.57}$$

Technique:

1. Make sure that $g(y)$ satisfies the conditions of Theorem 2.1. (It will if dg/dy is continuous.)

2. Solve $g(y) = 0$ to determine all the equilibrium solutions of (2.57), and mark them on the phase line. Remember these are solutions.

3. Decide on the sign of dy/dx — that is, $g(y)$ — by sketching $g(y)$ as a function of y. Between successive equilibrium solutions, mark the phase line with an arrow pointing left if $g(y) < 0$ and right if $g(y) > 0$.

4. Identify the type of equilibrium solution (stable, unstable, semistable) according to Figure 2.52.

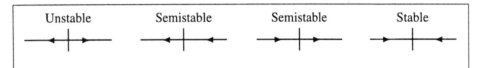

FIGURE 2.52 Types of equilbrium solutions

EXERCISES

1. This exercise deals with the differential equation

$$\frac{dy}{dx} = (y - a)(y - b)(y - c). \tag{2.58}$$

   (a) Does Theorem 2.1 apply to this equation?

   (b) Confirm that $y(x) = a$, $y(x) = b$, and $y(x) = c$ are the equilibrium solutions of (2.58).

   (c) Draw the phase line for each of the following cases i through iii. In each case identify all the equilibrium solutions as being stable, unstable, or semistable.

       i. $a = b = c = 1$

       ii. $a = b = 1, c = -1$

       iii. $a = 1, b = 0, c = -1$

   (a) Based on the information you obtained in part (c), describe qualitatively the behavior of the solutions in each of the three cases in part (c). Are your descriptions consistent with part (a), in particular as far as intersecting solutions are concerned?

   (b) Use technology to draw solution curves for each of the cases in part (c). Make sure the results you obtain here are consistent with the answers you gave to part (d).

2. Consider the differential equation

$$\frac{dy}{dx} = g(y)$$

where $g(y)$ has a continuous derivative. Show that between any two equilibrium solutions, the solution curves change concavity. [Hint: Sketch $g(y)$.]

3. During the 19th century there was a massive population of passenger pigeons in the United States. In order to breed successfully, a large number of pigeons had to be present. Successful breeding rarely occurred when small numbers were present. In the latter part of the century, passenger pigeons were hunted more heavily than in the past, and within 30 years they were extinct. It has been suggested that the population P of passenger pigeons, as a function of time t, could be modeled by the differential equation

$$\frac{dP}{dt} = aP(P - b)(c - P),$$

where a, b, and c are positive constants, and $b < c$. Without solving this differential equation, but by sketching the family of solutions after using a phase line analysis, decide whether this is a reasonable model. If it is, what are the physical interpretations of b and c?

4. In each of the following differential equations, explain how the behavior of the solution varies with the value of the constant a. Are there any values of a at which this behavior changes dramatically? (Do not attempt to solve these equations.)

(a) $\dfrac{dy}{dx} = a - y^2$

(b) $\dfrac{dy}{dx} = y(a - y^2)$

What Have We Learned?

MAIN IDEAS

- An AUTONOMOUS first order differential equation is of the form

$$\frac{dy}{dx} = g(y),$$

where $g(y)$ is a given function of y alone.

- If $dy/dx = g(y)$ remains unchanged after the interchange of y with $-y$ on both sides of the differential equation, then the family of solutions is symmetric about the x-axis.

- If $dy/dx = g(y)$ remains unchanged after the simultaneous interchange of y with $-y$ and x with $-x$ on both sides of the differential equation, then the family of solutions is symmetric about the origin.

- If the differential equation

$$\frac{dy}{dx} = g(x, y)$$

has a solution of the form $y(x) = constant,$ then this solution is called an EQUILIBRIUM SOLUTION.

- Let the differential equation

$$\frac{dy}{dx} = g(x, y)$$

have an equilibrium solution, that is, a solution of the form $y(x) = constant.$ An equilibrium solution is called STABLE if all other solutions of the differential equation that start near this equilibrium solution remain near this equilibrium solution as $x \to \infty$. An equilibrium solution is called UNSTABLE if all other solutions of the differential equation that start near this equilibrium solution move away from this equilibrium solution as $x \to \infty$. If an equilibrium solution is neither stable nor unstable, it is called SEMISTABLE.

- If $g(x, y)$ and $\partial g / \partial y$ are defined and continuous in a finite rectangular region containing the point (x_0, y_0) in its interior, then the differential equation

$$\frac{dy}{dx} = g(x, y)$$

has a unique solution passing through the point

$$y(x_0) = y_0.$$

This solution is valid for all x for which the solution remains inside the rectangle.

- A particular solution of

$$\frac{dy}{dx} = g(x, y)$$

that cannot be obtained from the family of solutions (containing the arbitrary constant C) by selecting a finite value for C is called a SINGULAR SOLUTION.

- A PHASE LINE is a line giving values of the dependent variable including the equilibrium solutions. Arrows on this line show regions where the solution is increasing or decreasing which is useful in determining the stability of equilibrium solutions.

---

*How to Analyze $dy/dx = g(y)$ **Graphically***

Purpose: To use the techniques of calculus, isoclines, and slope fields to sketch the solutions $y = y(x)$ of

$$\frac{dy}{dx} = g(y). \tag{2.59}$$

Technique: Use the following operations — the order is flexible.

1. Make sure a solution exists. See *How to See Whether Unique Solutions of $dy/dx = g(x, y)$ Exist* on page 74.

2. Analyze the phase line. See *How to Perform a Phase Line Analysis on $dy/dx = g(y)$* on page 89.

3. Determine regions in the xy-plane where the solution curves are increasing or decreasing.

   (a) Consider $g(y) = 0$. These will define regions in the xy-plane where solutions $y = y(x)$ have slope 0, and possible locations of local maxima and minima.

   (b) Consider $g(y) > 0$. These will define regions in the xy-plane where solutions $y = y(x)$ are increasing.

   (c) Consider $g(y) < 0$. These will define regions in the xy-plane where solutions $y = y(x)$ are decreasing.

4. Determine regions in the xy-plane in which the solution curves are concave up or concave down. Differentiate (2.59) with respect to x, and then use (2.59) to eliminate y' from the result. This leads to

$$\frac{d^2y}{dx^2} = \frac{dg}{dy}\frac{dy}{dx} = \frac{dg}{dy}g = f(y).$$

   (a) Consider $f(y) = 0$. These will define regions in the xy-plane where the solutions $y = y(x)$ may have inflection points (change concavity).

   (b) Consider $f(y) > 0$. These will define regions in the xy-plane where solutions $y = y(x)$ are concave up.

   (c) Consider $f(y) < 0$. These will define regions in the xy-plane where solutions $y = y(x)$ are concave down.

5. Determine the isoclines. Consider $g(y) = m$, where m is constant. Solve this equation for y in terms of m. These are horizontal lines, and they identify where all the solution curves have the same slope, namely, m.

6. Construct the slope field for (2.59), using technology.

7. Check whether the slope field has any symmetries.

8. Sketch several curves that are consistent with all the preceding information.

Comments about graphical analysis

- Operations 2 through 7 in the preceding technique need not be done in this order, depending on the problem. For example, we might first construct the slope field, and then check where the solution is increasing.

- Operations 2 through 5 usually require some algebra in order to solve equations like $g(y) = m$ and $f(y) = 0$ for y. Even if you have great algebraic skills, solving the equations is not always possible.

Words of Caution:

- It is a common oversight to mislay the equilibrium solutions.

- When analyzing the phase line, realize that if the uniqueness theorem is satisfied, no solutions cross equilibrium solutions.

- Graphical analysis will not show whether a solution goes to infinity for finite x. Solutions need not go on forever.

- If the existence and uniqueness theorem is satisfied, then solution curves cannot touch or cross.

- It is possible to miss nonunique solutions using graphical analysis.

- Be very careful when solving for $y(x)$. It is a common mistake to introduce functions that are not solutions.

- Software packages that find solutions of differential equations analytically are not infallible. They too can get the wrong answer. Slope fields are useful!

CHAPTER 3

General First Order Differential Equations

Where Are We Going?

In the first two chapters we considered two types of simple differential equations,

$$\frac{dy}{dx} = g(x),$$

with solution $y = f(x)$, and

$$\frac{dy}{dx} = g(y),$$

with solution $x = f(y)$.

In this chapter we consider the more general first order differential equation

$$\frac{dy}{dx} = g(x, y), \tag{3.1}$$

where the dependent variable y and the independent variable x are both present on the right-hand side of (3.1).

We show how the graphical techniques of Chapters 1 and 2 apply to equations like (3.1) with little or no modification. We also introduce a simple method of obtaining a numerical solution of (3.1), namely, Euler's method. Our primary goal is to fully understand the solution's behavior even though we do not necessarily have an explicit solution in terms of familiar functions.

3.1   Graphical Solutions

In this section we apply the techniques developed in Chapters 1 and 2 — graphical analysis and explicit solutions — to differential equations of the form (3.1) where both variables x and y are present on the right-hand side. We first analyze a simple example, and then make some general observations about what we have learned. We use these ideas to show how an implicit function may be graphed by converting it to a differential equation.

EXAMPLE 3.1

Consider the differential equation

$$\frac{dy}{dx} = x - y. \tag{3.2}$$

We want to sketch the solution curves of this differential equation using graphical techniques.

First we use the techniques of calculus to decide where the solutions are increasing and where they are decreasing. Solutions will have horizontal tangent lines when $x - y = 0$ — that is, when $y = x$. [Notice that although $y = x$ makes the right-hand side of (3.2) zero, the left-hand side would be 1, so $y = x$ is not a solution of (3.2).] Solutions will be increasing when $dy/dx = x - y > 0$ — that is, when $x > y$ — and decreasing when $dy/dx < 0$ — that is, when $x < y$. So the line $y = x$ divides the xy-plane into two regions. Solutions that lie above the line $y = x$ have the property that $x < y$, so they must be decreasing. Similarly, solutions that lie below the line $y = x$ must be increasing. These regions are sketched in Figure 3.1.

The line $y = x$ is the curve along which the solution curves will have slope 0; in other words, $y = x$ is the isocline corresponding to slope 0. Let's look at the isoclines corresponding to slope m — that is, the curves $x - y = m$, or $y = x - m$. These isoclines are straight lines with slope 1 and y-intercept given by $-m$. For example, the isocline $y = x - 1$ is where the solutions have slope 1. This isocline is parallel to but below the isocline $y = x$, which separates the increasing and decreasing solutions. Unlike the previous examples, these isoclines are neither vertical — as was the case for $dy/dx = g(x)$ — nor horizontal — as was the case for $dy/dx = g(y)$. Isoclines corresponding to $m = 0$, $m = \pm 2$, and $m = \pm 4$ are shown in Figure 3.2.

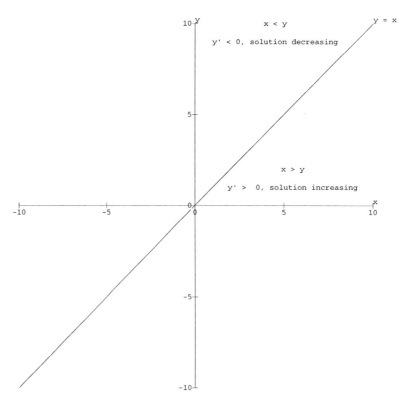

FIGURE 3.1   Regions where solutions of $dy/dx = x - y$ are increasing and decreasing

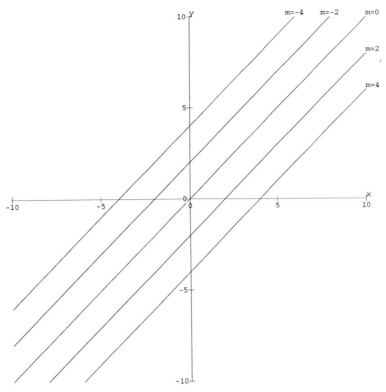

FIGURE 3.2 Isoclines of $dy/dx = x - y$ for $m = 0, m = \pm 2, m = \pm 4$

Now let's look at the concavity of the solutions of (3.2). To do this we need d^2y/dx^2, which we get from (3.2) by differentiating with respect to x,

$$\frac{d^2y}{dx^2} = 1 - \frac{dy}{dx}.$$

If we substitute for dy/dx from (3.2) into this last equation we find

$$\frac{d^2y}{dx^2} = 1 - x + y.$$

Solutions will be concave up when $d^2y/dx^2 > 0$ — that is, when $1 - x + y > 0$ — or equivalently when $y > x - 1$. Solutions will be concave down when $d^2y/dx^2 < 0$ — that is, when $y < x - 1$ — so points of inflection might occur along the line $y = x - 1$ where $d^2y/dx^2 = 0$. Thus, the line $y = x - 1$ divides the xy-plane into two regions. Solutions that lie above $y = x - 1$ have the property that $y > x - 1$, and so this will be where the solution curves are concave up. In the same way, the region $y < x - 1$ is where the solution curves will be concave down. These regions are sketched in Figure 3.3.

If we calculate the slopes corresponding to (3.2) for several points we obtain Table 3.1. Calculations such as these lead to Figure 3.4.

If we hand-draw a few solution curves on Figure 3.4, we find Figure 3.5. These are in complete agreement with our previous information concerning monotonicity, isoclines, and concavity.

However, there are two things to notice in Figure 3.5. First, the line $y = x - 1$ appears to be a solution curve. Second, all solution curves seem to tend to the line $y = x - 1$. If we substitute $y = x - 1$ into (3.2), we find that $y = x - 1$ is indeed a solution. The suggestion

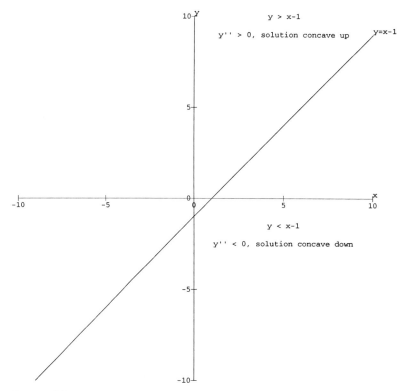

FIGURE 3.3   Regions where solutions of $dy/dx = x - y$ are concave up and concave down

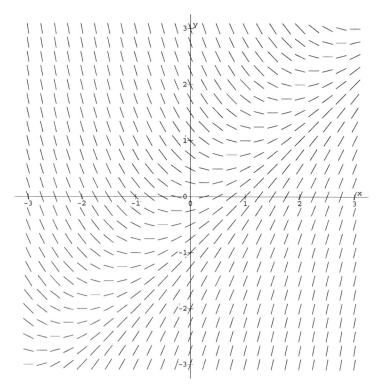

FIGURE 3.4   Slope field for $dy/dx = x - y$

TABLE 3.1
Slopes for
$dy/dx = x - y$

x	y	$x - y$
-2	-2	0
-2	0	-2
-2	2	-4
0	-2	2
0	0	0
0	2	-2
2	-2	4
2	0	2
2	2	0

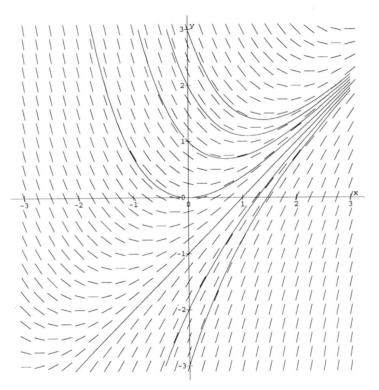

FIGURE 3.5 Some hand-drawn solution curves and slope field for $dy/dx = x - y$

that all solutions tend to $y = x - 1$ as $x \to \infty$ leads us to wonder whether solutions of (3.2) can be written in the form

$$y(x) = x - 1 + z(x), \tag{3.3}$$

where $z(x) \to 0$ as $x \to \infty$. If we substitute (3.3) and its derivative into (3.2), namely, $dy/dx = x - y$, we find that

$$1 + \frac{dz}{dx} = x - (x - 1 + z),$$

or

$$\frac{dz}{dx} = -z.$$

This equation is easily solved for $z(x)$, and we have

$$z(x) = Ce^{-x}, \tag{3.4}$$

where C is an arbitrary constant. Substituting (3.4) into (3.3) we find

$$y(x) = x - 1 + Ce^{-x}$$

is the solution of (3.2). Notice that, as expected, the term involving C goes to 0 as $x \to \infty$. Earlier we noticed that inflection points might occur along the line $y = x - 1$. They don't because $y = x - 1$ is a solution and by Theorem 2.1, page 72, no solution may cross this solution. [Explain why the differential equation (3.2) satisfies the hypothesis of Theorem 2.1.]   □

How to Analyze $dy/dx = g(x, y)$ *Graphically*

Purpose: To use the techniques of calculus, isoclines, and slope fields to sketch the solutions $y = y(x)$ of

$$\frac{dy}{dx} = g(x, y). \tag{3.5}$$

Technique: Use the following operations — the order is flexible.

1. Determine regions in the xy-plane where the solution curves are increasing or decreasing.

   (a) Consider $g(x, y) = 0$. These will define regions in the xy-plane where solutions $y = y(x)$ have slope zero. These are possible locations of local maxima and minima.

   (b) Consider $g(x, y) > 0$. These will define regions in the xy-plane where solutions $y = y(x)$ are increasing.

   (c) Consider $g(x, y) < 0$. These will define regions in the xy-plane where solutions $y = y(x)$ are decreasing.

2. Determine the isoclines. Consider $g(x, y) = m$, where m is a constant. Solve this equation, obtaining the isoclines as y in terms of x and m (or as x in terms of y and m). These isoclines identify where all the solution curves have the same slope, namely, m. Note that these isoclines need not be straight lines.

3. Determine regions in the xy-plane where the solution curves are concave up or concave down. Differentiate (3.5) with respect to x (be sure to use the chain rule), and then substitute (3.5) into the result to eliminate y'. This leads to

$$\frac{d^2y}{dx^2} = f(x, y).$$

   (a) Consider $f(x, y) = 0$. These will define regions in the xy-plane where the solutions $y = y(x)$ change concavity and where they may have inflection points.

(b) Consider $f(x, y) > 0$. These will define regions in the xy-plane where solutions $y = y(x)$ are concave up.

(c) Consider $f(x, y) < 0$. These will define regions in the xy-plane where solutions $y = y(x)$ are concave down.

4. Construct the slope field for (3.5), using technology.

5. Sketch several curves that are consistent with all the preceding information.

Comments about graphical analysis

- Operations 1 through 4 need not be done in this order. For example, we might first construct the slope field, and then check where the solution is increasing.

- Operations 1, 2, and 3 usually require some algebra in order to solve the equations $g(x, y) = m$ and $f(x, y) = 0$. Even if you have great algebraic skills, solving these equations is not always possible.

We will devote the rest of this section to these graphical techniques, applying them to various examples.

EXAMPLE 3.2: Graphing Implicit Functions

There are situations in mathematics in which having a simple expression does not mean we know much about this expression. For example, consider the simple equation

$$x^3 + 3y^2 = 27. \tag{3.6}$$

There does not appear to be an easy way to obtain the graph of this function, even though it is apparent that there is symmetry with respect to the x-axis [replace y by $-y$ in (3.6) and observe that the equation does not change].

One of the important uses of calculus is as an aid in drawing graphs of functions, mainly to find relative extreme values as well as intervals where the graphs are increasing or decreasing. Thus, we need the derivative of a function, but here we have no function. The procedure given in calculus is to assume that the equation (3.6) defines a function implicitly and then implicitly differentiate (3.6) with respect to x. Thus, we assume that y may be found as a function of x and differentiate (3.6) to obtain

$$3x^2 + 6y\frac{dy}{dx} = 0. \tag{3.7}$$

Solving (3.7) for the derivative, we find

$$\frac{dy}{dx} = -\frac{x^2}{2y}. \tag{3.8}$$

This is a differential equation whose family of solutions contains the curve we want. By looking at the sign of this derivative, we see that if $x \neq 0$, the curve we seek will be decreasing for $y > 0$, will be increasing for $y < 0$, and will have a vertical tangent when $y = 0$ (if $x \neq 0$). When $x = 0$ (and $y \neq 0$), the curve will have a horizontal tangent.

If we calculate the slopes corresponding to (3.8) for various values of x and y, we obtain Table 3.2. Plotting results of calculations such as these lead to Figure 3.6. Notice that the slope field is symmetric about the x-axis.

TABLE 3.2
Slopes for
$dy/dx = -x^2/2y$

x	y	$-x^2/2y$
± 2	-2	1
± 2	-1	2
± 2	0	∞
± 2	1	-2
± 2	2	-1
± 1	-2	$1/4$
± 1	-1	$1/2$
± 1	0	∞
± 1	1	$-1/2$
± 1	2	$-1/4$

We noted that the only place where the graph of our original equation has horizontal tangents is where $x = 0$ and $y \neq 0$. Vertical tangents could occur where $y = 0$ and $x \neq 0$. Other isoclines are given by

$$-\frac{x^2}{2y} = m,$$

that is,

$$y = -\frac{x^2}{2m}, \tag{3.9}$$

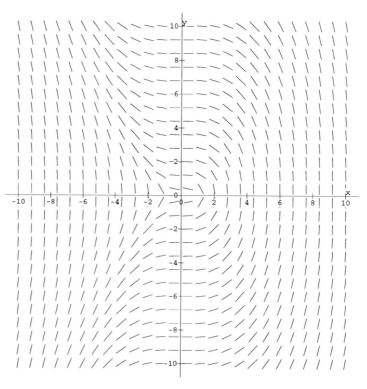

FIGURE 3.6   Slope field for $dy/dx = -x^2/2y$

where m is the specified slope. The isoclines (3.9) are parabolas through the origin, and it is apparent that the isoclines which correspond to steep slopes are very broad parabolas whereas those which correspond to very small slopes are very narrow parabolas. Furthermore, if $m > 0$, the parabolas are concave down, whereas if $m < 0$ they are concave up. In Figure 3.7 we have drawn the slope field corresponding to (3.8) and superimposed on it the isoclines (3.9) for $m = \pm 1, \pm 2, \pm 3, \pm 4$, and ± 5.

Because the slope field shows how the derivative of the solution behaves, we need a specific point on the curve to draw the graph of the original equation (3.6). One of the easy points to calculate on this graph is for $x = 0$, where $y = 3$, and this hand-drawn curve, that is,

$$x^3 + 3y^2 = 27, \tag{3.10}$$

is shown in Figure 3.8. Notice that this curve is symmetric about the x-axis, which it should be. Also notice that the hand-drawn curve is actually two particular solutions, one through $(0, 3)$ and the other through $(0, -3)$, joined at the point $(3, 0)$.

We confirm that Figure 3.8 is reasonable by considering the concavity of the solution curve. We differentiate (3.8) to find

$$\frac{d^2y}{dx^2} = -\frac{2x}{2y} + \frac{x^2}{2y^2}\frac{dy}{dx} = -\frac{x}{2y^2}\left(2y - x\frac{dy}{dx}\right).$$

If we use (3.8) in this last equation we find

$$\frac{d^2y}{dx^2} = -\frac{x}{2y^2}\left[2y - x\left(-\frac{x^2}{2y}\right)\right]$$

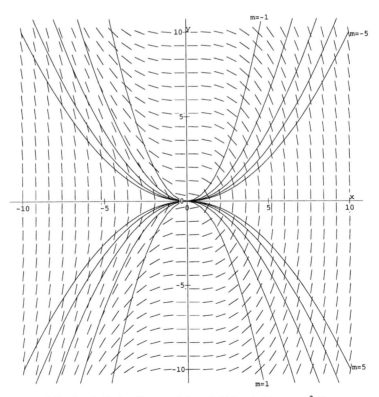

FIGURE 3.7 Parabolic isoclines and slope field for $dy/dx = -x^2/2y$

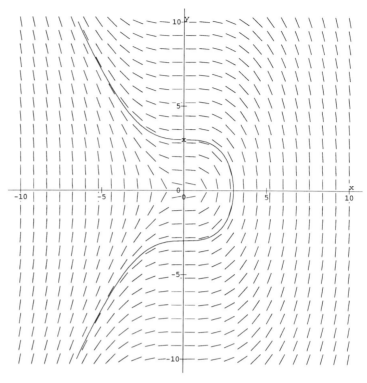

FIGURE 3.8   Hand-drawn graph of the equation $x^3 + 3y^2 = 27$

which can be written as

$$\frac{d^2y}{dx^2} = -\frac{x}{y}\frac{1}{4y^2}\left(4y^2 + x^3\right). \tag{3.11}$$

We may rewrite the expression in parentheses which appears on the right-hand side of (3.11) as $4y^2 + x^3 = y^2 + \left(x^3 + 3y^2\right)$ and use (3.10) to simplify this expression. Thus, we see that (3.11) becomes

$$\frac{d^2y}{dx^2} = -\frac{x}{y}\frac{1}{4y^2}\left(y^2 + 27\right).$$

Looking at the right-hand side of this equation we see that the solution curve is concave down if $x/y > 0$ — that is, in the first and third quadrants — and is concave up if $x/y < 0$ — that is, in the second and fourth quadrants. This is consistent with the graph in Figure 3.8.   □

To summarize, we started with an implicit equation relating x and y, and via implicit differentiation we obtained a differential equation as an aid to obtaining the graph of this equation. Thus, the original implicit equation is a solution of the differential equation; it is called an implicit solution of the differential equation. This leads to the following definition.

Definition 3.1:   **An IMPLICIT SOLUTION of a first order differential equation is any relationship between x and y that through implicit differentiation yields the original differential equation.**

Sometimes it is a simple task to obtain an explicit solution from an implicit solution; sometimes it is impossible. Note that in our last example, the implicit solution $x^3 + 3y^2 = 27$ yields two explicit solutions, namely,

$$y = \sqrt{9 - \frac{1}{3}x^3} \text{ and } y = -\sqrt{9 - \frac{1}{3}x^3},$$

both of which are needed to obtain the complete graph of our original equation.

EXAMPLE 3.3

Sometimes important information comes by considering specific isoclines, as we now show by investigating the differential equation

$$\frac{dy}{dx} = y^2 - xy + x^2 - 1. \tag{3.12}$$

The solution curve of this equation will be increasing whenever the derivative is positive — namely, when $y^2 - xy + x^2 - 1 > 0$ — and decreasing when $y^2 - xy + x^2 - 1 < 0$. Thus, the curve $y^2 - xy + x^2 - 1 = 0$ separates the xy-plane into regions of increasing and decreasing solutions. From your work with conics[1] you may know that the curve $y^2 - xy + x^2 - 1 = 0$ is an ellipse centered at the origin[2] with semiaxes $\sqrt{2}$ and $\sqrt{2/3}$ rotated through 45°. Because $y^2 - xy + x^2 - 1 > 0$ outside this ellipse, our solution curves are increasing, whereas inside this ellipse they are decreasing. These regions are shown in Figure 3.9.

The slope field for (3.12) is shown in Figure 3.10. This figure suggests that the slope field is symmetric about the origin, which can be confirmed by noting that replacing x with $-x$ and y with $-y$ on both sides of (3.12) leaves it unchanged. This figure also suggests that the behavior near the origin is different from that further from the origin. No horizontal tangents appear in the figure other than those in an oval-shaped region around the origin. It appears that the maximum negative slope occurs at the origin. This is made obvious in Figure 3.11, which shows both the slope field and the ellipse where the slopes are zero.

To explore this further we note that the isoclines for slope m occur when

$$y^2 - xy + x^2 - 1 = m. \tag{3.13}$$

If we select a specific value of x, say x_0, then the corresponding values of y may be obtained from (3.13) by moving m to the left-hand side and using the quadratic formula to solve for y, namely,

$$y = \frac{x_0 \pm \sqrt{x_0^2 - 4(x_0^2 - 1 - m)}}{2} = \frac{x_0 \pm \sqrt{4(1 + m) - 3x_0^2}}{2}. \tag{3.14}$$

This means that for the solution curve to have tangents with slope m, we must have

$$4(1 + m) - 3x_0^2 \geq 0. \tag{3.15}$$

---

[1] The quadratic $Ax^2 + Bxy + Cy^2 + Dx + Ey + F = 0$ can be put in the form $A'x'^2 + C'y'^2 + D'x' + E'y' + F' = 0$ by a rotation of axes through the angle θ, where $\cot 2\theta = (A - C)/B$.

[2] This curve could also be sketched using the techniques of Example 3.2 (see Exercise 16, page 112) or by plotting $y = \left(x \pm \sqrt{x^2 - 4(x^2 - 1)}\right)/2$, which is the solution of the quadratic $y^2 - xy + x^2 - 1 = 0$.

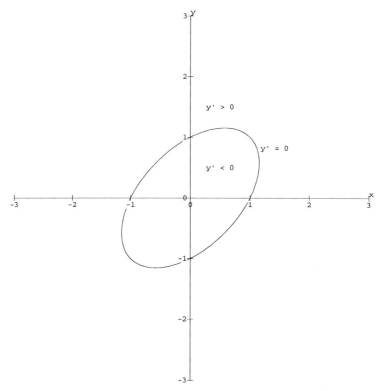

FIGURE 3.9   The ellipse $y^2 - xy + x^2 - 1 = 0$

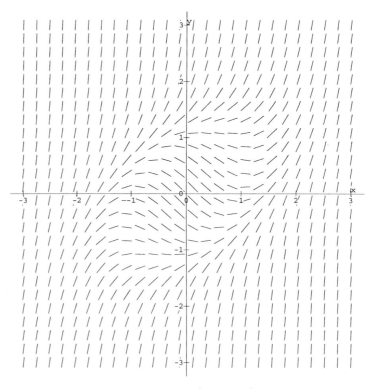

FIGURE 3.10   Slope field for $dy/dx = y^2 - xy + x^2 - 1$

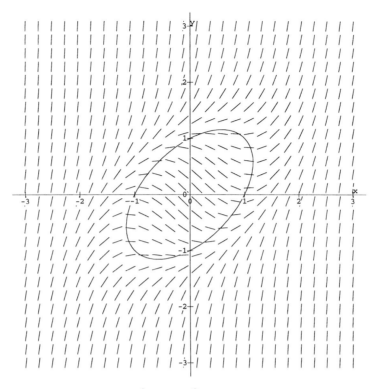

FIGURE 3.11 The ellipse $y^2 - xy + x^2 - 1 = 0$ and slope field for $dy/dx = y^2 - xy + x^2 - 1$

Clearly this inequality cannot be satisfied if $m < -1$. If $m = -1$, then the inequality (3.15) implies that $x_0 = 0$, in which case (3.14) gives $y_0 = 0$. Thus, the maximum negative slope that can occur is $m = -1$, and this occurs only at the origin.

Also note that from (3.15), we have only horizontal tangents ($m = 0$) when $4 - 3x_0^2 \geq 0$, that is, for values of x_0 satisfying the inequality

$$-\frac{2}{\sqrt{3}} \leq x_0 \leq \frac{2}{\sqrt{3}}.$$

By solving (3.13) for x, a similar analysis shows that horizontal tangents occur only for

$$-\frac{2}{\sqrt{3}} \leq y_0 \leq \frac{2}{\sqrt{3}}.$$

The isocline for zero slope is the ellipse $y^2 - xy + x^2 = 1$ as shown in Figure 3.11. Figure 3.12 shows some hand-drawn solution curves of (3.12) consistent with these facts. $\square$

EXERCISES

1. Sketch the slope field corresponding to

$$y\frac{dy}{dx} = x.$$

(a) What symmetry does the slope field have?

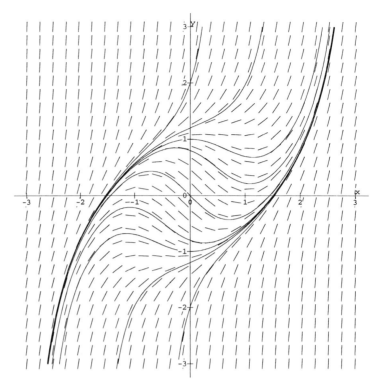

FIGURE 3.12 Hand-drawn solution curves and slope field for $dy/dx = y^2 - xy + x^2 - 1$

(b) What is the equation of the isocline where all solutions have zero slope?

(c) Draw isoclines for $m = 0, \pm 0.5, \pm 1,$ and ± 2.

(d) Decide where the solution curves are concave up; concave down.

(e) Draw the solution curve that passes through the point $x = 0, y = 1$.

(f) Show that $y^2 - x^2 = 1$ passes through $(0, 1)$ and satisfies $yy' = x$.

2. Sketch the slope field corresponding to

$$\frac{dy}{dx} = x^2 + y^2.$$

(a) What symmetry does the slope field have?

(b) When will the solution to this differential equation have a horizontal tangent?

(c) Draw isoclines for $m = 1/4, 1, 4,$ and 9.

(d) Draw the solution curve that passes through $(-1, 1)$.

3. Use graphical methods to sketch solution curves for

$$\frac{dy}{dx} = \sqrt{x^2 + y^2}.$$

4. Use graphical methods to sketch solution curves for

$$\frac{dy}{dx} = -2xy.$$

5. Look at the slope field for

$$\frac{dy}{dx} = e^{-x} - 2y$$

in the window $-3 < x < 3$, $-3 < y < 3$.

(a) Does it appear that solutions of this differential equation have a relative maximum? (You may want to consider the isocline that corresponds to a slope of zero.)

(b) Does it appear that all solutions of this differential equation have

i. a horizontal asymptote? (Use the differential equation to justify your answer.)

ii. a vertical asymptote? (Use the differential equation to justify your answer.)

(c) Confirm that

$$y(x) = e^{-x} + (y_0 - 1)e^{-2x}$$

satisfies $y' = e^{-x} - 2y$ subject to $y(0) = y_0$. Explain how this solution is consistent with your answers in parts (a) and (b).

6. Consider the following eight first order differential equations:

   i. $dy/dx = x + 1$

   ii. $dy/dx = x - 1$

   iii. $dy/dx = y + 1$

   iv. $dy/dx = y - 1$

   v. $dy/dx = y^2 - 1$

   vi. $dy/dx = y + x$

   vii. $dy/dx = y - x$

   viii. $dy/dx = y - x + 1$

(a) The slope fields for three of the preceding equations are given in Figures 3.13, 3.14, and 3.15. Match an equation with each of the three figures, carefully stating the reasons for your choices. Do not plot the slope fields for i through viii to answer this question.

(b) Briefly outline a general strategy for this matching process.

7. We wish to find all curves $y = y(x)$ with the property that the tangent line at each point on the curve is perpendicular to the line connecting that point to the origin (see Figure 3.16). Use the fact that if the line is perpendicular to the curve, then the product of the slopes of the line and the curve must be -1 to arrive at the differential equation

$$\frac{dy}{dx} = -\frac{x}{y}.$$

Use graphical analysis to sketch solutions of this equation. Based on what you see, guess the equation of the curve $y = y(x)$. Check that your guess satisfies the differential equation.

8. Figures 3.17 and 3.18 are mystery slope fields, believed to be the slope fields for two of the following differential equations:

$$\frac{dy}{dx} = \frac{x^2+1}{y^2+1}, \ \frac{dy}{dx} = -\frac{x^2+1}{y^2+1}, \ \frac{dy}{dx} = \frac{y^2+1}{x^2+1}, \ \frac{dy}{dx} = -\frac{y^2+1}{x^2+1}.$$

(a) Identify to which differential equation each of the mystery slope fields belongs. Confirm all the information in the slope fields using isoclines and concavity.

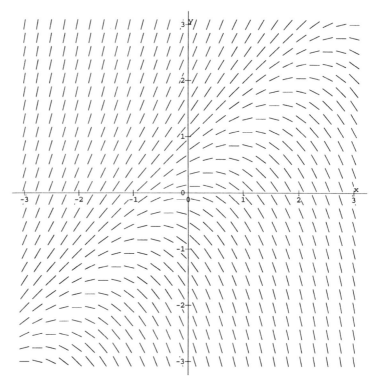

FIGURE 3.13   Slope field A

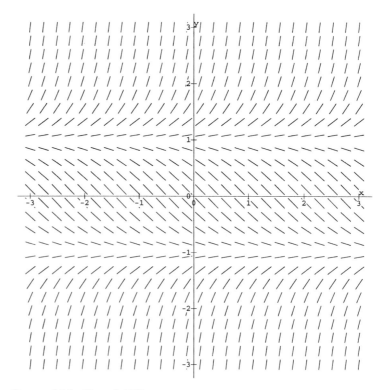

FIGURE 3.14   Slope field B

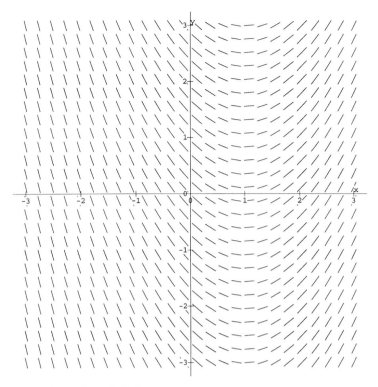

FIGURE 3.15 Slope field C

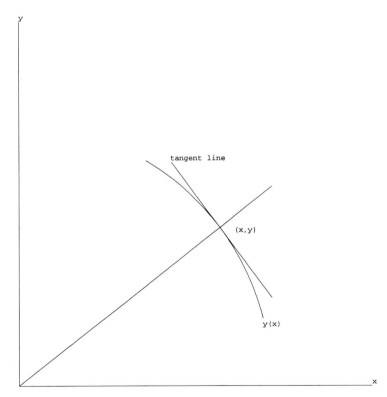

FIGURE 3.16 Tangent line problem

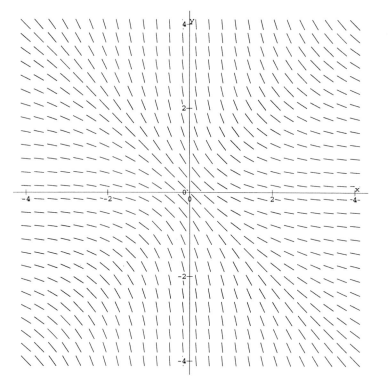

FIGURE 3.17   Mystery slope field 1

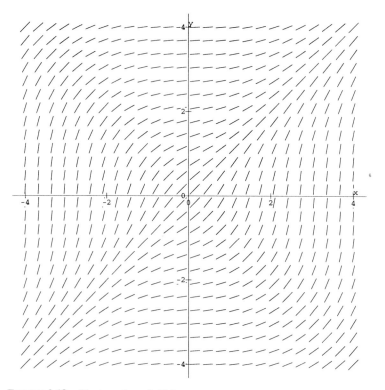

FIGURE 3.18   Mystery slope field 2

(b) Now superimpose the two mystery slope fields (perhaps by placing one on top of the other and holding both up to the light, or by plotting the slope fields for

$$\frac{dy}{dx} = a\frac{x^2 + 1}{y^2 + 1} + b\frac{y^2 + 1}{x^2 + 1}$$

with $a = \pm1$ and $b = 0$, and then $a = 0$ and $b = \pm1$). What do you notice? Would this have made your previous analysis in part (a) easier?

9. Substitute the given function $y(x)$ into the differential equation that follows, and determine n so $y(x)$ is an explicit solution.

(a) $y(x) = x^n$, $\qquad xy' - 3y = 0$

(b) $y(x) = nx + x^2$, $\qquad xy' - 2y + x = 0$

(c) $y(x) = 1 - e^{-nx}$, $\qquad y' + 32y - 32 = 0$

(d) $y(x) = (x - n)^2 e^x$, $\qquad (x - 2)y' - xy = 0$

(e) $y(x) = \exp(1 - ne^x)$, $\qquad y' + e^x y = 0$ (Note: $\exp u = e^u$.)

10. Show that the given function $y(x)$ is an explicit solution of the differential equation that follows for any choice of the constant C.

(a) $y(x) = Cx^4 - 1$, $\qquad xy' - 4y = 4$

(b) $y(x) = 1 - Ce^{-2x}$, $\qquad y' + 2y = 2$

(c) $y(x) = C(x - 2)^2 e^x$, $\qquad (x - 2)y' - xy = 0$

(d) $y(x) = C\exp(1 - e^x)$, $\qquad y' + e^x y = 0$ (Note: $\exp u = e^u$.)

11. For what value of the constant C does the graph of the solution for each part of Exercise 10 pass through the point $(0, -1)$?

12. Show that the following are implicit solutions of the given differential equations for any choice of the constant C.

(a) $x^2 + y^2 = C$, $\qquad y' = -x/y$

(b) $\sin xy + x^3 + xy^2 = C$, $\qquad (x\cos xy + 2xy)y' = -3x^2 - y^2 - y\cos xy$

(c) $xe^{-xy} + x^2 y = C$, $\qquad (x^2 - x^2 e^{-xy})y' = xye^{-xy} - e^{-xy} - 2xy$

(d) $\ln|ye^C| + e^{x^2} = y^3$, $\qquad (-3y^2 + 1/y)y' + 2xe^{x^2} = 0$

(e) $xy + (Cy)^{-1} = 3x$, $C \neq 0$, $\qquad (3x - 2xy)y' = y^2 - 3y$

13. Consider the implicitly defined function

$$x^2 + y^2 = C.$$

Show that, for this function,

$$\frac{dy}{dx} = -\frac{x}{y}.$$

Use slope fields to show that the curves characterized by the derivatives are circles.

14. Consider the implicitly defined function

$$x^2 - y^2 = C.$$

Show that, for this function,

$$\frac{dy}{dx} = \frac{x}{y}.$$

Use slope fields to show that the curves characterized by the derivatives are hyperbolae.

15. Consider the implicitly defined function

$$(x^2 + y^2)^2 = 12(x^2 - y^2).$$

Show that, for this function,

$$\frac{dy}{dx} = \frac{x(6 - x^2 - y^2)}{y(6 + x^2 + y^2)}.$$

Use slope fields to show the curves characterized by the derivatives. They are called lemniscates.[3] Notice that far from the origin, the curves look like circles, whereas near the origin they look like hyperbolae. Explain why this is the case by comparing this differential equation with those in Exercises 13 and 14.

16. Use the preceding techniques to sketch the implicit function

$$y^2 - xy + x^2 - 1 = 0.$$

Notice that the points $(0, \pm 1)$ lie on the curve.

17. Use the preceding techniques to sketch the implicit function

$$y^4 - 4(x + 1)y^3 + 3 = 0.$$

Notice that the point $(0, 1)$ lies on the curve.

18. Use the preceding techniques to sketch the implicit function

$$x^3 + y^3 = -8.$$

19. Use the preceding techniques to sketch the implicit function

$$x^3 + xy = 8.$$

3.2   Symmetry

In a number of previous examples we observed that the slope field is sometimes symmetric: about the x-axis, about the y-axis, or about the origin. For example, slope fields for $dy/dx = g(x)$ could be symmetric about the y-axis or the origin, while those for $dy/dx = g(y)$ could be symmetric about the x-axis or the origin. In this section we bring these ideas together in one place.

How to Test $dy/dx = g(x, y)$ for Symmetry

Purpose: To determine the symmetry of the slope field directly from the differential equation

$$\frac{dy}{dx} = g(x, y). \tag{3.16}$$

---

[3]From the Latin "lemniscata" meaning "adorned with ribbons".

> **Technique:**
>
> 1. Replace x with $-x$ in (3.16). If the resulting equation is identical to (3.16) — that is, (3.16) is unchanged — then the slope field is symmetric about the y-axis.
>
> 2. Replace y with $-y$ in (3.16). If the resulting equation is identical to (3.16), then the slope field is symmetric about the x-axis.
>
> 3. Replace x with $-x$ and y with $-y$ in (3.16). If the resulting equation is identical to (3.16), then the slope field is symmetric about the origin.

Comments about testing for symmetry

- This is a very useful result and is easily applied. Symmetry is one thing we should always check when analyzing a differential equation graphically.

- If a slope field is symmetric about any two of the x-axis, the y-axis, and the origin, it is automatically symmetric about the third (see Exercise 3).

- Symmetry of the slope field does not guarantee that a particular solution curve has the same symmetry (see Exercise 5).

- Symmetry of the slope field guarantees the same symmetry for the **family** of solutions. For example, consider a slope field that is symmetric about the x-axis (replacing y with $-y$ leaves the differential equation unchanged). If a particular solution curve passes through an initial point (x_0, y_0), then there is a solution curve (which could be the same curve) that passes through the point $(x_0, -y_0)$. This solution curve can be obtained by reflecting the original solution curve about the x-axis.

- The word mathematicians use in place of "unchanged" is "invariant". Thus, if the differential equation (3.16) is invariant under the interchange of x with $-x$, then the slope field is symmetric about the y-axis.

Words of Caution: Remember, when checking (3.16) for symmetry, make all substitutions on both sides of the equation, not just on the right-hand side of (3.16).

EXAMPLE 3.4

Identify the symmetries of the slope field for

$$\frac{dy}{dx} = \frac{y^2}{x}. \tag{3.17}$$

1. If we replace x with $-x$ on both sides of (3.17) we find

$$\frac{dy}{-dx} = \frac{y^2}{-x},$$

which reduces to

$$\frac{dy}{dx} = \frac{y^2}{x}.$$

Because this is identical to (3.17), the slope field is symmetric about the y-axis.

2. If we replace y with $-y$ on both sides of (3.17) we find

$$\frac{-dy}{dx} = \frac{(-y)^2}{x},$$

which reduces to

$$\frac{dy}{dx} = \frac{-y^2}{x}.$$

Because this is not identical to (3.17), the slope field is not symmetric about the x-axis.

3. If we replace x with $-x$, and y with $-y$ on both sides of (3.17) we find

$$\frac{-dy}{-dx} = \frac{(-y)^2}{-x},$$

which reduces to

$$\frac{dy}{dx} = \frac{-y^2}{x}.$$

Because this is not identical to (3.17), the slope field is not symmetric about the origin.

These observations are confirmed by Figure 3.19, which shows the slope field for (3.17). □

EXERCISES

1. Without sketching the slope field for

$$y^3 \frac{dy}{dx} = x^3,$$

decide what type of symmetry (if any) the slope field has. Sketch the slope field to confirm your analysis.

2. Without sketching the slope field for

$$\frac{dy}{dx} = x^2 + y^2,$$

decide what type of symmetry (if any) the slope field has. Sketch the slope field to confirm your analysis.

3. Show that
   (a) A slope field that is symmetric about both the x-axis and the y-axis is automatically symmetric about the origin.
   (b) A slope field that is symmetric about both the x-axis and the origin is automatically symmetric about the y-axis.
   (c) A slope field that is symmetric about both the origin and the y-axis is automatically symmetric about the x-axis.

4. (a) Give an example of a differential equation whose slope field is not symmetric about the x-axis, about the y-axis, or about the origin.
   (b) Give an example of a differential equation whose slope field is symmetric about the origin but is not symmetric about either the x-axis or the y-axis.
   (c) Give an example of a differential equation whose slope field is symmetric about the y-axis but is not symmetric about either the x-axis or the origin.
   (d) Give an example of a differential equation whose slope field is symmetric about the x-axis but is not symmetric about either the y-axis or the origin.

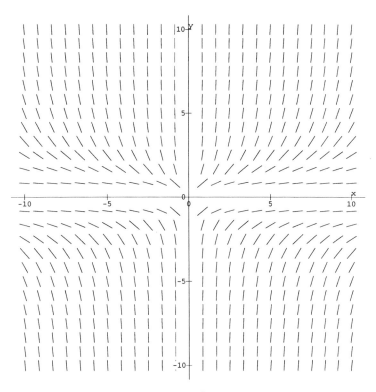

FIGURE 3.19 Slope field for $dy/dx = y^2/x$

5. This exercise concerns the differential equation

$$\frac{dy}{dx} = \frac{y}{2x}.$$

(a) Show that the slope field is symmetric about the x-axis, symmetric about the y-axis, and symmetric about the origin.

(b) Show that $y = \sqrt{x}$ $(x > 0)$ is a solution curve that passes through the point $(1, 1)$.

(c) Show that the solution curve in part (b) has no symmetries.

6. Find conditions on $f(y)$ and $g(x)$ such that the family of solutions of differential equations of the type $dy/dx = f(y)g(x)$ will be

(a) Symmetric with respect to the x-axis.

(b) Symmetric with respect to the y-axis.

(c) Symmetric with respect to the origin.

7. The slope field for $dy/dx = g(x, y)$ — with a particular $g(x, y)$ — is shown in Figure 3.20 for the window $0 < x < 5, 0 < y < 5$.

(a) Construct four members of the family of solutions on this figure by hand-drawing solution curves that pass through the points $(0, 1)$, $(0, 2)$, $(0, 3)$, and $(0, 4)$.

(b) Now suppose that you know that the slope field is symmetric about the y-axis. In which other quadrant(s) does this allow you to extend the slope field? Draw the four members of the family of solutions that are associated with those you drew in part (a).

(c) Repeat part (b) if the slope field is symmetric about the x-axis.

(d) Repeat part (b) if the slope field is symmetric about the origin.

(e) Repeat part (b) if the slope field is symmetric about both the x-axis and the y-axis.

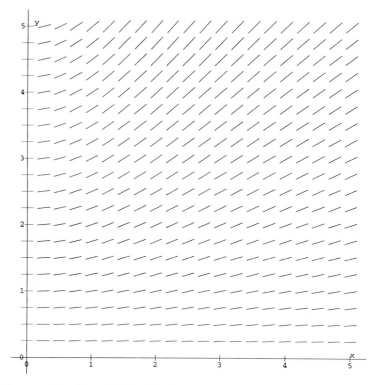

FIGURE 3.20   Slope field for Exercise 7

8. Repeat Exercise 7 for the slope field shown in Figure 3.21.

9. Show that differential equations of the type

$$\frac{dy}{dx} = g(x)$$

are invariant under the interchange of y with $y + C$ for every constant C. What does this tell you about the family of solutions of these differential equations?

10. Show that differential equations of the type

$$\frac{dy}{dx} = g(y)$$

are invariant under the interchange of x with $x + C$ for every constant C. What does this tell you about the family of solutions of autonomous differential equations?

11. A family of curves is symmetric about the line L if, whenever a member of the family passes through the point P, a member of the family also passes through the point Q, the mirror image of P through L. (See Figure 3.22.) Under these circumstances the line is L called a line of symmetry for the family of curves.

   (a) Show that if L is the line $y = x \tan A$, if P has coordinates (x_1, y_1), and if Q has coordinates (x_2, y_2), then

$$\begin{cases} x_2 = x_1 \cos 2A + y_1 \sin 2A, \\ y_2 = x_1 \sin 2A - y_1 \cos 2A. \end{cases}$$

   [Hint: Join the origin O to the point P. Let r be the distance OP. Let $A + B$ be the angle between OP and the x-axis. In the same way let $A - B$ be the angle between OQ and the

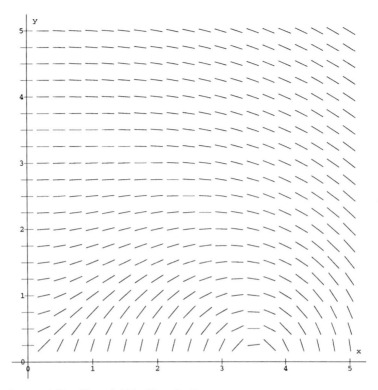

FIGURE 3.21 Slope field for Exercise 8

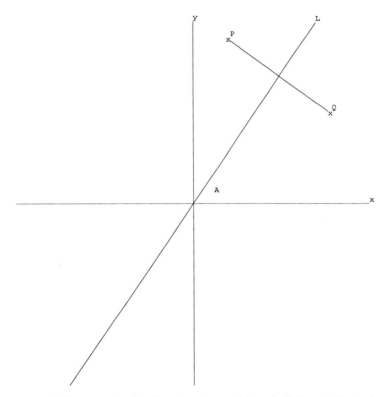

FIGURE 3.22 The point Q is the mirror image of the point P through the line L

x-axis. Express x_1, y_1, x_2, and y_2 in terms of r, $A + B$, and $A - B$. Now eliminate r and B from these four equations.]

(b) Show that if a family of curves is symmetric about the line

    i. $y = 0$, then the family of curves is unchanged if y is replaced with $-y$.

    ii. $x = 0$, then the family of curves is unchanged if x is replaced with $-x$.

    iii. $y = x$, then the family of curves is unchanged under the interchange of y with x.

    iv. $y = -x$, then the family of curves is unchanged under the interchange of y with $-x$.

(c) Show that the family of solutions of the differential equation

$$\frac{dy}{dx} = -\frac{x}{y}$$

is symmetric about all lines through the origin. Confirm this by constructing the slope field of this differential equation.

(d) Show that the family of solutions for the differential equation

$$\frac{dy}{dx} = \frac{x}{y}$$

is symmetric about the line $y = 0$ and the line $x = 0$. Does the family of solutions have any other lines of symmetry? Confirm your comments by constructing the slope field for this differential equation.

12. Find all lines of symmetry for the family of solutions for the following differential equations.

(a) $\dfrac{dy}{dx} = \dfrac{\sqrt{x^2 + y^2}}{x}$

(b) $\dfrac{dy}{dx} = \dfrac{\sqrt{x^2 + y^2}}{y}$

(c) $\dfrac{dy}{dx} = \dfrac{x\sqrt{x^2 + y^2}}{y}$

(d) $\dfrac{dy}{dx} = \sqrt{x^2 + y^2}$

(e) $\dfrac{dy}{dx} = \dfrac{x^2}{y^2}$

(f) $\dfrac{dy}{dx} = x + y$

13. This exercise deals with the differential equation

$$\frac{dy}{dx} = ax + by$$

where a and b are constants. What conditions must these constants satisfy if this differential equation has the following lines of symmetry? Confirm your answers by constructing the slope field for the appropriate differential equation.

(a) The x-axis

(b) The y-axis

(c) The line $y = x$

(d) The line $y = -x$

14. This exercise deals with the differential equation

$$\frac{dy}{dx} = \frac{ax + by}{cx + dy}$$

where a, b, c, and d are constants. What conditions must these constants satisfy if this differential equation has the following lines of symmetry? Confirm your answers by constructing the slope field for the appropriate differential equation.

(a) The x-axis

(b) The y-axis

(c) The line $y = x$

(d) The line $y = -x$.

3.3 Numerical Solutions: Euler's Method

The preceding techniques let us graphically sketch the solution curve $y = y(x)$ for the differential equation

$$\frac{dy}{dx} = g(x, y) \tag{3.18}$$

subject to the condition

$$y(x_0) = y_0 \tag{3.19}$$

without knowing the analytical solution. However, there are times when we need not only a sketch of the solution, but also fairly accurate numerical values of $y = y(x)$ for various values of x.

In Chapter 1 we found that the solution of $dy/dx = g(x)$ subject to (3.19) can be written in the form

$$y(x) = \int_{x_0}^{x} g(t)\, dt + y_0.$$

In this case, a numerical approximation for $y(x)$ can be computed for given values of x by any of the standard techniques used in the numerical evaluation of definite integrals, such as the left-hand rule, the right-hand rule, the midpoint rule, the trapezoidal rule, Simpson's rule, and so on.[4] Indeed, we used this technique to evaluate the Error Function in Chapter 1. Although we did not do so in Chapter 2, similar comments apply to finding $x = x(y)$ through standard numerical integration techniques. However, these numerical integration techniques fail on more general differential equations like (3.18).

We now introduce a method of approximating the solution of (3.18) subject to (3.19) that relies only on the form of the differential equation itself. This technique is known as EULER'S METHOD (Euler is pronounced "oiler") and is based on locally approximating a function by its tangent line.

In calculus we often find it convenient to approximate the graph of a differentiable function $y(x)$ near a point $x = a$ by its tangent line at that point, namely,

$$y(x) \approx y(a) + y'(a)(x - a).$$

---

[4]You should refer to your calculus text for a discussion of the pros and cons of these methods.

This approximation is called a "linear approximation" or "local linearity" and is frequently written as

$$y(a + h) \approx y(a) + y'(a)h, \tag{3.20}$$

where we have replaced $x - a$ with h. The geometric interpretation of (3.20) is shown in Figure 3.23.

EXAMPLE 3.5

Suppose we start with the differential equation

$$\frac{dy}{dx} = xy, \tag{3.21}$$

along with an initial condition

$$y(1) = 0.5,$$

where we want to obtain a numerical value for y when $x = 1.5$ — namely, $y(1.5)$. At present we have no analytical way of solving (3.21), but Figure 3.24 shows the slope field for (3.21) and the initial point, from which we could sketch a solution. From this hand-drawn solution curve we could estimate $y(1.5)$, but not very accurately.

We now use the ideas developed at the beginning of this section to find a numerical approximation at $x = 1.5$. We start at the initial point $(1, 0.5)$. We use the differential equation to discover that the slope of the solution curve at this point is $(1)(0.5) = 0.5$, from (3.21). We now draw a line from $(1, 0.5)$ to the right with slope 0.5 and assume that near

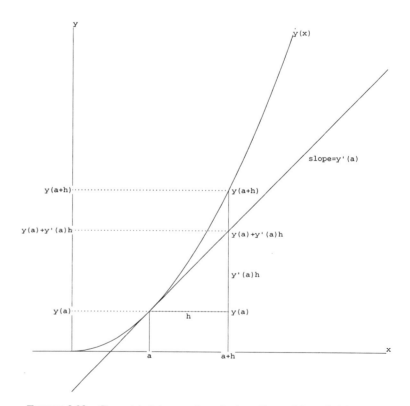

FIGURE 3.23   Geometric interpretation of $y(a + h) \approx y(a) + y'(a)h$

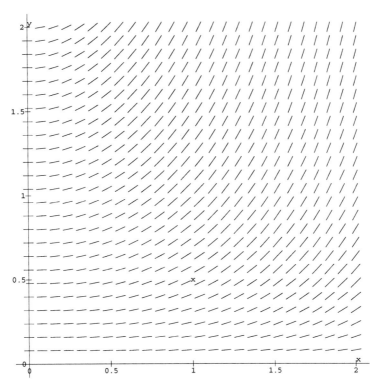

FIGURE 3.24 The initial point $(1, 0.5)$ and slope field for $dy/dx = xy$

the point $(1, 0.5)$ the graph of the function and its tangent line will be close together. If we use this tangent line to approximate the solution curve for $1 \leq x \leq 1.1$, then at $x = 1.1$ (3.20) gives an approximation for the solution as $0.5 + (0.5)(0.1)$, or 0.55. This is recorded at the extreme left-hand side of Table 3.3 where $h = 0.1$.

At the point $(1.1, 0.55)$, we use the differential equation to compute the slope of the tangent line as $(1.1)(0.55)$, or 0.605. We now can construct this new tangent line, starting at $(1.1, 0.55)$, with slope 0.605. We use this tangent line to approximate the solution of the differential equation for $x = 1.2$ as $0.55 + (0.605)(0.1)$, or 0.6105. A series of five steps using this pattern of moving along the tangent line, at each new point, to approximate the solution is shown in Table 3.3. This results in having $y(1.5) \approx 0.88082$. This point and the intermediate points are shown in Figure 3.25.

In the next section we will obtain the explicit solution of (3.21) subject to the condition $y(1) = 0.5$, as

$$y = \frac{1}{2} e^{\frac{1}{2}(x^2 - 1)}. \tag{3.22}$$

TABLE 3.3 Numerical approximation to the solution of $y' = xy$

x	y	$y' = xy$	h	$y'h$
1.0	0.5	(1)(0.5)=0.5	0.1	(0.5)(0.1)=0.05
1.1	0.5+0.05=0.55	(1.1)(0.55)=0.605	0.1	(0.605)(0.1)=0.0605
1.2	0.55+0.0605=0.6105	(1.2)(0.6105)=0.7326	0.1	(0.7326)(0.1)=0.07326
1.3	0.6105+0.07326=0.68376	(1.3)(0.68376)=0.88889	0.1	(0.88889)(0.1)=0.08889
1.4	0.68376+0.08889=0.77265	(1.4)(0.77265)=1.0817	0.1	(1.0817)(0.1)=0.10817
1.5	0.77265+0.10817=0.88082			

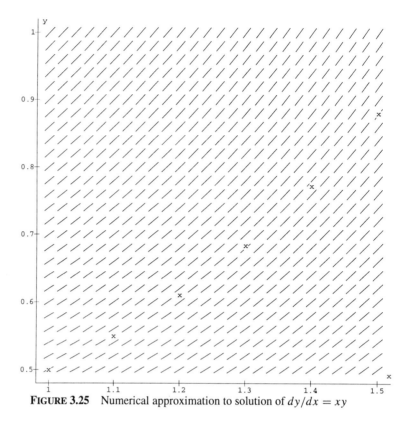

FIGURE 3.25   Numerical approximation to solution of $dy/dx = xy$

Because the accuracy of this procedure depends on the step size h, in Table 3.4 we compare the Euler approximation with the exact solution for different step sizes. In this table we can see how the accuracy of Euler's method depends on the step size of increments. Figure 3.26 shows the numerical approximation for $h = 0.1$ and the exact solution. Figure 3.27 shows the numerical approximation for $h = 0.001$ and the exact solution. Can you explain why, in this example, all the numerical values using Euler's method are smaller than the exact value for every value of $x > 1$? ☐

We can construct an algorithm for this process — namely, that the y value at a new value of x equals the y value at the old value of x plus the slope at the old value of x times the distance between these two values of x. This is perhaps more clearly stated in terms of an equation. We do it for a general first order differential equation,

$$\frac{dy}{dx} = g(x, y),$$

TABLE 3.4   **Comparison of solutions of $dy/dx = xy$, $y(1) = 0.5$**

x	Euler ($h = 0.1$)	Euler ($h = 0.01$)	Euler ($h = 0.001$)	Exact
1.0	0.5000	0.5000	0.5000	0.5000
1.1	0.5500	0.5548	0.5553	0.5554
1.2	0.6105	0.6217	0.6229	0.6230
1.3	0.6838	0.7036	0.7057	0.7060
1.4	0.7727	0.8041	0.8076	0.8080
1.5	0.8808	0.9282	0.9335	0.9341

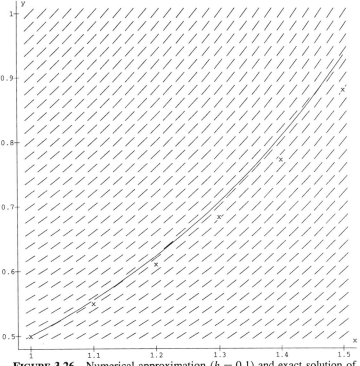

FIGURE 3.26 Numerical approximation ($h = 0.1$) and exact solution of $dy/dx = xy$

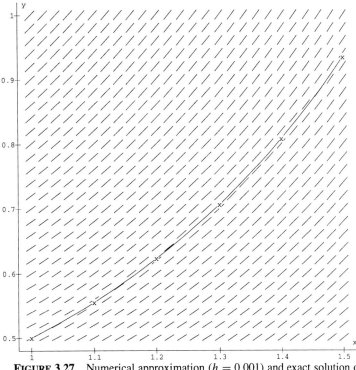

FIGURE 3.27 Numerical approximation ($h = 0.001$) and exact solution of $dy/dx = xy$

starting at the point (x_0, y_0). We first note that the slope of the tangent line to the solution curve at the point (x_0, y_0) is given by $g(x_0, y_0)$. This means the equation of the tangent line at this point is

$$y - y_0 = g(x_0, y_0)(x - x_0).$$

If we move to the right along this tangent line a horizontal distance h, we estimate the y value of the solution of the differential equation as

$$y(x_0 + h) \approx y(x_0) + g(x_0, y_0)h, \tag{3.23}$$

where $y(x_0) = y_0$.

Continuing this process gives

$$y(x_0 + 2h) \approx y(x_0 + h) + g(x_0 + h, y_0 + h)h. \tag{3.24}$$

Because the initial value of y was y_0, we call the first numerical approximation to the solution of the differential equation y_1, the second approximation y_2, and so on. This means we can rewrite (3.23) and (3.24) as

$$y_1 = y_0 + g(x_0, y_0)h$$

and

$$y_2 = y_1 + g(x_1, y_1)h, \text{ where } x_1 = x_0 + h,$$

or, after n steps,

$$y_n = y_{n-1} + g(x_{n-1}, y_{n-1})h$$

where

$$x_{n-1} = x_0 + (n-1)h.$$

This way of constructing a numerical approximation to a differential equation is called Euler's method.

How to Analyze $dy/dx = g(x, y)$ Numerically

Purpose: To use Euler's method to obtain a numerical approximation to

$$\frac{dy}{dx} = g(x, y), \tag{3.25}$$

starting at the point (x_0, y_0).

Technique:

1. Select a step size h.

2. Compute $y_1 = y(x_0) + g(x_0, y_0)h$, where $y(x_0) = y_0$. The value of y_1 is a numerical approximation to the solution of (3.25) at $x_1 = x_0 + h$, that is, $y(x_1) \approx y_1$.

3. Use x_1 and y_1 from step 2 to compute $y_2 = y_1 + g(x_1, y_1)h$. The value of y_2 is a numerical approximation to the solution of (3.25) at $x_2 = x_1 + h = x_0 + 2h$, that is, $y(x_2) \approx y_2$.

4. Continue in this way so that after n steps the value of y_n, where $y_n = y_{n-1} + g(x_{n-1}, y_{n-1})h$, is a numerical approximation to the solution of (3.25) at $x_n = x_0 + nh$, that is, $y(x_n) \approx y_n$.

Comments about analyzing $y' = g(x, y)$ numerically

- The usual advice in choosing the step size h is "the smaller the better."

- If you want to find a numerical solution over an interval $a \le x \le b$, starting at $(a, y(a))$ and ending when $x = b$, choose the step size h so that $h = (b - a)/n$, where n is the number of steps you will take along the x-axis.

- Euler's method also works going from larger values of x to smaller. In this case h is negative.

- Euler's method is one of many ways to obtain a numerical approximation to a solution of a differential equation, or, in short, a numerical solution. Some of these ways are described in the appendix.

- Every numerical method has its problems. Exercise 4 shows some of these for Euler's method.

- **From now on, we assume that either you have access to a computer/calculator program that performs Euler's method or you are willing to do these calculations by hand.**

EXERCISES

1. Complete steps i through iv for each of the following first order differential equations (a) through (k):

   i. Find the exact solution (see Exercise 1, Section 1.1), and evaluate it at x_1, x_2, x_3, and x_4.

   ii. Use Simpson's rule to find approximate values for $y(x_1)$, $y(x_2)$, $y(x_3)$, and $y(x_4)$. Each time start with the point $(x_0, y(x_0))$. Compare these results with the exact values.

   iii. Use Euler's method with $h = 0.1$ to find approximate values for $y(x_1)$, $y(x_2)$, $y(x_3)$, and $y(x_4)$. Compare these results with the exact values.

   iv. Repeat iii with $h = 0.05$, making the appropriate changes in $x_1, x_2, x_3, \cdots, x_8$.

   (a) $dy/dx = x^3$ subject to $y(1) = 1$. $x_0 = 1, x_1 = 1.1, x_2 = 1.2, x_3 = 1.3, x_4 = 1.4$

   (b) $dy/dx = x^4$ subject to $y(1) = 1$. $x_0 = 1, x_1 = 1.1, x_2 = 1.2, x_3 = 1.3, x_4 = 1.4$

   (c) $dy/dx = \cos x$ subject to $y(0) = 0$. $x_0 = 0, x_1 = 0.1, x_2 = 0.2, x_3 = 0.3, x_4 = 0.4$

   (d) $dy/dx = \sin x$ subject to $y(\pi) = 2$. $x_0 = \pi, x_1 = \pi + 0.1, x_2 = \pi + 0.2, x_3 = \pi + 0.3, x_4 = \pi + 0.4$

   (e) $dy/dx = e^{-x}$ subject to $y(0) = 1$. $x_0 = 0, x_1 = 0.1, x_2 = 0.2, x_3 = 0.3, x_4 = 0.4$

   (f) $dy/dx = 1/x^2$ subject to $y(1) = 1$. $x_0 = 1, x_1 = 1.1, x_2 = 1.2, x_3 = 1.3, x_4 = 1.4$

   (g) $dy/dx = 1/x$ subject to $y(-1) = 1$. $x_0 = -1, x_1 = -0.9, x_2 = -0.8, x_3 = -0.7, x_4 = -0.6$

   (h) $dy/dx = 1/[x(x - 1)]$ subject to $y(2) = 1$. $x_0 = 2, x_1 = 2.1, x_2 = 2.2, x_3 = 2.3, x_4 = 2.4$

   (i) $dy/dx = 1/(1 + x^2)$ subject to $y(1) = \pi/4$. $x_0 = 1, x_1 = 1.1, x_2 = 1.2, x_3 = 1.3, x_4 = 1.4$

   (j) $dy/dx = \ln x$ subject to $y(1) = 1$. $x_0 = 1, x_1 = 1.1, x_2 = 1.2, x_3 = 1.3, x_4 = 1.4$

   (k) $dy/dx = x^2 e^{-x}$ subject to $y(0) = 1$. $x_0 = 0, x_1 = 0.1, x_2 = 0.2, x_3 = 0.3, x_4 = 0.4$

2. Consider the differential equation

$$\frac{dy}{dx} = g(x)$$

subject to $y(0) = 0$. Explain why using Euler's method to approximate the solution gives the same result as using the left-hand rule for Riemann sums to approximate the definite integral $\int_0^x g(t)\,dt$. How would replacing the condition $y(0) = 0$ with $y(x_0) = y_0$ change this result? Does it matter if $g(x)$ is an increasing or decreasing function? Does it matter if $g(x)$ is positive or negative?

3. Solve the following problems numerically on the indicated interval using Euler's method with $h = 0.1$.

(a) $\qquad\qquad\qquad y' = 2xy, \qquad y(1) = 1, \qquad 1 \le x \le 1.5$

(b) $\qquad\qquad\qquad y' = 1 + y^2, \qquad y(0) = 0, \qquad 0 \le x \le 0.5$

(c) $\qquad\qquad\qquad y' = e^{-y}, \qquad y(0) = 0, \qquad 0 \le x \le 0.5$

(d) $\qquad\qquad\qquad y' = 2x - 3y, \qquad y(1) = 1, \qquad 1 \le x \le 1.5$

(e) $\qquad\qquad\qquad y' = x, \qquad y(0) = 1, \qquad 0 \le x \le 1$

4. Use Euler's method with $h = 0.18, 0.23, 0.25$, and 0.3 to obtain numerical solutions of the logistics equation

$$\frac{dy}{dx} = 10y(1 - y)$$

subject to $y(0) = 0.1$, from $x = 0$ to $x = 5$. Notice that in general these numerical solutions do not go to the stable equilibrium solution of $y(x) = 1$.

5. This exercise deals with the differential equation

$$\frac{dy}{dx} = (y - a)(y - b)(y - c).$$

(a) Draw direction fields for each of the following cases i through iii, and sketch by hand what you think the solution curves will look like.

     i. $a = b = c = 1$

     ii. $a = b = 1, c = -1$

     iii. $a = 1, b = 0, c = -1$

(b) Use technology — changing the step size if needed — to draw solution curves for each of the cases in part (a) for various initial conditions. Make sure the results you obtain here are consistent with the answers you gave to part (a).

3.4   Power Series Solutions

In the preceding section we concentrated on the differential equation

$$\frac{dy}{dx} = xy \qquad\qquad\qquad\qquad (3.26)$$

subject to the initial condition

$$y(1) = 0.5. \tag{3.27}$$

We stated, without justification, that its solution is

$$y(x) = \frac{1}{2} e^{\frac{1}{2}(x^2 - 1)}. \tag{3.28}$$

In this section we will justify this result. The method we introduce hinges on the fact that (3.26) satisfies the conditions of Theorem 2.1 on page 72 and so we know that there is a unique solution to this problem.

At the end of Chapter 1 we found that although we could not express the Error Function in terms of familiar functions, we could obtain some idea of its behavior by expressing it as a power series. We use the same idea here to find a solution $y(x)$ of (3.26) in the form of a power series

$$y(x) = \sum_{k=0}^{\infty} c_k x^k = c_0 + c_1 x + c_2 x^2 + c_3 x^3 + \cdots + c_n x^n + \cdots, \tag{3.29}$$

where c_0, c_1, c_2, $\cdots$, are constants. Once we find such a solution, then the uniqueness theorem guarantees that we have found the solution. So the idea is simple: we assume that (3.26) has a solution of the form (3.29) and then determine the constants c_0, c_1, c_2, $\cdots$. The crucial feature is not whether we are allowed to make such an assumption about the solution being a power series, but whether it allows us to find a solution.

To determine the constants c_0, c_1, c_2, $\cdots$, we substitute (3.29) and its derivative into (3.26), which gives

$$c_1 + 2c_2 x + 3c_3 x^2 + \cdots + nc_n x^{n-1} + \cdots = x(c_0 + c_1 x + c_2 x^2 + \cdots + c_m x^m + \cdots)$$
$$= c_0 x + c_1 x^2 + c_2 x^3 + \cdots + c_m x^{m+1} + \cdots.$$

This equation can be rewritten in the form

$$c_1 + \left(2c_2 - c_0\right) x + \left(3c_3 - c_1\right) x^2 + \left(4c_4 - c_2\right) x^3 + \left(5c_5 - c_3\right) x^4 + \cdots = 0. \tag{3.30}$$

We want this to be an identity in x — that is, valid for every x. Equating the coefficients of x^0, x^1, x^2, $\cdots$ to zero we get

$$\begin{cases} c_1 & = 0, \\ 2c_2 - c_0 & = 0, \\ 3c_3 - c_1 & = 0, \\ 4c_4 - c_2 & = 0, \\ 5c_5 - c_3 & = 0, \\ 6c_6 - c_4 & = 0, \end{cases} \tag{3.31}$$

and so on. From this we see that

$$c_1 = 0,$$

which, when used in the third equation of (3.31) implies that

$$c_3 = 0,$$

which, when used in the fifth equation of (3.31) implies that

$$c_5 = 0.$$

If we take more terms in (3.30) we find that all the odd terms are zero, that is,

$$c_{2n+1} = 0, \qquad n = 0, 1, 2, \cdots. \tag{3.32}$$

If we return to (3.31), we see that the second equation implies that

$$c_2 = \frac{1}{2} c_0,$$

which, when used in the fourth equation implies that

$$c_4 = \frac{1}{4} c_2 = \frac{1}{2^2} \frac{1}{2!} c_0,$$

which, when used in the sixth equation implies that

$$c_6 = \frac{1}{6} c_4 = \frac{1}{2^3} \frac{1}{3!} c_0.$$

If we take more terms in (3.30) we find all the even terms follow the same pattern, that is,

$$c_{2n} = \frac{1}{2^n} \frac{1}{n!} c_0, \qquad n = 1, 2, 3, \cdots. \tag{3.33}$$

If we substitute (3.32) and (3.33) into (3.29), we find

$$y(x) = c_0 + \frac{x^2}{2} c_0 + \frac{x^4}{2^2} \frac{1}{2!} c_0 + \frac{x^6}{2^3} \frac{1}{3!} c_0 + \cdots$$

which, if we define $z = x^2/2$, can be written in the form

$$y = c_0 \left[1 + z + \frac{1}{2!} z^2 + \frac{1}{3!} z^3 + \cdots \right].$$

We recognize the quantity in square brackets as the Taylor series expansion for e^z — convergent for all z — so we have

$$y(x) = c_0 e^z = c_0 e^{\frac{1}{2} x^2}, \tag{3.34}$$

where c_0 is an undetermined constant. This constant is determined by our initial condition $y(1) = 0.5$ which gives

$$c_0 = \frac{1}{2} e^{-\frac{1}{2}}.$$

When the latter is used in (3.34) we find that

$$y(x) = \frac{1}{2} e^{\frac{1}{2}(x^2 - 1)}$$

is the solution of $dy/dx = xy$ subject to $y(1) = 0.5$. Thus (3.28) is the solution of this initial value problem.

EXERCISES

1. Solve each of the following differential equations, by assuming that they each have a power series solution of the form (3.29). Try to express your series solution in terms of familiar functions.

   (a) $y' = -y$.

   (b) $y' = 2y$.

   (c) $y' = -y + x$.

   (d) $(1 + x)y' = 2y$.

What Have We Learned?

MAIN IDEAS

- An IMPLICIT SOLUTION of a first order differential equation is any relationship between x and y that through implicit differentiation yields the original differential equation.

- One method for obtaining a numerical approximation to $dy/dx = g(x, y)$ is Euler's method. It starts at a point (x_0, y_0) and computes subsequent values of x and y using $x_n = x_0 + nh$, $y_n = y_{n-1} + g(x_{n-1}, y_{n-1})h$, where h is the step size.

- We can obtain solutions of differential equations in the form of power series. Chapter 9 contains a detailed discussion of this method.

How to Fully Analyze $dy/dx = g(x, y)$

Purpose: To summarize the major steps required to fully analyze the differential equation

$$\frac{dy}{dx} = g(x, y).$$

Technique: Note that the order is flexible.

1. Check that a solution exists. See *How to See Whether Unique Solutions of* $dy/dx = g(x, y)$ *Exist* on page 74.

2. Perform a graphical analysis.

   (a) Check where solutions are increasing, decreasing, concave up, concave down, and have inflection points. See *How to Analyze* $dy/dx = g(x, y)$ *Graphically* on page 98.

   (b) Check whether the slope field has any symmetries. See *How to Test* $dy/dx = g(x, y)$ *for Symmetry* on page 112.

   (c) Use technology to plot the slope field. Confirm using isoclines.

   (d) If the equation is autonomous, analyze the phase line. See *How to Perform a Phase Line Analysis on* $dy/dx = g(y)$ on page 89.

3. Perform a numerical analysis using technology. See *How to Analyze* $dy/dx = g(x, y)$ *Numerically* on page 124.

4. Find an explicit or implicit solution. (Some methods are given in later chapters.)

CHAPTER 4

Separable Differential Equations and Applications

Where Are We Going?

In this chapter we continue the discussion of Chapter 3 by looking at a special type of first order differential equation, namely,

$$\frac{dy}{dx} = f(y)g(x),$$

where the right-hand side is the product of a function of y and a function of x. Such equations are called separable. In the first part of the chapter we focus on obtaining analytical solutions and developing a technique for graphing such solutions if we cannot solve explicitly for one of the variables. We devote the rest of the chapter to applications of separable differential equations, including a discussion on deriving differential equations from data.

4.1   Separable Equations

In this section we look at a simple application of differential equations to a mixture problem that leads to a type of differential equation we have not yet discussed — separable differential equations. We discover how to find analytical solutions of these equations and then apply the technique to a number of examples.

The mixture problem we consider now is to determine the concentration of solute in a container. To derive a differential equation that describes this process, we let y be a function that represents the amount of substance in a given container at time x, and we assume that the instantaneous rate of change of y with respect to x is given by

$$\frac{dy}{dx} = \begin{matrix} \text{rate at which substance} \\ \text{is added to the container} \end{matrix} - \begin{matrix} \text{rate at which substance} \\ \text{is leaving the container} \end{matrix}.$$

This equation is often called an EQUATION OF CONTINUITY or a CONSERVATION EQUATION.

EXAMPLE 4.1: Solute in a Container

A 300-gallon container is 2/3 full of water containing 50 pounds of salt. At time $x = 0$, valves are turned on so pure water is added to the container at a rate of 3 gallons per minute. If the well-stirred mixture is drained from the container at the rate of 2 gallons per minute, how many pounds of salt are in the container when it is full (and all valves are turned off)?

    We first note that more of the liquid is being added per minute than is being drained, so the number of gallons in the container is increasing. In fact we may use a continuity-type argument to note that the rate of change of volume V of liquid in the container equals the rate being added minus the rate being drained. Thus, we have that

$$\frac{dV}{dx} = 3 - 2 = 1 \text{ gallon per minute.}$$

Integration gives

$$V(x) = x + 200. \tag{4.1}$$

Why is the constant of integration 200?

    If y represents the number of pounds of salt in the container at time x, the concentration of salt is

$$\frac{y}{V} = \frac{y}{x + 200}.$$

Thus, the rate at which salt is leaving the container is $2y/(x + 200)$. Because the rate at which salt is arriving in the container is 0 — there is no salt in the incoming pure water — the continuity equation gives

$$\frac{dy}{dx} = -\frac{2y}{x + 200}, \tag{4.2}$$

which is valid for $0 < x < 100$. (Why don't we consider values of time greater than 100?) Because there are 50 pounds of salt in the container at $x = 0$, the proper initial condition is

$$y(0) = 50.$$

    If we look at the slope field for (4.2) in Figure 4.1, we observe that all solutions will be decreasing and concave up. [To fully convince yourself of these facts, consider (4.2) and the result of differentiating (4.2).] Note that the slope field ignores the condition that the container is full when $x = 100$. We have also hand-drawn the solution curve that passes through the inital point $(0, 50)$. From this curve we can estimate the value of $y(x)$ for $x = 100$ at about 25 pounds, which is an approximate answer to our original question. However, to obtain an exact answer we must obtain an explicit solution.

    Equation (4.2) is a type of differential equation that we have not seen before. However if we rewrite it in the form

$$\frac{1}{y}\frac{dy}{dx} = \frac{-2}{x + 200},$$

and then integrate with respect to x, we see that

$$\int \frac{1}{y}\frac{dy}{dx}\,dx = \int \frac{-2}{x + 200}\,dx,$$

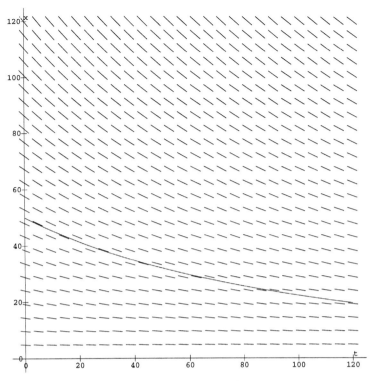

FIGURE 4.1 Hand-drawn solution curve and slope field for $dy/dx = -2y/(x + 200)$

or

$$\int \frac{1}{y} \, dy = \int \frac{-2}{x + 200} \, dx.$$

These integrals are easily performed and we find

$$\ln y + C_1 = -2 \ln (x + 200) + C_2,$$

where C_1 and C_2 are arbitrary constants. Notice we used the fact that x and y are positive. (Where do we use this fact?) This equation can be rewritten as

$$\ln y = -2 \ln (x + 200) + C,$$

where $C = C_2 - C_1$ is an arbitrary constant. Solving for y gives

$$y(x) = \frac{e^C}{(x + 200)^2}.$$

From the initial condition $y(0) = 50$, we find $e^C = 50 \left(200^2\right)$, so our final form of the solution is

$$y(x) = \frac{50 \left(200^2\right)}{(x + 200)^2}. \tag{4.3}$$

To answer the original question about how many pounds of salt are in the container when full, we note from (4.1) that the container will be full when $x = 100$, so $y(100)$ will be the amount of salt in the container at this time. From (4.3) we have that

$$y(100) = \frac{50\left(200^2\right)}{300^2} = 22\frac{2}{9} \text{ pounds of salt,}$$

which is close to our estimate of 25 pounds. $\square$

Equation (4.2) is an example of a special type of differential equation that may be solved using only integration techniques — SEPARABLE DIFFERENTIAL EQUATIONS .

Definition 4.1:   **A SEPARABLE DIFFERENTIAL EQUATION has the form**

$$\frac{dy}{dx} = f(y)g(x), \tag{4.4}$$

where the right-hand side is the product of a function of y and a function of x.

Another example of a separable differential equation is

$$\frac{dy}{dx} = xy,$$

discussed in Example 3.5 on page 120 of the last chapter.

Solutions of separable equations may be obtained by separating the y and x dependence as

$$\frac{1}{f(y)}\frac{dy}{dx} = g(x) \tag{4.5}$$

and then integrating both sides with respect to x, so that

$$\int \frac{1}{f(y)}\frac{dy}{dx}\,dx = \int g(x)\,dx,$$

or

$$\int \frac{1}{f(y)}\,dy = \int g(x)\,dx. \tag{4.6}$$

Observe that after isolating the two variables y and x, we are left with a calculus problem — namely, finding antiderivatives of both sides of an equation.

Comments about separable equations

- If we apply Theorem 2.1 on page 72 to (4.4) we see that we are guaranteed a unique solution through the point (x_0, y_0) if $g(x)$ and df/dy are continuous in the vicinity of (x_0, y_0). [Strictly speaking, Theorem 2.1 requires the condition that $f(y)$ is continuous in the vicinity of (x_0, y_0). Why is this condition unnecessary in the current situation?]

- Earlier we defined equilibrium solutions as solutions for which $y(x) = c$, for all x. From (4.4), we see that equilibrium solutions will occur at the roots $y(x) = c$ of

$$f(y) = 0. \tag{4.7}$$

Thus, to find equilibrium solutions of (4.4), we should solve (4.7) for $y = c$.

- If we look at the prescription we have just given to solve (4.4), we realize that we have implicitly assumed that $f(y) \neq 0$ in going from (4.4) to (4.5). Thus, to ensure that we do not inadvertently discard some solutions, we suggest the following procedure:

  (a) First, find all equilibrium solutions — that is, find all values of y that satisfy $f(y) = 0$.

  (b) Second, for $y \neq 0$, perform the indicated integrations in (4.6)

How to Solve Separable Differential Equations

Purpose: To find all solutions $y = y(x)$ of

$$\frac{dy}{dx} = f(y)g(x), \tag{4.8}$$

for given $f(y)$ and $g(x)$.

Technique:

1. Check that the uniqueness theorem is satisfied.

2. Find the equilibrium solutions. Solve $f(y) = 0$ for y. The solutions $y = c_1$, $y = c_2$, $\cdots$, are the equilibrium solutions.

3. Find the nonequilibrium solutions. Rewrite (4.8) in the form

$$\frac{1}{f(y)}\frac{dy}{dx} = g(x),$$

and integrate with respect to x:

$$\int \frac{1}{f(y)}\frac{dy}{dx}\,dx = \int g(x)\,dx,$$

or

$$\int \frac{1}{f(y)}\,dy = \int g(x)\,dx.$$

Comments about solving separable differential equations

- A common mistake in solving separable equations is to go straight to step 3, missing any equilibrium solutions.

- Frequently, the nonequilibrium solutions are implicitly defined. Sometimes, implicitly defined functions can be solved for $y = y(x)$, or for $x = x(y)$. If this can be done, do so.

- When writing down antiderivatives in step 3, we need only include an arbitrary constant on one side of the equation. (Why?) It is a common mistake to forget all constants of integration.

- The solutions obtained in steps 2 and 3 are all the solutions of (4.8) if the uniqueness theorem is satisfied.

EXAMPLE 4.2

We now consider the equation discussed in Example 3.5 on page 120, namely,

$$\frac{dy}{dx} = xy. \tag{4.9}$$

We merely want to go through the mechanics of finding the explicit solution, so we will suppress the graphical analysis of (4.9) in this example. The uniqueness theorem is satisfied everywhere because xy and $\partial(xy)/\partial y = x$ are both continuous everywhere. We notice that $y(x) = 0$ is an equilibrium solution.

Assuming $y(x) \neq 0$, we divide both sides by y to obtain

$$\frac{1}{y}\frac{dy}{dx} = x.$$

Integration gives

$$\ln|y| = \frac{1}{2}x^2 + c,$$

where c is an arbitrary constant. Taking exponentials we find

$$y(x) = Ce^{\frac{1}{2}x^2}, \tag{4.10}$$

where $C = \pm e^c$, which is never zero. If we allow C to be zero, the equilibrium solution $y(x) = 0$ can be absorbed into (4.10), so (4.10) contains all the explicit solutions of (4.9).

If we impose the same initial condition

$$y(1) = 0.5,$$

as we used in Example 3.5 on page 120, we see that

$$0.5 = Ce^{\frac{1}{2}},$$

so

$$C = \frac{1}{2}e^{-\frac{1}{2}}.$$

If we substitute this value for C into (4.10) we find

$$y(x) = \frac{1}{2}e^{\frac{1}{2}(x^2-1)},$$

as stated in (3.22) on page 121.   □

EXAMPLE 4.3

Another example of a separable equation is

$$\frac{dy}{dx} = \frac{2xy}{x^2+1}. \tag{4.11}$$

The uniqueness theorem is satisfied in the entire xy plane. Also, the slope field is symmetric about the x-axis, about the y-axis, and about the origin. Clearly, (4.11) has the equilibrium solution

$$y(x) = 0.$$

Assuming $y(x) \neq 0$, we write (4.11) in the form

$$\frac{1}{y}\frac{dy}{dx} = \frac{2x}{x^2+1}.$$

Integration of both sides using u-substitution yields

$$\ln|y| = \ln|x^2+1| + c. \tag{4.12}$$

Applying the exponential function to both sides of (4.12), we obtain

$$e^{\ln|y|} = e^{\ln|x^2+1|+c} = e^c e^{\ln|x^2+1|}$$

or

$$y(x) = C(x^2+1), \tag{4.13}$$

where $C = \pm e^c$, which is never zero. Note that we assumed that $y \neq 0$ in rearranging the equation. However, $y(x) = 0$ is an equilibrium solution, and it may be obtained by setting $C = 0$ in (4.13). Thus, (4.13) contains all the explicit solutions of (4.11) if we allow C to take on any value.

Before plotting (4.13), let's see what calculus tells us. From (4.11) we know the solution is increasing in the first and third quadrants and decreasing in the second and fourth. We now want to determine concavity, so we calculate the second derivative of y. Differentiating (4.11) gives

$$\frac{d^2y}{dx^2} = \frac{2\left[(y + xy')(x^2+1) - xy(2x)\right]}{(x^2+1)^2}$$

and substituting for y' from (4.11), we see that

$$\frac{d^2y}{dx^2} = \frac{2y}{x^2+1}.$$

This shows that the solution is concave up for $y > 0$ and concave down for $y < 0$. Finally we note that Theorem 2.1 is satisfied for all x and y, so solution curves will not intersect. Because $y(x) = 0$ (the x-axis) is a solution curve, no curves will cross it.

Figure 4.2 shows the slope field for (4.11) and some analytical solution curves, which are in good agreement with all the preceding information. □

EXAMPLE 4.4

As another example of a separable equation, let us look at

$$\frac{dy}{dx} = \frac{2xy}{x^2-1}, \text{ where } x \neq \pm 1. \tag{4.14}$$

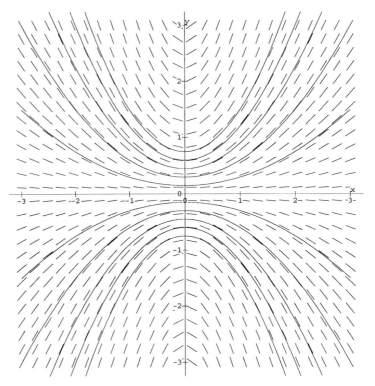

FIGURE 4.2 Analytical solution curves and slope field for $dy/dx = 2xy/(x^2 + 1)$

The slope field is symmetric about the x-axis, about the y-axis, and about the origin. Clearly (4.14) has the equilibrium solution

$$y(x) = 0.$$

Before continuing, we should realize that we are now dealing with three distinct regions in the xy-plane — namely, $x < -1$, $-1 < x < 1$, and $x > 1$. The conditions of Theorem 2.1 are satisfied for each of these regions separately, so within any region, solution curves will not cross. In particular they will not cross the x-axis, the solution curve $y(x) = 0$.

If we concentrate on $y \neq 0$, we can rewrite (4.14) as

$$\frac{1}{y}\frac{dy}{dx} = \frac{2x}{x^2 - 1},$$

and integration of both sides yields

$$\ln|y| = \ln|x^2 - 1| + c. \tag{4.15}$$

Applying the exponential function to both sides of (4.15), we obtain

$$e^{\ln|y|} = e^{\ln|x^2-1|+c} = e^c e^{\ln|x^2-1|}$$

or

$$y(x) = C(x^2 - 1). \tag{4.16}$$

Note that we assumed that $y \neq 0$ in rearranging the equation. However, $y = 0$ is an equilibrium solution, and it may be obtained by setting $C = 0$ in (4.16). Thus, (4.16) represents the explicit solution of (4.14) in each of the distinct regions.

Before plotting (4.16), let's see what calculus tells us. From (4.14) we know the solution is increasing in the first and third quadrants if $x^2 - 1 > 0$ (that is, $x < -1$ or $x > 1$) and decreasing if $x^2 - 1 < 0$ (that is, $-1 < x < 1$). In the second and fourth quadrants, we know the solution is decreasing if $x^2 - 1 > 0$ and increasing if $x^2 - 1 < 0$. Differentiating (4.14) and substituting for y' from (4.14), we see that

$$\frac{d^2y}{dx^2} = \frac{2y}{x^2 - 1},$$

so the solution is concave up for $y > 0$ and $x^2 - 1 > 0$, or $y < 0$ and $x^2 - 1 < 0$, and concave down for $y < 0$ and $x^2 - 1 > 0$, or $y > 0$ and $x^2 - 1 < 0$. Figure 4.3 shows this analysis with concave up and decreasing written in the upper left portion. In the other seven portions of the figure indicate the proper concavity and whether solutions are increasing or decreasing.

Figure 4.4 shows the slope field for (4.14) and some analytical solution curves, which are in good agreement with all the preceding information. However, we must stress that no solution passes from one region to another. In particular, in spite of appearances, no solution curves cross the x-axis at $x = \pm 1$. (Why?)   □

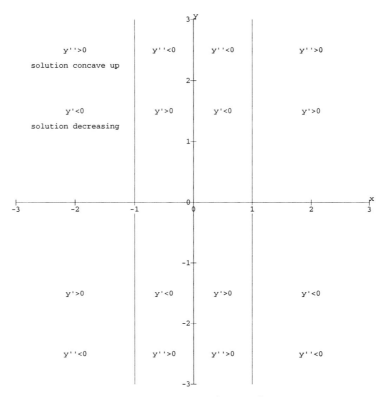

FIGURE 4.3 Graphical analysis of $dy/dx = 2xy/(x^2 - 1)$

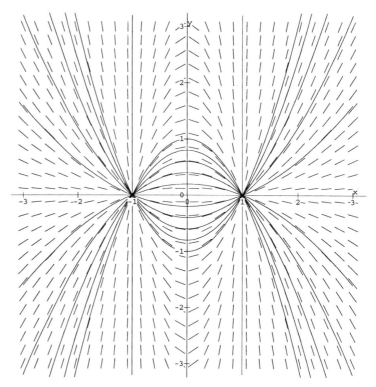

FIGURE 4.4 Analytical solution curves and slope field for $dy/dx = 2xy/(x^2 - 1)$

EXAMPLE 4.5

Now let's consider the equation

$$\frac{dy}{dx} = \frac{-y^5}{x\left(1 + y^4\right)} \tag{4.17}$$

for $x > 0$ and $y > 0$. Because $y > 0$ there are no equilibrium solutions, and because $x > 0$ and $y > 0$ there is no reason to look at symmetry.

Separation of the variables in (4.17) yields

$$\frac{1 + y^4}{y^5}\frac{dy}{dx} = -\frac{1}{x},$$

so that integration of both sides with respect to x gives

$$\int \frac{1 + y^4}{y^5}\frac{dy}{dx}\,dx = -\int \frac{1}{x}\,dx$$

or

$$\int \frac{1 + y^4}{y^5}\,dy = -\ln x + C. \tag{4.18}$$

The integral appearing on the left-hand side of (4.18) is easily computed if we realize that the integrand can be expressed in the form $y^{-5} + 1/y$, so we find the solution of (4.18) as

$$-\frac{1}{4y^4} + \ln y = -\ln x + C.$$

This is slightly different from our previous examples in that we are unable to solve this explicitly for y as a function of x. However, we can solve it for x as a function of y, obtaining

$$x(y) = \frac{c}{y}e^{1/(4y^4)}, \tag{4.19}$$

where $c = e^C$.

Before sketching this solution and the slope field, we will see what calculus has to tell us. Because the derivative (4.17) is negative in the first quadrant, the function y is decreasing. If we differentiate (4.17), and then use (4.17) to eliminate y' from the result, we find, after some messy algebra, that

$$\frac{d^2y}{dx^2} = \frac{2y^9(3 + y^4)}{x^2\left(1 + y^4\right)},$$

which is positive in the first quadrant. Thus, we know the solution is a decreasing, concave up function.

The graphs of the slope field and solution curves from (4.19), for six choices of c, are shown in Figure 4.5. ☐

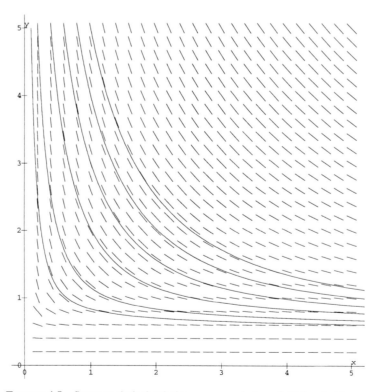

FIGURE 4.5 Some analytical solution curves and slope field for $dy/dx = -y^5/[x(1 + y^4)]$

EXAMPLE 4.6

As our final example on separation of variables, consider the differential equation

$$\frac{dy}{dx} = \frac{x^3 + 1}{y^3 + 1} \qquad\qquad (4.20)$$

in the first quadrant, so $x > 0$ and $y > 0$. Writing (4.20) in the separable form

$$\left(y^3 + 1\right)\frac{dy}{dx} = x^3 + 1$$

and integrating yields the implicit solution

$$\frac{1}{4}y^4 + y = \frac{1}{4}x^4 + x + C,$$

or

$$y^4 - x^4 + 4(y - x) = C. \qquad\qquad (4.21)$$

We cannot solve this either for $y(x)$ or for $x(y)$. So even though we have a family of implicit solutions, we have no graph of them. However, we can construct the slope field and draw in a few solutions by hand. These are shown in Figure 4.6. This figure suggests that $y = x$ is a solution. Clearly, $y = x$ satisfies (4.21) when $C = 0$ and is therefore a solution of (4.20). This figure also suggests that if $y > x$, the solution will be increasing and concave up,

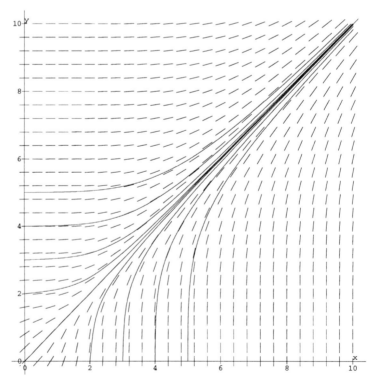

FIGURE 4.6   Some hand-drawn solution curves and slope field for $dy/dx = (x^3 + 1)/(y^3 + 1)$

whereas if $y < x$, the solution will be increasing and concave down. We will use calculus to see whether these suggestions are accurate.

The derivative given in (4.20) is positive for $x > 0$, $y > 0$, so the solutions will be increasing, as suggested by Figure 4.6. Also note that the slopes are larger than 1 if $x > y$ and smaller than 1 if $x < y$. To decide on the concavity, we differentiate (4.20) and substitute for y' from (4.20), giving, after some lengthy algebra,

$$\frac{d^2 y}{dx^2} = \frac{3}{\left(y^3 + 1\right)^3} \left(xy^3 + x + yx^3 + y\right) \left(xy^3 + x - yx^3 - y\right). \tag{4.22}$$

We are interested only in the first quadrant where $x > 0$ and $y > 0$. Here the sign of y'' is determined entirely by the sign of the last term on the right-hand side of (4.22) — namely, $xy^3 + x - yx^3 - y$ — which we will now analyze. Figure 4.6 suggests that the concavity changes at $y = x$, which suggests that $y - x$ will be a factor of $xy^3 + x - yx^3 - y$. It is, because

$$xy^3 + x - yx^3 - y = (y - x)(xy^2 + x^2 y - 1) = x(y - x)\left(y^2 + xy - \frac{1}{x}\right). \tag{4.23}$$

So the sign of y'' is determined by the product of these three terms. Figure 4.6 suggests that the only change of sign occurs at $y = x$, but perhaps we have missed something. If we think of the equation

$$y^2 + xy - \frac{1}{x} = 0$$

as a quadratic in y and solve it, we find

$$y = \frac{-x \pm \sqrt{x^2 + 4/x}}{2}.$$

Thus, (4.23) can be written as

$$xy^3 + x - yx^3 - y = x(y - x)\left(y + \frac{x - \sqrt{x^2 + 4/x}}{2}\right)\left(y + \frac{x + \sqrt{x^2 + 4/x}}{2}\right).$$

Because we are concerned with $x > 0$ and $y > 0$, the last term is always positive, but there is a possibility of y'' changing sign along the curve

$$y = -\frac{x - \sqrt{x^2 + 4/x}}{2}. \tag{4.24}$$

By multiplying the numerator and denominator of (4.24) by $x + \sqrt{x^2 + 4/x}$, we can express this curve in the form

$$y = \frac{2}{x\left(x + \sqrt{x^2 + 4/x}\right)}.$$

This curve behaves like $1/x^2$ for large x and like $1/\sqrt{x}$ for small x. We can ask where the curve (4.24) crosses the curve $y = x$, and answer that this will occur when

$$x = -\frac{x - \sqrt{x^2 + 4/x}}{2}.$$

Solving this equation gives $x = (1/2)^{1/3} \approx 0.7937$. If we look at Figure 4.6 again, we see that something unusual is happening near the origin. Figure 4.7 shows the slope field in the window $0 < x < 2$, $0 < y < 2$, together with the curve given by (4.24) and $y = x$, and some hand-drawn solution curves. The change in concavity near the origin is obvious. Without the concavity analysis, a cursory examination of Figure 4.6 would have led us to believe that an individual solution could not change concavity.

Figure 4.6 suggests that as $x \to \infty$, $y \to x$. Can you prove this from (4.21)? (See Exercise 5 on page 145.)   □

EXERCISES

1. Solve the differential equation

$$y \frac{dy}{dx} = x,$$

and compare your answer with the one you obtained in Section 3.1, Exercise 1.

2. Solve the differential equation

$$\frac{dy}{dx} = -2xy,$$

and compare your answer with the one you obtained in Section 3.1, Exercise 4.

3. Solve the following differential equations. Confirm your results by using slope fields.
   (a) $x^2 y' = y^2$

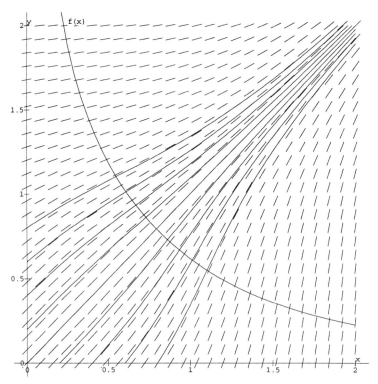

FIGURE 4.7 Some hand-drawn solution curves, the functions $f(x) = -[x - (x^2 + 4/x)^{1/2}]/2$, $y = x$, and slope field for $dy/dx = (x^3 + 1)/(y^3 + 1)$

(b) $e^{-x}y' - \sec y = 0$

(c) $yy' - x \sin x^2 = 0$

(d) $(4y + x^2 y)y' = 2x + xy^2$

(e) $x^2 y' + 4 + y^2 = 0$

(f) $(y^2 + 1)y' + y \tan x = 0$

(g) $y' = e^{x+y}$

(h) $yy' = e^{x+y}$

(i) $yy' - (1 + y)\cos^2 x = 0$

(j) $(1 + x^2)y' = 1 + y^2$

4. Solve the following differential equations subject to the given initial condition. Confirm your results by using graphical and numerical methods.

(a) $xy' = y,$ $\quad y(e) = 1$

(b) $y' + y^2 \cos x = 0,$ $\quad y(0) = 1/2$

(c) $xy' = (4 - y^2)^{1/2},$ $\quad y(8) = 1$

(d) $(1 - x^2)^{1/2} y' + y^3 = 0,$ $\quad y(1) = 1$

(e) $2(2 + y)y' + y(1 - x^2) = 0,$ $\quad y(0) = -1$

(f) $(4x^2 + x - 1)y' + (8x + 1)(y + 2) = 0,$ $\quad y(1) = 2$

(g) $y \sin x + (y^2 + 1)e^{\cos x}y' = 0,$ $\quad y(\pi/2) = 1$

(h) $3ye^{x^2}y' + 1 = 0,$ $\quad y(1) = 2$

(i) $3y^2 y' = \sin x^2,$ $\quad y(0) = 3$

(j) $y' = 2xy/(x^2 - 1),$ $\quad y(2) = -6$

(k) $y' = 2xy/(x^2 - 1),$ $\quad y(0) = 2$

(l) $y' = 2xy/(x^2 - 1),$ $\quad y(-2) = -6$

5. Show that all solutions of (4.20), namely,

$$y^4 - x^4 + 4(y - x) = C, \tag{4.25}$$

have the property that as $x \to \infty$, $y \to x$. [Hint: Use the fact that

$$y^4 - x^4 = (y - x)\left(y^3 + xy^2 + x^2 y + x^3\right)$$

to write (4.25) in the form

$$y - x = \frac{C}{\left(y^3 + xy^2 + x^2 y + x^3 + 4\right)},$$

and then let $x \to \infty$.]

6. Let a 200-gallon container of pure water have a salt concentration of 3 pounds per gallon added to the container at the rate of 4 gallons per minute.

(a) If the well-stirred mixture is drained from the container at the rate of 4 gallons per minute, find the number of pounds of salt in the container as a function of time.

(b) How many minutes does it take for the concentration in the container to reach 2 pounds per gallon?

(c) What does the concentration in the container approach for large values of time? Does this agree with your intuition?

7. Look in the textbooks of your other courses and find an example that uses a separable differential equation. Write a report about this example that includes the following items:

(a) A brief description of background material so a classmate will understand the origin of the differential equation.

(b) How the constants in the differential equation can be evaluated and what the initial condition means.

(c) The solution of this differential equation.

(d) An interpretation of this solution, and how it answers a question posed by the original discussion.

4.2 Graphing Separable Equations

When we integrate separable equations, our solutions will always have the form

$$G(y) = F(x), \tag{4.26}$$

that is, solutions are defined implicitly, with the dependence on x and y separated. As we have seen, sometimes it is possible to solve (4.26) for $y = y(x)$ or for $x = x(y)$, in which case we can sketch the graph of the solution directly. However, whether or not we can solve (4.26), it is always possible to sketch the solution curve relating x and y. We now demonstrate this technique.

EXAMPLE 4.7

Solve

$$\frac{dy}{dx} = \frac{18x^2}{-15y^2 - 6y - 4} \tag{4.27}$$

subject to

$$y(0) = 1. \tag{4.28}$$

Clearly (4.27) is separable, and so we are led to

$$-\int \left(15y^2 + 6y + 4\right) dy = 18 \int x^2 \, dx,$$

which is easily integrated to yield

$$-5y^3 - 3y^2 - 4y + C = 6x^3. \tag{4.29}$$

If we substitute the initial condition ($y(0) = 1$) into this equation, we find that $C = 12$, so (4.29) becomes

$$-5y^3 - 3y^2 - 4y + 12 = 6x^3. \tag{4.30}$$

Notice that this solution is of the form (4.26), where the left-hand side is a function of y and the right-hand side is a function of x.

To sketch the graph of the solution (4.30), we proceed as follows. We first divide the graph paper into four boxes. In the lower left-hand box we sketch the function of x — namely, $F(x) = 6x^3$ — and in the lower right-hand box we sketch the function of y — namely, $G(y) = -5y^3 - 3y^2 - 4y + 12$. In the upper right-hand box, we draw a diagonal line with slope 1 which connects the corners. In this way we create Figure 4.8. The scales used along the horizontal x- and y-axes in Figure 4.8 are independent of each other. However, it is critical in this construction that the scales used on the vertical axes for $F(x)$ and $G(y)$ are identical.

We proceed by selecting a value of x in the lower left-hand box, say x_0, and drawing a vertical line through the point $(x_0, F(x_0))$ into the upper left-hand box, which is where we will finally sketch the solution curve relating x and y. We now draw a horizontal line through the point $(x_0, F(x_0))$ until we cross the curve $G(y)$ in the lower right-hand box at the point $(y_0, G(y_0))$. At this point we have $F(x_0) = G(y_0)$ (remember the vertical axes have the same scale), so the point with numerical values $x = x_0$, $y = y_0$ must lie on the graph of $F(x) = G(y)$. What we need to do is plot this point in the upper left-hand box where we plot y versus x.

We have already drawn the vertical line $x = x_0$ in the upper left-hand box, so the next step is to draw a horizontal line corresponding to $y = y_0$ in this same box. This is where the line with slope 1 comes into play. We draw a vertical line through $(y_0, G(y_0))$ until we intersect the line with slope 1 in the upper right-hand box. We then draw a horizontal line through this point of intersection into the upper left-hand box. This is the line $y = y_0$. The point where this line meets the vertical line $x = x_0$ is the point (x_0, y_0), which lies on the graph of $F(x) = G(y)$. This is demonstrated in Figure 4.9. We repeat this process, starting with different values of x in the lower left-hand box, and then join the points in the upper left-hand box as is done in Figure 4.10. Finally, we remove the construction lines and have the graph of $F(x) = G(y)$ in the upper left-hand box, as shown in Figure 4.11.   □

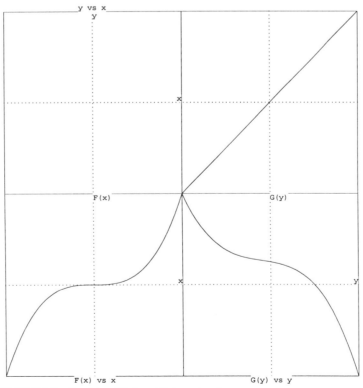

FIGURE 4.8   Graphing $G(y) = F(x)$ — step 1

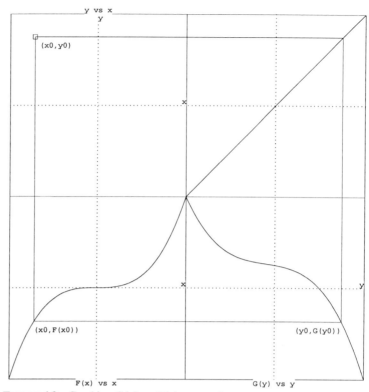

FIGURE 4.9   Graphing $G(y) = F(x)$ — step 2

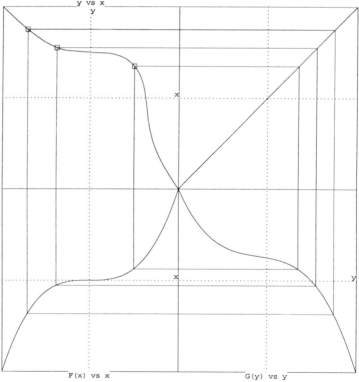

FIGURE 4.10   Graphing $G(y) = F(x)$ — step 3

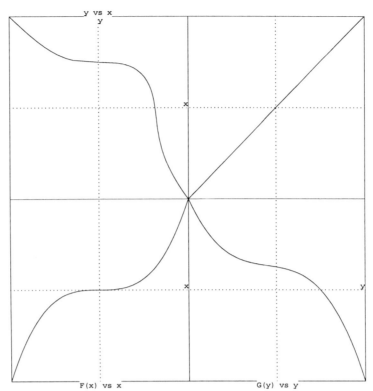

FIGURE 4.11 Graphing $G(y) = F(x)$ — step 4

How to Sketch Solutions of $G(y) = F(x)$

Purpose: To use graphical techniques to sketch the solution $y = y(x)$ of

$$G(y) = F(x). \tag{4.31}$$

Technique:

1. Divide the graph paper into four boxes. Label the boxes (see Figure 4.12).

   (a) The lower left-hand box is where we graph $F(x)$ versus x.

   (b) The lower right-hand box is where we graph $G(y)$ versus y.

   (c) The upper left-hand box is where we will eventually sketch $y = y(x)$.

   (d) The upper right-hand box is the auxiliary box.

2. Draw the graph of the function $F(x)$ in the lower left-hand box. Draw the graph of the function $G(y)$ in the lower right-hand box, making sure that the vertical scale is the same as the vertical scale used in the lower left-hand box. Draw the diagonal with slope 1 in the auxiliary box.

3. Select any point x_0 in the lower left-hand box. Draw the vertical line $x = x_0$ into the upper left-hand box. Draw a horizontal line through the point $(x_0, F(x_0))$ until it crosses the curve $G(y)$ in the lower right-hand box at the point $(y_0, G(y_0))$. [If it crosses the curve $G(y)$ at more than one point, consider every one of these points in the next step.]

4. Draw a vertical line through $(y, G(y_0))$ until it intersects the line in the auxiliary box. Draw a horizontal line through this point of intersection into the upper left-hand box.

The point where this line meets the vertical line $x = x_0$ is the point (x_0, y_0), which lies on the graph of $F(x) = G(y)$.

5. Repeat the last two steps for different values of x_0 obtaining many points in the upper left-hand box. Join them. The curve is the graph of the solution $y = y(x)$ of $G(y) = F(x)$.

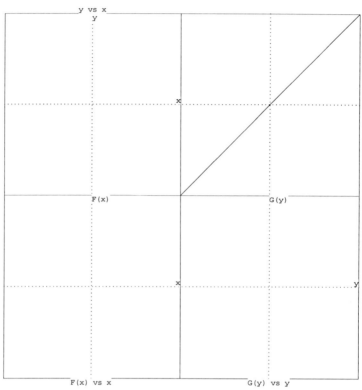

FIGURE 4.12   Graphing $G(y) = F(x)$ — the four quadrants

EXERCISES

1. Use the technique described in this section to sketch

$$x^2 = 1 - y^2.$$

Use Figure 4.13 for this purpose. It shows the functions x^2 versus x and $1 - y^2$ versus y.

2. Use the technique described in this section to sketch the following curves.
   (a) $x^2 = 10 - y^2$
   (b) $x^2 = 10 + y^2$
   (c) $x^3 + 3y^3 = 27$

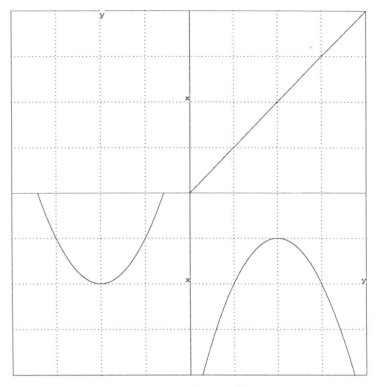

FIGURE 4.13 The figure for plotting $x^2 = 1 - y^2$

4.3 Applications — Deriving Differential Equations From Data

In this section we discuss two applications of differential equations that utilize real data sets. The main purpose is to illustrate how to obtain reasonable differential equations by analyzing the data set numerically.

EXAMPLE 4.8: The Heating of a Probe

A student[1] held a temperature probe firmly between her thumb and forefinger while the temperature of the probe was recorded. Table 4.1 shows the resulting data set where the temperature is in degrees centigrade and the time is in seconds. This data set is plotted in Figure 4.14. Our goal is to determine a formula that will allow us to find the temperature at other times.

Because the temperature is clearly changing with time we ask what is the relationship between the temperature of the probe T and the time t. To answer this question, we try to find a differential equation governing this process, which means we want to relate the derivative dT/dt to a function of t and T. We can obtain an approximate numerical value for dT/dt from Table 4.1 by using one of several approximate forms of the derivative. Here we list three such forms, the right-hand difference quotient

$$\frac{dT}{dt} \approx \frac{\Delta_R T}{\Delta t} = \frac{T(t+h) - T(t)}{h}, \tag{4.32}$$

---

[1]Melisa Enrico, University of Arizona.

TABLE 4.1 The heating of a temperature probe with time

Time	Temp
0	32.78
2	33.12
4	33.37
6	33.54
8	33.68
10	33.78
12	33.85
14	33.93
16	33.98
18	34.03
20	34.05

the left-hand difference quotient

$$\frac{dT}{dt} \approx \frac{\Delta_L T}{\Delta t} = \frac{T(t-h) - T(t)}{-h}, \tag{4.33}$$

and the average of the last two, the central difference quotient,

$$\frac{dT}{dt} \approx \frac{\Delta_C T}{\Delta t} = \frac{T(t+h) - T(t-h)}{2h}. \tag{4.34}$$

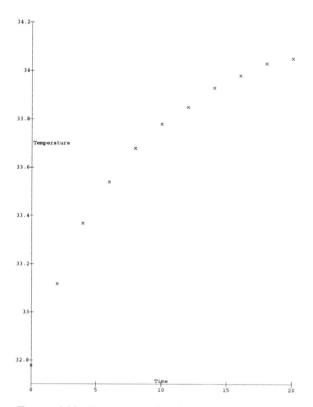

FIGURE 4.14   Temperature of probe versus time

TABLE 4.2 The heating of a temperature probe with time

Time	Temp	$[T(t+2) - T(t-2)]/4$
0	32.78	
2	33.12	0.1475
4	33.37	0.1050
6	33.54	0.0775
8	33.68	0.0600
10	33.78	0.0425
12	33.85	0.0375
14	33.93	0.0325
16	33.98	0.0250
18	34.03	0.0175
20	34.05	

Of these, the central difference quotient is the approximation most commonly used. Table 4.2 shows the central difference quotient calculations $\Delta_c T/\Delta t$ for Table 4.1.

In Figure 4.15 we plot this numerical approximation for dT/dt against the time t. We do not see any obvious relation between these variables. In Figure 4.16 we plot the numerical approximation for dT/dt against the temperature T, and it appears that the relationship is approximately linear with a negative slope. Figure 4.17 is Figure 4.16 with the addition of a straight line with slope -0.14 and vertical intercept 4.78. (Note the horizontal scale is from 33 to 34.)

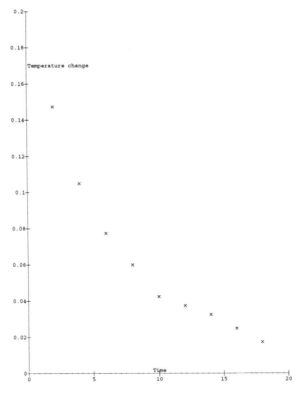

FIGURE 4.15 Numerical approximation for dT/dt versus time t

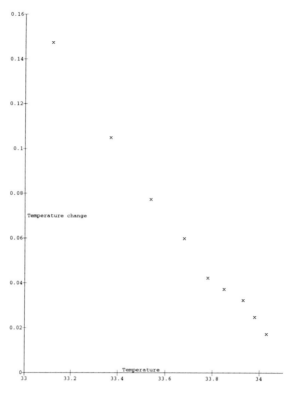

FIGURE 4.16   Numerical approximation for dT/dt versus temperature T

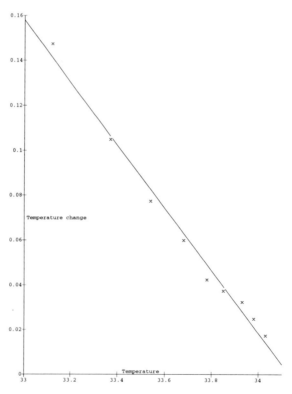

FIGURE 4.17   The line $4.78 - 0.14T$ and the numerical approximation for dT/dt versus temperature T

This suggests that the differential equation governing the temperature change might be of the form

$$\frac{dT}{dt} = a + kT,$$

where a and k are constants, with $k < 0$. This differential equation has the equilibrium solution $-a/k$, which we label T_a, so $T_a = -a/k$. We can then rewrite this differential equation in terms of T_a as

$$\frac{dT}{dt} = k(T - T_a). \tag{4.35}$$

Before trying to solve (4.35), we will use graphical arguments to analyze this model. For this to be a reasonable model for the heating of the probe, we expect the solution curves to be increasing, concave down, and tend to the exterior temperature of the finger as time increases. This agrees with the slope field for (4.35), as shown in Figure 4.18 for $T_a = 4$ and $k = -0.5$.

From (4.35) we see that if T, the temperature of the probe, is less than T_a, then $k(T - T_a)$ is positive (remember $k < 0$), so that T is an increasing function. Furthermore, the nearer T gets to T_a, the smaller the increase. If we take the derivative of (4.35) we find

$$\frac{d^2T}{dt^2} = k\frac{dT}{dt} = k^2\left(T - T_a\right).$$

This equation tells us that for $T < T_a$, the solution curves are concave down. All of this seems to suggest that (4.35) is a reasonable model.

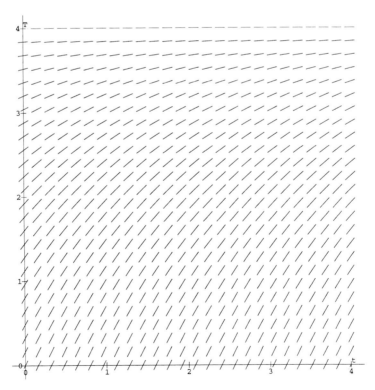

FIGURE 4.18   Slope field for $dT/dt = k(T - T_a)$

Because (4.35) is an autonomous differential equation (the right-hand side is independent of t), a phase line analysis can be performed on (4.35), and we find from Figure 4.19 that T_a is a stable equilibrium. This means that the temperature will tend to the quantity T_a as time increases, which means that T_a is the exterior temperature of the finger. The temperature T_a is often called the ambient temperature. (Note that this analysis also implies that if $T > T_a$, then T decreases to T_a.)

We now turn to finding the explicit solution of (4.35), which has the equilibrium solution

$$T(t) = T_a. \tag{4.36}$$

By rewriting (4.35) in the form

$$\frac{1}{T - T_a} \frac{dT}{dt} = k$$

and then integrating, we find

$$\ln|T - T_a| = kt + c.$$

We isolate the temperature by taking the exponential of this equation,

$$|T - T_a| = e^{kt+c},$$

which can be written as

$$T - T_a = Ce^{kt},$$

where the arbitrary constant C may be either positive or negative. (Why?) If we permit $C = 0$, we can absorb the equilibrium solution (4.36) into this solution. Rearranging gives

$$T(t) = T_a + Ce^{kt},$$

where the value of C is determined by requiring the value of the temperature at $t = 0$ to equal the initial temperature T_0

$$T_0 = T_a + C.$$

Thus, the final form of the solution is

$$T(t) = T_a + (T_0 - T_a)e^{kt}. \tag{4.37}$$

A brief look at this equation leads us to the conclusion that for large values of time, the exponential term will be insignificant (remember that $k < 0$) and the temperature of the probe will approach the ambient temperature, in this case the exterior temperature of the

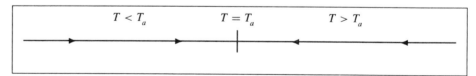

FIGURE 4.19   Phase line analysis for $dT/dt = k\left(T - T_a\right)$

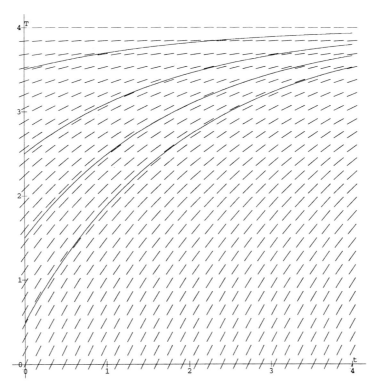

FIGURE 4.20 Some solution curves and slope field for $dT/dt = k(T - T_a)$

finger. This agrees with the slope field for (4.35) as shown in Figure 4.20, and it also agrees with our common sense.

We now return to the original experimental data set (Table 4.1) to see how well it compares to our mathematical model. It is clear that Figure 4.17 gives a crude estimate (based on the differential equation) of $k \approx -0.14$ and $kT_a = -a \approx 4.78$, from which we can estimate the ambient temperate as $T_a \approx 34.14$. However, we expect to get a more accurate estimate of the parameters by comparing the data with the exact solution (4.37). If we rewrite (4.37) in the form

$$T_a - T(t) = \left(T_a - T_0\right) e^{kt}$$

and take the logarithm of both sides, we find

$$\ln\left(T_a - T(t)\right) = \ln\left(T_a - T_0\right) + kt.$$

Thus, if we use the estimate $T_a \approx 34.14$ and plot $\ln[34.14 - T(t)]$ against t and then find a straight line, its slope will determine k, and its vertical intercept will determine $\ln\left(T_a - T_0\right)$. Figure 4.21 shows this plot, together with the straight line $0.308 - 0.135t$. Thus, we estimate $k \approx -0.135$ and $T_a - T_0 \approx e^{0.308}$, which gives $T_0 \approx 32.78$, in excellent agreement with the experimental value of T_0 and our crude initial estimate of $k \approx -0.14$. When these estimates, together with $T_a \approx 34.14$, are substituted in (4.37), we find

$$T(t) = 34.14 - e^{0.308 - 0.135t}.$$

The agreement between the graph of this function and the original data set may be seen in Figure 4.22.

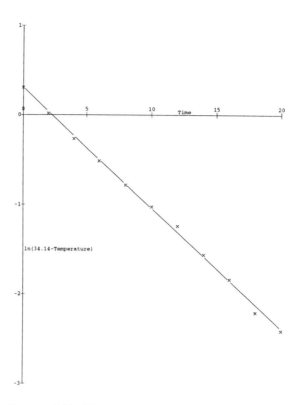

FIGURE 4.21 The line $0.308 - 0.135t$ and $\ln[34.14 - T(t)]$ versus time

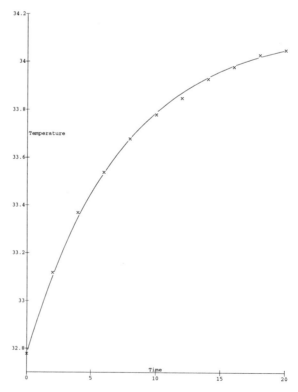

FIGURE 4.22 The function $34.14 - e^{0.308 - 0.135t}$ and temperature of probe versus time

Equation (4.35) on page 155 is called NEWTON'S LAW OF HEATING and is based on the assumption that the change in temperature of a body being heated is proportional to the difference between the temperature of the body and the temperature of the ambient medium. Equation (4.35) is also NEWTON'S LAW OF COOLING and is based on the assumption that the change in temperature of a body being cooled is proportional to the difference between the temperature of the body and the temperature of the ambient medium.   □

EXAMPLE 4.9: The Population of Kenya (1950–1990)

In recent years the population of Kenya[2] has grown rapidly, as can be seen from Table 4.3 and Figure 4.23. Can we find a differential equation which models this growth? Once we find this, its solution, subject to an appropriate initial condition, will give us a formula that can be used to predict the population at future times.

If we look carefully at Table 4.3, we see that the 1950 population of 6.265 million doubled somewhere between 1970 and 1975 (taking between 20 and 25 years), and then doubled again by 1990 (taking between 15 and 20 years). Thus, the doubling time is not constant, but is decreasing. This population growth therefore cannot satisfy the type of simple differential equation we used to model Botswana's population growth, denoted by P, namely,

$$\frac{dP}{dt} = kP,$$

because that differential equation results in a constant doubling time.

In modeling population growth, the common yardstick is the population growth per unit population — that is, the quantity $(1/P)(dP/dt)$. For example, in the case of Botswana,

$$\frac{1}{P}\frac{dP}{dt} = k. \tag{4.38}$$

We can discover what information the data for Kenya can give us about $(1/P)(dP/dt)$ by approximating this quantity by one of the numerical derivatives given on page 152, say the central difference approximation, in which case

$$\frac{1}{P}\frac{dP}{dt} \approx \frac{1}{P(t)}\frac{\Delta_C P}{\Delta t} = \frac{1}{P(t)}\frac{P(t+h) - P(t-h)}{2h}.$$

Table 4.4 shows the result of these numerical calculations.

TABLE 4.3   **Population of Kenya**

Year	Population (in millions)
1950	6.265
1955	7.189
1960	8.332
1965	9.749
1970	11.498
1975	13.741
1980	16.632
1985	20.353
1990	25.130

---

[2]"World Population Growth and Aging" by N. Keyfitz, University of Chicago Press, 1990, page 137.

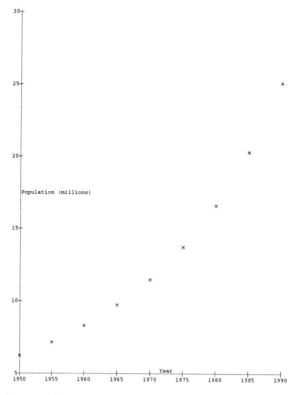

FIGURE 4.23   Population of Kenya

In Figure 4.24 we have plotted the numerical approximation for $(1/P)(dP/dt)$ against P, and we see no obvious straight line approximation. In Figure 4.25 we have plotted the numerical approximation for $(1/P)(dP/dt)$ against t, and we see a striking linear relationship, which we may characterize by the straight line $0.026227 + 0.000443t$.

Thus, our findings suggest that Kenya's population growth is similar to Botswana's, (4.38), except that the growth rate k is not constant but is a linearly increasing function of time, yielding the differential equation

$$\frac{1}{P}\frac{dP}{dt} = at + b,$$

TABLE 4.4   Estimating $(1/P)dP/dt$ **for Kenya**

t	$P(t)$	$[P(t+5) - P(t-5)]/(10P(t))$
1950	6.265	
1955	7.189	0.028752
1960	8.332	0.030724
1965	9.749	0.032475
1970	11.498	0.034719
1975	13.741	0.037363
1980	16.632	0.039755
1985	20.353	0.041753
1990	25.130	

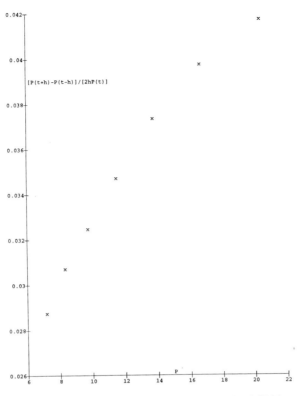

FIGURE 4.24   Numerical approximation of $(1/P)dP/dt$ versus P

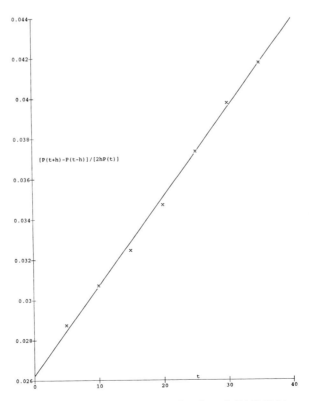

FIGURE 4.25   Numerical approximation of $(1/P)dP/dt$ versus t

where a and b are constants ($a > 0$). This is a separable differential equation, with implicit solution

$$\ln P(t) = \frac{1}{2}at^2 + bt + C. \tag{4.39}$$

If we let $t = 0$ correspond to the year 1950 and put $t = 0$ in (4.39), we find

$$C = \ln P(0),$$

so that

$$\ln P(t) = \frac{1}{2}at^2 + bt + \ln P(0). \tag{4.40}$$

From (4.40) we find

$$P(t) = P(0)e^{\frac{1}{2}at^2+bt}. \tag{4.41}$$

This is exponential growth, but here the exponent is quadratic in t rather than just linear in t, as in Botswana's case.

Now let's see how this fits the actual data for Kenya. Equation (4.41) contains three unknowns, $P(0)$, a, and b, which we have to estimate. We already have the initial estimates based on Figure 4.25, of $a \approx 0.000443$ and $b \approx 0.026227$, but we should be able to find better estimates now by comparing the data with the exact solution rather than with the differential equation. In order to obtain a linear equation we rewrite (4.40) in the form

$$\frac{\ln P(t) - \ln P(0)}{t} = \frac{1}{2}at + b.$$

Thus, plotting $[\ln P(t) - \ln P(0)]/t$ against t should yield a straight line from which we could estimate the slope, $a/2$, and the vertical intercept, b. However, this means we must first estimate $P(0)$. If, from Table 4.3, we accept $P(0) = 6.265$, then Figure 4.26 shows $[\ln P(t) - \ln P(0)]/t$ plotted against t together with the straight line

$$y(t) = 0.026434 + 0.00203t.$$

This gives $a \approx 2(0.00203) = 0.00406$ and $b \approx 0.026434$, which are consistent with our earlier crude estimates. So we have the population of Kenya growing according to

$$P(t) = 6.265\exp(0.000203t^2 + 0.026434t). \tag{4.42}$$

Figure 4.27 shows the population as well as the graph of the function (4.42) (adjusted so $t = 0$ corresponds to the year 1950), which seem in good agreement.

This model has the property that the population doubling time is not constant but is getting smaller as time progresses. Thus, in 1990 the population of 25 million is predicted to double in 15 years. A century later, in the year 2090, the population of Kenya is predicted to double in about 8 years. The population at that time is predicted to be about 13.5 billion, which is over twice the 1990 world population of 5.2 billion. In 2098 the population of Kenya is predicted to be 27 billion. In the year 2153 the population density of Kenya is predicted to be about one person per square foot. (The land areas of Kenya and Botswana are approximately the same, about the area of Texas.)   □

There are a variety of reasons why we would not expect a simple population model for any country to predict accurately far into the future. Among these are the inability to predict

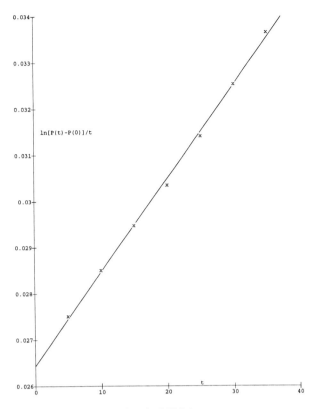

FIGURE 4.26 $[\ln P(t) - \ln P(0)]/t$ versus t

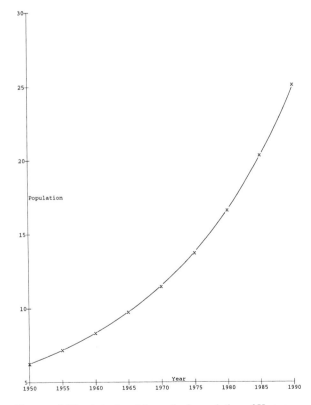

FIGURE 4.27 Actual and theoretical population of Kenya

future discoveries (for example, starting in 1970 the widespread use of "the pill", a birth control device, drastically reduced the birthrate in England); the inability to predict future diseases (for example, the AIDS epidemic has not yet had a major impact on published population figures, but will definitely change the deathrate); the inability to predict wars (for example, the Second World War raised the deathrate throughout Europe); and the effect of feedback, whereby the prediction of a population explosion changes the attitudes, habits, or behavior of a society, thereby lowering its birthrate (for example, China's decision to limit each couple to having only one child had a definite impact on the birthrate).

EXERCISES

1. Some students measured the temperature of coffee to see how rapidly it cooled in a room that was 24°C. The coffee temperature in degrees centigrade, taken at 1 minute intervals, is given in Table 4.5. This data set is plotted in Figure 4.28.

TABLE 4.5 The temperature of coffee as a function of time

Time	Temp
0	82.3
1	79.7
2	77.4
3	75.1
4	73.2
5	70.8
6	69.3
7	67.0
8	65.4
9	63.4
10	62.1

(a) Plot the approximate numerical value for dT/dt (by the central difference approximation) against the time t. This data set appears approximately linear. If this is accurate, it would mean that the cooling of coffee is governed by a differential equation of the type

$$\frac{dT}{dt} = \alpha t + \beta,$$

where α and β are constants. What is the sign of α? Solve this differential equation and describe what happens to T as $t \to \infty$. Is this a reasonable model for the cooling of coffee?

(b) Plot the approximate numerical value for dT/dt (by the central difference approximation) against the temperature T. This data set appears approximately linear. If this is accurate, it would mean that the cooling of coffee is governed by a differential equation of the type

$$\frac{dT}{dt} = aT + b,$$

where a and b are constants. What is the sign of a? Solve this differential equation and describe what happens to T as $t \to \infty$. Is this a reasonable model for the cooling of coffee?

(c) Expand the explicit solution you obtained in part (b) in a Taylor series about the origin. Explain how this Taylor series is related to the explicit solution you obtained in part (a).

2. Perform the following experiments related to Newton's law of cooling.

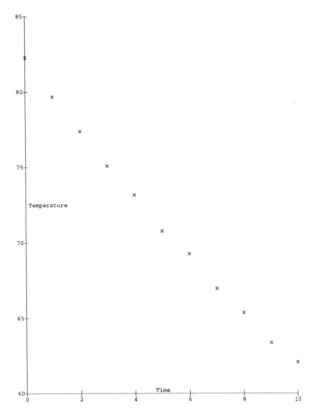

FIGURE 4.28   Temperature of coffee as a function of time

(a) Place a hot liquid in a cup, and measure the temperature of the liquid as a function of time. The appropriate time interval will depend on the insulting properties of your container. Plot this data set as in Exercise 1 (a) and (b) to see which gives a better fit.

(b) Repeat this experiment with hot liquid at the same initial temperature as in your experiment in part (a), only this time cover the top of the liquid with whipping cream (or a substitute). Compare the two data sets and explain any differences.

3. A cup of coffee at 180°F is placed in a room where the air temperature is 70°F. After 30 minutes the temperature of the coffee is 120°F. At this time the coffee is placed in an oven, preheated to 400°F, until the coffee is once more at 180°F. Assume that Newton's law of cooling (heating) applies to both time periods.

(a) Construct a separate phase line for each of these two time periods and then sketch the temperature of the coffee as a function of time for the entire time period.

(b) Find the analytical solution to this problem, and compare this solution with your sketch.

(c) At what time does the temperature of the coffee in the oven reach 180°F?

4. A rule of thumb used during homicide investigations is that if the temperature of the body of an average-sized person is within 1% of the room temperature, the person has been dead for at least 24 hours. Assume that at the time of death the person's body temperature was 98.6°F and that Newton's law of cooling (4.35) applies, that is,

$$\frac{dT}{dt} = k(T - T_a),$$

where $k < 0$. If the room temperature was a constant 72°F, what can you say about the limitations of k for an average-sized person?

5. Observations on the growth of animal tumors indicate that the size y obeys the differential equation

$$\frac{dy}{dt} = me^{-ht}y,$$

where h and m are positive constants. This differential equation is sometimes called the Gompertz growth law. Note that physical considerations require $y > 0$ and $t > 0$.

(a) Just by looking at the differential equation, what do you expect to happen as $t \to \infty$?

(b) Look at the slope field for $y > 0$, $t > 0$ for this equation to obtain some idea on how solution curves behave. What sort of behavior does this remind you of?

(c) Determine regions where the solutions of the differential equation are increasing, decreasing, concave up, or concave down.

(d) Does this differential equation have an equilibrium solution? If so, what is it? If not, explain why not.

(e) Solve this differential equation subject to the initial condition $y(0) = y_0$, $y_0 > 0$.

(f) Find a relationship between y_0, h, and m such that the graph of y versus t will have no inflection point.

(g) Show that the carrying capacity b for the Gompertz growth law depends on the initial condition. Is it possible to find an initial condition $y(0)$ for which $y(0) > b$?

(h) The Gompertz growth law is sometimes written

$$\frac{dy}{dt} = -ky \ln \frac{y}{a},$$

where k and a are positive constants. (See Exercise 7 on page 63.) What is the relationship between h, m, k and a?

6. The Chanter growth equation is

$$\frac{dy}{dt} = \alpha y (b - y) e^{-\beta t},$$

where α, b, and β are positive constants.

(a) It is claimed that this equation is a hybrid of the logistics equation and the Gompertz growth law described in Exercise 5. Justify this claim.

(b) Look at the slope field for this equation to obtain some idea on how solution curves behave. What sort of behavior does this remind you of?

(c) Determine regions where the solutions of the differential equation are increasing, decreasing, concave up, or concave down.

(d) Solve this differential equation subject to the initial condition $y(0) = y_0$, $y_0 > 0$.

7. Show that if $T(t)$ is a linear function of t, then

$$\frac{dT}{dt} = \frac{T(t + h) - T(t)}{h}$$

and

$$\frac{dT}{dt} = \frac{T(t + h) - T(t - h)}{2h}$$

for all $h \neq 0$. What does this suggest about accuracy if you use the right-hand or central difference quotient to estimate the derivative of a nonlinear function?

8. Show that if $T(t)$ is an exponential function of t, then

$$\frac{1}{T}\frac{dT}{dt} = \frac{\ln T(t + h) - \ln T(t)}{h}$$

and

$$\frac{1}{T}\frac{dT}{dt} = \frac{\ln T(t+h) - \ln T(t-h)}{2h}$$

for all $h \neq 0$. What does this suggest about accuracy if you use the right-hand or central difference quotient to estimate the derivative of a linear function?

9. A metal weight is dropped and, after some time, the distance fallen is recorded every $1/50$ second starting from $t = 0$. The following are the distances d the weight fell, measured in cm from the point at which $t = 0$:

$$0.00, \ 1.98, 4.45, 7.17, 10.44, 13.64 \ , 17.61, 21.71, 26.45, 31.77, 37.74.$$

By plotting d/t against t for $t > 0$, decide whether this data set is approximated by the function

$$d(t) = at^2 + bt,$$

where a and b are constants. What values did you get for a and b? It is believed that the weight is falling with constant acceleration g, so that $d''(t) = g$. Solve this differential equation for $d(t)$. Is this in good agreement with the data? What numerical value did you find for g, based on your values for a and b? What was the velocity of the weight at time $t = 0$?

4.4   Applications — Mechanics

In this section we discuss two applications of differential equations from mechanics that use real data sets.

EXAMPLE 4.10: Motion in the Presence of Velocity-Dependent Friction

In calculus you most likely saw the differential equation that modeled an object falling under the influence of gravity, obtained by using Newton's second law of motion (mass $\times$ acceleration = applied force) as

$$m\frac{dv}{dt} = mg, \tag{4.43}$$

where m is the mass of the object, v is velocity (positive velocities, or increasing distances, associated with downward movements), and g is acceleration due to gravity.

Table 4.6 and Figure 4.29 show data obtained from the United States Parachute Association that correspond to a sky diver in free fall in a stable spread-eagle position when falling from rest. It shows the sky diver's distance fallen (in feet) as a function of the time (in seconds). The third column in Table 4.6 contains the central difference quotients of the distance with respect to time, which is a numerical approximation of the velocity.

From the third column of Table 4.6 (or from the data points in Figure 4.29 , which seem to lie along a straight line as you move to the right), we conclude that the velocity, the rate of change of distance x, is approaching a constant value as t increases. Thus, the rate of change of velocity with respect to time approaches zero for larger times, giving rise to what is commonly called a TERMINAL VELOCITY . However, if we look at the slope field for (4.43) (or use isoclines, see *How to Analyze $dy/dx = g(x, y)$ Graphically* on page 98), we observe that the rate of change of the velocity in this differential equation $mdv/dt = mg$ is always a nonzero constant. Thus, (4.43) does not model this data. Missing in the differential equation is a term to account for air resistance.

TABLE 4.6 Distance fallen (in feet) against time (in seconds)

Time	Distance	Difference Quotient
0	0	
1	16	31.0
2	62	61.0
3	138	90.0
4	242	114.0
5	366	131.0
6	504	143.0
7	652	152.0
8	808	159.5
9	971	165.0
10	1138	169.0
11	1309	172.5
12	1483	

To obtain a better model, we consider a more general situation in which the object of mass m, besides moving under the influence of an external force mg, is subjected to a frictional force, the air resistance. A widely quoted rule of thumb concerning this frictional force is that for a slow-moving object (like a ball bearing falling in honey), the resistance is proportional to the velocity, whereas for a fast-moving object (like a ball bearing falling in

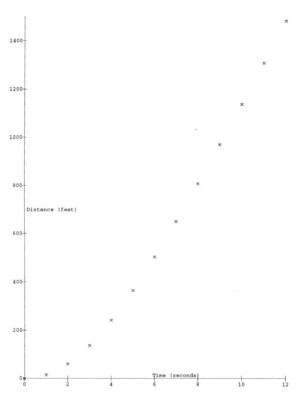

FIGURE 4.29   Free fall distance versus time

air), resistance is proportional to the square of the velocity. The second case is appropriate for skydiving, and so Newton's second law of motion gives

$$m\frac{dv}{dt} = mg - kv|v|, \tag{4.44}$$

where $-kv|v|$ represents this frictional force ($k > 0$). The reason for the term $v|v|$ rather than v^2 is that the air resistance always opposes the motion. If $v < 0$, a term like v^2 would be in the same direction as the motion and therefore aid it. We will concentrate on the case in which the sky diver falls downward from the aircraft — that is, $v \geq 0$. In this case (4.44) reduces to

$$\frac{dv}{dt} = g - \frac{kv^2}{m}, \tag{4.45}$$

where $k > 0$.

We observe that (4.45) has equilibrium solutions — namely, when $g = kv^2/m$. Because we are concerned only with positive values of v, we denote the positive equilibrium solution by V, so we have

$$V = \sqrt{\frac{mg}{k}}. \tag{4.46}$$

Because V is a quantity that can be measured, we replace m/k in (4.45) by V^2/g which results in

$$\frac{V^2}{g}\frac{dv}{dt} = V^2 - v^2. \tag{4.47}$$

(This will also simplify future algebraic manipulations.)

Because this is an autonomous differential equation — the right-hand side is independent of t — we do a phase line analysis, giving Figure 4.30. Here we see clearly that V is a stable equilibrium. Thus, for the usual situation where the sky diver's initial velocity v_0 is less than V, we see that the sky diver's velocity increases with time but never reaches the velocity V in finite time. The velocity V is the terminal velocity.

The data set in Table 4.6 was obtained for a sky diver falling from rest. If we measure the time t from the initial time $t = 0$, we have the initial condition

$$v(0) = 0.$$

To solve (4.47) we notice that it is separable and can be written in the form

$$\frac{1}{V^2 - v^2}\frac{dv}{dt} = \frac{g}{V^2}. \tag{4.48}$$

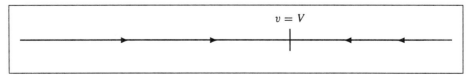

FIGURE 4.30 Phase line analysis for $\left(V^2/g\right)dv/dt = V^2 - v^2$, $v \geq 0$

We use partial fractions on the left-hand side of (4.48) and integrate to obtain

$$\frac{1}{2V} \int \left(\frac{1}{V-v} + \frac{1}{V+v} \right) dv = \int \frac{g}{V^2} \, dt.$$

This leads to

$$\frac{1}{2V} \ln \left(\frac{V+v}{V-v} \right) = \frac{gt}{V^2} + C.$$

From the initial condition, $v(0) = 0$, we have that $C = 0$ and our solution may be written as

$$\frac{V+v}{V-v} = e^{2gt/V}. \tag{4.49}$$

Because $2g/V$ is a constant, to simplify the expression we replace $2g/V$ by another constant α,

$$\alpha = \frac{2g}{V},$$

so (4.49) can be written

$$\frac{V+v}{V-v} = e^{\alpha t}.$$

Solving this for v, we find

$$v(t) = V \frac{e^{\alpha t} - 1}{e^{\alpha t} + 1}. \tag{4.50}$$

Notice that $v(t) \to V$ as $t \to \infty$, in complete agreement with our understanding that V is the terminal velocity. Figure 4.31 shows (4.50) for the case $V = 2g = 1$ (so $\alpha = 1$).

If we denote the distance that the body has fallen from its initial position by x (positive being downward), we see that (4.50) is

$$\frac{dx}{dt} = V \frac{e^{\alpha t} - 1}{e^{\alpha t} + 1}. \tag{4.51}$$

By using the substitution $u = e^{\alpha t} + 1$, we can integrate (4.51), subject to the initial condition

$$x(0) = 0.$$

Doing so gives

$$x(t) = \frac{2V}{\alpha} \ln \left[\frac{1}{2} \left(e^{\alpha t} + 1 \right) \right] - tV \tag{4.52}$$

as the distance fallen as a function of the time t.

If we now return to the original data set, there is only one constant we need to estimate — namely, V — because we know $g \approx 32.2$ ft/sec^2. Table 4.6 suggests a terminal velocity in excess of 172.5 ft/sec. Figure 4.32 shows the data set and the function (4.52) for one such choice of V — namely, $V = 182$ ft/sec.   □

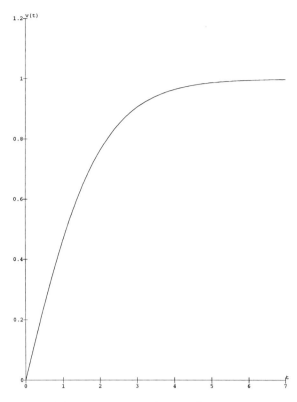

FIGURE 4.31   The velocity $(e^t - 1)/(e^t + 1)$

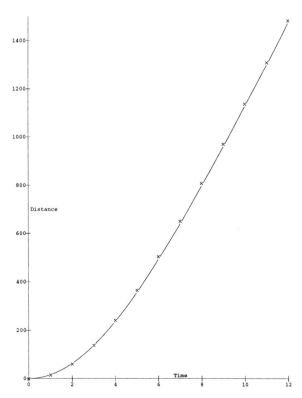

FIGURE 4.32   The data set and the theoretical function

EXAMPLE 4.11: Stopping Distances

A number of experiments have been performed that measure the distance a car travels when braking in an emergency stop as a function of the velocity of the car at the time the brakes are applied. Typical results are shown in Table 4.7 and presented graphically in Figure 4.33. Someone claims that these results are in good agreement with the model of a vehicle coming to rest under constant deceleration. Is that person right?

If we let $x(t)$ represent the distance in feet the vehicle traveled from the time when the brakes are applied [say at $t = 0$, so $x(0) = 0$], then the velocity, $v(t)$, and the acceleration, $a(t)$, of the vehicle are given by

$$v(t) = \frac{dx}{dt}, \text{ and } a(t) = \frac{dv}{dt}. \tag{4.53}$$

The assumption of constant deceleration is the statement that

$$a(t) = -k, \tag{4.54}$$

where k is a positive constant. If we take the units of time as seconds, then the units of k will be ft/sec^2. In view of (4.53), we see that (4.54) can be written as the differential equation

$$\frac{dv}{dt} = -k, \tag{4.55}$$

which has the solution

$$v(t) = -kt + b, \tag{4.56}$$

where b is a constant. If we let V_0 be the velocity of the vehicle when the brakes are applied at $t = 0$, then (4.56) implies that $b = V_0$. Thus, (4.56) becomes

$$v(t) = -kt + V_0. \tag{4.57}$$

TABLE 4.7 Observed car stopping distances

Velocity (mph)	Stopping distance (feet)
10	5.6
15	12.8
20	21.0
25	32.2
30	48.8
35	63.4
40	83.9
45	113.5
50	133.0
55	166.2
60	185.7
65	233.5

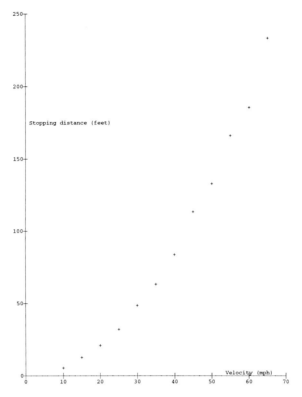

FIGURE 4.33 Stopping distance versus velocity

If we let T be the time at which the vehicle comes to rest — namely, the time T at which $v(T) = 0$ — then (4.57) tells us that

$$T = \frac{V_0}{k}. \tag{4.58}$$

So we have found a relationship between V_0, the initial velocity of the vehicle, and T, the time it takes for the vehicle to come to rest. What we want is the distance the vehicle has traveled in this time, $x(T) - x(0)$. In view of the fact that $v = dx/dt$, we see that (4.57) can be written as the differential equation

$$\frac{dx}{dt} = -kt + V_0,$$

which is immediately integrated, giving

$$x(t) = -\frac{1}{2}kt^2 + V_0 t + c, \tag{4.59}$$

where c is a constant. Because we specify that $x(0) = 0$, (4.59) implies that $c = 0$, in which case (4.59) becomes

$$x(t) = -\frac{1}{2}kt^2 + V_0 t. \tag{4.60}$$

If we denote the distance traveled by the vehicle from $t = 0$ to $t = T$ as d, so that $d = x(T)$, (4.60) tells us that

$$d = -\frac{1}{2}kT^2 + V_0 T. \tag{4.61}$$

If we substitute (4.58) into (4.61) and simplify, we find

$$d = \frac{1}{2k}V_0^2. \tag{4.62}$$

This tells us that the assumption of constant deceleration leads to a quadratic relationship between V_0, the initial velocity of the vehicle, and d, the distance it takes for the vehicle to stop. The constant k presumably has something to do with the vehicle. The only thing left is to see whether this agrees with Table 4.7.

With the ideas from earlier in this chapter, there are at least two different ways we can see whether the data in Table 4.7 satisfy a quadratic equation like

$$y = hx^2 \tag{4.63}$$

where h is a constant.

1. Plot y/x on the vertical axis against x on the horizontal axis. If the data set is quadratic, the resulting graph will approximate a straight line through the origin with slope h. Figure 4.34 shows that the line with slope 0.054 seems to give a good fit to the data set.

2. Plot $\ln y$ on the vertical axis against $\ln x$ on the horizontal axis. If the data set is quadratic, the resulting graph will approximate a straight line with slope 2 and with vertical intercept $\ln h$, because (4.63) can be written as

$$\ln y = \ln h + 2 \ln x.$$

Figure 4.35 shows that the line with slope 2 and intercept -2.92 seems to give a good fit to the data set. Note that $\ln 0.054 \approx -2.92$. How does this relate to Figure 4.34?

Noticing our units, the quantity V_0 is measured in feet per second, whereas the quantity x is measured in miles per hour, so $h \neq 1/(2k)$ but $h = (15/22)^2/(2k)$. Figure 4.36 shows the data set (Table 4.7) and the quadratic function (4.63) with $h = 0.054$. $\square$

EXERCISES

1. A sky diver falls from rest subject to the differential equation (4.48), namely,

$$\frac{1}{V^2 - v^2}\frac{dv}{dt} = \frac{g}{V^2},$$

where the terminal velocity V is 180 ft/sec. How long does it take for the sky diver to reach 90% of the terminal velocity? How far has she fallen when this happens?

2. Show that if (4.48), namely,

$$\frac{1}{V^2 - v^2}\frac{dv}{dt} = \frac{g}{V^2},$$

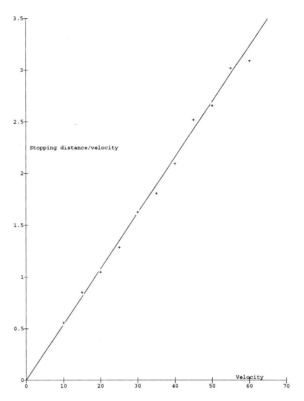

FIGURE 4.34   Stopping distance/velocity versus velocity

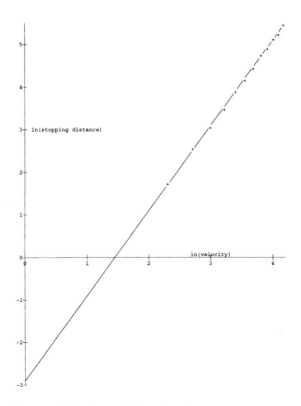

FIGURE 4.35   Plot of ln(stopping distance) versus ln(velocity)

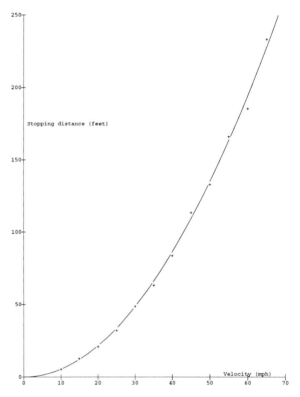

FIGURE 4.36   Stopping distance versus velocity and the quadratic function $d = 0.054v^2$

is solved subject to the initial conditions

$$v(t_0) = v_0, \qquad x(t_0) = x_0$$

then

$$x(t) = x_0 + \frac{V^2}{g} \ln \left| \frac{e^{\alpha(t-t_0)} + C}{1 + C} \right| - (t - t_0)V,$$

where α and C are defined by

$$\alpha = \frac{2g}{V}, \qquad C = \frac{V - v_0}{V + v_0}.$$

3. A sky diver falls from rest from 10,000 feet above the earth subject to the differential equation (4.48), namely,

$$\frac{1}{V^2 - v^2} \frac{dv}{dt} = \frac{g}{V^2},$$

where the terminal velocity V is 180 ft/sec. When he reaches 1,000 feet, his parachute opens instantaneously and his terminal velocity is now 22 ft/sec.[3] Assume that the differential equation (4.48) still applies but now $V = 22$.

(a) Construct a separate phase line for each of these two events and then sketch the velocity of the sky diver as a function of time for the entire time period.

---

[3]The velocity of 22 ft/sec is approximately your velocity on impact if you step off a 10 ft wall.

(b) Find the exact solution for $v(t)$, and compare this solution with your sketch. You may use the results from Exercise 2.

(c) By using your result from part (b) and the fact that $v = dx/dt$, find an exact solution for $x(t)$. Use this result to determine how fast he is travelling when he hits the ground.

4. In place of (4.45), some people have suggested that an object falling from an airplane experiences an air resistance that is linear in the velocity, namely,

$$m\frac{dv}{dt} = mg - kv.$$

Find a formula that represents the terminal velocity for this equation. Also find the velocity and the distance fallen as a function of time. By using different values for the terminal velocity, try to match the distance as a function of time with the data set in Table 4.6. Is this model a good one?

5. Use of the chain rule $(dv/dt = dv/dx\, dx/dt)$ allows us to write $dv/dt = -k$ in the form

$$\frac{dv}{dx}\frac{dx}{dt} = -k,$$

that is,

$$\frac{dv}{dx}v = -k.$$

Integrate this to find

$$\frac{1}{2}v^2(t) = -kx(t) + C.$$

and from this last equation derive (4.62) directly.

6. A student[4] shot a 120 grain hollow point bullet from a 300 Winchester Magnum rifle. He measured the velocity, v, as a function of distance, x, and compiled the data shown in Table 4.8. This data set is plotted in Figure 4.37. We are going to construct mathematical models and analyze this data set in various ways.

(a) It is claimed that this data set looks linear, so it obeys the law

$$v(x) = ax + b,$$

where a and b are constants. Find values for a and b so this linear model gives a good fit to the data set. With these values, estimate how far the bullet travels before coming to rest. Remembering that $v = dx/dt$, estimate when the bullet comes to rest.

TABLE 4.8   Velocity of bullet as a function of distance

Distance (ft)	Velocity (ft/sec)
0	3290
300	2951
600	2636
900	2342
1200	2068
1500	1813

[4]Richard Zimmer, Canyon del Oro High School, Tucson, AZ.

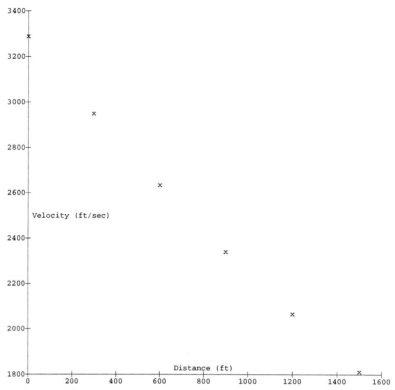

FIGURE 4.37   Velocity of bullet versus distance

(b) It is claimed that this data set looks quadratic, so it obeys the law

$$v(x) = ax^2 + bx + c,$$

where a, b, and c are constants. Find values for a, b, and c so this quadratic model gives a good fit to the data set. With these values estimate how far the bullet travels before coming to rest. Remembering that $v = dx/dt$, estimate when the bullet comes to rest.

(c) It is claimed that the bullet, when fired horizontally, is subject only to air resistance, which is proportional to a power of the velocity, that is

$$\frac{dv}{dt} = -kv^n,$$

where k and n are positive constants. Because

$$\frac{dv}{dt} = \frac{dv}{dx}\frac{dx}{dt} = \frac{dv}{dx}v,$$

we have

$$\frac{dv}{dx} = -kv^{n-1},$$

which can be written

$$\ln\left(-\frac{dv}{dx}\right) = (n-1)\ln v + \ln k.$$

Thus, if we plot $\ln(-dv/dx)$ versus $\ln v$, we should see a straight line with slope $n-1$ and y-intercept $\ln k$, from which we can estimate n and k. From Table 4.8 construct the central difference quotient $\Delta_C v/\Delta x$, which is an approximation to the derivative dv/dx. Now plot

$\ln(\Delta_c v/\Delta x)$ versus $\ln v$. Find values for k and n so this model gives a good fit to the data set. With these values, solve the differential equation and compare the exact answer to the actual data set. Estimate how far the bullet travels before coming to rest. Remembering that $v = dx/dt$, estimate when the bullet comes to rest.

(d) If we plot $\Delta_c v/\Delta x$ versus v we see a nearly straight line. This would mean that the bullet's velocity obeys the differential equation

$$\frac{dv}{dx} = av + b,$$

where a and b are constants. Find values for a and b so this model gives a good fit to the data set. With these values, solve the differential equation and compare the exact answer to the actual data set. Estimate how far the bullet travels before coming to rest. Remembering that $v = dx/dt$, estimate when the bullet comes to rest. Notice that this differential equation can be recast in terms of Newton's law, namely

$$\frac{dv}{dt} = \frac{dv}{dx}v = av^2 + bv.$$

(e) If we plot $\Delta_c v/\Delta x$ versus x we see a nearly straight line. This would mean that the bullet's velocity obeys the differential equation

$$\frac{dv}{dx} = ax + b,$$

where a and b are constants. Find values for a and b so this model gives a good fit to the data set. With these values, solve the differential equation and compare the exact answer to the actual data set. Estimate how far the bullet travels before coming to rest. Remembering that $v = dx/dt$, estimate when the bullet comes to rest. Notice that this differential equation can be recast in terms of Newton's law, namely

$$\frac{dv}{dt} = \frac{dv}{dx}v = axv + bv.$$

How is this model related to the model in part (b)?

4.5   Applications — Orthogonal Trajectories

The coordinate systems with which we are most familiar are the rectangular coordinate system and the polar coordinate system in two dimensions. Both have the property that setting one variable equal to a constant produces a family of curves (sometimes lines) that are perpendicular to the family obtained by setting the other variable equal to a constant.

A natural reason for the name of the rectangular coordinate system is that a rectangle is formed by any pair of vertical lines (given by $x = c_1$ and $x = c_2$) and any pair of horizontal lines (given by $y = k_1$ and $y = k_2$). It is also true in rectangular coordinate systems that lines given by $x = a$ are perpendicular to those given by $y = b$. Such lines are said to be orthogonal[5] to each other, and the lines are examples of ORTHOGONAL TRAJECTORIES . This situation could also occur for nonhorizontal and nonvertical lines or for families of curves. In this section we show how differential equations are used to construct orthogonal trajectories. This technique is used later in the text.

We start with two definitions.

---

[5]From the Greek "orthogōnion" meaning "right angle."

Definition 4.2:   **A family of curves is said to be PARAMETERIZED BY λ if the family is given by a relation $f(x, y, \lambda) = 0$, where the Greek letter λ (lambda) is a parameter that may take on various values.**

Thus we see that $y = 2x + \lambda$ gives a family of lines with slope 2 parameterized by λ. The equation $x^2 + y^2 = \lambda$ gives a family of circles with radius $\sqrt{\lambda}$. Notice that in the first case λ may be any real number, while in the second case λ may have only non-negative values.

Definition 4.3:   **Suppose we have a family of curves given by $f(x, y, \lambda) = 0$. The ORTHOGONAL TRAJECTORIES to this family is again a family of curves which intersect those given by $f(x, y, \lambda) = 0$ at right angles.**

When using this definition, we need to remember that two curves intersect at right angles, that is, are orthogonal, if their tangent lines at the point of intersection are perpendicular. We next consider two examples which start with families of nonhorizontal lines, and end with very different types of orthogonal trajectories.

EXAMPLE 4.12

Consider the family of straight lines with slope 3 given by

$$y = 3x + \lambda, \qquad\qquad (4.64)$$

where λ is a parameter that gives the y-intercept. Now two lines, with slope m_1 and m_2, are orthogonal if $m_1 = -1/m_2$, so the product of their slopes is equal to -1 ($m_1 m_2 = -1$). Thus, a family of lines orthogonal to those of (4.64) is given by

$$y = -\frac{1}{3}x + c,$$

where c is an arbitrary constant.

We could derive this equation for the orthogonal family in the following manner: The parameter λ in (4.64) is eliminated if we differentiate the equation to obtain

$$\frac{dy}{dx} = 3. \qquad\qquad (4.65)$$

This indicates that the slope of the lines tangent to all members of this family is 3. The differential equation that governs the family of lines that is orthogonal to those of (4.64) can be obtained from (4.65) by replacing dy/dx with $-1/(dy/dx)$ — namely,

$$\frac{1}{-dy/dx} = 3.$$

Thus, the family of orthogonal trajectories must satisfy the differential equation

$$\frac{dy}{dx} = -\frac{1}{3}.$$

Upon integration this yields

$$y = -\frac{1}{3}x + c.$$

See Figure 4.38. $\square$

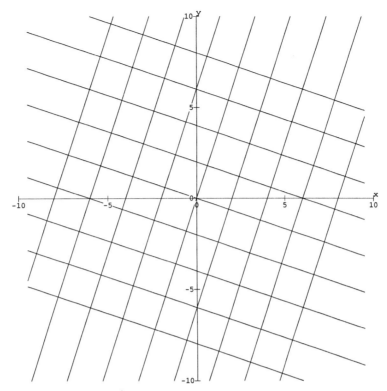

FIGURE 4.38   Orthogonal trajectories $y = 3x + \lambda$, $y = -x/3 + c$

How to Find Orthogonal Trajectories

Purpose: To construct the differential equation of the trajectories orthogonal to the family of curves

$$f(x, y, \lambda) = 0, \tag{4.66}$$

where a particular value of λ gives a particular curve.

Technique:

1. If possible solve (4.66) for λ, to get

$$\lambda = g(x, y). \tag{4.67}$$

2. Differentiate (4.67) with respect to x, using the chain rule,

$$0 = \frac{\partial g}{\partial x} + \frac{\partial g}{\partial y}\frac{dy}{dx}. \tag{4.68}$$

This is the differential equation for the original family of curves where g, $\partial g/\partial x$, and $\partial g/\partial y$ are known functions.

3. To set up the differential equation for the orthogonal trajectories, replace dy/dx with $-1/(dy/dx)$ in (4.68) to get

$$0 = \frac{\partial g}{\partial x} - \frac{\partial g}{\partial y} \frac{1}{dy/dx}. \tag{4.69}$$

Solve this differential equation for $y(x)$.

Comments about orthogonal trajectories

- Most people do not try to remember (4.69). It is more common to go through the technique described to reach (4.69).

- There is a different way of getting (4.68) from (4.66) without having to solve for λ. Differentiate (4.66) with respect to x to obtain

$$\frac{\partial f}{\partial x} + \frac{\partial f}{\partial y} \frac{dy}{dx} = 0. \tag{4.70}$$

Now in this case, the left-hand side of (4.70) may still contain the parameter λ. We can eliminate λ from (4.70) by using (4.66). If we do this we will have equation (4.68). We use this technique in the next example and in the first example of the following chapter.

- Although a particular curve may have no symmetry, if the original family of curves has a symmetry about the x-axis, the y-axis, or the origin, then so will the family of orthogonal trajectories. We can see that this is the case for the slope fields because (4.68) and (4.69) will have the same symmetries. Because we are concerned with all solutions of these equations, then the set of all solutions will have the same symmetries.

EXAMPLE 4.13

Now consider the family of lines through the origin parameterized by their slope, λ,

$$y = \lambda x. \tag{4.71}$$

This family of curves, when taken as a whole, is symmetrical about the x-axis, the y-axis, and the origin. (Why?) The orthogonal trajectories will have the same symmetries.

We now use the preceding technique to find the orthogonal trajectories. If we solve (4.71) for λ we find

$$\lambda = \frac{y}{x} \tag{4.72}$$

for $x \neq 0$. Differentiating (4.72) with respect to x, we find

$$0 = \frac{1}{x^2} \left(x \frac{dy}{dx} - y \right),$$

or

$$x \frac{dy}{dx} = y.$$

This is the differential equation for the family of curves (4.71).

An alternative way of obtaining this differential equation is to simply differentiate (4.71) to obtain

$$\frac{dy}{dx} = \lambda$$

We then eliminate λ by substituting its value from (4.71) — namely, $\lambda = y/x$ into this expression to obtain $dy/dx = y/x$, as we found by the other method.

To find the family of trajectories orthogonal to these straight lines, we must solve the differential equation

$$x\frac{-1}{dy/dx} = y,$$

or

$$\frac{dy}{dx} = -\frac{x}{y}.$$

Separating the variables and integrating we find

$$\frac{y^2}{2} = -\frac{x^2}{2} + c, \text{ or } y^2 + x^2 = 2c.$$

Thus, we see that the family of circles

$$y^2 + x^2 = 2c$$

is orthogonal to the family of straight lines

$$y = \lambda x.$$

As anticipated, the family of circles is symmetric about the x-axis, the y-axis, and the origin. □

Orthogonal trajectories occur in many situations, including weather maps, fluid flow, and electrostatics. On weather maps the wind direction from areas of high pressure to those of low pressure is orthogonal to the curves of constant pressure (often called isobars). A similar situation occurs in planar fluid flow when the fluid motion is along curves that are the orthogonal trajectories of the curves of equipotential. In the third instance, the curves of force in an electrostatic field are the orthogonal trajectories of the curves of constant potential.

EXAMPLE 4.14

Suppose that curves of constant electrostatic potential are given by the family of ellipses

$$\frac{x^2}{2} + y^2 = \lambda \tag{4.73}$$

and that we wish to find the curves of force associated with this potential. This family of curves is symmetrical about the x-axis, the y-axis, and the origin. The orthogonal trajectories will have the same symmetries.

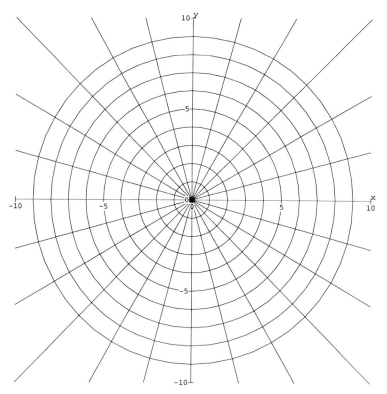

FIGURE 4.39   Orthogonal trajectories $y = \lambda x$, $y^2 + x^2 = 2c$

To find the family of curves orthogonal to these ellipses, we differentiate (4.73) with respect to x and obtain

$$x + 2y \frac{dy}{dx} = 0.$$

The differential equation for the family of curves orthogonal to these ellipses must satisfy

$$x + 2y \frac{-1}{dy/dx} = 0,$$

or

$$\frac{dy}{dx} = \frac{2y}{x}.$$

This is a separable differential equation with equilibrium solution $y(x) = 0$, and nonequilibrium solution

$$\ln |y| = 2 \ln |x| + C.$$

If we solve for y we get

$$y(x) = cx^2,$$

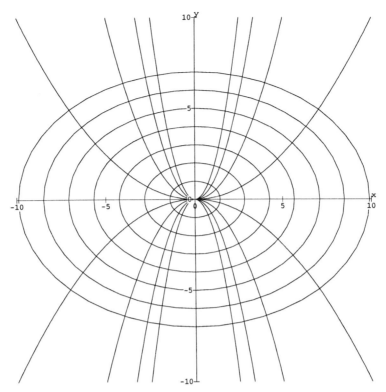

FIGURE 4.40 Orthogonal trajectories $x^2/2 + y^2 = \lambda$, $y = cx^2$

in which we have included the equilibrium solution. Thus, we have a family of parabolas that give the curves of force that are orthogonal to the elliptical level curves of the electrostatic potential field (see Figure 4.40). □

EXERCISES

1. Find the orthogonal trajectories of the families of curves given by

   (a) $y = -2x + \lambda$

   (b) $y = \lambda x + 4$

   (c) $y^2 = \lambda x$

   (d) $y^2 = \lambda x^3$

   (e) $x^2 - \lambda^2 y^2 = 16$

   (f) $x^2 + \lambda^2 y^2 = 16$

   (g) $x^2 + ay^2 = \lambda^2$ (a fixed)

   (h) $y = \lambda/x$

   (i) $y = \lambda e^x$

   (j) $y = \lambda e^{-x}$

   (k) $y = \lambda e^{ax}$ (a fixed)

   (l) $y = 1 + \lambda e^{-2x}$

   (m) $y = \ln(x^3 + \lambda)$

   (n) $x^{2/3} - y^{2/3} = \lambda^{2/3}$

   (o) $e^x \cos y = \lambda$

2. Find the value of the constant a such that the family of curves in Exercise 1g is orthogonal to those in Exercise 1d.

3. Use a plotting package to observe graphs of $y^2 = (x + \lambda)^2$ for several values of λ. (Note that you may need to plot $y = \pm|x + \lambda|$ on your machine.)

   (a) Record what you observe.

   (b) The curves such as you observed in part (a) are called "self-orthogonal." Give reasons why this name seems to be appropriate.

   (c) Differentiate the original equation as a first step in finding the differential equation for the orthogonal trajectories, and use the results to amplify your answer to part (b).

4. Parabolic cylinder coordinates are given by $x = \left(u^2 - v^2\right)/2$, $y = uv$.

   (a) Eliminate u from the preceding equations, and from the result show that putting $v = \lambda$ gives parabolas in the xy-plane. Which way do these parabolas open?

   (b) Find the differential equation for the family of curves orthogonal to these parabolas, and show that the solution of the differential equation also gives parabolas. Which way do these parabolas open?

5. The streamlines for irrotational, incompressible flow in the region bounded by two perpendicular lines — say, the positive x- and positive y-axes — are given by the family of hyperbolas $xy = b$. (This represents the flow around the inside of a corner, as shown in Figure 4.41.) The orthogonal trajectories to the streamlines give lines of constant velocity potential. Show that these orthogonal trajectories are also hyperbolas.

6. The velocity potential for a certain flow is given by $(x + y)^2 = \lambda$. Find the streamlines for this situation — that is, the orthogonal trajectories. Describe a possible flow consistent with these streamlines.

7. Oblique (or isogonal) trajectories occur when one family of curves intersects another family of curves at some angle θ, $\theta \neq \pi/2$. If one family is given by the solution of the differential equation

$$\frac{dy}{dx} = f(x, y),$$

the other family, which intersects these curves at an angle of θ, is given by the solution of the differential equation

$$\frac{dy}{dx} = \frac{f(x, y) + \tan \theta}{1 - f(x, y) \tan \theta}.$$

   (a) Derive this formula. [Hint: Use the identity for the tangent of the sum of two angles.]

   (b) Find the family of lines that intersect $y = x + 3$ at an angle of $\pi/3$.

   (c) Find the family of curves that intersect the family of hyperbolas $y^2 = 2x^2 + \lambda$ at an angle of $\arctan(1/2)$.

   (d) Find the family of curves that intersect the circle $x^2 + y^2 = \lambda^2$ at an angle of $\pi/4$.

   (e) Find the family of curves that intersect the parabolas $y^2 = \lambda x$ at an angle of $\pi/4$.

8. On the graph of the function $r = f(\theta)$, in polar coordinates, the angle ψ_1 between the intersection of extended radius and the tangent line is given by

$$\tan \psi_1 = \frac{r}{dr/d\theta} = \frac{f}{f'}.$$

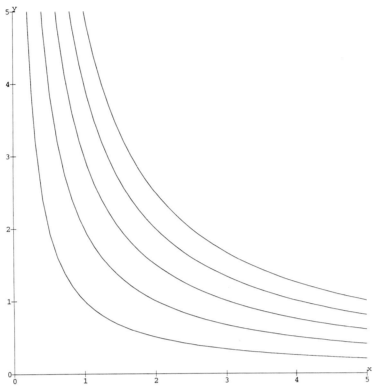

FIGURE 4.41 Flow inside a corner

If another graph in polar coordinates is to be orthogonal to this graph, the corresponding angle ψ_2 will be larger by adding $\pi/2$ or smaller by subtracting $\pi/2$. Thus, ψ_2 will satisfy the equation

$$\tan\psi_2 = \tan(\psi_1 \pm \frac{\pi}{2}) = -\frac{1}{\tan\psi_1}$$

(justify this last step). In terms of the variables (r, θ), we have the equation for the orthogonal trajectories to the curve $r = f(\theta)$ given by

$$\tan\psi_2 = \frac{r}{dr/d\theta} = -\frac{f'}{f},$$

or

$$\frac{1}{r}\frac{dr}{d\theta} = -\frac{f}{f'}.$$

It is therefore evident that in polar coordinates, the negative reciprocal also comes into play at the start of finding orthogonal trajectories.

(a) Show that the differential equation for the orthogonal trajectories to the family of circles $r = \lambda\cos\theta$ is

$$\frac{1}{r}\frac{dr}{d\theta} = \frac{\cos\theta}{\sin\theta}.$$

(b) Integrate this equation to obtain the family

$$r = c\sin\theta,$$

and graph several members of each family.

9. Find the orthogonal trajectories to the family of cardioids[6] given by

$$r = \lambda(1 - \cos\theta).$$

Compare your results with that of Exercise 3. Is the terminology of that exercise appropriate here? Why or why not?

10. Find the orthogonal trajectories to the family of cardioids given by

$$r = \lambda(1 - \sin\theta).$$

11. Find the orthogonal trajectories of the spirals given by

$$r = \lambda\theta$$

(the spirals of Archimedes). Graph some of these spirals and some of the orthogonal trajectories. Were you surprised at the results? Why or why not?

12. Find the orthogonal trajectories to those of $r = \lambda/\sqrt{\theta}$.

13. The flow within a two-dimensional wedge of angle α has streamlines given by $r^{\pi/\alpha}\sin(\theta\pi/\alpha) = \lambda$. (See Figure 4.42.) Use the information from Exercise 8 to find the associated velocity potential for this situation. Now let $\alpha = \pi/2$, and compare with the results of Exercise 5.

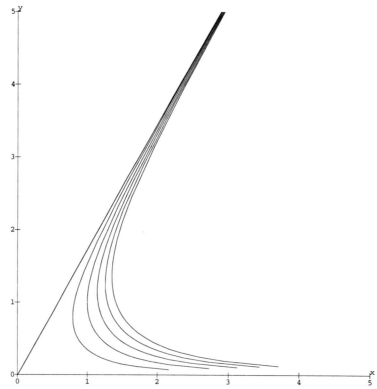

FIGURE 4.42   Flow within a wedge

---

[6]From Greek, meaning "heart-shaped."

14. Show that the family of curves given by $x^2/(m^2 + \lambda) + y^2/(n^2 + \lambda) = 1$ (where m and n are specified constants) is self-orthogonal. [Hint: Show that the given family of curves satisfies the differential equation

$$\left(m^2 - n^2\right) \frac{dy}{dx} = \left(x\frac{dy}{dx} - y\right)\left(y\frac{dy}{dx} + x\right)$$

and that this equation is unchanged if dy/dx is replaced with $-1/(dy/dx)$.]

4.6   Applications — Functional Equations

In this section we show how differential equations can be used to solve functional equations.

EXAMPLE 4.15: A Functional Equation

We now look at a type of problem that arises in various branches of mathematics — namely, solving functional equations. A simple example of this problem is to find all differentiable functions $f(x)$ that satisfy the condition

$$f(u + v) = f(u) + f(v) \text{ for all } u, v. \tag{4.74}$$

At first sight this problem appears unrelated to differential equations.

If we try $v = 0$ in (4.74) we see that

$$f(u) = f(u) + f(0),$$

which tells us that $f(0) = 0$.

If (4.74) is rewritten in the form

$$\frac{f(u + v) - f(u)}{v} = \frac{f(v)}{v}$$

or, because $f(0) = 0$,

$$\frac{f(u + v) - f(u)}{v} = \frac{f(v) - f(0)}{v}.$$

If we let $v \to 0$, the left-hand side becomes $f'(u)$ and the right-hand side becomes $f'(0)$, which is independent of u, and so is a constant, say, m. Thus, we find the differential equation

$$f'(u) = m,$$

subject to $f(0) = 0$, with solution

$$f(u) = mu. \tag{4.75}$$

We see that (4.75) satisfies (4.74). Thus, *the only differentiable function $f(u)$ satisfying (4.74) is the linear function mu.* $\square$

EXAMPLE 4.16: Another Functional Equation

The function $\tan x$ satisfies the identity

$$\tan(x + y) = \frac{\tan x + \tan y}{1 - \tan x \tan y},$$

which is the cornerstone of many applications of the tangent function. Is this the only function which satisfies this identity? To answer this question we seek the most general differentiable function, defined at $x = 0$, that satisfies the functional equation

$$f(x + y) = \frac{f(x) + f(y)}{1 - f(x)f(y)}. \tag{4.76}$$

If we try $x = y = 0$ in (4.76), we obtain

$$f(0) = \frac{2f(0)}{1 - [f(0)]^2},$$

or

$$f(0)\left[f^2(0) + 1\right] = 0.$$

This is true if $[f(0)]^2 = -1$ or $f(0) = 0$, so because $f(0)$ is a real number, we have

$$f(0) = 0.$$

If we consider the definition of $f'(0)$ and use this last fact, we have

$$f'(0) = \lim_{y \to 0} \frac{f(y) - f(0)}{y} = \lim_{y \to 0} \frac{f(y)}{y}. \tag{4.77}$$

We now want to manipulate (4.76) to obtain a derivative. If we subtract $f(x)$ from both sides of (4.76) and obtain a common denominator in the result, we see that

$$f(x + y) - f(x) = \frac{f(x) + f(y)}{1 - f(x)f(y)} - f(x) = \frac{\left[1 + f^2(x)\right]f(y)}{1 - f(x)f(y)}.$$

We then divide both sides by y and take the limit of the result as $y \to 0$. This gives

$$\lim_{y \to 0} \frac{f(x + y) - f(x)}{y} = \lim_{y \to 0} \left\{ \frac{\left[1 + f^2(x)\right]}{[1 - f(x)f(y)]} \right\} \frac{f(y)}{y},$$

where we notice that the left-hand side is the definition of df/dx. Now we use the fact that $f(y) \to 0$ as $y \to 0$ along with (4.77) to obtain the differential equation

$$\frac{df}{dx} = \left[1 + f^2(x)\right] f'(0).$$

This equation can be written in separated form

$$\int \frac{df}{1 + f^2(x)} = \int f'(0)\, dx,$$

giving

$$\tan^{-1}\left(f(x)\right) = f'(0)x + C.$$

Using the initial condition $f(0) = 0$, gives $C = 0$, and

$$f(x) = \tan\left(f'(0)x\right).$$

Thus, the only function satisfying (4.76) is not just $\tan x$, but $\tan ax$, where a is any number. If it is further specified that the derivative of the function we seek has the value 1 at the origin, then the only function is $\tan x$. □

EXERCISES

1. Imagine we conduct the following experiment using a spring scale that measures weight (that is, force). We want to see how the distance the spring pointer moves down is related to the weight of the object hung on the scale. With nothing on the scale it reads 0, so the distance displaced is 0 when there is 0 weight on the scale. Put a weight w_1 on the scale and measure the displacement of the pointer, namely, $d(w_1)$. (See Figure 4.43.) Now we replace the weight w_1 with a weight w_2 and measure $d(w_2)$. Finally we put both w_1 and w_2 on the scale and measure $d(w_1 + w_2)$. We discover that

$$d(w_1 + w_2) = d(w_1) + d(w_2). \tag{4.78}$$

We repeat this for various w_1 and w_2 and discover that (4.78) is always valid (unless of course we stretch the spring too much by hanging a very heavy weight on the scale thereby losing its spring properties!). What does (4.78) tell us about the relationship between d and w? [Hint: Look at (4.74).] What does this tell us about the relationship between w and d? This relationship is called HOOKE'S LAW .

2. Show that the only functions $f(x)$ that satisfy the condition

$$f(u + v) = f(u)$$

for all u, v, are $f(x) = C$, where C is a constant.

   (a) Use this result to show that if

$$\frac{dy}{dx} = g(x, y)$$

is invariant under the interchange of x with $x + a$ for every constant a — so that if x is replaced by $x + a$ in $y' = g(x, y)$ the resulting equation is identical to $y' = g(x, y)$ — then $g(x, y)$ is a function of y alone.

   (b) Show that if

$$\frac{dy}{dx} = g(x, y)$$

is invariant under the interchange of y with $y + b$ for every constant b, then $g(x, y)$ is a function of x alone.

   (c) Show that if

$$\frac{dy}{dx} = g(x, y)$$

is invariant under the interchange of x with $x + a$ and y with $y + b$, for every a and b, then $g(x, y)$ is constant.

3. The purpose of this exercise is to find all differentiable functions $f(x)$ that satisfy the condition

$$f(u + v) = f(u)f(v) \text{ for all } u, v,$$

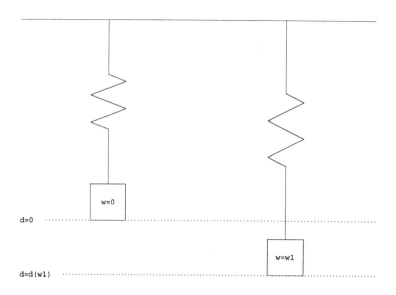

FIGURE 4.43   Spring with and without weight

where $f(x) \neq 0$.

(a) Show that

$$f(0) = 1$$

and

$$f'(u) = af(u),$$

where a is constant, namely, $a = f'(0)$.

(b) For the function $f(u)$ and the constant a obtained in part (a), define

$$g(u) = f(u)e^{-au}.$$

Prove that

$$g(0) = 1$$

and that

$$g'(u) = 0.$$

(c) Integrate the last equation in part (b) subject to $g(0) = 1$, and explain how this shows that $f(x) = e^{ax}$ is the only nonzero differentiable function that has the property that

$$f(u + v) = f(u)f(v) \text{ for all } u, v.$$

4. Repeat the technique you used in Exercise 3 to show that the only function $f(x)$ that satisfies

$$f(u + v) = \frac{1}{k}f(u)f(v) \text{ for all } u, v,$$

is $f(x) = ke^{ax}$.

5. This problem deals with two functions, $f(x)$ and $g(x)$, which have the following properties:

    i. $f(0) = 0$, and $g(0) = 1$.

    ii. $f(x)$ and $g(x)$ are differentiable for all x.

    iii. $f'(x) = g(x)$, and $g'(x) = -f(x)$.

(a) Construct the function

$$h(x) = f^2(x) + g^2(x),$$

and show that $h'(x) = 0$; deduce that

$$f^2(x) + g^2(x) = 1$$

for all x.

(b) Let $F(x)$ and $G(x)$ be another two functions that have exactly the same properties as $f(x)$ and $g(x)$ (that is, properties i through iii). Construct the function

$$k(x) = [F(x) - f(x)]^2 + [G(x) - g(x)]^2,$$

and show that $k'(x) = 0$. Now show that $F(x) = f(x)$, and $G(x) = g(x)$, for all x.

(c) Can you think of any functions $f(x)$ and $g(x)$ that satisfy all of the preceding conditions?

(d) Explain how parts (b) and (c) guarantee that $f(x) = \sin x$ and $g(x) = \cos x$.

(e) Use the preceding ideas to show that $\sinh x$ and $\cosh x$ are the only functions that satisfy the following conditions:

    i. $f(0) = 0$, and $g(0) = 1$.

    ii. $f(x)$ and $g(x)$ are differentiable for all x.

    iii. $f'(x) = g(x)$, and $g'(x) = f(x)$.

4.7   Applications — Calculus

In this section we show how differential equations arise when answering questions from elementary calculus.

EXAMPLE 4.17: An Application Based on the Mean Value Theorem

We now discuss an application based on the MEAN VALUE THEOREM, as follows:

Theorem 4.1:   *If $f(x)$ is continuous for $u \le x \le v$ and is differentiable for $u < x < v$, then there is at least one value c between u and v, $u < c < v$, for which*

$$f'(c) = \frac{f(v) - f(u)}{v - u}.$$

    The geometrical interpretation of this result — that there is always a tangent line to $f(x)$ parallel to the line joining the endpoints of the interval — is shown in Figure 4.44.

    In general, for an arbitrary function $f(x)$, it is impossible to find c explicitly in terms of the end points u and v. However, if $f(x)$ is quadratic, that is,

$$f(x) = a_0 + a_1 x + a_2 x^2,$$

then c is always the average of u and v — that is,

$$c = \frac{u + v}{2}.$$

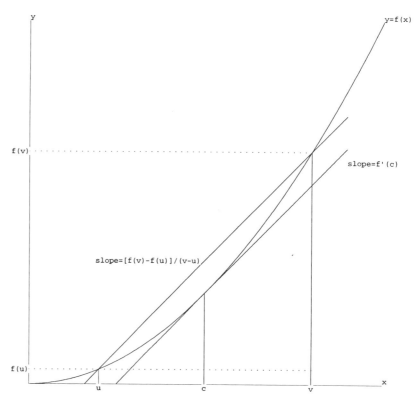

FIGURE 4.44   Geometrical interpretation of the mean value theorem

In other words the value c in the conclusion of the mean value theorem when applied to a quadratic function always occurs at the average of the endpoints of the interval. (See Exercise 1, page 199.)

A natural question to ask is what other, non-quadratic functions have this property — that is, what functions satisfy

$$f'\left(\frac{u+v}{2}\right) = \frac{f(v) - f(u)}{v - u} \tag{4.79}$$

for every u and v? For our purposes we will assume that the function $f(u)$ we seek is defined for all u and is twice differentiable.

We know that a general quadratic satisfies this condition, so let us exploit this by defining the function $g(u)$ as the difference between $f(u)$ and the first three terms in its Taylor series, namely,

$$g(u) = f(u) - f(0) - f'(0)u - \frac{1}{2}f''(0)u^2. \tag{4.80}$$

Thus, if we can show that

$$g(u) = 0,$$

then the general function $f(u)$ is quadratic, otherwise it is not.

If we use (4.80) to calculate $g'(u) = f'(u) - f'(0) - f'(0)u$ and $g''(u) = f''(u) - f''(0)$, and then evaluate $g(u)$, $g'(u)$, and $g''(u)$ at $u = 0$ we obtain

$$g(0) = g'(0) = g''(0) = 0. \tag{4.81}$$

We now use the fact that $g'(u) = f'(u) - f'(0) - f''(0)u$ to write

$$g'\left(\frac{u+v}{2}\right) = f'\left(\frac{u+v}{2}\right) - f'(0) - f''(0)\left(\frac{u+v}{2}\right).$$

Using (4.79) in this last equation changes its form to

$$g'\left(\frac{u+v}{2}\right) = \frac{f(v) - f(u)}{v - u} - f'(0) - f''(0)\left(\frac{u+v}{2}\right).$$

By collecting terms on the right-hand side of this equation over a common denominator and rearranging we obtain

$$g'\left(\frac{u+v}{2}\right) = \frac{f(v) - f(u) - f'(0)(v - u) - f''(0)\left(v^2 - u^2\right)/2}{v - u},$$

or

$$g'\left(\frac{u+v}{2}\right) = \frac{f(v) - f'(0)v - f''(0)v^2/2 - \left[f(u) - f'(0)u - f''(0)u^2/2\right]}{v - u}.$$

The terms on the right-hand side of this equation are close to the definitions of $g(v)$ and $g(u)$. What is missing is the term $f(0)$, so we add and subtract such a term so our last expression becomes

$$g'\left(\frac{u+v}{2}\right) = \frac{f(v) - f(0) - f'(0)v - f''(0)v^2/2 - \left[f(u) - f(0) - f'(0)u - f''(0)u^2/2\right]}{v - u}.$$

We now use (4.80) to finally obtain

$$g'\left(\frac{u+v}{2}\right) = \frac{g(v) - g(u)}{v - u} \tag{4.82}$$

for every u and v. If we can find the most general $g(u)$ satisfying (4.81) and (4.82), then we can substitute it into (4.80) to find $f(u)$. So we now turn to (4.81) and (4.82).

If we try $v = -u$ into (4.82) and use (4.81), we find that

$$g(u) = g(-u) \tag{4.83}$$

for every u. If we differentiate (4.82) with respect to u (keeping v constant) we find

$$\frac{1}{2}g''\left(\frac{u+v}{2}\right) = \frac{-g'(u)(v - u) + [g(v) - g(u)]}{(v - u)^2} \tag{4.84}$$

for every u and v. If we try $v = -u$ in (4.84) and use (4.81) and (4.83), we find the differential equation

$$g'(u) = 0,$$

which implies that

$$g(u) = a,$$

where a is an arbitrary constant. But from (4.81), $g(0) = 0$, which means that $a = 0$, so

$$g(u) = 0.$$

From this and (4.80) we conclude that

$$f(u) = f(0) + f'(0)u + \frac{1}{2}f''(0)u^2.$$

In other words, *the only (twice differentiable) function that satisfies (4.79) is a quadratic.*

$\square$

EXAMPLE 4.18: The Failure of Newton's Method

Newton's method finds the roots of a specified function $f(x)$ by an iterative procedure. In essence, we make an initial approximation, x_0, for the root of $f(x)$, and obtain the next approximation, x_1, from

$$x_1 = x_0 - \frac{f(x_0)}{f'(x_0)}.$$

Then we take x_1 as the new approximation and use it in place of x_0 in the preceding expression, to produce

$$x_2 = x_1 - \frac{f(x_1)}{f'(x_1)}.$$

The whole process is repeated until either a root, c, is found or the technique fails. The geometrical interpretation of Newton's method is shown in Figure 4.45.

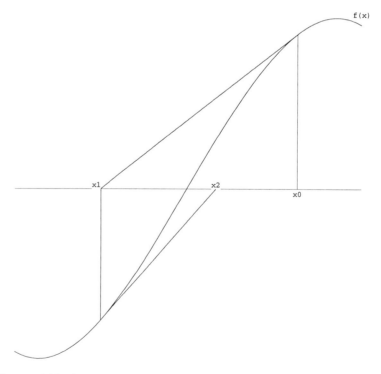

FIGURE 4.45 Geometrical interpretation of Newton's method

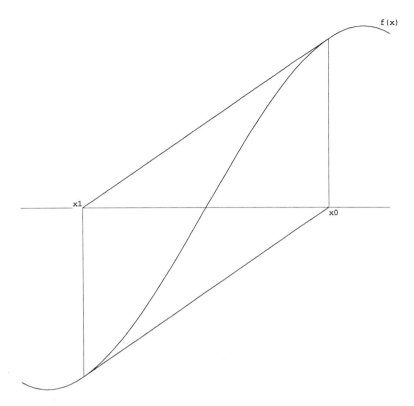

FIGURE 4.46 Failure of Newton's method

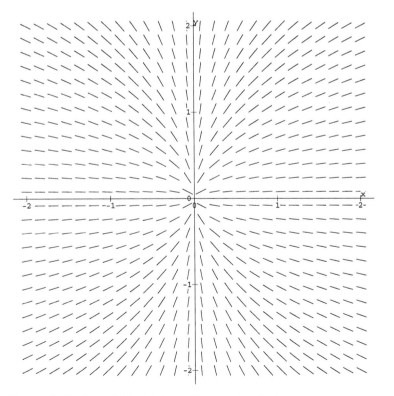

FIGURE 4.47 Slope field for failure of Newton's method

One way in which this technique fails [assuming $f'(x_0) \neq 0$] is with a poor choice for x_0, which leads to $x_1 = c - x_0$, and then, using x_1, we end up back at x_0. Thus, the process oscillates indefinitely between x_1 and x_0. This is illustrated in Figure 4.46. Usually, a better choice for the initial approximation x_0 avoids this problems.

However, it can happen that no matter what initial approximation x_0 is selected, the process oscillates indefinitely between $c - x_0$ and x_0. This leads to the following problem (where, without loss of generality, we assume $c = 0$).

Find all functions with the following property: For all x ($x \neq 0$), $f(x)$ satisfies

$$-x = x - \frac{f(x)}{f'(x)}, \tag{4.85}$$

where $f'(x) \neq 0$.

If we put $y = f(x)$ and $dy/dx = f'(x)$ in (4.85) and solve the resulting equation for dy/dx we obtain

$$\frac{dy}{dx} = \frac{y}{2x}. \tag{4.86}$$

Before solving this separable differential equation, let's use our graphical techniques to try to identify properties of the graph of f. The differential equation (4.86) implies that the slope field is symmetric about the x-axis, about the y-axis, and about the origin. The slope field for (4.86) is shown in Figure 4.47 and has these properties. The isoclines with slope m are characterized by the straight lines

$$y = 2mx,$$

which is consistent with Figure 4.47. The concavity is obtained from d^2y/dx^2, which in this case is

$$\frac{d^2y}{dx^2} = \frac{1}{2}\left(\frac{xy' - y}{x^2}\right) = -\frac{1}{4}\left(\frac{y}{x^2}\right).$$

Thus, the function we seek will be concave down for $y > 0$ and concave up for $y < 0$. This is consistent with Figure 4.47.

Now let's turn to solving (4.86). We first note that $y(x) = 0$ is the equilibrium solution. If we rewrite (4.86) as

$$\frac{1}{y}\frac{dy}{dx} = \frac{1}{2x},$$

we may integrate to find

$$\ln|y| = \frac{1}{2}\ln|x| + C.$$

In our usual manner, we write this solution in the form

$$y(x) = A\sqrt{|x|},$$

which includes the equilibrium solution. This is the only function for which Newton's method fails in the manner just described. The graph with the $A = 1$ is shown in Figure 4.48. Notice that this solution is not symmetric about the x-axis, even though the family of solutions associated with this slope field is.   □

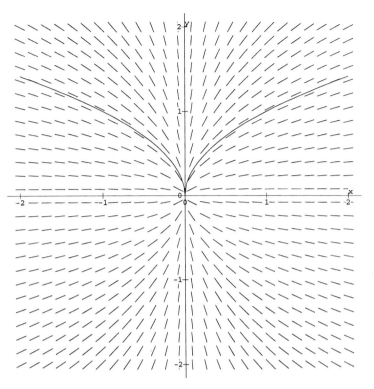

FIGURE 4.48 Solution and slope field for failure of Newton's method

EXERCISES

1. Show that if $f(x)$ is a quadratic function — that is, $f(x) = a_0 + a_1 x + a_2 x^2$ — then solving

$$f'(c) = \frac{f(u) - f(v)}{u - v}$$

for c (for every u and v) gives

$$c = \frac{u + v}{2}.$$

2. Show that if

$$\frac{dT}{dt} = \frac{T(t + h) - T(t)}{h},$$

then $T(t)$ is a linear function of t.

3. Show that if

$$\frac{dT}{dt} = \frac{T(t + h) - T(t - h)}{2h},$$

then $T(t)$ is a linear function of t.

4. Show that if

$$\frac{1}{T} \frac{dT}{dt} = \frac{\ln T(t + h) - \ln T(t)}{h},$$

then $T(t)$ is an exponential function of t.

5. Show that if

$$\frac{1}{T}\frac{dT}{dt} = \frac{\ln T(t+h) - \ln T(t-h)}{2h},$$

then $T(t)$ is an exponential function of t.

4.8   Other Applications

In this section we apply the techniques we have learned to problems that do not fit into the previous sections.

EXAMPLE 4.19: The Water-Skier Problem

Consider the situation in which a boat and water-skier are alongside a dock, connected by a tightly stretched rope. At time $t = 0$, the boat moves with a constant velocity v away from the dock in a direction perpendicular to the dock. We construct a coordinate system so that initially the skier is at $(a, 0)$ and the boat at $(0, 0)$. Later, at time t, the boat is at $(0, vt)$, and the skier at $(x(t), y(t))$. Thus, the rope has length a, and the boat is traveling up the y-axis with velocity v (Figure 4.49).

We want to determine what curve the skier follows. There are two key assumptions that we make to tackle this problem: (i) the rope is always tangent to the skier's trajectory;

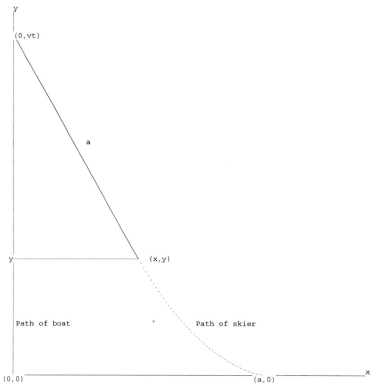

FIGURE 4.49   The water-skier problem

(ii) the rope has a constant length. We can see from Figure 4.49 that the requirement that the rope is tangent to the trajectory leads to

$$\frac{dy}{dx} = -\frac{vt - y}{x}, \tag{4.87}$$

whereas requiring the rope to have constant length gives

$$(vt - y)^2 + x^2 = a^2. \tag{4.88}$$

Eliminating $(vt - y)$ between (4.87) and (4.88) gives the equation

$$\frac{dy}{dx} = -\frac{\sqrt{a^2 - x^2}}{x}. \tag{4.89}$$

Because the problem requires that x always be positive, from (4.89) we see that the slope of the y versus x trajectory will always be negative, so y will be a decreasing function of x. We also note that the only place the trajectory can have a horizontal tangent is $x = a$.

If we differentiate (4.89) we obtain

$$\frac{d^2y}{dx^2} = \frac{a^2}{x^2\sqrt{a^2 - x^2}},$$

which is never negative, so the trajectory will always be concave up. (Does this agree with your intuition about what will happen to the skier?)

If we look at the differential equation (4.89), we see that the isoclines are given by

$$-\sqrt{a^2 - x^2}/x = m,$$

which can be rewritten as

$$x = \frac{a}{\sqrt{1 + m^2}}. \tag{4.90}$$

Because a and m are constants, this is a vertical line. Note from (4.90) that the slopes are steeper at points on the curve that are closer to $x = 0$ (that is, the y-axis).

Figure 4.50 shows the slope field as well as the hand-drawn solution curve for the case $a = 1$. Notice that we have obtained all the information about the solution of (4.89) from its slope field and an analysis of the differential equation by means of isoclines and concavity, without obtaining the explicit solution.

To find an explicit form of the trajectory we must integrate (4.89), which requires some skill. To find an antiderivative of $-\sqrt{a^2 - x^2}/x$, we use the substitution $u = \sqrt{a^2 - x^2}$, so $u^2 = a^2 - x^2$. We then integrate $u^2/(a^2 - u^2) = -1 + a^2/(a^2 - u^2)$. The result is

$$y = -\sqrt{a^2 - x^2} + \frac{a}{2}\ln\left(\frac{a + \sqrt{a^2 - x^2}}{a - \sqrt{a^2 - x^2}}\right) + C.$$

Because $y = 0$ when $x = a$, then $C = 0$. (This curve is called a tractrix, from the Latin "trahere," to pull.) □

EXAMPLE 4.20: Tape Deck Counters

Conduct the following experiment using either a VCR or a tape deck equipped with a digital tape-counter. Put a tape in the unit, set the tapecounter to zero, and, as you start the tape playing, note the time. Then record the time every 5 to 10 minutes along with the number

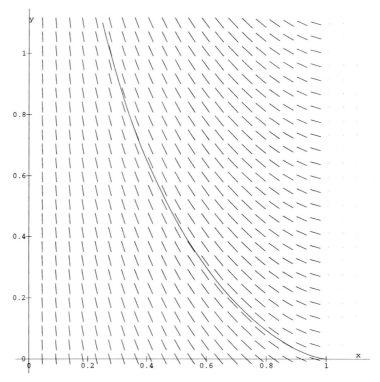

FIGURE 4.50 Slope field and hand-drawn solution curve for $dy/dx = -(1 - x^2)^{1/2}/x$, $y(1) = 0$

on the tapecounter. A data set obtained in this way from the tape "Flashdance" is shown in Table 4.9, which shows the number of revolutions n, the number on the tapecounter, as a function of time t. This is presented graphically in Figure 4.51. Can we model this situation? To do so requires finding a relationship between the number of revolutions n and the time t. The number of revolutions is related to the angular velocity of either the take-up reel or the feeder reel because the tape passes with constant velocity past the read-write head.

Let's concentrate on the take-up reel as sketched from above in Figure 4.52. Imagine that the reel starts running at time $t = 0$ and that at any time t, the radius of the tape on

TABLE 4.9   Tapecounter versus time

Time (secs)	Counter
0	0
360	525
720	985
1080	1395
1380	1709
1680	2003
1980	2278
2280	2540
2520	2741
2820	2982
3120	3214
3420	3436

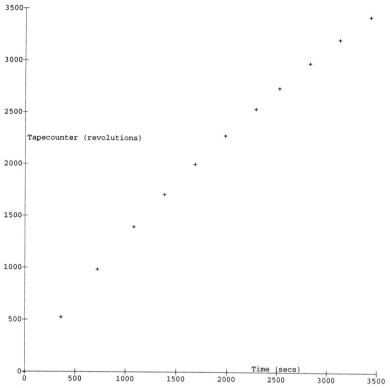

FIGURE 4.51   Tapecounter versus time

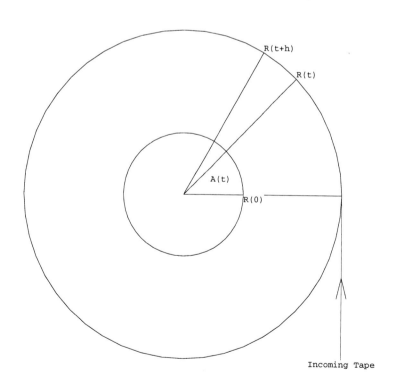

FIGURE 4.52   The take-up reel (seen from above)

the reel is $R(t)$. Thus, $R(0)$ is the radius of the take-up reel when empty. The area of the tape on the reel (as seen from above) at time t is that of the washer-shaped region in Figure 4.52, namely, $\pi \left[R^2(t) - R^2(0) \right]$. If c is the thickness of the tape, the amount of tape on the reel at time t is $\pi \left[R^2(t) - R^2(0) \right] / c$. On the other hand, if v is the constant velocity of the tape past the read-write head, then after the tape has been running for time t the amount of tape on the take-up reel will be vt. From these two expressions for the amount of tape on the reel at time t, we have

$$vt = \frac{\pi}{c} \left[R^2(t) - R^2(0) \right],$$

from which

$$R(t) = R(0)\sqrt{bt + 1}, \tag{4.91}$$

where b is the constant

$$b = \frac{cv}{\pi R^2(0)}.$$

If $A(t)$ is the angle (in radians) of the take-up reel at time t measured from an initial angle of zero at time $t = 0$, we have

$$A(0) = 0. \tag{4.92}$$

Now let's see how A depends on t by looking at the motion of the reel. Consider what happens as the time increases by a small amount from t to $t + h$. The change in the angle will be $A(t + h) - A(t)$. When $A(t + h) - A(t)$ is multiplied by an appropriate radius $\mathcal{R}$, this is the amount of tape added to the reel in this time interval — namely, vh — so we will have

$$[A(t + h) - A(t)]\,\mathcal{R} = vh,$$

which can be written as

$$\frac{A(t + h) - A(t)}{h}\mathcal{R} = v. \tag{4.93}$$

There are a variety of functions we could use for the radius $\mathcal{R}$, such as

- the radius at the beginning of the time interval $R(t)$,
- the radius at the end of the time interval $R(t + h)$,
- the mean of the radius at the beginning and end $\frac{1}{2}\left[R(t + h) + R(t) \right]$,
- the average value of the radius from t to $t + h$, namely, $\frac{1}{h}\left(\int_t^{t+h} R(u)du \right)$.

However, from (4.93) it is obvious that we are interested in what happens to $\mathcal{R}$ as $h \to 0$. It is easy to see that all of the suggested possibilities for $\mathcal{R}$ have the common property that

$$\lim_{h \to 0} \mathcal{R} = R(t). \tag{4.94}$$

From (4.93) and (4.94) we thus find

$$\frac{dA}{dt} = \frac{v}{R(t)}. \tag{4.95}$$

From (4.91) and (4.95) we have the differential equation

$$\frac{dA}{dt} = \frac{v}{R(0)\sqrt{bt+1}},$$ (4.96)

subject to (4.92). The solution of (4.96) subject to (4.92) is

$$A(t) = \frac{2v}{bR(0)}\left(\sqrt{bt+1} - 1\right),$$

which relates the angle $A(t)$ to the time t.

If we assume that the number of revolutions (the number on the tapecounter) is proportional to the angle $A(t)$ — that is, $n(t) = kA(t)$ — we find

$$n(t) = \frac{2vk}{bR(0)}\left(\sqrt{bt+1} - 1\right).$$ (4.97)

If we introduce the constant a by

$$a = \frac{2vk}{bR(0)},$$

(4.97) can be rewritten in the final form

$$n(t) = a\left(\sqrt{bt+1} - 1\right),$$ (4.98)

where a and b are constants depending on the tape and the machine used. This is the relation for which we have been looking.

The question now is whether this equation fits the experimental data in Table 4.9. One way to test this is to rewrite (4.98) by solving it for t, to find

$$t = \frac{1}{ba^2}n^2 + \frac{2}{ba}n.$$

This equation implies that t is a quadratic function of n. If we rewrite it (for $n \neq 0$) as

$$\frac{t}{n} = \frac{1}{ba^2}n + \frac{2}{ba}$$ (4.99)

and plot t/n along the vertical axis and n along the horizontal axis, the result will be a straight line with slope $1/ba^2$ and vertical intercept $2/(ba)$. We can then see whether the data set (Table 4.9) is consistent with (4.98). Figure 4.53 shows the data set plotted in this manner along with a straight line with slope 0.000107 and t/n-intercept 0.626179, which, from (4.99), suggests that $a = 2926$ and $b = 0.00109$. Figure 4.54 shows the original data set (Table 4.9) plotted with the function (4.98), where $a = 2926$ and $b = 0.00109$. $\square$

EXERCISES

1. The experiment described in Example 4.20 used a 120-minute tape. I have used the tape to record some programs, and the present tapecounter reading is 4000. What is the longest program I can still record, if initially the tapecounter was set to 0?

2. The analysis of the tape deck described in Example 4.20 assumed that the digital tapecounter measured the rotation of the take-up reel. However, some tape decks are designed so that the

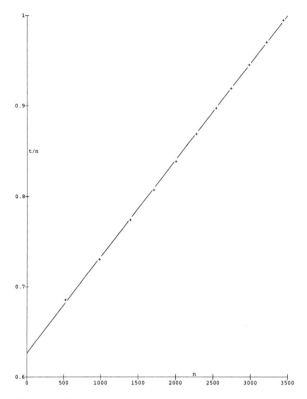

FIGURE 4.53 t/n versus n

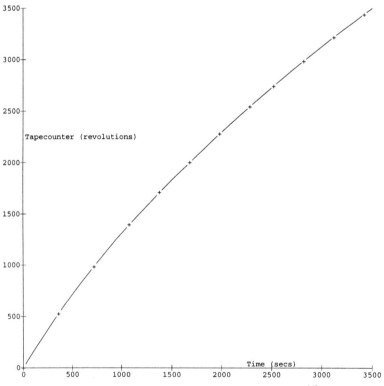

FIGURE 4.54   Tapecounter versus time and $n(t) = a[(bt + 1)^{1/2} - 1]$

tapecounter measures the rotation of the feeder reel. How does this change the analysis and graphs presented in Example 4.20?

3. Perform the experiment described in Example 4.20. Use either a VCR or a tape deck equipped with a digital tapecounter. Put a tape in the unit, set the tape-counter to zero, and, as you start the tape playing, note the time. Then record the time every 5 to 10 minutes along with the number on the tapecounter. Compare the data set obtained in this way with (4.98).

What Have We Learned?

MAIN IDEAS

- A first order differential equation is separable if it can be written in the form where the derivative equals a product of two functions, each of which contains only one of the variables. If x and y are the variables, separable differential equations have the form

$$\frac{dy}{dx} = f(y)g(x).$$

- To solve a separable differential equation, see *How to Solve Separable Differential Equations* on page 135.

- To graph implicit equations, see *How to Sketch Solutions of $G(y) = F(x)$* on page 149.

- Differential equations may be obtained by analyzing data sets. See page 151.

- Differential equations are found in many different applications. See the following sections for reference.

  — Mechanics, page 167.
  — Orthogonal Trajectories, page 179.
  — Functional Equations, page 189.
  — Calculus, page 193.
  — Other Applications, page 200.

CHAPTER 5

Applications Leading to New Techniques

Where Are We Going?

In this chapter we continue the discussion of Chapters 3 and 4 on the general first order differential equation

$$\frac{dy}{dx} = g(x, y).$$

We focus on applications that lead us to develop new techniques of solution. The types of differential equations we consider will be classified as equations with homogeneous coefficients, linear equations, Bernoulli equations, and Clairaut equations. We devote Section 5.5 to certain special second order differential equations that can be solved by the techniques of this chapter. We will need some of these results in Chapter 7.

5.1 Differential Equations with Homogeneous Coefficients

In this section we start with another example from orthogonal trajectories. However, the differential equation that arises in this case is not one that we have encountered previously. We develop a technique for solving this differential equation and show how it is used in other examples. This technique is used later in the text.

EXAMPLE 5.1

Consider the problem of finding the trajectories orthogonal to the family of "offset" circles given by

$$x^2 + y^2 = 2\lambda y, \tag{5.1}$$

as shown in Figure 5.1.

This family of curves is symmetrical about the x-axis, the y-axis, and the origin. (Why?) The orthogonal trajectories will have the same symmetries. To determine the

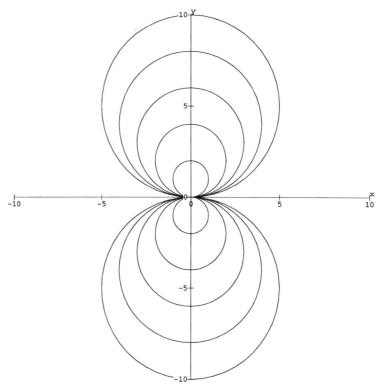

FIGURE 5.1 Circles $x^2 + y^2 = 2\lambda y$

differential equation satisfied by this family of circles, we differentiate (5.1) implicitly to obtain

$$2x + 2y\frac{dy}{dx} = 2\lambda\frac{dy}{dx}.$$
(5.2)

Because we did not solve for λ before differentiating, we must now eliminate λ from (5.2) by replacing it by its value from (5.1). This gives

$$2x + 2y\frac{dy}{dx} = \frac{x^2 + y^2}{y}\frac{dy}{dx},$$

or, collecting coefficients of dy/dx,

$$\frac{x^2 - y^2}{y}\frac{dy}{dx} = 2x.$$

This means that the differential equation satisfied by the orthogonal trajectories to these circles is

$$-\frac{x^2 - y^2}{y}\frac{1}{dy/dx} = 2x,$$

which can be written as

$$\frac{dy}{dx} = \frac{y^2 - x^2}{2xy} = \frac{1}{2}\left(\frac{y}{x} - \frac{x}{y}\right).$$
(5.3)

Although (5.3) is not a separable differential equation, we do notice the presence of the ratios y/x and x/y on the right-hand side. Because x and y occur only in these ratios, let us try a change of the dependent variable from $y(x)$ to $z(x)$ by using one of the ratios, say,

$$z = \frac{y}{x}.$$

This means that

$$y = xz,$$

and differentiating this equation gives us

$$\frac{dy}{dx} = z + x\frac{dz}{dx}.$$

Making these two substitutions in (5.3) gives the transformed differential equation as

$$z + x\frac{dz}{dx} = \frac{1}{2}\left(z - \frac{1}{z}\right).$$

This equation is separable and may be rearranged as

$$\frac{2z}{z^2 + 1}\frac{dz}{dx} = -\frac{1}{x},$$

so integration gives

$$\ln(z^2 + 1) = -\ln|x| + C,$$

or

$$x(z^2 + 1) = 2c.$$

Returning to the original variables and completing the square, we observe that the family of offset circles

$$x^2 + y^2 = 2cx, \ \text{ or } \ (x - c)^2 + y^2 = c^2,$$

is orthogonal to the original family given by (5.1). The two families are shown in Figure 5.2.   □

It was no accident that the differential equation (5.3) was transformed to a separable one after the change of variable $y = xz$. The following few steps show that this would happen to any equation of the form

$$\frac{dy}{dx} = g\left(\frac{y}{x}\right). \tag{5.4}$$

If we make the change of variables $z = y/x$ in equation (5.4), it becomes

$$x\frac{dz}{dx} + z = g(z),$$

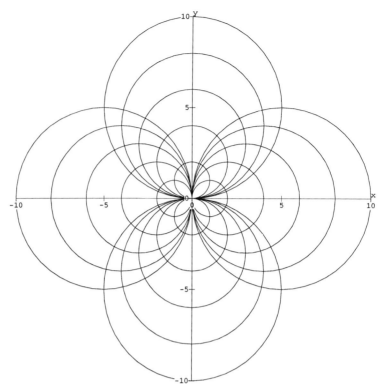

FIGURE 5.2   Orthogonal trajectories $x^2 + y^2 = 2\lambda y$, $x^2 + y^2 = 2cx$

which may be seen to be separable by rearranging it as

$$\frac{1}{g(z) - z} \frac{dz}{dx} = \frac{1}{x}.$$

So a first order differential equation of the type (5.4) can always be converted to a separable differential equation by a change of variables. We now formalize this discussion.

Definition 5.1:   **A DIFFERENTIAL EQUATION WITH HOMOGENEOUS COEFFICIENTS is one which may be put in the form**

$$\frac{dy}{dx} = g\left(\frac{y}{x}\right).$$

How to Solve Differential Equations with Homogeneous Coefficients

Purpose: To solve first order differential equations of the form

$$\frac{dy}{dx} = g\left(\frac{y}{x}\right). \tag{5.5}$$

Technique:

1. Make the change of variable

$$z = \frac{y}{x},$$

so that

$$y = xz \tag{5.6}$$

and

$$\frac{dy}{dx} = x\frac{dz}{dx} + z. \tag{5.7}$$

2. Substitute (5.6) and (5.7) into (5.5) to eliminate all y dependence.

3. Rearrange the result so terms involving z are separated from terms involving x.

4. Integrate this separable equation to find either an explicit or implicit solution, $z = z(x)$.

5. If the last step yields an explicit solution for $z = z(x)$, use (5.6) to find the explicit solution of (5.5) as $y = xz(x)$. If, on the other hand, there is an implicit solution relating z with x, replace z in this implicit solution with y/x to obtain an implicit solution of (5.5).

Comments about differential equations with homogeneous coefficients

- Sometimes it is quite difficult to recognize that a differential equation is of the type (5.5). One way to test whether a differential equation

$$\frac{dy}{dx} = g(x, y) \tag{5.8}$$

is of the type (5.5) is to substitute $y = xz$ into $g(x, y)$ obtaining $g(x, xz)$. If the resulting function is independent of x, then (5.8) is of the type (5.5). This shows that $g(x, y)$ is homogeneous of degree 0.

- A common mistake is to omit the z on the right-hand side of (5.7).

- Sometimes the resulting separable equation is easier to integrate if we make the change of variables $v = x/y$ instead of $z = y/x$.

- The slope fields of differential equations with homogeneous coefficients, (5.5), are always symmetric about the origin. (Why?)

- This type of differential equation is named because functions of the form $g\left(y/x\right)$ are homogeneous[1] of degree 0 in x and y.

EXAMPLE 5.2

We show how this technique works in practice by considering

$$\frac{dy}{dx} = \frac{x^3 y}{x^4 + y^4}, \qquad (x, y) \neq (0, 0). \tag{5.9}$$

We notice this is not separable, but is it an equation with homogeneous coefficients?
If we divide the numerator and denominator by x^4, we find

$$\frac{dy}{dx} = \frac{y/x}{1 + (y/x)^4}.$$

---

[1] A function $F(x, y)$ is homogeneous of degree n if $F(tx, ty) = t^n F(x, y)$.

The right-hand side is a function of y/x, so it is a differential equation with homogeneous coefficients. [Alternatively, we could test whether the right-hand side of (5.9) is of the form $g(y/x)$ by substituting $y = xz$ to find

$$\frac{x^3 y}{x^4 + y^4} = \frac{x^3 xz}{x^4 + x^4 z^4} = \frac{z}{1 + z^4}.$$

Seeing this is independent of x, we have a differential equation with homogeneous coefficients.]

We make the substitution $z = y/x$ and obtain

$$x\frac{dz}{dx} + z = \frac{z}{1 + z^4},$$

which can be rewritten as

$$x\frac{dz}{dx} = \frac{z}{1 + z^4} - z = \frac{-z^5}{1 + z^4}.$$

This is separable, and it can be written as

$$\frac{1 + z^4}{z^5}\frac{dz}{dx} = -\frac{1}{x},$$

or

$$\left(\frac{1}{z^5} + \frac{1}{z}\right)\frac{dz}{dx} = -\frac{1}{x}.$$

Integration and combination of logarithmic terms yield the implicit solution

$$4\ln|zx| = z^{-4} + C.$$

Replacing z with y/x, we find the implicit solution of (5.9) in the form

$$4\ln|y| = \left(\frac{x}{y}\right)^4 + C.$$

$\square$

We now look at another application of differential equations with homogeneous coefficients.

EXAMPLE 5.3: The Focal Property of the Parabolic Reflector

An important property of the parabola is that rays coming from the focus are reflected parallel to the axis. Or, equivalently, parallel rays are focused at a single point after reflection. This focal property (shown in Figure 5.3) is why parabolas are used in the design of car headlights and radio telescopes. However, constructing a parabolic telescope is very expensive, and so a natural question to ask is whether any other shapes have this focal property.

To be specific, let rays be projected parallel to the x-axis from infinity (see Figure 5.4). They strike the mirror with shape $f(x)$ at the point with coordinates (x_0, y_0) and are then focused at the origin O. (The rays and mirror continue into the lower half plane, but they are not drawn in Figure 5.4.) When a ray strikes the mirror, it is reflected so that the angle between the incoming ray and the mirror equals the angle between the reflected ray

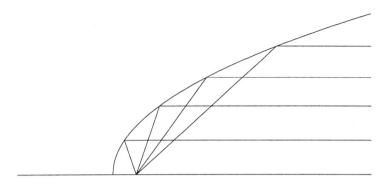

FIGURE 5.3 Focal property

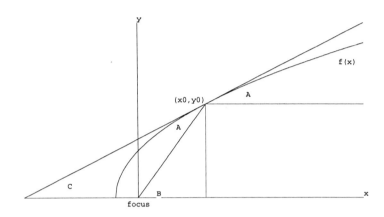

FIGURE 5.4 Geometry for focal property

and the mirror. At the place where the ray strikes the mirror, the mirror acts as if it were a straight line with slope $f'(x_0)$. Thus, the angle between the ray and the mirror is actually the angle A between the ray and the straight line with slope $f'(x_0)$.

From the geometry of Figure 5.4, where A, B, and C are all angles, we see that

$$A = C,$$

so

$$B = A + C = 2A. \tag{5.10}$$

Also we see that

$$\tan B = \frac{y_0}{x_0} \tag{5.11}$$

and

$$\tan A = f'(x_0). \tag{5.12}$$

From (5.10) we have

$$\tan B = \tan 2A = \frac{2 \tan A}{1 - \tan^2 A},$$

which, from (5.11) and (5.12), gives rise to

$$\frac{y_0}{x_0} = \frac{2f'(x_0)}{1 - [f'(x_0)]^2}. \tag{5.13}$$

If we solve (5.13) for $f'(x_0)$ we find

$$f'(x_0) = \frac{-x_0 \pm \sqrt{x_0^2 + y_0^2}}{y_0}.$$

This is the slope of the tangent line to the curve $f(x)$ at the point (x_0, y_0) that causes a line parallel to the x-axis to be reflected through the origin. Because we want this to happen for every point (x_0, y_0) — namely, (x, y) — we have the differential equation

$$\frac{dy}{dx} = \frac{-x \pm \sqrt{x^2 + y^2}}{y} = \frac{-1 \pm \sqrt{1 + (y/x)^2}}{y/x}, \tag{5.14}$$

where $y = f(x)$. Clearly, there will be two solutions of this differential equation — one corresponding to $+$ and the other corresponding to $-$. Thus, there is some hope of a solution in addition to the parabola.

Before trying to solve this analytically, let's look at the slope field for (5.14) to see first whether this differential equation is reasonable, and second what the ambiguity in sign implies. Figure 5.5 shows the slope field and a few hand-drawn solution curves for (5.14) with the plus sign. Notice we appear to have parabolas opening to the right. Figure 5.6 shows the slope field and a few hand-drawn solution curves for (5.14) with the minus sign. Notice, in this case, we appear to have parabolas opening to the left. So the ambiguity in sign seems to dictate the orientation of the figure rather than a different type of solution.

Now let us turn to finding the explicit solution of (5.14) to confirm these observations. Equation (5.14) is a differential equation with homogeneous coefficients, so we use the substitution

$$y = xz$$

to find, after some simplification, the separable differential equation

$$x\frac{dz}{dx} = \frac{-1 - z^2 \pm \sqrt{1 + z^2}}{z}.$$

Multiplying both sides of this equation by -1 and separating variables leads to

$$\int \frac{z\,dz}{1 + z^2 \mp \sqrt{1 + z^2}} = -\int \frac{dx}{x}.$$

By making the substitution $u^2 = 1 + z^2$ in the left-hand integral we find that

$$\int \frac{u\,du}{u^2 \mp u} = \int \frac{du}{u \mp 1}.$$

Now we integrate to obtain

$$\ln|u \mp 1| = -\ln|x| + C,$$

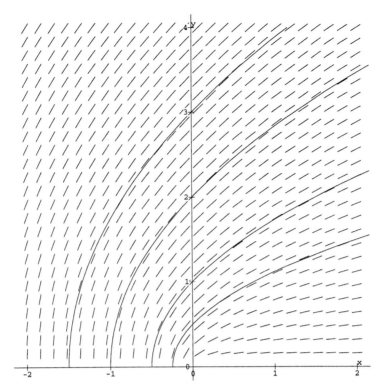

FIGURE 5.5 Hand-drawn solution curves and slope field for $dy/dx = [-x + (x^2 + y^2)^{1/2}]/y$

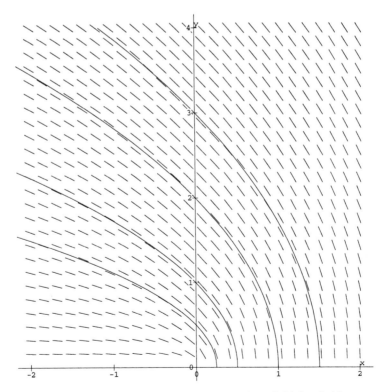

FIGURE 5.6 Hand-drawn solution curves and slope field for $dy/dx = [-x - (x^2 + y^2)^{1/2}]/y$

or

$$x(u \mp 1) = c.$$

This can be rewritten as

$$xu = c \pm x,$$

which, when squared, leads to

$$x^2 u^2 = c^2 \pm 2cx + x^2.$$

To put this expression in terms of our original variables, we use the fact that

$$u^2 = 1 + z^2 = 1 + \frac{y^2}{x^2},$$

so our solution becomes

$$y^2 = c^2 \pm 2cx.$$

As expected, this represents two parabolas, one opening to the left and the other to the right. Thus, *the parabola is the only curve with the focal property.*   □

EXERCISES

1. First confirm that the following are differential equations with homogeneous coefficients, and then solve them analytically. For explicit solutions, graph and compare with its slope field.

(a) $(x^2 + 3y^2)y' + 2xy = 0$

(b) $3xyy' + x^2 + y^2 = 0$

(c) $xy' + \sqrt{x^2 + y^2} - y = 0$

(d) $(x^2 - xy + y^2)y' + y^2 = 0$

(e) $[y \tan(x/y) - x] y' + y = 0$

(f) $(xy - x^2)y' - y^2 = 0$

(g) $x^2 y' + y^2 - xy = 0$

(h) $2xyy' + x^2 + y^2 = 0$

(i) $3xyy' + x^2 + y^2 = 0$

(j) $xy' - y \ln(y/x) - y = 0$

(k) $(x + 2y)y' + x + y = 0$

2. A function $F(x, y)$ is homogeneous of degree n if $F(tx, ty) = t^n F(x, y)$. For example,

$$F(x, y) = \frac{x^4 + y^4}{x}$$

is homogeneous of degree 3 because

$$F(tx, ty) = \frac{(tx)^4 + (ty)^4}{tx} = t^3 \frac{x^4 + y^4}{x} = t^3 F(x, y).$$

Check to see if the following functions are homogeneous. If so, give the degree.

(a) $F(x, y) = x^2 + 3xy - x^3(x + y)^{-1}$

(b) $F(x, y) = x^2 + 3xy - x^2(x + y)$

(c) $F(x, y) = x \sin(x/y) + x^2/y$

(d) $F(x, y) = \sqrt{x^4 + 4x^2y^2 + x^2}$

(e) $F(x, y) = \ln x - \ln y + e^{x/y}$

3. Show that the numerators and denominators of the differential equations in the examples for this section are homogeneous of the same degree.

(a) $\dfrac{dy}{dx} = \dfrac{y^2 - x^2}{2xy}$

(b) $\dfrac{dy}{dx} = \dfrac{x^3 y}{x^4 + y^4}$

(c) $\dfrac{dy}{dx} = \dfrac{-x \pm \sqrt{x^2 + y^2}}{y}$

4. This question deals with solving the differential equation

$$\frac{dy}{dx} = f\left(\frac{a_1 x + b_1 y + c_1}{a_2 x + b_2 y + c_2}\right),$$

where a_1, b_1, c_1, a_2, b_2, and c_2 are constants.

(a) If

$$a_1 b_2 - a_2 b_1 \neq 0,$$

show that the transformation

$$x = u + \alpha, \ y = v + \beta,$$

through careful choice of the constants α and β, reduces the differential equation to the differential equation with homogeneous coefficients

$$\frac{dv}{du} = f\left(\frac{a_1 u + b_1 v}{a_2 u + b_2 v}\right).$$

(b) If

$$a_1 b_2 - a_2 b_1 = 0,$$

show that the transformation

$$z = a_1 x + b_1 y$$

reduces the differential equation to a separable differential equation involving z and x.

5. Using the technique described in Exercise 4, solve the following differential equations.

(a) $\dfrac{dy}{dx} = -\dfrac{x + y - 2}{x - y + 4}$

(b) $\dfrac{dy}{dx} = -\dfrac{x + y + 1}{2x + 2y - 4}$

(c) $\dfrac{dy}{dx} = -\dfrac{-x + 2y - 1}{2x - 4y + 3}$

5.2   Linear Differential Equations

In this section we start with a variation on a preceding example involving the population of Botswana where now emigration is assumed to occur. However, the differential equation that arises in this case is not one we have encountered previously. We develop a technique for solving this differential equation. Then by looking at the solution we are able to see a different but preferable method to obtain the same solution. We show how this method is used in other examples. This method is used heavily later in the text.

EXAMPLE 5.4: The Population of Botswana with Emigration

Recall in Chapter 2 we discovered that the population of Botswana has been growing exponentially with a doubling time of about 20 years. Suppose that in 1990, residents of that country worry about this growth rate and think about emigrating to other countries. It seems reasonable that any emigration for this reason would increase gradually over time, perhaps linearly. The integral of this emigration rate over the next 20 years equals the total emigration during that time period. If we hypothesize that 1/4 of the 1990 population of 1.285 million will leave the country over the next two decades in a linear manner at the rate of at million people per year, we have

$$\int_0^{20} at \, dt = \frac{1.285}{4},$$

from which we find

$$\frac{1}{2} a \, (20)^2 = \frac{1.285}{4},$$

so

$$a = \frac{2}{(20)^2} \frac{1.285}{4} = 1.60625 \times 10^{-3}.$$

Thus, we have

$$\text{emigration rate in millions per year} = at = 1.60625 \times 10^{-3} t.$$

If we use the growth rate for Botswana from Chapter 2 and its population for 1990 as our initial condition, we obtain the differential equation

$$\frac{dP}{dt} = kP - at \tag{5.15}$$

subject to

$$P(0) = 1.285 \text{ (million)}, \tag{5.16}$$

where $k = 0.0355$ and $a = 1.60625 \times 10^{-3}$.

In (5.15) we have an example of a linear differential equation. The word "linear" is used because the right-hand side of (5.15) is a linear function of P. In terms of these variables other linear differential equations could be expressed by

$$\frac{dP}{dt} = f(t)P + g(t),$$

where f and g are functions of t.

Figure 5.7 shows the slope field and the hand-drawn solution curve for (5.15) subject to (5.16). It suggests that $P(20) \approx 2.2$, that is, the population after 20 years will be about 2.2 million.

[The software you are using may have trouble imitating the slope field in Figure 5.7, due to the fact that the horizontal and vertical scales are so different — in this case 20 units horizontally and 3 units vertically. If this happens, you should rescale either the dependent or the independent variable so the new scales are approximately equal. For example, in this case you could either introduce a new dependent variable y, where $P = y/10$ — so the new units will be 20 units horizontally and 30 units vertically — or introduce a new independent variable x, where $t = 10x$ — so the new units will be 2 units horizontally and 3 units vertically.]

Let us now move away from this specific initial condition and consider other initial values of the population. If we look more carefully at Figure 5.7, particularly in the region between 0 and 1 on the vertical axis and 0 to 20 on the horizontal axis, we see that it appears possible for the population to die out [that is, there is a time when $P(t) = 0$] depending on the initial value $P(0)$. Figure 5.8 shows a number of hand-drawn solution curves, from which it is difficult to decide how many solutions will eventually die out.

We will use calculus to try to make sense of Figure 5.8. From (5.15), we see that $P(t)$ is increasing if $kP - at$ is positive and decreasing if $kP - at$ is negative. The line $P = at/k$ is where $P' = 0$, which is the isocline corresponding to horizontal tangents. So the line $P = at/k$ separates the plane into two regions: Solution curves that lie above this line will be increasing, whereas below the line they will be decreasing.

If we differentiate (5.15) and use (5.15) to eliminate P' from the result, we find

$$\frac{d^2 P}{dt^2} = k^2 \left(P - \frac{a}{k}t - \frac{a}{k^2} \right).$$

The line $P = at/k + a/k^2$ is where $P'' = 0$, so it separates the plane into two regions: Solution curves that lie above this line will be concave up and those that lie below this line will be concave down. Notice that this line is parallel to the line $P = at/k$ but is displaced a distance a/k^2 up the vertical axis. Figure 5.9 shows these regions.

Figure 5.10 shows the line $P = at/k$ corresponding to zero slope, the line $P = at/k + a/k^2$ corresponding to zero second derivatives, and some hand-drawn solution curves to (5.15). This suggests that if a solution curve has its initial value $P(0)$ between 0 and a/k^2 the population will eventually die out. If $P(0) \geq a/k^2$, then the population will continue to grow.

Now let's try to find an analytical solution of (5.15). Note that (5.15) is not separable nor is it an equation with homogeneous coefficients.[2] To solve (5.15) we recall that in the previous section we transformed a differential equation we could not solve to one we could solve by making a change in the dependent variable y of the form $y(x) = xz(x)$, where x was the independent variable. We now make a similar change of variable here (where P is the dependent variable, and t the independent variable). However, because we do not know in advance what the multiplicative factor should be, we simply let it be an unknown function $u(t)$. That is, we let

$$P(t) = u(t)z(t), \tag{5.17}$$

---

[2] Note also that $P(t) = at/k$ is in no sense an equilibrium solution of (5.15) because it does not satisfy the differential equation and its derivative is not zero.

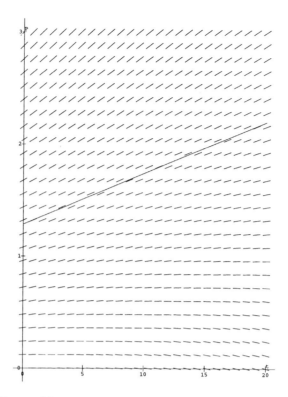

FIGURE 5.7   Hand-drawn solution curve and slope field for
$dP/dt = kP - at$

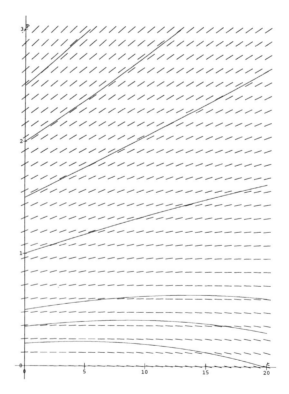

FIGURE 5.8   Hand-drawn solution curves and slope field for
$dP/dt = kP - at$

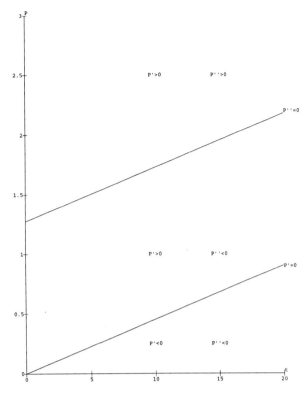

FIGURE 5.9 The lines $P = at/k$ and $P = at/k + a/k^2$

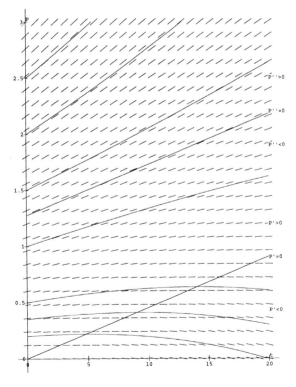

FIGURE 5.10 The lines $P = at/k$ and $P = at/k + a/k^2$, hand-drawn solution curves, and slope field for $dP/dt = kP - at$

where z is the new dependent variable and $u(t)$ is a specific function chosen to make the resulting differential equation solvable. If we differentiate (5.17) we obtain

$$\frac{dP}{dt} = u\frac{dz}{dt} + \frac{du}{dt}z, \qquad (5.18)$$

and substituting (5.17) and (5.18) into (5.15) yields

$$u\frac{dz}{dt} + \frac{du}{dt}z = kuz - at.$$

Because we want to choose a $u(t)$ that simplifies this differential equation, we put all terms involving u on the left-hand side. This yields

$$u\frac{dz}{dt} + \left(\frac{du}{dt} - ku\right)z = -at.$$

If we choose u so that the term in parenthesis is zero, then we would have

$$u\frac{dz}{dt} + 0z = -at,$$

or

$$\frac{dz}{dt} = -\frac{at}{u(t)}. \qquad (5.19)$$

Because $u(t)$ will be a known function of time, (5.19) may be integrated directly. The required function $u(t)$ that lets this happen satisfies

$$\frac{du}{dt} - ku = 0,$$

or

$$\frac{1}{u}\frac{du}{dt} = k. \qquad (5.20)$$

We need only one function $u(t)$ to make this change of variable work, so we set to zero the arbitrary constant that results from integrating (5.20) and write a solution of (5.20) as

$$u(t) = e^{kt}. \qquad (5.21)$$

Now we may integrate (5.19) directly, using integration by parts, to obtain

$$z(t) = \int -\frac{at}{u(t)}\,dt = \int -ate^{-kt}\,dt$$

as

$$z(t) = a\left(\frac{t}{k} + \frac{1}{k^2}\right)e^{-kt} + C. \qquad (5.22)$$

Equations (5.17) and (5.21) let the solution $P(t)$ of the original problem be expressed as

$$P(t) = e^{kt}z(t)$$

which, by (5.22), can be written as

$$P(t) = a\left(\frac{t}{k} + \frac{1}{k^2}\right) + Ce^{kt}. \qquad (5.23)$$

Since, from (5.23),

$$P(0) = \frac{a}{k^2} + C,$$

then

$$C = P(0) - \frac{a}{k^2},$$

and we can rewrite (5.23) as

$$P(t) = a\left(\frac{t}{k} + \frac{1}{k^2}\right) + \left(P(0) - \frac{a}{k^2}\right)e^{kt}. \qquad (5.24)$$

Substituting the values for a, k, and $P(0)$ — chosen to satisfy the initial condition given in (5.16) — lets us write our solution (5.24) in final form as

$$P(t) \approx 0.0452t + 1.275 + 0.01045e^{0.0355t}. \qquad (5.25)$$

This means that after 20 years the population is

$$P(20) \approx 2.2,$$

which is about a 71% increase in population instead of about the 100% increase we would have without any emigration. This is in excellent agreement with the information we obtained from Figure 5.7.

Note that our solution in (5.25) is an increasing function for all values of t. If we look at (5.24), we see that if $P(0) - a/k^2 > 0$, the population will increase exponentially; if $P(0) - a/k^2 = 0$, the population will increase linearly; and if $P(0) - a/k^2 < 0$, the population will eventually die out. In our case $P(0) - a/k^2 = 0.01045$, which is just positive.

One other comment needs to be made. The solution with $P(0) - a/k^2 = 0$ [namely, $P(t) = at/k + a/k^2$] is unstable, because for $P(0) - a/k^2$ slightly positive or slightly negative, the associated solutions recede from the solution for $P(0) - a/k^2 = 0$ as t increases. Thus, a small change in $P(0)$ will make a large change in the eventual population, either going to exponential growth or dying out. This is consistent with Figure 5.10.

Finally, we can rewrite (5.24) in a more illuminating form, namely,

$$P(t) = P(0)e^{kt} - \frac{a}{k^2}\left(e^{kt} - 1 - kt\right).$$

In this form we may see the impact of emigration. The first term on the right-hand side corresponds to no emigration, and the second to the impact of emigration. □

The technique we used to reduce the differential equation (5.15) to one that is separable will work with any first order differential equation of the form

$$\frac{dy}{dx} + p(x)y = q(x), \qquad (5.26)$$

where $p(x)$ and $q(x)$ are given functions of x.

Definition 5.2:   **Differential equations that have the form (5.26) are called LINEAR DIF-FERENTIAL EQUATIONS. Note that there is at most one factor involving y or dy/dx in each term, and it occurs to the first power.**

To show that the general linear differential equation (5.26) can always be reduced to a separable equation, we again seek a change of the dependent variable of the form

$$y(x) = u(x)z(x). \tag{5.27}$$

Here z is the new dependent variable, and we will look for a judicious choice of the function $u(x)$ so the transformed equation will be separable. Making this substitution — using the product rule to differentiate $u(x)z(x)$ — changes the form of (5.26) to

$$u\frac{dz}{dx} + \frac{du}{dx}z + p(x)uz = q(x).$$

We rearrange this equation as

$$u\frac{dz}{dx} + \left(\frac{du}{dx} + p(x)u\right)z = q(x). \tag{5.28}$$

We now choose $u(x)$ to satisfy the differential equation

$$\frac{du}{dx} + p(x)u = 0, \tag{5.29}$$

or, in separated form,

$$\frac{1}{u}\frac{du}{dx} = -p(x). \tag{5.30}$$

Choosing u in this manner reduces (5.28) to the separable differential equation

$$u\frac{dz}{dx} = q(x),$$

and by dividing this equation by u and integrating, we get

$$z(x) = \int \frac{q(x)}{u(x)}\,dx + C. \tag{5.31}$$

The function u in the denominator of this equation is found by integrating (5.30) as

$$\ln|u(x)| = \int -p(x)\,dx. \tag{5.32}$$

Again note that because we are seeking only one specific function, $u(x)$, we may set the arbitrary constant of integration at this step to zero and choose the positive solution of $u(x)$ from (5.32) as

$$u(x) = e^{-\int p(x)\,dx}. \tag{5.33}$$

Combining (5.31) and (5.27) gives the solution, $y(x)$, of our original differential equation as

$$y(x) = u(x) \int \frac{q(x)}{u(x)} \, dx + Cu(x), \tag{5.34}$$

where $u(x)$ is given by (5.33) and C is an arbitrary constant.

Most people find that memorizing (5.34), with $u(x)$ given by (5.33), as the solution to a general first order linear differential equation is not worth the effort. In fact, further analysis of the structure of this solution, (5.34), leads to the following simple way for its development. If we rewrite (5.34) in the form

$$\frac{1}{u(x)} y(x) = \int \frac{q(x)}{u(x)} \, dx + C$$

and differentiate, we find

$$\frac{d}{dx} \left(\frac{1}{u(x)} y(x) \right) = \frac{q(x)}{u(x)}. \tag{5.35}$$

This suggests that we might be able to multiply (5.26) by $1/u(x) = e^{\int p(x)\,dx}$ and write it in the form (5.35), which can be integrated immediately to yield (5.34). This observation leads to the following three-step process.

How to Solve Linear Differential Equations

Purpose: To find $y = y(x)$ that satisfies the linear differential equation

$$a_1(x) \frac{dy}{dx} + a_0(x) y = f(x),$$

where $a_1(x)$ is not zero.

Technique:

1. Put the linear differential equation in the standard form of

$$\frac{dy}{dx} + p(x) y = q(x), \tag{5.36}$$

by dividing by $a_1(x)$. Compute the function

$$\mu(x) = e^{\int p(x)\,dx},$$

which has the property that

$$\frac{d\mu}{dx} = \mu(x) p(x). \tag{5.37}$$

2. Multiply both sides of the differential equation (5.36) by this function $\mu(x)$ to get

$$\mu(x) \frac{dy}{dx} + \mu(x) p(x) y = \mu(x) q(x). \tag{5.38}$$

Notice that the term $\mu(x)p(x)$ on the left-hand side of (5.38) is equal to $d\mu/dx$ by (5.37). If we make this replacement in (5.38) we obtain

$$\mu(x)\frac{dy}{dx} + \frac{d\mu}{dx}y = \mu(x)q(x).$$

Here we see that the left-hand side of this equation may be written as the derivative of the product $\mu(x)y$. Thus

$$\frac{d}{dx}[\mu(x)y] = \mu(x)\frac{dy}{dx} + \frac{d\mu}{dx}y = \mu(x)\frac{dy}{dx} + \mu(x)p(x)y,$$

so (5.38) is equivalent to

$$\frac{d}{dx}[\mu(x)y] = \mu(x)q(x). \tag{5.39}$$

You should always check your work at this point by differentiating the left-hand side of (5.39). The resulting equation should then be the same as what you obtain by multiplying the original differential equation by $\mu(x)$.

3. Integrate both sides of (5.39) to obtain

$$\mu(x)y = \int \mu(x)q(x)\,dx + C.$$

Finally, divide this last equation by the function $\mu(x) = e^{\int p(x)\,dx}$ to obtain an explicit solution of (5.36), that is,

$$y(x) = \frac{1}{\mu(x)} \int \mu(x)q(x)\,dx + \frac{C}{\mu(x)}, \tag{5.40}$$

which agrees with (5.34).

Comments about linear differential equations

- The function $\mu = e^{\int p(x)\,dx}$ is called an INTEGRATING FACTOR of (5.36) because if we multiply (5.36) by this factor, we can integrate both sides of the resulting equation.

- Because (5.40) contains every solution of (5.36), the function in (5.40) is called the GENERAL SOLUTION of (5.36). See Exercise 23.

- A common mistake when going from (5.36) to (5.38) is to forget to multiply the right-hand side of (5.36) by the integrating factor. This can lead to some unexpected integrals when we reach (5.40).

- Another common mistake when going from (5.39) to (5.40) is to is forget C, the constant of integration. As a result, only a particular solution of (5.36) is found.

- Solutions of first order linear differential equations are automatically explicit solutions.

- Linear differential equations are very important, because they occur frequently in applications. The next chapter is devoted to their applications and properties.

EXAMPLE 5.5

As an example consider the linear differential equation

$$(x^2 + 1)\frac{dy}{dx} - 2xy = -e^{-x}(x^2 + 1)^2, \tag{5.41}$$

where we apply the preceding three steps.

1. Divide by $x^2 + 1$ to change the form of (5.41) to

$$\frac{dy}{dx} - \frac{2x}{(x^2 + 1)} y = -e^{-x} (x^2 + 1). \qquad (5.42)$$

Here $p(x) = -2x/(x^2 + 1)$ and $q(x) = -e^{-x}(x^2 + 1)$. The integrating factor is

$$\mu(x) = e^{\int p(x)\,dx} = e^{\int \frac{-2x}{x^2+1}\,dx} = e^{-\ln(x^2+1)} = e^{\ln(x^2+1)^{-1}} = (x^2 + 1)^{-1} = \frac{1}{x^2 + 1}.$$

2. Multiply (5.42) by the integrating factor $\mu(x) = 1/(x^2 + 1)$ to obtain

$$\frac{1}{x^2 + 1} \frac{dy}{dx} - \frac{2x}{(x^2 + 1)^2} y = -e^{-x} (x^2 + 1) \frac{1}{x^2 + 1}$$

or

$$\frac{1}{x^2 + 1} \frac{dy}{dx} - \frac{2x}{(x^2 + 1)^2} y = -e^{-x}.$$

Here we take advantage of the fact that the left-hand side of this equation is the derivative of a product to write it as

$$\frac{d}{dx} \left(\frac{1}{x^2 + 1} y \right) = -e^{-x}. \qquad (5.43)$$

[You should differentiate the left-hand side of (5.43) to verify this.]

3. Integrate (5.43) to obtain

$$\frac{1}{x^2 + 1} y = \int -e^{-x} dx = e^{-x} + C,$$

and multiply by $x^2 + 1$ to obtain the explicit solution of (5.41) as

$$y(x) = (e^{-x} + C) (x^2 + 1),$$

or

$$y(x) = e^{-x} (x^2 + 1) + C (x^2 + 1). \qquad (5.44)$$

Note that the term $C(x^2 + 1)$ is the general solution of $(x^2 + 1)\,dy/dx - 2xy = 0$, whereas the term $e^{-x}(x^2 + 1)$ is a particular solution of (5.41).

Figure 5.11 shows the slope field and various solution curves of (5.41). We notice that the solution that passes through the point $(0, 1)$ appears to divide the family of solutions into two groups, those that tend to $+\infty$ and those that tend to $-\infty$ as $x \to \infty$. Confirm this observation directly from (5.44). $\qquad \square$

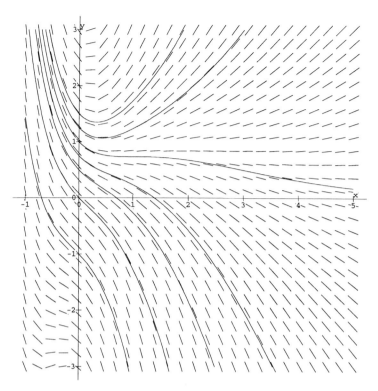

FIGURE 5.11   Analytical solution curves and slope field for $(x^2 + 1)\, dy/dx - 2xy = -e^{-x}\left(x^2 + 1\right)^2$

EXAMPLE 5.6

Finally, let us also use this three-step process on the first differential equation of this section,

$$\frac{dP}{dt} = kP - at. \tag{5.45}$$

1. Write the differential equation as

$$\frac{dP}{dt} - kP = -at. \tag{5.46}$$

Here $p(t) = -k$ and $q(t) = -at$, and the integrating factor is

$$\mu(t) = e^{\int -k\, dt} = e^{-kt}.$$

2. Multiply (5.46) by this integrating factor to obtain

$$e^{-kt}\frac{dP}{dt} - ke^{-kt}P = -ate^{-kt},$$

or combining the left-hand side as the derivative of a product gives

$$\frac{d}{dt}\left(e^{-kt}P\right) = -ate^{-kt}. \tag{5.47}$$

(Differentiate this expression to verify this.)

3. Integrate (5.47) to obtain

$$e^{-kt} P = a \left(\frac{t}{k} + \frac{1}{k^2} \right) e^{-kt} + C,$$

and divide by e^{-kt} to give the explicit solution of (5.45) as

$$P(t) = a \left(\frac{t}{k} + \frac{1}{k^2} \right) + Ce^{kt},$$

which is what we obtained previously. Note that the term Ce^{kt} is the general solution of $dP/dt - kP = 0$, whereas the first term is a particular solution of (5.45).

We have already pointed out — based on the discussion following (5.25) — that the solution corresponding to $C = 0$ is a straight line and unstable. Remember that equilibrium solutions are solutions of $dy/dx = g(x, y)$ of the type $y = c$ — that is, they are horizontal lines. Equation (5.45) is an example of a differential equation that has a line as a solution, but the line is not horizontal, so the solution is not an equilibrium solution. □

EXERCISES

Solve the following differential equations. Make sure that the analytical solution you obtain is supported by the slope field, increasing and decreasing regions, and concavity issues.

1. $\dfrac{dy}{dx} + y = x$

2. $\dfrac{dy}{dx} - 2y = xe^{2x}$

3. $\dfrac{dy}{dx} + y = e^x$

4. $y' - 6y = e^x$

5. $y' + 3y = 3x^2 e^{-3x}$

6. $\dfrac{dy}{dx} + \dfrac{1}{x \ln x} y = 3x^2$

7. $x \dfrac{dy}{dx} + 2y - \dfrac{x}{x^2 + 2} = 0$

8. $\dfrac{dy}{dx} + \dfrac{1}{x + 1} y = \dfrac{\cos x}{x + 1}$

9. $\dfrac{dy}{dx} + xy = 3x$

10. $\dfrac{1}{x} \dfrac{dy}{dx} + 2y = 2x^2$

11. $\cos x \dfrac{dy}{dx} + y \sin x = 3 \sin x \cos^2 x$

12. $x \dfrac{dy}{dx} - y = x^2 \cos x$

13. $\dfrac{dy}{dx} - \dfrac{1}{2x}y = 2$

14. $y' - \dfrac{2}{x^2}y = \dfrac{1}{x^2}$

15. $y' + 2xy = 2x$

16. $y' + (2 + x^{-1})y = 2e^{-2x}$

17. $(x^2 + 1)y' + 4xy = x$

18. $\dfrac{dy}{dx} + \dfrac{4x}{x^2 + 1}y = 3x$

19. $x\dfrac{dy}{dx} + \dfrac{2x + 1}{x + 1}y = x - 1$

20. $x^4 y' + 2x^3 y = 1$

21. Compare the form of your solutions to Exercises 1 through 5. State what is similar and what is different about the terms in your solutions.

22. Show that the solution of $y' + 2xy = 1$, $\qquad y(0) = 0$ is $y(x) = e^{-x^2} \int_0^x e^{t^2}\, dt$.

   (a) Can this solution vanish for $x > 0$?

   (b) Show that this solution has an extreme value at the point $(x_m, 1/(2x_m))$ where x_m satisfies $y(x_m) = 1/(2x_m)$. Show that at this extreme value we have $y''(x_m) = -2y(x_m)$. Explain why this implies that there is a maximum at $(x_m, 1/(2x_m))$. Can this solution have a minimum? Can it have a second maximum? Find a numerical value for x_m.

   (c) Show that this solution has a point of inflection at $\left(x_i, x_i/\left(2x_i^2 - 1\right)\right)$ where x_i satisfies $y(x_i) = x_i/\left(2x_i^2 - 1\right)$. Can this solution have a second point of inflection? Find a numerical value for x_i.

   (d) Use L'Hôpital's rule[3] to show that this solution has the property that $y(x) \to 0$ as $x \to \infty$.

   (e) Use L'Hôpital's rule to show that this solution has the property that $2xy(x) \to 1$ as $x \to \infty$, so that $y(x) \to 1/(2x)$ as $x \to \infty$.

   (f) Use the information from parts (a) through (e) to sketch this solution for $x > 0$.

   (g) Show that the solution satisfies $y(x) = -y(-x)$. Use this information to sketch the solution for all x.

23. Show that if $dy/dx + p(x)y = q(x)$ has an equilibrium solution, say $y = a$, then $q(x)$ must have the form $q(x) = ap(x)$. This means that linear differential equations that have an equilibrium solution have the form

$$\frac{dy}{dx} = p(x)(a - y).$$

Find the explicit solution of this equation and show that it is obtained from (5.40) by setting $C = 0$. Since this shows that all solutions of linear differential equations have the form (5.40), the expression (5.40) is called the general solution.

24. Look at the slope field for the differential equation

$$\left(x^2 + 1\right) y' + xy = \sqrt{x^2 + 1}.$$

---

[3]One version of L'Hôpital's rule is, if $\lim_{x \to a} f(x) = 0$, $\lim_{x \to a} g(x) = 0$, and $\lim_{x \to a} \left[f'(x)/g'(x)\right] = L$, then $\lim_{x \to a} \left[f(x)/g(x)\right] = L$.

Make a conjecture about the behavior of solutions $y(x)$ as $x \to \infty$. Try to prove this conjecture by analyzing the differential equation together with its isoclines. Now obtain the explicit family of solutions of this differential equation. Look at what happens to this family as $x \to \infty$, and confirm this previous conjecture.

25. Suppose that over the next 20 years, P_0/α (that is, $1/\alpha$ of the initial population) left some specific country whose current rate of change of population is proportional to the population. If this number of people left the country at a constant rate over these 20 years, the differential equation that modeled this situation would be

$$\frac{dP}{dt} = kP - \frac{1}{20\alpha}P_0,$$

subject to the initial condition

$$P(0) = P_0.$$

Show that if $k > 1/(20\alpha)$, all solutions of this initial value problem are increasing and concave up. (Do this using the differential equation directly and also using the explicit solution.)

26. Use the solution of the preceding exercise to compare the population of Botswana after 20 years of linear emigration to that of 20 years of constant emigration. (Assume that the total emigration is the same in the two models.)

27. Consider the following initial value problem, which models some population as a function of time.

$$\frac{dP}{dt} = kP - at,$$

$$P(0) = P_0.$$

(a) State in words what assumptions are made in writing down this model, and give interpretations for the three constants, k, a, and P_0.

(b) Use the differential equation to discover where the solution to this initial value problem is concave up or down, increasing or decreasing.

(c) Solve this initial value problem and find conditions on the parameters which will guarantee that the population will (i) die out, (ii) not die out. How much of your answer could have you determined from your results of part (b)?

28. The differential equation for a barrel sinking in a body of water under the influence of gravity and water resistance is

$$m\frac{dv}{dt} = mg - kv,$$

where m is the mass, g is the gravitational constant, and kv is the frictional force.

(a) By using the slope field or the differential equation directly, determine whether there is a limiting velocity. If so, what is it? If not, why isn't there one?

(b) If the barrel has an initial velocity at the surface of the water of v_0, find the explicit solution of the associated initial value problem. Show that this explicit solution verifies your answer in part (a).

5.3   Bernoulli's Equation

In this section we look at an example involving the weight of fish as a function of time. The differential equation that arises in this case is not one we have encountered previously. However, a simple substitution converts this differential equation into a linear differential equation. This permits us to develop a technique for solving the original differential equation. We then show how it is used in other examples.

EXAMPLE 5.7: Weight of Fish

In developing mathematical models of fish population dynamics, the growth of a fish is considered to be primarily a function of interaction with the environment. The assumption is that growth during a given time period equals the difference between what goes into the fish during that time period and what leaves the fish. If we consider a fish that weighs $w(t)$ pounds at time t, measured in years, a common metabolic growth model has

$$\frac{dw}{dt} = Hw^m - kw^n,$$

where H, k, m, and n are positive constants specific to the type of fish under consideration and the environment in which the fish is located. Here we use $m = 2/3$ and $n = 1$, the values used in early studies.[4] The slope field and some solution curves for this equation are shown in Figure 5.12 for $H = 1$, $k = 0.6$. Note the similarities with the slope field associated with the logistics equation — namely, an apparent equilibrium solution with solutions increasing below this equilibrium and decreasing above. This can be seen if we factor the differential equation as

$$\frac{dw}{dt} = Hw^{2/3} - kw = w^{2/3}\left(H - kw^{1/3}\right), \tag{5.48}$$

from which it is apparent that equilibrium solutions occur when

$$w = 0 \text{ and } w = \left(\frac{H}{k}\right)^3.$$

The equilibrium solution $w = 0$ is of no interest in this example, and $w = (H/k)^3$ corresponds to the limiting weight of the fish. For the case shown in Figure 5.12, namely, $H = 1$, $k = 0.6$, this limiting weight occurs when $w = (5/3)^3 \approx 4.63$, as shown in Figure 5.12.

We also note from Figure 5.12 that inflection points seem to occur in the lower left corner where we have small values of w. We can obtain more information about these inflection points from the differential equation. If we differentiate (5.48), we obtain

$$\frac{d^2w}{dt^2} = \left(\frac{2}{3}Hw^{-1/3} - k\right)\frac{dw}{dt},$$

from which it is clear that inflection points occur when

$$w = \left(\frac{2H}{3k}\right)^3,$$

---

[4]R.J.H. Beverton and S.J. Holt, "On the Dynamics of Exploited Fish Populations," Fishery Invest. Ser 2, **19**, 1957, pages 1–533.

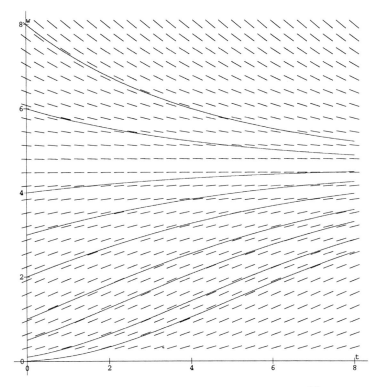

FIGURE 5.12   Solution curves and slope field for $dw/dt = Hw^{2/3} - kw$

that is, somewhere between $w = 0$ and the equilibrium solution at $w = (H/k)^3$. (Note that the places where $dw/dt = 0$ give equilibrium solutions and are not considered when looking for inflection points.) For the case shown in Figure 5.12 — namely, $H = 1$, $k = 0.6$ — this occurs when $w = (10/9)^3 \approx 1.37$, in agreement with Figure 5.12. Note that the times at which inflection points occur depend on the initial values.

Let us now seek an explicit solution of (5.48) for the case shown in Figure 5.12, where $H = 1$, $k = 0.6$:

$$\frac{dw}{dt} = w^{2/3} - 0.6w. \tag{5.49}$$

Although this equation is separable (and, in fact, autonomous), the technique to allow simple integration is obscure. (See Exercise 11 on page 243.) We were successful in using the substitution

$$w = uz, \tag{5.50}$$

for solving linear differential equations, so let's try this again. Using (5.50) in (5.49), we find

$$\frac{du}{dt}z + u\frac{dz}{dt} = (uz)^{2/3} - 0.6uz,$$

which can be rewritten as

$$\left(\frac{du}{dt} + 0.6u\right)z + u\frac{dz}{dt} = (uz)^{2/3}. \tag{5.51}$$

If we choose u so that

$$\frac{du}{dt} + 0.6u = 0,$$

namely,

$$u = e^{-0.6t}, \tag{5.52}$$

then (5.51) reduces to

$$\frac{dz}{dt} = u^{-1/3}z^{2/3}. \tag{5.53}$$

Substituting (5.52) into (5.53) and rearranging results in

$$z^{-2/3}\frac{dz}{dt} = e^{0.2t}.$$

Integration yields

$$z = \left(\frac{5}{3}e^{0.2t} - C\right)^3. \tag{5.54}$$

When (5.52) and (5.54) are substituted into (5.50), we have, in terms of our original variable,

$$w(t) = uz = \left(e^{-0.2t}\right)^3 \left(\frac{5}{3}e^{0.2t} - C\right)^3,$$

or

$$w(t) = \left(\frac{5}{3}(1 - Ce^{-0.2t})\right)^3, \tag{5.55}$$

where

$$C = 1 - 0.6w(0)^{1/3} \tag{5.56}$$

and $w(0)$ is the initial weight of the fish. From (5.55) we see that the equilibrium solution is $(5/3)^3$, in agreement with our earlier observation. Note that (5.55) shows that this equilibrium is completely independent of the choice of the initial weight of the fish.

The explicit solution (5.55) can now be used to calculate the times at which inflection points occur — namely, the times at which $w(t) = (10/9)^3$. From (5.55), this is equivalent to asking for the times at which

$$\left(\frac{10}{9}\right)^3 = \left(\frac{5}{3}(1 - Ce^{-0.2t})\right)^3.$$

Taking the cube root of both sides of this equation and solving for t gives

$$t = 5\ln(3C),$$

which, in view of (5.56), can be written as

$$t = 5\ln\left[3\left(1 - 0.6w(0)^{1/3}\right)\right] = 5\ln\left(3 - 1.8w(0)^{1/3}\right).$$

As expected, this time depends on the initial value $w(0)$. To guarantee that $t > 0$ in this last expression we must have $3 - 1.8w(0)^{1/3} > 1$, which implies that $w(0) < (10/9)^3$. Thus, only solution curves with initial weights less than $(10/9)^3$ will have an inflection point.

The solution of (5.48) for other choices of H and k is possible by the same technique, the solution being

$$w(t) = \left[\frac{H}{k} \left(1 - Ce^{-kt/3} \right) \right]^3. \tag{5.57}$$

Figure 5.13 shows the graph of (5.57) for $H = 4.049$, $k = 0.285$, along with data on the growth of North Sea plaice. [5] $\square$

Equation (5.48) belongs to a class of differential equations known as BERNOULLI'S DIFFERENTIAL EQUATIONS.

Definition 5.3: **A BERNOULLI DIFFERENTIAL EQUATION has the form**

$$\frac{dy}{dx} + p(x)y = q(x)y^n,$$

where n is a specified constant.

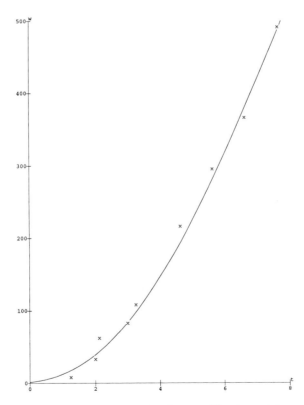

FIGURE 5.13 Solution of $dw/dt = Hw^{2/3} - kw$ and data

---

[5]"Modelling Exploited Marine Fish Stocks" by J. Beyer and P. Sparre, Applications of Ecological Modelling in Environmental Management, Part A, 1983, page 505.

Note in the case of

$$\frac{dw}{dt} = Hw^{2/3} - kw,$$

$p(x) = k, q(x) = H$, and $n = 2/3$.

How to Solve Bernoulli's Differential Equation

Purpose: To solve Bernoulli's differential equation

$$\frac{dy}{dx} + p(x)y = q(x)y^n \tag{5.58}$$

for $y = y(x)$.

Technique:

1. Make the substitution

$$y = uz$$

in (5.58) to find

$$\frac{du}{dx}z + u\frac{dz}{dx} + puz = qu^n z^n,$$

which can be written as

$$\left(\frac{du}{dx} + pu\right)z + u\left(\frac{dz}{dx} - qu^{n-1}z^n\right) = 0. \tag{5.59}$$

2. Solve the separable equation

$$\frac{du}{dx} + pu = 0$$

for $u = u(x)$, and substitute the result in (5.59). This leads to the separable equation

$$\frac{dz}{dx} - qu^{n-1}z^n = 0,$$

which we solve for $z = z(x)$.

3. The solution of (5.58) is $y(x) = u(x)z(x)$.

Comments about Bernoulli's differential equation

- There is another way to solve Bernoulli's equation by converting it to a linear differential equation. See Exercise 15 on page 244.

In previous examples we looked for equilibrium solutions — say, $y = 0$ — and then proceeded to divide by y assuming that $y \neq 0$. The next example considers the possibility of y being zero at a specific point.

EXAMPLE 5.8

Consider the differential equation

$$\frac{dy}{dx} + \frac{2x}{x^2 + 1} y = \frac{1}{\left(x^2 + 1\right)^2 y^2}, \tag{5.60}$$

which has no equilibrium solutions. The first thing to notice is that if there exists an x_0 for which $y(x_0) = 0$, then dy/dx at x_0 will be infinite, so that the curve $y = y(x)$ will approach the point $y(x_0) = 0$ (the x-axis) vertically. Then notice that (5.60) is a Bernoulli differential equation with $n = -2$. Thus, we change the dependent variable from y to z by

$$y = uz$$

so

$$\frac{du}{dx}z + u\frac{dz}{dx} + \frac{2x}{x^2 + 1}uz = \frac{1}{(x^2 + 1)^2 u^2 z^2}. \tag{5.61}$$

Selecting u so that

$$\frac{du}{dx} + \frac{2x}{x^2 + 1}u = 0,$$

namely, $u = 1/\left(x^2 + 1\right)$, reduces (5.61) to

$$\frac{1}{x^2 + 1}\frac{dz}{dx} = \frac{1}{z^2},$$

which has the solution

$$z = \left(x^3 + 3x + C\right)^{1/3}.$$

Returning to the original dependent variable gives the solution

$$y(x) = \frac{\left(x^3 + 3x + C\right)^{1/3}}{x^2 + 1}. \tag{5.62}$$

For any given C it can be shown that $x^3 + 3x + C$ has exactly one real root.[6] Thus, for any choice of C (that is, for any solution curve) there is exactly one point x_0 for which $y(x_0) = 0$. So we know that every solution curve will have exactly one place where it approaches the x-axis vertically.

Several other things are evident at this point:

- From (5.60) we see that the isocline for zero slope is given by $y = 1/[2x(x^2 + 1)]^{1/3}$. Thus, solution curves will have relative extrema at places that are at larger and larger y values as x approaches the origin from the right, and smaller and smaller y values as x approaches positive infinity.

- Also from (5.60) we see that the slope field has symmetry with respect to the origin.

- From (5.62) we see that the solution curves will be single-valued functions of x that are bounded for all finite values of x . The y-intercept is given by $C^{1/3}$.

---

[6]The root is $\left(a + \sqrt{1 + a^2}\right)^{1/3} + \left(a - \sqrt{1 + a^2}\right)^{1/3}$ where $a = -C/2$.

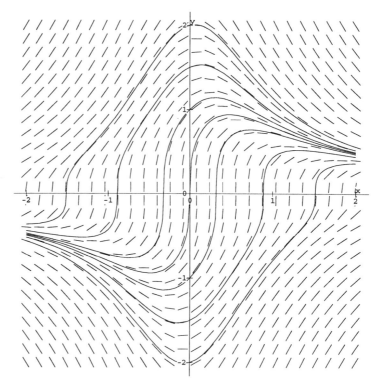

FIGURE 5.14 Analytical solution curves and slope field for $dy/dx + 2xy/(x^2 + 1) = 1/[(x^2 + 1)^2 y^2]$

Figure 5.14 gives the slope field and several solution curves for this situation. [In Figure 5.14, there are 14 solution curves, not 7, because the derivative is undefined on the x-axis.] Note the agreement with the preceding points.   □

The logistics equation has been used in Chapter 2 to model human populations. In the next example, we apply it to a population of animals in nature that are confined to a specific region. A convenient form for writing the logistics model is

$$\frac{dP}{dt} = kP \left(1 - \frac{P}{b} \right),$$

where P is the number of animals (in thousands), t is time (in years), k is an effective reproduction rate, and b is the carrying capacity. We have already seen that $P = b$ is a stable equilibrium solution. Thus, for any positive initial condition, all solutions will approach the value b as t becomes large. So by this model, b will be the number of animals in this region when this system is in equilibrium; that is, it is the carrying capacity of the region. (Recall that a stable equilibrium solution is one in which solutions are increasing just below this equilibrium and decreasing just above.)

EXAMPLE 5.9: Animal Population Growth

Often the carrying capacity of a region is dependent on rainfall or other seasonal events. Such an effect can be incorporated into this model by letting the carrying capacity have an oscillatory component. Thus, we consider the differential equation

$$\frac{dP}{dt} = kP \left(1 - \frac{P}{b + c \sin \omega t} \right),$$

where $b > c$, usually by a large amount.

To simplify subsequent calculations, we will consider the special case where $k = 1$, $b = 7$, $c = 1$, and $\omega = 6$, so the differential equation becomes

$$\frac{dP}{dt} = P\left(1 - \frac{P}{7 + \sin 6t}\right). \tag{5.63}$$

The slope field for this situation is shown in Figure 5.15 along with numerical solution curves for several different initial populations. Note that regardless of the initial condition, the population soon settles into a periodic pattern. Equation (5.63) is a Bernoulli equation,

$$\frac{dP}{dt} - P = -\frac{1}{7 + \sin 6t}P^2, \tag{5.64}$$

so we make the substitution

$$P = uz,$$

and (5.64) transforms into

$$\left(\frac{du}{dt} - u\right)z + u\frac{dz}{dt} = -\frac{1}{7 + \sin 6t}u^2z^2. \tag{5.65}$$

We select u so that $du/dt - u = 0$ — namely, $u = e^t$ — so (5.65) reduces to

$$\frac{dz}{dt} = -\frac{e^t}{7 + \sin 6t}z^2. \tag{5.66}$$

Integration of (5.66) gives

$$\frac{1}{z} = \int_0^t \frac{e^s}{7 + \sin 6s}\,ds + C.$$

The explicit solution is thus

$$P(t) = uz = e^t\left(\int_0^t \frac{e^s}{7 + \sin 6s}\,ds + C\right)^{-1},$$

where $C = 1/P(0)$.

Even though we cannot find a simple expression for this integral, we can discover some facts about the long-term behavior of this solution. The function $7 + \sin 6s$ in the integrand is bounded below by 6 and above by 8. Thus, we have that

$$\left(\int_0^t \frac{e^s}{6}\,ds + C\right)^{-1} \leq \left(\int_0^t \frac{e^s}{7 + \sin 6s}\,ds + C\right)^{-1} \leq \left(\int_0^t \frac{e^s}{8}\,ds + C\right)^{-1},$$

so

$$\left(\frac{e^t - 1}{6} + C\right)^{-1} \leq \left(\int_0^t \frac{e^s}{7 + \sin 6s}\,ds + C\right)^{-1} \leq \left(\frac{e^t - 1}{8} + C\right)^{-1},$$

and our explicit solution has the bounds

$$e^t\left(\frac{e^t - 1}{6} + C\right)^{-1} \leq P(t) \leq e^t\left(\frac{e^t - 1}{8} + C\right)^{-1},$$

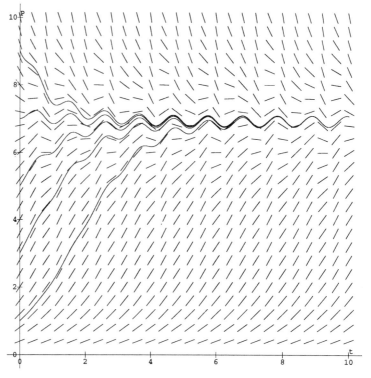

FIGURE 5.15 Numerical solution curves and slope field for $dP/dt = P[1 - P/(7 + \sin 6t)]$

or

$$\frac{6}{1 + (6C - 1)e^{-t}} \leq P(t) \leq \frac{8}{1 + (8C - 1)e^{-t}}. \tag{5.67}$$

From (5.67) we see that the long-term behavior for this population is bounded by 6 and 8. This oscillating behavior is indicated in Figure 5.15. □

EXERCISES

Solve the following Bernoulli differential equations.

1. $\dfrac{dy}{dx} + \sqrt{x}\,y = \dfrac{2}{3}\sqrt{\dfrac{x}{y}}$

2. $\dfrac{dy}{dx} + \dfrac{y}{x} = \dfrac{1}{x^3 y^3}$

3. $x^2 \dfrac{dy}{dx} + y^2 - xy = 0$

4. $x \dfrac{dy}{dx} - (1 + x)y - y^2 = 0$

5. $\dfrac{dy}{dx} + \dfrac{2y}{x} = -x^9 y^5$

6. $\dfrac{dy}{dx} + y = 2x^2 y^2$

7. $y\dfrac{dy}{dx} + xy^2 - x = 0$

8. $2y\dfrac{dy}{dx} + y^2 \sin x - \sin x = 0$

9. $x\dfrac{dy}{dx} - \dfrac{y}{2\ln x} - y^2 = 0$

10. $\dfrac{dy}{dx} - y + xe^{-2x}y^3 = 0$

11. Show that the nonequilibrium solutions of the differential equation

$$\frac{dw}{dt} = Hw^m - kw$$

can be obtained by using the fact that

$$\frac{1}{Hw^m - kw} = \frac{1}{w\left(Hw^{m-1} - k\right)} = \frac{1}{k}\left(\frac{Hw^{m-2}}{Hw^{m-1} - k} - \frac{1}{w}\right).$$

Confirm that using this approach, the solution of (5.49) is (5.55).

12. If $n > m > 0$, find two nonnegative equilibrium solutions of the differential equation modeling the growth of an individual fish,

$$\frac{dw}{dt} = Hw^m - kw^n,$$

and show that any solution curve will be increasing between the two. The constants H and k are both positive.

(a) Show that for any size of fish at birth, the limiting value of its weight is $(H/k)^{1/(n-m)}$.

(b) Using this model, find the weight corresponding to the greatest growth rate.

(c) What would happen if you would use the initial condition $w(0) = 0$?

13. Consider a Bernoulli equation of the form

$$\frac{dy}{dx} + y = m^{n-1}y^n,$$

where m and n are positive constants.

(a) For the case $n = 2$, find all the equilibrium solutions and determine their stability.

(b) Repeat part (a) for the case $n = 3$.

(c) Repeat part (a) for the case $n = 5$.

(d) Can you find values of m and n such that an equilibrium solution of this equation given by $y = 0$ will be stable? If so, what are they? If not, explain fully.

14. Consider the case in which the $p(x)$ and $q(x)$ in the Bernoulli equation

$$\frac{dy}{dx} + p(x)y = q(x)y^n$$

are bounded for all finite values of x.

(a) Find conditions on n and q such that all solutions of the Bernoulli equation will approach the x-axis vertically.

(b) What condition on n will guarantee that 0 will be an equilibrium solution? Is it possible for this equilibrium solution ever to be stable? If so, give these conditions. If not, explain fully.

15. Show that the Bernoulli equation

$$\frac{dy}{dx} + p(x)y = q(x)y^n$$

can be converted to the linear differential equation

$$\frac{1}{1-n}\frac{du}{dx} + p(x)u = q(x)$$

by the substitution

$$u = y^{1-n}.$$

16. A differential equation of the form

$$\frac{dy}{dx} = a_0(x) + a_1(x)y + a_2(x)y^2,$$

with $a_0(x)$ and $a_2(x)$ not identically zero, is called a RICCATI DIFFERENTIAL EQUATION .

(a) Show that if we know a solution $y_1(x)$ of a Riccati differential equation, then the change of variable

$$y = y_1(x) - \frac{1}{u}$$

results in the following linear differential equation for u:

$$\frac{du}{dx} + \left[a_1(x) + 2a_2(x)y_1(x)\right]u = a_2(x).$$

(b) To find this first solution of a Riccati differential equation, we often try simple functions like ax^b or ae^{bx}. Find values of a and b such that ax^b is a solution of

$$\frac{dy}{dx} = 1 + x + 2x^2\cos x - (1 + 4x\cos x)\,y + 2\,(\cos x)\,y^2.$$

(c) Make the change of variable given in part (a) to obtain the differential equation

$$\frac{du}{dx} - u = 2\cos x.$$

(d) Solve the differential equation in part (c), and show that the general solution of our original differential equation in part (b) is

$$y = x - (\sin x - \cos x + Ce^x)^{-1},$$

where C is an arbitrary constant.

17. Solve the following Riccati differential equations.

(a) $\dfrac{dy}{dx} = 5 + \dfrac{4}{x}y + \dfrac{1}{4x^2}y^2$

(b) $\dfrac{dy}{dx} = (1 - y)\left(\dfrac{1}{x} - \dfrac{1}{10} + \dfrac{y}{10}\right)$

(c) $\dfrac{dy}{dx} = \dfrac{6}{x^2} - 2y^2$

(d) $\dfrac{dy}{dx} = -1 + 2y - y^2$

(e) $\dfrac{dy}{dx} = 3e^{4x} + 2y - 12y^2$

(f) $\dfrac{dy}{dx} = e^{2x} + \left(1 + \dfrac{5}{2}e^x\right)y + y^2$

(g) $\dfrac{dy}{dx} = \dfrac{4}{x^2} - \dfrac{y}{x} - y^2$

18. Show that if y is a solution of the Riccati differential equation (in the form from Exercise 16), then $u = 1/y$ is a solution of the differential equation

$$\frac{du}{dx} = -a_2(x) - a_1(x)u - a_0(x)u^2.$$

19. Show that if we know one solution, $y_1(x)$, of a Riccati equation, the substitution

$$y = y_1(x) + v$$

converts the general Riccati equation to the Bernoulli equation

$$\frac{dv}{dx} - \left[a_1(x) + 2a_2(x)y_1(x)\right]v = a_2(x)v^2.$$

Explain the relationship between this result and that of Exercise 16.

20. The logistics differential equation

$$\frac{dy}{dx} = ay(b - y)$$

can be classified either as a Bernoulli or as a Riccati differential equation.

(a) Use the method in Exercise 16 to solve it as a Riccati equation.

(b) Use the appropriate change of dependent variable to solve it as a Bernoulli equation.

21. In Chapter 2 we considered the logistics equation as a model of population growth. Here we derive a differential equation that includes the effect of births, deaths, immigration, emigration, and crowding. If we write

$$\begin{array}{c} \text{rate of change} \\ \text{of population} \end{array} = \begin{array}{c} \text{rate of additions} \\ \text{to population} \end{array} - \begin{array}{c} \text{rate of subtractions} \\ \text{from population} \end{array}$$

and let y be the population at time t, we would have

$$\frac{dy}{dt} = (By + I) - (Dy + E + cy^2) = a + by - cy^2,$$

where $b = B - D =$ birthrate $-$ deathrate (per unit population), $a = I - E =$ immigration $-$ emigration (per unit time), and c accounts for competition or inhibition of large populations. Note that we could think of the right-hand side of the preceding equation as the first three terms in a Taylor series of a general function that describes the population growth.

(a) Look at the slope field for this differential equation and explain the effect on its solutions by independently changing a, b, and c. Pay particular attention to the existence (or nonexistence) of equilibrium solutions and their behavior as you change the parameters.

(b) Solve the equation for the case $a = c = 0$.

(c) Solve the equation for the case $b = c = 0$.

(d) Solve the equation for the case $c = 0$.

(e) Solve the equation for the case $a = 0$.

(f) Solve the equation for the case $c \neq 0$. (Note this is a Riccati equation.)

5.4 Clairaut's Equation

In this section we look at an example from string art. The differential equation that arises in this case is not one we have encountered previously. We develop a technique for solving it and show how it is used in other examples.

EXAMPLE 5.10: String Art

Imagine we put nails along the x- and y-axes at the points 0, 1, 2 , $\cdots$, 10. We tie one end of a piece of string to the nail at the point $(0, 0)$ and the other end to the nail at the point $(0, 10)$. Now we join the nail at $(1, 0)$ to the one at $(9, 0)$, the nail at $(2, 0)$ to the one at $(0, 8)$, and so on until we have joined $(10, 0)$ to $(0, 0)$. Figure 5.16 illustrates the end result. If we look at this figure we see that the straight lines have created the illusion of a curve. The question we want to answer is, what is the equation of this curve?[7]

Let's denote the function we are looking for by $y = f(x)$. A typical string will be tangent to $y = f(x)$ at a point (x_0, y_0), say. This string is tied to a nail on the x-axis, at, say, $(u, 0)$, and its other end is tied to the point $(0, 10 - u)$ on the y-axis, as shown in Figure 5.17.

The relationship between the string and the curve is that the slope of the string is the slope of the curve — namely, $f'(x_0)$. The equation for a straight line with slope $m = f'(x_0)$ is

$$y - y_0 = m(x - x_0).$$

Because this straight line passes through the point $(u, 0)$, we have

$$-y_0 = m(u - x_0). \tag{5.68}$$

However, from Figure 5.17 we clearly have

$$m = \frac{u - 10}{u},$$

which, when solved for u, gives

$$u = \frac{10}{1 - m}. \tag{5.69}$$

If we substitute (5.69) into (5.68) and rearrange, we find

$$y_0 - x_0 m = \frac{10m}{m - 1}, \tag{5.70}$$

where $m = f'(x_0)$. However, we want (5.70) to be valid at all points (x, y) on the curve, so (5.70) gives the differential equation

$$y - xy' = \frac{10y'}{y' - 1}. \tag{5.71}$$

---

[7]Based on an idea in "Calculus in a Real and Complex World" by F. Wattenberg, PWS 1995, page 232.

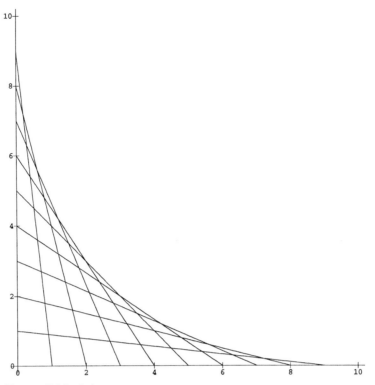

FIGURE 5.16 String art

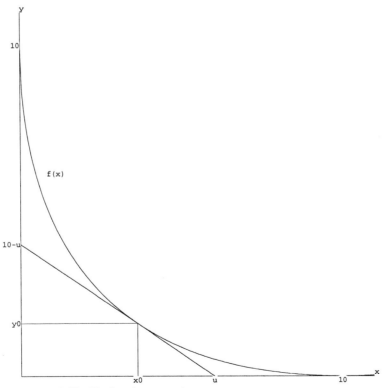

FIGURE 5.17 The function we seek

Furthermore, the curve we seek must pass through the point $(10, 0)$, so (5.71) is subject to the initial condition

$$y(10) = 0. \tag{5.72}$$

If we multiply equation (5.71) by $y' - 1$ and simplify, we find

$$x \left(y'\right)^2 + (10 - x - y)\, y' + y = 0.$$

This is a quadratic equation for y', which can be solved to find

$$y' = \frac{1}{2x} \left[-(10 - x - y) \pm \sqrt{(10 - x - y)^2 - 4xy} \right]. \tag{5.73}$$

This differential equation is not one we have seen before, and so far we have no way to solve it.

However, if we write (5.71) in the form

$$y = xy' + \frac{10y'}{y' - 1}, \tag{5.74}$$

we see that if we could find y' as a function of x, then we could substitute it into the right-hand side of (5.74) and read off $y = y(x)$ immediately. So the problem becomes finding y' as a function of x. If we define

$$p = y',$$

then we want to find $p(x)$. One way to do this is to find a differential equation that p must satisfy and solve it for $p(x)$. This suggests that we should differentiate (5.71) with respect to x, which gives

$$y' - xy'' - y' = 10\frac{y''\left(y' - 1\right) - y'y''}{\left(y' - 1\right)^2},$$

or, upon simplification,

$$-xy'' = 10\frac{-y''}{\left(y' - 1\right)^2}.$$

We can rewrite this last equation in the form

$$y'' \left(x - \frac{10}{\left(y' - 1\right)^2} \right) = 0,$$

from which we conclude that either

$$y'' = 0 \tag{5.75}$$

or

$$x - \frac{10}{\left(y' - 1\right)^2} = 0. \tag{5.76}$$

We treat each of these equations in turn.

In the case of (5.75), we see that an integration yields $y' = C$. Substituting this into (5.74) gives

$$y = Cx + \frac{10C}{C-1},\qquad\qquad (5.77)$$

which is a family of straight lines.

In the case of (5.76), we see that we can write it in the form

$$\left(y' - 1\right)^2 = \frac{10}{x},$$

or

$$y' = 1 \pm \sqrt{\frac{10}{x}}.$$

When the latter is substituted into (5.74), we find the solution

$$y = 10 + x \pm 2\sqrt{10x}.\qquad\qquad (5.78)$$

These are singular solutions[8], because they are not contained in (5.77). (Why?)

Thus, all the solutions of (5.71) have the form of (5.77) or (5.78).

The solution (5.77) satisfies the initial condition (5.72) only if $C = 0$, which leads to the solution $y(x) = 0$, which is of no interest. In fact, the solution (5.77) is exactly the family of straight lines that was used to construct the string art. Part of the singular solution (5.78), the part with the minus sign, does satisfy our initial condition (5.72). Thus, the shape generated by this string art is given by

$$y = 10 + x - 2\sqrt{10x}.$$

This is part of a parabola with axis $y = x$ and vertex at $(5/2, 5/2)$ — see Exercise 3 on page 252. What shape is generated by the part of (5.78) with the plus sign?   □

Definition 5.4: **A differential equation of the form**

$$y - xy' = f(y')\qquad\qquad (5.79)$$

is known as CLAIRAUT'S EQUATION.

Clairaut's equation arises in a number of different problems and is easy to solve. In developing its solution we simplify the notation by introducing the variable p, where

$$p = \frac{dy}{dx}.\qquad\qquad (5.80)$$

---

[8]Recall that a singular solution of

$$\frac{dy}{dx} = g(x, y)$$

is a particular solution that cannot be obtained from the family of solutions (containing the arbitrary constant C) by selecting a finite value for C.

If we use this notation in (5.79) and differentiate with respect to x we find

$$-xp' = \frac{df}{dp}p',$$

or equivalently

$$\left(\frac{df}{dp} + x\right)p' = 0.$$

This equation may be satisfied in two ways; either

$$p' = 0 \tag{5.81}$$

or

$$\frac{df}{dp} = -x. \tag{5.82}$$

Equations (5.81) and (5.80) imply that

$$y' = p = C. \tag{5.83}$$

If we substitute (5.83) into (5.79), we find

$$y = Cx + f(C),$$

which is a family of straight lines. If we solve (5.82) subject to (5.79), we obtain the singular solution of (5.79).

How to Solve Clairaut's Equation

Purpose: To find the explicit and singular solutions of Clairaut's equation

$$y - xy' = f(y'). \tag{5.84}$$

Technique:

1. To make the notation less confusing, set $y' = p$ to get

$$y - xp = f(p).$$

2. Differentiate this last equation with respect to x to obtain

$$-xp' = \frac{df}{dp}p',$$

or

$$\left(\frac{df}{dp} + x\right)p' = 0.$$

3. There are two cases to consider.

(a) $p' = 0$. This implies that $y' = p = C$. Substitute this into (5.84) to find

$$y = xC + f(C).$$

This gives the family of explicit solutions of (5.84).

(b) $df/dp = -x$. We try to solve this for $p = p(x)$ and then use the fact that $y' = p$ to obtain

$$y = xp(x) + f(p(x)).$$

This is the singular solution of (5.84).

Comments about Clairaut's differential equation

- It is sometimes quite difficult to decide whether an equation is of the Clairaut type. Solving for y and seeing whether the result may be expressed in the form

$$y = xy' + f(y')$$

  is usually very successful.

- The family of explicit solutions can be written down immediately from Clairaut's equation by replacing y' with C. The family of explicit solutions is always a family of straight lines.

- Usually the singular solution cannot be obtained from the family of explicit solutions.

- If we are unable to solve $df/dp = -x$ for $p = p(x)$, we can treat p as a parameter, and the singular solution will be given parametrically by

$$x = -df/dp,$$
$$y = -p\,df/dp + f(p).$$

EXAMPLE 5.11

Solve the differential equation

$$y - xy' = \frac{1}{4}\left(y'\right)^2. \tag{5.85}$$

We introduce the variable $p = y'$ and differentiate (5.85) to find

$$-xp' = \frac{1}{2}pp',$$

from which we have either

$$p' = 0$$

or

$$p = -2x.$$

If $p' = 0$, then $y' = p = C$, which, when substituted into (5.85), yields the family of explicit solutions

$$y = xC + \frac{1}{4}C^2.$$

If $y' = p = -2x$, then when we substitute it into (5.85) we find the singular solution

$$y = -2x^2 + \frac{1}{4}(2x)^2 = -x^2.$$

Notice that this singular solution cannot be obtained from the family of explicit solutions for any choice of the constant C. □

EXERCISES

1. Solve the following differential equations.

   (a) $y - xy' = \left(y'\right)^2$

   (b) $y - xy' = -\left(y'\right)^3$

   (c) $y - xy' = \sqrt{1 + \left(y'\right)^2}$

   (d) $y\left(y'\right)^2 - x\left(y'\right)^3 - 1 = 0$

   (e) $x\left(y'\right)^2 - (1 + y)y' + 1 = 0$

2. Use technology to plot the slope field for (5.73). Notice that the curve we seek is the boundary of the region in the xy plane where the derivative exists. From this information, find the equation of this curve.

3. Confirm that the change of variables from (x, y) to (X, Y), where

   $$\begin{aligned} X &= (x - y)/\sqrt{2}, \\ Y &= (x + y)/\sqrt{2}, \end{aligned}$$

   is a clockwise rotation through $\pi/4$ of the xy-plane. Show that under this change of variables, the equation

   $$y = 10 + x - 2\sqrt{10x}$$

   becomes

   $$Y = \frac{1}{10\sqrt{2}}X^2 + \frac{5}{\sqrt{2}},$$

   which is a parabola symmetric about the Y-axis and vertex at $X = 0$, $Y = 5/\sqrt{2}$ — that is, a parabola with axis of symmetry $y = x$ and vertex at $x = 5/2$, $y = 5/2$.

4. Consider the string art that is constructed by joining nails on the positive x- and y-axes, which are placed so that the area under the string is always 1. Show that the differential equation that characterizes the curve created by these straight lines is

   $$y - xy' = \sqrt{-2y'}.$$

   (a) What is the equation of this curve?

   (b) Use the idea suggested in Exercise 2 (that is, find an explicit formula for y' and see where it is undefined) to confirm the equation you found in part (a) for the curve.

5. Consider the string art that is constructed by joining nails on the positive x- and y-axes which are placed so that the length of the string is always 1. Show that the differential equation that characterizes the curve created by these straight lines is

$$y - xy' = \frac{-y'}{\sqrt{1 + (y')^2}}.$$

What is the equation of this curve?

6. By changing from the unknown variable y to a new unknown variable v where $v = y^3$, solve

$$y - 3xy' = 6y^2 (y')^2.$$

7. Show that the differential equation (5.14) (which gives the focal property of a reflector) can be written in the form

$$y - 2xy' = y (y')^2.$$

By changing from the unknown variable y to a new unknown variable v where $v = y^2$, solve (5.14).

8. By changing from the variables x and y to the new variables u and v, where $u = x^2$ and $v = y^2$, show that the differential equation

$$xy (y')^2 + (x^2 - a^2 - y^2) y' - xy = 0$$

reduces to

$$u \left(\frac{dv}{du}\right)^2 + (u - a^2 - v) \frac{dv}{du} - v = 0.$$

Solve this equation for v in terms of u and dv/dx and then show that it is a Clairaut equation. Now solve the Clairaut equation for $v = v(u)$, and then find the solution of the original equation.

9. The ellipse has the property that all light rays emitted from one focus pass through the other focus after reflection off the ellipse. This exercise is designed to show that the ellipse is the only curve with this reflective property. If you choose the coordinate system so the light that is emitted from the point $(a, 0)$ is received at the point $(-a, 0)$, show that the differential equation that the curve must satisfy is the one in Exercise 8.

10. Identify the following differential equations as separable equations, those with homogeneous coefficients, linear equations, Bernoulli equations, Clairaut equations, or none of these. (Some may fall into more than one category; see Table 5.1.) Do not attempt to solve any of these equations.

(a) $(2 - y)\dfrac{dy}{dx} = x^2 + 1$

(b) $\tan \dfrac{dy}{dx} - x\dfrac{dy}{dx} + y = 0$

(c) $(x + y)\dfrac{dy}{dx} + y - x = 0$

(d) $2xy + (x^2 + \cos y)\dfrac{dy}{dx} = 0$

(e) $y \sin x + \dfrac{1}{y} - \dfrac{dy}{dx} = 0$

(f) $3x(y^2 + 1) + y(x^2 + 1)\dfrac{dy}{dx} = 0$

(g) $x\dfrac{dy}{dx} + 3y - x^2 = 0$

(h) $x^3 + y^3 - xy^2\dfrac{dy}{dx} = 0$

(i) $\dfrac{dy}{dx} = \dfrac{xy^2 - 1}{1 - x^2 y}$

(j) $y \sin x + e^x - \dfrac{dy}{dx} = 0$

(k) $y + \left(\dfrac{dy}{dx}\right)^2 - x\dfrac{dy}{dx} = 0$

(l) $xy^2 - y + \dfrac{dy}{dx} = 0$

TABLE 5.1 Solving First Order Differential Equations

Differential Equation	Response		Type
$y' = f(y)g(x)$?	Yes	$\longrightarrow$	Separable
$y' = g(y/x)$?	Yes	$\longrightarrow$	Homogeneous coefficients
$y' + p(x)y = q(x)$?	Yes	$\longrightarrow$	Linear
$y' + p(x)y = q(x)y^n$?	Yes	$\longrightarrow$	Bernoulli
$y' = a_0(x) + a_1(x)y + a_2(x)y^2$?	Yes	$\longrightarrow$	Riccati
$y - xy' = f(y')$?	Yes	$\longrightarrow$	Clairaut

5.5 Equations Reducible to First Order

In this section we look at differentiable equations of order greater than 1, a topic to be studied extensively in Chapters 7 though 13. Under very special circumstances it is possible to reduce some higher order differential equations to first order equations, that then can be solved. The results of this section form the backbone of the later treatment.

In the last section we converted a first order differential equation into one with second derivatives. Sometimes we can go the other way, when an equation that contains second derivatives can be reduced to a first order differential equation. In fact we used this idea when we dealt with the problem of the sky diver in free fall, because (4.45), namely,

$$m\frac{dv}{dt} = mg - kv^2, \tag{5.86}$$

could be written

$$m\frac{d^2x}{dt^2} = mg - k\left(\frac{dx}{dt}\right)^2. \tag{5.87}$$

Consequently, we actually had an equation with a second derivative, called a second order differential equation. The reason that (5.87) reduced to an equation we could solve, namely, (5.86), was because x was not explicitly present in (5.87), only its derivatives occur. The substitution

$$v = \frac{dx}{dt}, \qquad \frac{dv}{dt} = \frac{d^2x}{dt^2},$$

resulted in a differential equation containing only v and t (and not x).

We illustrate this technique in the following example.

EXAMPLE 5.12

Solve

$$\frac{d^2y}{dx^2} - \frac{dy}{dx} - e^{2x} = 0. \tag{5.88}$$

If we let

$$v = \frac{dy}{dx},$$

then (5.88) becomes the first order equation

$$\frac{dv}{dx} - v - e^{2x} = 0.$$

This is a linear differential equation for v, with integrating factor $\mu = e^{-x}$ and solution

$$v = e^{2x} + C_1 e^x.$$

Because $v = dy/dx$, this equation can be written

$$\frac{dy}{dx} = e^{2x} + C_1 e^x.$$

We can integrate this equation to find

$$y(x) = \frac{1}{2} e^{2x} + C_1 e^x + C_2,$$

which is the solution of (5.88). $\qquad\qquad\square$

How to Solve $F\left(\frac{d^2y}{dx^2}, \frac{dy}{dx}, x\right) = 0$ ***for*** $y = y(x)$

Purpose: Find $y = y(x)$ that satisfies

$$F\left(\frac{d^2y}{dx^2}, \frac{dy}{dx}, x\right) = 0, \tag{5.89}$$

that is, where the **dependent variable y is missing**.

Technique:

1. Make the substitution

$$v = \frac{dy}{dx}, \tag{5.90}$$

so

$$\frac{dv}{dx} = \frac{d^2y}{dx^2}. \tag{5.91}$$

2. The differential equation (5.89) becomes

$$F\left(\frac{dv}{dx}, v, x\right) = 0. \tag{5.92}$$

This is a first order differential equation for $v = v(x)$. If possible, solve it for $v = v(x)$. The solution may be explicit or implicit.

3. If it is explicit, we solve

$$\frac{dy}{dx} = v(x)$$

for $y = y(x)$.

Comments about solving $F(d^2y/dx^2, dy/dx, x) = 0$ **for** $y = y(x)$

- If the solution of (5.92) is implicit, we may be stuck at this stage.

EXAMPLE 5.13: Curves of Constant Curvature

In calculus you may have learned about the curvature, $\kappa(x)$, of a curve $y = f(x)$ at the point (x, y) — namely,

$$\kappa(x) = \frac{y''}{\left(1 + (y')^2\right)^{3/2}}. \tag{5.93}$$

For any given curve, $\kappa(x)$ measures how much the curve arches at the point (x, y). Values of $\kappa(x)$ near zero indicate the curve is nearly straight, whereas large values correspond to a more arched curve. For example, the parabola $y = x^2$ has curvature

$$\kappa(x) = \frac{2}{\left(1 + 4x^2\right)^{3/2}},$$

which has a maximum when $x = 0$ and goes to zero as $x \to \infty$. Thus, the greatest arching of a parabola is at the origin, and as x goes to infinity the parabola becomes straighter, in complete agreement with our knowledge of a parabola. (It is common to believe mistakenly that y'' measures how much a curve arches. It does not. In the case of the parabola $y = x^2$ we clearly have $y'' = 2$, which is a constant for all points. However, the parabola certainly does not have the same arch at each point; it arches more at the origin than the sides. See Figure 5.18.)

A natural question to ask is, "What curves have constant curvature?" In view of (5.93), this is clearly equivalent to solving

$$y'' = \kappa \left[1 + (y')^2\right]^{3/2} \tag{5.94}$$

for $y = y(x)$ where κ is a nonzero constant. At first sight (5.94) does not appear to fall into the category of a first order differential equation, because the second derivative y'' is present. However, if, as previously suggested, we make the substitution

$$v = y', \tag{5.95}$$

then (5.94) becomes

$$v' = \kappa \left(1 + v^2\right)^{3/2}, \tag{5.96}$$

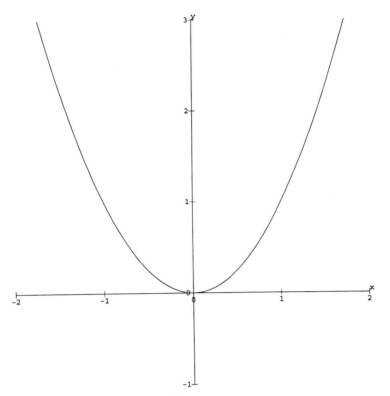

FIGURE 5.18 The parabola $y = x^2$

which is a first order differential equation in v.

The problem of solving (5.96) reduces to

$$\int \frac{dv}{\left(1 + v^2\right)^{3/2}} = \kappa \int dx. \tag{5.97}$$

The integral on the left-hand side of (5.97) can be evaluated by making the substitution $v = \tan\theta$ — or by using integral tables — to give

$$\frac{v}{\sqrt{1 + v^2}} = \kappa x + C. \tag{5.98}$$

Solving (5.98) for v yields

$$v = \frac{\pm(\kappa x + C)}{\sqrt{1 - (\kappa x + C)^2}},$$

which, in view of (5.95), gives the first order equation

$$\frac{dy}{dx} = \frac{\pm(\kappa x + C)}{\sqrt{1 - (\kappa x + C)^2}}. \tag{5.99}$$

An elementary integration leads to the solution of (5.99), namely,

$$y(x) = \mp\frac{1}{\kappa}\sqrt{1 - (\kappa x + C)^2} + a, \tag{5.100}$$

where a is the constant of integration. Equation (5.100) can be rewritten in the form

$$(y - a)^2 + (x - b)^2 = \frac{1}{\kappa^2},$$ (5.101)

where we have put $b = -C/\kappa$. Equation (5.101) is a circle with center (b, a) and radius $1/\kappa$. Thus, *the circle is the only curve with constant curvature.*   □

EXAMPLE 5.14

Solve

$$\frac{d^2y}{dx^2} + a^2 y = 0,$$ (5.102)

where a is a positive constant.

If we use the substitution $v = dy/dx$ in this differential equation and calculate the second derivative as $d^2y/dx^2 = dv/dx$, we obtain

$$\frac{dv}{dx} + a^2 y = 0.$$

Although this is a first order differential equation, it has two dependent variables present, both v and y, and we have no means of solving it. One way of overcoming this dilemma is to think of v as a function of y. Thus, when we compute dv/dx, we can use the chain rule and write

$$\frac{dv}{dx} = \frac{dv}{dy}\frac{dy}{dx}.$$

Because $dy/dx = v$, we have

$$\frac{dv}{dx} = \frac{dv}{dy}v,$$

and (5.102) can be written in the form

$$v\frac{dv}{dy} + a^2 y = 0.$$

The solution of this separable equation is

$$v^2 = C - a^2 y^2,$$

which, since $v = dy/dx$, becomes

$$\frac{dy}{dx} = \pm\sqrt{C - a^2 y^2}.$$ (5.103)

The solution of (5.103) is seen to be

$$\frac{1}{a}\sin^{-1}\left(\frac{ay}{\sqrt{C}}\right) = \pm x + b,$$

where b is the constant of integration. This equation leads to

$$y(x) = \left(\sqrt{C}/a\right)\sin(\pm ax + ab) = \left(\sqrt{C}/a\right)(\sin ab\cos ax \pm \cos ab\sin ax),$$

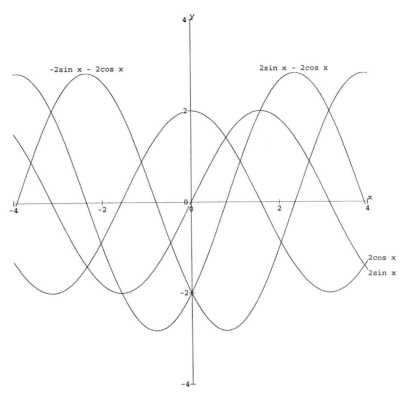

FIGURE 5.19   Some solutions of $y'' + y = 0$

which can be rewritten in the form

$$y(x) = C_1 \cos ax + C_2 \sin ax, \tag{5.104}$$

where $C_1 = \left(\sqrt{C}/a\right) \sin ab$ and $C_2 = \pm\left(\sqrt{C}/a\right) \cos ab$. Thus, (5.104) is the solution of (5.102), where C_1 and C_2 are arbitrary constants. Figure 5.19 shows some of the solutions of (5.102) for the case $a = 1$. □

   Thus, there is another case where a change of variable reduces an equation involving second derivatives to a first order differential equation — namely, when the independent variable is missing

$$F\left(\frac{d^2y}{dx^2}, \frac{dy}{dx}, y\right) = 0. \tag{5.105}$$

How to Solve $F\left(\frac{d^2y}{dx^2}, \frac{dy}{dx}, y\right) = 0$ ***for*** $y = y(x)$

Purpose: To find $y = y(x)$ that satisfies

$$F\left(\frac{d^2y}{dx^2}, \frac{dy}{dx}, y\right) = 0, \tag{5.106}$$

that is, where the **independent variable x is missing**.

Technique:

1. Make the substitution

$$v = \frac{dy}{dx}.$$ (5.107)

In this case we compute y'' in a different manner to eliminate x as the variable. Here we use the chain rule to write

$$\frac{d^2y}{dx^2} = \frac{d}{dx}\left(\frac{dy}{dx}\right) = \frac{d}{dx}(v) = \frac{dv}{dy}\frac{dy}{dx} = \frac{dv}{dy}v.$$ (5.108)

Notice this is different from the form of the second derivative used in (5.91).

2. The differential equation (5.106) now becomes

$$F\left(v\frac{dv}{dy}, v, y\right) = 0.$$ (5.109)

This is a first order differential equation for $v = v(y)$. If possible, solve it for $v = v(y)$. The solution may be explicit or implicit.

3. If it is explicit, we solve

$$\frac{dy}{dx} = v(y)$$

for $y = y(x)$.

Comments about solving $F(d^2y/dx^2, dy/dx, y) = 0$ **for** $y = y(x)$

- Again if the solution of (5.109) is implicit, we may be stuck at this stage.

EXAMPLE 5.15

Solve

$$\frac{d^2y}{dx^2} - a^2y = 0,$$ (5.110)

where a is a positive constant, for $y = y(x)$.

From (5.107) and (5.108), (5.110) can be written in the form

$$v\frac{dv}{dy} - a^2y = 0.$$

The solution of this separable equation is

$$v^2 = C + a^2y^2,$$

which by (5.107) becomes

$$\frac{dy}{dx} = \pm\sqrt{C + a^2y^2}.$$ (5.111)

Using a table of integrals, we may put the solution of (5.111) in the form

$$\frac{1}{a}\ln\left(ay + \sqrt{C + a^2 y^2}\right) = \pm x + b,$$

where b is the constant of integration. This equation leads to

$$ay + \sqrt{C + a^2 y^2} = A e^{\pm ax},$$

where $A = e^{ab}$. We move the ay term from the left-hand side to the right-hand side and square to find

$$y(x) = \frac{A}{2a} e^{\pm ax} - \frac{C}{2aA} e^{\mp ax},$$

which can be rewritten in the form

$$y(x) = C_1 e^{ax} + C_2 e^{-ax}, \tag{5.112}$$

where $C_1 = A/(2a)$ and $C_2 = -C/(2aA)$ if the plus sign is used, and $C_2 = A/(2a)$ and $C_1 = -C/(2aA)$ if the minus sign is used. Thus, (5.112) is the solution of (5.110), where C_1 and C_2 are arbitrary constants. Figure 5.20 shows some of the solutions of (5.110) for the case $a = 1$.   □

The differential equations in Examples 5.14 and 5.15 will be seen again in Chapter 7.

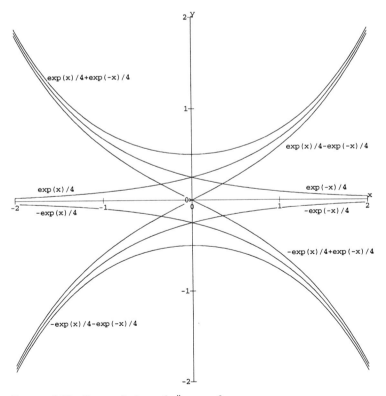

FIGURE 5.20   Some solutions of $y'' - y = 0$

EXERCISES

Solve the following differential equations.

1. $xy'' + y' = 0$

2. $2xy'' + (y')^2 - 1 = 0$

3. $yy'' + (y')^2 = 0$

4. $xy'' + y' + x = 0$

5. $xy'' - y' = (y')^3$

6. $y'' + (y')^2 = 1$

7. $x^2 y'' + (y')^2 - 2xy' = 0$

8. $yy'' + (y')^2 + 4 = 0$

9. $(y^2 + 1) y'' - 2y (y')^2 = 0$

10. Find the equation of a curve whose curvature is always equal to 0 — that is, $\kappa = 0$ in (5.93).

11. If a projectile of mass m is fired vertically from the earth, the projectile's distance from the center of the earth, x, is governed by the differential equation

$$m \frac{d^2 x}{dt^2} = -\frac{mg R^2}{x^2},$$

where g is the gravitational constant and R is the radius of the earth.

(a) Letting

$$\frac{dx}{dt} = v \text{ and } \frac{d^2 x}{dt^2} = v \frac{dv}{dx},$$

solve the foregoing differential equation for v, subject to $v(R) = v_0$, obtaining

$$v = \sqrt{v_0^2 + 2g R^2 \left(\frac{1}{x} - \frac{1}{R} \right)}.$$

(b) Show that if $v_0^2 - 2g R > 0$, the velocity will always be nonzero. (This critical value of v_0, $v_0 = \sqrt{2g R}$, is called the *escape velocity* of the earth.)

12. The differential equation

$$\frac{d^2 y}{dx^2} = g(y)$$

is a special case of (5.105), so the substitution $v = dy/dx$ may be used to solve this equation. Use the following method to find the solution. Multiply both sides of the differential equation $y'' = g(y)$ by dy/dx and integrate to find that

$$\left(\frac{dy}{dx} \right)^2 = 2G(y) + C,$$

where $G(y)$ is any antiderivative of $g(y)$. Integrate once more to find that

$$\int \frac{dy}{\sqrt{2G(y) + C}} = x + K,$$

where K is also an arbitrary constant.

13. Use the technique described in Exercise 12 to solve the differential equation (5.102), namely,

$$\frac{d^2y}{dx^2} + a^2y = 0.$$

Compare your answer with (5.104).

14. Use the technique described in Exercise 12 to solve the differential equation (5.110), namely,

$$\frac{d^2y}{dx^2} - a^2y = 0.$$

Compare your answer with (5.112).

15. In Section 3.4 we showed that sometimes it is possible to solve first order differential equations by assuming a power series solution of the form

$$y(x) = \sum_{k=0}^{\infty} c_k x^k = c_0 + c_1 x + c_2 x^2 + c_3 x^3 + \cdots + c_n x^n + \cdots.$$

(a) Assume that

$$\frac{d^2y}{dx^2} - a^2y = 0$$

has a power series solution.

    i. Substitute the power series into the differential equation, and find $c_2, c_3, c_4, \cdots$ in terms of c_0 and c_1.

    ii. Identify the power series in terms of familiar functions, and compare your answer with (5.112).

(b) Assume that

$$\frac{d^2y}{dx^2} + a^2y = 0$$

has a power series solution.

    i. Substitute the power series into the differential equation, and find $c_2, c_3, c_4, \cdots$ in terms of c_0 and c_1.

    ii. Identify the power series in terms of familiar functions, and compare your answer with (5.104).

16. This exercise is designed to show that if $y = y_1(x)$ and $y = y_2(x)$ are two functions which satisfy

$$\frac{d^2y}{dx^2} + a^2y = 0$$

subject to $y(x_0) = y_0$ and $y'(x_0) = y_0'$, then $y_1(x) = y_2(x)$.

(a) Construct the function

$$D(x) = a^2 \left[y_1(x) - y_2(x) \right]^2 + \left[y_1'(x) - y_2'(x) \right]^2.$$

Show that if $D(x) = 0$ then $y_1(x) = y_2(x)$.

(b) Show that

$$\frac{dD}{dx} = 0$$

and

$$D(x_0) = 0.$$

Explain why this implies that

$$D(x) = 0.$$

(c) Use parts (a) and (b) to show that there is only one solution of

$$\frac{d^2 y}{dx^2} + a^2 y = 0$$

subject to $y(x_0) = y_0$ and $y'(x_0) = y'_0$.

17. Let a cable of uniformly distributed weight be suspended between two supports (see Figure 5.21). If we put the origin of our coordinate system the distance a below the lowest point of the cable and the coordinates of any point on the cable are (x, y), the differential equation relating x and y is

$$a\frac{d^2 y}{dx^2} = \left[1 + \left(\frac{dy}{dx} \right)^2 \right]^{1/2}$$

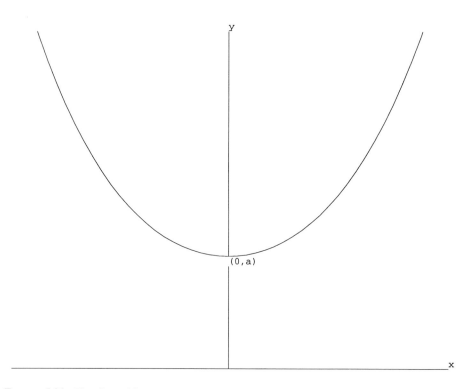

FIGURE 5.21   Hanging cable

The constant a in this equation is the ratio of the tension in the cable at the lowest point to the density of the cable. Show that the solution of this differential equation, subject to the condition $y(0) = a$, $y'(0) = 0$, may be expressed as

$$y = a \cosh \frac{x}{a}.$$

(The graph of this equation is called a "catenary," which is derived from the Latin word for "chain.")

What Have We Learned?

MAIN IDEAS

How to Identify and Solve $dy/dx = g(x, y)$

Purpose: To summarize the major steps required to identify and solve the differential equation

$$\frac{dy}{dx} = g(x, y).$$

Technique:

1. If

$$\frac{dy}{dx} = f(y)g(x)$$

then the equation is **separable**. See *How to Solve Separable Differential Equations* on page 135.

2. If

$$\frac{dy}{dx} = g\left(\frac{y}{x}\right)$$

then the equation has **homogeneous coefficients**. See *How to Solve Differential Equations with Homogeneous Coefficients* on page 212.

3. If

$$a_1(x)\frac{dy}{dx} + a_0(x)y = f(x)$$

then we have a **linear** equation. See *How to Solve Linear Differential Equations* on page 227.

4. If

$$\frac{dy}{dx} + p(x)y = q(x)y^n$$

then we have a **Bernoulli** equation. See *How to Solve Bernoulli's Differential Equation* on page 238.

5. If

$$\frac{dy}{dx} = a_0(x) + a_1(x)y + a_2(x)y^2$$

then we have a **Riccati** equation. See page 244.

6. If

$$y - x\frac{dy}{dx} = f\left(\frac{dy}{dx}\right)$$

then we have a **Clairaut** equation. See *How to Solve Clairaut's Equation* on page 250.

Comment: Table 5.2 summarizes the types of differential equations we have covered.

How to Solve $F(y'', y', y, x) = 0$

1. If F is a function of y'', y', and x, see *How to Solve* $F\left(y'', y', x\right) = 0$ *for* $y = y(x)$ on page 255.

2. If F is a function of y'', y', and y, see *How to Solve* $F\left(y'', y', y\right) = 0$ *for* $y = y(x)$ on page 259.

3. More general cases will be treated in Chapters 7 through 13.

USEFUL RESULTS

- The general solution of

$$\frac{d^2y}{dx^2} + a^2y = 0,$$

where a is a specified constant, is

$$y(x) = C_1 \cos ax + C_2 \sin ax.$$

- The general solution of

$$\frac{d^2y}{dx^2} - a^2y = 0,$$

where a is a specified constant, is

$$y(x) = C_1 e^{ax} + C_2 e^{-ax}.$$

TABLE 5.2 Solving First Order Differential Equations

Differential Equation	Response		Type
$y' = f(y)g(x)$?	Yes	$\longrightarrow$	Separable
$y' = g(y/x)$?	Yes	$\longrightarrow$	Homogeneous coefficients
$y' + p(x)y = q(x)$?	Yes	$\longrightarrow$	Linear
$y' + p(x)y = q(x)y^n$?	Yes	$\longrightarrow$	Bernoulli
$y' = a_0(x) + a_1(x)y + a_2(x)y^2$?	Yes	$\longrightarrow$	Riccati
$y - xy' = f(y')$?	Yes	$\longrightarrow$	Clairaut

CHAPTER 6

First Order Linear Differential Equations and Applications

Where Are We Going?

First order linear differential equations occur so often in applications that we now spend an entire chapter analyzing them, building on the techniques developed in Chapter 5. An important result for linear differential equations is the principle of linear superposition, which we discuss and use in several ways. We introduce terms such as "transient" and "steady state" to describe certain types of solutions. We revisit previous examples to illustrate the new techniques developed here. Some of these new techniques are precursors for developing solutions in future chapters.

6.1 Further Analysis of Linear Equations

We start this section by restating *How to Solve Linear Differential Equations* from page 227 and give two examples to illustrate its use. We conclude this section with two applications, one dealing with solute in a container and the other with fish harvesting. The second application foreshadows a technique used in the next section and again in Chapter 11.

To remind ourselves how to solve the first order linear differential equation

$$a_1(x)\frac{dy}{dx} + a_0(x)y = f(x),$$

we repeat the technique from the last chapter.

How to Solve $a_1(x)dy/dx + a_0(x)y = f(x)$

Purpose: To find $y = y(x)$ that satisfies the linear differential equation

$$a_1(x)\frac{dy}{dx} + a_0(x)y = f(x),$$

where $a_1(x)$ is not zero.

Technique:

1. Put the linear differential equation in the standard form

$$\frac{dy}{dx} + p(x)y = q(x) \tag{6.1}$$

by dividing by $a_1(x)$. Compute the integrating factor $\mu = e^{\int p(x)\,dx}$.

2. Multiply both sides of the differential equation (6.1) by the integrating factor μ, noting that the left-hand side may now be written as a derivative of a product,

$$\mu \frac{dy}{dx} + \mu p(x)y = \frac{d}{dx}(\mu y).$$

This yields

$$\frac{d}{dx}(\mu y) = \mu q(x). \tag{6.2}$$

3. Integrate both sides of (6.2) to find

$$\mu y(x) = \int \mu q(x)\,dx + C,$$

and divide by μ to obtain the explicit solution

$$y(x) = \frac{1}{\mu} \int \mu q(x)\,dx + \frac{C}{\mu}. \tag{6.3}$$

Comments on first order linear differential equations

- Solutions of first order linear differential equations are automatically explicit solutions.
- Because the explicit solution (6.3) contains all solutions of (6.1), it is called the GENERAL SOLUTION of (6.1).
- The term on the right-hand side of (6.1) — namely $q(x)$ — is frequently called the FORCING FUNCTION.
- If we apply the existence and uniqueness theorem, Theorem 2.1, to the linear differential equation (6.1), we find that the conditions of that theorem are satisfied if $p(x)$ and $q(x)$ are continuous in some interval $a < x < b$. However, if these conditions are satisfied, we have the following stronger theorem, which guarantees the existence and uniqueness of solutions for **linear** differential equations.

Theorem 6.1: *Consider the general first order linear differential equation*

$$\frac{dy}{dx} + p(x)y = q(x), \tag{6.4}$$

subject to the initial condition

$$y(x_0) = y_0.$$

If $p(x)$ and $q(x)$ are continuous for $a < x < b$, and x_0 is in this interval, then there exists a unique solution of (6.4) that is continuous for $a < x < b$.

Comments on the existence and uniqueness theorem for first order linear differential equations

- This is a stronger theorem than our previous result. In the previous result a unique solution is guaranteed only in the vicinity of the initial point. Here a unique solution is guaranteed in the same interval that $p(x)$ and $q(x)$ are continuous, no matter how large. Thus, we need look only at $p(x)$ and $q(x)$ to find intervals in which a unique solution will exist.

EXAMPLE 6.1

As an example consider the differential equation

$$\left(x^2 + 1\right) \frac{dy}{dx} + xy = \sqrt{x^2 + 1}.$$ (6.5)

We apply the three steps given in the preceding technique.

1. Divide by $x^2 + 1$ to change the form of (6.5) to

$$\frac{dy}{dx} + \frac{x}{\left(x^2 + 1\right)} y = \frac{1}{\sqrt{x^2 + 1}}.$$ (6.6)

Here $p(x) = x/\left(x^2 + 1\right)$ and $q(x) = 1/\sqrt{x^2 + 1}$, which are continuous everywhere, so, by Theorem 6.1, we are assured that solutions exist everywhere. The integrating factor is

$$\mu = e^{\int p(x)\,dx} = e^{\int \frac{x}{x^2+1}\,dx} = e^{\frac{1}{2}\ln(x^2+1)} = e^{\ln\sqrt{x^2+1}} = \sqrt{x^2 + 1}.$$

2. Multiply (6.6) by the integrating factor $\mu = \sqrt{x^2 + 1}$ to obtain

$$\sqrt{x^2 + 1}\frac{dy}{dx} + \frac{x}{\sqrt{x^2 + 1}} y = 1,$$

or equivalently

$$\frac{d}{dx}\left(\sqrt{x^2 + 1}\, y\right) = 1.$$ (6.7)

3. Integrate (6.7) to obtain

$$\sqrt{x^2 + 1}\, y = \int 1\,dx + C = x + C,$$

and divide by $\sqrt{x^2 + 1}$ to obtain the general solution of (6.5) as

$$y(x) = \frac{x + C}{\sqrt{x^2 + 1}}.$$ (6.8)

Figure 6.1 shows the slope field and analytical solution curves of (6.5) through the points $(0, -9)$, $(0, -5)$, $(0, 0)$, $(0, 5)$, and $(0, 9)$. We notice that all these solution curves seem to tend to 1 as $x \to \infty$. This observation can be confirmed from (6.8), because $y \to 1$ as $x \to \infty$ for all values of C. We also see that the slope field is symmetric about the origin, which is confirmed by the invariance of (6.5) under the transformation $(x, y) \to (-x, -y)$. However, the only solution curve that is symmetric with respect to the origin is the one in which $C = 0$ in (6.8). $\square$

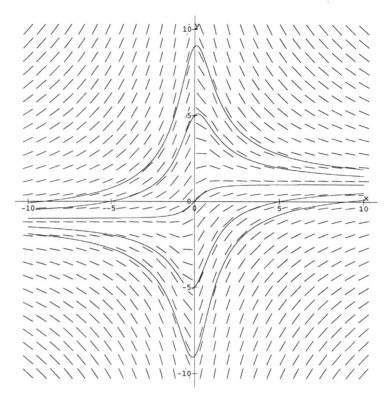

FIGURE 6.1 Analytical solution curves and slope field for $(x^2 + 1)\, dy/dx + xy = (x^2 + 1)^{1/2}$

EXAMPLE 6.2

As another example consider the differential equation

$$x\frac{dy}{dx} + 2y = 4x^2. \tag{6.9}$$

Apply the three steps previously given.

1. Divide by x to change the form of (6.9) to

$$\frac{dy}{dx} + \frac{2}{x}y = 4x. \tag{6.10}$$

Here $p(x) = 2/x$ and $q(x) = 4x$, and because $p(x)$ is not defined at $x = 0$, Theorem 6.1 implies that our solutions could be divided into two regions — namely, $x < 0$ and $x > 0$. The integrating factor is

$$\mu = e^{\int \frac{2}{x}\, dx} = e^{2\ln|x|} = e^{\ln x^2} = x^2.$$

2. Multiply (6.10) by $\mu = x^2$ to obtain

$$x^2\frac{dy}{dx} + 2xy = 4x^3,$$

or equivalently

$$\frac{d}{dx}\left(x^2 y\right) = 4x^3.$$ (6.11)

3. Integrate (6.11) to obtain

$$x^2 y = \int 4x^3\,dx + C = x^4 + C,$$

and divide by x^2 to obtain the general solution of (6.9) as

$$y(x) = \frac{x^4}{x^2} + \frac{C}{x^2} = x^2 + Cx^{-2}.$$ (6.12)

We see that (6.12) contains two type of solutions, those in which y is defined at $x = 0$ (namely, when $C = 0$, so that $y = x^2$) and those in which y is not defined at $x = 0$ (namely, when $C \neq 0$).

Figure 6.2 shows the slope field and various solution curves of (6.9), including $y = x^2$. We see that the slope field is symmetric about the y-axis. (How would you confirm this?)

We also notice from Figure 6.2 that as $x \to 0$ from the left, all solution curves (except $y(x) = x^2$) go to $+\infty$ if the solution curve starts above $y = x^2$ and go to $-\infty$ if the solution curve starts below $y = x^2$. So $y(x) = x^2$ is an UNSTABLE solution as x approaches 0 from the left.

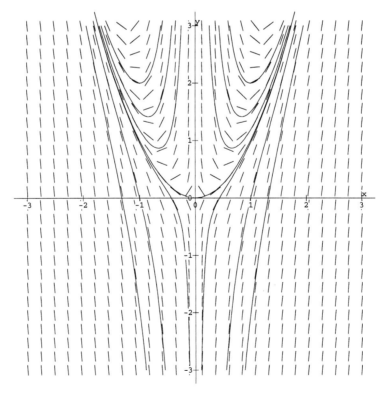

FIGURE 6.2 Analytical solution curves and slope field for $x\,dy/dx + 2y = 4x^2$

For $x > 0$ all solution curves in Figure 6.2 appear to go to ∞ as $x \to \infty$; furthermore, they all seem to approach $y = x^2$ as $x \to \infty$. For this reason we call x^2 a STABLE solution[1] when $x > 0$ and $x \to \infty$.

These observations that were made by looking at Figure 6.2 can be confirmed directly from the solution (6.12). Finally, we note from (6.12) that the only solution that is valid for all x is the one where $C = 0$, namely, $y(x) = x^2$. $\square$

In Example 4.1 on page 132, we discussed a simple mixture problem that lead to a separable differential equation. Now we look at a more complicated example. Recall that to derive the differential equation that describes mixture processes, we let x be a function that represents the amount of substance in a given container at time t, and we assume that the instantaneous rate of change of x with respect to t is governed by the equation of continuity — also called the conservation equation — given by

$$\frac{dx}{dt} = \begin{matrix} \text{rate at which substance} \\ \text{is added to the container} \end{matrix} - \begin{matrix} \text{rate at which substance} \\ \text{is leaving the container} \end{matrix}.$$

EXAMPLE 6.3: Solute in a Container Again

A 300-gallon container is 2/3 full of water containing 50 pounds of salt. At time $x = 0$, valves are turned on so a salt solution of concentration 1/3 pounds per gallon is added to the container at a rate of 3 gallons per minute. If the well-stirred mixture is drained from the container at the rate of 2 gallons per minute, how many pounds of salt are in the container when it is full (and all valves are turned off)?

Notice that the only difference between this example and Example 4.1 is that the water being added is not pure, but here contains 1/3 pounds of salt per gallon. We mimic the same analysis used in Example 4.1.

We first note that more of the salt solution is being added per minute than is being drained, so the number of gallons in the container is increasing. In fact we may use a continuity-type argument to note that the rate of change of volume V of liquid in the container equals the rate being added minus the rate being drained. Thus, we have that

$$\frac{dV}{dt} = 3 - 2 = 1 \text{ gallon per minute.}$$

Integration gives

$$V(t) = t + 200. \tag{6.13}$$

(Why is the constant of integration set equal to 200?)

If x represents the number of pounds of salt in the container at time t, the concentration of salt is

$$\frac{x}{V} = \frac{x}{t + 200}.$$

Thus, the rate at which salt is leaving the container is $2x/(t + 200)$, and from the continuity equation we have

$$\frac{dx}{dt} = (3)(\frac{1}{3}) - \frac{2x}{t + 200} = 1 - \frac{2x}{t + 200}, \tag{6.14}$$

---

[1]This terminology is consistent with that used for stable and unstable equilibrium solutions.

which is valid for $0 < t < 100$. (Why don't we consider values of time greater than 100?) Because there are 50 pounds of salt in the container at $t = 0$, the proper initial condition is

$$x(0) = 50.$$

If we look at the slope field for (6.14) in Figure 6.3, we observe that all solutions will be increasing and concave down. [To fully convince yourself of these facts, consider (6.14) and the result of differentiating (6.14).] Note that the slope field ignores the condition that the container is full when $t = 100$. We have also hand-drawn the solution curve that passes through the initial point $(0, 50)$. From this curve we can estimate the value of $x(t)$ for $t = 100$ at about 90 pounds, which is an approximate answer to our original question. However, to obtain an exact answer we must obtain an explicit solution.

The only difference between (6.14) and the differential equation in Example 4.1, namely, $dx/dt = -2x/(t + 200)$, is the presence of the constant 1 on the right-hand side. However, that constant is enough to prevent (6.14) from being a separable differential equation. However, if we rearrange (6.14) as

$$\frac{dx}{dt} + \frac{2x}{t + 200} = 1, \tag{6.15}$$

we see that it is a linear differential equation. Using our usual techniques, we first find the integrating factor as

$$\mu = e^{\int \frac{2}{t+200}\, dt} = (t + 200)^2.$$

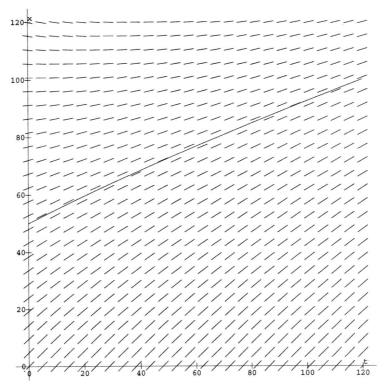

FIGURE 6.3 Hand-drawn solution curve and slope field for $dx/dt = 1 - 2x/(t + 200)$

If we multiply both sides of (6.15) by this integrating factor and combine terms, we obtain

$$(t+200)^2 \frac{dx}{dt} + 2(t+200)x = (t+200)^2,$$

or

$$\frac{d}{dt}\left[(t+200)^2 x\right] = (t+200)^2. \tag{6.16}$$

Integration of (6.16) yields

$$(t+200)^2 x = \frac{1}{3}(t+200)^3 + C,$$

from which we find x as

$$x(t) = \frac{1}{3}(t+200) + \frac{C}{(t+200)^2}.$$

The choice of C as $-50(200^2)/3$ will satisfy the initial condition $x(0) = 50$, so our final form for the solution is

$$x(t) = \frac{1}{3}(t+200) - \frac{50(200)^2}{3(t+200)^2}. \tag{6.17}$$

To answer the original question — how many pounds of salt are in the container when it is full — we note from (6.13) that the container will be full when $t = 100$, so $x(100)$ will be the amount of salt in the container at this time. From (6.17) we have that

$$x(100) = \frac{1}{3}(300) - \frac{50(200)^2}{3(300)^2} = \frac{2500}{27} = 92\frac{16}{27} \text{ pounds of salt,}$$

which is close to our estimate of 90 pounds.   □

EXAMPLE 6.4: Fish Harvesting

We started Section 5.2 with a mathematical model of human population growth that included emigration. Now we consider a model of the population of fish in a lake in which there are no predators but there is an abundant supply of food for the fish (for example, a fish farm). Fish are harvested in a periodic manner described by the function $h(t) = a + b\sin 2\pi t$, where a and b are constants, $a > b$, a and b are given in thousands, and t is given in years. Note that $h(t)$ is always positive, it oscillates between $a + b$ and $a - b$, and

$$\int_0^1 h(t)\,dt = a,$$

the number of fish harvested per year.

If we assume that fish reproduce at a rate proportional to their population, the appropriate differential equation that models this situation is

$$\frac{dP}{dt} = kP - (a + b\sin 2\pi t). \tag{6.18}$$

Here P is the fish population (in thousands) and $k > 0$ is the net growth rate. Figure 6.4 gives the slope field for this situation for $k = 0.5$, $a = 3$, $b = 1$, along with numerical solutions for five different initial conditions. It is apparent that in these five cases, the fish population either dies out or grows quite rapidly. There does not seem to be a simple

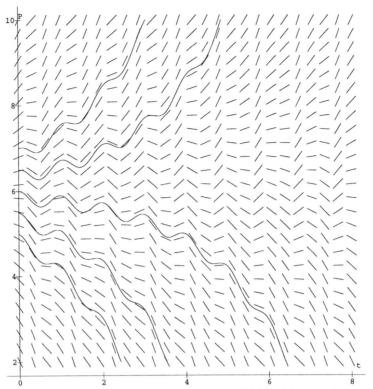

FIGURE 6.4 Numerical solution curves and slope field for fish harvesting

periodic solution for this mathematical model. Let us find the general solution of this linear differential equation to investigate this further.

We put (6.18) in the standard form

$$\frac{dP}{dt} - kP = -(a + b\sin 2\pi t) \tag{6.19}$$

and compute the integrating factor as e^{-kt}. We then multiply (6.19) by this integrating factor to obtain

$$e^{-kt}\frac{dP}{dt} - e^{-kt}kP = -e^{-kt}(a + b\sin 2\pi t)$$

and rewrite this equation as

$$\frac{d}{dt}\left(e^{-kt}P\right) = -e^{-kt}(a + b\sin 2\pi t). \tag{6.20}$$

Finally, we integrate (6.20) — using integration by parts or, preferably, a table of integrals — to obtain

$$e^{-kt}P = e^{-kt}\left(\frac{a}{k} + b\frac{k\sin 2\pi t + 2\pi \cos 2\pi t}{k^2 + 4\pi^2}\right) + C$$

and divide by e^{-kt} to obtain the general solution

$$P(t) = \left(\frac{a}{k} + b\frac{k\sin 2\pi t + 2\pi \cos 2\pi t}{k^2 + 4\pi^2}\right) + Ce^{kt}. \tag{6.21}$$

The term Ce^{kt} is the general solution of $dP/dt - kP = 0$, whereas the other term is a particular solution of (6.19). The choice of the constant of integration as

$$C = P_0 - \frac{a}{k} - \frac{2\pi b}{k^2 + 4\pi^2}$$

gives the initial population in (6.21) as P_0.

From (6.21) we see that for any positive value of the growth rate, k, the exponential part of the solution will dominate the trigonometric part for large values of t. If $C > 0$, the solution will grow without bound, and if $C < 0$, it will become unbounded in the negative direction. The only way there will be a bounded solution is for C to equal zero. This solution is unstable, because for C slightly positive or slightly negative, the associated solutions do not stay close to the solution for $C = 0$. The solution curve corresponding to the initial population, P_0, that makes $C = 0$ in (6.21) and the solution curves for $C = -0.05$ and $C = 0.05$ are shown in Figure 6.5. □

EXERCISES

Solve the following differential equations.

1. $\dfrac{dy}{dt} + 2y = 2e^{-t}$

2. $\dfrac{dy}{dt} + 2y = 20e^{3t}$

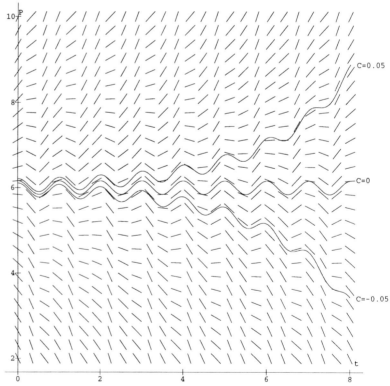

FIGURE 6.5 Analytical solution curves and slope field for fish harvesting with $C = -0.05, 0,$ and 0.05

3. $\dfrac{dy}{dt} + 2ty = 4t$

4. $\dfrac{dy}{dt} + \dfrac{2}{t}y = 6$

5. $\dfrac{dy}{dt} + \dfrac{2}{t}y = \dfrac{\sin t}{t^2}$

6. $t\dfrac{dy}{dt} + 2y = e^t$

7. $\dfrac{dT}{dt} - kT = \alpha,$ where k and α are positive constants

8. $\dfrac{dT}{dt} - kT = \beta t,$ where k and β are positive constants

9. $t\dfrac{dP}{dt} + 3P = \dfrac{\ln t}{t}$

10. $t\dfrac{dP}{dt} + 2P = 6\sin t$

11. $\dfrac{dP}{dt} + (\sin t)\,P = 4\cos t \sin t$

12. $\dfrac{dy}{dt} - 3y = 6,\ \ y(0) = 1$

13. $\dfrac{dy}{dt} - y = \sin 2t,\ \ y(0) = 0$

14. $\dfrac{dy}{dt} - 7y = 14t,\ \ y(0) = 0$

15. $\dfrac{dy}{dt} + \dfrac{2}{t}y = t,\ \ y(1) = 1$

16. $\dfrac{dy}{dt} + 2ty = t,\ \ y(0) = 2$

17. Consider the use of Newton's law of cooling to model the effect of temperature oscillations outside a building on the temperature within. Suppose you are leaving for four days and wonder if you can shut off your heating system during your absence. (There are plants inside that cannot tolerate temperatures below 40^oF.) Assume that the outside temperature varies sinusoidally from a mean of 45^oF, with a 10^oF oscillation up and down. If when you leave in the morning the building is 70^oF, and the outside temperature is 45^oF, are your plants safe? (The surface area of the building, type of construction, insulation, and heat energy of the house are taken into account if you use the value $k = -0.2$ in Newton's law of cooling.)

18. Let a 200-gallon container of pure water have a salt concentration of 3 pounds per gallon added to the container at the rate of 4 gallons per minute.

  (a) If the well-stirred mixture is drained from the container at the rate of 5 gallons per minute, find the number of pounds of salt in the container as a function of time. For how many minutes is this solution valid?

  (b) How many minutes does it take for the concentration in the container to reach 2 pounds per gallon?

19. A container is filled with 10 gallons of water containing 5 pounds of salt. A salt solution of concentration 3 pounds per gallon is pumped into the container at a rate of 2 gallons per minute, and the well-stirred mixture drains at the same rate. How much salt is in the container after 15 minutes?

20. An open container has 5 pounds of impurities dissolved in 150 liters of water. Pure water is pumped into the container at a rate of 3 liters per minute, and the well-stirred mixture is drained at the rate of 2 liters per minute.

(a) How many pounds of impurities remain after 10 minutes?

(b) If the container holds a maximum of 200 liters, what is the concentration of impurities just before the solution overflows?

21. Look in the textbooks for your other courses and find an example that uses a linear differential equation. Write a report about this example that includes the following items:

(a) A brief description of background material so a classmate will understand the origin of the differential equation.

(b) How the constants in the differential equation can be evaluated and the meaning of the initial condition.

(c) The details of the solution of this differential equation.

(d) An interpretation of this solution, and how it answers a question posed by the original discussion.

6.2   The Principle of Linear Superposition

In the last example in the previous section, the function on the right-hand side of the differential equation (6.19) took the form of the sum of two functions, $-a - b\sin 2\pi t$, and the particular solution in (6.21) had a similar form. Repeating the differential equation and its solution we have

$$\frac{dP}{dt} - kP = -(a + b\sin 2\pi t), \tag{6.22}$$

$$P(t) = \left(\frac{a}{k} + b\frac{k\sin 2\pi t + 2\pi \cos 2\pi t}{k^2 + 4\pi^2}\right) + Ce^{kt}. \tag{6.23}$$

Because the constants a and b in (6.22) occur separately in (6.23), we observe that we could have obtained the part of the solution in (6.23) that is in parentheses by solving two separate problems. That is, solving

$$\frac{dP}{dt} - kP = -a$$

gives

$$P = \frac{a}{k} + C_1 e^{kt},$$

and solving

$$\frac{dP}{dt} - kP = -b\sin 2\pi t$$

gives

$$P = b\frac{k\sin 2\pi t + 2\pi \cos 2\pi t}{k^2 + 4\pi^2} + C_2 e^{kt}.$$

Adding the two gives our former solution (6.23) if we replace the sum of two arbitrary constants with a single arbitrary constant, that is, let $C = C_1 + C_2$.

This process of adding two solutions of a linear differential equation and obtaining a solution of an associated problem is called LINEAR SUPERPOSITION. This property is true in general for linear differential equations as we now show. If $y_1(t)$ satisfies

$$\frac{dy_1}{dt} + p(t)y_1 = q_1(t)$$

and $y_2(t)$ satisfies

$$\frac{dy_2}{dt} + p(t)y_2 = q_2(t),$$

then adding these two equations and using properties of derivatives gives

$$\frac{d(y_1 + y_2)}{dt} + p(t)(y_1 + y_2) = q_1(t) + q_2(t).$$

Thus, their sum, $y(t) = y_1(t) + y_2(t)$, satisfies the linear differential equation

$$\frac{dy}{dt} + p(t)y = q_1(t) + q_2(t).$$

A related property of linear differential equations is that if $y_1(t)$ satisfies

$$\frac{dy}{dt} + p(t)y = q(t)$$

then $y(t) = ay_1(t)$, where a is a constant, satisfies the linear differential equation

$$\frac{dy}{dt} + p(t)y = aq(t).$$

(See Exercise 12 on page 286).

These two results can be combined, and form the PRINCIPLE OF LINEAR SUPER-POSITION.

Theorem 6.2:   *If $y_1(t)$ and $y_2(t)$ are solutions of*

$$\frac{dy}{dt} + p(t)y = q_1(t)$$

and

$$\frac{dy}{dt} + p(t)y = q_2(t),$$

respectively, then their linear combination, $ay_1(t) + by_2(t)$, is a solution of

$$\frac{dy}{dt} + p(t)y = aq_1(t) + bq_2(t),$$

where a and b are specified constants.

Comments about linear superposition

- It is important that the left-hand sides of the differential equations satisfied by $y_1(t)$ and $y_2(t)$ be exactly the same — namely, $y' + p(t)y$.

- The principle of linear superposition is not generally true for nonlinear differential equations. (See Exercise 13 on page 286.)
- The principle of linear superposition is not generally true if a and b are functions of t. For example, if $y_1(t)$, $y_2(t)$, and $y_3(t)$ are the solutions of

$$\frac{dy}{dt} + y = 1,$$

$$\frac{dy}{dt} + y = e^t,$$

and

$$\frac{dy}{dt} + y = 1 + te^t,$$

respectively, $y_3(t) \neq y_1(t) + ty_2(t)$. (See Exercise 14 on page 286.)

As an example of how we might use the principle of linear superposition, consider Exercises 4 and 5 on page 277 of the previous section. There we found that the general solution of

$$\frac{dy}{dt} + \frac{2}{t}y = 6$$

was

$$y(t) = C_1 t^2 + 2t,$$

and the general solution of

$$\frac{dy}{dt} + \frac{2}{t}y = \frac{\sin t}{t^2}$$

was

$$y(t) = C_2 t^2 - \frac{\cos t}{t^2}.$$

Thus we may use the principle to write down the general solutions of

$$\frac{dy}{dt} + \frac{2}{t}y = 6 + \frac{\sin t}{t^2},$$

$$\frac{dy}{dt} + \frac{2}{t}y = -1,$$

and

$$\frac{dy}{dt} + \frac{2}{t}y = 12 - 3\frac{\sin t}{t^2},$$

as

$$y(t) = Ct^2 + 2t - \frac{\cos t}{t^2},$$

$$y(t) = Ct^2 - \frac{1}{3}t,$$

and

$$y(t) = Ct^2 + 4t + 3\frac{\cos t}{t^2},$$

respectively.

EXAMPLE 6.5: High Temperature Furnace

Now we consider the situation in which we have a high temperature furnace (like those used in factories that produce steel or glass) where the temperature in the furnace increases in a linear manner throughout the time period of interest. If this linear increase in temperature is denoted by $at + b$, then the appropriate initial value problem for the temperature $T(t)$ of a cold object placed in such a furnace is

$$\frac{dT}{dt} = k(T - at - b), \text{ where } k < 0, \tag{6.24}$$

$$T(0) = T_0.$$

Here b is the initial temperature of the furnace, a is the rate of change of the furnace temperature with time, and T_0 is the initial temperature of the object. Note that (6.24) is Newton's law of heating with the ambient temperature changing according to $at + b$.

Figure 6.6 gives the slope field for (6.24) (for $a = 10$, $b = 70$, and $k = -0.1$) and conveys the distinct impression that as time increases, all solutions of this differential equation approach a straight line. Figure 6.7 gives numerical solutions of (6.24) for several different initial conditions and strengthens this belief. Both figures indicate that the entire region is divided into two sections, one in which all solutions are concave up (above some straight line) and one in which they are concave down. They also indicate that all solutions that start in a region where the solution is concave down will always be increasing, whereas for initial conditions in the upper-left part of the figures, the solution will start out decreasing but will always end up increasing. Using isoclines to find where we have horizontal tangents is informative. From (6.24) we see that the equation for isoclines with horizontal tangents is

$$T(t) = at + b,$$

and it is clear from Figures 6.6 and 6.7 that this corresponds to the locations where solutions may have only relative minimum values.

To examine concavity, we differentiate (6.24) to obtain

$$\frac{d^2T}{dt^2} = k\left(\frac{dT}{dt} - a\right) = k\left[k\left(T - at - b\right) - a\right].$$

Thus, $T'' = 0$ for $T(t) = at + b + a/k$. Above this line all solution curves will be concave up, and below this line all solution curves will be concave down. (Why don't we see any inflection points in Figures 6.6 and 6.7?) Because all the numerical solutions in Figure 6.7 seem to approach a linear asymptote, we try a solution of the form

$$T(t) = mt + c. \tag{6.25}$$

Substituting this function into the differential equation (6.24), we find that it will be satisfied if m and c are chosen so

$$m = k(mt + c - at - b) = k(m - a)t + k(c - b)$$

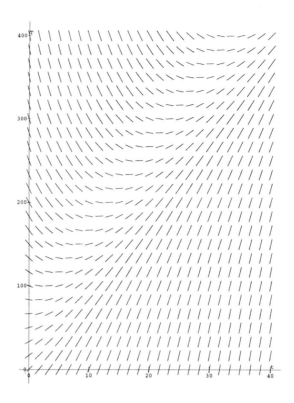

FIGURE 6.6   Slope field for furnace

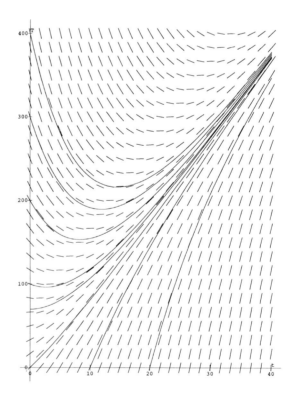

FIGURE 6.7   Numerical solution curves and slope field for furnace

is valid for all values of t. This is possible if we choose $m = a$ and $c = b + a/k$, so our solution of (6.24) is

$$T(t) = at + b + \frac{a}{k}.$$

From the slope field, we see that this is also our solution for large values of time. Thus, the line dividing regions of concavity is actually a particular solution.

To further investigate general solutions of the linear equation (6.24), we write the differential equation in the standard form as

$$\frac{dT}{dt} - kT = -kat - kb. \tag{6.26}$$

In Exercises 7 and 8 on page 277 of the previous section, we found that the general solution of

$$\frac{dT}{dt} - kT = \alpha$$

was

$$T(t) = C_1 e^{kt} - \frac{\alpha}{k},$$

and the general solution of

$$\frac{dT}{dt} - kT = \beta t$$

was

$$T(t) = C_2 e^{kt} - \beta t - \frac{\beta}{k}.$$

Now we look at (6.26), make the association $\alpha = -kb$ and $\beta = -ka$, and use the principle of linear superposition to write the general solution of (6.26) as

$$T(t) = Ce^{kt} + at + b + \frac{a}{k}. \tag{6.27}$$

We choose the arbitrary constant of integration as $C = T_0 - b - a/k$ so (6.27) will satisfy the initial condition $T(0) = T_0$. With $C = 0$ we have the solution

$$T(t) = at + b + \frac{a}{k},$$

which is the line we found earlier.

Differentiating (6.27) will verify the accuracy of our earlier predictions regarding increasing or decreasing regions as well as our considerations about concavity.

One last comment pertaining to this example is necessary. Notice that if we think of the right-hand side of (6.26), $-k(at + b)$, as an input to the system (the furnace), then the input is a linear function, as is the steady state output, $at + b + a/k$. This is clear from equations (6.25) and (6.27) as well as from Figure 6.8, which shows these two linear functions and the graph of (6.27) for $a = 10$, $b = 70$, $T(0) = 70$, and $k = -0.1$. $\qquad\square$

We noted earlier that in Figure 6.7 it appeared that solutions, regardless of the initial condition, approached a straight line as $t \to \infty$. This is also apparent from (6.27), where,

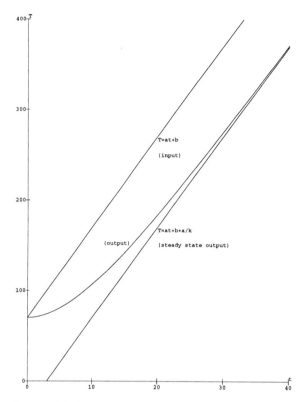

FIGURE 6.8   Linear input and output

because $k < 0$, the term $Ce^{kt} \to 0$ as $t \to \infty$. Thus, as $t \to \infty$, all solutions approach the straight line solution $T(t) = at + b + a/k$. This gives rise to the following definition.

Definition 6.1: **Consider the explicit solution of a linear differential equation. If there is a portion of this solution that approaches zero as the independent variable approaches infinity it is called the TRANSIENT part of the solution. For this situation, the portion of this solution that does not approach zero as the independent variable approaches infinity is called the STEADY STATE part of the solution.**

Comments on transient and steady state parts of the solution

- Often these parts of the solution are referred to as the "transient solution" and the "steady state solution".

- The words steady state refer to the fact that this part accurately describes the solution's behavior after a long period of time, not the fact that it is "steady".

In the previous example where the general solution is $Ce^{kt} + at + b + a/k$, the transient solution is Ce^{kt} because $k < 0$, and the steady state solution is $at + b + a/k$.

EXERCISES

Solve the following differential equations. Identify the transient and steady state solutions, if they exist.

1. $\dfrac{dy}{dt} + y = 3$

2. $\dfrac{dy}{dt} + y = t$

3. $\dfrac{dy}{dt} + y = t^2$

4. $\dfrac{dy}{dt} + y = \cos t$

5. $\dfrac{dy}{dt} + y = \sin t$

6. $\dfrac{dy}{dt} + y = \sin 5t$

7. $\dfrac{dy}{dt} + y = e^t$

8. Use the principle of linear superposition and your results from Exercises 1 through 7 to find the general solutions of the following differential equations. Identify the transient and steady state solutions, if they exist.

   (a) $\dfrac{dy}{dt} + y = 3 + t$

   (b) $\dfrac{dy}{dt} + y = 3 \sin t + \sin 5t$

   (c) $\dfrac{dy}{dt} + y = 3 + t + \sin t$

9. For the equation $y' + y = f(t)$, use the principle of linear superposition and your results from Exercises 1 through 7 to find the general solution of this differential equation, for the following forcing functions. Identify the transient and steady state solutions, if they exist.

   (a) $f(t) = 2t + 1$

   (b) $f(t) = t + \cos t$

   (c) $f(t) = 6 \cos t + 5 \sin t$

   (d) $f(t) = 2t^2 + t + 3$

   (e) $f(t) = 5 + 2e^t$

   (f) $f(t) = 2 \sin \left(t + \dfrac{\pi}{6} \right)$

10. Solve the following differential equations. Identify the transient and steady state solutions, if they exist.

   (a) $\dfrac{dy}{dt} - \dfrac{2}{t} y = t^2 \cos t$

   (b) $\dfrac{dy}{dt} - \dfrac{2}{t} y = t^2 e^{2t}$

   (c) $\dfrac{dy}{dt} - \dfrac{2}{t} y = t^3 e^{3t}$

11. For the equation

$$\frac{dy}{dt} - \frac{2}{t} y = f(t)$$

use the principle of linear superposition and your results from Exercise 10 to find the general solution of this differential equation for the following forcing functions. Identify the transient and steady state solutions, if they exist.

(a) $t^2 \cos t + t^3 e^{3t}$

(b) $\pi t^2 e^{2t} - 13t^3 e^{3t}$

(c) $t^2 \left(3 \cos t + 4e^{2t} \right)$

12. Show that if $y(t) = f(t)$ satisfies the linear differential equation

$$\frac{dy}{dt} + p(t)y = q(t),$$

then $y = af(t)$ satisfies

$$\frac{dy}{dt} + p(t)y = aq(t).$$

13. Consider the situation in which $y = y_1(x)$ satisfies the differential equation

$$\frac{dy}{dx} = f(x, y) + q_1(x)$$

and in which $y = y_2(x)$ satisfies the differential equation

$$\frac{dy}{dx} = f(x, y) + q_2(x).$$

If $y = y_1(x) + y_2(x)$ satisfies

$$\frac{dy}{dx} = f(x, y) + q_1(x) + q_2(x),$$

show that

$$f(x, y_1 + y_2) = f(x, y_1) + f(x, y_2).$$

By using the result from Example 4.15, show that $f(x, y) = p(x)y$. What does this tell you about the original differential equations satisfied by y_1 and y_2?

14. Let $y_1(t)$, $y_2(t)$, and $y_3(t)$ be the solutions of

$$\frac{dy}{dt} + y = 1,$$

$$\frac{dy}{dt} + y = e^t,$$

and

$$\frac{dy}{dt} + y = 1 + te^t,$$

respectively. Show that $y_3(t) \neq y_1(t) + ty_2(t)$.

6.3 Solving Linear Differential Equations with Constant Coefficients

At this point we have a procedure for finding general solutions of all linear differential equations. In this section we consider some applications where the two terms involving the dependent variable have constant coefficients — namely,

$$\frac{dy}{dt} + ky = q(t)$$

where k is constant. For this situation we develop an alternative method of solution which foreshadows a technique — given in Chapter 11 — for solving higher order differential equations with constant coefficients.

Our first application considers the situation in which human cells are susceptible to infection by a virus such that once a specific cell is infected, it can never again act as a host for the virus. To form a mathematical model in which cells are forming, becoming infected, and being eliminated (excreted), we let $f(t)$ be the total number of cells at time t and $P(t)$ be the number of infected cells. We could form a continuity equation to account for all the infected cells as

$$\begin{array}{ccc} \text{Rate of change of} \\ \text{infected cells} \end{array} = \begin{array}{ccc} \text{Rate of change of} \\ \text{newly infected cells} \end{array} - \begin{array}{ccc} \text{Rate of change of} \\ \text{eliminated cells} \end{array}.$$

Because the number of healthy cells is $f - P$, we may write this equation in terms of f and P as

$$\frac{dP}{dt} = r(f - P) - sP,$$

where r is the constant rate of infection of healthy cells per unit population and s is the constant elimination (excretion) rate per unit population. This equation can be rearranged in the form

$$\frac{dP}{dt} + (r + s)P = rf,$$

where $f(t)$ is a specified function. This is a linear differential equation, and because the coefficients of dP/dt and P are constants — namely, 1 and $r + s$ — it is called a LINEAR DIFFERENTIAL EQUATION WITH CONSTANT COEFFICIENTS.

In the following examples we concentrate on the specific case in which $r = s = 1/2$, that is,

$$\frac{dP}{dt} + P = \frac{1}{2}f, \tag{6.28}$$

and look at the effect for different choices of the forcing function $\frac{1}{2}f(t)$.

EXAMPLE 6.6: Infected Cells

In this first example the total number of cells is taken to be a linear function of time — namely, $f(t) = a + bt$ — so the differential equation (6.28) becomes

$$\frac{dP}{dt} + P = \frac{1}{2}(a + bt). \tag{6.29}$$

The general solution of this linear differential equation may be found by the principle of linear superposition from the solutions of

$$\frac{dP}{dt} + P = \frac{1}{2}a,$$

and

$$\frac{dP}{dt} + P = \frac{1}{2}bt.$$

Using the results of Exercises 7 and 8 on page 277, we find

$$P(t) = \frac{a}{2} + \frac{b}{2}(t - 1) + Ce^{-t}.$$

The graph of this solution is given in Figure 6.9 for $a = 4$, $b = 0$, and several values of C, whereas Figure 6.10 shows the results for $a = 4$ and $b = 2$. Notice that in both of these figures, the solution approaches a straight line as time increases. In Figure 6.9 the line is horizontal (the equilibrium solution), whereas in Figure 6.10, the solution approaches a straight line with slope $b/2$. In both cases the straight line solution (given by $C = 0$) is stable and is the steady state solution.   □

Exercise 6 on page 301 asks for the general solution of a similar cell-virus problem in which the total cell population was experiencing oscillations, with the appropriate differential equation being

$$\frac{dP}{dt} + P = \frac{1}{2}(a + 2b \sin t). \tag{6.30}$$

One form of the answer to this exercise is

$$P(t) = \frac{1}{2}a + \frac{1}{2}b(\sin t - \cos t) + Ce^{-t}. \tag{6.31}$$

Here the first two terms on the right-hand side constitute the steady state solution while Ce^{-t} is the transient solution. The graph of this solution is shown in Figure 6.11 for $a = 4$ and $b = 2$ and for several different initial conditions. Notice that with a sinusoidal forcing function, the graph of the steady state solution contains oscillations which are close to the period of the function on the right-hand side of (6.30).

We collect these results together in Table 6.1. Notice that the general solutions of these linear equations consist of two parts, a particular solution of the differential equation plus Ce^{-t}, where Ce^{-t} is the general solution of the ASSOCIATED HOMOGENEOUS DIFFERENTIAL EQUATION, namely,

$$\frac{dP}{dt} + P = 0.$$

Definition 6.2:   The ASSOCIATED HOMOGENEOUS DIFFERENTIAL EQUATION is obtained from the original nonhomogeneous differential equation by setting the forcing function to zero. Thus, if

$$\frac{dy}{dt} + p(t)y = q(t)$$

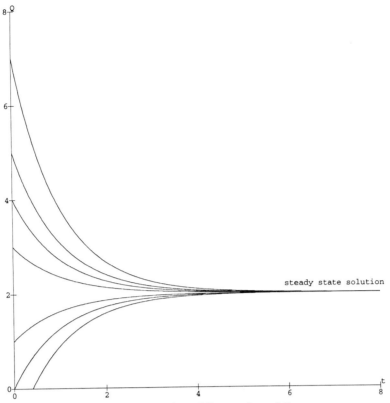

FIGURE 6.9   Graph of $a/2 + Ce^{-t}$ for different values of C

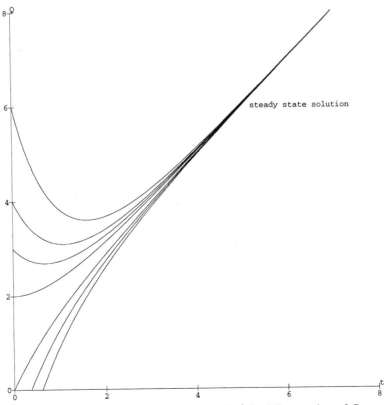

FIGURE 6.10   Graph of $a/2 + b(t-1)/2 + Ce^{-t}$ for different values of C

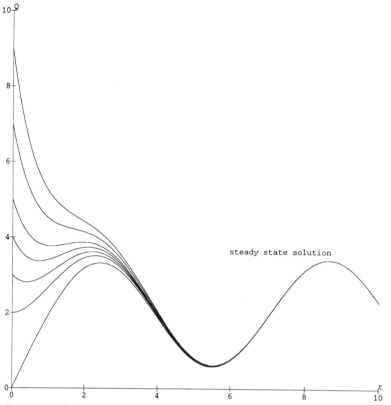

FIGURE 6.11 Graph of $a/2 + b(\sin t - \cos t)/2 + Ce^{-t}$ for different values of C

TABLE 6.1 **Forcing functions and general solutions for** $dP/dt + P = f(t)$

$f(t)$	General solution $P(t)$
$a/2$	$a/2 + Ce^{-t}$
$bt/2$	$b(t-1)/2 + Ce^{-t}$
$b \sin t$	$b(\sin t - \cos t)/2 + Ce^{-t}$

is the original equation, the associated homogeneous differential equation is

$$\frac{dy}{dt} + p(t)y = 0.$$

Comments about associated homogeneous differential equations

- Even though this section is devoted to linear differential equations with constant coefficients, this definition applies to all linear differential equations.

What we want to do in this section is to consider a method that avoids integrating factors for obtaining general solutions of first order linear equations with constant coefficients. Before doing this, we introduce some new notation to simplify our discussion.

1. The general solution of the associated homogeneous differential equation is identified by adding the subscript h to the dependent variable. Thus, for the preceding cases,

$$\frac{dP}{dt} + P = 0$$

has solution

$$P_h = Ce^{-t},$$

where C is an arbitrary constant.

2. Recall that any function that satisfies a differential equation is called a particular solution. Here we identify a particular solution by adding a subscript p to the dependent variable. Normally this function has no arbitrary constants. Thus, in Table 6.1 the forcing functions are $a/2$, $bt/2$, and $b \sin t$, and the particular solutions are $a/2$, $b(t-1)/2$, and $b(\sin t - \cos t)/2$; that is, they are designated as

$$P_p = \frac{a}{2},$$

$$P_p = \frac{b}{2}(t-1),$$

and

$$P_p = \frac{b}{2}(\sin t - \cos t).$$

The sum of P_h and the appropriate formula for P_p gives the general solution of the original differential equation.

In Table 6.1 we notice that a particular solution for a constant forcing function was a constant, while that for $bt/2$ was $b(t-1)/2$. If the right-hand side of our equation were a general polynomial, what type of particular solution would be appropriate? Because the derivative of a polynomial is again a polynomial of one degree less, it seems reasonable to try a polynomial. We will use this reasoning in our next example.

EXAMPLE 6.7

Consider the differential equation

$$\frac{dT}{dt} - 2T = 6t - 4t^2. \tag{6.32}$$

The general solution of the associated homogeneous equation of (6.32) is

$$T_h(t) = Ce^{2t},$$

while, in accordance with our previous discussion, we will try a particular solution of the form

$$T_p(t) = A + Bt + Ct^2. \tag{6.33}$$

Substituting this expression into the differential equation gives

$$B + 2Ct - 2(A + Bt + Ct^2) = 6t - 4t^2. \tag{6.34}$$

We want this to be an identity in t, which means it has to be true for all values of t. Therefore we can choose any convenient values for t to find conditions on A, B, and C.

If we try $t = 0$ in (6.34), we find

$$B - 2A = 0. \tag{6.35}$$

If we try $t = 1$ in (6.34) and use (6.35), we find

$$-2B = 2$$

or

$$B = -1, \tag{6.36}$$

giving

$$A = -\frac{1}{2}. \tag{6.37}$$

If we try $t = -1$ in (6.34) and use (6.36) and (6.37), we find

$$-4C - 2 = -10$$

or

$$C = 2. \tag{6.38}$$

Because we have

$$C = 2, \ B = -1, \ \text{and} \ A = -\frac{1}{2},$$

the particular solution (6.33) is

$$T_p(t) = -\frac{1}{2} - t + 2t^2.$$

Thus, the general solution of (6.32) is

$$T(t) = T_p(t) + T_h(t) = -\frac{1}{2} - t + 2t^2 + Ce^{2t}.$$

To notice the mathematical techniques we have avoided by using this method, you should solve (6.32) using an integrating factor.   □

There is another way of solving (6.34) — namely, by equating coefficients of like powers of t on the two sides of the equation to obtain

$$B - 2A = 0,$$
$$2C - 2B = 6,$$

and

$$-2C = -4.$$

The solution of this set of equations is again

$$C = 2, \ B = -1, \ \text{and} \ A = -\frac{1}{2}.$$

EXAMPLE 6.8: Fish Harvesting

We now focus on differential equation (6.19) from Section 6.1, which dealt with fish harvesting, namely,

$$\frac{dP}{dt} - kP = -a - b \sin 2\pi t, \tag{6.39}$$

and ask what types of functions we could use for P such that a linear combination of P and its derivative would result in the right-hand side. Notice that we have the sum of a constant and a sine function so the principle of linear superposition can be used.

We first consider a particular solution for the constant. From our previous reasoning (and common sense) letting P equal a constant (say A) results in the equation

$$-kA = -a,$$

so we choose $A = a/k$ and obtain the first part of the particular solution, namely a/k.

Turning our attention to the other term on the right-hand side of (6.39), we ask what functions can we multiply by a constant and add to its derivative to give a sine function. The derivative of a sine function is a cosine function, so choosing only a sine function will not work. However, because the derivative of a linear combination[2] of sine and cosine functions gives a linear combination of sine and cosine functions, that is what we will try. (Or, note in Table 6.1 that the particular solution corresponding to the sine function had both sine and cosine functions.) Thus, we substitute

$$P_p(t) = \frac{a}{k} + A \sin 2\pi t + B \cos 2\pi t$$

into (6.39) and obtain

$$2A\pi \cos 2\pi t - 2B\pi \sin 2\pi t - kA \sin 2\pi t - kB \cos 2\pi t = -b \sin 2\pi t,$$

or

$$(2A\pi - kB) \cos 2\pi t - (2B\pi + kA) \sin 2\pi t = -b \sin 2\pi t.$$

Because we want this to be an identity, the coefficient of $\cos 2\pi t$ on the left-hand side of the preceding equation must be zero and the coefficient of $\sin 2\pi t$ must be $-b$. This gives the pair of algebraic equations

$$2A\pi - kB = 0$$

$$2B\pi + kA = b,$$

with solution

$$A = \frac{bk}{k^2 + 4\pi^2}$$

---

[2]Recall that a linear combination of $f(t)$ and $g(t)$ is $C_1 f(t) + C_2 g(t)$, where C_1 and C_2 are constants.

$$B = \frac{2b\pi}{k^2 + 4\pi^2},$$

in agreement with (6.21).   □

EXAMPLE 6.9: Population with Immigration

Thus far we have considered first order linear differential equations with constant coefficients in which the forcing functions (that is, the right-hand sides) are either oscillatory or polynomials in the independent variable. Consider now a simple population model in which immigration begins at time $t = 0$ and approaches a — the limit for large values of t — according to $a(1 - e^{-bt})$. That is, consider a population P that satisfies

$$\frac{dP}{dt} = kP + a\left(1 - e^{-bt}\right), \tag{6.40}$$

subject to the initial condition

$$P(0) = P_0.$$

Before finding the general solution, we notice that the right-hand side of (6.40) is always positive so P will always be an increasing function. Taking the derivative of both sides of (6.40) gives

$$\frac{d^2 P}{dt^2} = k\frac{dP}{dt} + abe^{-bt}.$$

This shows that all solutions will be concave up for positive values of k, a, and b. A typical slope field — with $k = a = b = 1$ — is shown in Figure 6.12.

We want to find a particular solution and write down the general solution of this differential equation. To do this we rewrite (6.40) in the standard form of a linear equation, namely,

$$\frac{dP}{dt} - kP = a - ae^{-bt}.$$

Here the forcing function is the sum of two terms, so we use the principle of linear superposition to examine each part separately.

The particular solution corresponding to

$$\frac{dP}{dt} - kP = a$$

is

$$P_{p_1}(t) = -\frac{a}{k}.$$

The second part of the forcing function is the exponential $-ae^{-bt}$, and because the derivative of an exponential function is again an exponential function, it seems reasonable to assume a particular solution of the form Ae^{-bt}. Substituting this function into the differential equation

$$\frac{dP}{dt} - kP = -ae^{-bt},$$

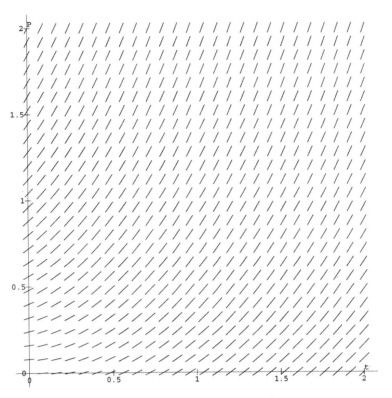

FIGURE 6.12   Slope field for $dP/dt = kP + a(1 - e^{-bt})$

gives

$$-bAe^{-bt} - kAe^{-bt} = -ae^{-bt},$$

so $A = a/(b + k)$, and

$$P_{p_2}(t) = \frac{a}{b + k}e^{-bt}.$$

The entire particular solution is

$$P_p(t) = P_{p_1}(t) + P_{p_2}(t) = -\frac{a}{k} + \frac{a}{b + k}e^{-bt}.$$

Before we satisfy our initial condition, we need to obtain the general solution of our original differential equation as the sum of this particular solution and the general solution of the associated homogenous equation. Thus, we have

$$P(t) = Ce^{kt} - \frac{a}{k} + \frac{a}{b + k}e^{-bt}.$$

Our initial condition $P(0) = P_0$ is satisfied if we choose $C = P_0 + a/k - a/(b + k)$. For $k = a = 1$, the top curve in Figure 6.13 shows the effect of immigration with $b = 1$. The bottom curve corresponds to $b = 0$.

Often the steady state solution of a first order equation is the particular solution and the transient solution is the general solution of the associated homogeneous equation. The preceding example shows that this is not always the case.

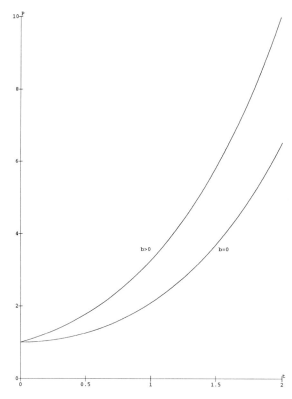

FIGURE 6.13   Graph of $Ce^{kt} - a/k + [a/(b+k)]e^{-bt}$

We should note that our analysis for finding a particular solution in the preceding example fails if $b + k = 0$ — that is, if $b = -k$. To decide what to do in this case, we consider a differential equation of the form

$$\frac{dP}{dt} - kP = 5e^{kt}. \tag{6.41}$$

In accordance with our previous examples, we should use a trial solution of the form Ae^{kt}. Doing so yields

$$kAe^{kt} - kAe^{kt} = 5e^{kt},$$

or

$$0 = 5e^{kt}.$$

There is no choice of A that will allow this equation to be satisfied! This happened because the trial particular solution is also a solution of the associated homogeneous equation, so Ae^{kt} will never be a particular solution of (6.41) for any choice of A.

We then ask ourselves if there are other functions which, when differentiated, yield an exponential function. Because the derivative of products of exponential functions times other functions can yield exponential functions (using the product rule), we try an unknown function times an exponential. Doing so gives $P_p(t) = z(t)e^{kt}$, and substituting this expression into the differential equation gives

$$z'(t)e^{kt} + kz(t)e^{kt} - kz(t)e^{kt} = 5e^{kt}.$$

This requires that $z'(t) = 5$, so the particular solution is $5te^{kt}$. This suggests a trial solution of the form Ate^{kt}, that is, t times the initial trial solution.

As another example we note that an appropriate form to try for a particular solution of

$$\frac{dP}{dt} - kP = 7$$

is simply a constant, A, so straightforward substitution gives $A = -7/k$. However, this is not valid if $k = 0$. Now in this case our differential equation is $dP/dt = 7$, and integration gives $P = 7t + C$. Thus, the particular solution is $7t$, and the appropriate trial solution is again t times the usual form of the trial solution.   □

We now collect our discoveries so far in this section.

- One discovery is that the general solution of a first order linear differential equation may be written as the sum of a particular solution and the general solution of the associated homogeneous differential equation.

- A second discovery is that we know how to choose the appropriate form for a particular solution of a linear differential equation with constant coefficients if the forcing function is in the form of a polynomial, exponential, or sinusoidal function. These forms are shown in Table 6.2. In this table the constants $a_0, a_1, a_2, \cdots, a_n, a, b, c,$ and ω occur in the forcing function for the differential equation and the constants that appear in the trial solution are to be determined so the differential equation is satisfied.

- If any part of the trial solution also satisfies the associated homogenous differential equation, then use t times the usual function on the right-hand side of Table 6.2. In the exercises more forcing functions are considered, and we can augment Table 6.2 with Table 6.3, which contains products of the three types of functions occurring in the left-hand side of Table 6.2.

- If any part of the suggested trial solution in Table 6.3 satisfies the associated homogeneous differential equation, then use t times the usual function indicated on the right-hand side of Table 6.3. Because of the principle of linear superposition, we multiply only the trial solution pertaining to this particular solution.

TABLE 6.2 Appropriate particular solutions of $a_1 dP/dt + a_0 P = f(t)$

$f(t)$	Trial Solution
$a_0 + a_1 t + a_2 t^2 + \cdots + a_n t^n$	$A_0 + A_1 t + A_2 t^2 + \cdots + A_n t^n$
ae^{ct}	Ae^{ct}
$a \sin wt + b \cos wt$	$A \sin wt + B \cos wt$

TABLE 6.3 More particular solutions of $a_1 dP/dt + a_0 P = f(t)$

$f(t)$	Trial Solution
$e^{ct}(a \sin wt + b \cos wt)$	$e^{ct}(A \sin wt + B \cos wt)$
$e^{ct}\sum_{k=0}^{k=n} a_k t^k$	$e^{ct}\sum_{k=0}^{k=n} A_k t^k$
$\sum_{k=0}^{k=n} a_k t^k (a \sin wt + b \cos wt)$	$\sum_{k=0}^{k=n} A_k t^k \sin wt + \sum_{k=0}^{k=n} B_k t^k \cos wt$
$e^{ct}\sum_{k=0}^{k=n} a_k t^k (a \sin wt + b \cos wt)$	$e^{ct}(\sum_{k=0}^{k=n} A_k t^k \sin wt + \sum_{k=0}^{k=n} B_k t^k \cos wt)$

Comments about solving linear differential equations with constant coefficients

- This method for solving linear differential equations with constant coefficients is called the METHOD OF UNDETERMINED COEFFICIENTS .

- The reason we know that we have found the solution using this method, even though we are in effect guessing particular solutions, is that the uniqueness theorem guarantees that no matter how we find the solution, it is *the* solution. So judicious guessing is a valid technique for finding solutions for this situation.

- This technique replaces integration with algebraic manipulation.

We illustrate the process of finding the general solution for a first order linear differential equation with constant coefficients using these two tables with two simple examples.

EXAMPLE 6.10

Because we wish to illustrate use of the previous tables by using a three-step method, we consider only the task of finding the general solution of the initial value problem,

$$\frac{dy}{dt} + 20y = 25 \sin 2t, \quad y(0) = 0,$$

and do not consider any graphical analysis.

1. Find the general solution of the associated homogeneous differential equation

$$\frac{dy}{dt} + 20y = 0$$

as Ce^{-20t}.

2. Give the proper form for a particular solution corresponding to $25 \sin 2t$ using Tables 6.2 and 6.3, which in this case is $A \sin 2t + B \cos 2t$, and substitute this trial form into the original differential equation to evaluate the unknown constants. This operation yields

$$(2A + 20B) \cos 2t + (-2B + 20A) \sin 2t = 0 \cos 2t + 25 \sin 2t,$$

so A and B must satisfy the algebraic equations

$$2A + 20B = 0 \text{ and } -2B + 20A = 25.$$

Solve these equations to find $B = -25/202$ and $A = 625/202$, so the particular solution is

$$(625/202) \sin 2t - (25/202) \cos 2t.$$

3. The general solution of our original differential equation is the sum of the two solutions from steps 1 and 2, namely,

$$y(t) = Ce^{-20t} + (625/202) \sin 2t - (25/202) \cos 2t.$$

Choosing the arbitrary constant C as $25/202$ lets us satisfy the initial condition. Notice from the form of this general solution that the first term decays to zero while the other two terms oscillate forever. Thus, the first term is a transient solution and the other two terms comprise the steady state solution. Notice that although this steady state solution is not steady, its character does not change with time.   □

EXAMPLE 6.11

Here we consider the initial value problem

$$\frac{dy}{dt} + 20y = 40t + 25te^{-20t}, \quad y(0) = 0,$$

and we find the general solution by the method of our previous example.

1. We already have the general solution of the associated homogeneous differential equation as Ce^{-20t}, so we continue to step 2.

2. Since the right-hand side of the differential equation is composed of two parts — namely, $40t$ and $25te^{-20t}$ — we treat them separately and use the principle of linear superposition.

   (a) Table 6.3 gives the proper form for a particular solution corresponding to $40t$ as $(A_0 + A_1 t)$. We substitute this trial solution into

   $$\frac{dy}{dt} + 20y = 40t$$

   and obtain

   $$A_1 + 20(A_0 + A_1 t) = 40t.$$

   Thus, we need $A_1 = 2$ and $A_0 = -1/10$, and the particular solution for this part of the forcing function is $-1/10 + 2t$.

   (b) Table 6.3 gives the proper form for a particular solution corresponding to $25te^{-20t}$ as $(A_0 + A_1 t)e^{-20t}$, but because $A_0 e^{-20t}$ is a solution of the associated homogeneous differential equation, we multiply[3] this expression by t and substitute the result, $e^{-20t}(A_0 t + A_1 t^2)$, into

   $$\frac{dy}{dt} + 20y = 25te^{-20t}$$

   to obtain

   $$-20(A_0 t + A_1 t^2)e^{-20t} + (A_0 + 2A_1 t)e^{-20t} + 20(A_0 t + A_1 t^2)e^{-20t} = 25te^{-20t}.$$

   Combining terms gives $A_0 + 2A_1 t = 25t$, so $A_0 = 0$ and $A_1 = 25/2$, and our particular solution is $(25/2) t^2 e^{-20t}$.

3. The general solution of our original differential equation will be the sum of the solution from step 1 plus the particular solutions from step 2, giving

$$y(t) = Ce^{-20t} - \frac{1}{10} + 2t + \frac{25}{2}t^2 e^{-20t}.$$

Choosing the arbitrary constant C as $1/10$ lets us satisfy the initial condition. Notice here that the entire solution approaches the line $-1/10 + 2t$ with increasing t. Thus, $-1/10 + 2t$ is the steady state solution whereas the other two terms are the transient solution. $\square$

---

[3] See the discussion at the end of Example 6.9.

Comments about trial solutions

- If the forcing function consists of different functions added together, repeat step 2 for each different part of $f(t)$ before proceeding to step 3.

- Note that t multiplied the trial solution associated with $25te^{-20t}$ and not the one associated with $40t$. Finding a solution for each type of forcing function separately and using the principle of linear superposition is a good idea.

- If the forcing function is not listed in Tables 6.2 and 6.3, this method fails. For example, consider the linear differential equation with constant coefficients

$$\frac{dy}{dt} + y = \frac{1}{2\sqrt{t}} e^{\sqrt{t}-t},$$

which has an integrating factor of e^t and a general solution of

$$y(t) = e^{\sqrt{t}-t} + Ce^{-t}.$$

There are no simple extensions of the trial solutions in Tables 6.2 and 6.3 to develop an appropriate trial solution for this example.

- The second type of linear equation for which this method cannot be generally applied is an equation that has nonconstant coefficients. For example, consider

$$x^3 \frac{dy}{dx} + 3x^2 y = \cos x,$$

with the solution

$$y(x) = \frac{\sin x}{x^3} + \frac{C}{x^3}.$$

Again, there is no obvious way of selecting a form for the particular solution as we could for the forcing functions in the tables.

EXERCISES

1. Redo Exercises 1 through 7 on page 284 by using Tables 6.2 and 6.3. Compare the answers you obtain to those you found when solving Exercises 1 through 7.

2. Find the general solution of the following differential equations by using Tables 6.2 and 6.3.

   (a) $\dfrac{dy}{dt} + 4y = 6e^t$

   (b) $\dfrac{dy}{dt} + 4y = 6t$

   (c) $\dfrac{dy}{dt} + 4y = 3\cos 4t$

   (d) $\dfrac{dy}{dt} + 4y = 2e^{-4t}$

   (e) $\dfrac{dy}{dt} + 4y = 6te^t$

   (f) $\dfrac{dy}{dt} + 4y = 6t^2 - 2$

   (g) $\dfrac{dy}{dt} + 4y = 3\cos t - 2\sin t$

(h) $\dfrac{dy}{dt} + 4y = 2te^{-4t}$

3. Write the proper form for the particular solution of the following differential equations.

(a) $\dfrac{dy}{dt} - 2y = t^3$

(b) $\dfrac{dy}{dt} - 2y = t\cos 3t + 2$

(c) $\dfrac{dy}{dt} - 2y = e^{2t} + e^{3t}$

(d) $\dfrac{dy}{dt} - 2y = te^{2t} + 8\sin t$

(e) $\dfrac{dy}{dt} - 2y = 4t\sin 3t$

(f) $\dfrac{dy}{dt} - 2y = e^t\cos 3t + 7t$

(g) $\dfrac{dy}{dt} - 2y = te^{2t}\sin t$

(h) $\dfrac{dy}{dt} - 2y = t^3 e^{-2t} + \sin t$

4. Use the techniques of Section 6.1 to find the general solution of the differential equations in Exercise 2.

5. After developing a special technique to solve some class of differential equations, we have sometimes given a *How to Solve* $\cdots$ summary with headings *Purpose, Technique,* and *Comments.* Create a *How to Solve Linear Differential Equations with Constant Coefficients* by adding statements under Purpose, Technique, and Comments that summarize what is discussed in this section.

6. The first example in this section considered the differential equation

$$\frac{dP}{dt} + P = \frac{1}{2}f$$

as a model of human cells affected by a virus, where $P(t)$ gives the number of human cells infected and $f(t)$ gives the total number of cells at time t. Consider now the situation in which the total number of cells is governed by the oscillatory function

$$f(t) = a + 2b\sin t,$$

with a and b positive constants, and $a > b$. Write down the proper form for the particular solution for this differential equation, and show that the general solution has the form of (6.31).

7. We want to show that the general solution of

$$a_1\frac{dP}{dt} + a_0 P = f(t), \tag{6.42}$$

where a_0 and a_1 ($\neq 0$) are constants, may be expressed as

$$P(t) = Ce^{-(a_0/a_1)t} + P_p(t). \tag{6.43}$$

Here C is an arbitrary constant and $P_p(t)$ is any function that satisfies (6.42).

(a) Show that (6.43) satisfies (6.42).

(b) Show that if $f(t)$ is continuous for some interval containing t_0, then (6.42) satisfies the conditions of Theorem 6.1, so we can conclude that it has a unique solution of (6.42) that satisfies the initial condition $P(t_0) = P_0$.

(c) Show that letting $C = \left[P_0 - P_p(t_0) \right] e^{(a_0/a_1)t_0}$ allows the solution in (6.43) to satisfy the initial condition and that since t_0 and P_0 are chosen arbitrarily, all solutions of (6.42) may be written in the form of (6.43).

6.4   More Applications

In this section we consider more examples of first order differential equations to illustrate the wide variety of applications.

EXAMPLE 6.12: Air Quality

A conference room with volume 2000 cubic meters contains air with 0.002% carbon monoxide. At time $t = 0$, the ventilation system starts blowing in air containing $2 + \sin(t/5)$ cubic meters of carbon monoxide per cubic meter of air. If the ventilation system inputs and extracts air at a rate of 0.2 cubic meters per minute, how long before the air in the room contains 0.015% carbon monoxide?

If x represents the volume of carbon monoxide in cubic meters at time t, using a conservation equation gives

$$\frac{dx}{dt} = \text{inflow} - \text{outflow} = \left(0.02 + 0.01 \sin \frac{t}{5} \right)(0.2) - \left(\frac{x}{2000} \right)(0.2),$$

which becomes

$$\frac{dx}{dt} = 0.004 + 0.002 \sin \frac{t}{5} - 0.0001x. \tag{6.44}$$

The proper initial condition is

$$x(0) = (0.00002)(2000) = 0.04.$$

To solve (6.44) we first rewrite it as

$$\frac{dx}{dt} + kx = a + b \sin \frac{t}{5} \tag{6.45}$$

where $k = 0.0001, a = 0.004$, and $b = 0.002$, so the arithmetic will not be so messy.

This linear differential equation has constant coefficients, so we use the method of undetermined coefficients. The solution of the associated homogeneous equation

$$\frac{dx}{dt} + kx = 0$$

is Ce^{-kt}. The particular solution to try for a forcing function of $a + b \sin(t/5)$ is $A + B \sin(t/5) + D \cos(t/5)$, which when substituted in (6.45) yields

$$\frac{1}{5} B \cos \frac{t}{5} - \frac{1}{5} D \sin \frac{t}{5} + k \left(A + B \sin \frac{t}{5} + D \cos \frac{t}{5} \right) = a + b \sin \frac{t}{5}.$$

This implies that

$$kA = a, \qquad \frac{1}{5}B + kD = 0, \qquad -\frac{1}{5}D + kB = b,$$

which gives

$$A = \frac{a}{k}, \qquad B = \frac{bk}{k^2 + (1/5)^2}, \qquad D = -\frac{1}{5}\frac{b}{k^2 + (1/5)^2}.$$

Thus the particular solution is

$$\frac{a}{k} + \frac{b}{k^2 + (1/5)^2}\left(k \sin\frac{t}{5} - \frac{1}{5}\cos\frac{t}{5}\right),$$

and so our general solution is

$$x(t) = \frac{a}{k} + \frac{b}{k^2 + (1/5)^2}\left(k \sin\frac{t}{5} - \frac{1}{5}\cos\frac{t}{5}\right) + Ce^{-kt}.$$

For the specific parameters above, this solution becomes

$$x(t) = 40 + 0.000005\cos\frac{t}{5} - 0.01\sin\frac{t}{5} + Ce^{-0.0001t}, \qquad \textbf{(6.46)}$$

where $C = 0.04 - 40 - 0.000005$.

To answer the above question about when the conference room will contain 0.015% carbon monoxide, we need to find the value of t for which $x(t) = (0.00015)(2000) = 0.3$. Figure 6.14, shows the graph of (6.46) along with that of the line $x = 0.3$. Zooming gives the time for this amount of carbon monoxide as $t \approx 67.3$. (Since the coefficient of the cosine term in (6.46) is so small in comparison with the other terms, you could set it to zero and find this value of t using your calculator.) $\qquad \square$

EXAMPLE 6.13: An Electrical RL Circuit

Consider a simple electrical circuit consisting of wires connecting a resistor, inductor, switch, and voltage generator, as shown in Figure 6.15.[4] If the switch is closed, the resulting current — the rate of flow of charged particles — will be described by the differential equation

$$L\frac{dI}{dt} + RI = E(t), \qquad \textbf{(6.47)}$$

where L is the inductance (units of henries), R is the resistance (units of ohms), I is the current (units of amperes), and $E(t)$ is the output of the voltage generator (units of volts). Such a circuit is called an RL CIRCUIT. Consider a specific circuit where the inductance is 1 henry, the resistance is 20 ohms and the voltage generator has the form of $25e^{-20t}$, which eventually decays to zero. We consider $t = 0$ as the time when the switch is closed. A natural initial condition for this situation is that there be zero current when the switch is closed, so $I(0) = 0$. We wish to find the behavior of the current in this circuit as a function of time.

---

[4]On a humorous note, this figure was constructed — as were all figures in this book — by using mathematical equations to describe each of the components. What equation would you write down to draw the spiral that represents the inductance?

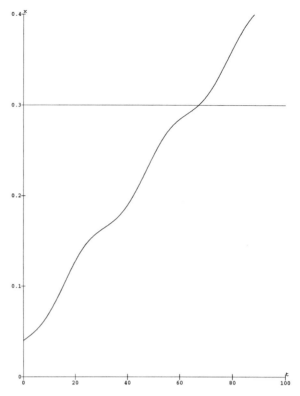

FIGURE 6.14   The functions $x = 40 + 0.000005 \cos(t/5) - 0.01 \sin(t/5) + Ce^{-0.0001t}$ and $x = 0.3$

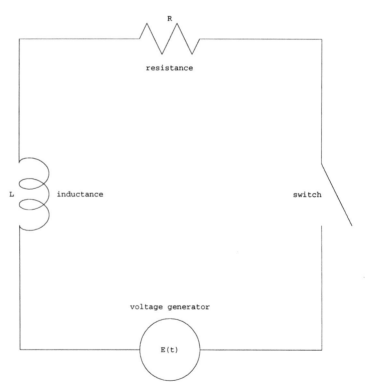

FIGURE 6.15   RL electrical circuit

We find an explicit solution by using a three step-method on the resulting initial value problem,

$$\frac{dI}{dt} + 20I = 25e^{-20t}, \quad I(0) = 0. \tag{6.48}$$

1. Find the general solution of the associated homogeneous differential equation,

$$\frac{dI}{dt} + 20I = 0$$

as

$$I_h(t) = Ce^{-20t}.$$

2. Give the proper form for a particular solution corresponding to $25e^{-20t}$ using Tables 6.2 and 6.3, which in this case is Ate^{-20t}. (Why isn't it Ae^{-20t}?) Substitute this trial form into the original differential equation to evaluate the unknown constant A. This substitution yields

$$A = 25,$$

so a particular solution is

$$I_p(t) = 25te^{-20t}.$$

3. The general solution of our original differential equation will be the sum of the solution from step 1 and the particular solution from step 2, giving

$$I(t) = Ce^{-20t} + 25te^{-20t}.$$

Choosing the arbitrary constant C as 0 lets us satisfy the initial condition of $I(0) = 0$, giving

$$I(t) = 25te^{-20t}. \tag{6.49}$$

This function is plotted in Figure 6.16. Notice that the current increases to a maximum and then decays. We could ask when the current will be zero again. The answer is never. We can show this from (6.49) where we see that the only time when $I(t) = 0$ is $t = 0$. From (6.49) we see that $I \to 0$ as $t \to \infty$. Thus the current never reaches zero for a finite value of the time. However, from a practical point view, there will be a time after which we are unable to detect the current.

A useful quantity to define is the SETTLING TIME, which is the time after which a response is no larger than 1% of its maximum value. What is the settling time for the current in our example? For this we first need to find when the maximum current, I_{max}, occurs. This will be the time, T, when $I' = 0$. From (6.48) we see that this will occur when

$$20I(T) = 25e^{-20T},$$

which, by (6.49), can be written

$$20\left(25Te^{-20T}\right) = 25e^{-20T},$$

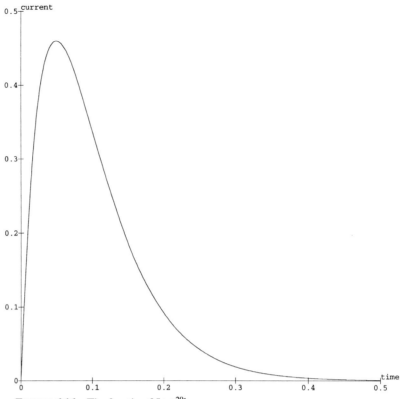

FIGURE 6.16   The function $25te^{-20t}$

so that

$$T = \frac{1}{20}.$$

Thus the maximum current is

$$I_{\max} = I(T) = \frac{25}{20e} \approx 0.46.$$

Both of these values are consistent with Figure 6.16. The settling time, t_s, will be the time after which the current is no larger than $I_{\max}/100$. Thus the settling time satisfies

$$25t_s e^{-20t_s} = \frac{25}{20e}\frac{1}{100},$$

or,

$$t_s e^{-20t_s} = \frac{1}{2000e}.$$

This equation cannot be solved using analytical techniques, but can be solved by any standard numerical root-finding method. We find two approximate solutions, namely, 0.00018 and 0.389192. The first occurs before the time when $I_{\max}$ occurs. The second is the settling time, so

$$t_s \approx 0.389192 \text{ secs.}$$

Figure 6.17 shows the current and the horizontal line $I = I_{\max}/100 \approx 0.0046$. $\square$

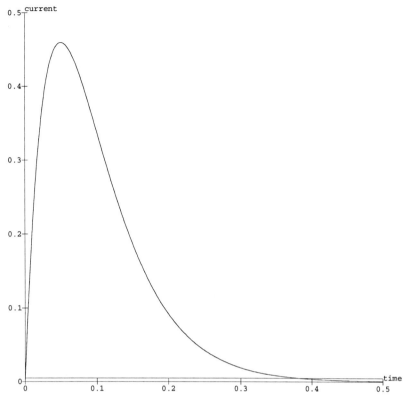

FIGURE 6.17 The function $25te^{-20t}$ and the line $I_{max}/100$

EXAMPLE 6.14: Yam in the Oven

Consider the situation in which a cook places a yam in an oven (at room temperature, 70^oF) and simultaneously turns the oven on to 400^oF. If it takes the oven 5 minutes to reach 400^oF and if it does so in a linear manner, the temperature in the oven, $T_a(t)$, is described by

$$T_a(t) = 70 + \frac{400 - 70}{5}t$$

for the first 5 minutes, after which the temperature remains at a constant 400^oF. Thus,

$$T_a(t) = \begin{cases} 70 + 66t & \text{if } 0 \le t \le 5, \\ 400 & \text{if } t > 5. \end{cases} \tag{6.50}$$

Figure 6.18 shows the function the function $T_a(t)$. It is continuous at $t = 5$. (Why?)

If the temperature of the yam is given by $T(t)$ and it obeys Newton's law of heating with $T_a(t)$ the ambient temperature, we have

$$\frac{dT}{dt} = k\left[T - T_a(t)\right], \; k < 0. \tag{6.51}$$

If the yam is initially at room temperature, the appropriate initial condition is

$$T(0) = 70.$$

The slope field associated with (6.51) for $k = -0.04$, is shown in Figure 6.19, and it appears that regardless of the initial temperature of the yam, its temperature approaches a limiting value. Also, there seems to be an equilibrium solution somewhere about 400^oF. It

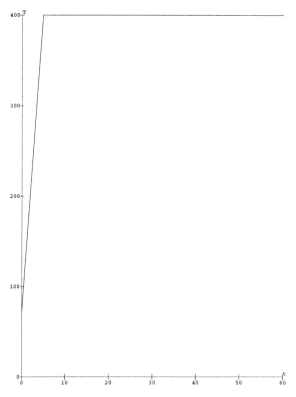

FIGURE 6.18   The function $T_a(t)$

appears that for $t > 5$ the solution curve is increasing, but we must look at the numerical solution curve (given in Figure 6.20 for the initial condition of 70^oF) to discover that this solution curve starts out concave up and then changes to concave down for larger values of t. From equation (6.51) it is clear that if we consider only values of t greater than 5, then $T = 400$ is an equilibrium solution. From (6.51), we also see that the solution is increasing for all t. But to analyze concavity in more detail, we need to differentiate (6.51) to obtain

$$\frac{d^2T}{dt^2} = k\left(\frac{dT}{dt} - \frac{dT_a(t)}{dt}\right),$$

which, by (6.50), reduces to

$$\frac{d^2T}{dt^2} = \begin{cases} k^2\left[T(t) - 70 - 66t - 66/k\right] & \text{if } 0 \le t < 5, \\ k^2\left[T(t) - 400\right] & \text{if } t > 5. \end{cases} \tag{6.52}$$

Notice that $T_a(t)$ is not differentiable at $t = 5$, because it has a "corner" there — see Figure 6.18. Thus, d^2T/dt^2 does not exist at $t = 5$, which is why the value $t = 5$ is absent from (6.52).

Equation (6.52) shows that although initially all solutions are concave up [$k < 0$ and when $t = 0$, $T(0) = 70$], they are concave down for $t > 5$ and $T(t) < 400$. From this formulation, an inflection point can occur only if $T(t) - T_a(t) = 66/k$ at some point during the time period interval $0 < t < 5$ or if d^2T/dt^2 does not exist (namely, $t = 5$), but we cannot tell from (6.52) which will happen. However, we can discover this from the explicit solution.

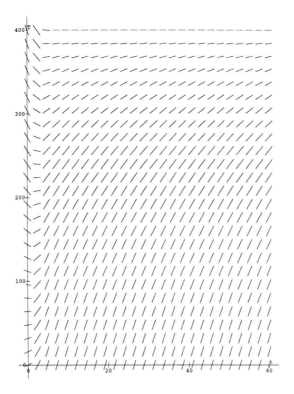

FIGURE 6.19   Slope field for yam heating up

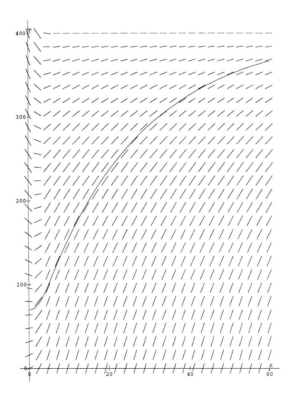

FIGURE 6.20   Numerical solution curve and slope field for yam heating up

Equation (6.51) is a linear differential equation, so we find the explicit solution by writing (6.51) as

$$\frac{dT}{dt} - kT = -kT_a(t) \qquad\qquad (6.53)$$

and computing the integrating factor as

$$e^{\int -k\,dt} = e^{-kt}.$$

We then multiply (6.53) by this integrating factor to obtain

$$e^{-kt}\frac{dT}{dt} - ke^{-kt}T = -kT_a(t)e^{-kt},$$

or, equivalently,

$$\frac{d}{dt}\left(e^{-kt}T\right) = -kT_a(t)e^{-kt}, \qquad\qquad (6.54)$$

and integrate (6.54) to obtain

$$e^{-kt}T = -k\int T_a(t)e^{-kt}\,dt + C. \qquad\qquad (6.55)$$

Writing the indefinite integral as one from 0 to t, we can choose the arbitrary constant to satisfy the initial condition $T(0) = 70$, divide by the integrating factor, and write the explicit solution as

$$T(t) = -ke^{kt}\int_0^t T_a(u)e^{-ku}\,du + 70e^{kt}. \qquad\qquad (6.56)$$

Equation (6.50) lets us expand (6.56) as

$$T(t) = \begin{cases} 70 + 66t + 66\left(1 - e^{kt}\right)/k & \text{if } 0 \le t \le 5, \\ 400 + \left[66(1 - e^{5k})/k\right]e^{k(t-5)} & \text{if } t > 5. \end{cases} \qquad\qquad (6.57)$$

From (6.57) we may confirm our earlier finding (as well as our common sense) that the temperature of the yam approaches 400^oF for large values of time. The graph of this function coincides with the numerical solution given in Figure 6.20.

Let's return to the location of the inflection point. If we substitute (6.57) into (6.52) for $t < 5$, we find

$$\frac{d^2T}{dt^2} = -66ke^{kt},$$

which is always positive. Thus, the inflection point occurs at exactly $t = 5$. $\qquad\qquad \square$

We note that the solution of (6.53) for any ambient temperature, namely,

$$\frac{dT}{dt} - kT = -kT_a(t),$$

subject to the initial condition

$$T(t_0) = T_0,$$

follows immediately from (6.55). It is

$$T(t) = -ke^{kt} \int_{t_0}^{t} T_a(u)e^{-ku} \, du + T_0 e^{kt}.$$

EXERCISES

1. A conference room with volume 2000 cubic meters contains air with 0.002% carbon monoxide. At time $t = 0$, the ventilation system starts blowing in air that contains carbon monoxide amounting to $2 + \sin(t/5)$ (percent by volume). If the ventilation system inputs (and extracts) air at a rate of 0.2 cubic meters per minute, how long before the air in the room contains 0.015% carbon monoxide?

2. A conference room contains 3000 cubic meters of air that is free of carbon monoxide. The ventilation system blows in air, free of carbon monoxide, at a rate of 0.3 cubic meters per minute and extracts air at the same rate. If at time $t = 0$, people in the room start smoking and add carbon monoxide to the room at a rate of 0.02 cubic meters per minute, how long before the air in the room contains 0.015% carbon monoxide?

3. If the simple RL circuit described by (6.47) has $L = 1$ henry and $R = 60$ ohms and the voltage source is a 12 volt battery, find the explicit solution that describes what happens following the closure of the switch with $I(0) = 0$. What is the steady state solution in this case. Are there any equilibrium solutions to this problem? Please explain.

4. Let the circuit of Exercise 3 have a voltage source equal to $6 \sin 2t$. If $I(0) = 0$, find the explicit solution to this initial value problem. What is the steady state solution in this exercise? Are there any equilibrium solutions to this problem? Please explain.

5. Let the circuit of Exercise 3 have a voltage source equal to $e^{-60t} + \sin t$. If $I(0) = 0$, find the explicit solution to this initial value problem. What is the steady state solution in this exercise? Are there any equilibrium solutions to this problem? Please explain.

6. Table 6.4 records the voltage as a function of time for a charging capacitor, and Figure 6.21 is a graph of this voltage as a function of time.

   (a) How would you describe what happens to the voltage as time increases?

TABLE 6.4   Voltage versus time

Time	Voltage	$\Delta_C V/\Delta t$
0	0.00	
2	1.95	0.781
4	3.25	0.550
6	4.15	0.385
8	4.79	
10	5.19	
12	5.45	
14	5.62	
16	5.75	
18	5.83	
20	5.89	
22	5.92	
24	5.95	
26	5.97	

(b) The third column of Table 6.4 shows the change in voltage as a function of time. The changes between the first and second, the second and third, and the third and fourth time periods have already been entered. Complete the rest of this column, and give your observation about the change of voltage as time increases.

(c) The rate of increase of the voltage with time appears to decrease as the voltage approaches an apparent upper limit of 6, suggesting that the rate of change of voltage with time may be proportional to $(6 - \text{voltage})$. To test this hypothesis, plot $\Delta_C V / \Delta t$ of the third column in the table on the vertical axis and $(6 - V)$ on the horizontal axis (subtract the second column of the table from 6). Describe the behavior of the resulting graph.

(d) Assuming that your answer to part (c) was that the graph was close to a straight line, estimate the slope m of this line. Because this slope gives the proportionality constant between the rate of increase of voltage and $(6 - \text{the voltage})$, we can conclude that the differential equation that models this process is

$$\frac{dV}{dt} = m(6 - V). \tag{6.58}$$

(e) Figure 6.22 shows the slope field for (6.58) with a positive value for m. Can you predict the long-term behavior of the voltage? How does this depend on the initial voltage (that is, when $t = 0$)?

(f) Is there an equilibrium solution? If so, what is it? If not, why not?

(g) Solve (6.58) subject to the initial condition $V(0) = 0$, and graph your result. Compare this graph with Figure 6.21, describing any similarities and differences.

7. An RC circuit consists of a resistor, capacitor, switch, and voltage generator, as shown in Figure 6.23. The differential equation that models this circuit is

$$R\frac{dq}{dt} + \frac{1}{C}q = E(t), \tag{6.59}$$

where q is the charge on the capacitor. (The units of C are farads, called its capacitance, and the units of charge are coulombs.) Find the explicit solution of (6.59) when the resistance is 10 ohms, the capacitance is $1/100$ farads, the initial charge on the capacitor is 5 coulombs, and the voltage generator is a 12 volt battery. What is the steady state solution in this exercise? Are there any equilibrium solutions to this problem? Please explain.

8. Repeat Exercise 7 if the voltage generator now has the form $E(t) = 12 \sin 4t$.

9. Solve the equation

$$\frac{dy}{dt} + y = f(t),$$

where

(a)
$$f(t) = \begin{cases} 2, & 0 \le t < 1 \\ 1, & t \ge 1 \end{cases}, \quad y(0) = 0.$$

(b)
$$f(t) = \begin{cases} 2, & 0 \le t < 1 \\ 0, & t \ge 1 \end{cases}, \quad y(0) = 0.$$

(c)
$$f(t) = \begin{cases} 5, & 0 \le t < 10 \\ 1, & t \ge 10 \end{cases}, \quad y(0) = 6.$$

(d)
$$f(t) = \begin{cases} e^{-t}, & 0 \le t < 2 \\ e^{-2}, & t \ge 2 \end{cases}, \quad y(0) = y_0.$$

10. A pumpkin pie recipe says to place the ingredients in a preheated oven at 425^oF for 15 minutes, then turn the thermostat to 350^oF and continue baking for 45 minutes. Assume that the temperature

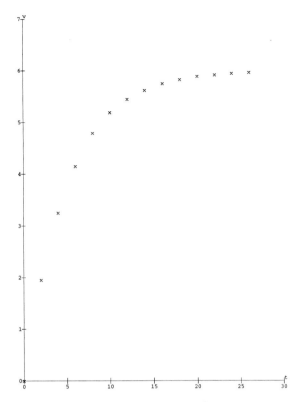

FIGURE 6.21 Graph of voltage versus time

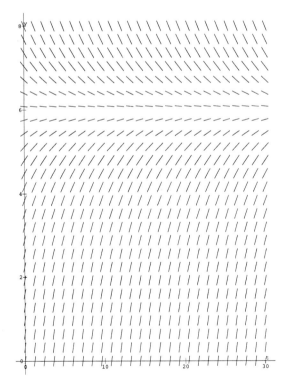

FIGURE 6.22 Direction field for $dV/dt = m(6 - V)$ for $m > 0$

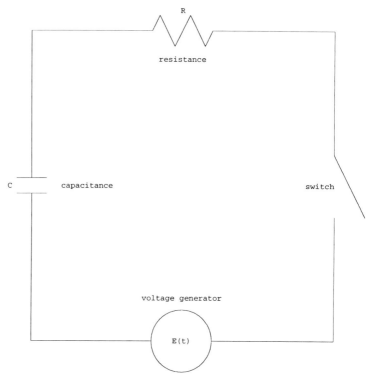

FIGURE 6.23   RC electrical circuit

of the oven changes instantaneously when we change the thermostat from 425^oF to 350^oF. Use Newton's law of heating, namely,

$$\frac{dT}{dt} = k \left[T - T_a(t) \right],$$

where $T_a(t)$ is

$$T_a(t) = \begin{cases} 425 & \text{if } 0 \leq t < 15, \\ 350 & \text{if } t > 15. \end{cases}$$

If the initial temperature of the uncooked pie is 70^oF, find the temperature T of the pie at time t. Compare the graph of this solution to the graph of a similar problem with $T_a(t) = 350$ for all time. From this comparison, explain the advantage of using a hotter oven for the first 15 minutes.

11. A differential equation that arises in the study of traffic flow is

$$\frac{dx}{dt} = \frac{1}{2}V + \frac{x}{2t},$$

where V is the maximum velocity of the car in traffic flow and x is the directed distance of a car from a traffic light (t is time). If the car starts from rest, then from the differential equation we have that $x = -x_0$ at $t = x_0/V$.

(a) Find the solution of this initial value problem, and show that V is in fact the maximum velocity for the car.

(b) Find the time it takes the car to reach the traffic light, and compare it with the time it would take if the car were going at its maximum speed the entire distance from its starting place to the traffic light. Does your answer surprise you?

12. Consider a glass thermometer filled with mercury. If T is the temperature inside the thermometer (that is, of the mercury) and T_a is the temperature on the exterior, by Newton's law of cooling we have that

$$\frac{dT}{dt} = k(T - T_a).$$

Properties of mercury and the glass are included in the constant k. Consider the situation in which time t is measured in minutes and the thermometer is characterized by the value $k = -0.2$.

(a) If this thermometer is taken from a room at $70°F$ to a porch outside where the temperature is $50°F$, how long before the thermometer reading will be within 10 percent of the correct temperature?

(b) If this thermometer is kept outside, where the diurnal temperature varies plus or minus $12°F$ around a mean of $50°F$, find the phase lag between the outside temperature and the temperature reading of the thermometer. (Hint: Model the outside temperature as a sinusoidal function, and look for the steady state solution of the problem.)

13. A simple model of the cardiovascular system represents arteries as a reservoir between the heart and the arterioles (smaller arteries). The output from the heart is the input to the reservoir, and the output from the reservoir is what flows into the arterioles. If we consider $P(t)$ as the pressure in the reservoir, R as the resistance to flow into the arterioles, and $f(t)$ the output from the heart, the appropriate differential equation is

$$C\frac{dP}{dt} + \frac{1}{R}P = f(t),$$

where C is the compliance of the reservoir. [The units of $f(t)$ are volume per unit time.]

(a) Consider the case in which the interest is in the steady state solution and the phase shift between the output from the heart and the input from the arterioles. If $f(t)$ is represented by $a + b\sin\omega t$, find the effect of C and R on this phase shift. (You can get a feeling of what is happening by observing the slope field and some numerical solutions before verifying your findings analytically.)

(b) Now consider the case in which the output from the heart is represented by the following periodic function with period $2\pi/\omega$

$$f(t) = \begin{cases} a\sin\omega t & 0 < t < \pi/\omega, \\ 0 & \pi/\omega < t < 2\pi/\omega. \end{cases}$$

Is there a steady state solution now? [The suggestion for part (a) also applies here.]

14. Consider an empty cylindrical container whose radius is 5 feet and height is 10 feet. At time $t = 0$, a tap is turned on, letting water enter the container at the constant rate of 3 cubic feet per minute. An exit valve is open, and it lets water leave the container at a rate proportional to the depth of water in the container (let the proportionality constant be k). Set up the appropriate differential equation to describe this process. If the exit valve remains open, how long will it take to fill the tank as a function of k? Can you determine values of k so the tank will never be filled? If so, what are they? If not, why not?

15. Repeat Exercise 14 with the water leaving the container at a rate equal to $k \times \sqrt{depth}$.

16. Repeat Exercise 14 with a conical container of height 10 feet and radius at the top of 5 feet. Can you determine values of k so the tank will never be filled? If so, what are they? If not, why not?

17. Repeat Exercise 14 after adding the effect of loss due to evaporation, which is proportional to the surface area (have the proportionality constant be m). Does this problem have an equilibrium solution?

18. Consider an open cylindrical container that is initially half full of pure water. At $t = 0$, two valves are turned on: One lets in a saline solution of 3 lb/gal at a rate of 4 gal/min, and the other

lets out the well-stirred mixture at a rate equal to $0.2 \times depth$. If the container initially contains 200 gallons, how long before the concentration of the solution is $1/2$ of the input concentration? Does this happen before the tank overflows or is emptied?

19. A simple dynamical model in neurobiology considers a neuron as having three regions: dendrites, a cell body, and an axon. Neurons are connected to other neurons by synapses. The nerve cell body functions as an input-output device, responding to electrical stimuli. One of the differential equations that occurs in this model is

$$C\frac{du}{dt} + \frac{1}{R}(u - u_0) = I(t),$$

where C, R, and u_0 are constants, u is the cell potential (that is, the voltage difference between the inside and outside of this simple neuron), and $I(t)$ is the electrical current injected into the cell. A recent article[5] describing such a model includes the following two statements about this model. (i) "For a typical neuron, $u_0 < u_{thresh}$, so u will decay to u_0 when the injected current vanishes." (ii) "If $I(t)$ is a large constant current I_c, the cell potential will change in an almost linear fashion between u_0 and u_{thresh}."

   (a) Explain these two statements by just analyzing the differential equation. (You may wish to use some specific values of the constants and look at a typical slope field.)

   (b) Find the explicit solution of the differential equation when $I(t)$ equals a constant, and explain these two statements using your solution.

   (c) Find an explicit formula for u_{thresh} in terms of the other constants.

20. One of the standard glucose tolerance tests infuses glucose continuously into the bloodstream at the known rate of G mg/min. Blood concentrations are measured at subsequent time intervals until steady state is reached. It is assumed that the glucose is used by the body according to the differential equation

$$\frac{dx}{dt} = -kx + G,$$

where x is the amount of glucose, t is time, G is the infusion rate of glucose, and k is the "turnover rate" (what is checked to see if the body is processing glucose normally).

   (a) Explain the relationship between this turnover rate and the equilibrium solution of the differential equation.

   (b) Data from a typical glucose tolerance test are given as Table 6.5. Note that the measurements stopped before equilibrium was reached. We will now find the turnover rate by two methods. First, use the data from the table to obtain difference quotients, and plot the numerical derivative versus the concentration. If you now find the best straight line through these points, you will note that its vertical intercept is G and its slope is the turnover rate. Second, show that the explicit solution of the differential equation has the form

$$x(t) = ae^{-kt} + (85 - a),$$

and find values of a and k that best fit the data. The value of k will be the turnover rate.

21. Suppose that we input a dye of concentration C at a flow rate of r into a compartment, with volume V, in the body that removes this substance at the known rate R. Then, the concentration of the substance x in this compartment is modelled by the differential equation

$$\frac{dx}{dt} = -kx + Cr - R,$$

where k is the "reaction rate". Use the equilibrium solution to derive an expression for the reaction rate in terms of the limiting solution.

---

[5]J. Hopfield "Neurons, Dynamics, and Computation", Physics Today **47** No. 2, February, 1994, pages 40–46.

TABLE 6.5 Data from glucose tolerance test

Time (minutes)	$x(t)$ (mg/dl)
0	85.0
10	105.4
20	120.1
30	127.3
40	131.9
50	134.4

What Have We Learned?

MAIN IDEAS

- The principle of linear superposition (see page 279) is sometimes very helpful in solving linear differential equations.

- "Steady state" and "Transient" solutions refer to the long-term and short-term behavior of solutions of differential equations (see page 284).

- Steady state solutions are often stable solutions.

- A special technique can be used for solving linear differential equations with constant coefficients (see Tables 6.2 and 6.3 on page 297) and Exercise 5 on page 301.

- The aforementioned technique [see also *How to Solve* $a_1(x)dy/dx + a_0(x)y = f(x)$ on page 267] can be used to solve initial value problems for linear differential equations occurring in a wide variety of applications.

CHAPTER 7

Second Order Differential Equations with Constant Coefficients

---

Where Are We Going?

In this chapter we extend the techniques we have developed for first order differential equations by establishing a relationship between systems of first order equations and second order linear differential equations. We show that for second order differential equations with constant coefficients, with the introduction of a phase plane, many of our graphical techniques carry over from the previous chapters. We describe how a trajectory in the phase plane can be constructed from a numerical, graphical, or analytical point of view. For the important case of constant coefficients, we develop the characteristic equation method to obtain the general solution of second order linear differential equations. We also state an existence theorem and use it to discuss the properties of solutions for initial value problems.

7.1 Examples of Second Order Equations with Constant Coefficients

In this section we consider three examples of relationships that lead to a set of coupled first order differential equations. We show how our previous graphical techniques apply to this situation and how explicit solutions may be obtained by combining them into a single second order differential equation. We will see this last technique again in Chapter 12.

EXAMPLE 7.1: Sea Battles

It is the year 1805. You are the commander of a fleet of warships about to do battle with an enemy. Your ships and those of your enemy are equal in quality. The only difference is in numbers: You have 37 ships, and your opponent, 12. You engage the enemy in battle. Who will win the battle? How many of the winner's ships will survive? How long will the battle last? The first question does not require much intuition. You should win the battle, because you have a larger force and the ships are evenly matched. The second question could be answered with the guess that 25 ships will survive, because $37 - 12 = 25$. The third question might receive the response, "Who knows?"

We will try to answer these questions by looking at a simple model, in which we assume that the rate of change of the number of ships is proportional to the number of enemy ships. If x is the number of ships in your fleet and y the number of ships in the enemy's fleet — so x and y are nonnegative — then the appropriate differential equations are

$$\frac{dx}{dt} = -ay, \qquad \frac{dy}{dt} = -bx, \tag{7.1}$$

where a and b measure the effectiveness of the two fleets and are both positive. [Why is there a negative sign on the right-hand sides of the equations in (7.1)?] If the fleets are equally matched in quality, then $a = b$, so (7.1) becomes

$$\frac{dx}{dt} = -ay, \qquad \frac{dy}{dt} = -ax. \tag{7.2}$$

This is a coupled set of first order differential equations, in which a solution consists of x and y given as functions of the time t. From (7.2) we observe that although it is apparent that x and y will be decreasing functions of time (as is to be expected in a battle without reinforcements), we currently have no means of solving (7.2). We also have one more variable than we usually have when we draw a slope field. However, because x and y are both functions of the time t, we could (in principle) solve for t in terms of one of the dependent variables — say x — and substitute the result into the expression for y. This would eliminate the variable t and give an equation relating x and y. For such situations the graph of y versus x is called a TRAJECTORY (or ORBIT), and the xy-plane is called the PHASE PLANE associated with (7.2). We proceed to analyze (7.2) in this light.

Because the right-hand sides of the two equations in (7.2) do not involve the variable t explicitly, we may use the chain rule to combine these two equations into one. If we treat y as a function of x, and x as a function of t, using the chain rule gives

$$\frac{dy}{dx} = \left(\frac{dy}{dt}\right)\left(\frac{dt}{dx}\right) = \frac{dy/dt}{dx/dt} = \frac{-ax}{-ay},$$

or

$$\frac{dy}{dx} = \frac{x}{y}. \tag{7.3}$$

Notice that because x and y are both nonnegative, the slope field for (7.3) (shown in Figure 7.1) has positive slopes. This means that both x and y must increase or decrease together in time, but the slope field does not tell us which of these happens. However, the original differential equations (7.2) show that x and y must both decrease with time at rate a. We could indicate this on the slope field by adding arrows to the slopes to indicate the direction the trajectories follow as time increases, creating a field of vectors. Such a slope field is known as a DIRECTION FIELD and is shown in Figure 7.2, where the arrows indicate the direction of travel. The length of each vector has been scaled so that those vectors with small values of $(dx/dt, dy/dt)$ are short, and those with large values are long.

We see that $y = x$ is a solution of (7.3). So if initially x and y have equal values, they will continue to do so, and as time increases this solution approaches the origin — namely, $x = y = 0$, which is when the two fleets are eliminated simultaneously.

The direction field also suggests that, for the case $a = b$, the side that starts with the greater number of ships will always have ships surviving when the other fleet is destroyed completely. That is, if the initial value of x is greater than the initial value of y, the initial point on the direction field picture will be placed below the line $y = x$, and because (by the uniqueness theorem) other solution curves cannot cross the solution curve $y = x$, the x-intercept will be positive. That means that x is positive when y is zero, so x wins the

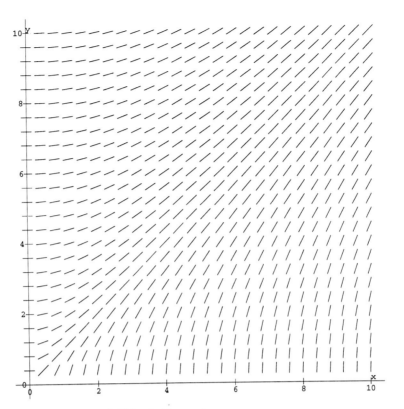

FIGURE 7.1 Slope field for $dy/dx = x/y$

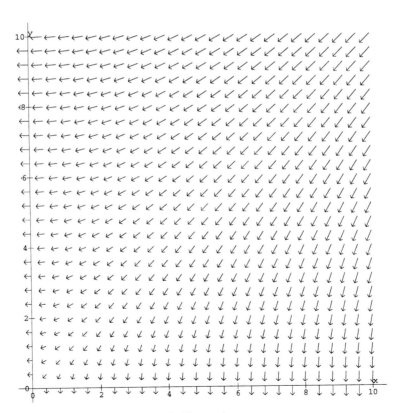

FIGURE 7.2 Direction field for $dy/dx = x/y$

battle. Figure 7.3 shows a few hand-drawn trajectories for this situation, including $y = x$. Each trajectory corresponds to a battle that started with different numbers of ships — that is, different initial conditions. Similarly, if the initial value of y is greater than the initial value of x, the initial point on the direction field picture will be placed above the line $y = x$, and the y-intercept will be positive. That means that y is positive when x is zero, so y wins the battle (see Figure 7.4).

The regions where solutions are concave up or concave down are evident in both of these figures — namely, above or below the line $y = x$. We confirm this by differentiating (7.3) to find

$$\frac{d^2y}{dx^2} = \left[\frac{y - x\,(dy/dx)}{y^2}\right] = \left[\frac{y - x\,(x/y)}{y^2}\right],$$

or

$$\frac{d^2y}{dx^2} = \frac{1}{y^3}(y - x)(y + x).$$

This result means that $y = x$ is the dividing line between trajectories that are concave up and those that are concave down. Thus, initial values of x and y that place the initial point below this dividing line will result in y going to zero before x, whereas if the initial point is above this line, the opposite will happen.

We can discover how many ships survive if we can find the trajectories in the phase plane. This we can do by integrating the separable equation (7.3). We find

$$x^2(t) - y^2(t) = C, \tag{7.4}$$

where C is the constant of integration. We can find C by evaluating (7.4) at $t = 0$, giving

$$x^2(0) - y^2(0) = C,$$

so (7.4) becomes

$$x^2(t) - y^2(t) = x^2(0) - y^2(0), \tag{7.5}$$

valid for all $t \geq 0$. Thus, if we start at $t = 0$ with x_0 and y_0, respectively, that is,

$$x(0) = x_0,$$
$$y(0) = y_0,$$

then (7.5) implies that

$$x^2(t) - y^2(t) = x_0^2 - y_0^2 \tag{7.6}$$

for all values of t, $t \geq 0$. Thus, for the case where $x_0 > y_0$, y will die out first. If we denote the value of x when $y = 0$ with x_r and call it the residual value of x, then from (7.6) we have that

$$x_r^2 = x_0^2 - y_0^2. \tag{7.7}$$

That is, we can predict who wins and that the number of survivors will be

$$x_r = \sqrt{x_0^2 - y_0^2}.$$

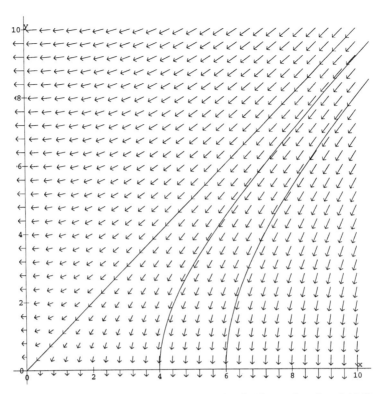

FIGURE 7.3 Trajectories and direction field for $dy/dx = x/y$, where initially $x \geq y$

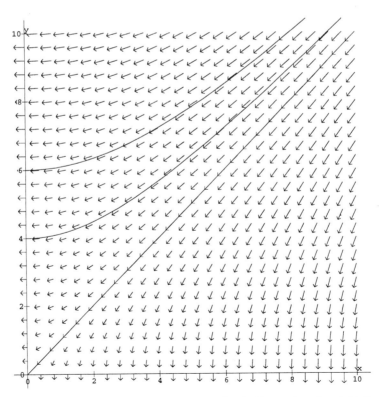

FIGURE 7.4 Trajectories and direction field for $dy/dx = x/y$, where initially $x \leq y$

In the case of 37 ships versus 12, the number of surviving ships is 35, a long way from our guess of 25.

Surprisingly, we have answered the first two questions without knowing the exact time dependence of each variable.

We now turn to the third question. When will the battle end? This requires an analysis in which the time is explicitly involved, so we proceed to determine the time behavior of x and y. If we solve the second equation in (7.2) for x we have

$$x = -\frac{1}{a}\frac{dy}{dt},$$

which, when differentiated, gives

$$\frac{dx}{dt} = -\frac{1}{a}\frac{d^2y}{dt^2}.$$

If we substitute this result into the first equation in (7.2), we obtain

$$\frac{d^2y}{dt^2} = a^2y,$$

or

$$\frac{d^2y}{dt^2} - a^2y = 0. \tag{7.8}$$

This is a SECOND ORDER LINEAR DIFFERENTIAL EQUATION. It is SECOND ORDER because it contains the second derivative as its highest derivative. It is LINEAR because it is a linear combination of d^2y/dt^2, dy/dt, and y with coefficients 1, 0, and $-a^2$. A precise definition will be given later in this section.

We solved this differential equation in Section 5.5 by reducing it to a first order differential equation, with an explicit solution given by

$$y(t) = C_1e^{at} + C_2e^{-at}, \tag{7.9}$$

where C_1 and C_2 are arbitrary constants. The fact that this is a solution may be seen by substituting (7.9) into (7.8).

To find the explicit solution for $x(t)$, we substitute (7.9) into the second equation in (7.2) to obtain

$$x(t) = -\frac{1}{a}\frac{dy}{dt} = -\frac{1}{a}\left(aC_1e^{at} - aC_2e^{-at}\right) = -C_1e^{at} + C_2e^{-at}. \tag{7.10}$$

Let us now use these explicit solutions to determine their time history, starting with the initial values $y(0) = y_0$ and $x(0) = x_0$. Setting $t = 0$ in (7.9) and (7.10) leads to the algebraic equations

$$y_0 = C_1 + C_2 \text{ and } x_0 = -C_1 + C_2,$$

with solution $C_1 = -(x_0 - y_0)/2$ and $C_2 = (x_0 + y_0)/2$. This gives our solution of the system of equations (7.2) as

$$\begin{aligned} x(t) &= \frac{1}{2}\left(x_0 - y_0\right)e^{at} + \frac{1}{2}\left(x_0 + y_0\right)e^{-at}, \\ y(t) &= -\frac{1}{2}\left(x_0 - y_0\right)e^{at} + \frac{1}{2}\left(x_0 + y_0\right)e^{-at}. \end{aligned} \tag{7.11}$$

From these expressions we see that if $x_0 > y_0$, then x will always be positive, and there will be a time $t = T$ when y vanishes, that is — $y(T) = 0$. At that time the value of x will be $x(T)$, and this represents the number of surviving ships. From (7.11) we have that $y(T) = 0$ when

$$e^{2aT} = \frac{x_0 + y_0}{x_0 - y_0}.$$

Thus, the battle ends at time

$$T = \frac{1}{2a} \ln \frac{x_0 + y_0}{x_0 - y_0}.$$

This time T depends not only on the initial values x_0 and y_0 but also on a, the efficiency of the ships. (The value of a could be determined experimentally. How? See Exercise 1 on page 334.)

Using this expression for T in (7.11) gives the number of surviving ships as

$$x_r = x(T) = e^{-aT} \left[\frac{1}{2} (x_0 - y_0) e^{2aT} + \frac{1}{2} (x_0 + y_0) \right]$$

or

$$x_r = x(T) = \sqrt{\frac{x_0 - y_0}{x_0 + y_0}} (x_0 + y_0) = \sqrt{x_0^2 - y_0^2}, \tag{7.12}$$

which is what we discovered earlier in (7.7). Figure 7.5 shows the solutions (7.11) for the case $x_0 = 37$ and $y_0 = 12$. From this figure we see that both x and y decrease steadily, and when the number y is zero the residual value of x (the number of survivors) is 35, which can be confirmed by using (7.12). □

During the First World War, these ideas were used to design a successful strategy for airplane dogfights. There is also evidence that in the Napoleonic Wars, the British naval commander used a strategy similar to this at the Battle of Trafalgar.

EXAMPLE 7.2: Denise and Chad's Relationship

A situation that sometimes leads to war (at least locally) is the romantic attraction of two individuals ill-suited to each other. Here we try to model this situation for Denise and her boyfriend Chad. We observed that Denise's affection for Chad increases when her affection is reciprocated. However, Chad's affection for Denise increases when his affection is not reciprocated. If x represents Denise's affection and y represents Chad's, then

$$\frac{dx}{dt} = ay, \qquad \frac{dy}{dt} = -bx, \tag{7.13}$$

where a and b are positive constants, is a model consistent with this information. Positive values of affection represent a liking for the other individual, and negative values express dislike. We want to analyze their stormy relationship.

We follow the previous example and use the chain rule to combine these two equations into one. Thus, if we treat y as an implicit function of x, and x as a function of t, we have

$$\frac{dy}{dx} = \left(\frac{dy}{dt} \right) \left(\frac{dt}{dx} \right) = \frac{dy/dt}{dx/dt} = -\frac{bx}{ay},$$

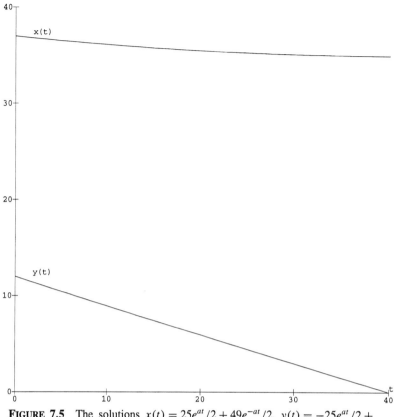

FIGURE 7.5 The solutions $x(t) = 25e^{at}/2 + 49e^{-at}/2$, $y(t) = -25e^{at}/2 + 49e^{-at}/2$

or

$$\frac{dy}{dx} = -\frac{bx}{ay}. \tag{7.14}$$

Notice that for this model a and b are both positive, whereas x and y can have either sign, giving positive slopes in the second and fourth quadrants and negative slopes in the first and third quadrants. [Regions of concavity may be determined by differentiating (7.14) and then using (7.14) in the result to eliminate y'.]

A typical slope field for (7.14), with $a = 1$ and $b = 2$, is shown in Figure 7.6. The slope field suggests that solutions of (7.14) will have some symmetry. Symmetry with respect to the x-axis follows from the fact that (7.14) is unchanged if y is replaced with $-y$. Symmetry with respect to the y-axis follows from the fact that (7.14) is unchanged if x is replaced with $-x$. This may also be verified by solving the separable differential equation in (7.14) as

$$ay^2 + bx^2 = C, \tag{7.15}$$

where C is a constant of integration. This gives the shape of the trajectories in the phase plane as ellipses.

Note that even though we do not know the explicit time dependence of x and y, we can determine the direction of travel as time increases by looking at (7.13). For example, in the first quadrant, where x and y are both positive, x will be an increasing function of t, and y will be a decreasing function. This means that time, as a parameter, is proceeding in a clockwise manner in the first quadrant. We analyze the other three quadrants in a similar

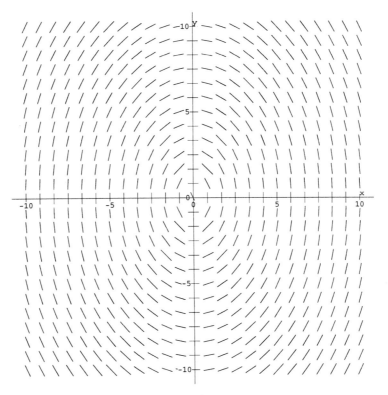

FIGURE 7.6   Slope field for $dy/dx = -2x/y$

way and find that this clockwise rotation around the origin persists. Figure 7.7 shows the direction field of (7.14).

   If we superimpose several of the elliptical trajectories (7.15) on Figure 7.7, we find Figure 7.8. Time, as a parameter, is proceeding clockwise around the ellipse, allowing us to tell the story of this relationship as time progresses from a fixed starting point. For example, in the first quadrant we see that as x increases, y decreases. Thus, as Denise (x) becomes more and more enamored with Chad, Chad's affection for Denise is decreasing. We can also determine from Figure 7.8 that there is no time when both are ecstatic over each other. That would occur if they each experienced maximum affection at the same time. They don't.

   Now let's try to obtain the explicit dependence of x and y on t. We return to the original system of equations (7.13) and solve it in a manner similar to that used in the previous example. Thus, we solve the second equation in (7.13) for x as $x = (-1/b)dy/dt$ and substitute this expression into the first equation to obtain

$$\frac{dx}{dt} = -\frac{1}{b}\frac{d^2y}{dt^2} = ay,$$

or

$$\frac{d^2y}{dt^2} = -aby. \tag{7.16}$$

This second order linear differential equation was also solved in Section 5.5, the solution being

$$y(t) = C_1 \cos \sqrt{ab}t + C_2 \sin \sqrt{ab}t. \tag{7.17}$$

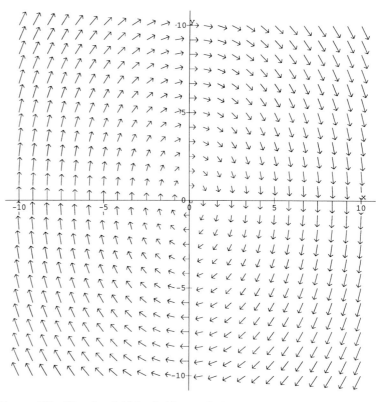

FIGURE 7.7   Direction field for $dy/dx = -2x/y$

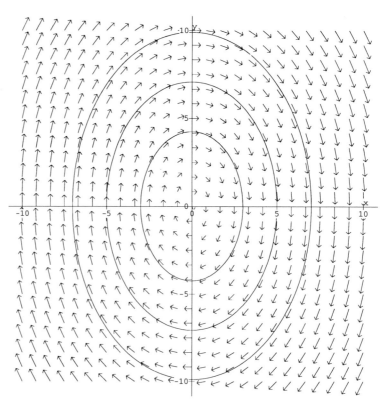

FIGURE 7.8   Trajectories and direction field for $dy/dx = -2x/y$

[Verify that (7.17) satisfies (7.16).] If we substitute (7.17), into the second equation in (7.13) we see that

$$x(t) = \sqrt{\frac{a}{b}} \left(C_1 \sin \sqrt{ab}\, t - C_2 \cos \sqrt{ab}\, t \right)$$

gives the rest of the solution of our original system of equations.

If we start counting time when[1] y is very attracted to x and x is feeling neutral toward y, we might impose initial conditions $x(0) = 0$, $y(0) = 2.5$. This requires that $C_1 = 2.5$ and $C_2 = 0$, so for this case our solution is

$$x(t) = 2.5 \sqrt{\frac{a}{b}} \sin \sqrt{ab}\, t, \tag{7.18}$$

$$y(t) = 2.5 \cos \sqrt{ab}\, t. \tag{7.19}$$

It is evident from (7.18) and (7.19) that the relative size of a and b (actually $\sqrt{a/b}$) determines which individual has the greater emotional range. If $a < b$, then the amplitude of Denise's affection is less than 2.5, the amplitude of Chad's affection. Then Chad will have the greater emotional range.

Figure 7.9 shows the solutions (7.18) and (7.19) for the case $a = 1$, $b = 2$, which shows that y has the greater emotional range. We can see that between $t = 0$ and $t = t_1$, x and y have affection for each other, although y is becoming less attracted to x while x is becoming more attracted to y. From $t = t_1$ to $t = t_2$, x has affection for y, but that affection is not reciprocated. In fact the least attraction that y has for x occurs just as x is neutral toward y — that is, when $x = 0$ at $t = t_2$. At this stage y is more attracted to x, but x is less and less attracted to y. Between $t = t_2$ and $t = t_3$, x and y dislike each other. The next time interval when x and y are both positive is between $t = t_4$ and $t = t_5$, when they are both attracted to each other again. The emotional cycle from $t = 0$ to $t = t_4$ is repeated forever. For only a quarter of the time do Denise and Chad simultaneously like each other. This is consistent with Figure 7.8. Only in the first quadrant do they simultaneously have affection for each other. $\qquad\square$

EXAMPLE 7.3: Parental Interference

Suppose that the differential equation that modeled Denise's feelings was of the form

$$\frac{dx}{dt} = ay - cx,$$

where c is a positive constant. Here we see that her change in affection for Chad is diminished by a term that is proportional to her affection. What could this mean? Well, if her parents were worried that she would become so enamored with Chad that she would leave home, then the more affection she had for Chad, the more pressure the parents might exert to have them break up.

---

[1]Henceforth, instead of referring to Denise and Chad by name, we will use x and y to identify them, although, in fact, x represents Denise's feelings for Chad, and y represents Chad's feelings for Denise.

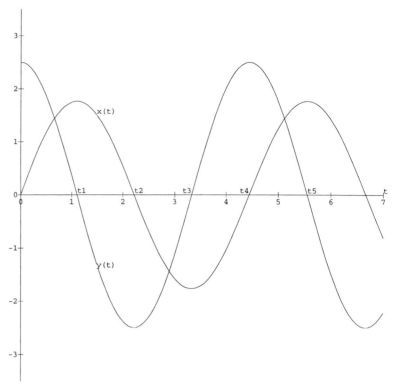

FIGURE 7.9   The solutions $x(t) = 2.5 \cdot 2^{-1/2} \sin 2^{1/2}t$, $y(t) = 2.5 \cos 2^{1/2}t$

Let's assume that the differential equation that modeled Chad's feelings remained the same, so we have

$$\frac{dx}{dt} = ay - cx, \qquad \frac{dy}{dt} = -bx. \tag{7.20}$$

Combining these equations as before, we obtain

$$\frac{dy}{dx} = -\frac{bx}{ay - cx}, \tag{7.21}$$

where a, b, and c are positive constants.

The effect of the additional term $-cx$ on the direction field is shown on Figure 7.10 with $a = 1, b = 2, c = 0.4$. Note that horizontal tangents occur only along the vertical axis, while the isocline for vertical tangents is given by the line $y = 0.4x$. Because dy/dt is negative in the first quadrant, the direction of travel will be clockwise. The slope is positive along the x-axis, so it is evident that all solution curves will spiral toward the origin. This is even clearer in Figure 7.11, where we have drawn the numerical solutions for three different initial conditions x_0, y_0 — namely, $x_0 = 0, y_0 = 2.5$; $x_0 = 0, y_0 = 5$; and $x_0 = 0$, $y_0 = 10$. This suggests that all solution curves eventually end at the origin $x = 0, y = 0$, which means that eventually Denise and Chad will have no affection for each other.

Another way to see the effect of $-cx$ is by contrasting the direction fields and trajectories of Figures 7.7 and 7.8 with those of Figures 7.10 and 7.11.

Equation (7.21) is a differential equation with homogeneous coefficients (see Section 5.1), which may be solved by a change in the dependent variable (see Exercise 6, page 335). If we do that, we obtain the exact equations for the trajectories drawn in Figure 7.11.

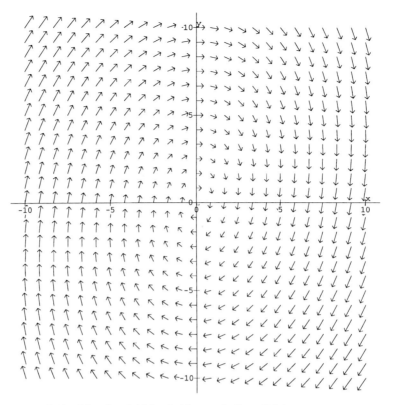

FIGURE 7.10 Direction field for $dy/dx = -2x/(y - 0.4x)$

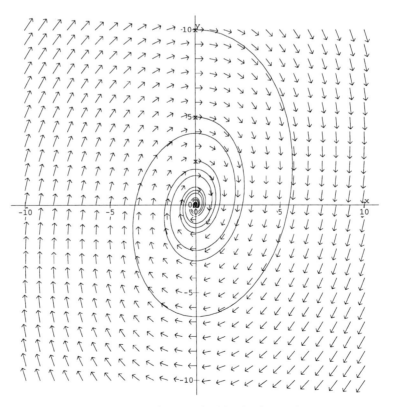

FIGURE 7.11 Trajectories and direction field for $dy/dx = -2x/(y - 0.4x)$

We can find the solutions for $y(t)$ and $x(t)$ by substituting the second equation in (7.20), $x = -y'/b$, into the first (twice), giving

$$\frac{d^2 y}{dt^2} + c\frac{dy}{dt} + aby = y'' + cy' + aby = 0. \tag{7.22}$$

This equation looks like (7.16) with the addition of the term cy', but it is one we have not yet solved. However, it is a second order linear differential equation.

In keeping with previous attempts to solve new differential equations, we define the dependent variable z by the product

$$y = u(t)z(t), \tag{7.23}$$

where we want to choose $u(t)$ to simplify the differential equation. If we substitute (7.23) into (7.22), we find

$$uz'' + \left(2u' + cu\right)z' + \left(u'' + cu' + abu\right)z = 0. \tag{7.24}$$

The obvious choice to force one of the coefficients to zero is

$$2u' + cu = 0,$$

which would eliminate z'. This means we should choose $u = e^{-ct/2}$, in which case (7.24) reduces to

$$e^{-ct/2}z'' + \left(\frac{1}{4}c^2 - \frac{1}{2}c^2 + ab\right)e^{-ct/2}z = 0,$$

or

$$z'' + \left(-\frac{1}{4}c^2 + ab\right)z = 0. \tag{7.25}$$

For realistic situations the parental influence c is small compared with their mutual affection, so the constant $c^2/4$ is small in comparison with the product ab (speaking as fathers, this is unfortunate, but true). The term in parentheses, $-c^2/4 + ab$, is therefore positive. This equation is just like (7.16) with solution (7.17). Therefore, the solution of (7.25) is

$$z(t) = C_1 \cos \beta t + C_2 \sin \beta t,$$

where $\beta = \sqrt{ab - c^2/4}$, and C_1 and C_2 are arbitrary constants. If we substitute this expression for $z(t)$ into (7.23) we find

$$y(t) = e^{-ct/2}\left(C_1 \cos \beta t + C_2 \sin \beta t\right), \tag{7.26}$$

which, from the second equation in (7.20), gives

$$x(t) = \frac{1}{2b}e^{-ct/2}\left[\left(cC_1 - 2\beta C_2\right)\cos \beta t + \left(2\beta C_1 + cC_2\right)\sin \beta t\right], \tag{7.27}$$

where $\beta = \sqrt{ab - c^2/4}$. We see from (7.26) and (7.27) that as $t \to \infty$, both x and y go to zero. So eventually, neither Denise nor Chad will have any affection for each other. (Let us hope they didn't get married.) This confirms the earlier suggestion from the phase plane analysis that all solution curves eventually end at the origin.

If we use the same initial conditions in (7.26) and (7.27) as we used in (7.18) and (7.19) — namely, $x(0) = 0$, $y(0) = 2.5$ — we find

$$y(t) = \frac{5}{4\beta} e^{-ct/2} \left(2\beta \cos \beta t + c \sin \beta t\right), \tag{7.28}$$

$$x(t) = \frac{5a}{2\beta} e^{-ct/2} \sin \beta t. \tag{7.29}$$

Figure 7.12 shows (7.28) and (7.29) for the special case shown in Figure 7.10 — namely, $a = 1$, $b = 2$, and $c = 0.4$. Comparing Figures 7.9 and 7.12 shows the impact of the parental interference parameter c. Because both amplitudes decrease with time, they have less affection for each other, until their affection ultimately dies out.[2] □

In these examples we moved back and forth between (7.2) and (7.8), between (7.13) and (7.16), and between (7.20) and (7.22), all of which are special cases of more general types of differential equations.

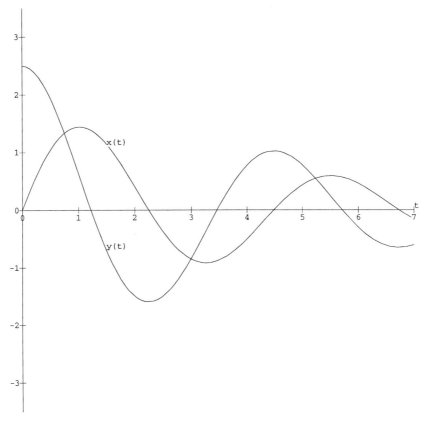

FIGURE 7.12 The functions $x(t) = (2.5/\beta)e^{-0.2t} \sin \beta t$ and $y(t) = (1.25/\beta)e^{-0.2t} \left(2\beta \cos \beta t + 0.4 \sin \beta t\right)$ with $\beta = 7/5$

---

[2]Because this is in the best interests of both Denise and Chad, some might say that parents know best.

Definition 7.1:   **The set of differential equations**

$$\frac{dx}{dt} = P(x, y, t), \qquad \frac{dy}{dt} = Q(x, y, t), \tag{7.30}$$

where $P(x, y, t)$ and $Q(x, y, t)$ are given functions of the three variables x, y, t is called a SYSTEM OF FIRST ORDER DIFFERENTIAL EQUATIONS. If P and Q are independent of t, the system is called AUTONOMOUS.[3]

Thus, the differential equations modelling the sea battles and Denise and Chad are autonomous systems of first order differential equations.

Definition 7.2:   **The differential equation**

$$a_2(t)\frac{d^2y}{dt^2} + a_1(t)\frac{dy}{dt} + a_0(t)y = h(t), \tag{7.31}$$

where $a_2(t)$, $a_1(t)$, and $a_0(t)$ are given functions of t and $a_2(t) \neq 0$, is called a SECOND ORDER LINEAR DIFFERENTIAL EQUATION. If a_2, a_1, and a_0 are independent of t, (7.31) is called a SECOND ORDER LINEAR DIFFERENTIAL EQUATION WITH CONSTANT COEFFICIENTS. If $h(t)$ is identically zero — that is, zero for all values of t — the differential equation (7.31) is called HOMOGENEOUS; otherwise, it is called NONHOMOGENEOUS.

Thus, (7.8), (7.16), and (7.22) are homogeneous second order linear differential equations with constant coefficients.

We could ask if it is always possible to write the system of first order differential equations (7.30) as a second order linear differential equation (7.31). The answer is no (see Exercise 9, page 335).

We could ask if it is always possible to write the second order linear differential equation (7.31) as a system of first order differential equations (7.30). The answer is yes; by introducing a new variable $v(t) = y'$ we could rewrite (7.31) in the form

$$\frac{dv}{dt} = -\frac{a_1}{a_2}v - \frac{a_0}{a_2}y + h(t), \qquad \frac{dy}{dt} = v. \tag{7.32}$$

In this case the phase plane variables will be y and v.

If we have a situation where a_2, a_1, a_0, and $h(t)$ are all constants, then (7.32) is an autonomous system. Thus,

- a homogeneous second order linear differential equation with constant coefficients can always be converted to an autonomous system of first order differential equations. The phase plane variables will be y and $v = dy/dt$.

In a later section we will exploit this capability.

EXERCISES

1. If the sea battle characterized by (7.2) began with $x_0 = 37$ and $y_0 = 12$ and lasted for 40 hours, what is the value of a?

---

[3]The use of the word "autonomous" is consistent with the previous usage in Chapter 2, because in both cases the independent variable is absent from the right-hand side.

2. Find the time-dependent behavior of $x(t)$ and $y(t)$ for Example 7.1 when $a \neq b$. That is, first eliminate the variable x from the system of equations in (7.1) to obtain $y'' - aby = 0$, and then solve for $y(t)$ and $x(t)$, assuming that $y(0) = y_0$ and $x(0) = x_0$. If $b > a$, and $y_0 = x_0$, who will win the battle? How many ships will remain after the battle is over?

3. Comment on the emotional ranges of Denise and Chad by using both the explicit solutions in (7.18) and (7.19) and the equations for trajectories in the phase plane given by (7.15).

4. This exercise uses the following terminology: You "like" someone when you have affection for him or her. You "dislike" someone when you have negative feelings for that person. A couple is "happy" when they both like each other. A couple is "unhappy" when they both dislike each other. An individual is "happy" if he or she is liked. An individual is "unhappy" if he or she is disliked. Assume that the units of time in Figure 7.9 are weeks.

   (a) When were Denise and Chad happy as individuals? As a couple?

   (b) When were Denise and Chad unhappy as individuals? As a couple?

   (c) Assuming this relationship lasted the seven weeks shown in this figure, estimate the proportion of the time Denise and Chad were happy as a couple.

   (d) During this seven-week period, were Denise and Chad happy as a couple for longer periods of time than they were unhappy as a couple? Explain your reasoning.

   (e) If you were a parent who was happy if either Denise was happy or Chad was unhappy, what proportion of the seven weeks were you happy? Explain.

   (f) At what point during the seven weeks would you liked to have double-dated with Denise and Chad? Explain.

5. Concavity of the trajectories in the phase plane may be determined by differentiating (7.21). Do so, and show that if $ab - c^2/4 > 0$, then concavity changes on either side of the line $y = cx/a$.

6. Equation (7.21) is of the form $dy/dx = g(y/x)$, which was discussed in Section 5.1. Use the change of variables suggested in that section to find the solution curves of (7.21) (that is, find the equation for the trajectories in the phase plane). Differentiate your solution twice, and compare the resulting concavity regions with those from Exercise 5.

7. Show that for the differential equation $y'' - aby = 0$, if $y_1(t)$ and $y_2(t)$ are solutions, so is $C_1 y_1(t) + C_2 y_2(t)$, where C_1 and C_2 are arbitrary constants.

8. Show that for the differential equation $a_2 y'' + a_1 y' + a_0 y = 0$, if $y_1(t)$ and $y_2(t)$ are solutions, so is $C_1 y_1(t) + C_2 y_2(t)$, where C_1 and C_2 are arbitrary constants.

9. Show that, by eliminating all x dependence, the system of equations

$$\frac{dx}{dt} = -\sin y, \qquad \frac{dy}{dt} = x$$

gives rise to the second order differential equation

$$\frac{d^2 y}{dt^2} + \sin y = 0.$$

Explain why this is not a linear differential equation.

10. Show that any system of first order linear differential equations of the form

$$x' = a(t)x + b(t)y + f(t)$$

$$y' = c(t)x + k(t)y + g(t)$$

(7.33)

gives rise to a single second order linear differential equation for y'' ($x' = dx/dt$, $y' = dy/dt$).

11. For the following cases, use the techniques from Exercise 10 to find a second order linear differential equation obtained by eliminating one of the dependent variables ($x' = dx/dt$, $y' = dy/dt$).

(a)
$$x' = x - y$$
$$y' = -x + 2y$$

(b)
$$x' = -2x + y$$
$$y' = -3x + 2y + \cos t$$

(c)
$$x' = x - y$$
$$y' = 2x - y - t$$

(d)
$$x' = 3x - 2y$$
$$y' = 2x - 2y$$

(e)
$$x' = 5x - 2y + 3$$
$$y' = 6x - 2y$$

(f)
$$x' = -4x - 6y + 6e^{2t}$$
$$y' = x + y + 3e^{2t}$$

12. The method of elimination used in the previous exercise will also work for first order linear differential equations that are not in the form given by (7.33). However, in this case, you may have to differentiate one of the equations before making a useful substitution. Find a second order differential equation by eliminating one of the dependent variables in the following systems of differential equations ($x' = dx/dt$, $y' = dy/dt$).

(a)
$$x' + y' + y = 0$$
$$3x' + 2y' = 7$$

(b)
$$x' + y' = -e^t$$
$$2x' - 2y' - y = 0$$

(c)
$$x' - 3y' = e^{2t}$$
$$2x' - 2y' - y = 8$$

(d)
$$x' + 2x + y' = e^{-t}$$
$$2tx' + 3y' + y = t - 3$$

(e)
$$tx' + x + y' = \cos t$$
$$2x' - y' - y = -\sin t$$

(f)
$$x' + 2tx + y' = \sin t$$
$$2tx' + 2y' + 5y = 0$$

7.2 General Second Order Linear Differential Equations with Constant Coefficients

Let us start by summarizing what we discovered in Section 7.1 regarding the solutions of three specific second order linear differential equations — namely, (7.8), (7.16), and (7.22), here relabeled (7.34), (7.35), and (7.36)

$$\frac{d^2y}{dt^2} - a^2y = 0, \tag{7.34}$$

$$\frac{d^2y}{dt^2} + aby = 0, \tag{7.35}$$

and

$$\frac{d^2y}{dt^2} + c\frac{dy}{dt} + aby = 0 \quad \text{where } ab - c^2/4 > 0, \tag{7.36}$$

respectively. Their solutions are (7.9), (7.17), and (7.26) — that is,

$$y(t) = C_1 e^{at} + C_2 e^{-at}, \tag{7.37}$$
$$y(t) = C_1 \cos\sqrt{ab}\,t + C_2 \sin\sqrt{ab}\,t, \tag{7.38}$$

and

$$y(t) = e^{-ct/2}\left(C_1 \cos\sqrt{ab - c^2/4}\,t + C_2 \sin\sqrt{ab - c^2/4}\,t\right). \tag{7.39}$$

Because the coefficients of d^2y/dt^2, dy/dt, and y are constants in (7.34), (7.35), and (7.36), these are examples of second order linear differential equations with constant coefficients.

At first sight there does not appear to be a pattern to the solutions (7.37), (7.38), and (7.39). However, if we recall EULER'S FORMULA

$$e^{i\theta} = \cos\theta + i\sin\theta$$

and its consequences

$$\sin\theta = \frac{e^{i\theta} - e^{-i\theta}}{2i}$$

and

$$\cos\theta = \frac{e^{i\theta} + e^{-i\theta}}{2},$$

we observe that the solutions (7.37), (7.38), and (7.39) can all be written in the form of a linear combination of exponential functions of the type $y = e^{rt}$, where r may be either a real or a complex number.[4] This raises the question as to whether we can find the solutions

---

[4] The appendix contains a discussion of complex numbers. We use the letter i to denote $\sqrt{-1}$. Some disciplines use j for $\sqrt{-1}$, usually to avoid confusion when i already has a well-established common usage within that discipline. For example, in electrical engineering i usually represents current.

(7.37), (7.38), and (7.39) without having to go through the lengthy process described in Sections 5.5 and 7.1.

Let's first look at (7.34) and try an exponential-type function $y(t) = e^{rt}$, where r may be either a real or a complex number in the hope of finding a value of r that makes e^{rt} a solution . We compute the second derivative[5] as $y'' = r^2 e^{rt}$ and substitute these expressions into (7.34) to find

$$y'' - a^2 y = \left(r^2 - a^2\right) e^{rt} = 0.$$

Thus r must satisfy the quadratic equation

$$r^2 - a^2 = 0.$$

This quadratic equation has two real solutions

$$r = \pm a.$$

Thus, both $y(t) = e^{at}$ and $y(t) = e^{-at}$ satisfy (7.34). Now, a linear combination of its solutions gives $y(t) = C_1 e^{at} + C_2 e^{-at}$, which is the solution we already found in (7.37).

We now turn to (7.35) and (7.36). Because (7.35) is just a special case of (7.36) with $c = 0$, we include both cases by considering (7.36) with $ab - c^2/4 > 0$. As in our previous development, we substitute $y(t) = e^{rt}$ in the differential equation (7.36) giving

$$r^2 + cr + ab = 0.$$

Again this is a quadratic equation. The quadratic formula yields two roots, r_1 and r_2, which are defined by

$$r_1 = \frac{1}{2}\left(-c - \sqrt{c^2 - 4ab}\right)$$

and

$$r_2 = \frac{1}{2}\left(-c + \sqrt{c^2 - 4ab}\right).$$

Because $c^2 - 4ab < 0$, we write

$$r_1 = \frac{1}{2}\left(-c - i\sqrt{4ab - c^2}\right) = -\frac{c}{2} - i\sqrt{ab - c^2/4} = -\frac{c}{2} - i\beta$$

and

$$r_2 = \frac{1}{2}\left(-c + i\sqrt{4ab - c^2}\right) = -\frac{c}{2} + i\sqrt{ab - c^2/4} = -\frac{c}{2} + i\beta,$$

where $\beta = \sqrt{ab - c^2/4}$. Thus, we may write our solutions of (7.36) in different forms:

$$\begin{aligned}
y(t) &= C_1 e^{r_1 t} + C_2 e^{r_2 t} \\
&= C_1 e^{(-c/2 - i\beta)t} + C_2 e^{(-c/2 + i\beta)t} \\
&= C_1 e^{-ct/2} e^{-i\beta t} + C_2 e^{-ct/2} e^{i\beta t} \\
&= e^{-ct/2}\left(C_1 e^{-i\beta t} + C_2 e^{i\beta t}\right),
\end{aligned}$$

---

[5]The justification for $de^{rt}/dt = re^{rt}$, where r is a complex number, is given in Exercise 18, page 346.

or, using Euler's formula,

$$y(t) = e^{-ct/2} \left(K_1 \cos \beta t + K_2 \sin \beta t \right),$$

where $K_1 = C_1 + C_2$, $K_2 = i(C_2 - C_1)$.

In a similar manner, we now attempt to determine exponential solutions for any second order linear differential equation with constant coefficients. We try a solution of the form e^{rt} in the general equation

$$a_2 \frac{d^2 y}{dt^2} + a_1 \frac{dy}{dt} + a_0 y = 0 \tag{7.40}$$

(where $a_2 \neq 0$, a_1, and a_0 are constants) and obtain

$$\left(a_2 r^2 + a_1 r + a_0 \right) e^{rt} = 0.$$

Because the exponential function is never zero, this equation will be satisfied only if the unknown parameter r is a solution of the quadratic equation

$$a_2 r^2 + a_1 r + a_0 = 0. \tag{7.41}$$

This equation is central to finding solutions of (7.40), and is called the CHARACTERISTIC EQUATION associated with (7.40). This name is used because the behavior of the solutions of (7.40) are characterized by the roots of this quadratic equation.

Solutions of the characteristic equation (7.41) have one of three possible forms: two real distinct solutions, two complex solutions, or one real repeated solution. We have just seen examples of the first two cases and will now cover each of the three cases in turn.

1. **Two real distinct roots.** If in the linear differential equation with constant coefficients of the form (7.40) we have

$$a_1^2 - 4a_2 a_0 > 0,$$

then the roots of (7.41) will be real numbers given by

$$r_1 = \frac{-a_1 - \sqrt{a_1^2 - 4a_2 a_0}}{2a_2}$$

and

$$r_2 = \frac{-a_1 + \sqrt{a_1^2 - 4a_2 a_0}}{2a_2}.$$

Thus, our solution of the original differential equation (7.40) will be

$$y(t) = C_1 e^{r_1 t} + C_2 e^{r_2 t}. \tag{7.42}$$

2. **Two complex distinct roots.** If

$$a_1^2 - 4a_2 a_0 < 0,$$

the roots of (7.41) will be complex numbers given by

$$r_1 = \frac{-a_1 - i\sqrt{4a_2 a_0 - a_1^2}}{2a_2} = \alpha - i\beta$$

and

$$r_2 = \frac{-a_1 + i\sqrt{4a_2 a_0 - a_1^2}}{2a_2} = \alpha + i\beta.$$

Notice that r_1 and r_2 are complex conjugates. The solution could still have the form of (7.42), where the arbitrary constants will be complex numbers. However, usually it is advantageous to use Euler's formula to write the solution in terms of real arbitrary constants and real valued functions as

$$y(t) = e^{\alpha t}\left(C_1 \cos \beta t + C_2 \sin \beta t\right). \tag{7.43}$$

3. One real repeated root. If

$$a_1^2 - 4a_2 a_0 = 0,$$

we obtain only one root α of the characteristic equation, namely,

$$\alpha = r_1 = r_2 = -\frac{a_1}{2a_2}.$$

This means we have found only one solution of the second order differential equation, namely, $e^{\alpha t}$. To find another solution, we try our customary substitution of the form

$$y(t) = e^{\alpha t} z(t). \tag{7.44}$$

Substituting (7.44) into (7.40) we find

$$e^{\alpha t}\left[a_2\left(z'' + 2\alpha z' + \alpha^2 z\right) + a_1\left(z' + \alpha z\right) + a_0 z\right] = 0.$$

The term $e^{\alpha t}$ is never zero, so we may divide through by that term and rearrange what is left to obtain

$$a_2 z'' + \left(a_1 + 2a_2 \alpha\right) z' + \left(a_2 \alpha^2 + a_1 \alpha + a_0\right) z = 0. \tag{7.45}$$

Now the expression multiplying the z term in (7.45) has the form of the left-hand side of the characteristic equation, and because α is its solution, that term is zero. Also, because $\alpha = -a_1/(2a_2)$, we have $a_1 + 2a_2 \alpha = 0$, and so (7.45) reduces to

$$a_2 z'' = 0. \tag{7.46}$$

The solution of (7.46) is

$$z(t) = C_1 + C_2 t,$$

so, from (7.44), we have

$$y(t) = C_1 e^{\alpha t} + C_2 t e^{\alpha t}. \tag{7.47}$$

Equation (7.47) gives the form of our solution for Case 3.

The solutions (7.42), (7.43), and (7.47) will have a simpler form if $a_2 = 1$, $a_1 = 2a$, and $a_0 = b$, which we can always achieve by dividing (7.40) by a_2 and letting $a_1/a_2 = 2a$ and $a_0/a_2 = b$. In this way we can rewrite (7.40) in the form

$$\frac{d^2y}{dt^2} + 2a\frac{dy}{dt} + by = 0, \tag{7.48}$$

where a and b are constants.

We summarize the solutions we have found for this equation with the preceding recipe.

1. If $a^2 - b > 0$, then the solution of (7.48) is

$$y(t) = C_1 e^{r_1 t} + C_2 e^{r_2 t}, \tag{7.49}$$

where $r_1 = -a - \sqrt{a^2 - b}$, and $r_2 = -a + \sqrt{a^2 - b}$.

2. If $a^2 - b = 0$, then the solution of (7.48) is

$$y(t) = \left(C_1 + C_2 t\right) e^{-at}. \tag{7.50}$$

3. If $a^2 - b < 0$, then the solution of (7.48) is

$$y(t) = e^{-at}\left(C_1 \cos\sqrt{b - a^2}\,t + C_2 \sin\sqrt{b - a^2}\,t\right). \tag{7.51}$$

Because these are all the possible solutions of (7.48), we will look at them a little more closely.

- Solution (7.49) contains exponential growth ($r_1 > 0$ and $r_2 > 0$), exponential decay ($r_1 < 0$ and $r_2 < 0$), and exponential growth and decay (r_1 and r_2 have opposite signs) as well as the possibility of either of r_1 or r_2 being zero, giving a constant plus an exponential term.

- Solution (7.50) contains exponential growth ($a < 0$), exponential decay ($a > 0$), or a term linear in time ($a = 0$).

- Solution (7.51) contains exponential growing oscillations ($a < 0$), exponential decaying oscillations ($a > 0$), and pure oscillations ($a = 0$).

- The solutions given by (7.49), (7.50), and (7.51) are general solutions of (7.48).

Thus, combinations of exponential growth or decay, linear growth or decay, and oscillations are the only possible types of solutions for second order linear differential equation with constant coefficients, (7.48).

If in (7.48) we fix b and gradually reduce a from values where $a^2 > b$, through $a^2 = b$, until $a^2 < b$, we progress from (7.49), through (7.50), to (7.51). For example, if we consider the initial values $y(0) = 1$ and $y'(0) = 0$ and fix $b = 64$, then Figure 7.13 shows (7.49) with $a = 40$, 20, and 10. The dotted curve is (7.50) with $a = 8$. Figure 7.14 shows (7.51) with $a = 6, 4, 2$, and 1. The dotted curve is (7.50) with $a = 8$. Figure 7.15 combines Figures 7.13 and 7.14. What do you expect to happen when $a = 0$? What do you expect when $a = -1$?

A quick way of determining the characteristic equation associated with a second order linear differential equation with constant coefficients uses operator notation. With operator notation, capital letters denote differentiation. Thus, we may write $y'(t) = dy/dt = Dy$, where the D stands for differentiation with respect to t. With this notation, the usual

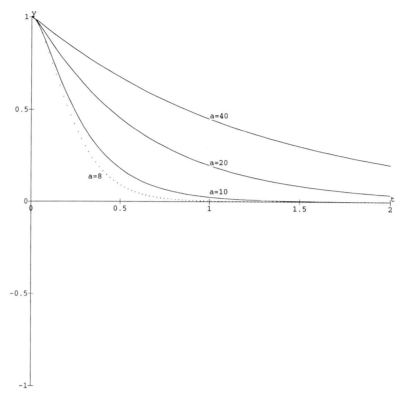

FIGURE 7.13   The functions $y = C_1 e^{r_1 t} + C_2 e^{r_2 t}$ and $y = (C_1 + C_2 t) e^{-at}$

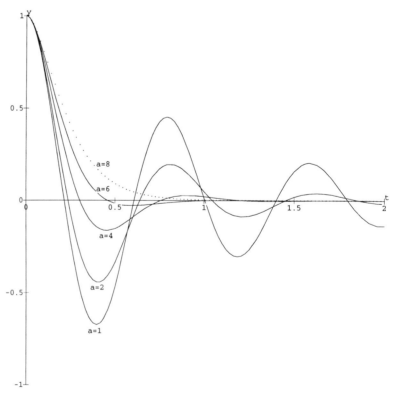

FIGURE 7.14   The   functions $y = (C_1 + C_2 t) e^{-at}$ and $y = e^{-at} (C_1 \cos \beta t + C_2 \sin \beta t)$, where $\beta^2 = b - a^2$

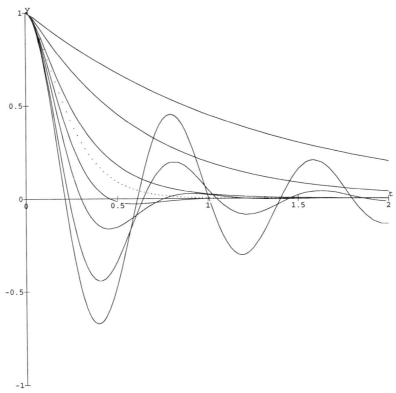

FIGURE 7.15   The functions $y = C_1 e^{r_1 t} + C_2 e^{r_2 t}$, $y = \left(C_1 + C_2 t \right) e^{-at}$, and $y = e^{-at} \left(C_1 \cos \beta t + C_2 \sin \beta t \right)$, where $\beta^2 = b - a^2$

algebraic operations of differentiation are slightly contracted. Thus, the standard rule for differentiating a sum would be written as

$$D\left[f(t) + g(t) \right] = Df(t) + Dg(t),$$

so our basic second order differential equation becomes

$$a_2 D^2 y + a_1 Dy + a_0 y = (a_2 D^2 + a_1 D + a_0)y = 0.$$

The associated characteristic equation is

$$a_2 r^2 + a_1 r + a_0 = 0,$$

so it is apparent that rather than taking derivatives each time, we may determine the characteristic equation by simply replacing the operator D in the differential equation with the parameter r and setting the result to zero. After finding the roots of the characteristic equation, we write our solution in terms of exponentials involving these roots. In fact we could list a three-step method as follows.

> ### *How to Solve Second Order Linear Differential Equations with Constant Coefficients*
>
> **Purpose:** To find the general solution of the second order linear differential equation with constant coefficients,
>
> $$a_2 y'' + a_1 y' + a_0 y = (a_2 D^2 + a_1 D + a_0)y = 0.$$
>
> **Technique:**
>
> **1.** Write the characteristic equation as
>
> $$a_2 r^2 + a_1 r + a_0 = 0.$$
>
> **2.** Solve the characteristic equation, calling the solutions r_1 and r_2. There are three possibilities:
>
> (a) r_1 and r_2 are real and distinct.
>
> (b) $r_1 = r_2$ (there is a double root).
>
> (c) r_1 and r_2 are complex conjugates of each other, written as $\alpha \pm i\beta$.
>
> **3.** Write the general solution corresponding to the possibilities (a), (b), and (c) of step 2 as
>
> (a) $y(t) = C_1 e^{r_1 t} + C_2 e^{r_2 t}$.
>
> (b) $y(t) = C_1 e^{r_1 t} + C_2 t e^{r_1 t}$.
>
> (c) $y(t) = C_1 e^{\alpha t} \cos \beta t + C_2 e^{\alpha t} \sin \beta t$ [or the form from (a); either is correct].

EXAMPLE 7.4

Find the general solution of the following differential equations.

i.
$$y'' + y' - 6y = 0 \qquad (7.52)$$

ii.
$$y'' + 2y' + y = 0 \qquad (7.53)$$

iii.
$$y'' + 2y' + 5y = 0 \qquad (7.54)$$

The procedure for each case is to use a solution of the form $y = e^{rt}$ and find values of r that solve the resulting characteristic equation.

i. Assuming $y = e^{rt}$ in (7.52) results in the characteristic equation

$$r^2 + r - 6 = 0,$$

which may be factored as

$$(r - 2)(r + 3) = 0.$$

The roots of this characteristic equation are 2 and -3, so the general solution of (7.52) is

$$y(t) = C_1 e^{2t} + C_2 e^{-3t}. \qquad (7.55)$$

ii.   Assuming $y = e^{rt}$ in (7.53) results in the characteristic equation

$$r^2 + 2r + 1 = 0,$$

which may be factored as

$$(r + 1)(r + 1) = 0.$$

Because we have a double root of this characteristic equation, we write the general solution of (7.53) as

$$y(t) = C_1 e^{-t} + C_2 t e^{-t}.$$

iii.   Assuming $y = e^{rt}$ in (7.54) results in the characteristic equation

$$r^2 + 2r + 5 = 0,$$

which has the complex roots

$$r_1 = -1 - 2i, \ r_2 = -1 + 2i.$$

Here we have a choice: We may write the general solution as

$$y(t) = K_1 e^{r_1 t} + K_2 e^{r_2 t},$$

where r_1 and r_2 are $-1 - 2i$ and $-1 + 2i$ respectively, and the constants K_1 and K_2 will be complex arbitrary numbers, or we may write it as

$$y(t) = e^{-t} \left(C_1 \cos 2t + C_2 \sin 2t \right),$$

where the constants C_1 and C_2 will be real arbitrary numbers.   □

We must stress that what we have presented here is a recipe for solving second order linear differential equations with constant coefficients. We have not justified that the solution we get from this recipe contains all solutions. It does, but we will justify this later.

EXERCISES

Use the recipe just outlined to find the general solutions of the following differential equations.

1. $y'' - y' - 6y = 0$

2. $y'' - y' - 2y = 0$

3. $y'' - 2y' + y = 0$

4. $y'' - 2y' - 8y = 0$

5. $y'' + 2y' - 8y = 0$

6. $y'' - 6y' + 9y = 0$

7. $y'' - y' - 12y = 0$

8. $y'' - 2y' + 10y = 0$

9. $y'' - 5y' + 6y = 0$

10. $y'' - 6y' + 10y = 0$

11. $y'' + 16y = 0$

12. $y'' - 9y = 0$

13. $6y'' + y' - y = 0$

14. $9y'' - 6y' + y = 0$

15. $y'' + 2y' + 4y = 0$

16. $y'' + 4y' + 2y = 0$

17. Solve $y'' + 2ay' + 225y = 0$ for each of the following cases.
   (a) $a = 9$
   (b) $a = 15$
   (c) $a = 25$

18. The derivative of a complex function

$$F(t) = u(t) + iv(t),$$

where $u(t)$ and $v(t)$ are real functions, is defined by

$$F'(t) = u'(t) + iv'(t).$$

   (a) Prove that $d\left(e^{it}\right)/dt = ie^{it}$.
   (b) Prove that $d\left(e^{rt}\right)/dt = re^{rt}$ for $r = \alpha + i\beta$, where α and β are real numbers.
   (c) Prove that the usual rules for differentiating apply to complex valued functions, that is, if $F_1(t) = u_1(t) + iv_1(t)$ and $F_2(t) = u_2(t) + iv_2(t)$, then
      i. $\left[F_1(t) \pm F_2(t)\right]' = F_1'(t) \pm F_2'(t)$,
      ii. $\left[F_1(t)F_2(t)\right]' = F_1'(t)F_2(t) + F_1(t)F_2'(t)$.

19. The indefinite integral of a complex function

$$F(t) = u(t) + iv(t),$$

where $u(t)$ and $v(t)$ are real functions, is defined by

$$\int F(t)\,dt = \int u(t)\,dt + i\int v(t)\,dt.$$

   (a) Prove that $\int e^{it}\,dt = e^{it}/i + C$.
   (b) Prove that $\int e^{rt}\,dt = e^{rt}/r + C$ for $r = \alpha + i\beta$, where α and β are real numbers.
   (c) Using $\sin t = (e^{it} - e^{-it})/(2i)$, evaluate $\int e^t \sin t\,dt$.
   (d) Using $\cos t = (e^{it} + e^{-it})/2$, evaluate $\int e^t \cos t\,dt$.

7.3 Initial Value Problems and an Existence Theorem

Our solutions of the second order differential equations so far in this chapter contained two arbitrary constants. Thus, we could impose two conditions to determine the values of these two constants. It frequently happens that these two conditions are imposed at the same value of the independent variable. Such conditions are called INITIAL CONDITIONS. The combination of a differential equation together with initial conditions is called an INITIAL VALUE PROBLEM. Our examples in Section 7.1 are initial value problems. Changing the initial conditions usually changes the solution.

EXAMPLE 7.5: Denise and Chad's Relationship

We return to a special case of the mathematical model of affection with Denise, x, and Chad, y,

$$\frac{dx}{dt} = ay, \qquad \frac{dy}{dt} = -bx,$$

for the case when $a = 1$ and $b = 2$, so that

$$\frac{dx}{dt} = y, \qquad \frac{dy}{dt} = -2x.$$

If we start counting time when Chad, y, had maximum affection and Denise, x, had none, we would have the initial conditions

$$x(0) = 0, \ y(0) = y_{max},$$

or

$$y(0) = y_{max}, \ y'(0) = 0. \tag{7.56}$$

In Section 7.1 we combined these two first order equations into the second order one,

$$y'' + 2y = 0,$$

with solution

$$y(t) = C_1 \cos \sqrt{2}t + C_2 \sin \sqrt{2}t. \tag{7.57}$$

In order to find values for C_1 and C_2 from the initial conditions (7.56), we need $y'(t)$, which we get by differentiating (7.57):

$$y'(t) = \sqrt{2}\left(-C_1 \sin \sqrt{2}t + C_2 \cos \sqrt{2}t\right). \tag{7.58}$$

If we put $t = 0$ in (7.57) and (7.58), we find

$$y(0) = C_1,$$
$$y'(0) = \sqrt{2}C_2.$$

Choosing the arbitrary constants C_1 and C_2 as $C_1 = y_{max}$ and $C_2 = 0$ will satisfy the given initial conditions. This means that our explicit solution of the original problem is

$$x(t) = \frac{1}{\sqrt{2}} y_{max} \sin \sqrt{2} t, \qquad (7.59)$$

$$y(t) = y_{max} \cos \sqrt{2} t.$$

In Figure 7.16 we show (7.59) for various values of y_{max}. Notice that for all these curves $x(0) = 0$, so the initial x value is the same for each of the solutions. In the first order case this would have been enough to guarantee a unique solution. Not so for second order — notice in Figure 7.16 that the solution curves are not unique because they intersect. So, what do we need to guarantee a unique solution? □

In Chapter 6 we gave a theorem (Theorem 6.1) regarding first order linear initial value problems. Here we state the counterpart for second order linear initial value problems.

Theorem 7.1: *Consider the general second order linear differential equation*

$$a_2(t) \frac{d^2 y}{dt^2} + a_1(t) \frac{dy}{dt} + a_0(t) y = 0, \qquad (7.60)$$

where $a_2(t)$, $a_1(t)$, and $a_0(t)$ are given functions of t, subject to the initial conditions

$$y(t_0) = y_0, \qquad y'(t_0) = y_0^*,$$

where y_0 and y_0^* are constants. If $a_2(t)$, $a_1(t)$, and $a_0(t)$ are continuous, $a_2(t) \neq 0$ for $a < t < b$, and t_0 is contained in this interval, then for the interval $a < t < b$ there exists a unique solution of (7.60) with a continuous first derivative.

Comments about second order linear initial value problems

- In the first order case we exploited the fact that solutions do not intersect by using the uniqueness property. Here a unique solution is specified by two conditions, the initial position and the initial slope, so this theorem implies that a solution that starts from a specified position and with a specified slope is unique. Thus, we should not be surprised to find that solutions that start from the same point, but with different slopes, intersect. This is what happens in Figure 7.16.

- If $y(t_0) = 0$ and $y'(t_0) = 0$, then the function $y(t) = 0$ satisfies both (7.60) and these initial conditions. Consequently, $y(t) = 0$ is the only solution that passes through $y(t_0) = 0$ with zero slope. No other solution can have the property that $y(t_0) = 0$ and $y'(t_0) = 0$.

- Notice that this theorem applies to the coefficients $a_2(t)$, $a_1(t)$, and $a_0(t)$ being variable (that is, they may depend on t). If a_2, a_1, and a_0 are independent of t (that is, constants), then they are continuous for all t, so the theorem guarantees that second order linear differential equations with constant coefficients always have unique solutions valid for all t.

- The conclusions of this theorem can be extended to differential equations of the type

$$a_2(t) \frac{d^2 y}{dt^2} + a_1(t) \frac{dy}{dt} + a_0(t) y = h(t)$$

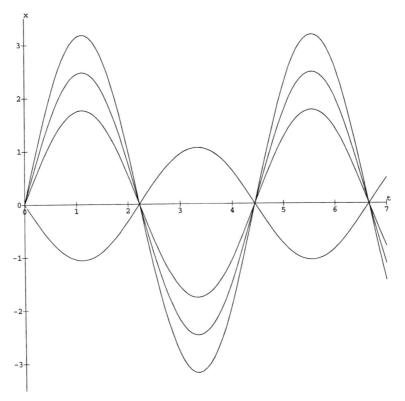

FIGURE 7.16   The solution $(y_{max}/2^{1/2}) \sin (2^{1/2}t)$ for different y_{max}

if $h(t)$ is continuous in the interval $a < t < b$. In this nonhomogeneous differential equation, $h(t)$ is called a FORCING FUNCTION.

EXAMPLE 7.6

Solve

$$y'' + y' - 6y = 0 \tag{7.61}$$

subject to the condition that

$$y(0) = 1, \qquad y'(0) = 0. \tag{7.62}$$

We have already found the solution of (7.61), given by (7.55), namely,

$$y(t) = C_1 e^{2t} + C_2 e^{-3t}. \tag{7.63}$$

If we differentiate (7.63), we find

$$y'(t) = 2C_1 e^{2t} - 3C_2 e^{-3t}. \tag{7.64}$$

When (7.63) and (7.64) are evaluated at the initial time $t = 0$, we have

$$y(0) = C_1 + C_2$$

and

$$y'(0) = 2C_1 - 3C_2.$$

In view of (7.62), these last two equations give a system of two equations in the two unknowns C_1 and C_2, namely,

$$C_1 + C_2 = 1,$$
$$2C_1 - 3C_2 = 0,$$

with solution

$$C_1 = \frac{3}{5}, \qquad C_2 = \frac{2}{5}.$$

We substitute these values into (7.63) to find the solution of (7.61) subject to (7.62) as

$$y(t) = \frac{3}{5}e^{2t} + \frac{2}{5}e^{-3t}.$$

$\square$

EXAMPLE 7.7: A System of Functional Equations

The functions $\sin t$ and $\cos t$ are differentiable functions that satisfy the identities

$$\sin(t_1 + t_2) = \sin t_1 \cos t_2 + \cos t_1 \sin t_2$$

and

$$\cos(t_1 + t_2) = \cos t_1 \cos t_2 - \sin t_1 \sin t_2,$$

for all t_1, t_2. This leads to the problem of finding all differentiable functions $x(t)$, and $y(t)$, that satisfy the identities

$$x(t_1 + t_2) = x(t_1)y(t_2) + y(t_1)x(t_2) \tag{7.65}$$

and

$$y(t_1 + t_2) = y(t_1)y(t_2) - x(t_1)x(t_2) \tag{7.66}$$

for all t_1, t_2. We see that $x(t) = 0$ and $y(t) = 0$ satisfy (7.65) and (7.66), so we seek functions that are not identically zero.

At first sight, finding $x(t)$ and $y(t)$ that satisfy (7.65) and (7.66) does not appear to be related to differential equations. However, we have already seen a number of functional equation problems that can be converted to differential equations that can be solved.

We construct the derivatives of x and y with respect to t,

$$x'(t) = \lim_{h \to 0} \frac{x(t+h) - x(t)}{h} \tag{7.67}$$

and

$$y'(t) = \lim_{h \to 0} \frac{y(t+h) - y(t)}{h}. \tag{7.68}$$

If we use (7.65) in (7.67), we find

$$x'(t) = \lim_{h \to 0} \frac{x(t)y(h) + y(t)x(h) - x(t)}{h} = \lim_{h \to 0} \left[x(t)\frac{y(h) - 1}{h} + y(t)\frac{x(h)}{h} \right]. \qquad (7.69)$$

The first term on the right-hand side would contain $y'(0)$ if $y(0) = 1$, and the second term would contain $x'(0)$ if $x(0) = 0$. Let's return to (7.65) and (7.66) to see what they tell us about $x(0)$ and $y(0)$.

Putting $t_1 = t$ and $t_2 = 0$ in (7.65) and (7.66), we find

$$x(t) = x(t)y(0) + y(t)x(0) \qquad (7.70)$$

and

$$y(t) = y(t)y(0) - x(t)x(0). \qquad (7.71)$$

Now we put $t = 0$ in (7.70) and (7.71) and find

$$x(0) = 2x(0)y(0) \qquad (7.72)$$

and

$$y(0) = y^2(0) - x^2(0). \qquad (7.73)$$

Equation (7.72) gives rise to two cases: $y(0) = 1/2$ or $x(0) = 0$. If we substitute $y(0) = 1/2$ in (7.73), we find $x^2(0) = -1/4$, which is not possible, so we are forced to conclude that $x(0) = 0$. If we substitute $x(0) = 0$ in (7.71), we find $y(t) = 0$, or $y(0) = 1$. From (7.65) we see that $y(t) = 0$ implies that $x(t) = 0$, which is not what we are looking for. So the only possibilities are $x(0) = 0$ and $y(0) = 1$.

Using these facts we can rewrite (7.69) as

$$x'(t) = \lim_{h \to 0} \left[x(t)\frac{y(h) - 1}{h} + y(t)\frac{x(h)}{h} \right] = \lim_{h \to 0} \left[x(t)\frac{y(h) - y(0)}{h} + y(t)\frac{x(h) - x(0)}{h} \right],$$

or

$$x'(t) = ax(t) + by(t), \qquad (7.74)$$

where $a = y'(0)$ and $b = x'(0)$. In the same way, we can use (7.68) to find

$$y'(t) = -bx(t) + ay(t). \qquad (7.75)$$

Thus, our problem is to solve the system of first order linear differential equations (7.74) and (7.75), subject to the initial conditions

$$x(0) = 0, \qquad y(0) = 1. \qquad (7.76)$$

Notice that we can obtain no more information about the values of a and b from (7.74) or (7.75), because putting $t = 0$ in either (7.74) or (7.75) leads to identities. However, we also notice that if $b = 0$, then (7.74) and (7.75) subject to (7.76), can be solved immediately to yield

$$x(t) = 0 \text{ and } y(t) = e^{at}, \qquad (7.77)$$

which do satisfy (7.65) and (7.66). But, when we posed the original question, we were thinking of functions that were not identically zero, unlike $x(t) = 0$. Thus, from now on, we concentrate on $b \neq 0$. Nevertheless, it did come as a surprise that (7.77) satisfied (7.65) and (7.66) — we were expecting the trigonometric functions.

If we differentiate (7.75) with respect to t, and then substitute from (7.74) for $x'(t)$, we find

$$y'' = -bx' + ay' = -b(ax + by) + ay'.$$

If we now solve (7.75) for $x = (ay - y')/b$ (remember $b \neq 0$) and then substitute this in the last equation, we get

$$y'' = -b(ax + by) + ay' = -b\left(\frac{a}{b}(ay - y') + by\right) + ay',$$

or

$$y'' - 2ay' + (a^2 + b^2)y = 0.$$

This is a second order linear differential equation with constant coefficients, with solution

$$y(t) = e^{at}(C_1 \cos bt + C_2 \sin bt).$$

The condition $y(0) = 1$ implies that $C_1 = 1$, so that the last equation becomes

$$y(t) = e^{at}(\cos bt + C_2 \sin bt). \tag{7.78}$$

We substitute (7.78) into (7.75) and solve for $x(t)$ to find

$$x(t) = e^{at}(-C_2 \cos bt + \sin bt). \tag{7.79}$$

The condition $x(0) = 0$ implies that $C_2 = 0$, so that (7.78) and (7.79) become

$$x(t) = e^{at} \sin bt, \qquad y(t) = e^{at} \cos bt. \tag{7.80}$$

We see that the functions in (7.80) satisfy (7.65) and (7.66), and so, contrary to expectation, $\sin t$ and $\cos t$ are not the unique solutions of (7.65) and (7.66). Rather $e^{at} \sin bt$ and $e^{at} \cos bt$ are.   □

EXERCISES

1. Solve the following initial value problems.

   (a) $(D^2 + D - 2)y = 0$, subject to $y(0) = 0$ and $y'(0) = 3$

   (b) $(D^2 + 2D - 10)y = 0$, subject to $y(0) = 0$ and $y'(0) = 4$

   (c) $y'' + 6y' + 9y = 0$, subject to $y(0) = 2$ and $y'(0) = 0$

   (d) $(12D^2 + D - 1)y = 0$, subject to $y(0) = 4$ and $y'(0) = 0$

   (e) $y'' + 3y' = 0$, subject to $y(0) = 4$ and $y'(0) = 3$

   (f) $y'' - 2\pi y' + 2\pi^2 y = 0$, subject to $y(0) = 0$ and $y'(0) = -2\pi$

   (g) $y'' + 10y' + 100y = 0$, subject to $y(0) = 15$ and $y'(0) = 4$

   (h) $y'' + 10y' + 100y = 0$, subject to $y(1) = 0$ and $y'(1) = 0$

2. Show that $y = A \cos(\eta x + \phi)$ satisfies

$$y'' + \eta^2 y = 0$$

for any choice of A and ϕ. Explain how this function fits with Theorem 7.1.

3. Give the solution of

$$y'' + 16y = 0, \quad y(0) = 2\sqrt{2}, \quad y'(0) = 8\sqrt{2}$$

in the form of Exercise 2.

4. Find the second order differential equation with real constant coefficients that has the given general solution.
   (a) $y = C_1 e^{-x} + C_2 e^{3x}$
   (b) $y = C_1 e^{-2x} \cos 3x + C_2 e^{-2x} \sin 3x$
   (c) $y = C_1 e^{7x} + C_2 x e^{7x}$
   (d) $y = A \sin(3x + \phi)$
   (e) $y = C_1 e^{ax} \cos bx + C_2 e^{ax} \sin bx$

5. For what values of η does

$$y'' + \eta^2 y = 0, \quad y(0) = 0, \, y'(\pi) = 0$$

have a nonzero solution? Give these solutions.

6. The functions $\sinh t$ and $\cosh t$ are defined by

$$\sinh t = \frac{1}{2}\left(e^t - e^{-t}\right) \text{ and } \cosh t = \frac{1}{2}\left(e^t + e^{-t}\right)$$

and satisfy the identities

$$\sinh(t_1 + t_2) = \sinh t_1 \cosh t_2 + \cosh t_1 \sinh t_2$$

and

$$\cosh(t_1 + t_2) = \cosh t_1 \cosh t_2 + \sinh t_1 \sinh t_2$$

for all t_1, t_2. Find the most general differentiable functions $x(t)$ and $y(t)$ that satisfy the identities

$$x(t_1 + t_2) = x(t_1)y(t_2) + y(t_1)x(t_2)$$

and

$$y(t_1 + t_2) = y(t_1)y(t_2) + x(t_1)x(t_2)$$

for all t_1, t_2.

7. Find the most general differentiable functions $x(t)$ and $y(t)$ that satisfy the identities

$$x(t_1 + t_2) = x(t_1)y(t_2) + y(t_1)x(t_2)$$

and

$$y(t_1 + t_2) = y(t_1)y(t_2)$$

for all t_1, t_2.

8. Look in the textbooks of your other courses and find an example that uses a second order linear differential equation with constant coefficients. Write a report about this example that includes the following items:

   (a) A brief description of background material so a classmate will understand the origin of the differential equation.

   (b) How the constants in the differential equation can be evaluated and what the initial condition means.

   (c) The solution of this differential equation.

   (d) An interpretation of this solution, and how it answers a question posed by the original discussion.

7.4   Constructing the Phase Plane Trajectory from the Explicit Solution

In this section we will show how trajectories in the phase plane and solutions of second order differential equations are related.

EXAMPLE 7.8: Denise and Chad's Relationship

We return to the first example discussed in the last section, in which, with $y_{max} = 5/2$, we found that

$$x(t) = \frac{5}{2\sqrt{2}} \sin \sqrt{2}t$$

and

$$y(t) = \frac{5}{2} \cos \sqrt{2}t$$

were the solutions of

$$\frac{dx}{dt} = y, \qquad \frac{dy}{dt} = -2x$$

subject to

$$x(0) = 0, \quad y(0) = 5/2.$$

Now that we have explicit functions to represent the affections of Denise and Chad, we can add time to the trajectory of y versus x in Figure 7.8. First we construct a table of values of x and y as functions of t (Table 7.1). The variable t plays the role of a parameter, and the pairs of (x, y) values from Table 7.1 can be plotted. Figure 7.17 shows these pairs of points added to the graph from Figure 7.8.

However, we do not need to compute a table before obtaining the graph of Figure 7.8. It is possible to do this from the graphs of $x(t)$ and $y(t)$ directly. In Figure 7.18 we have plotted y versus t in the upper right-hand box and x versus t in the lower right-hand box. The upper left-hand box is where we will plot the trajectory in the phase plane, and it is labeled y versus x. In the lower left-hand box we have drawn a line at a 45° angle.

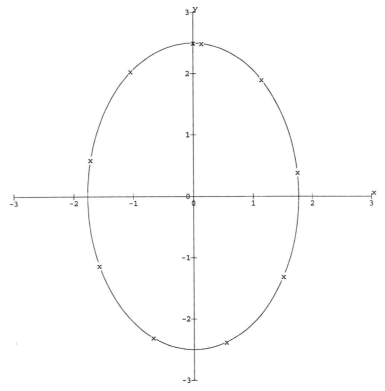

FIGURE 7.17   Phase plane trajectory of $dx/dt = y$, $dy/dt = -2x$, $x(0) = 0$, $y(0) = 5/2$

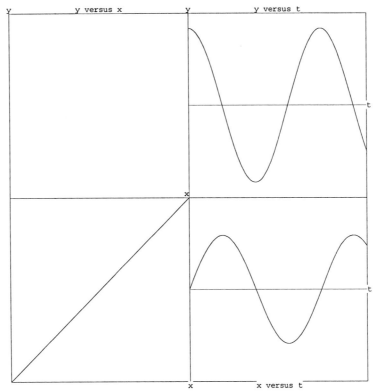

FIGURE 7.18   Constructing the phase plane trajectory for Denise and Chad — step 1

TABLE 7.1 **Values of x and y for Denise and Chad**

t	x	y
0.0	0.00	2.50
0.5	1.15	1.90
1.0	1.75	0.39
1.5	1.51	-1.31
2.0	0.54	-2.38
2.5	-0.68	-2.31
3.0	-1.58	-1.13
3.5	-1.72	0.59
4.0	-1.04	2.03
4.5	0.14	2.49

How to Construct the Trajectory in the Phase Plane

Purpose: To construct the phase plane trajectory from the solution curves $x = x(t)$, $y = y(t)$.

Technique:

1. Select any point on the t-axis in the y versus t box. Let's call it $t = a$.

2. Draw a vertical line through $t = a$ so that it crosses both the curve $y = y(t)$ in the y versus t box and the curve $x = x(t)$ in the x versus t box. This identifies the values of $y(a)$ and $x(a)$. We want to transfer these values to the y versus x box.

3. Draw a horizontal line through $y(a)$ into the y versus x box. $y(a)$ lies on this line.

4. Draw a horizontal line through $x(a)$ into the lower left box until it intersects the 45° line. Then draw a vertical line through this point of intersection into the y versus x box. Where this line intersects the horizontal line from part 3 gives the point $(x(a), y(a))$ in the y versus x box.

5. Repeat all the preceding steps for different values of a, and then join the points in the y versus x box. Notice that the direction of the trajectory in the phase plane is determined by the order of the points plotted for increasing t.

   This process is not as complicated as it sounds. Look at Figure 7.19 and reread the steps. Some more lines and a completed trajectory are shown in Figure 7.20.

We have an additional way of looking at the phase plane. If we consider time as an axis perpendicular to the xy-plane, then the three-dimensional graph of the affections of Denise and Chad as a function of time would give a helixlike curve. If we then take the projection into the xy-plane of this curve, we obtain the phase plane. This is shown in Figure 7.21.   □

   This view gives us a way to picture the uniqueness theorem in terms of nonintersecting curves. Imagine a three-dimensional space with coordinates (y, y', t). Projection down the t-axis gives the phase plane. Consider the plane corresponding to $t = t_0$. A point in this plane represents the initial condition $(y(t_0), y'(t_0))$. Now imagine the curve $y = y(t)$ progressing in time in this three-dimensional space. It will start from $(y(t_0), y'(t_0))$ and follow $y = y(t)$ and $y' = y'(t)$. The theorem says no other curve in this space can pass through the point $(y(t_0), y'(t_0))$, so no other solution can intersect the solution through $(y(t_0), y'(t_0))$.

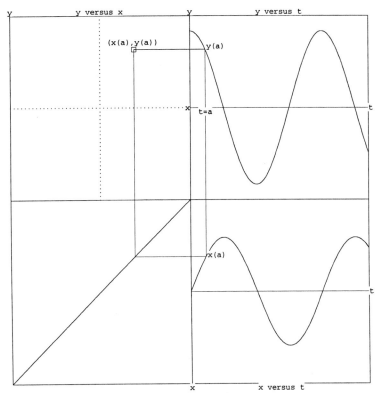

FIGURE 7.19 Constructing the phase plane trajectory for Denise and Chad — step 2

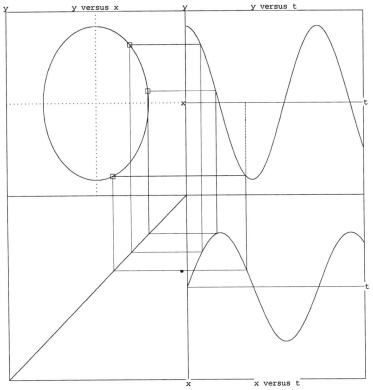

FIGURE 7.20 Constructing the phase plane trajectory for Denise and Chad — step 3

FIGURE 7.21   Helix in three-dimensional space (x, y, t)

EXAMPLE 7.9: Sea Battles

In the model of sea battles in Section 7.1, we solved

$$\frac{dx}{dt} = -ay, \qquad \frac{dy}{dt} = -ax$$

subject to

$$y(0) = y_0, \ x(0) = x_0,$$

the solution being

$$y(t) = \frac{1}{2}\left(y_0 - x_0\right)e^{at} + \frac{1}{2}\left(y_0 + x_0\right)e^{-at}, \tag{7.81}$$

$$x(t) = \frac{1}{2}\left(-y_0 + x_0\right)e^{at} + \frac{1}{2}\left(y_0 + x_0\right)e^{-at}. \tag{7.82}$$

If we plot the graphs of these two functions in the same manner as we did for the model of Denise and Chad, we obtain Figure 7.22. If we take a short straight edge and construct the graph of y versus x from the graphs of y versus t and x versus t using the same process we used in the previous example, we could obtain several points on the trajectory corresponding to (7.81) and (7.82) in the phase plane. This is shown in Figure 7.23.   □

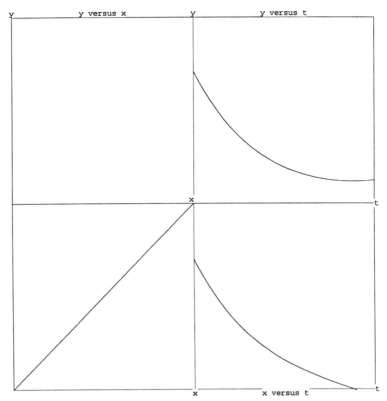

FIGURE 7.22 Constructing the phase plane trajectory for sea battles — step 1

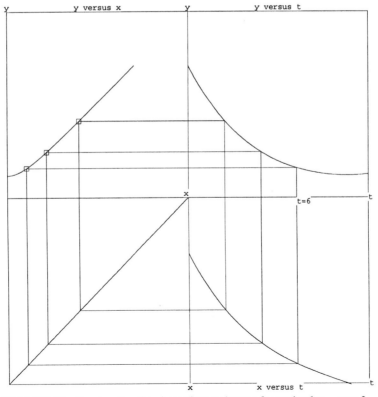

FIGURE 7.23 Constructing the phase plane trajectory for sea battles — step 2

EXAMPLE 7.10: Parental Interference

Finally we look at the last example in Section 7.1, in which we had the system of differential equations

$$\frac{dy}{dt} = -2x, \qquad \frac{dx}{dt} = y - 0.4x. \tag{7.83}$$

If we solve this subject to

$$x(0) = 2.5, \qquad y(0) = 0$$

we find

$$y(t) = -\frac{1}{\beta} e^{-ct/2} \sin \beta t, \tag{7.84}$$

$$x(t) = 2.5 e^{-ct/2} \left[\cos \beta t - c/(2\beta) \sin \beta t\right], \tag{7.85}$$

where $\beta = 7/5$ and $c = 0.4$.

Figure 7.24 shows a graph of (7.84) and (7.85) as functions of t in the two right-hand boxes and shows the 45° line in the lower left-hand box. The graph of the y versus x trajectory in the phase plane may now be constructed by the graphical method given in the previous two examples. This is done in Figure 7.25.

Notice that all three graphs in Figure 7.25 suggest that as time increases, both the x and y values approach zero as a limit. In fact, from the general solution of this problem as given in (7.26) and (7.27), we see that both x and y approach zero as t approaches infinity, regardless of the choice of the arbitrary constants C_1 and C_2. Thus, in the phase plane the trajectory will spiral in toward the origin. Notice that the point $x = 0$, $y = 0$, satisfies the two differential equations in (7.83) and is therefore classified as an EQUILIBRIUM POINT (which no other trajectory can intersect). If the trajectory for every initial condition is bounded and approaches an equilibrium point as t approaches infinity, as it does in this case, we say that the equilibrium point is ASYMPTOTICALLY STABLE.   □

Definition 7.3: **The system of autonomous differential equations**

$$x' = P(x, y)$$
$$y' = Q(x, y)$$

has an EQUILIBRIUM POINT at (x_0, y_0) if

$$P(x_0, y_0) = 0,$$
$$Q(x_0, y_0) = 0.$$

Equilibrium points of a system of autonomous differential equations play the same role as that of the equilibrium solutions of an autonomous first order equation.

For the previous example in which there was no parental involvement in the affection model for Denise and Chad, the point $(0, 0)$ was an equilibrium point. There we found that trajectories in the phase plane were all ellipses, but different initial conditions could lead to different ellipses. Thus, we could say that all the trajectories are bounded but do not approach the equilibrium point as t approaches infinity. An equilibrium point is NEUTRALLY STABLE if every trajectory in the phase plane is bounded but does not approach the equilibrium point as t approaches infinity. It should come as no surprise to you that if an equilibrium point is neither asymptotically stable nor neutrally stable, it is called UNSTABLE.

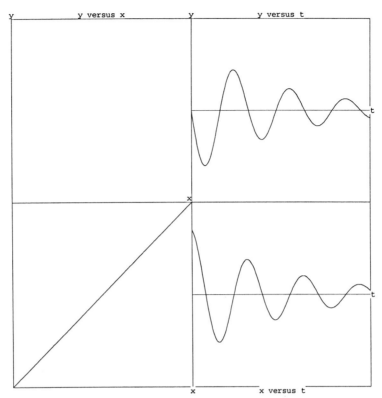

FIGURE 7.24   Constructing the phase plane trajectory for parental interference —
step 1

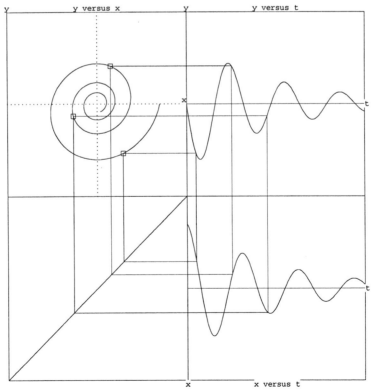

FIGURE 7.25   Constructing the phase plane trajectory for parental interference —
step 2

EXERCISES

1. Which systems of equations in Exercise 11, Section 7.1, have equilibrium points? Find the equilibrium points for those systems that have them, and explain why the others don't.

2. Give the equilibrium points for the following systems of differential equations.

   (a) $\begin{aligned} x' + 2y &= 6 \\ 2y' - 2x + 7y &= 9 \end{aligned}$

   (b) $\begin{aligned} x' + 3x - 2y &= 0 \\ 3y' - 7x + 4y &= 0 \end{aligned}$

   (c) $\begin{aligned} x' + x - 3y &= 2 \\ y' - 2x + 6y &= -4 \end{aligned}$

3. In Section 7.2 we considered the second order differential equation

   $$y'' + 2ay' + by = 0, \tag{7.86}$$

   along with relations between a and b that yielded different types of solutions. Change the form of (7.86) to that of a first order system (with dependent variables v and y) by the change of variables $v = y'$. Examine the direction field in the phase plane, and describe what conclusions you can draw regarding the stability of the equilibrium point at $v = y = 0$ for several specific values a and b. In particular, include the following situations:

   (a) $a^2 - b > 0$

   (b) $a^2 - b = 0$

   (c) $a^2 - b < 0$, and $a > 0$

   (d) $a^2 - b < 0$, and $a < 0$

4. The differential equation for the trajectories in the phase plane in Exercise 2(b) has homogeneous coefficients. Solve this differential equation to obtain an explicit solution for the trajectories in the phase plane.

5. Exercise 11, Section 7.1, contains a number of systems of first order linear differential equations. The instructions in that exercise call for finding a second order differential equation by eliminating one of the dependent variables.

   i. For the following systems of differential equations, use that technique to find the associated second order differential equation and its general solution.

   ii. Use this solution to solve for the other dependent variable, and choose the arbitrary constants to satisfy the given initial conditions.

   iii. Use the method of this section to obtain the trajectory in the phase plane from the solution of this initial value problem. Note any equilibrium points, and classify them as neutrally stable, asymptotically stable, or unstable. Examine the direction field in the phase plane, and make sure it supports your classifications.

   (a)
   $$\begin{aligned} x' &= 2x - y \\ y' &= -x + 2y \end{aligned}$$

   with $x(0) = 2$, $y(0) = 0$.

   (b)
   $$\begin{aligned} x' &= 2x - 2y \\ y' &= 3x - 2y \end{aligned}$$

   with $x(0) = 0$, $y(0) = 2$.

   (c)
   $$x' = 3x - 2y$$

$$y' = 2x - 2y$$

with $x(0) = 2$, $y(0) = 1$.

(d)
$$x' = 5x - 2y$$
$$y' = 4x - 2y$$

with $x(0) = 0$, $y(0) = 6$.

6. Exercise 12 Section 7.1, on page 336, contains a number of systems of first order linear differential equations. The instructions in that exercise call for finding a second order differential equation by eliminating one of the dependent variables. These equations were not in the standard form

$$\frac{dx}{dt} = P(x, y, t), \qquad \frac{dy}{dt} = Q(x, y, t),$$

but could be put in that form by appropriate operations on the two differential equations. Consider, for example, Exercise 12 (a),

$$\begin{aligned} x' + y' + y &= 0, \\ 3x' + 2y' &= 7. \end{aligned} \qquad (7.87)$$

Multiplying the top equation by -3 and adding the result to the bottom equation results in a differential equation containing x', but not y'. In a similar manner, substituting the value of x' from the top equation into the bottom one gives an equation involving y', but not x'. These two new equations have the form

$$\frac{dx}{dt} = P(x, y), \qquad \frac{dy}{dt} = Q(x, y). \qquad (7.88)$$

    i. For each of the following systems of equation, obtain the equivalent system of differential equations in the form (7.88).

    ii. Find the associated second order differential equation for the resulting system and its general solution.

    iii. Use this solution to solve for the other dependent variable, and choose the arbitrary constants to satisfy the initial conditions $x(0) = 0$, $y(0) = 1$.

    iv. Use the method of this section to obtain the trajectory in the phase plane from the solution of this initial value problem. Note any equilibrium points, and classify them as neutrally stable, asymptotically stable, or unstable. Examine the direction field in the phase plane and make sure it supports your classifications.

(a) The system in (7.87).

(b) The system

$$\begin{aligned} x' + y' &= 5 \\ 2x' - 2y' - y &= 0. \end{aligned}$$

(c) The system

$$\begin{aligned} x' + y' - 3y &= 0 \\ x' - 3x - 6y &= 0. \end{aligned}$$

What Have We Learned?

MAIN IDEAS

- A single second order differential equation may be obtained from a system of two first order differential equations.

- An autonomous system of two first order linear differential equations results from an appropriate substitution in any second order linear differential equation of the form

$$a_2 y'' + a_1 y' + a_0 y = 0,$$

  where a_2, a_1, and a_0 are constants. This allows our graphical analysis from earlier chapters to be used in analyzing properties of solutions of these second order equations.

- The characteristic equation may be used to find explicit solutions of all second order linear differential equations with constant coefficients. See *How to Solve Second Order Linear Differential Equations with Constant Coefficients* on page 344.

- If all the coefficients in a second order linear differential equation are continuous (and the one multiplying the second derivative is never zero), then all initial value problems associated with this equation will have a unique solution. See Theorem 7.1 on page 348.

- Trajectories in the phase plane may be constructed from an explicit solution of a system of differential equations using a geometrical technique. See *How to Construct the Trajectory in the Phase Plane* on page 356.

CHAPTER 8

Applications

Where Are We Going?

We start this chapter with a detailed discussion of the oscillatory motion of a pendulum, where we consider both a linear and nonlinear model. We then move on to discover properties of spring-mass systems, including overdamped, critically damped, and underdamped situations. We also discuss similar motions in simple electric circuits. We conclude this chapter by analyzing two point boundary value problems, which leads to a discussion of eigenvalues, eigenfunctions, and orthogonality.

8.1 The Simple Pendulum

In science and engineering there are many situations in which the dependent variable that models some phenomena oscillates. Such motion is called OSCILLATORY. This section deals with the motion of a simple pendulum.

We begin our discussion with some groundwork. The simple pendulum consists of a light rigid rod of length h, hinged at one end, with a mass m attached to the other. When the pendulum is at rest in a vertical position with the weight directly below the hinge, we have the configuration from which measurements are made. We start our discussion of oscillatory motion by considering a simple pendulum in motion, where the position of the pendulum is characterized by the angle x (in radians) between the rod and the rest position. See Figure 8.1, in which positive angles are measured to the right, and $dx/dt > 0$ represents counterclockwise motion.

If the weight of the rod is negligible, the hinge is frictionless, and there is no air resistance, the differential equation governing the pendulum in motion is[1]

$$m\frac{d^2x}{dt^2} = -\frac{mg}{h}\sin x,$$

or

$$x'' + \lambda^2 \sin x = 0, \tag{8.1}$$

---

[1] The derivation of this equation can be found in "Physics" by R. Resnick, D. Halliday, and K.S. Krane, Volume 1, 4th edition, Wiley 1992, page 323.

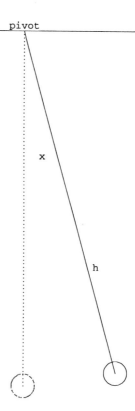

FIGURE 8.1   Simple pendulum

where

$$\lambda = \sqrt{\frac{g}{h}}. \tag{8.2}$$

Equation (8.1) is a second order nonlinear differential equation and cannot be solved in terms of familiar functions. However, there is an existence theorem for general differential equations of the form

$$\frac{d^2x}{dt^2} = f\left(t, x, x'\right), \tag{8.3}$$

which is very similar to Theorem 2.1. It essentially says that if f, $\partial f/\partial x$, and $\partial f/\partial x'$ are continuous in the vicinity of t_0, x_0, and x_0^*, then the differential equation (8.3) has a unique solution subject to the initial conditions $x(t_0) = x_0$, $x'(t_0) = x_0^*$.

Because the differential equation (8.1) satisfies the conditions of this theorem for every $x(t_0) = x_0$, $x'(t_0) = x_0^*$, we are guaranteed that a solution curve exists and is unique. For example, the functions $x(t) = n\pi$ ($n =$ any integer) satisfy (8.1) given the initial conditions $x(t_0) = n\pi$, $x'(t_0) = 0$. The functions $x(t) = n\pi$ are thus solutions, and no other solution can have the property that $x(t_0) = n\pi$, $x'(t_0) = 0$. Physically these are solutions that correspond to the pendulum hanging at rest, with the mass either directly below the hinge ($x = 0$, $x = \pm 2\pi$, $\cdots$) or directly above the hinge ($x = \pm\pi$, $x = \pm 3\pi$, $\cdots$). We are not able to tell the difference between the solutions $x(t) = 0$ and $x(t) = 2\pi$, unless we know that someone had first rotated the pendulum through 2π and then let it hang there. Similar comments apply to the other solutions.

EXAMPLE 8.1: Simple Frictionless Pendulum, Linearized

If x is small (that is, the pendulum's displacement is always near the rest position), we can approximate $\sin x$ by the first term in its Taylor series,

$$\sin x \approx x,$$

in which case (8.1) becomes

$$\frac{d^2 x}{dt^2} + \lambda^2 x = 0. \tag{8.4}$$

We have seen this differential equation many times, with its general solution as

$$x(t) = C_1 \cos \lambda t + C_2 \sin \lambda t, \tag{8.5}$$

where C_1 and C_2 are arbitrary constants. If initially we have

$$x(0) = x_0, \qquad \frac{dx}{dt}(0) = v_0,$$

then (8.5) becomes

$$x(t) = x_0 \cos \lambda t + \frac{v_0}{\lambda} \sin \lambda t. \tag{8.6}$$

If we introduce the angular velocity $v(t) = dx/dt$ we can write (8.4) as the system of equations

$$\frac{dx}{dt} = v, \qquad \frac{dv}{dt} = -\lambda^2 x, \tag{8.7}$$

Notice that the equilibrium point of (8.32) and (8.33) is $x = 0$, $v = 0$, which is when the pendulum is at rest, hanging vertically. Notice also that the uniqueness theorem guarantees that there is only one solution with initial conditions $x = 0$, $v = 0$ [the solution $x(t) = 0$], so no other trajectory can pass through the equilibrium point in the phase plane.

The equation for the direction field in the phase plane is

$$\frac{dv}{dx} = \left(\frac{dv}{dt}\right) / \left(\frac{dx}{dt}\right) = -\lambda^2 \frac{x}{v},$$

which can be integrated to give the trajectories

$$v^2(t) + \lambda^2 x^2(t) = C. \tag{8.8}$$

These trajectories are ellipses in the phase plane. Utilizing our initial conditions allows us to write this equation as

$$v^2(t) + \lambda^2 x^2(t) = v^2(0) + \lambda^2 x^2(0) = v_0^2 + \lambda^2 x_0^2. \tag{8.9}$$

The direction field and some trajectories with different initial conditions are shown in Figure 8.2 for $\lambda = 3$.

Looking at (8.7) we see that in the first quadrant where $x > 0$ and $v > 0$, x is an increasing function of t, and v is a decreasing function of t. Similar comments apply to the remaining three quadrants. Thus, motion along any trajectory is clockwise. This is consistent with the direction of the arrows in Figure 8.2.

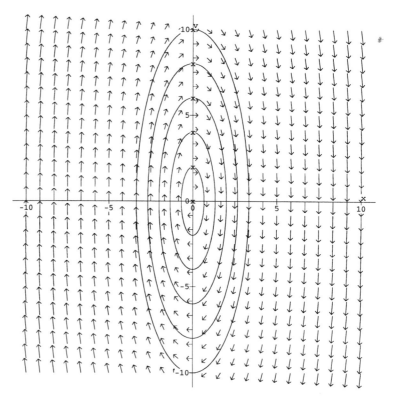

FIGURE 8.2   Trajectories and direction field for $dv/dx = -\lambda^2 x/v$

Let's interpret one of these trajectories in terms of the motion of the pendulum — say, the trajectory which corresponds to $C = 16$ — which has the equation

$$v^2 + 9x^2 = 16.$$

This ellipse crosses the x-axis at $x = \pm 4/3$, and the v-axis at $v = \pm 4$, as can be seen in Figure 8.2. We will start at the point $(4/3, 0)$ and proceed clockwise once around the ellipse until we return to $(4/3, 0)$. The point $(4/3, 0)$ corresponds to an angle of $4/3$ with a velocity of zero. This is when the pendulum is at rest at its extreme right-hand position. The part of the trajectory in the fourth quadrant corresponds to the pendulum swinging back towards the vertical position, which it reaches at $(0, -4)$, when the velocity is -4. In the third quadrant the pendulum is moving to the left, until it reaches its extreme left-hand position at an angle of $-4/3$, the point $(-4/3, 0)$ in the phase plane. The second quadrant characterizes the pendulum swinging back to the vertical position $(0, 4)$, and the first quadrant is where the pendulum swings past vertical back to its original starting angle on the right-hand side. Following the ellipse around for a second time repeats the process. In other words, the ellipse corresponds to a pendulum swinging backwards and forwards forever. Different ellipses correspond to different initial conditions. The smaller the ellipse the smaller the pendulum's maximum angle, or the slower the pendulum.

From (8.9) we see that v will have its extreme values when $x = 0$ and that x will have its extreme values when $v = 0$. In other words, the fastest the pendulum will move is when it is passing its vertical position $x = 0$, and the pendulum will have zero velocity when it is at a maximum angle from its vertical position.

For (8.4) to model a real pendulum the angle x must be small (although mathematically there is no restriction on the size of x). If we think of the situation where we let the pendulum hang at rest and then give it an initial velocity — so $x(0) = 0$ and $v(0) = v_0$ — this model predicts that the pendulum will oscillate backwards and forwards forever, as is indicated both by the direction field in the phase plane, and by the corresponding time-dependent solution

$$x(t) = \frac{v_0}{\lambda} \sin \lambda t. \tag{8.10}$$

This is also seen in Figure 8.3 where we used the technique described under *How to Construct the Trajectory in the Phase Plane* on page 356. We show the graph of the angle $x(t)$ given by (8.10) (for $\lambda = 3$ and $v_0 = 3$) in the lower right-hand box, the graph of the angular velocity $v(t) = x' = v_0 \cos \lambda t$ in the upper right-hand box, and the trajectory in the phase plane in the upper left-hand box.

These trajectory predicts that no matter how large the initial velocity v_0, the motion is always bounded and oscillatory. This is clearly unrealistic, because if we have a sufficiently large initial velocity we would expect the pendulum to go over the top, and, because there is no friction, continue going over the top forever. In this case the motion would be unbounded. This reinforces our earlier statement that for the model to be realistic, x must be restricted to small values.

If the pendulum is released from rest, its subsequent motion is

$$x(t) = x_0 \cos \lambda t.$$

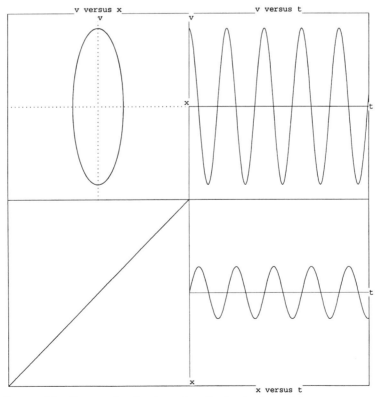

FIGURE 8.3   Constructing the phase plane for the simple pendulum

The period T of the pendulum is the smallest time it takes to return to its maximum displacement of x_0. Thus, the period is the smallest T for which $x(T) = x_0$, so T must satisfy $x_0 = x_0 \cos \lambda T$. Solving for the smallest $T > 0$, we find

$$T = \frac{2\pi}{\lambda} = 2\pi \sqrt{\frac{h}{g}}. \tag{8.11}$$

Notice that the period is independent of the mass and the initial angle and that it varies as the square root of the length of the pendulum. The validity of (8.11) can be checked by simple experiments (see Exercise 1 on page 381).

We can use (8.9) to obtain the period T in a different way, without solving (8.4), as follows. If the pendulum is released from rest, then (8.9) can be written

$$\left(\frac{dx}{dt}\right)^2 = \lambda^2 \left(x_0^2 - x^2\right). \tag{8.12}$$

This results in

$$\frac{dx}{dt} = \pm \lambda \sqrt{x_0^2 - x^2},$$

or

$$\frac{dt}{dx} = \frac{1}{\pm \lambda \sqrt{x_0^2 - x^2}}, \tag{8.13}$$

where $+$ applies to the case where x is increasing (moving to the right) and $-$ applies to the case where x is decreasing (moving to the left). Thus, one period would correspond to four different motions, from the initial (maximum) value to $x = 0$, from $x = 0$ to the minimum value of x, and then back again. The extreme values of x will occur when $dx/dt = 0$ — namely, $x(t) = \pm x_0$. Thus, the pendulum will swing an equal distance in both directions and will return to its original angle in one period. This period will be the sum of the time the pendulum takes to move from x_0 to 0, then from 0 to $-x_0$, then from $-x_0$ to 0, and finally back to x_0. The time the pendulum takes to move from x_0 to 0 is

$$\int_{x=x_0}^{x=0} dt,$$

with similar integrals for the other three time intervals. Thus the period T will be

$$T = \int_{x=x_0}^{x=0} dt + \int_{x=0}^{x=-x_0} dt + \int_{x=-x_0}^{x=0} dt + \int_{x=0}^{x=x_0} dt,$$

corresponding to the four successive motions, or

$$T = \int_{x=x_0}^{x=0} \frac{dt}{dx} dx + \int_{x=0}^{x=-x_0} \frac{dt}{dx} dx + \int_{x=-x_0}^{x=0} \frac{dt}{dx} dx + \int_{x=0}^{x=x_0} \frac{dt}{dx} dx.$$

However, we must make sure we substitute the correct dt/dx from (8.13) into the integrals in this last equation. Remember that $-$ corresponds to the angle decreasing (the first and second integrals), whereas $+$ corresponds to the angle increasing (the third and fourth integrals). We thus have

$$T = \int_{x_0}^0 \frac{1}{-\lambda\sqrt{x_0^2 - x^2}}\,dx + \int_0^{-x_0} \frac{1}{-\lambda\sqrt{x_0^2 - x^2}}\,dx +$$

$$+ \int_{-x_0}^0 \frac{1}{\lambda\sqrt{x_0^2 - x^2}}\,dx + \int_0^{x_0} \frac{1}{\lambda\sqrt{x_0^2 - x^2}}\,dx. \qquad (8.14)$$

If we make use of $\int_a^b f(x)\,dx = -\int_{-a}^{-b} f(-x)\,dx$ and $\int_a^b f(x)\,dx = -\int_b^a f(x)\,dx$, we see that each of the four integrals in (8.14) is equivalent to the first one, so the full period is four times the time it takes for the pendulum to swing from $x = x_0$ to $x = 0$ (in complete agreement with our intuition). That is,

$$T = -4 \int_{x_0}^0 \frac{1}{\lambda\sqrt{x_0^2 - x^2}}\,dx,$$

or

$$T = 4\sqrt{\frac{h}{g}} \int_0^{x_0} \frac{1}{\sqrt{x_0^2 - x^2}}\,dx. \qquad (8.15)$$

This last integral has two unusual features: it appears to depend on x_0 (the initial angle), and it is an improper integral (why?). We could check to see whether this integral actually depends on x_0 by computing T for different values of x_0 by a standard numerical integration technique, such as Simpson's rule. Unfortunately, such numerical techniques are not reliable when applied to improper integrals. One way to get around this problem is to use a change of variable that converts this improper integral into a proper integral. A typical change of variables for this type of integral would be

$$x = x_0 \sin u, \qquad (8.16)$$

in which case

$$
\begin{aligned}
T &= 4\sqrt{\frac{h}{g}} \int_0^{x_0} \frac{1}{\sqrt{x_0^2 - x^2}}\,dx \\
&= 4\sqrt{\frac{h}{g}} \int_0^{\pi/2} \frac{1}{\sqrt{x_0^2 - x_0^2 \sin^2 u}}\, x_0 \cos u\,du \\
&= 4\sqrt{\frac{h}{g}} \int_0^{\pi/2} du.
\end{aligned}
\qquad (8.17)
$$

In this case the integral reduces to a proper integral, which we can evaluate directly without recourse to numerical integration. Also notice that the dependence on x_0 vanished, so in fact we find

$$T = 2\pi\sqrt{\frac{h}{g}},$$

in agreement with (8.11).   □

EXAMPLE 8.2

Consider the specific situation where $g/h = 1$, so $\lambda = 1$ and the differential equation becomes

$$x'' + x = 0, \tag{8.18}$$

with initial conditions

$$x(0) = 0.2, \qquad x'(0) = 0.15. \tag{8.19}$$

These conditions correspond to the situation in which the pendulum is released at an angle of 0.2 radian with an initial angular velocity of 0.15 radian per second, which means that initially the angle is going to increase. We want to find the maximum angle that the pendulum attains (before swinging back), when this first occurs, and when the pendulum is vertical the first time.

Using (8.6) the solution of (8.18) subject to (8.19) is given by

$$x(t) = 0.2 \cos t + 0.15 \sin t.$$

The maximum angle will occur when $dx/dt = 0$ and $x > 0$. Because $dx/dt = -0.2 \sin t + 0.15 \cos t$, the times when $dx/dt = 0$ satisfy the equation

$$\tan t = \frac{0.15}{0.2} = \frac{3}{4},$$

that is,

$$t = \arctan \frac{3}{4} + n\pi \approx 0.6435 + n\pi, \qquad n = 0, \pm 1, \pm 2, \cdots.$$

The maximum angle will occur when $t = t_1$ secs, where

$$t_1 = \arctan \frac{3}{4} \approx 0.6435.$$

At this time $\cos t_1 = 4/5$ and $\sin t_1 = 3/5$, so the maximum angle $x_1 = x(t_1)$ is

$$x_1 = x(t_1) = 0.2 \cos t_1 + 0.15 \sin t_1 = 0.2 \cdot \frac{4}{5} + 0.15 \cdot \frac{3}{5} = 0.25.$$

The first time $t_2 > 0$ when the pendulum is vertical is when $x(t_2) = 0$ — that is, when

$$0.2 \cos t_2 + 0.15 \sin t_2 = 0,$$

or

$$t_2 = \pi - \arctan \frac{0.2}{0.15} \approx 2.2143.$$

Notice that we used the explicit solution to find x_1, t_1, and t_2. We will be returning to this problem in this chapter and the next.   □

EXAMPLE 8.3: Simple Frictionless Pendulum

We are unable to solve the nonlinear differential equation for the frictionless pendulum

$$\frac{d^2x}{dt^2} + \lambda^2 \sin x = 0,$$

where $\lambda = \sqrt{g/h}$, in terms of familiar functions. However, we can use it to obtain useful information by following steps similar to the linearized case. We introduce the angular velocity, $v = dx/dt$, and write (8.1) as the system of equations

$$\frac{dx}{dt} = v, \qquad \frac{dv}{dt} = -\lambda^2 \sin x.$$

The trajectories in the phase plane will satisfy

$$\frac{dv}{dx} = -\lambda^2 \frac{\sin x}{v},$$

and Figure 8.4 shows the corresponding direction field. (Why is the motion clockwise?)
This differential equation may be integrated to yield

$$v^2(t) - 2\lambda^2 \cos x(t) = c, \tag{8.20}$$

where c is an arbitrary constant. This is the counterpart of our earlier model (8.8). Note that we found (8.8) by approximating $\sin x$ with x, for small x. Thus, for small values of

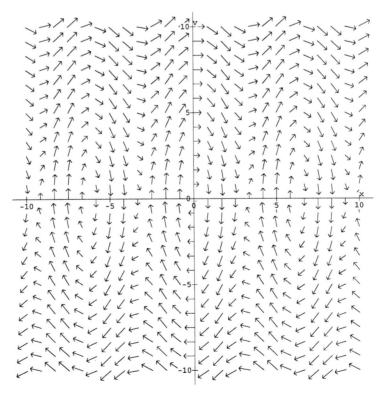

FIGURE 8.4   Direction field for $dv/dx = -(\lambda^2 \sin x)/v$

x, (8.20) should be equivalent to (8.8). As stated (8.20) is not too revealing; to explore this further, we use the trigonometric identity

$$\cos x = 1 - 2\sin^2 \frac{x}{2}$$

to rewrite (8.20) in the form

$$v^2(t) + 4\lambda^2 \sin^2 \frac{x(t)}{2} = C,$$

where $C = c + 2\lambda^2$. With the initial conditions $x(0) = x_0$, $v(0) = v_0$, this conservation of energy equation becomes

$$v^2(t) + 4\lambda^2 \sin^2 \frac{x(t)}{2} = v_0^2 + 4\lambda^2 \sin^2 \frac{x_0}{2}. \tag{8.21}$$

In this form, the approximation $\sin x/2 \approx x/2$ reduces (8.21) to (8.9) as we expected.

If we consider the pendulum starting at $x(0) = 0$ with prescribed initial velocity $v(0) = v_0$, we have the trajectory in the phase plane given by

$$v^2(t) + 4\lambda^2 \sin^2 \frac{x(t)}{2} = v_0^2.$$

Solving this for $v(t)$ gives

$$v(t) = \pm 2\lambda \sqrt{\mu^2 - \sin^2 \frac{x(t)}{2}}, \tag{8.22}$$

where

$$\mu = \frac{v_0}{2\lambda}.$$

Here $+$ applies to the case in which x is increasing, and $-$ applies when x is decreasing. For $v(t)$ to be real, x must satisfy

$$\mu^2 - \sin^2 \frac{x(t)}{2} \geq 0. \tag{8.23}$$

This gives rise to four types of trajectories depending on the magnitude of μ.

1. If $\mu^2 > 1$, then (8.23) is always satisfied in the sense that $\sin^2 x/2 \leq 1$ for all x. In this case the trajectories (8.22) will be defined for every x, and the velocity $v(t)$ can never be zero. The condition $\mu^2 > 1$ is equivalent to $v_0 > 2\lambda$ or $v_0 < -2\lambda$. So for large initial velocities, the trajectory is defined for all x. Physically this corresponds to a large enough initial velocity to propel the pendulum over the top. Because we are dealing with a frictionless pendulum this will result in the mass circling the pivot forever. Examples of these trajectories are shown in Figure 8.5.

2. If $\mu^2 = 1$, then (8.23) is satisfied for every x. However, (8.22) implies that the velocity is zero whenever $\sin x/2 = \pm 1$; that is, at $x = \pm\pi, \pm 3\pi, \pm 5\pi \cdots = (2n + 1)\pi$, where n is any integer. So if the trajectory continues for all x, it would have the property that there is a time t_0 for which $x(t_0) = \pi$ and $v(t_0) = 0$. However, we have already pointed out that the only solution for which $x(t_0) = \pi$ and $v(t_0) = 0$ is the one for which $x(t) = \pi$ for all time. Consequently, our trajectory never reaches this point in finite time, so the trajectory does not continue for all x but approaches the upper limit of π. Physically this corresponds to an initial velocity that is exactly right to make the pendulum come

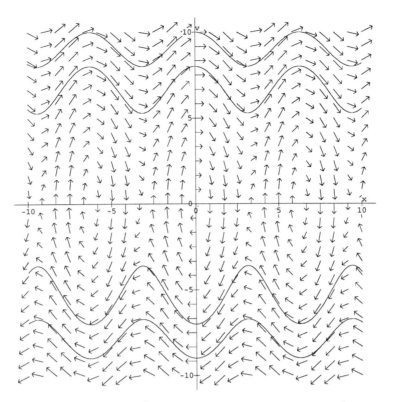

FIGURE 8.5 Trajectories ($\mu^2 > 1$) and direction field for $dv/dx = -(\lambda^2 \sin x)/v$

to rest in a vertical position with the mass balancing above the hinge. However, it will take the pendulum an infinite time to reach this position. (This same result is proved in a different way in Exercise 6, page 381.) An example of this trajectory is shown in Figure 8.6.

3. If $0 < \mu^2 < 1$, then (8.23) will be satisfied for some values of x. For example, if $\mu^2 = 1/2$, then the only values of x for which $\sin^2 x/2 \leq 1/2$ are $(4n - 1)\pi/2 \leq x \leq (4n + 1)\pi/2$, where n is any integer. For the initial condition $x(0) = 0$ to be included in this set, we must have $n = 0$, so this implies that $-\pi/2 \leq x \leq \pi/2$ for $\mu^2 = 1/2$. The condition $0 < \mu^2 < 1$ is equivalent to $0 < v_0 < 2\lambda$ or $-2\lambda < v_0 < 0$. So for moderate initial velocities, the trajectory is defined for these values of x. Physically this corresponds to a pendulum swinging backwards and forwards, and for small x it corresponds to the linearized motion described in the previous example. The velocity will be zero when x has its extreme values. Examples of these trajectories are shown in Figure 8.7.

4. If $0 = \mu^2$, then $\sin^2 x/2 \leq 0$, which is true only when $\sin x/2 = 0$ — that is, when $x = 2n\pi$. This corresponds to $v_0 = 0$, meaning that for zero initial velocity, we have only discrete values of x. For our initial condition, $n = 0$, so this corresponds to having the pendulum hang vertically for all time.

To summarize our previous discussion we combine all the cases in a single picture of the phase plane as shown in Figure 8.8 for $\lambda = 3$. Also shown are trajectories with $x(0) = 0$ and $v(0) = 10, 8, -7,$ and -9 (which go on forever); $v(0) = 6$ (which takes an infinite time to reach the point $x = \pi$); $v(0) = 4, 2,$ and -5 (which oscillate forever); and $v(0) = 0$, where the pendulum is stationary for all time. If we consider the trajectory through $x(0) = 0$, $v(0) = 10$, the direction of travel of the pendulum starts from left to right. The

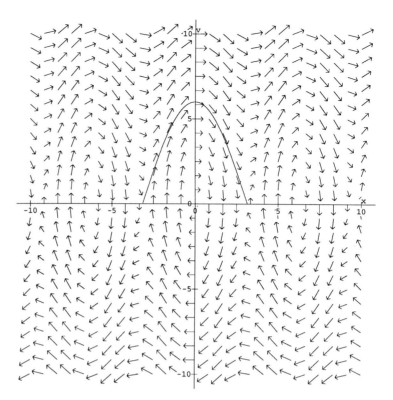

FIGURE 8.6   Trajectory ($\mu^2 = 1$) and direction field for $dv/dx = -(\lambda^2 \sin x)/v$

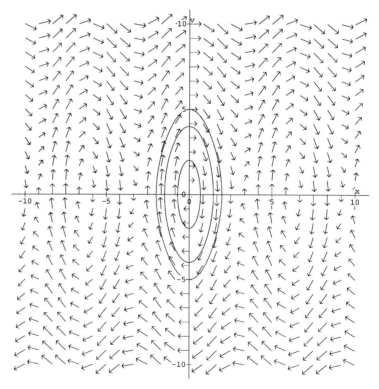

FIGURE 8.7   Trajectories ($0 < \mu^2 < 1$) and direction field for $dv/dx = -(\lambda^2 \sin x)/v$

trajectory for $x < 0$ is the trajectory corresponding to the motion of the pendulum prior to our consideration of it at $x(0) = 0$, $v(0) = 10$. In the same way, the motion of the pendulum corresponding to the trajectory through $x(0) = 0$, $v(0) = -9$, starts from right to left. The trajectory for $x > 0$ is the trajectory corresponding to the motion of the pendulum prior to our consideration of it at $x(0) = 0$, $v(0) = -9$.

It should be noted that while we cannot solve $x'' + \lambda^2 \sin x = 0$ in terms of familiar functions, we have discovered a great deal about its solutions by analyzing its trajectories in the phase plane.

Let us now further explore the notion of period. We pull the pendulum to one side to an initial angle of x_0 $(0 < x_0 < \pi)$ and release the pendulum from rest, giving us the initial conditions

$$x(0) = x_0, \qquad v(0) = 0. \tag{8.24}$$

To compute the period, we want the time the pendulum takes to return to the original angle x_0. Looking at the various trajectories in the phase plane (Figure 8.8), we see that the only trajectories for which the notion of period makes sense would be closed, ellipse-like trajectories. (Why?)

If we use $x(0) = x_0$, $v(0) = 0$ in (8.21), we find

$$\left(\frac{dx}{dt} \right)^2 = 4\lambda^2 \left(\sin^2 \frac{x_0}{2} - \sin^2 \frac{x(t)}{2} \right),$$

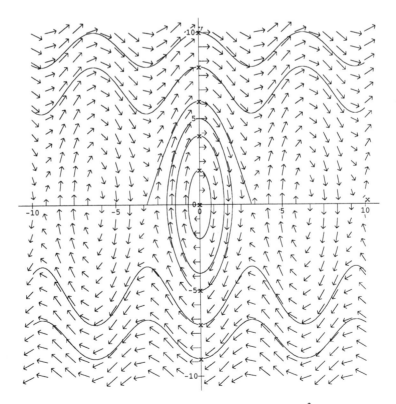

FIGURE 8.8 Trajectories and direction field for $dv/dx = -(\lambda^2 \sin x)/v$

which is the counterpart of (8.12), the linearized case. From this we see that the velocity will be zero whenever $\sin^2(x_0/2) - \sin^2(x(t)/2) = 0$ — that is, when $x(t) = \pm x_0$. The counterpart of (8.13) is

$$\frac{dt}{dx} = \frac{1}{\pm 2\lambda\sqrt{\sin^2(x_0/2) - \sin^2(x/2)}},$$

where $+$ applies to the case in which x is increasing (moving to the right) and $-$ applies when x is decreasing (moving to the left). Thus, one period would correspond to four different motions, from the initial (maximum) value of x_0 to $x = 0$, from $x = 0$ to the minimum value of x, $-x_0$, and then back again. The pendulum will therefore swing an equal distance in both directions and will return to its original angle in one period. Using arguments similar to those used in the previous example, we see that this period will also be four times the time it takes for the pendulum to swing from $x = x_0$ to $x = 0$, that is,

$$T = 4\int_{x=x_0}^{x=0} dt = 4\int_{x_0}^{0} \frac{dt}{dx} dx = -4\int_{x_0}^{0} \frac{1}{2\lambda\sqrt{\sin^2(x_0/2) - \sin^2(x/2)}} dx.$$

In view of (8.2), this last integral can be rewritten as

$$T = 2\sqrt{\frac{h}{g}}\int_{0}^{x_0} \frac{1}{\sqrt{k^2 - \sin^2(x/2)}} dx, \tag{8.25}$$

where

$$k = \sin\frac{x_0}{2}.$$

This integral is the counterpart of (8.15). Again the period appears to depend on x_0 and again we have an improper integral. We follow the pattern suggested by the linearized case and make a substitution similar to (8.16) to try to convert the improper integral into a proper one, namely,

$$\sin\frac{x(t)}{2} = k\sin u.$$

From this we see that

$$\frac{1}{2}\cos\frac{x(t)}{2} dx = k\cos u\, du,$$

or

$$dx = \frac{2k\cos u}{\cos x/2} du = \frac{2k\cos u}{\sqrt{1 - \sin^2 x/2}} du = \frac{2k\cos u}{\sqrt{1 - k^2\sin^2 u}} du.$$

Because $u = 0$ when $x = 0$, and $u = \pi/2$ when $x = x_0$, the integral in (8.25) can be written

$$\int_{0}^{x_0} \frac{1}{\sqrt{k^2 - \sin^2(x/2)}} dx$$

$$= \int_{0}^{\pi/2} \frac{1}{\sqrt{k^2 - k^2\sin^2 u}} \frac{2k\cos u}{\sqrt{1 - k^2\sin^2 u}} du = 2\int_{0}^{\pi/2} \frac{du}{\sqrt{1 - k^2\sin^2 u}}.$$

Thus, (8.25) becomes

$$T = 4\sqrt{\frac{h}{g}} \int_0^{\pi/2} \frac{du}{\sqrt{1 - k^2 \sin^2 u}}, \tag{8.26}$$

where

$$k = \sin \frac{x_0}{2}. \tag{8.27}$$

This is a proper integral provided $k^2 < 1$, which is always satisfied for the trajectories we are considering ($-\pi < x_0 < \pi$). There are two major differences between (8.26) and its counterpart (8.17) in the linearized case. First, T depends on the initial angle x_0 (through k), and second, the integral in (8.26) cannot be evaluated in terms of familiar functions.

We can always evaluate the integral numerically for given x_0 (and hence given k) by any of the standard numerical integration techniques. For example, using Simpson's rule, we can construct Table 8.1, which shows that the period depends on the initial angle x_0.

The integral in (8.26) occurs in a number of different applications, and it is known as the COMPLETE ELLIPTIC INTEGRAL OF THE FIRST KIND. ☐

TABLE 8.1 Simpson's rule for period

x_0	T
$\pi/180$ (1°)	$6.2833\sqrt{h/g}$
$\pi/36$ (5°)	$6.2862\sqrt{h/g}$
$\pi/6$ (30°)	$6.3926\sqrt{h/g}$
$\pi/4$ (45°)	$6.5343\sqrt{h/g}$
$\pi/3$ (60°)	$6.7430\sqrt{h/g}$

EXAMPLE 8.4

Consider the situation with $\lambda = 1$ where our differential equation,

$$x'' + \sin x = 0,$$

is subject to the same conditions as Example 8.2, namely,

$$x(0) = 0.2, \qquad x'(0) = 0.15.$$

We wish to find the maximum angle that the pendulum attains (before swinging back), when this first occurs, and also when the pendulum is vertical the first time.

Contrary to the situation in Example 8.2, we do not have an explicit solution in terms of familiar functions. However, we can still find these values by using (8.21), which in this case is

$$v^2(t) + 4 \sin^2 \frac{x(t)}{2} = \alpha^2, \tag{8.28}$$

where

$$\alpha^2 = (0.15)^2 + 4 \sin^2 0.1 \approx 0.06237.$$

The maximum angle will occur when $v = 0$ and $x > 0$, which, by (8.28), is at time t_1 when the angle x_1 is

$$x_1 = x(t_1) = 2 \arcsin \frac{\alpha}{2} \approx 0.2504.$$

(The corresponding angle in Example 8.2 is 0.25.) So this is the angle, but what is the time t_1? From (8.28) we have

$$\frac{dt}{dx} = \frac{1}{\pm 2\sqrt{(\alpha/2)^2 - \sin^2(x(t)/2)}}, \tag{8.29}$$

so

$$t_1 = \int_{x=x_0}^{x=x_1} dt = \int_{x_0}^{x_1} \frac{dt}{dx} dx = \int_{x_0}^{x_1} \frac{1}{2\sqrt{(\alpha/2)^2 - \sin^2(x/2)}} dx,$$

or

$$t_1 = \int_{0.2}^{2 \arcsin \alpha/2} \frac{1}{2\sqrt{(\alpha/2)^2 - \sin^2(x/2)}} dx.$$

[Why did we use $+$ instead of $-$ from (8.29)?] We could try to evaluate this integral numerically by Simpson's rule, for example, but it is an improper integral, so numerical techniques are suspect. As in Example 8.3, we try to convert this into a proper integral using the transformation

$$\sin \frac{x(t)}{2} = \frac{\alpha}{2} \sin u,$$

obtaining

$$t_1 = \int_{0.2}^{2 \arcsin \alpha/2} \frac{1}{2\sqrt{(\alpha/2)^2 - \sin^2(x(t)/2)}} dx = \int_{u_1}^{u_2} \frac{du}{\sqrt{1 - (\alpha/2)^2 \sin^2 u}},$$

where $u_1 = \arcsin[(2/\alpha) \sin(0.2/2)] \approx 0.9265$ and $u_2 = \arcsin[(2/\alpha) \sin(\arcsin \alpha/2)] = \arcsin 1 = \pi/2$. Evaluating this integral numerically, using Simpson's rule, gives

$$t_1 \approx 0.6487.$$

(The corresponding time in Example 8.2 is 0.6435.)

The first time $t_2 > 0$ when the pendulum is vertical is when $x(t_2) = 0$, which is the time taken for the pendulum to swing out to x_1 (that is, t_1) and then return to $x = 0$, namely,

$$t_2 = t_1 + \int_{x=x_1}^{x=0} dt = t_1 + \int_{x_1}^{0} \frac{dt}{dx} dx.$$

However, because the pendulum is swinging to the left when it goes from x_1 to 0, we must use $-$ in (8.29), so

$$t_2 = t_1 + \int_{x_1}^{0} \frac{1}{-2\sqrt{(\alpha/2)^2 - \sin^2(x/2)}} dx = t_1 - \int_{2 \arcsin \alpha/2}^{0} \frac{1}{2\sqrt{(\alpha/2)^2 - \sin^2(x/2)}} dx.$$

This improper integral can be converted to a proper integral as before, and we find

$$t_2 = t_1 - \int_{\pi/2}^{0} \frac{du}{\sqrt{1 - (\alpha/2)^2 \sin^2 u}}.$$

Simpson's rule gives

$$t_2 \approx 2.257.$$

(The corresponding time in Example 8.2 is 2.2143.)

Notice that although we had no explicit solution, we were still able to find x_1, t_1, and t_2. We will return to this problem in the next chapter. $\square$

EXERCISES

1. Perform the following experiment. Take a piece of string and fix a weight on one end. Anchor the other end, and measure the length of the pendulum. Displace the pendulum through a small angle, release it from rest, and measure the time it takes to return to the initial angle.

   (a) Repeat this a few times with different initial angles. Is the resulting period independent of the small initial angle?

   (b) Repeat this a few times with different masses. Is the resulting period independent of the mass?

   (c) Repeat the experiment for pendulums of different lengths. Do your results agree with (8.11)?

2. Repeat the experiment described in Exercise 1 for larger initial angles. Show that the model is less reliable as the initial angle increases.

3. On Figure 8.8, sketch the trajectories corresponding to the following physical situations.

   (a) A pendulum hangs so that the mass is directly below the hinge for all time. How many such trajectories are there?

   (b) A pendulum is carefully suspended so that the mass is directly above the hinge for all time. How many such trajectories are there?

4. Find the solution of the linearized equation for the simple pendulum (8.4) if the pendulum is released from rest at an angle of $1/10$ radians. The relative difference between x and $\sin x$ is $(x - \sin x)/x$. Show that it satisfies

   $$\frac{x - \sin x}{x} < \frac{1}{3!}x^2.$$

   What is the maximum relative difference between x and $\sin x$ for this motion?

5. Find the solution of the linearized equation (8.4) if the pendulum starts from its equilibrium position with an initial velocity of 2 radians per second. What is the maximum relative difference between x and $\sin x$ for this motion?

6. From (8.22) show that the time it takes for the pendulum that starts at $x(0) = 0$ with velocity $v(0) = v_0$ to reach the angle α is given by

   $$t = \int_{0}^{\alpha} \frac{1}{2\lambda\sqrt{\mu^2 - \sin^2(x(t)/2)}}\, dx,$$

   where

   $$\mu = \frac{v_0}{2\lambda}.$$

If $v_0 = 2\lambda$, show that as $\alpha \to \pi$, $t \to \infty$. Discuss the physical interpretation of this result.

7. Show that if $-\pi < x_0 < \pi$, then

$$\int_0^{x_0} \frac{1}{\sqrt{\sin^2(x_0/2) - \sin^2(x(t)/2)}} \, dx > \int_0^{x_0} \frac{1}{\sqrt{1 - \sin^2(x(t)/2)}} \, dx.$$

Use this result to show that

$$\lim_{x_0 \to \pi} \int_0^{x_0} \frac{1}{\sqrt{\sin^2(x_0/2) - \sin^2(x(t)/2)}} \, dx = \infty.$$

Explain how this relates to (8.25).

8. Expand the integrand in (8.26) as a power series in $\sin^2 u$, and use the fact that

$$\int_0^{\pi/2} \sin^{2n} u \, du = \frac{1 \cdot 3 \cdot 5 \cdots (2n-1)}{2 \cdot 4 \cdot 6 \cdots 2n} \frac{\pi}{2}$$

to find

$$T = 2\pi \sqrt{\frac{h}{g}} \left[1 + \left(\frac{1}{2}\right)^2 k^2 + \left(\frac{1 \cdot 3}{2 \cdot 4}\right)^2 k^4 + \left(\frac{1 \cdot 3 \cdot 5}{2 \cdot 4 \cdot 6}\right)^3 k^6 + \cdots \right],$$

where $k = \sin(x_0/2)$. Use this result with $x_0 = \pi/180$ [so that $k = \sin(\pi/360)$] to estimate T. Compare your result with the linearized period of $T = 2\pi\sqrt{h/g} \approx 6.2832$. As x_0 increases, do you expect T to get closer to or farther from the linearized period of $T = 2\pi\sqrt{h/g}$? As $x_0 \to \pi$, what do you expect to happen to T? Is it possible for the same pendulum to have every period from 0 to ∞, just by adjusting the initial angle?

9. A SECONDS PENDULUM CLOCK is one that ticks every second at the end of each swing of the pendulum, so it has a period of 2 seconds. We want to construct such a frictionless clock.

(a) We decide to use the linearized model for the pendulum. How long should we make the pendulum, if $g = 32.2$ feet/sec^2? What has this to do with the size of grandfather clocks?

(b) We build the clock according to the results from part (a) and start it from rest at an initial angle of 10°. We know that the correct frictionless model is not the linearized model but is governed by (8.1). Will our clock run fast or slow? Over a 24-hour period, how inaccurate is our clock? How often should we restart the clock so it is never more than a minute wrong?

(c) I have a pendulum clock that has a pendulum 9 inches long. At what angle should I start it from rest for it to tick every second in the nonlinear case?

10. By analyzing the trajectory in the phase plane for the nonlinear pendulum with initial conditions $x(0) = 0$, $x'(0) = \beta$, answer the following questions.

(a) For what values of β will x be bounded for all times?

(b) For what value of β will the pendulum approach an unstable equilibrium in the limit as $t \to \infty$?

(c) If the initial conditions are changed to $x(0) = \alpha$, $x'(0) = 0$, can you choose α so that in the resulting motion x becomes unbounded? Explain fully.

11. Recall that the Taylor series for $\sin x$ (about the origin) is

$$\sin x = x - \frac{1}{3!}x^3 + \frac{1}{5!}x^5 - \cdots.$$

For oscillations where x is small, we might approximate $\sin x$ by $x - x^3/3!$ and get the differential equation for a simple pendulum

$$\frac{d^2x}{dt^2} + \lambda^2 \left(x - \frac{1}{3!}x^3 \right) = 0 \qquad (8.30)$$

in place of the linearized pendulum (8.4).

(a) Write (8.30) in terms of the phase plane variables (x and $v = dx/dt$), and compare the direction field of your result with the one we obtained for the linearized pendulum, Figure 8.2. Explain fully any similarities and differences.

(b) Find the equation for the trajectories in the phase plane, and compare it with the equation for the trajectories in the phase plane for $x'' + \lambda^2 x = 0$, given by (8.8).

(c) Write (8.30) in terms of the phase plane variables (x and $v = dx/dt$) and compare the direction field of your result with the one we obtained for the nonlinear pendulum, Figure 8.8. Explain fully any similarities and differences.

(d) Compare the equation for the trajectories in the phase plane for $x'' + \lambda^2 \sin x = 0$, given by (8.20), with the equation for the trajectories in the phase plane from part (b).

12. You might think that all trajectories in the phase plane that are closed — such as ellipses — correspond to periodic motion. To investigate this, consider

$$x(t) = \sin t^2 \qquad \text{and} \qquad y(t) = \cos t^2.$$

(a) Find the equation of the corresponding trajectory in the phase plane (the xy-plane) and draw its graph. [Hint: Consider $x^2(t) + y^2(t)$.] Is this trajectory closed?

(b) Graph the functions $x(t)$ and $y(t)$, and comment on the possibility of periodic motion.

13. The ELLIPTIC INTEGRAL OF THE FIRST KIND is defined by

$$F(z, k) = \int_0^z \frac{du}{\sqrt{1 - k^2 \sin^2 u}},$$

for $k^2 < 1$.

(a) Prove that $F(z, k) = F(z, k')$ where $k'^2 = 1 - k^2$.

(b) Prove that

$$\int_0^{\pi/2} \frac{dx}{\sqrt{\sin x}} = 2\sqrt{2} F(\frac{\pi}{2}, \frac{1}{\sqrt{2}}).$$

(Hint: Let $x = \pi/2 - y$ and then let $\cos y = \cos^2 u$.)

14. The ELLIPTIC INTEGRAL OF THE SECOND KIND is defined by

$$E(z, k) = \int_0^z \sqrt{1 - k^2 \sin^2 u} \, du,$$

for $k^2 < 1$.

(a) Show that the perimeter of the ellipse $x^2/a^2 + y^2/b^2 = 1$, $(a > b)$ in the first quadrant, is

$$\int_0^a \sqrt{\frac{a^2 - e^2 x^2}{a^2 - x^2}} \, dx,$$

where $e = \sqrt{a^2 - b^2}/a$. By letting $x = a \sin u$, show that the perimeter of an ellipse is $4aE(\pi/2, e)$.

(b) What is the length of $\sin x$, from $x = 0$ to $x = 2\pi$?

15. Consider the differential equation

$$\left(\frac{dx}{dt}\right)^2 = (1 - x^2)(1 - k^2x^2),$$

where $k^2 < 1$. Prove that this differential equation has periodic solutions with period

$$4 \int_0^1 \frac{1}{\sqrt{(1 - x^2)(1 - k^2x^2)}} \, dx.$$

8.2   Spring-Mass System

This section deals with the motion of a mass that is attached to a spring. This is another example of oscillatory motion.

EXAMPLE 8.5: Simple Harmonic Motion

For example, suppose we have a mass attached to a vertical spring, as shown in Figure 8.9. If the mass is not moving, we say it is in equilibrium. In this situation the force on the spring due to the hanging mass is exactly balanced by the tension in the spring. Now consider a situation in which the mass is extended a distance x_0 beyond its equilibrium position $x = 0$ (the position at which the mass will hang without undergoing oscillations) and released. We take positive values of the displacement x to correspond to positions where the mass is below this equilibrium position.

If the spring obeys Hooke's law (where the restoring force is proportional to the distance x), then Newton's law of motion gives

$$m\frac{d^2x}{dt^2} = -kx, \tag{8.31}$$

where $k > 0$ is the proportionality constant and $m > 0$ is the mass. The negative sign on the right-hand side of (8.31) indicates that the restoring force is in opposition to the acceleration.

If we represent (8.31) as a system of differential equations in terms of the displacement and velocity, we have

$$\frac{dx}{dt} = v \tag{8.32}$$

and

$$\frac{dv}{dt} = -\frac{k}{m}x, \tag{8.33}$$

so the first order differential equation satisfied by the trajectories in the phase plane becomes

$$\frac{dv}{dx} = -\frac{k}{m}\frac{x}{v}. \tag{8.34}$$

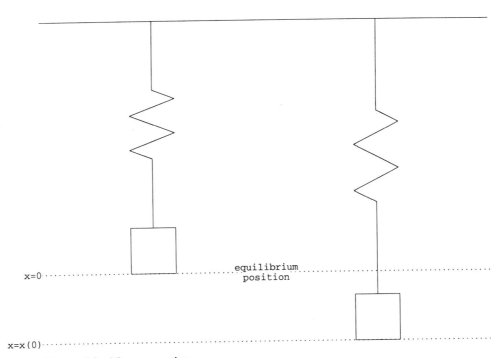

FIGURE 8.9   Mass on a spring

Notice that the equilibrium point of (8.32) and (8.33) is $x = 0$, $v = 0$, which is when the mass is in its equilibrium position.[2] Notice also that the uniqueness theorem guarantees that there is only one solution with initial conditions $x = 0$, $v = 0$ [the solution $x(t) = 0$], so no other trajectory can pass through the equilibrium point in the phase plane.

Figure 8.10 gives the direction field for (8.34) (with[3] $k/m = 2.6$), which indicates regions where the trajectories in the phase plane will be increasing, decreasing, concave up, and concave down. It appears that the trajectories will have horizontal tangents when $x = 0$ and vertical tangents when $v = 0$. This behavior may be easily verified by analyzing (8.34) and its derivative

$$\frac{d^2v}{dx^2} = -\frac{k}{m^2}\frac{mv^2 + kx^2}{v^3}.\tag{8.35}$$

Looking back at (8.33) we see that in the first quadrant, x is an increasing function of t, and v is a decreasing function of t. Thus, motion along any trajectory is clockwise. This is consistent with the direction of the arrows in Figure 8.10.

We may also solve the separable differential equation in (8.34) and determine that the trajectories are the ellipses

$$v^2 + \frac{k}{m}x^2 = C.\tag{8.36}$$

From (8.36) we see that v will have its extreme values when $x = 0$ and that x will have its extreme values when $v = 0$. In other words, the fastest the mass will move is when

---

[2]We must be careful to distinguish between the equilibrium *point* $x = 0$, $v = 0$, and the equilbrium *position* $x = 0$.

[3]This choice for the ratio of k to m was made so that the figures look good!

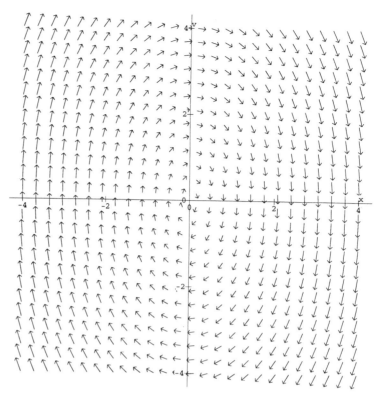

FIGURE 8.10   Direction field for $dv/dx = -(k/m)(x/v)$

it is passing its equilibrium position $x = 0$, and the mass will have zero velocity when it is a maximum distance from its equilibrium position.

The arbitrary constant C in (8.36) may be determined for the situation previously mentioned if we know the initial values of x and v. In this problem we pulled the mass down a distance x_0 before releasing it, so if we assume that it is simply released from this stationary position, the initial velocity will be zero and our initial conditions become

$$x(0) = x_0, \qquad x'(0) = v(0) = 0. \tag{8.37}$$

For these initial conditions, the value of the constant C is $(k/m)x_0^2$, and motion in the phase plane will start along the positive x-axis and proceed into the fourth quadrant (where $v < 0$). The velocity starts at zero and then takes on negative values, because in Figure 8.9 the mass is moving upwards, in the direction opposite to the positive x-axis.

Also notice that the point $(0, 0)$ is an equilibrium point of the system given in (8.35) and that all other trajectories are ellipses. Thus, the equilibrium point $(0, 0)$ is neutrally stable .

If we now examine the time-dependent behavior of the displacement, $x(t)$, the concavity in the xt-plane will be easy to determine, because the original differential equation (8.31) is in terms of the second derivative. Here we see that if $x > 0$, the curve will be concave down, and if $x < 0$, the curve will be concave up. The change from concave up to concave down (and vice versa) takes place when $x = 0$, so we may also conclude that these places will be points of inflection. This information allows us to state that the graph of $x(t)$ will consist of "right side up" humps and "upside down" humps between the coordinate intercepts. What is missing is the height of the humps and the time scale that will determine these intercepts. To obtain this information, we write (8.31) as

$$\frac{d^2x}{dt^2} + \frac{k}{m}x = 0 \tag{8.38}$$

and recognize this as a familiar differential equation (see Sections 5.5, 7.1, and 7.2) with a general solution of the form

$$x(t) = C_1 \cos\sqrt{\frac{k}{m}}t + C_2 \sin\sqrt{\frac{k}{m}}t. \tag{8.39}$$

If we impose the initial conditions (8.37) on (8.39), we obtain the explicit solution of (8.38) as

$$x(t) = x_0 \cos\sqrt{\frac{k}{m}}t, \tag{8.40}$$

from which we may calculate

$$v(t) = -x_0\sqrt{\frac{k}{m}} \sin\sqrt{\frac{k}{m}}t.$$

The graph of (8.40) is shown in Figure 8.11 (with $k/m = 2.6$ and $x_0 = 3$). Notice that the regions of concavity and the position of the inflection points are as we determined earlier, but now we have the added information about the horizontal and vertical scales as well as the values for t where the solution is increasing or decreasing.

Figure 8.12 shows the graph of x as a function of t, the graph of v as a function of t, and the graph of the corresponding trajectory in the phase plane. Draw the appropriate lines like we did in Section 7.3 to show how the trajectory in the phase plane can be obtained from the time-dependent graphs giving the displacement and velocity.

In Figure 8.12 we see that the time it takes to make one complete trip around a trajectory is the same as the time it takes for either the displacement or the velocity curve to finish a complete cycle. This time is called the PERIOD of this oscillatory motion and is given by $2\pi/\sqrt{k/m} = 2\pi/\omega$, defining

$$\omega = \sqrt{\frac{k}{m}}$$

to simplify the form of some of our future equations. Thus, systems with a greater mass (and the same spring constant k) will have longer periods (oscillate more slowly), whereas stiffer springs with larger values of k (and the same mass) will have shorter periods (oscillate faster). The value of $\omega/(2\pi)$ is called the NATURAL FREQUENCY of the system.

If we consider the general initial value problem for the spring-mass system, we would have solutions of (8.31) satisfy

$$x(0) = x_0, \qquad x'(0) = v(0) = v_0.$$

Using these initial conditions in our general solution, given by (8.39), we obtain

$$x(t) = x_0 \cos\omega t + \frac{v_0}{\omega} \sin\omega t. \tag{8.41}$$

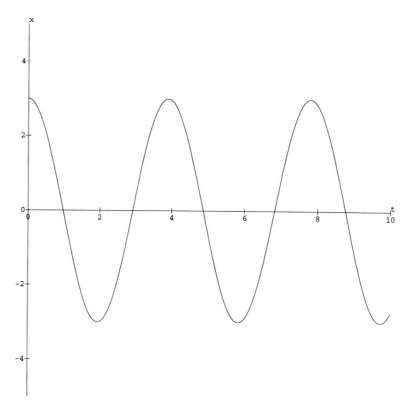

FIGURE 8.11 Graph of $x(t) = x_0 \cos[(k/m)^{1/2}t]$

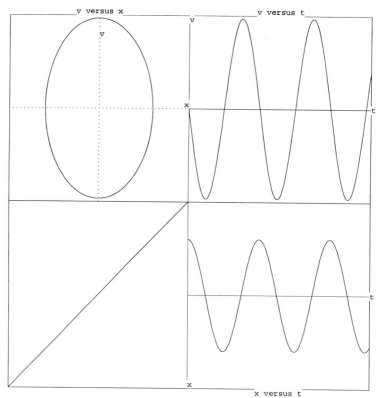

FIGURE 8.12 Constructing the phase plane for simple harmonic motion

Exercise 2 on page 400 gives an alternative form of the solution by expressing (8.41) as

$$x(t) = A \cos(\omega t + \phi),$$

where

$$A = \sqrt{x_0^2 + \left(\frac{v_0}{\omega}\right)^2} \tag{8.42}$$

and

$$\phi = \arctan\left(-\frac{v_0}{\omega x_0}\right). \tag{8.43}$$

Here A is the amplitude of the oscillations associated with this initial value problem. The angle ϕ is often called the PHASE ANGLE[4] or PHASE SHIFT, and these types of oscillations are called SIMPLE HARMONIC MOTION.

The trajectory in the phase plane associated with this general initial value problem is given [from (8.36)] by

$$v^2 + \frac{k}{m}x^2 = v_0^2 + \frac{k}{m}x_0^2.$$

Because $k/m = \omega^2$ and $v_0^2 + \omega^2 x_0^2 = \omega^2 A^2$, this equation can be written in the form

$$\frac{1}{\omega^2 A^2}v^2 + \frac{1}{A^2}x^2 = 1. \tag{8.44}$$

From (8.44) it is apparent that the semiaxes of the ellipse in the phase plane are given by ωA for the velocity and A for the displacement. Thus, for values of $\omega > 1$, the velocity undergoes larger oscillations in the phase plane than the displacement, whereas for $\omega < 1$, the reverse is true. The initial point (where $t = 0$ and the motion starts) on the phase plane trajectory is given by (x_0, v_0). Because the point $(0, 0)$ is an equilibrium point [associated with the initial conditions $x(0) = v(0) = 0$] of this system and all the solutions are ellipses, we see that $(0, 0)$ is neutrally stable. □

In the next three examples we consider the impact of a simple friction force on a spring-mass system, where we gradually increase the magnitude of the force. One way to picture this is to think of the motion of a spring-mass in a medium in which the density is gradually increased — starting with a vacuum, then to air, to water, to oil, and to molasses.

EXAMPLE 8.6: Underdamped Motion

To work on a slightly more complicated example, we consider the case in which a friction force also acts on the spring-mass system. We assume this friction force is proportional to the velocity and dampens the motion. To include this friction force, we need to add the term $-b\,dx/dt$, where b is a small positive constant, to the right-hand side of (8.31). This gives

$$m\frac{d^2x}{dt^2} = -kx - b\frac{dx}{dt}, \tag{8.45}$$

---

[4]Some authors define the phase angle as the value of $-\phi$.

or, with

$$v = \frac{dx}{dt},$$

$$m\frac{dv}{dt} + bv + kx = 0,$$

so in the phase plane,

$$\frac{dx}{dt} = v \tag{8.46}$$

and

$$\frac{dv}{dt} = -\frac{b}{m}v - \frac{k}{m}x. \tag{8.47}$$

We note that $x = 0$, $v = 0$, is again an equilibrium point. The uniqueness theorem guarantees that there is only one solution with initial conditions $x = 0$, $v = 0$, so no other solution can pass through this equilibrium point.

The differential equation of the trajectories in the phase plane is now

$$\frac{dv}{dx} = \frac{-bv - kx}{mv}. \tag{8.48}$$

Our first observation is that the only symmetry of the family of trajectories is about the origin. We note that trajectory slopes will still be negative in the first quadrant, and, looking at (8.46) and (8.47), we see that motion will again be clockwise. However, it is clear from (8.48) that places where the trajectories have horizontal tangents will no longer be on the vertical axis, but instead along the line

$$v = -\frac{k}{b}x.$$

These facts are reinforced by the direction field in Figure 8.13 (for $b = 3$, $k = 13$, $m = 4$), and the swirling pattern of the field suggests that the trajectories spiral in toward the origin. We can examine the concavity in the phase plane by considering the derivative of (8.48), which may be put in the form

$$\frac{d^2v}{dx^2} = -\frac{k}{m^2v^3}\left(mv^2 + bxv + kx^2\right). \tag{8.49}$$

Thus, additional changes in concavity (besides when $v = 0$) can occur along the lines given by

$$v = \frac{1}{2m}x\left(-b \pm \sqrt{b^2 - 4km}\right). \tag{8.50}$$

If the argument of the square root is a negative number, these lines will not exist in the phase plane. In this case the quantity in parentheses in (8.49) is positive, and the concavity is determined solely by the sign of v. For this situation we see that as the value of b is increased from 0 (its value for simple harmonic motion), all trajectories will be concave up for $v < 0$ and concave down for $v > 0$ as long as $b^2 - 4km$ is negative. Note the first example in this section had $b = 0$, so these statements also apply there.

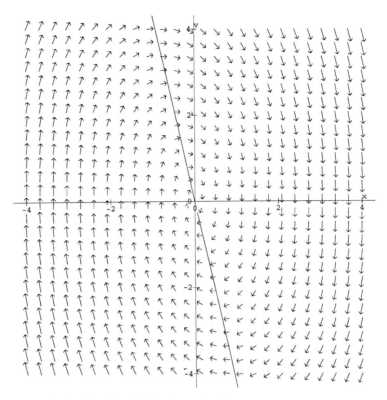

FIGURE 8.13 Direction field for damped motion

To analyze the situation further, we solve the differential equation in the phase plane. We note that (8.48) has homogeneous coefficients and may be solved (see Exercise 17 on page 404) to yield the implicit solution

$$\ln\left(\frac{x^2}{v^2} + \frac{b}{m}\frac{x}{v} + \frac{k}{m}\right) - \frac{2b}{mq}\arctan\left[\frac{1}{q}\left(2\frac{x}{v} + \frac{b}{m}\right)\right] = C, \tag{8.51}$$

where $q = \sqrt{4k/m - (b/m)^2} \neq 0$. These are the trajectories in the phase plane. After all this work, we have found an equation that is algebraically correct, but nearly useless!

We now seek the explicit solutions for the displacements from (8.45). This second order differential equation with constant coefficients will have solutions of the form e^{rx}, so if we substitute this exponential into (8.45) we obtain the characteristic equation

$$mr^2 + br + k = 0. \tag{8.52}$$

If we assume the damping effect is small, $b^2 - 4mk$ will be negative, and we write the expression for the roots of (8.52) as

$$-\frac{b}{2m} \pm i\frac{1}{2m}\sqrt{4km - b^2} = \alpha \pm i\beta,$$

where $\alpha = -b/(2m)$ and $\beta = \sqrt{4km - b^2}/(2m)$. This means that the general solution of (8.45) may be written as

$$x(t) = C_1 e^{\alpha t}\cos\beta t + C_2 e^{\alpha t}\sin\beta t. \tag{8.53}$$

To evaluate the arbitrary constants C_1 and C_2, we need some initial conditions. If we assume that at time $t = 0$, the mass is at its equilibrium position with a velocity of v_0, our initial conditions would be

$$x(0) = 0, \;\; x'(0) = v_0,$$

and the arbitrary constants would have the values $C_1 = 0$, $C_2 = v_0/\beta$. This gives our explicit solution of (8.45) as

$$x(t) = \frac{v_0}{\beta} e^{\alpha t} \sin \beta t, \tag{8.54}$$

from which we may calculate

$$v(t) = e^{\alpha t} \left(v_0 \cos \beta t + \frac{v_0 \alpha}{\beta} \sin \beta t \right). \tag{8.55}$$

Figure 8.14 shows the graph of the explicit solutions in (8.54) and (8.55) along with the trajectory in the phase plane corresponding to these initial conditions. Thus, with these initial conditions, the mass oscillates with decreasing amplitude until it finally comes to rest at the origin in infinite time.

If we return to (8.53), the general solution for $x(t)$ valid for any initial conditions, we can rewrite it as

$$x(t) = e^{\alpha t} \left(C_1 \cos \beta t + C_2 \sin \beta t \right).$$

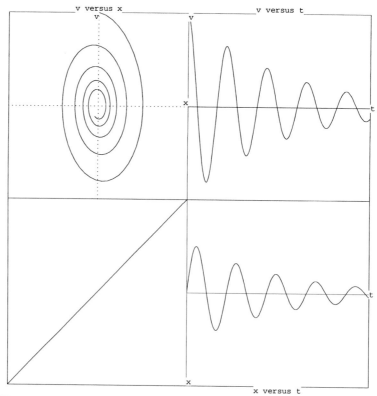

FIGURE 8.14   Constructing the phase plane for damped motion

Because $\alpha < 0$, then $x(t) \to 0$ as $t \to \infty$. If we compute the velocity, we find

$$v(t) = e^{\alpha t}\left(\alpha C_1 \cos \beta t + \alpha C_2 \sin \beta t - \beta C_1 \sin \beta t + \beta C_2 \cos \beta t\right),$$

which also has the property that $v(t) \to 0$ as $t \to \infty$. Thus, regardless of the initial conditions, all solutions will approach $(0, 0)$ in the phase plane as $t \to \infty$. In other words, regardless of the initial conditions, the mass oscillates with decreasing amplitude until it finally comes to rest at the origin in infinite time.

However, the point $(0, 0)$ is the equilibrium point corresponding to zero initial displacement and zero initial velocity. Consequently, by the uniqueness theorem, no other solution can reach $(0, 0)$ in finite time. Because all solutions spiral to this origin of the phase plane, regardless of initial conditions, the equilibrium point is asymptotically stable. □

EXAMPLE 8.7: Critically Damped Motion

Now we consider the case where the damping force acting on the spring-mass system is very large, so the only difference between this situation and our previous example is the size of b. There were two places where the size of b mattered in our analysis in the previous example, both of which involved the quantity $b^2/4 - km$. If that quantity is negative, all solutions oscillate. Furthermore, if $b \neq 0$, they approach zero as t approaches infinity. In the phase plane, $(0, 0)$ is asymptotically stable, because all solutions spiral in toward the origin as time increases. However, if b is such that $b^2/4 - km$ is not negative, we need further analysis. In particular recall that when this quantity was not negative, we had a possible change of concavity at places other than along the line $v = 0$.

Let us start by seeing what happens if $b^2/4 - km = 0$. We look at a typical direction field for this situation, as shown in Figure 8.15 (for $b = 4$, $k = 1$, $m = 4$). This direction field is quite different from the previous two (Figures 8.10 and 8.13) in that there is far less variety in the slopes. The only variation in the pattern occurs for points near the x-axis.

The direction field suggests that there might be a trajectory that is a straight line through the origin. If we substitute $v = ax$ into (8.48) we obtain

$$a = -\frac{b}{m} - \frac{k}{ma},$$

so a must satisfy the quadratic equation

$$ma^2 + ba + k = 0.$$

This gives

$$a = \frac{1}{2m}\left(-b \pm \sqrt{b^2 - 4km}\right),$$

and because we are considering the case where $b^2 - 4km = 0$, we find

$$a = -\frac{b}{2m}.$$

Thus,

$$v = -\frac{b}{2m}x$$

is a trajectory in the phase plane. We draw this trajectory on the previous direction field (see Figure 8.16) and note that because of our uniqueness theorem, no other trajectory may

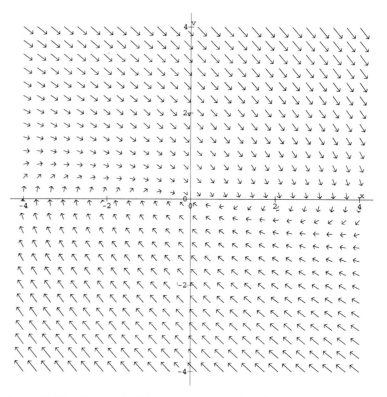

FIGURE 8.15 Direction field for critically damped motion

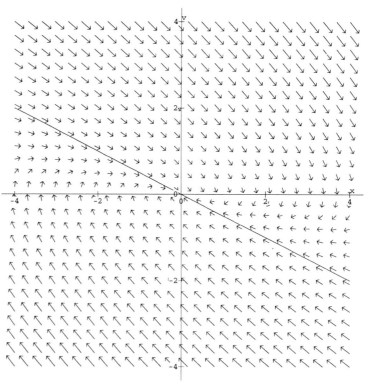

FIGURE 8.16 Trajectory $v = -bx/(2m)$ and direction field for critically damped motion

cross this line. Because the line goes through the origin, solutions for this situation cannot circle the origin, so the motion will not be oscillatory. Because this situation corresponds to the smallest value of the damping coefficient where oscillations cease, it is called CRITICAL DAMPING.

Figure 8.17 shows trajectories for several different initial conditions. Notice that with the exception of points on the line $v = -bx/(2m)$, the trajectories are concave up in the lower half plane and concave down in the upper half plane, as was true in the previous two examples. If we start with an initial condition on the positive x-axis (that is, a positive initial displacement and a zero initial velocity), the mass will never pass through the equilibrium point, but only approach it as a limit. If we start with an initial condition on the positive v-axis, (so the mass is initially at the equilibrium position $x = 0$ with a positive initial velocity) the displacement will increase to a maximum, where the velocity changes sign. It then will approach the origin in the limit. The scenarios for situations along the other axes are left for the exercises (see Exercise 16 on page 404).

To obtain the explicit solution for the displacement as a function of time for the case $b^2 - 4km = 0$, we return to (8.45),

$$m\frac{d^2x}{dt^2} + b\frac{dx}{dt} + kx = 0, \tag{8.56}$$

and to the characteristic equation resulting from a solution of the form e^{rt}, namely,

$$mr^2 + br + k = 0. \tag{8.57}$$

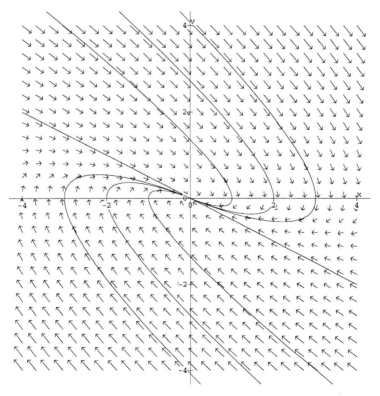

FIGURE 8.17   Various trajectories and direction field for critically damped motion

Because we are considering the case where $b^2 - 4km = 0$, (8.57) has a double root of $r = -b/(2m)$, and the general solution of (8.56) may be written as

$$x(t) = C_1 e^{-bt/(2m)} + C_2 t e^{-bt/(2m)}. \tag{8.58}$$

From (8.58) we see that $x(t) \to 0$ as $t \to \infty$, and there is no oscillatory behavior, which reinforces our earlier conclusions.

If we impose the general initial conditions

$$x(0) = x_0, \ x'(0) = v_0,$$

we may determine the arbitrary constants in (8.58) and write the solution of this initial value problem as

$$x(t) = x_0 e^{-bt/(2m)} + \left(\frac{bx_0}{2m} + v_0\right) t e^{-bt/(2m)}. \tag{8.59}$$

The graph of (8.59) [and that of $v(t) = x'(t)$] is shown in Figure 8.18 for typical values of the parameters and positive values for x_0 and v_0. Also shown is the corresponding trajectory in the phase plane. It is possible to draw appropriate lines like we did in Section 7.3 to show how the trajectory in the phase plane can be obtained from the time-dependent graphs giving the displacement and velocity.   □

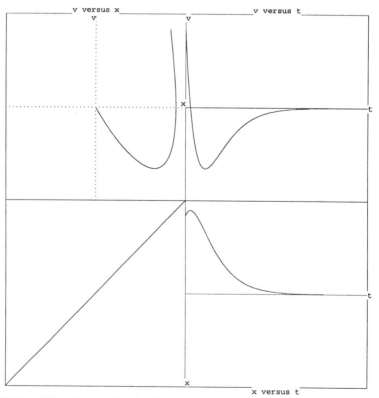

FIGURE 8.18   Constructing the phase plane for critically damped motion

EXAMPLE 8.8: Overdamped Motion

To complete our analysis of damped motion, considering the remaining case $b^2 - 4km > 0$, we return to the original differential equation

$$m\frac{d^2x}{dt^2} + b\frac{dx}{dt} + kx = 0 \tag{8.60}$$

and to the corresponding equations in the phase plane

$$\frac{dx}{dt} = v,$$

$$\frac{dv}{dt} = -\frac{b}{m}v - \frac{k}{m}x,$$

and

$$\frac{dv}{dx} = \frac{-bv - kx}{mv}. \tag{8.61}$$

In Example 8.6 we examined concavity in the phase plane with the aid of the second derivative (8.49), given as

$$\frac{d^2v}{dx^2} = -\frac{k}{m^2v^3}\left(mv^2 + bxv + kx^2\right).$$

We had noted — see (8.50) — that additional changes in concavity could occur along the lines given by

$$v = -w_1x \text{ and } v = -w_2x,$$

where

$$w_1 = \frac{1}{2m}\left(b + \sqrt{b^2 - 4km}\right), \qquad w_2 = \frac{1}{2m}\left(b - \sqrt{b^2 - 4km}\right).$$

So far we have considered cases where $b^2 - 4km \leq 0$. If $b^2 - 4km > 0$, then we may write

$$mv^2 + bxv + kx^2 = m\left(v + w_1x\right)\left(v + w_2x\right).$$

Thus, the expression for the second derivative in the phase plane is

$$\frac{d^2v}{dx^2} = -\frac{1}{mv^3}k\left(v + w_1x\right)\left(v + w_2x\right).$$

This expression shows that the phase plane is divided into six regions by the three lines through the origin given by

$$v = 0, \quad v = -w_1x, \text{ and } v = -w_2x.$$

Substituting $v = ax$ into (8.61) shows that the lines $v = -w_1x$ and $v = -w_2x$ are also trajectories.

Figure 8.19 shows these regions for typical values of m, b, and k ($b = 6$, $k = 1$, $m = 4$). Because the two nonhorizontal lines dividing regions of concavity are trajectories, and trajectories cannot cross, any specific trajectory will remain in the region bounded by

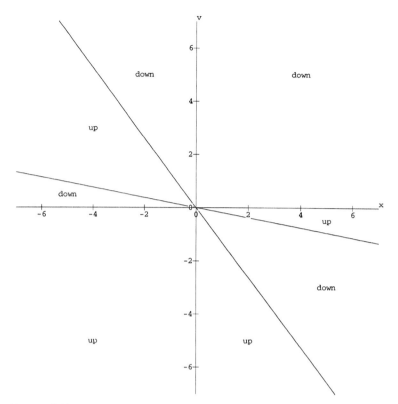

FIGURE 8.19   Concavity regions for overdamped motion

these two lines. Figure 8.20 shows the direction field for this situation, along with trajectories for different initial conditions and the two straight line trajectories that also divide regions of different concavity. Notice that in the limit as t increases, all trajectories approach the origin from either the second or fourth quadrants. That is, the mass can approach the equilibrium position from below with negative values of velocity when $x > 0$, or from above with positive values of velocity when $x < 0$. Does this agree with your intuition?

To obtain the explicit solution for the displacement as a function of time from (8.60), we use the characteristic equation (8.57), where now the roots will be real numbers given by

$$\frac{1}{2m}\left(-b \pm \sqrt{b^2 - 4km}\right). \tag{8.62}$$

If we label the smaller root r_1 and the larger root r_2 (note that they both will be negative), the general solution of (8.60) may be written as

$$x(t) = C_1 e^{r_1 t} + C_2 e^{r_2 t}. \tag{8.63}$$

Choosing $C_1 = (x_0 r_2 - v_0)/(r_2 - r_1)$ and $C_2 = (v_0 - x_0 r_1)/(r_2 - r_1)$ means that (8.63) will satisfy the initial conditions

$$x(0) = v_0, \, x'(0) = v_0,$$

so the solution of this initial value problem is

$$x(t) = \frac{x_0 r_2 - v_0}{r_2 - r_1} e^{r_1 t} + \frac{v_0 - x_0 r_1}{r_2 - r_1} e^{r_2 t}. \tag{8.64}$$

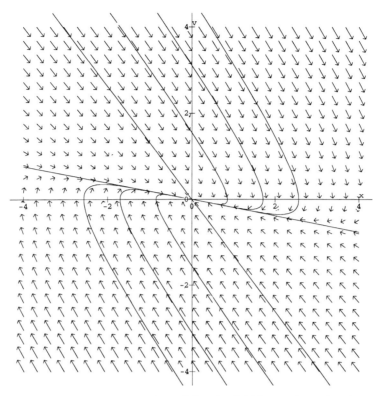

FIGURE 8.20 Various trajectories and direction field for overdamped motion

Here we obtain confirmation of our earlier observation that all solutions decay to zero in the limit as t approaches infinity. Taking the derivative of (8.64) to find $v(t)$ shows that $v(t)$ will also approach zero in the limit. Thus, all trajectories will approach the equilibrium point in the phase plane at $(0, 0)$, so the origin is asymptotically stable. $\square$

We can summarize the motion of a spring-mass system, which is governed by the initial value problem

$$m\frac{d^2x}{dt^2} + b\frac{dx}{dt} + kx = 0 \tag{8.65}$$

subject to

$$x(0) = x_0, \ x'(0) = v_0,$$

in the following manner. We look at the characteristic equation obtained by using a solution of (8.65) in the form e^{rt}, namely,

$$mr^2 + br + k = 0. \tag{8.66}$$

We think of m and k as being fixed and we are allowed to vary the damping factor b.

- If $b^2 - 4mk < 0$, the roots of (8.66) are complex numbers, oscillations occur, and the resulting motion is called UNDERDAMPED.

- If $b^2 - 4mk = 0$, the roots of (8.66) are real, but repeated. Oscillations do not occur. The resulting motion is called CRITICALLY DAMPED.

- If $b^2 - 4mk > 0$, there are two (negative) real roots of (8.66). Oscillations do not occur. The resulting motion is called OVERDAMPED.

Figures 8.21, 8.22, and 8.23 show the graphs of solutions of (8.65) where x_0 and v_0 are both positive, m and k are fixed, and b is given different values, corresponding to the underdamped, critically damped, and overdamped cases.

EXERCISES

1. Solve the following initial value problems for $x(t)$, and find the equation of the corresponding trajectory in the phase plane.

   (a)
   $$\frac{d^2x}{dt^2} + 16x = 0, \quad x(0) = 3, \quad x'(0) = 12$$

   (b)
   $$9\frac{d^2x}{dt^2} + x = 0, \quad x(0) = -5, \quad x'(0) = \frac{1}{3}$$

   (c)
   $$4\frac{d^2x}{dt^2} + 9x = 0, \quad x(0) = 0, \quad x'(0) = 0$$

   (d)
   $$\frac{d^2x}{dt^2} + 100x = 0, \quad x(0) = 0, \quad x'(0) = 0.2$$

2. (a) Show that the function $A\cos(\omega t + \phi)$, where $w = \sqrt{k/m}$, satisfies (8.31). Show that choosing A and ϕ as in (8.42) and (8.43) allows this function to satisfy the initial conditions $x(0) = x_0, x'(0) = v_0$.

   (b) Expand your solution in part (a), and obtain the form of the solution given in (8.41).

3. Consider a spring-mass system that is governed by

   $$4\frac{d^2x}{dt^2} + x = 0,$$

   where motion is started by stretching the spring 2 units from its equilibrium position and releasing it from rest.

   (a) Write the corresponding differential equation for trajectories in the phase plane (in terms of x and $v = dx/dt$).

   (b) Give the appropriate initial conditions for the statement of this problem both for the original second order differential equation and for the trajectories in the phase plane.

   (c) Find the equation of the trajectory in the phase plane associated with this initial value problem.

   (d) Find an explicit solution for the displacement as a function of time.

   (e) Construct the phase plane trajectory from your solution in part (d) following the technique described on page 356, then plot the trajectory from your answer to part (c) in the phase plane to see if they agree.

4. Consider a spring-mass system that is governed by

   $$\frac{d^2x}{dt^2} + 8x = 0,$$

   where motion is started from the equilibrium position by giving the mass an initial velocity v_0.

   (a) Write the corresponding differential equation for the trajectories in the phase plane (in terms of x and $v = dx/dt$).

   (b) What are the appropriate initial conditions for the statement of this problem in terms of both the original second order differential equation and the one for the trajectories in the phase plane?

   (c) Find the equation of the trajectory in the phase plane associated with this initial value problem.

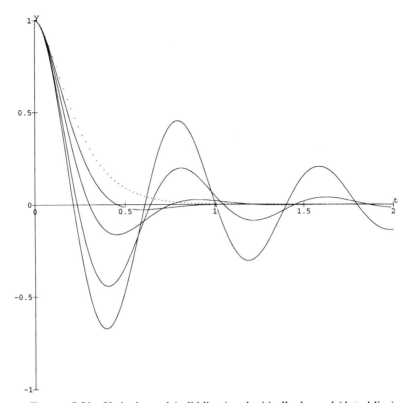

FIGURE 8.21 Underdamped (solid lines) and critically damped (dotted line) motions

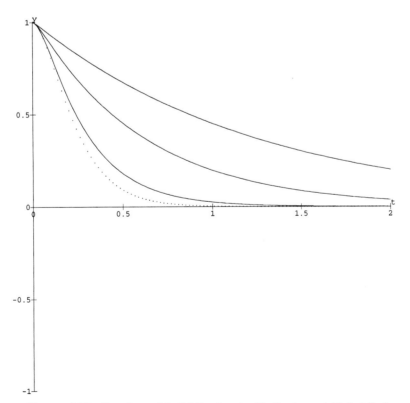

FIGURE 8.22 Overdamped (solid lines) and critically damped (dotted line) motions

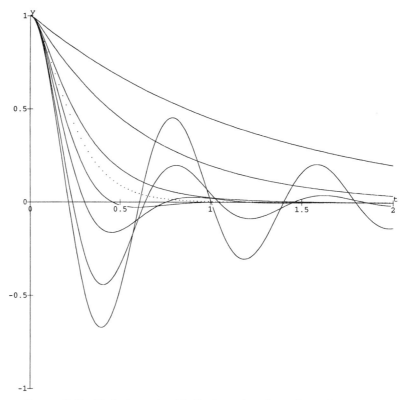

FIGURE 8.23   Underdamped, critically damped, and overdamped motions

(d) Find an explicit solution for the displacement as a function of time.

(e) Construct the phase plane trajectory from your solution in part (d) following the technique described on page 356, then plot the trajectory from your answer to part (c) in the phase plane to see if they agree.

5. Consider a spring-mass system that is governed by

$$16\frac{d^2x}{dt^2} + x = 0$$

subject to an initial displacement of -2 and an initial velocity of 3. What are the amplitude and period of the resulting motion? Find the equation of the trajectory in the phase plane corresponding to these initial conditions.

6. Find the amplitude, period, and phase angle for motion described by

$$x'' + 9x = 0, \quad x(0) = -2, \quad x'(0) = -6.$$

7. Find the amplitude, period, and phase angle for motion described by

$$x'' + \pi^2 x = 0, \quad x(0) = 1, \quad x'(0) = \pi\sqrt{3}.$$

8. Solve

$$m\frac{d^2x}{dt^2} + b\frac{dx}{dt} + kx = 0$$

for the following situations.

(a) $m = 1, \quad b = 1/8, \quad k = 1, \quad x(0) = 0, \quad x'(0) = 1/2$

(b) $m = 1,$ $b = 8,$ $k = 16,$ $x(0) = 0,$ $x'(0) = -3$

(c) $m = 64,$ $b = 16,$ $k = 17,$ $x(0) = 1,$ $x'(0) = 0$

(d) $m = 9,$ $b = 6,$ $k = 37,$ $x(0) = 1,$ $x'(0) = 1$

(e) $m = 9,$ $b = 6,$ $k = 37,$ $x(0) = 0,$ $x'(0) = 0$

9. For what values of b are the motions governed by

$$4x'' + bx' + 9x = 0$$

(a) Overdamped?

(b) Underdamped?

(c) Critically damped?

(d) Show that

$$f(t) = \frac{e^{r_2 t} - e^{r_1 t}}{r_2 - r_1}$$

satisfies

$$m\frac{d^2x}{dt^2} + b\frac{dx}{dt} + kx = 0$$

if r_1 and r_2 satisfy

$$mr^2 + br + k = 0.$$

(e) Set $r_2 = r_1 + \varepsilon$ in the formula for $f(t)$ in part (d). Show that, for the underdamped case,

$$\lim_{\varepsilon \to 0} f(t) = te^{r_1 t}.$$

(Note that you can use L'Hôpital's rule, or the Taylor series expansion $e^z = 1 + z + z^2/2! + \cdots$.)

10. Show that if $m > 0$ and $b > 0$, the general solution of

$$m\frac{d^2x}{dt^2} + b\frac{dx}{dt} + kx = 0$$

approaches 0 as $t \to \infty$ for all nonnegative values of k.

11. Show that the solution of

$$m\frac{d^2x}{dt^2} + b\frac{dx}{dt} + kx = 0$$

for the overdamped case with $x(0) = x_0$ and $x'(0) = v_0$, may have at most one horizontal tangent in the phase plane. (See Figure 8.20.)

12. Repeat Exercise 11 for the critically damped case.

13. If we multiply the differential equation

$$m\frac{d^2x}{dt^2} + kx = 0$$

by dx/dt, we should observe that each term can be integrated separately. We integrate

$$m\frac{dx}{dt}\frac{d^2x}{dt^2} + kx\frac{dx}{dt} = 0$$

to obtain

$$\frac{1}{2}m\left(\frac{dx}{dt}\right)^2 + \frac{1}{2}kx^2 = E.$$

The first term on the left-hand side of this equation is the kinetic energy of the object; the second term represents the potential energy. For example, with the spring-mass system, $kx^2/2$ represents the energy stored in the spring. This equation is a statement of conservation of energy, $E =$ constant.

(a) Write down the conservation of energy equation for the system in Exercise 5, and evaluate the constant E.

(b) Show that the kinetic energy of the system is a maximum when the potential energy is a minimum. Where does this happen in the phase plane?

14. If the total energy of the damped spring-mass system described by (8.60) is defined as

$$E = \frac{1}{2}m\left(\frac{dx}{dt}\right)^2 + \frac{1}{2}kx^2,$$

show that E satisfies

$$\frac{dE}{dt} = -b\left(\frac{dx}{dt}\right)^2$$

and is therefore a decreasing function of time.

15. An object in liquid is kept afloat when the upward force due to buoyancy exceeds the downward pull of gravity. If such an object is displaced slightly from its equilibrium position and released, the subsequent motion is described by

$$x'' + ax = 0,$$

where a is the ratio of the gravitational constant to the displacement of the body in equilibrium. x is positive in the downward direction.

(a) An oil drum in the form of a right circular cylinder is in equilibrium when it lies upright and half submerged in a lake. If the cylinder is 4 feet long and the gravitational constant is taken as 32 feet per second per second, the governing differential equation is

$$x'' + 16x = 0,$$

where x is measured in feet and t in seconds. Find the resulting motion if the cylinder is further submerged so that it extends 1 foot above the water, and then is released from rest.

(b) If the initial conditions in part (a) are changed to give the cylinder an initial velocity of 2 feet per second in its equilibrium position, find the resulting solution.

16. In Figure 8.16, explain the trajectories corresponding to initial conditions that start

(a) along the negative x-axis, and

(b) along the negative v-axis.

17. Use a change of variables $w = x/v$ or $u = v/x$ to derive the implicit solution of (8.48) as (8.51).

18. The linearized differential equation describing the oscillations of a simple pendulum including a damping term is

$$\frac{d^2x}{dt^2} + 2\alpha\frac{dx}{dt} + \lambda^2 x = 0,$$

where $\lambda^2 = g/h$ and α are constants.

(a) Consider a simple pendulum with $\lambda^2 = 4$.

    i. For what values of α will the resulting motion of a pendulum with $x(0) = 0.5$ and $x'(0) = 0$ be underdamped?

    ii. For what values of α will the pendulum in part i always stay on the same side of the vertical line on which it was released?

(b) If $\alpha = 1/10$, $\lambda^2 = 4.01$, and the pendulum is set in motion with initial conditions $x(0) = 0$, $x'(0) = 1$, find the maximum value of $|x|$ for the resulting motion.

(c) If the initial conditions of part (b) are changed to $x(0) = 0$, $x'(0) = \beta$, for what value of β will the maximum value of $|x|$ be 0.2 radian?

19. The linearized differential equation describing the oscillations of a simple pendulum including a damping term is

$$\frac{d^2x}{dt^2} + 2\alpha\frac{dx}{dt} + \lambda^2 x = 0,$$

where α is a constant. Construct the phase plane for different values of α and fixed λ^2. Compare and contrast this with the phase plane analysis when $\alpha = 0$.

20. The differential equation describing the oscillations of a simple pendulum including a damping term is

$$\frac{d^2x}{dt^2} + 2\alpha\frac{dx}{dt} + \lambda^2 \sin x = 0,$$

where α is a constant. Construct the phase plane for different values of α and fixed λ^2.

(a) Compare and contrast this with the phase plane analysis when $\alpha = 0$.

(b) Compare and contrast this with the phase plane analysis for the linearized pendulum from Exercise 19.

21. The linearized differential equation for a damped pendulum in Exercise 18 is similar to those of spring-mass systems and electric circuits. Make a table giving the equivalent terms in these models.

8.3   Electrical Circuits

In Example 6.14 we considered a simple electrical circuit involving an inductor and a resistor, and in the exercises in Section 6.4 we studied a circuit with a capacitor and a resistor. In this section we will consider electrical circuits with other combinations of these elements.

EXAMPLE 8.9: Ideal Oscillator

As our first example in this section, we consider the problem of the discharge of a charged capacitor in a circuit with an inductor. We may think of a capacitor as consisting of two parallel plates with a dielectric (an insulating material that could simply be air) between the two plates. If the capacitor is charged, there is an excess of electrons on one of the plates, and the voltage (or potential difference) across the plates is proportional to this charge. If the charge is q (units of coulombs), the proportionality constant is $1/C$ (C has units of farads), and V is the voltage (units of volts), we have that

$$V = \frac{1}{C}q.$$

As concerns the other element in this circuit, we can think of an inductor as a wire wound around a cylindrical core. An inductor has the property that a changing current passing through this coil initiates a magnetic field that opposes the change in the current. The voltage across such an inductor is proportional to the rate of change of the current, giving the equation

$$V = L\frac{dI}{dt},$$

where L is the inductance (units of henries) and I is the current (units of amperes). Because the letters L and C are used to quantify the circuit's inductance and capacitance, respectively, such a circuit is called an LC circuit.

We now need to use a law attributed to Kirchhoff, which says that the algebraic sum of the voltages around a closed circuit must be zero. Thus, for our LC circuit we have that

$$L\frac{dI}{dt} + \frac{q}{C} = 0.$$

Because the current in an electrical circuit is the rate of change of charge, our LC circuit is described by the system of first order differential equations

$$\frac{dq}{dt} = I, \qquad \frac{dI}{dt} = -\frac{1}{LC}q. \tag{8.67}$$

If we consider the phase plane associated with (8.67), with q on the horizontal axis and I on the vertical, we see that the differential equation for the trajectories is given by

$$\frac{dI}{dq} = \frac{dI/dt}{dq/dt} = -\frac{q}{LCI}. \tag{8.68}$$

This separable differential equation is easily integrated to yield

$$LCI^2 + q^2 = constant \tag{8.69}$$

as the equation for the trajectories in the phase plane.[5] The direction field for the phase plane and typical trajectories are shown in Figure 8.24 for the special case $C = 1/2, L = 4$.[6] Notice from (8.67) that because I is decreasing when q is positive and q is increasing when I is positive, motion will be clockwise. The trajectories are closed, so we know that we have oscillatory motion. Also, we see from the form of the trajectories in (8.69) that the equilibrium point of (8.67) at the origin is neutrally stable.

To discover the time-dependent behavior of this system, we eliminate I from the differential equations in (8.67) and obtain

$$\frac{d^2q}{dt^2} + \frac{1}{LC}q = 0.$$

---

[5]Note that if we divide (8.69) by $2C$, it becomes

$$(L/2)I^2 + q^2/(2C) = constant.$$

Because the first term on the left-hand side of this equation gives the magnetic energy stored in the inductor and the second term gives the electric energy stored in the capacitor, this is a type of conservation of energy equation. Note also that the extreme values of one type of energy occur when the other type of energy is zero.

[6]Unless specified otherwise, henceforth numerical values specified for q, C, V, L, and I are assumed to be in units of coulombs, farads, volts, henries, and amperes, respectively.

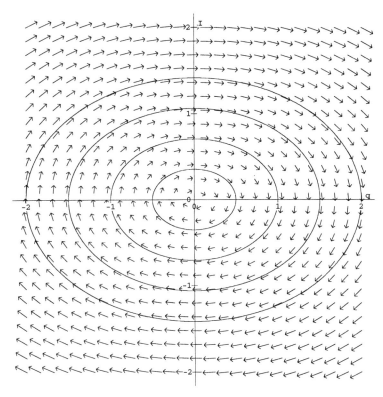

FIGURE 8.24 Direction field and trajectories for an LC circuit

This is a familiar equation, with a general solution of the form

$$q(t) = C_1 \cos \omega t + C_2 \sin \omega t, \tag{8.70}$$

where $\omega = 1/\sqrt{LC}$. Because we are analyzing the discharging of a charged capacitor, we have initial conditions of

$$q(0) = q_0, \qquad q'(0) = 0.$$

This means we may evaluate the arbitrary constants in (8.70) as $C_1 = q_0$, $C_2 = 0$, giving our solution of this initial value problem as

$$q(t) = q_0 \cos \omega t. \tag{8.71}$$

From (8.71) we see that the oscillations in the charge are periodic, with a period of $2\pi\sqrt{LC}$. The amplitude of the voltage drop across the capacitor is q_0/C. The current in this circuit is given by

$$I = \frac{dq}{dt} = -\omega q_0 \sin \omega t,$$

so both the current in the circuit and the charge on the capacitor oscillate with time. For the special case $C = 1/2$, $L = 4$, $q_0 = 1$, Figure 8.25 contains the graph of the charge on the capacitor as a function of time in the lower right-hand box, the current in the circuit as a function of time in the upper right-hand box, and the corresponding trajectory in the phase plane in the upper left-hand box. A 45° line is shown in the lower left-hand box to

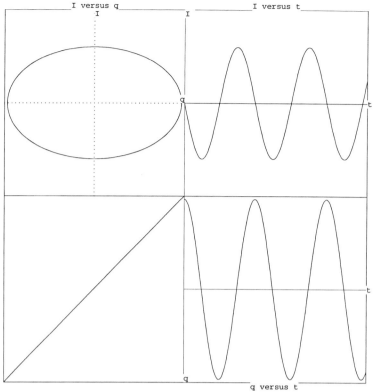

FIGURE 8.25   Charge, current, and phase plane trajectory for an LC circuit

allow you to draw the appropriate lines to show how the trajectory in the phase plane can be obtained from the explicit solutions for the charge and current.   □

Of course the ideal oscillator in the previous example was so called because no resistance was included, but there is in fact some resistance in all actual circuits. If we conduct an experiment where we pass an oscillatory current through a resistor and plot the voltage across the resistor and the current on the same graph, we obtain Figure 8.26. It is clear from this graph that the voltage and current have the same period; the only difference is the amplitude of oscillation. Thus, we see that Ohm's law — which states that voltage and current are proportional, with the proportionality constant as the resistance — also applies here. If we let R be the resistance (units of ohms) in our former circuit, we have that the voltage drop across the resistor is

$$V = RI.$$

If we now apply Kirchhoff's voltage law to a circuit consisting of a capacitor, resistor, and inductor connected in series, as shown in Figure 8.27, we obtain

$$L\frac{dI}{dt} + RI + \frac{1}{C}q = 0.$$

If we use the relationship between q and I, we obtain the coupled differential equations

$$\frac{dq}{dt} = I, \qquad \frac{dI}{dt} = -\frac{R}{L}I - \frac{1}{LC}q. \tag{8.72}$$

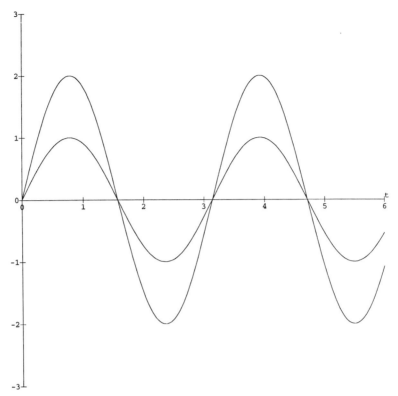

FIGURE 8.26 Voltage and current for a resistor

The corresponding differential equation for trajectories in the phase plane is

$$\frac{dI}{dq} = \frac{dI/dt}{dq/dt} = -\frac{RCI + q}{LCI}.$$ (8.73)

It is clear from (8.72) that motion is still clockwise, whereas from (8.73) we see that horizontal tangents now occur along the line

$$I = -\frac{1}{RC}q.$$

The trajectories still have vertical tangents along the horizontal axis. The direction field for (8.73) with a low resistance is shown in Figure 8.28, where the swirling pattern strongly suggests that all motion spirals in toward the origin, giving the origin as a stable equilibrium point. We could examine (8.73) some more by looking at concavity or obtaining an explicit solution (because it is a first order differential equation with homogeneous coefficients), but instead we move to the explicit time-dependent solution for q to conclude our analysis.

If we eliminate I from the two differential equations in (8.72), we obtain the second order differential equation

$$\frac{d^2q}{dt^2} + \frac{R}{L}\frac{dq}{dt} + \frac{1}{LC}q = 0,$$

or, equivalently,

$$L\frac{d^2q}{dt^2} + R\frac{dq}{dt} + \frac{1}{C}q = 0.$$ (8.74)

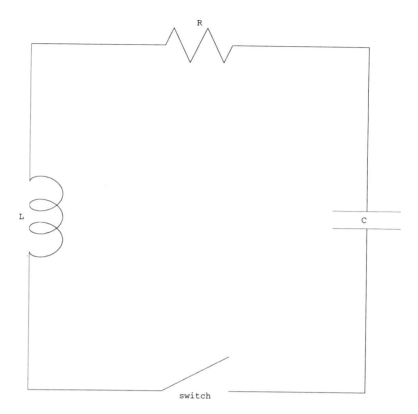

FIGURE 8.27   Series RLC circuit

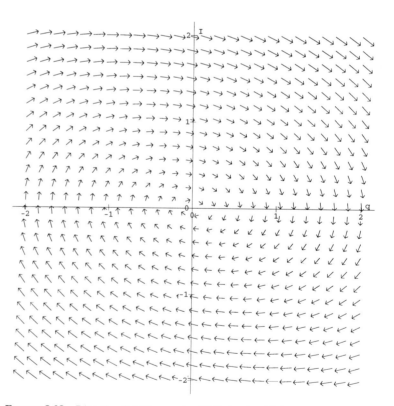

FIGURE 8.28   Direction field for a series RLC circuit with low resistance

Notice that the differential equation for this simple electric circuit has the same form as that of a damped spring-mass system if we replace L with the mass, m, R with the damping factor b, and $1/C$ with the spring constant k. In fact, because it is easier to change the resistance in a circuit than to change the value of b in an experiment involving the oscillations of a mass on a spring, people use experiments with circuits to model damped motion of a spring-mass system.

Equation (8.74) is a familiar equation, which we know has solutions of the form e^{rt}, where r satisfies the characteristic equation

$$Lr^2 + Rr + \frac{1}{C} = 0. \tag{8.75}$$

The roots of (8.75) are given by

$$r = \frac{-R \pm \sqrt{R^2 - 4L/C}}{2L}. \tag{8.76}$$

For small values of resistance, the quantity under the square root will be negative, so we may write our general solution of (8.74) as

$$q(t) = C_1 e^{\alpha t} \cos \beta t + C_2 e^{\alpha t} \sin \beta t,$$

where $\alpha = -R/(2L)$ and $\beta = \sqrt{4L/C - R^2}/(2L)$. Because $\alpha < 0$, we see that regardless of the values of C_1 and C_2, all solutions for $q(t)$ and $I(t) = q'(t)$ approach $(0, 0)$ as $t \to \infty$, so the origin is a stable equilibrium point.

If we consider the situation in which we have a charged capacitor in the circuit of Figure 8.27 and at time $t = 0$ the switch is closed, we have the initial conditions of (8.71) — namely, $q(0) = q_0$, $q'(0) = 0$. This means that in our general solution, we must choose $C_1 = q_0$ and $C_2 = -q_0\alpha/\beta$, so our solution for these initial conditions is

$$q(t) = q_0 e^{\alpha t} \left(\cos \beta t - \alpha/\beta \sin \beta t \right). \tag{8.77}$$

For the special case $C = 1/2$, $L = 4$, $R = 1$, $q_0 = 1$, in Figure 8.29 we plot the expression for $q(t)$ from (8.77) in the lower right-hand box, the corresponding expression for $I(t) = q'(t)$ in the upper right-hand box, and the trajectory in the phase plane corresponding to these two explicit solutions in the upper left-hand box. The 45° line is also included in the lower left-hand box to allow you to relate these three graphs. This situation corresponds to the underdamped situation for a damped spring-mass system for a mass of 4, spring constant of 2, and damping factor of 1.

EXAMPLE 8.10: Series RLC Circuits with Large Values of Resistance

We now focus on the situation in which the resistance in the series RLC circuit is large enough that the roots of the characteristic equation associated with our second order differential equation for $q(t)$ are real numbers. This means that the general solution of (8.74) is given by

$$q(t) = C_1 e^{r_1 t} + C_2 e^{r_2 t}, \tag{8.78}$$

where $r_1 = \left(-R - \sqrt{R^2 - 4L/C}\right)/(2L)$ and $r_2 = \left(-R + \sqrt{R^2 - 4L/C}\right)/(2L)$. Our first observation is that because $\sqrt{R^2 - 4L/C} < R$, both r_1 and r_2 are negative, and again

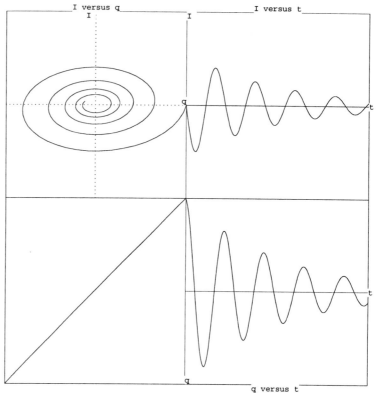

FIGURE 8.29   Charge, current, and phase plane trajectory for a series RLC circuit with low resistance

$(0, 0)$ is a stable equilibrium point in the phase plane. If we use the initial conditions for the discharge of a capacitor from before, (8.71), we see that

$$C_1 + C_2 = q_0 \text{ and } C_1 r_1 + C_2 r_2 = 0.$$

Thus, we must take

$$C_1 = \frac{r_2 q_0}{r_2 - r_1} \text{ and } C_2 = -\frac{r_1 q_0}{r_2 - r_1},$$

so (8.78) becomes

$$q(t) = \frac{r_2 q_0}{r_2 - r_1} e^{r_1 t} - \frac{r_1 q_0}{r_2 - r_1} e^{r_2 t}.$$

The corresponding value for the current is obtained from $I(t) = q'(t)$. For the special case $C = 1/2, L = 4, R = 10, q_0 = 1$, Figure 8.30 gives the graph of $q(t)$ in the lower right-hand box, that of $I(t)$ in the upper right-hand box, and the corresponding trajectory in the phase plane in the upper left-hand box. As usual, the 45° line in the lower left-hand box allows the construction of lines to develop the phase plane from the two explicit solutions. This situation corresponds to the overdamped situation in the mathematical model describing the damped spring-mass system.   □

The case where $R = \sqrt{4L/C}$, corresponding to the critically damped case for a series RLC circuit, is left for the exercises.

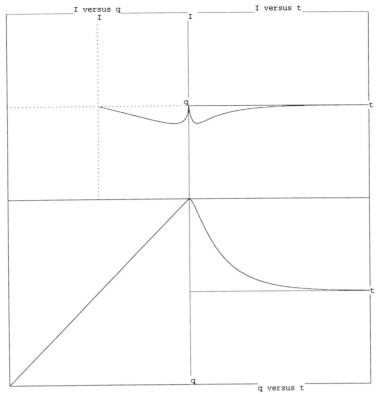

FIGURE 8.30   Charge, current, and phase plane trajectory for a series RLC circuit with large resistance

EXAMPLE 8.11: Parallel RLC Circuit

We now consider the situation in which the resistor, inductor, and capacitor are connected in parallel, as shown in Figure 8.31. Here we use Kirchhoff's current law, which states that the algebraic sum of currents at any point must equal zero. To obtain the current for each of the elements, we recall that the voltage drop across the three individual elements is

$$V = q/C \quad \text{(capacitor)},$$
$$V = RI \quad \text{(resistor)},$$
$$V = L \, dI/dt \quad \text{(inductor)}.$$

Using the fact that $I = dq/dt$, we obtain the currents in each of the individual elements as

$$I = C \, dV/dt \quad \text{(capacitor)},$$
$$I = V/R \quad \text{(resistor)},$$
$$I = (1/L) \int V \, dt \quad \text{(inductor)}.$$

Thus, for the circuit of Figure 8.31, we can conclude that

$$C \frac{dV}{dt} + \frac{1}{R} V + \frac{1}{L} \int V \, dt = 0.$$

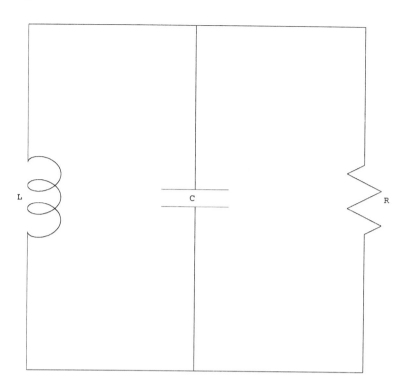

FIGURE 8.31 Parallel RLC circuit

If we differentiate this equation with respect to t, we obtain

$$C\frac{d^2V}{dt^2} + \frac{1}{R}\frac{dV}{dt} + \frac{1}{L}V = 0, \tag{8.79}$$

a second order differential equation with constant coefficients. We now obtain the explicit solution for V corresponding to the initial conditions

$$V(0) = 1, \qquad V'(0) = 0. \tag{8.80}$$

Again we know that solutions of (8.79) have the form of e^{rt}, where r satisfies the characteristic equation

$$Cr^2 + \frac{1}{R}r + \frac{1}{L} = 0.$$

This quadratic equation has solutions given by

$$r_1, r_2 = \frac{-1/R \pm \sqrt{1/R^2 - 4C/L}}{2C}, \tag{8.81}$$

where r_1 is the root obtained by taking the minus sign in (8.81). If $1/R^2 > 4C/L$, then both of these roots will be real numbers, giving the general solution of the differential equation as

$$V(t) = C_1 e^{r_1 t} + C_2 e^{r_2 t}, \tag{8.82}$$

where C_1 and C_2 are arbitrary constants. If we choose these constants so our solution satisfies the initial conditions in (8.80), we have

$$C_1 + C_2 = 1 \text{ and } C_1 r_1 + C_2 r_2 = 0.$$

Thus, we must take

$$C_1 = \frac{r_2}{r_2 - r_1} \text{ and } C_2 = -\frac{r_1}{r_2 - r_1},$$

so (8.82) becomes

$$V(t) = \frac{r_2}{r_2 - r_1} e^{r_1 t} - \frac{r_1}{r_2 - r_1} e^{r_2 t}. \tag{8.83}$$

Because r_1 and r_2 are both negative in (8.83), both $V(t)$ and $V'(t)$ will go to zero in the limit as $t \to \infty$, so the origin is a stable equilibrium point in the $V'(t)$ versus $V(t)$ phase plane.

One thing that is different in the two circuits (parallel and series) is the response to changes in the value of the resistance. In the series circuit, from (8.76), we have that the charge will not be oscillatory if $R^2 - 4L/C \geq 0$, with critical damping occurring with the equal sign.[7] For any less resistance than this, the response will oscillate. In the parallel RLC circuit, from (8.81), we see that voltage will not be oscillatory if $1/R^2 - 4C/L \geq 0$, with critical damping associated with the equal sign. Here, if R is slightly larger than that which gives critical damping, the voltage in this circuit will oscillate in an underdamped mode! This agrees with our intuition because increasing the resistance in a series RLC circuit makes it more difficult for current to flow, whereas increasing the resistance in a parallel RLC circuit encourages the current to flow in the other two components. If the resistance were infinite, we would have the pure oscillatory LC circuit from before. Notice also that if we think of L and C as being fixed, with R allowed to vary, for critical damping in the series circuit we have $R = 2\sqrt{L/C}$. In the parallel circuit, $R = (1/2)\sqrt{L/C}$.

To further explore the behavior of parallel RLC circuits, we consider the specific case where $L = 1/5$ henries, $C = 1/4$ farads, and $R = 1/3$ ohms. This means the solution in (8.83) is

$$V(t) = 1.25 \, e^{-2t} - 0.25 \, e^{-10t}. \tag{8.84}$$

Although we have already noted that $V(t) \to 0$ as $t \to \infty$, the voltage never reaches zero for a finite value of time.

In Section 6.4 we defined the SETTLING TIME, which is the time it takes for a response to fall to 1% of its maximum value. What is the settling time for this example? For the initial value problem for which (8.84) is our explicit solution, we see 1 is the maximum value of V, so that if we call the settling time t_s, we have that

$$0.01 = 1.25 \, e^{-2t_s} - 0.25 \, e^{-10t_s}. \tag{8.85}$$

This equation cannot be solved by analytically, but can be solved by any standard root-

---

[7]If we want to compare the charge on the capacitor in the two circuits, recall that $q = CV$. Thus, the changes in charge are proportional to the changes in voltage.

finding method. We find[8]

$$t_s \approx 2.414 \text{ sec.}$$

If we now keep the same values of $L = 1/5$ henries and $C = 1/4$ farad from our last example, and give R the value required for critical damping, we have that $R = (1/2)\sqrt{L/C} = 1/\sqrt{5}$. Using these values back in the characteristic equation (8.81), we have that its double root is given by $r = -1/(2RC) = -2\sqrt{5}$. This gives our explicit solution as

$$V(t) = (1 - rt)e^{rt}, \tag{8.86}$$

from which we again see that 1 is the maximum value of V. Thus, the settling time t_s for the critically damped case is given by the solution to

$$0.01 = (1 - rt_s)e^{rt_s}.$$

This may be solved by any standard root-finding method, yielding $t_s \approx 1.484$ sec. As expected, the critical damping case has a shorter settling time than did our previous over-damped case. The graphs of the explicit solutions for the voltage as given in (8.84) and (8.86) are shown in Figure 8.32. Also shown in this figure is the explicit solution for an underdamped case ($R = 1$), given by

$$V(t) = e^{-2t}\left(\cos 4t + 0.5 \sin 4t\right).$$

(See Exercise 6 on page 419.)

Now in the underdamped case, the voltage in the parallel circuit will oscillate forever, with its maximum and minimum values continually decreasing. The question may arise if there is some resistance at which the underdamped case will result in having a smaller t_s than that for the critically damped case. To investigate this situation, we find the explicit solution of (8.79) for $L = 1/5$, $C = 1/4$, and R unspecified as

$$V(t) = e^{-2t/R}\left(\cos \beta t + \frac{2}{\beta R} \sin \beta t\right), \tag{8.87}$$

where $\beta = 2\sqrt{5 - 1/R^2}$. Notice that relative extrema for $V(t)$ in (8.87) will occur at positive values of t for which $V'(t) = 0$ (see the underdamped case in Figure 8.32). Differentiating (8.87) we see that

$$V'(t) = e^{-2t/R}\left(-\beta - \frac{4}{R^2\beta}\right)\sin \beta t,$$

so this occurs when $\beta t = n\pi$. The value of the voltage at these times is

$$V\left(\frac{n\pi}{\beta}\right) = (-1)^n e^{-2n\pi/(R\beta)}, \tag{8.88}$$

so we need to see if we can find values of R and n such that the absolute value of the expression in (8.88) is less than 0.01. This inequality may be solved for $n = 1, 2, 3, \cdots$,

---

[8]Because the last term on the right-hand side of (8.85) decays much faster than the first term, for most accuracy requirements we may delete the last term and solve the resulting equation $0.01 = 1.25e^{-2t_s}$, which gives $t_s = -1/2 \ln(0.008) \approx 2.414$.

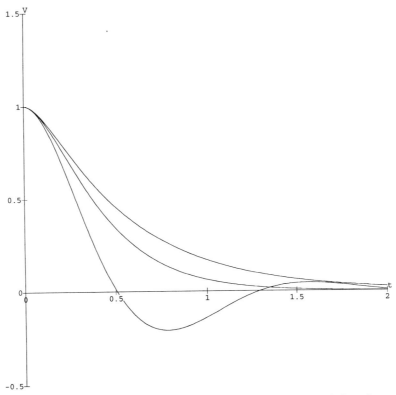

FIGURE 8.32 Overdamped, critically damped, and underdamped solutions for a parallel RLC circuit

in many different ways, and we find that with $n = 1$, any value of R less than 0.541 will work. The time (corresponding to $R = 0.541$ ohms) at which $V(t)$ attains this first minimum of 0.01 is $\pi/\beta \approx 1.247$ sec. Figure 8.33 compares the critically damped case to this underdamped case. Because we needed R greater than $1/\sqrt{5}$, any value of R between $1/\sqrt{5} \approx 0.447$ and 0.541 will give a smaller value of t_s than that obtained from the R from critical damping (namely, $1/\sqrt{5}$). It is interesting to note that the settling time for the underdamped case corresponding to $R = 0.541$ is less than 1.247 sec, the time at which the first minimum occurs. This can be seen in Figure 8.34, which shows the settling time of about 0.938 sec. □

EXERCISES

1. Differentiate (8.68) to show that the concavity of solutions in the phase plane agrees with that given by the trajectories of (8.69).

2. Find the general solution of the differential equation that models the series RLC circuit, namely, (8.74), or

$$L\frac{d^2q}{dt^2} + R\frac{dq}{dt} + \frac{1}{C}q = 0,$$

for the case when $R^2 = 4L/C$.

(a) Evaluate the arbitrary constants in your general solution using the initial conditions corresponding to the discharging of a charged capacitor.

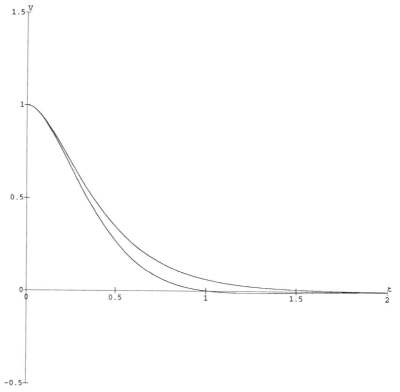

FIGURE 8.33   Critically damped and underdamped parallel RLC circuit

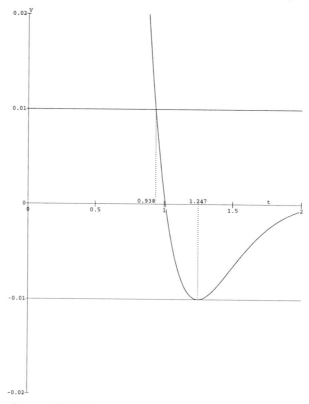

FIGURE 8.34   Settling time for underdamped parallel RLC circuit

(b) Plot the explicit solutions for $q(t)$ and $q'(t)$ and use them to obtain the graph of the trajectories in the phase plane. Show that the origin is an equilibrium point for these trajectories in the phase plane.

(c) If we have $R = 4$, $L = 1$, and $C = 1/4$, find the value of t such that the solution in part (a) reaches 1% of its maximum value; that is, find the settling time.

3. Solve the following initial value problems associated with the differential equation for a series RLC circuit

$$L\frac{d^2q}{dt^2} + R\frac{dq}{dt} + \frac{1}{C}q = 0.$$

(a) $L = 1$, $R = 1/8$, $C = 1$, $q(0) = 6$, $q'(0) = 0$

(b) $L = 1$, $R = 1/8$, $C = 1$, $q(0) = 0$, $q'(0) = 6$

(c) $L = 9$, $R = 6$, $C = 1/37$, $q(0) = 6$, $q'(0) = 0$

(d) $L = 9$, $R = 6$, $C = 1/37$, $q(0) = 0$, $q'(0) = 6$

4. For what values of the resistance R is the charge in a series RLC circuit (governed by $4q'' + Rq' + 9q = 0$)

(a) overdamped?

(b) underdamped?

(c) critically damped?

5. For what values of the resistance R is the voltage in a parallel RLC circuit [governed by $(1/9)V'' + (1/R)V' + (1/4)V = 0$]

(a) overdamped?

(b) underdamped?

(c) critically damped?

6. Show that the solution of the initial value problem

$$(1/4)V'' + V' + 5V = 0, \qquad V(0) = V_0, \qquad V'(0) = 0$$

is given by $V(t) = V_0 e^{-2t} (\cos 4t + 0.5 \sin 4t)$.

7. Obtain (8.87); that is, show that the solution of the initial value problem

$$(1/4)V'' + (1/R)V' + 5V = 0, \qquad V(0) = 1, \qquad V'(0) = 0$$

is given by

$$V(t) = e^{-2t/R} \left(\cos \beta t + \frac{2}{\beta R} \sin \beta t \right),$$

where $\beta = 2\sqrt{5 - 1/R^2}$.

8. Plot your solutions from Exercise 7 for $R = 0.1, 0.5, 1.0, 1.5$, and 2.0 on the same graph. Explain what you see and why it is the way it is.

9. Find the settling time for the voltage that satisfies

$$(1/4)V'' + V' + 5V = 0, \qquad V(0) = 0, \qquad V'(0) = 1.$$

10. A resistor cannot store energy, so we define the total energy of the series RLC circuit system described by (8.74) as

$$E = \frac{1}{2}L\left(\frac{dq}{dt}\right)^2 + \frac{1}{2C}q^2.$$

Show that E satisfies

$$\frac{dE}{dt} = -R\left(\frac{dq}{dt}\right)^2$$

and is therefore a decreasing function of time.

8.4   Boundary Value Problems

Initial value problems have conditions on the dependent variable given at a single value of the independent variable, such as the case for

$$\frac{d^2y}{dt^2} + 3\frac{dy}{dt} - 4y = 0, \qquad y(0) = 0, \qquad y'(0) = 2.$$

With boundary value problems, conditions are given at two different values of the independent variable, usually at the endpoints of its domain.

EXAMPLE 8.12

An example of a boundary value problem would be to find a function satisfying the differential equation

$$\frac{d^2y}{dt^2} + 3\frac{dy}{dt} - 4y = 0, \qquad 0 < t < 1, \tag{8.89}$$

if

$$y(0) = 0, \qquad y(1) = 2. \tag{8.90}$$

The procedure for solving boundary value problems is similar to that of solving initial value problems — namely, finding a general solution of the differential equation and then assigning the arbitrary constants in this general solution so the boundary conditions are satisfied. (If the differential equation has a forcing function, then we must find a particular solution before the arbitrary constants are determined.)

The general solution of (8.89) is obtained by using a solution of the form e^{rt}. This leads to the characteristic equation

$$r^2 + 3r - 4 = 0,$$

with solutions $r = -4$ and $r = 1$, and gives our general solution as

$$y(t) = C_1 e^{-4t} + C_2 e^t,$$

where C_1 and C_2 are arbitrary constants. From (8.90) we see that the boundary conditions will be satisfied if C_1 and C_2 satisfy

$$C_1 e^0 + C_2 e^0 = 0,$$

$$C_1 e^{-4} + C_2 e^1 = 2.$$

Solving these equations gives

$$C_1 = -C_2 = \frac{2}{e^{-4} - e},$$

and the solution of our boundary value problem becomes

$$y(t) = \frac{2}{e^{-4} - e} \left(e^{-4t} - e^t \right).$$

Now let us look at a similar boundary value problem where the differential equation is

$$y'' + 4y = 0 \tag{8.91}$$

and the boundary conditions are

$$y(0) = 0, \qquad y(b) = 2. \tag{8.92}$$

The general solution of (8.91) is

$$y(t) = C_1 \sin 2t + C_2 \cos 2t.$$

The first of the boundary conditions is satisfied if we choose $C_2 = 0$, and with that choice the second requires that

$$C_1 \sin 2b = 2.$$

If $\sin 2b \neq 0$, we can determine a value for C_1 such that this boundary condition is fulfilled — namely, $C_1 = 2/\sin 2b$. However, if $\sin 2b = 0$, there is no way to choose C_1 such that the second boundary condition in (8.92) is satisfied, so this boundary value problem has no solution.

Notice that if we change the second boundary condition to $y(\pi/2) = 0$, from our general solution, with $C_2 = 0$, we see that, regardless of the choice of C_1,

$$y\left(\frac{\pi}{2}\right) = C_1 \sin \pi = 0.$$

This gives an infinite number of solutions, one for each choice of the arbitrary constant C_1.

Thus, we have examples of boundary value problems that have a unique solution, no solution, or an infinite number of solutions. Lest we start thinking that we are doing something wrong, we should recall that all of our existence and uniqueness theorems to this point apply to initial value problems, not to boundary value problems. $\square$

EXAMPLE 8.13: Eigenvalues and Eigenfunctions for Boundary Value Problems

One of the important occurrences of boundary value problems is in mathematical models of vibrating systems. For example, a problem that arises in the vibration of an elastic string (such as a violin or guitar string) is to find values of the parameter λ such that the following boundary value problem has a nontrivial solution:

$$\frac{d^2y}{dx^2} + \lambda y = 0, \tag{8.93}$$

$$y(0) = 0, \qquad y(4) = 0. \tag{8.94}$$

These boundary conditions are understood to mean that the string is 4 units in length and that there is no displacement at either end.

From our previous experience, we see that the types of solutions of (8.93) depend upon the sign of λ. We now cover the three possible choices for λ in turn.

1. $\lambda > 0$. For convenience in notation we let $\lambda = \mu^2$. For this case the general solution of (8.93) is given by

$$y(x) = C_1 \sin \mu x + C_2 \cos \mu x,$$

and choosing $C_2 = 0$ satisfies the first boundary condition at $x = 0$. This leaves the other boundary condition as

$$y(4) = C_1 \sin 4\mu = 0,$$

which could be satisfied if either $C_1 = 0$ (giving the trivial solution) or

$$\sin 4\mu = 0.$$

Because we are seeking nontrivial solutions, choosing any μ satisfying

$$4\mu = n\pi, \qquad n = 1, 2, 3, \cdots$$

will provide a solution to the boundary value problem. Because $\lambda = \mu^2$ this gives an infinite number of possibilities for λ — namely, $n^2\pi^2/16$ — which are collectively referred to as the EIGENVALUES for this boundary value problem. ("Eigen" is the German word for "characteristic", so the name emphasizes the special nature of λ characteristic to this problem.) For each of these eigenvalues, $n^2\pi^2/16$, we have the solution (8.93), namely, $C_1 \sin \mu x = C_1 \sin(n\pi x/4)$. The function $\sin(n\pi x/4)$ is called an EIGENFUNCTION corresponding to the eigenvalue $n^2\pi^2/16$. It is common to label the eigenvalues and eigenfunctions associated with a boundary value problem with subscripts. In this case we would say that the eigenvalues and eigenfunctions for the boundary value problem given by (8.93) and (8.94) are

$$\lambda_n = \frac{1}{16}n^2\pi^2 \qquad \text{and} \qquad y_n(x) = \sin\frac{n\pi x}{4}, \qquad n = 1, 2, 3, \cdots.$$

Note that a constant times an eigenfunction will still satisfy both the differential equation in (8.93) and the boundary conditions in (8.94). Thus, when $\lambda > 0$, $C_1 \sin(\pi x/4)$, $C_2 \sin(2\pi x/4)$, $C_3 \sin(3\pi x/4)$, $\cdots$, form an infinite set of solutions with an infinite number of arbitrary constants.

2. $\lambda = 0$. In this case our general solution is

$$y(x) = C_1 + C_2 x,$$

and the two boundary conditions require that

$$y(0) = C_1 = 0,$$

$$y(4) = C_1 + 4C_2 = 0.$$

These two algebraic equations have the unique trivial solution $C_1 = C_2 = 0$, and we obtain no eigenvalues from this case.

3. $\lambda < 0$. For convenience in notation we let $\lambda = -\mu^2$, so the differential equation becomes

$$y'' - \mu^2 y = 0,$$

with a general solution given by[9]

$$y(x) = C_1 \sinh \mu x + C_2 \cosh \mu x.$$

Setting $C_2 = 0$ satisfies the first boundary condition in (8.94) at $x = 0$, leaving the other boundary condition as

$$C_1 \sinh 4\mu = 0.$$

Because $\sinh 4\mu = \left(e^{4\mu} - e^{-4\mu}\right)/2$ is 0 only for $\mu = 0$ (and here $\mu^2 > 0$), C_1 must equal zero, and we are left without any nontrivial solutions for this case. Thus, all our eigenvalues and eigenfunctions only come from the case where $\lambda > 0$.   □

EXAMPLE 8.14

The buckling of a long shaft, or column, of length L under an axial compressive force is modeled by the differential equation

$$\frac{d^2}{dx^2}\left(EI\frac{d^2 y}{dx^2}\right) + P\frac{d^2 y}{dx^2} = 0. \tag{8.95}$$

Here y is the displacement from equilibrium, x is the distance along the shaft, EI is the flexural rigidity, and P is the axial compressive force. For this example, we assume the shaft is supported at the ends in a manner such that there is zero deflection and zero moment. Zero deflection means that

$$y(0) = y(L) = 0,$$

and zero moment means that

$$y''(0) = y''(L) = 0.$$

When EI and P are both constant, we may divide both sides of (8.95) by EI, and because

---

[9]Recall that $\sinh x = (e^x - e^{-x})/2$ and $\cosh x = (e^x + e^{-x})/2$.

we know that $P/(EI)$ is positive, we define λ by $\lambda^2 = P/(EI)$. This gives the resulting equation as

$$\frac{d^4 y}{dx^4} + \lambda^2 \frac{d^2 y}{dx^2} = 0. \tag{8.96}$$

One of the ways to solve this problem is to mimic a technique from Section 5.5 and let $w = d^2 y/dx^2$. We then consider the resulting second order differential equation

$$\frac{d^2 w}{dx^2} + \lambda^2 w = 0.$$

This differential equation has its general solution given by

$$w(x) = C_1 \cos \lambda x + C_2 \sin \lambda x,$$

and because $d^2 y/dx^2 = w$, we can integrate twice to obtain

$$y(x) = -\frac{C_1}{\lambda^2} \cos \lambda x - \frac{C_2}{\lambda^2} \sin \lambda x + C_3 x + C_4$$

as the general solution of (8.96). If we apply the boundary conditions at $x = 0$, we have

$$y(0) = -\frac{C_1}{\lambda^2} + C_4 = 0$$
$$y''(0) = C_1 = 0.$$

This gives

$$C_1 = C_4 = 0.$$

We now apply our remaining boundary conditions at $x = L$ and obtain

$$y(L) = -\frac{C_2}{\lambda^2} \sin \lambda L + C_3 L = 0,$$
$$y''(L) = C_2 \sin \lambda L = 0.$$

Here we see that C_3 must equal zero and $C_2 \sin \lambda L = 0$. Now if $C_2 = 0$, all of the arbitrary constants are zero, and we simply obtain the trivial equilibrium solution $y(x) = 0$. If we are to have any other solutions, then λ must be such that

$$\sin \lambda L = 0.$$

This equation is satisfied for values of λ such that

$$\lambda L = n\pi, \qquad n = 1, 2, 3, \cdots;$$

thus, the eigenvalues for this example are

$$\lambda^2 = \frac{n^2 \pi^2}{L^2}, \qquad n = 1, 2, 3, \cdots.$$

Because the length of the column and its physical characteristics are all prescribed constants — that is, L, P, EI, and therefore λ^2 are all constants — we have discovered that there are an infinite number of applied compressive loads P_n, that satisfy all the conditions of this example. These discrete values,

$$P_n = \frac{n^2 \pi^2 E I}{L^2}, \qquad n = 1, 2, 3, \cdots,$$

are called the critical buckling loads. The smallest of these loads is P_1, given by

$$P_1 = \frac{\pi^2 E I}{L^2}.$$

For axial compressive loads smaller than this value, the column is stable in the unbent position. In other words, the column will remain straight when such a load is applied. The critical value P_1 is also called the Euler load and is the upper limit of stability for the unbent column.

The eigenfunctions corresponding to these eigenvalues are given by

$$y_n(x) = \sin \frac{n \pi x}{L}, \qquad n = 1, 2, 3, \cdots. \tag{8.97}$$

Of course in giving the solution to the boundary value problem, each of these eigenfunctions is multiplied by an arbitrary constant. The solution corresponding to the smallest critical buckling load,

$$y_1(x) = C_1 \sin \frac{\pi x}{L},$$

gives the corresponding deflection.

In Example 8.13 we found eigenfunctions for a boundary value problem as

$$y_n(x) = \sin \frac{n \pi x}{4}, \qquad n = 1, 2, 3, \cdots.$$

The domain of interest for these eigenfunctions was $0 < x < 4$. Note that this set of functions has the property that for $n \neq m$,

$$\int_0^4 y_n(x) y_m(x)\, dx = \int_0^4 \sin(n \pi x / 4) \sin(m \pi x / 4)\, dx$$

$$= \frac{1}{2} \int_0^4 \cos\left((n - m)\,\pi x / 4\right) - \cos\left((n + m)\,\pi x / 4\right)\, dx$$

$$= \frac{1}{2} [\sin\left((n - m)\,\pi x / 4\right) / \left((n - m)\,\pi / 4\right) +$$

$$- \sin\left((n + m)\,\pi x / 4\right) / \left((n + m)\,\pi / 4\right)]\big|_0^4$$

$$= 0.$$

This property is of great importance in the development of series representations of functions over a finite interval, so we give the following relevant definitions:

A set of functions $\{y_n(x), n = 1, 2, 3, \cdots\}$ is said to be ORTHOGONAL WITH RESPECT TO THE WEIGHT FUNCTION $p(x)$ on the interval $[a, b]$ if

$$\int_a^b y_n(x) y_m(x) p(x)\, dx = 0, \qquad n \neq m. \tag{8.98}$$

The NORM of the nth element in such a set of orthogonal functions is defined by

$$\|y_n(x)\| = \left(\int_a^b \left[y_n(x) \right]^2 p(x)\,dx \right)^{1/2}, \qquad n = 1, 2, 3, \cdots.$$

- If $p(x) = 1$ in (8.98), the phrase "with respect to the weight function $p(x)$" is deleted from the definition.

Now for our set of functions in Example 8.13, we compute the integral

$$\int_0^4 \sin^2 \left(\frac{n\pi}{4} x \right) dx = 2, \qquad n = 1, 2, 3, \cdots,$$

and note that we would say that $\{ y_n = \sin(n\pi x/4), n = 1, 2, 3, \cdots \}$ is orthogonal on the interval $[0, 4]$ with the norm of every function the same, namely,

$$\left\| \sin \left(\frac{n\pi}{4} x \right) \right\| = \sqrt{2}, \qquad n = 1, 2, 3, \cdots.$$

The example involving the axial loading of a beam also produced an orthogonal set of functions, namely, $\{ y_n = \sin(n\pi x/L), n = 1, 2, 3, \cdots \}$. The appropriate interval in this example is $0 \le x \le L$, and the norm of these functions is easily obtained by integration as

$$\left\| \sin \left(\frac{n\pi}{L} x \right) \right\| = \sqrt{\frac{L}{2}}, \qquad n = 1, 2, 3, \cdots. \qquad \square$$

EXAMPLE 8.15

The boundary value problem that we solved in Example 8.13 comes from consideration of a vibrating string. There the string is considered to be stretched between the places it was fastened at $x = 0$ and $x = 4$. The ordinary differential equation we solved was obtained from a partial differential equation that describes the time-varying motion of the string. Although we will not consider partial differential equations at this point, we will analyze the eigenfunctions in this context. A typical situation is for the string to be displaced from its equilibrium position and released. If the initial shape is that of a constant times one of the eigenfunctions, an ideal string will oscillate with a constant frequency (corresponding to that eigenfunction), with a time-dependent amplitude. This will be true for any of the eigenfunctions, so these eigenfunctions describe what are called the normal (or natural) modes of vibration. The eigenfunction corresponding to the smallest eigenvalue is called the FUNDAMENTAL MODE. Because the frequency of oscillation is proportional to $\sqrt{\lambda}$, the smallest eigenvalue corresponds to the lowest frequency of vibration. For this situation the fundamental mode is given by

$$y_1(x) = \sin \left(\frac{\pi}{4} x \right)$$

and is graphed in Figure 8.35. Notice that this mode is symmetric about the midpoint of the string. The graph of the next lowest mode

$$y_2(x) = \sin \left(\frac{\pi}{2} x \right)$$

is given in Figure 8.36 and is obviously antisymmetric about the midpoint of the string. To describe the motion of the string caused by an arbitrary initial displacement requires consideration of an infinite sum of constants times eigenfunctions. This leads to the topic of Fourier series and will not be covered in this book. $\qquad \square$

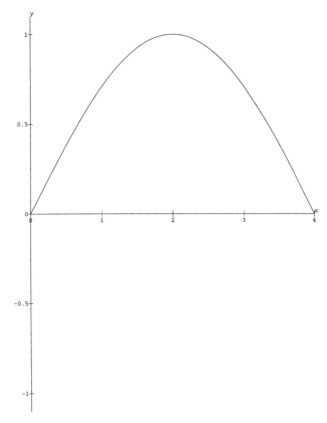

FIGURE 8.35   The fundamental mode, $\sin(\pi x/4)$

EXERCISES

1. Solve the following problems.

(a) $y'' + 100y = 0,$ $y(0) = 4,$ $y'(0) = 500$

(b) $y'' + 10y' + 26y = 0,$ $y(0) = 1,$ $y'(0) = 10$

(c) $y'' + 3y' + 14y = 0,$ $y(0) = 0,$ $y'(0) = 5$

2. For what value of b will the boundary value problem

$$y'' + 16y = 0,\qquad y(0) = 0,\qquad y'(b) = 1$$

(a) Have no solutions?

(b) Have exactly one solution?

(c) Have an infinite number of solutions?

3. For what value of b will the boundary value problem

$$y'' + 16y = 0,\qquad y(0) = 0,\qquad y'(b) = 0$$

(a) Have no solutions?

(b) Have exactly one solution?

(c) Have an infinite number of solutions?

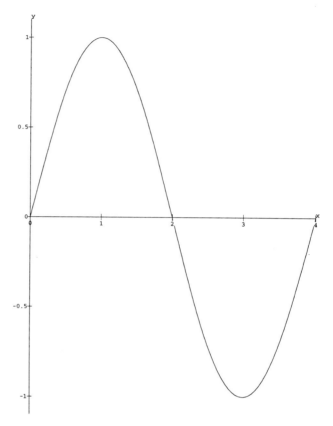

FIGURE 8.36 The mode for second lowest frequency, $\sin(\pi x/2)$

4. Show that if $y_1(t)$ and $y_2(t)$ satisfy a general second order differential equation

$$a_2(t)\frac{d^2y}{dt^2} + a_1(t)\frac{dy}{dt} + a_0(t)y = 0 \qquad \textbf{(8.99)}$$

and

$$y_1(0) = A_1, \qquad y_1(L) = B_1, \qquad y_2(0) = A_2, \qquad y_2(0) = B_2,$$

then $\alpha y_1(t) + \beta y_2(t)$ satisfies (8.99) along with the boundary conditions

$$y(0) = \alpha A_1 + \beta A_2, \qquad y(L) = \alpha B_1 + \beta B_2.$$

5. Find the eigenvalues and corresponding eigenfunctions for the following boundary value problems, where $\lambda^2 > 0$.

   (a) $y'' + \lambda^2 y = 0,$ $\qquad y(0) = y(b) = 0$

   (b) $y'' + \lambda^2 y = 0,$ $\qquad y'(0) = y(b) = 0$

   (c) $y'' + \lambda^2 y = 0,$ $\qquad y(0) = y'(b) = 0$

   (d) $y'' + \lambda^2 y = 0,$ $\qquad y'(0) = y'(b) = 0$

6. Ignoring damping, a spring-mass system is described by

$$\frac{d^2x}{dt^2} + \lambda^2 x = 0, \qquad \lambda^2 = \frac{k}{m}.$$

Find the values of the ratio k/m such that any motion that starts at its equilibrium position is at the same position at $t = 3$. [That is, if $x(0) = 0$ then $x(3) = 0$ also.]

7. Show that the eigenfunctions from

$$y'' + \lambda^2 y = 0, \qquad y'(0) = y(b) = 0$$

form an orthogonal set on $[0, b]$, and find the norm of the functions in this set.

8. Show that the eigenfunctions from

$$y'' + \lambda^2 y = 0, \qquad y(0) = y'(b) = 0$$

form an orthogonal set on $[0, b]$, and find the norm of the functions in this set.

9. Find an equation that λ must satisfy in order that the boundary value problem

$$y'' + \lambda^2 y = 0, \qquad y(0) = 0, \qquad y'(3) + 2y(3) = 0$$

will have a nontrivial solution.

(a) Show that the number of eigenvalues is infinite. If these eigenvalues are labeled λ_1, λ_2, λ_3, ..., what are the associated eigenfunctions?

(b) Show that the set of eigenfunctions is orthogonal, and compute the norm of the functions in this set.

What Have We Learned?

MAIN IDEAS

- We considered three applications which used similar linear second order differential equations with constant coefficients, namely,
  - the damped linear pendulum

  $$x'' + 2\alpha x' + \lambda^2 x = 0$$

  - the damped spring-mass

  $$mx'' + bx' + kx = 0$$

  - the series RLC circuit

  $$Lq'' + Rq' + \frac{1}{C}q = 0$$

- For such systems
  - undamped motion oscillates indefinitely. The oscillation does not change. The motion does not approach zero with increasing time.
  - underdamped motion oscillates indefinitely. The oscillation decays to zero with increasing time, and the motion approaches zero.
  - overdamped motion does not oscillate. The motion approaches zero with increasing time.
  - critically damped motion is the motion between underdamped motion and overdamped motion.

- For the nonlinear pendulum

$$x'' + \lambda^2 \sin x = 0,$$

  much of the solution behavior may be discovered without the aid of an explicit solution.

- Boundary value problems consist of differential equations with conditions given at two points in the domain (usually at the boundaries of the domain).

- Boundary value problems with a parameter in the differential equation may lead to orthogonal sets of functions.

CHAPTER 9

Second Order Linear Differential Equations

Where Are We Going?

In this chapter we look at the general second order linear differential equation

$$a_2(x)\frac{d^2y}{dx^2} + a_1(x)\frac{dy}{dx} + a_0(x)y = 0, \tag{9.1}$$

where $a_2(x)$, $a_1(x)$, and $a_0(x)$ are continuous functions of x and $a_2(x) \neq 0$ for $a < x < b$.

We will define what we mean by a general solution of (9.1) and develop several techniques for finding such solutions. We need linearly independent functions in one of these techniques, so they are introduced here.

We state theorems which show when we have solutions of (9.1) in the form of Taylor series. Because we often cannot find a general expression for the coefficients in a series solution, we introduce several theorems to help determine properties of solutions such as regions of convergence and whether or not solutions oscillate.

9.1 General Properties

We begin this chapter with a discussion of the qualitative nature of explicit solutions of the general second order linear differential equation given in (9.1). In Chapter 7 we introduced an existence and uniqueness theorem (Theorem 7.1), dealing with such differential equations, namely,

Theorem 9.1: *Consider the second order linear differential equation*

$$a_2(x)\frac{d^2y}{dx^2} + a_1(x)\frac{dy}{dx} + a_0(x)y = 0, \tag{9.2}$$

subject to the initial conditions

$$y(x_0) = y_0, \quad y'(x_0) = y_0^*.$$

If $a_2(x)$, $a_1(x)$, and $a_0(x)$ are continuous, with $a_2(x) \neq 0$ for $a < x < b$, and x_0 is contained in this interval, then for the interval $a < x < b$ there exists a unique solution of (9.2) with a continuous first derivative.

We have seen that if a_2, a_1, and a_0 are all constants, then we can always analyze the qualitative behavior of the solutions of (9.2) by means of the phase plane. Also, for this constant coefficient case, we can always find the explicit solution of (9.2).

However, if a_2, a_1, and a_0 are not all constants, the phase plane analysis fails because the associated system of first order differential equations is not autonomous. Furthermore, if a_2, a_1, and a_0 are not all constants, obtaining a solution of (9.2) in terms of familiar functions is the exception rather than the rule. For example, the innocent-looking differential equation

$$\frac{d^2y}{dx^2} - xy = 0, \tag{9.3}$$

which is known as Airy's equation and is important in aerodynamics, cannot be solved in terms of familiar functions.

In the subsequent sections we will look at various techniques for solving special cases of (9.2). However, situations exist where we can predict the behavior of solutions of (9.2) regardless of our finding an explicit solution.

We introduce these ideas by looking at the behavior of solutions of

$$\frac{d^2y}{dx^2} + ay = 0, \tag{9.4}$$

where a is constant. If $a < 0$ we may put $a = -k^2$, and we know the general solution of (9.4) is $y(x) = C_1 e^{kx} + C_2 e^{-kx}$, whereas if $a > 0$ we may put $a = k^2$, and the general solution is $y(x) = C_1 \cos kx + C_2 \sin kx$.

There are some major behavioral differences between these two solutions. In the case $a < 0$, where $y(x) = C_1 e^{kx} + C_2 e^{-kx}$ any nontrivial solution[1] can be zero at most once [at $x = \ln(-C_2/C_1)/(2k)$ if $-C_2/C_1 > 0$, and nowhere if $-C_2/C_1 \leq 0$]. In the case $a > 0$, where $y(x) = C_1 \cos kx + C_2 \sin kx$, any nontrivial solution can be zero an infinite number of times, because all solutions oscillate about zero.[2] Also in this case, if we take any two solutions that are not proportional to each other — say $\cos kx$ and $\sin kx$ — then between every pair of zeros of one solution, there is a zero of the other solution. See Figure 9.1. (This isn't true if we select, say, $\cos kx$ and $2 \cos kx$ as our solutions, which is why we restricted the observation to solutions that are not proportional to each other. See Figure 9.2.)

These ideas can be generalized to the case where a is not constant by the following results, which are proved and extended in the appendix.

Theorem 9.2: *Consider the differential equation*

$$\frac{d^2y}{dx^2} + Q(x)y = 0, \tag{9.5}$$

where $Q(x)$ is continuous for $x \geq x_0$ and where $y = y_1(x)$ is a nontrivial solution of (9.5).

(a) If $Q(x) \leq 0$ for $x \geq x_0$, then $y_1(x)$ has at most one zero for $x > x_0$. Thus the maximum number of times that $y_1(x)$ can cross the x-axis to the right of x_0 is once.

---

[1] Recall that $y(x) = 0$ is the TRIVIAL solution of (9.2). A NONTRIVIAL solution is a solution that is not identically zero. Unless indicated otherwise, we will assume the word solution refers to a nontrivial solution.

[2] When a function $f(x)$ is zero at $x = x_0$, mathematicians frequently say $f(x)$ *vanishes* at $x = x_0$. So a function that is zero an infinite number of times is said to vanish an infinite number of times.

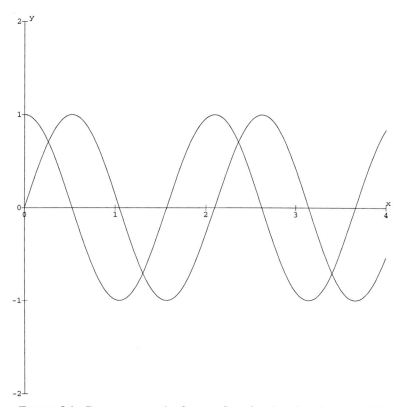

FIGURE 9.1 Betwen every pair of zeros of one function, there is a zero of the other function

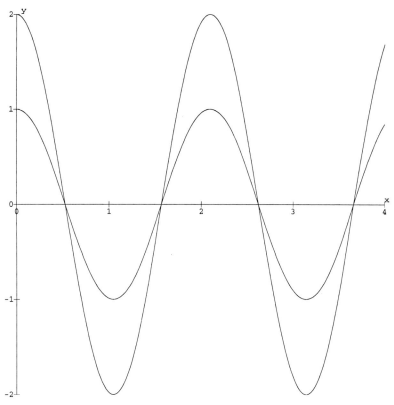

FIGURE 9.2 The functions $\cos kx$ and $2\cos kx$

(b) *If $Q(x) \geq k^2 > 0$ (k a real constant) for $x \geq x_0$, then $y_1(x)$ has an infinite number of zeros for $x \geq x_0$.*

(c) *If $y = y_2(x)$ is another nontrivial solution of (9.5) and there is no constant c for which $y_1(x) = cy_2(x)$, then between two consecutive zeros[3] of $y_1(x)$ there is exactly one zero of $y_2(x)$, and between two consecutive zeros of $y_2(x)$ there is exactly one zero of $y_1(x)$. So if $y_1(x)$ has an infinite number of zeros, so does $y_2(x)$; if $y_1(x)$ has a finite number of zeros, so does $y_2(x)$.*

Comments about Theorem 9.2

- Nontrivial functions that have an infinite number of zeros are said to OSCILLATE.

- Zeros of two functions are called INTERLACED if between any two zeros of one function, there is a zero of the other function, and vice versa.

- We might suspect that part (b) of the theorem could be replaced by "If $Q(x) > 0$ for $x \geq x_0$ then $y_1(x)$ has an infinite number of zeros for $x \geq x_0$" but it cannot. For example the differential equation $y'' + \left(a^2/x^2\right) y = 0$ has oscillating solutions if $a^2 > 1/4$ and nonoscillating solutions if $a^2 \leq 1/4$. (See Exercise 7, page 438.)

EXAMPLE 9.1

What can be said about the solutions of

$$\frac{d^2y}{dx^2} - \frac{2}{x^2}y = 0 \tag{9.6}$$

for $x > 0$?

We use part (a) of Theorem 9.2. Because $Q(x) = -2/x^2 \leq 0$ for $x \geq a$, where a is any positive constant, each solution of (9.6) can have at most one zero to the right of a, no matter how small a. Thus, the maximum number of times that a nontrivial solution of (9.6) can cross the positive x-axis is once. □

EXAMPLE 9.2

What can be said about the solutions of

$$\frac{d^2y}{dx^2} + \left(1 + \frac{1}{4x^2}\right) y = 0 \tag{9.7}$$

for $x > 0$?

We use part (b) of Theorem 9.2. Because $Q(x) = 1 + 1/(4x^2) \geq 1 > 0$, every nontrivial solution of (9.7) oscillates. A typical numerical solution, for $y(1) = 0$, $y'(1) = 1$, is shown in Figure 9.3. □

EXAMPLE 9.3: Airy's Equation

What can be said about the solutions of Airy's equation (9.3), that is,

$$\frac{d^2y}{dx^2} - xy = 0? \tag{9.8}$$

---

[3]It can be shown that a nontrivial solution of (9.5) can have only a finite number of zeros on a closed interval. Thus, the zeros of a solution are all distinct.

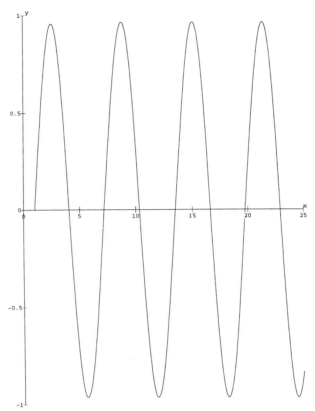

FIGURE 9.3 A numerical solution of $y'' + \left[1 + 1/(4x^2)\right] y = 0$

For $x > 0$ we use part (a) of Theorem 9.2. Because $Q(x) = -x \le 0$, every nontrivial solution of (9.8) has at most one zero for $x > 0$.

For $x < 0$ we make the change of variables $x = -t$, which transforms (9.8) into

$$\frac{d^2 y}{dt^2} + ty = 0.$$

Now consider $x \le -a$ for any $a > 0$, which means that $t \ge a$. Because $Q(t) = t \ge a > 0$, part (b) of Theorem 9.2 shows that every nontrivial solution of (9.8) oscillates for $x \le -a$.

Thus, every nontrivial solution of (9.8) oscillates for $x < 0$ and has at most one zero for $x > 0$. $\qquad \square$

Theorem 9.2 gives us a powerful technique to determine qualitative information about solutions of differential equations of the form $y'' + Q(x)y = 0$. However, we often encounter differential equations of the form

$$\frac{d^2 y}{dx^2} + p(x)\frac{dy}{dx} + q(x)y = 0. \qquad \textbf{(9.9)}$$

Thus, a natural question to ask is how we can relate the solutions of (9.9) to the solutions of

$$\frac{d^2 y}{dx^2} + Q(x)y = 0.$$

In the past we have found the substitution

$$y(x) = u(x)z(x) \tag{9.10}$$

very useful, so let us try it again here, where the object is to select u in a clever way to make the differential equation for z easier to solve. Substituting (9.10) into (9.9) and collecting together the terms in z, z', and z'' leads to

$$y'' + py' + qy = uz'' + \left(2u' + pu\right)z' + \left(u'' + pu' + qu\right)z = 0. \tag{9.11}$$

We can eliminate the z' term by choosing $u(x)$ to satisfy

$$2u' + pu = 0,$$

or

$$\frac{u'}{u} = -\frac{1}{2}p,$$

which means

$$u(x) = e^{-\frac{1}{2}\int p(x)\,dx}. \tag{9.12}$$

If we substitute (9.12) and its derivatives into (9.11), we have

$$u\left[z'' + \left(q - \frac{1}{2}p' - \frac{1}{4}p^2\right)z\right] = 0.$$

Thus, since from (9.12) $u(x) > 0$, we have established the following result.

Theorem 9.3:   *The solutions of the differential equation*

$$\frac{d^2y}{dx^2} + p(x)\frac{dy}{dx} + q(x)y = 0$$

are related to the solutions of

$$\frac{d^2z}{dx^2} + Q(x)z = 0,$$

where

$$Q(x) = q(x) - \frac{1}{2}p'(x) - \frac{1}{4}p^2(x), \tag{9.13}$$

by

$$y(x) = e^{-\frac{1}{2}\int p(x)\,dx}z(x). \tag{9.14}$$

Comments about Theorem 9.3

- Because the coefficient of $z(x)$ in (9.14) is always positive, $y(x)$ and $z(x)$ have the same zeros. Thus, the number of zeros of $y(x)$ and $z(x)$ and their distances apart are the same.
- The results of Theorem 9.2 now may be applied to the differential equation (9.9) by using $Q(x)$ defined by (9.13).

EXAMPLE 9.4

What can be said about the solutions of

$$y'' + \frac{2}{x}y' - \frac{2}{x^2}y = 0 \tag{9.15}$$

for $x > 0$?

Here $p(x) = 2/x$ and $q(x) = -2/x^2$, so by (9.13), we have

$$Q(x) = q(x) - \frac{1}{2}p'(x) - \frac{1}{4}p^2(x) = -\frac{2}{x^2} - \frac{1}{2}\left(-\frac{2}{x^2}\right) - \frac{1}{4}\left(\frac{2}{x}\right)^2,$$

which gives

$$Q(x) = -\frac{2}{x^2}.$$

Because $Q(x) \leq 0$, the nontrivial solutions of (9.15) have at most one zero for $x > 0$. $\square$

EXAMPLE 9.5

We are told that the differential equation

$$y'' + 2xy' + 2y = 0 \tag{9.16}$$

has a solution $y(x) = e^{-x^2}$. Is it possible for any nontrivial solution of (9.16) to oscillate?

We use part (c) of Theorem 9.2. It is not possible for any solution of (9.16) to oscillate, because e^{-x^2} has no zeros. If another solution were to oscillate, then between every zero of that other solution, the solution e^{-x^2} would have to equal zero. $\square$

EXERCISES

1. What can be said about the nontrivial solutions of the differential equation

$$y'' + (1 + e^x)y = 0?$$

2. Show that all nontrivial solutions of

$$y'' + \frac{1}{x}y' + y = 0$$

   oscillate for $x > 0$. (This equation is known as BESSEL'S EQUATION OF ORDER ZERO.)

3. Show that all nontrivial solutions of

$$y'' + \frac{1}{x}y' + \left(1 - \frac{1}{4x^2}\right)y = 0$$

   oscillate for $x > 0$. (This equation is known as BESSEL'S EQUATION OF ORDER $1/2$.) By making the substitution (9.10), find the general solution of this differential equation, and confirm that its solutions oscillate.

4. Show that if μ is a constant, all nontrivial solutions of

$$y'' + \frac{1}{x}y' + \left(1 - \frac{\mu^2}{x^2}\right)y = 0$$

oscillate for $x \geq a > 0$. (This equation is known as BESSEL'S EQUATION OF ORDER μ.)

5. If n is a positive integer, it can be shown that the differential equation

$$y'' - 2xy' + 2ny = 0$$

has a solution that is a polynomial of degree n. Show that no nontrivial solution of this differential equation can oscillate. (This equation is known as HERMITE'S EQUATION OF ORDER n.)

6. If n is a positive integer, it can be shown that the differential equation

$$y'' + \left(\frac{1}{x} - 1\right)y' + \frac{n}{x}y = 0$$

has a solution that is a polynomial of degree n. Show that no nontrivial solution of this differential equation can oscillate. (This equation is known as LAGUERRE'S EQUATION OF ORDER n.)

7. Show that for $x > 0$, the differential equation

$$y'' + \frac{a^2}{x^2}y = 0$$

is satisfied by

(a) $y(x) = C_1 x^{r_1} + C_2 x^{r_2}$, where $r_1 = \left(1 + \sqrt{1 - 4a^2}\right)/2$, $r_2 = \left(1 - \sqrt{1 - 4a^2}\right)/2$ if $a^2 < 1/4$.

(b) $y(x) = C_1\sqrt{x} + C_2\sqrt{x}\ln x$ if $a^2 = 1/4$.

(c) $y(x) = C_1\sqrt{x}\cos(\omega \ln x) + C_2\sqrt{x}\sin(\omega \ln x)$ where $\omega = \left(\sqrt{4a^2 - 1}\right)/2$, if $a^2 > 1/4$.

(d) How many zeros (for $x > 0$) can each of the solutions in parts (a), (b), and (c) have?

8. Show that if a function $f(x)$ oscillates and $f'(x)$ is continuous, then the function $f'(x)$ oscillates.

9. Use the result from Exercise 8 to show that if the solutions of

$$y'' + p(x)y' + y = 0$$

oscillate, then the solutions of

$$z'' + p(x)z' + \left(1 + p'(x)\right)z = 0$$

also oscillate. [Hint: Differentiate the first equation, and set $z = y'$.] By setting $p(x) = 1/x$, show that the fact that solutions of Bessel's equation of order one oscillate is a consequence of the oscillation of solutions of Bessel's equation of order zero.

10. Figure 9.4 is the graph of a solution of one of the following second order differential equations. Choose the differential equation that could give rise to this figure, and give reasons for your choice. (Hint: Remember Theorem 9.2.)

(a) $y'' + \left(1 + e^{-x}\right)y = 0$

(b) $y'' - \left(1 + e^{-x}\right)y = 0$

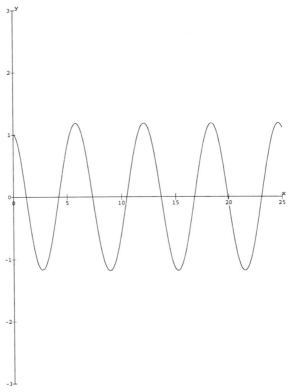

FIGURE 9.4 Mystery solution A

9.2 Reduction of Order

In Chapter 7, when looking at the differential equation

$$a\frac{d^2y}{dx^2} + b\frac{dy}{dx} + cy = 0,$$

where a, b, and c are constants and $b^2 = 4ac$, we found that the trial solution e^{rx} led to the characteristic equation

$$ar^2 + br + c = 0$$

with equal roots, so we found only one solution of the type e^{rx}. The way we found the second solution, xe^{rx}, was by using the fact that e^{rx} was a solution and then looking for solutions of the form $e^{rx}z(x)$.

This technique works in general on

$$a_2(x)\frac{d^2y}{dx^2} + a_1(x)\frac{dy}{dx} + a_0(x)y = 0$$

and requires that we know one (nontrivial) solution — say, $y = y_1(x)$ — and then try to find the general solution by using the well-tried change of dependent variable

$$y(x) = y_1(x)z(x).$$

Before looking at this method in general, let's look at an example to see how the technique works.

EXAMPLE 9.6

Solve

$$x^2 \frac{d^2 y}{dx^2} + 2x \frac{dy}{dx} - 2y = 0 \tag{9.17}$$

by using the fact that $y(x) = x$ is a solution.

We first note that if we substitute $y(x) = x$ into the left-hand side of (9.17), it reduces to zero, so indeed $y(x) = x$ is a solution of (9.17). Thus, $y_1(x) = x$, and we try

$$y(x) = y_1(x)z(x) = xz(x). \tag{9.18}$$

Differentiating (9.18) twice, and substituting the results into (9.17) gives

$$x^2 \left(xz'' + 2z' \right) + 2x \left(xz' + z \right) - 2xz = 0,$$

or

$$xz'' + 4z' = 0.$$

Notice that the term explicitly involving the dependent variable z is missing. Thus, according to our results from Section 5.5, we can reduce this second order equation to a first order equation if we define $v = z'$. The second order equation then becomes

$$xv' + 4v = 0, \tag{9.19}$$

which is a separable first order differential equation for $v(x)$. Notice we have reduced the order of the differential equation we are trying to solve from two to one. Solving for $v(x)$ gives

$$z' = v(x) = C_1 x^{-4}$$

(where C_1 is an arbitrary constant), which when integrated once more, gives

$$z(x) = -\frac{1}{3} C_1 x^{-3} + C_2 \tag{9.20}$$

(where C_2 is an arbitrary constant). If we substitute this into (9.18), we find

$$y(x) = x \left(-\frac{1}{3} C_1 x^{-3} + C_2 \right) = -\frac{1}{3} C_1 \frac{1}{x^2} + C_2 x$$

as the solution of (9.17). This is in agreement with our findings from Example 9.4 where we discovered that nontrivial solutions of (9.17) could have no more than one zero for $x > 0$.

$\square$

Now let's go back to the general situation

$$a_2(x)y'' + a_1(x)y' + a_0(x)y = 0, \tag{9.21}$$

where we know one solution — say, $y = y_1(x)$ — so that

$$a_2 y_1'' + a_1 y_1' + a_0 y_1 = 0. \tag{9.22}$$

We try to find the general solution by using the change of dependent variable

$$y(x) = y_1(x) z(x). \tag{9.23}$$

Substituting (9.23) into (9.21), we find

$$a_2 \left(y_1 z'' + 2 y_1' z' + y_1'' z \right) + a_1 \left(y_1 z' + y_1' z \right) + a_0 y_1 z = 0.$$

In this last equation we collect together like terms in z, z', and z'' to find

$$a_2 y_1 z'' + \left(2 a_2 y_1' + a_1 y_1 \right) z' + \left(a_2 y_1'' + a_1 y_1' + a_0 y_1 \right) z = 0.$$

Because y_1 is a known solution of (9.21) — that is, (9.22) is true — the last equation reduces to

$$a_2 y_1 z'' + \left(2 a_2 y_1' + a_1 y_1 \right) z' = 0.$$

If we define $v = z'$, then the last equation is

$$a_2 y_1 v' + \left(2 a_2 y_1' + a_1 y_1 \right) v = 0,$$

which is the counterpart of (9.19). Remember that in this equation, $a_2(x)$, $a_1(x)$, and $y_1(x)$ are all known functions. This is a first order separable equation with solution

$$z'(x) = v(x) = C_1 \exp \left(- \int \frac{2 a_2 y_1' + a_1 y_1}{a_2 y_1} \, dx \right) = C_1 \exp \left[- \int \left(2 \frac{y_1'}{y_1} + \frac{a_1}{a_2} \right) dx \right].$$

Because

$$\exp \left(- \int 2 \frac{y_1'}{y_1} \, dx \right) = \exp \left(-2 \int \frac{dy_1}{y_1} \right) = \exp \left(-2 \ln |y_1| \right) = \frac{1}{y_1^2},$$

we find

$$z'(x) = v(x) = C_1 \frac{1}{y_1^2} \exp \left(- \int \frac{a_1}{a_2} \, dx \right).$$

Integrating once more gives

$$z(x) = C_1 \int \left[\frac{1}{y_1^2} \exp \left(- \int \frac{a_1}{a_2} \, dx \right) \right] dx + C_2.$$

This is the counterpart of (9.20). Finally, in view of (9.23), we have the solution for the general case

$$y(x) = C_1 y_1 \int \left[\frac{1}{y_1^2} \exp \left(- \int \frac{a_1}{a_2} \, dx \right) \right] dx + C_2 y_1. \tag{9.24}$$

How to Reduce the Order

Purpose: To solve

$$a_2(x)y'' + a_1(x)y' + a_0(x)y = 0 \tag{9.25}$$

if we already have one (nontrivial) solution $y = y_1(x)$.

Technique:

1. Check that $y = y_1(x)$ is a solution of (9.25).

2. Introduce the new dependent variable $z(x)$ by

$$y(x) = y_1(x)z(x). \tag{9.26}$$

3. Substitute (9.26) into (9.25) and collect together like terms in z, z', and z''. If everything has been done correctly, the coefficient of z should be zero.

4. Put $v = z'$ in the equation obtained in the last step. This should yield a first order separable equation, which we solve for $v = v(x)$.

5. Solve for $z = z(x)$ by integrating $z' = v(x)$, where $v(x)$ has been obtained from step 4.

6. Substitute this $z(x)$ into (9.26) to get the final solution.

Comments about reduction of order

- In steps 4 or 5 we may be unable to integrate and obtain familiar functions.
- We should always get two constants of integration in the final solution. It is a common mistake to lose one or both.
- The general solution should always be a linear combination of the original known solution $y_1(x)$ and another solution.

EXERCISES

1. Show that $y = e^{2x}$ is a solution of

$$y'' - 4y' + 4y = 0,$$

and find the general solution. Confirm that your answer is consistent with Theorems 9.2 and 9.3.

2. Show that $y = x^2$ is a solution of

$$x^2 y'' + xy' - 4y = 0,$$

and find the general solution. Confirm that your answer is consistent with Theorems 9.2 and 9.3.

3. Show that $y = x$ is a solution of

$$x^2 y'' - xy' + y = 0,$$

and find the general solution for $x > 0$. Do Theorems 9.2 and 9.3 apply in this case? Explain fully.

4. Show that $y = e^x$ is a solution of

$$xy'' - 2(x + 1)y' + (x + 2)y = 0,$$

and find the general solution. Confirm that your answer is consistent with Theorems 9.2 and 9.3.

5. Show that $y = x \sin x$ is a solution of

$$x^2 y'' - 2xy' + (x^2 + 2)y = 0,$$

and find the general solution. Confirm that your answer is consistent with Theorems 9.2 and 9.3.

6. Show that $y = \sin x / \sqrt{x}$ $(x > 0)$ is a solution of

$$4x^2 y'' + 4xy' + (4x^2 - 1)y = 0,$$

and find the general solution for $x > 0$. Confirm that your answer is consistent with Theorems 9.2 and 9.3.

7. Show that $y = x$ is a solution of

$$x^2 y'' - x(x + 2)y' + (x + 2)y = 0,$$

and find the general solution. Confirm that your answer is consistent with Theorems 9.2 and 9.3.

8. Show that $y = x^2$ is a solution of

$$x^2 y'' - 3xy' + 4y = 0,$$

and find the general solution. Do Theorems 9.2 and 9.3 apply in this case? Explain fully.

9.3 Linear Independence and Dependence

A common feature of the general solutions for the specific cases of

$$a_2(x)\frac{d^2 y}{dx^2} + a_1(x)\frac{dy}{dx} + a_0(x)y = 0, \tag{9.27}$$

that we have obtained in Sections 5.5, 7.2, 8.1 through 8.4, and 9.2, is that they consist of the sum of two terms, each being an arbitrary constant multiplying a particular solution. We might wonder if all solutions of (9.27) have this same structure. That is, given two solutions $y_1(x)$ and $y_2(x)$ of (9.27), is it possible to give the general solution by simply forming a linear combination of the two solutions, namely, $C_1 y_1(x) + C_2 y_2(x)$.

For example, consider the differential equation

$$y'' - 4y = 0,$$

which has the general solution

$$y(x) = C_1 e^{2x} + C_2 e^{-2x}.$$

Thus, $e^{2x}, 3e^{2x}, 5e^{2x}, 2e^{-2x}$, and $7e^{-2x}$ are all particular solutions (along with many others). If we select two of these particular solutions — say, $y_1(x) = e^{2x}$ and $y_2(x) = 3e^{2x}$ — is their linear combination $C_1 e^{2x} + C_2 3 e^{2x}$ the general solution of our differential equation. Using a little algebra gives us $C_1 e^{2x} + C_2 3 e^{2x} = (C_1 + C_2 3) e^{2x}$, so this is not equivalent to our previously obtained general solution. Thus, we must take more care when choosing two particular solutions.

In this section we will introduce the topic of linear independence of functions, which we will use to be assured that the linear combination we form with two particular solutions will indeed be the general solution of our second order linear differential equation.

The words *independence* and *dependence* in mathematics have meanings very similar to their common usage. If one function may be written as a constant times a second function we could write

$$f(x) = cg(x),$$

where c is a nonzero constant. Here we see that to find the value of f at some point x, we compute $g(x)$ and then multiply the result by the constant c. For this reason we say that $f(x)$ depends on $g(x)$. Equivalently, if we divide by c we find

$$g(x) = \frac{1}{c} f(x),$$

and say that $g(x)$ depends on $f(x)$. An alternative way of expressing this relationship is as follows: Two functions f and g are linearly dependent on an interval $a < x < b$ in the common domain of f and g if there exist two constants, b_1 and b_2 (not both zero), such that

$$b_1 f(x) + b_2 g(x) = 0, \tag{9.28}$$

for all x in this interval. However, in differential equations what we really need is a method for making sure we do *not* have dependent functions. Thus, we look at (9.28) and give the following definition.

Definition 9.1: **Two functions $f(x)$ and $g(x)$ are LINEARLY INDEPENDENT if the only way for**

$$b_1 f(x) + b_2 g(x) = 0$$

to be true for $a < x < b$ in some common domain of f and g is to have $b_1 = b_2 = 0$. If two functions are not linearly independent, they are LINEARLY DEPENDENT.

Comments about linear independence and linear dependence

- If either f or g is identically zero, f and g are dependent. All we need to do is choose the coefficient of the identically zero function to be nonzero and the other coefficient to be zero.

- If one function is a constant multiple of another, the two functions are linearly dependent. We show this by noting that if $f = Cg$, then $f - Cg = 0$, and we have the form of (9.28) with $b_1 = 1$ and $b_2 = -C$. Thus, the functions e^{2x} and $3e^{2x}$ from our earlier discussion are linearly dependent.

- The two functions $\sin \beta x$ and $\cos \beta x$ are linearly independent for all values of x if $\beta \neq 0$. To show this we start with the equation

$$b_1 \sin \beta x + b_2 \cos \beta x = 0 \tag{9.29}$$

and see if we can find nonzero values of b_1 and b_2 such that the equation is true for all x. Because (9.29) must be true for all values of x, it must also be true for the two specific values of 0 and $\pi/(2\beta)$. Choosing $x = 0$ gives $b_1 \sin 0 + b_2 \cos 0 = 0$, so $b_2 = 0$. Choosing $x = \pi/(2\beta)$ gives $b_1 = 0$, and, because both b_1 and b_2 are zero, $\sin \beta x$ and $\cos \beta x$ are linearly independent functions for any nonzero β and all values of x.

- If two functions are linearly independent for all intervals in the common domain of the functions, we often omit mention of the domain. Thus, we simply say that $\sin \beta x$ and $\cos \beta x$ are linearly independent for $\beta \neq 0$.

- If $r_1 \neq r_2$, then $e^{r_1 x}$ and $e^{r_2 x}$ are linearly independent for all values of x. Here we need to determine if there exist constants, not both zero, such that

$$b_1 e^{r_1 x} + b_2 e^{r_2 x} = 0 \tag{9.30}$$

for all x in some interval. In other words we want (9.30) to be an identity. If (9.30) is true for x in some interval, we may differentiate this equation and obtain

$$r_1 b_1 e^{r_1 x} + r_2 b_2 e^{r_2 x} = 0. \tag{9.31}$$

Now we look upon (9.30) and (9.31) as two equations in the two unknowns, b_1 and b_2, for which we seek nonzero values that make these equations true. Solving these two equations, we find

$$b_1 (r_2 - r_1) e^{r_1 x} = 0$$

and

$$b_2 (r_2 - r_1) e^{r_2 x} = 0.$$

Now because $(r_2 - r_1) e^{r_1 x} e^{r_2 x} \neq 0$, then both b_1 and b_2 are zero, and the functions $e^{r_1 x}$ and $e^{r_2 x}$ are linearly independent for all values of x.

We can use a similar argument to develop a test for linear independence of any two functions — say, f and g — that are differentiable on the interval $a < x < b$. If we form the equation

$$b_1 f(x) + b_2 g(x) = 0$$

and also write down its derivative,

$$b_1 f'(x) + b_2 g'(x) = 0,$$

we have two equations in the two unknowns b_1 and b_2. Solving gives

$$b_1 \left[f(x) g'(x) - f'(x) g(x) \right] = 0$$

and

$$b_2 \left[f(x) g'(x) - f'(x) g(x) \right] = 0.$$

If the quantity in brackets in these latter two expressions is not zero, then b_1 and b_2 must equal zero, so f and g are linearly independent functions. Because this quantity occurs so often in considerations of linear independence and dependence it is given the name WRONSKIAN.

Definition 9.2: **The Wronskian of two functions *f* and *g*, labeled *W[f, g]*, is defined as**

$$W[f, g] = f(x)g'(x) - f'(x)g(x).$$

We now summarize this discussion with the following theorem.

Theorem 9.4: *If the Wronskian of two functions is not identically zero on some nonzero interval, then the two functions are linearly independent on that interval.*

Comments about the Wronskian

- Notice that the Wronskian is given by the value of the determinant

$$\begin{vmatrix} f(x) & g(x) \\ f'(x) & g'(x) \end{vmatrix}.$$

- Note that this theorem does *not* say that if the Wronskian of two functions is zero, then the two functions are linearly dependent. The theorem tells us nothing about linear independence or dependence if the Wronskian is identically zero. For example, the Wronskian of the two functions x and $|x|$ is identically zero for all values of x. However, these functions are linearly independent on any interval that includes the origin as an interior point and are linearly dependent on any interval that does not.

We can use this theorem to prove a result that puts the method of solving differential equations that we developed in Chapter 7 on solid footing. It also foreshadows what we will need for linear differential equations of order greater than two.

Theorem 9.5: *Consider the general second order linear differential equation*

$$a_2(x)\frac{d^2y}{dx^2} + a_1(x)\frac{dy}{dx} + a_0(x)y = 0, \tag{9.32}$$

where $a_2(x)$, $a_1(x)$, and $a_0(x)$ are continuous, with $a_2(x) \neq 0$ for $a < x < b$. If $f(x)$ and $g(x)$ are two solutions of (9.32) and the functions f and g are linearly independent for $a < x < b$, then all solutions of (9.32) may be written as

$$y(x) = C_1 f(x) + C_2 g(x). \tag{9.33}$$

Proof Choose general initial values for y and y' at $x = x_0$ (where $a < x_0 < b$) as

$$y(x_0) = y_0, \quad y'(x_0) = y_0^*. \tag{9.34}$$

The coefficients in (9.32) are such that the hypothesis of Theorem 9.1 is satisfied. Thus, we can use the conclusion of that theorem which states that there is a unique solution of (9.32) which satisfies the initial conditions in (9.34). If we impose the initial conditions of (9.34) on our solution in (9.33), we obtain the two simultaneous algebraic equations

$$C_1 f(x_0) + C_2 g(x_0) = y_0,$$
$$C_1 f'(x_0) + C_2 g'(x_0) = y_0^*.$$

Eliminating first C_2 and then C_1 from these equations gives

$$C_1 \left[f(x_0)g'(x_0) - f'(x_0)g(x_0) \right] = y_0 g'(x_0) - y_0^* g(x_0),$$
$$C_2 \left[f(x_0)g'(x_0) - f'(x_0)g(x_0) \right] = y_0^* f(x_0) - y_0 f'(x_0).$$

From here we see that if the term $\left[f(x_0)g'(x_0) - f'(x_0)g(x_0)\right] \neq 0$, then we obtain a unique solution for C_1 and C_2. However, this expression is simply the Wronskian of f and g, which, because $f(x)$ and $g(x)$ are both solutions of (9.32), is never zero if f and g are linearly independent (see Exercise 10 on page 449). Thus, C_1 and C_2 are uniquely determined, and we have obtained the unique solution of our initial value problem. Because x_0, y_0, and y_0^* were chosen arbitrarily, we see that all solutions of (9.32) may be written as a linear combination of two linearly independent solutions.

Comments about Theorem 9.5

- Our task will be to find linearly independent solutions. This form of our solution, (9.33), where C_1 and C_2 are arbitrary constants is called the GENERAL SOLUTION of (9.32) because all solutions may be written in this form.

- The significance of this theorem is that no matter how we obtain $f(x)$ and $g(x)$, if they are linearly independent and satisfy (9.32), then the general solution is (9.33). Thus, the recipe used earlier in Chapter 7 is justified.

EXAMPLE 9.7

We found earlier that $e^{\alpha x}$ and $xe^{\alpha x}$ were both solutions of the second order differential equation with constant coefficients

$$a_2 \frac{d^2 y}{dx^2} + a_1 \frac{dy}{dx} + a_0 y = 0 \tag{9.35}$$

for the case $a_1^2 - 4a_2 a_0 = 0$. Taking the Wronskian of these two solutions gives

$$W[e^{\alpha x}, xe^{\alpha x}] = \begin{vmatrix} e^{\alpha x} & xe^{\alpha x} \\ \alpha e^{\alpha x} & (1 + \alpha x)\,e^{\alpha x} \end{vmatrix} = e^{2\alpha x}.$$

Because this Wronskian is not zero, we may conclude that these two functions are linearly independent and that the general solution of (9.35) may be written as

$$y(x) = C_1 e^{\alpha x} + C_2 xe^{\alpha x}.$$

$\square$

EXAMPLE 9.8

We now show that in the reduction of order technique of the last section, the two solutions $y_1(x)$ and

$$y_2(x) = y_1(x)z(x),$$

where

$$z(x) = \int \left[\frac{1}{y_1^2} \exp\left(-\int \frac{a_1}{a_2}\,dx\right) \right] dx$$

[see (9.24)] are linearly independent. The Wronskian $W[y_1(x), y_2(x)]$ in this case becomes

$$W[y_1(x), y_2(x)] = W[y_1(x), y_1(x)z(x)] = \begin{vmatrix} y_1 & y_1 z \\ y_1' & y_1 z' + y_1' z \end{vmatrix} = y_1^2 z'.$$

But, from

$$z' = \frac{1}{y_1^2} \exp\left(-\int \frac{a_1}{a_2} dx\right),$$

we have

$$W[y_1(x), y_2(x)] = \exp\left(-\int \frac{a_1}{a_2} dx\right).$$

Because the exponential function can never be zero, the Wronskian is never zero, proving that $y_1(x)$ and $y_2(x)$ are linearly independent. □

EXERCISES

1. Show that the following sets of functions are linearly independent for all values of x by computing their Wronskian.
   (a) x, x^2
   (b) $x, 4 + x$
   (c) $1, e^x$
   (d) $x, 1/x, x > 0$
   (e) $\sinh 2x, \cosh 2x$

2. Determine all values of the constant k that make the following sets of functions linearly independent.
   (a) $x, k + x$
   (b) $x, 1 + kx$
   (c) $1 + 2x, k + x$
   (d) $1 + 2x, 1 + kx$
   (e) $3, 2 + kx$
   (f) $3x, kx$
   (g) $x^2, (x + k)^2$
   (h) $9 \sin 2x, k \sin x \cos x$
   (i) $\sin^2 kx + \cos^2 kx, 6$
   (j) $3e^{2x}, 6e^{(k-2)x}$

3. Sketch graphs of the functions in each part of Exercise 2, and discuss the graphical interpretation of the various situations that lead to linear independence.

4. Show that the functions from each of the three cases in (7.42), (7.43), and (7.47) are linearly independent.

5. Show that the following sets of functions are linearly independent.
   (a) e^{2x}, e^{-3x}
   (b) e^{ix}, e^{-ix}
   (c) e^{-x+5ix}, e^{-x-5ix}
   (d) e^{-x}, xe^{-x}
   (e) e^{rx}, xe^{rx}
   (f) $e^{ix}, \sin x$
   (g) $\sin \beta x, \cos \beta x, \beta \neq 0$
   (h) $\sin x, \sin^2 x$

(i) $e^{-x}\cos 5x$, $e^{-x}\sin 5x$

6. Let $f(x)$ be an odd function and $g(x)$ be an even function.

   (a) Prove that f and g are linearly independent on any interval that includes the origin as an interior point.

   (b) Draw graphs of typical functions for f and g, and explain part (a) from the graph.

   (c) Draw the graph of a situation in which f and g are linearly dependent for $x > 0$, and linearly independent for $x < 0$.

7. Show that $e^{\alpha x}\cos \beta x$ and $e^{\alpha x}\sin \beta x$, $\beta \neq 0$, are linearly independent functions.

8. Show that $x^a \cos(b \ln x)$ and $x^a \sin(b \ln x)$, $b \neq 0$, are linearly independent functions.

9. Show that x^a and $x^a \ln x$ are linearly independent functions.

10. The purpose of this exercise is to show that two solutions of $a_2(x)y'' + a_1(x)y' + a_0(x)y = 0$ are linearly independent if and only if their Wronskian is nonzero.

   (a) Show that if y_1 and y_2 both satisfy the differential equation

$$a_2 y'' + a_1 y' + a_0 y = 0,$$

   then

$$a_2\left(y_1 y_2'' - y_2 y_1''\right) + a_1\left(y_1 y_2' - y_2 y_1'\right) = 0.$$

   (b) Because the Wronskian $W[y_1, y_2] = y_1 y_2' - y_1' y_2$, show that W satisfies the differential equation

$$a_2 W' + a_1 W = 0,$$

   with solution $W = C \exp(-\int a_1(x)/a_2(x)\,dx)$.

   (c) Show that the Wronskian of the two solutions of $a_2 y'' + a_1 y' + a_0 y = 0$ is either always zero or never zero.

   (d) Show that any two solutions of $a_2 y'' + a_1 y' + a_0 y = 0$ are linearly independent if and only if their Wronskian is nonzero.

11. For any two linearly independent functions, $y_1(x)$ and $y_2(x)$, construct the 3×3 determinant

$$A = \begin{vmatrix} y_1 & y_2 & y \\ y_1' & y_2' & y' \\ y_1'' & y_2'' & y'' \end{vmatrix}.$$

   (a) Prove that if $A = 0$, then $y(x)$ satisfies a second order linear differential equation of the type $a_2(x)y'' + a_1(x)y' + a_0(x)y = 0$. What are $a_2(x)$, $a_1(x)$, and $a_0(x)$ in terms of $y_1(x)$, $y_2(x)$, and their derivatives?

   (b) Prove that $y_1(x)$ and $y_2(x)$ both satisfy the differential equation obtained in part (a).

   (c) Explain how the results obtained in parts (a) and (b) can be used to construct a second order linear differential equation $a_2(x)y'' + a_1(x)y' + a_0(x)y = 0$ from any two of its linearly independent solutions $y_1(x)$ and $y_2(x)$.

   (d) What happens if $y_1(x)$ and $y_2(x)$ are chosen as linearly dependent functions?

12. Use the technique outlined in Exercise 11 to construct a second order linear differential equation $a_2(x)y'' + a_1(x)y' + a_0(x)y = 0$ that has the general solution $y(x) = C_1 y_1(x) + C_2 y_2(x)$ where $y_1(x)$ and $y_2(x)$ are two linearly independent functions, given as follows.

   (a) x, x^2

(b) x, x^{-1}

(c) x, e^x

(d) e^{2x}, e^{-3x}

(e) e^{-x}, xe^{-x}

(f) e^{rx}, xe^{rx}

(g) $\sin \beta x, \cos \beta x$

(h) $e^{-x} \cos 5x, e^{-x} \sin 5x$

(i) $\sin x, \sin^2 x$

(j) $e^{1/x}, e^{-1/x}$

(k) $e^x, \sin x$

9.4 Cauchy-Euler Equation

All of the second order differential equations that we covered in Chapter 7 were linear ones with constant coefficients. These occur very often in applications, but they are not the only type of second order differential equations for which we can develop a systematic method of obtaining explicit solutions. In this section we find explicit solutions of the CAUCHY-EULER DIFFERENTIAL EQUATION, namely,

$$b_2 x^2 \frac{d^2 y}{dx^2} + b_1 x \frac{dy}{dx} + b_0 y = 0$$

where b_2, b_1, and b_0 are constants.

EXAMPLE 9.9: Electric Potential of a Charged Spherical Shell

A differential equation that occurs in a mathematical model describing the electric potential due to a charged spherical shell is given by

$$x^2 \frac{d^2 y}{dx^2} + 2x \frac{dy}{dx} - n(n+1)y = 0, \tag{9.36}$$

where x is the distance from the center of the spherical shell, y is the potential, and n is a positive constant. This differential equation does not have constant coefficients, and is one we have not yet considered. However, Theorems 9.2 and 9.3 can be used to show that nontrivial solutions of (9.36) cannot vanish more than once for $x > 0$ and cannot vanish more than once for $x < 0$ (see Exercise 6, page 455).

In Chapter 7 we discovered a simple method for solving second order linear differential equations with constant coefficients. Thus, we look for ways to replace xy' and $x^2 y''$ by derivatives that have constant coefficients, by using a change of variables from the independent variable x to a new independent variable t, where

$$\frac{dy}{dt} = x \frac{dy}{dx}.$$

The fact that $dy/dt = (dy/dx)(dx/dt)$ suggests we relate x and t by the differential equation

$$\frac{dx}{dt} = x,$$

which has the solution

$$x = Ae^t.$$

Thus, we could use the change of variables

$$x = e^t, \qquad \text{if } x > 0, \qquad \text{and} \qquad x = -e^t, \qquad \text{if } x < 0.$$

In our case $x > 0$, because distance is positive, so we use the change of variables

$$x = e^t, \tag{9.37}$$

which is equivalent to

$$t = \ln x, \tag{9.38}$$

so differentiation with respect to x gives

$$\frac{dt}{dx} = \frac{1}{x}. \tag{9.39}$$

By using the chain rule, along with (9.37), we find

$$\frac{dy}{dt} = \frac{dy}{dx}\frac{dx}{dt} = \frac{dy}{dx}e^t = x\frac{dy}{dx},$$

or

$$x\frac{dy}{dx} = \frac{dy}{dt}, \tag{9.40}$$

as desired. So, in (9.36) we can replace the $2x\,dy/dx$ term by $2\,dy/dt$. But what about the $x^2 d^2y/dx^2$ term?

If we differentiate (9.40) with respect to x using the product rule on the left-hand side, we find

$$x\frac{d^2y}{dx^2} + \frac{dy}{dx} = \frac{d}{dx}\left(\frac{dy}{dt}\right) = \frac{d^2y}{dt^2}\frac{dt}{dx}.$$

If we multiply this equation by x and use (9.39), we find

$$x^2\frac{d^2y}{dx^2} + x\frac{dy}{dx} = \frac{d^2y}{dt^2},$$

which, from (9.40), can be written as

$$x^2\frac{d^2y}{dx^2} = \frac{d^2y}{dt^2} - \frac{dy}{dt}. \tag{9.41}$$

If we substitute (9.40) and (9.41) into (9.36), our original differential equation becomes

$$\frac{d^2y}{dt^2} - \frac{dy}{dt} + 2\frac{dy}{dt} - n(n+1)y = 0,$$

or

$$\frac{d^2 y}{dt^2} + \frac{dy}{dt} - n(n+1)y = 0. \tag{9.42}$$

Because (9.42) is a linear differential equation with constant coefficients, we know we have solutions in the form $y = e^{rt}$, which gives the characteristic equation

$$r^2 + r - n(n+1) = (r-n)(r+n+1) = 0.$$

This has two solutions, $r = n$ and $r = -n - 1$, so our solution of (9.42) is

$$y(t) = C_1 e^{nt} + C_2 e^{-(n+1)t}.$$

We can use either (9.37) or (9.38) to express this solution in terms of our original variable x. In this way we get the general solution of our original differential equation as

$$y(x) = C_1 x^n + C_2 x^{-(n+1)}.$$

This solution is in agreement with Theorems 9.2 and 9.3, which state that nontrivial solutions of (9.36) cannot vanish more than once for $x > 0$ and cannot vanish more than once for $x < 0$ (see Exercise 6 on page 455).

If we are considering the potential inside the spherical shell, from symmetry considerations we expect the potential to be zero at the exact center $x = 0$, which we have if we choose $C_2 = 0$. If we are dealing with the potential outside the spherical shell, we expect the potential to go to 0 as $x \to \infty$, so we would choose $C_1 = 0$. Thus, for either situation we may obtain bounded solutions of our differential equation. $\square$

How to Solve a Cauchy-Euler Differential Equation

Purpose: To solve differential equations of the form

$$b_2 x^2 \frac{d^2 y}{dx^2} + b_1 x \frac{dy}{dx} + b_0 y = 0, \tag{9.43}$$

where b_2, b_1, and b_0 are constants.

Technique:

1. If $x > 0$, transform the independent variable from x to t, where

$$x = e^t, \tag{9.44}$$

so that

$$t = \ln x, \tag{9.45}$$

$$x \frac{dy}{dx} = \frac{dy}{dt}, \tag{9.46}$$

and

$$x^2\frac{d^2y}{dx^2} = \frac{d^2y}{dt^2} - \frac{dy}{dt}. \tag{9.47}$$

[If $x < 0$ use the transformation $x = -e^t$, so (9.45) becomes $t = \ln(-x)$, and (9.46) and (9.47) remain the same.]

2. Substitute (9.46) and (9.47) into (9.43) and combine like terms, obtaining

$$b_2\frac{d^2y}{dt^2} + (b_1 - b_2)\frac{dy}{dt} + b_0y = 0, \tag{9.48}$$

which is a second order linear equation with constant coefficients.

3. Solve (9.48) for $y = y(t)$.

4. Use (9.44) or (9.45) and the solution $y = y(t)$ obtained in step 3 to find the final solution $y = y(x)$.

Comments about solutions of the Cauchy-Euler differential equation

- The solution of (9.43) consists of a linear combination of two linearly independent solutions (see Exercise 3, page 454).

- The only solutions of (9.43) are linear combinations of functions of the form x^n, $x^\alpha \cos(\beta \ln x)$, $x^\alpha \sin(\beta \ln x)$, and $x^m \ln x$. (Why?)

EXAMPLE 9.10

Solve

$$x^2\frac{d^2y}{dx^2} + 4x\frac{dy}{dx} - 4y = 0. \tag{9.49}$$

We transform from x to t where $x = e^t$. Substituting (9.46) and (9.47) into (9.49) yields

$$\frac{d^2y}{dt^2} + 3\frac{dy}{dt} - 4y = 0.$$

This is a linear differential equation with constant coefficients and has the general solution

$$y(t) = C_1e^{-4t} + C_2e^t.$$

Because $x = e^t$, this last equation gives the general solution of (9.49) as

$$y(x) = C_1\frac{1}{x^4} + C_2x.$$

$\square$

EXAMPLE 9.11

Solve the initial value problem

$$x^2\frac{d^2y}{dx^2} + 3x\frac{dy}{dx} + 5y = 0$$

subject to

$$y(1) = 0, \qquad y'(1) = 6.$$

We change the independent variable from x to t by $x = e^t$. Substituting (9.46) and (9.47) into our differential equation above yields

$$\frac{d^2 y}{dt^2} + 2\frac{dy}{dt} + 5y = 0.$$

This linear differential equation with constant coefficients has the characteristic equation

$$r^2 + 2r + 5 = (r+1)^2 + 4 = 0,$$

so its general solution is

$$y(t) = C_1 e^{-t} \cos 2t + C_2 e^{-t} \sin 2t.$$

In terms of the original variables, this becomes

$$y(x) = C_1 x^{-1} \cos(2 \ln x) + C_2 x^{-1} \sin(2 \ln x).$$

Applying the initial conditions gives $C_1 = 0$, $C_2 = 3$, so our solution becomes

$$y(x) = 3x^{-1} \sin(2 \ln x). \qquad\qquad \square$$

EXERCISES

1. Find the general solution of the following differential equations, assuming $x > 0$. (Note: $y' = dy/dx$, $y'' = d^2 y/dx^2$.) Wherever possible, confirm the solutions are consistent with Theorems 9.2 and 9.3.

(a) $x^2 y'' - 2y = 0$

(b) $x^2 y'' + 2xy' - 6y = 0$

(c) $x^2 y'' + xy' = 0$

(d) $x^2 y'' + 9xy' + 2y = 0$

(e) $x^2 y'' + xy' + 9y = 0$

(f) $x^2 y'' - 5xy' + 25y = 0$

(g) $x^2 y'' + 5xy' + 4y = 0$

(h) $2x^2 y'' - xy' + y = 0$

(i) $x^2 y'' - 5xy' + 5y = 0$

(j) $2x^2 y'' + xy' - y = 0$

(k) $x^2 y'' - xy' + y = 0$

(l) $x^2 y'' - 3xy' + 4y = 0$

2. Show that $x^{-1} \cos(2 \ln x)$ and $x^{-1} \sin(2 \ln x)$ are linearly independent for $x > 0$.

3. Show that if $y_1(t)$ and $y_2(t)$ are linearly independent functions of t and satisfy a second order linear differential equation with constant coefficients, then $Y_1(x) = y_1(\ln x)$ and $Y_2(x) = y_2(\ln x)$ are linearly independent functions of x for $x \neq 0$. (Hint: Obtain the Wronskian of Y_1 and Y_2 in terms of the Wronskian of y_1 and y_2.)

4. Make the change of variable $x = \alpha t$, where α is a constant, in (9.43), and show that (9.43) transforms into an equation independent of α. Because we may think of $x = \alpha t$ as a change in scale (or dimension), (9.43) is sometimes called the EQUIDIMENSIONAL DIFFERENTIAL EQUATION.

5. Show that, for $x > 0$, the differential equation

$$y'' + \frac{a^2}{x^2} y = 0$$

has the general solutions that follow.

(a) $y(x) = C_1 x^{r_1} + C_2 x^{r_2}$, where $r_1 = \left(1 + \sqrt{1 - 4a^2}\right)/2$, $r_2 = \left(1 - \sqrt{1 - 4a^2}\right)/2$ if $a^2 < 1/4$

(b) $y(x) = C_1 \sqrt{x} + C_2 \sqrt{x} \ln x$ if $a^2 = 1/4$

(c) $y(x) = C_1 \sqrt{x} \cos(\omega \ln x) + C_2 \sqrt{x} \sin(\omega \ln x)$, where $\omega = \left(\sqrt{4a^2 - 1}\right)/2$ if $a^2 > 1/4$

6. Show that the nontrivial solutions of

$$x^2 \frac{d^2 y}{dx^2} + 2x \frac{dy}{dx} - n(n+1)y = 0,$$

where n is a positive constant, cannot vanish more than once for $x > 0$ and cannot vanish more than once for $x < 0$. (Hint: See Theorems 9.2 and 9.3.)

7. Show that trying a solution of the form x^r for

$$b_2 x^2 \frac{d^2 y}{dx^2} + b_1 x \frac{dy}{dx} + b_0 y = 0, \tag{9.50}$$

where b_2, b_1, and b_0 are constants, leads to the quadratic equation

$$b_2 r^2 + (b_1 - b_2)r + b_0 = 0.$$

Show that the form of the solution depends on the sign of $(b_1 - b_2)^2 - 4b_2 b_0$, and find the general solution of (9.50) for the three possible cases.

9.5 Taylor Series

Thus far we have mapped out a strategy for solving two types of second order differential equations — namely, linear equations with constant coefficients and Cauchy-Euler equations. We want to develop another technique that will apply to first or second order differential equations in general. Recall that on several previous occasions we had no choice other than leaving our solution in terms of indefinite or definite integrals. At times we then used a Taylor series expansion of the known function that occurred in the integrand. This finally resulted in our solution taking the form of an infinite series. We now see if we can obtain such solutions directly.

EXAMPLE 9.12: The Linearized Pendulum Again

To consider a specific example we return to the solution of the linearized pendulum equation

$$y'' + y = 0, \tag{9.51}$$

subject to

$$y(0) = 0.2, \qquad y'(0) = 0.15. \tag{9.52}$$

(Here y is the angle and x is the time. We have chosen the gravitation constant divided by the pendulum length as 1 to keep the algebra simple.) Note that these initial conditions correspond to the situation where the pendulum is released at an angle of 0.2 radian with an initial angular velocity of 0.15 radian per second, which means that initially the angle is going to increase. Although we already know the solution for this problem is

$$y(x) = 0.2 \cos x + 0.15 \sin x, \tag{9.53}$$

let's look at the same problem again through different eyes and act as though we were unaware of this solution.

We note that the initial conditions in (9.52) give us the first two terms in a Taylor series expansion of our solution about $x = 0$,

$$y(x) = \sum_{n=0}^{\infty} \frac{1}{n!} y^{(n)}(0) x^n = y(0) + y'(0)x + \frac{1}{2!} y''(0)x^2 + \frac{1}{3!} y'''(0)x^3 + \cdots.$$

Thus, an approximation to our solution using just the first two terms of the Taylor series would be the Taylor polynomial of degree 1,

$$y(x) = 0.2 + 0.15x.$$

(Notice that this agrees with (9.53) if we use one term approximations for $\cos x \approx 1$ and $\sin x \approx x$.) If we could find more terms in this Taylor series, then we would expect to get a better approximation to our solution. This means we should try to find $y''(0)$, $y'''(0)$, $y^{iv}(0)$, $\cdots$. How can we do this?

If we write (9.51) in the form

$$y''(x) = -y(x), \tag{9.54}$$

we can obtain the value of the second derivative at $x = 0$ as

$$y''(0) = -y(0) = -0.2.$$

Thus, an approximation to our solution using the first three terms of the Taylor series would be the Taylor polynomial of degree 2,

$$y(x) = 0.2 + 0.15x + \frac{1}{2!}(-0.2)x^2 = 0.2 + 0.15x - 0.1x^2.$$

More terms in this expansion can be obtained after differentiating (9.54): For example,

$$y''' = -y' \tag{9.55}$$

gives

$$y'''(0) = -y'(0) = -0.15,$$

and the first four terms in a Taylor series for the solution of (9.51) subject to (9.52) is the Taylor polynomial of degree 3,

$$y(x) = 0.2 + 0.15x + \frac{1}{2!}(-0.2)x^2 + \frac{1}{3!}(-0.15)\,x^3 = 0.2 + 0.15x - 0.1x^2 - 0.025x^3.$$

We can continue with this process: For example, in the next step we differentiate (9.55) and evaluate the result at $x = 0$ to find the Taylor polynomial of degree 4 as

$$y(x) = 0.2 + 0.15x - 0.1x^2 - 0.025x^3 + \frac{1}{120}x^4. \tag{9.56}$$

Following this same routine allows us to compute as many terms in a Taylor series expansion of a solution of such initial value problems as needed to achieve the desired accuracy. However, unless we can obtain the explicit form of the expression for the nth derivative of our solution evaluated at 0 from this procedure, we cannot determine for which values of x (if any) the series converges. In general we cannot obtain this form for the nth derivative, although in the particular case of (9.51) we can (see Exercise 7 on page 463).

We can compare (9.56) with the Taylor series of the exact solution (9.53) by substituting the Taylor series for the $\cos x$ and $\sin x$ terms and keeping all terms up to x^4. This gives

$$y(x) = 0.2[1 - x^2/2 + x^4/24 - \cdots] + 0.15[x - x^3/6 + \cdots]$$
$$= 0.2 + 0.15x - 0.1x^2 - 0.025x^3 + (1/120)x^4 + \cdots, \tag{9.57}$$

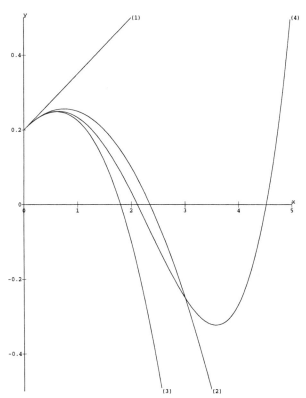

FIGURE 9.5   Taylor polynomials of degrees 1 through 4 for $y'' = -y$, $y(0) = 0.2$, $y'(0) = 0.15$

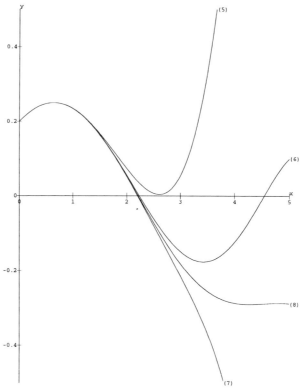

FIGURE 9.6 Taylor polynomials of degrees 5 through 8 for $y'' = -y$, $y(0) = 0.2$, $y'(0) = 0.15$

which is in exact agreement with (9.56). Figure 9.5 shows the Taylor polynomials of degrees 1 through 4 (obtained above). Figure 9.6 shows the Taylor polynomials of degrees 5 through 8. Notice how we can begin to see the solution emerging as the common points between the seventh and eighth degree polynomials.

From the Taylor series in Figure 9.7 for the eighth degree polynomial, we can predict the maximum angle that the pendulum attains (before swinging back) and when this first occurs, given approximately by $y_1 = 0.25$, $x_1 = 0.64$. We can also find the first time that the pendulum is vertical, given approximately by $x_2 = 2.21$. We can compare these with the values we got in Example 8.2, namely, $y_1 = 0.25$, $x_1 = 0.6435$, and $x_2 = 2.2143$.

For comparison Figure 9.8 shows the Taylor polynomials of degrees 5 through 8 as well as the exact solution. □

EXAMPLE 9.13: The Pendulum Again

Here we want to compare the Taylor series of the solution of the initial value problem that governs the nonlinear motion of a pendulum,

$$y'' + \sin y = 0, \tag{9.58}$$

$$y(0) = 0.2, \qquad y'(0) = 0.15, \tag{9.59}$$

with that of the linearized problem. Remember that we cannot find a solution for (9.58) in terms of familiar functions.

The values of the solution and its first derivative are given by the initial conditions (9.59), and the value of the second derivative at $x = 0$ is obtained from (9.58) as $y''(0) =$

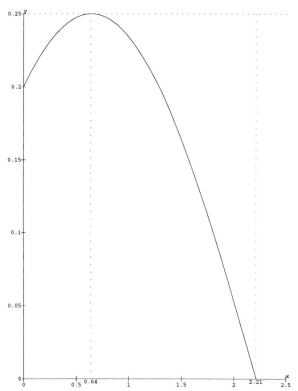

FIGURE 9.7 Taylor polynomial of degree 8 for $y'' = -y$, $y(0) = 0.2$, $y'(0) = 0.15$

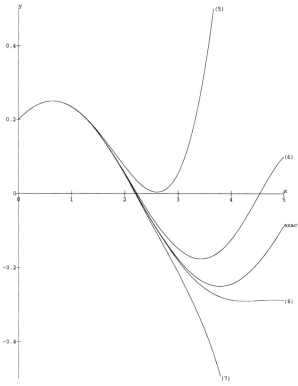

FIGURE 9.8 The exact solution and the Taylor polynomials of degrees 5 through 8 for $y'' = -y$, $y(0) = 0.2$, $y'(0) = 0.15$

$-\sin 0.2 \approx -0.1987$. To find values of the third and fourth derivatives at $x = 0$, we return to (9.58) and differentiate to obtain

$$y''' = -(\cos y)y' \qquad \text{and} \qquad y'''' = (\sin y)(y')^2 - (\cos y)y''.$$

Using the previously obtained values of y and its derivatives at $x = 0$ gives $y'''(0) = -(\cos 0.2)(0.15) \approx -0.1470$ and $y''''(0) = (\sin 0.2)(0.15)^2 - (\cos 0.2)(-0.1987) \approx 0.1992$. Thus, we have the first five terms in a Taylor expansion of our original initial value problem (9.58) and (9.59) as

$$y(x) \approx 0.2 + 0.15x - (0.1987/2!)x^2 - (0.1470/3!)x^3 + (0.1992/4!)x^4$$

$$\approx 0.2 + 0.15x - 0.0993x^2 - 0.0245x^3 + 0.0083x^4.$$

(9.60)

Note that there is very little difference in the two expressions in (9.60) and (9.57) ($1/120 \approx 0.0083$). Figure 9.9 shows the Taylor polynomials of degrees 1 through 4, from (9.60), and Figure 9.10 compares the polynomials of degree 4 for the linearized and nonlinear pendulums.

Figure 9.11 shows the Taylor polynomials of degrees 5 through 8 for the solution of (9.58). Notice how we can begin to see the solution emerging as the common points between the seventh and eighth degree polynomials.

From the Taylor series in Figure 9.12 for the eighth degree polynomial, we can predict the maximum angle that the pendulum attains (before swinging back) and when this first occurs, given approximately by $y = 0.25$, $x = 0.65$. We can also find the first time that the pendulum is vertical, given approximately by $x = 2.24$. We can compare these with the

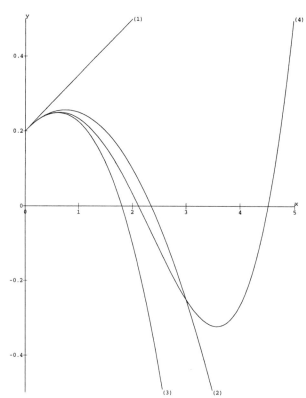

FIGURE 9.9   Taylor polynomials of degrees 1 through 4 for $y'' + \sin y = 0$, $y(0) = 0.2$, $y'(0) = 0.15$

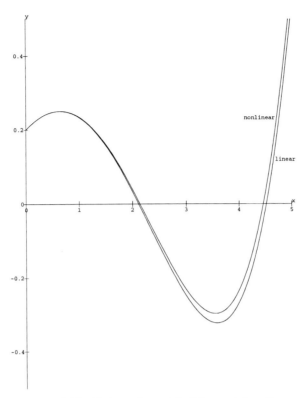

FIGURE 9.10 Taylor polynomial of degree 4 for $y'' = -\sin y$ and $y'' = -y$

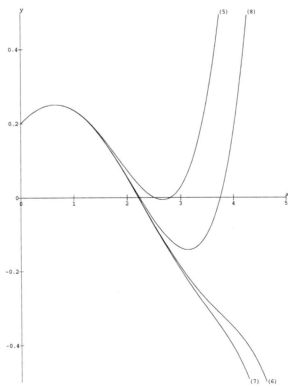

FIGURE 9.11 Taylor polynomials of degrees 5 through 8 for $y'' = -\sin y$, $y(0) = 0.2$, $y'(0) = 0.15$

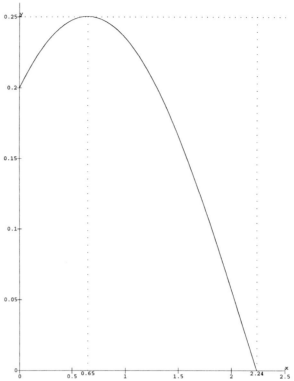

FIGURE 9.12　Taylor polynomial of degree 8 for $y'' = -\sin y$, $y(0) = 0.2$, $y'(0) = 0.15$

values we obtained in Example 8.4 — namely, $y_1 = 0.2504$, $x_1 = 0.6487$, and $x_2 = 2.257$.

$\square$

EXAMPLE 9.14

For our last example in this section we consider the second order linear equation

$$y'' + 4xy' + 2y = 0 \tag{9.61}$$

subject to the initial conditions

$$y(0) = y_0, \qquad y'(0) = y_0^*. \tag{9.62}$$

Because all the coefficients in (9.61) are continuous on all finite intervals and the coefficient of the second derivative is never zero, Theorem 9.1 applies. Thus, (9.61) has a unique solution.

If we solve (9.61) for the second derivative, we have

$$y''(x) = -4xy'(x) - 2y(x), \tag{9.63}$$

from which we can obtain the value of the second derivative at $x = 0$ as

$$y''(0) = -4 \cdot 0 \cdot y'(0) - 2y(0) = -2y_0.$$

More terms in this expansion can be obtained by differentiating (9.63): For example,

$$y'''(x) = -4xy'' - 4y' - 2y' = -4xy''(x) - 6y'(x),$$

giving

$$y'''(0) = -4 \cdot 0 \cdot y''(0) - 6y'(0) = -6y_0^*.$$

This gives us the first four terms in a Taylor series for the solution of (9.61) subject to (9.62) as

$$y(x) = y_0 + y_0^* x - \frac{2}{2!} y_0 x^2 + \frac{1}{3!} \left(-6y_0^* \right) x^3. \tag{9.64}$$

Following this same routine allows us to compute as many terms in a Taylor series solution of such initial value problems as we desire. However, again, if we cannot obtain the form of the expression for the nth derivative evaluated at 0 from this procedure, we cannot determine the interval over which our series converges. However, we may use Theorems 9.2 and 9.3 to discover whether this solution oscillates. Does it? □

EXERCISES

1. Find the first four terms in the Taylor series expansion of the solution to

$$y' = \sqrt{x^2 + y^2}, \qquad y(0) = 1.$$

2. Find the first four terms in the Taylor series expansion of the solution to

$$y' = e^x y^2 + 3 \sin x, \qquad y(0) = 1.$$

3. Find the first five terms in the Taylor series expansion of the solution to

$$y'' = -\sin y, \qquad y(0) = 1, \qquad y'(0) = 0.$$

4. Find the first five terms in the Taylor series expansion of the solution to

$$y' = x^2 + y^2, \qquad y(-1) = -1.$$

5. Find the Taylor series solution to

$$y' = 2xy, \qquad y(0) = 1,$$

and show that it may be expressed as e^{x^2}.

6. Develop a *How to Find a Taylor Series Solution for the Initial Value Problem* $y'' = f(x, y)$, $y(0) = y_0$, $y'(0) = y_0^*$ page by adding statements under Purpose, Technique, and Comments that summarize what is discussed in this section.

7. Solve $y'' + y = 0$, subject to $y(0) = y_0$, $y'(0) = y_0^*$, using Taylor series. Compare the answer with the one obtained by treating the equation as a linear differential equation with constant coefficients.

8. Solve $y'' - y = 0$, subject to $y(0) = y_0$, $y'(0) = y_0^*$, using Taylor series. Compare the answer with the one obtained by treating the equation as a linear differential equation with constant coefficients.

9.6   Power Series and Ordinary Points

There is an alternative way of finding coefficients in a Taylor series expansion of a solution of a differential equation. We can simply substitute the power series

$$y(x) = \sum_{n=0}^{\infty} c_n x^n = c_0 + c_1 x + c_2 x^2 + c_3 x^3 + c_4 x^4 + \cdots \qquad (9.65)$$

into the differential equation and try to evaluate $c_0, c_1, c_2, \cdots, c_n$. For linear differential equations, this operation yields linear algebraic equations. For nonlinear differential equations it leads to nonlinear algebraic equations.

EXAMPLE 9.15: The Linearized Pendulum Revisited Once More

To illustrate this technique, we return to the linearized pendulum equation

$$y'' + y = 0. \qquad (9.66)$$

If we substitute (9.65) into (9.66), we find

$$\left[2c_2 + 6c_3 x + 12c_4 x^2 + 20c_5 x^3 + 30c_6 x^4 + \cdots \right] + \\ + \left[c_0 + c_1 x + c_2 x^2 + c_3 x^3 + c_4 x^4 + \cdots \right] = 0.$$

If we rearrange[4] these terms in increasing powers of x, we obtain

$$\left(2c_2 + c_0 \right) + (6c_3 + c_1)x + (12c_4 + c_2)x^2 + (20c_5 + c_3)x^3 + (30c_6 + c_4)x^4 + \cdots = 0.$$

Because we want this equation to be satisfied for all values of x, we set the coefficient of each power of x to zero to obtain

$$2c_2 + c_0 = 0, \quad 6c_3 + c_1 = 0, \quad 12c_4 + c_2 = 0, \quad 20c_5 + c_3 = 0, \quad 30c_6 + c_4 = 0,$$

from which we find

$$c_2 = -\frac{1}{2}c_0, \quad c_3 = -\frac{1}{6}c_1, \quad c_4 = -\frac{1}{12}c_2, \quad c_5 = -\frac{1}{20}c_3, \quad c_6 = -\frac{1}{30}c_4.$$

The first gives us c_2 in terms of c_0, whereas the second gives c_3 in terms of c_1. The third gives c_4 in terms of c_2, but we already know c_2 in terms of c_0. Thus, we find

$$c_4 = -\frac{1}{12}c_2 = -\frac{1}{12}\left(-\frac{1}{2}c_0 \right) = \frac{1}{24}c_0.$$

In the same way we have

$$c_5 = \frac{1}{120}c_1, \quad c_6 = -\frac{1}{720}c_0.$$

If we substitute these values for c_2 to c_6 into (9.65) and collect together terms involving c_0

---

[4]Recall that absolutely convergent series may be rearranged without changing their interval of convergence.

and c_1, we find

$$y(x) = c_0 \left(1 - \frac{1}{2}x^2 + \frac{1}{24}x^4 - \frac{1}{720}x^6 + \cdots \right) + c_1 \left(x - \frac{1}{6}x^3 + \frac{1}{120}x^5 + \cdots \right).$$

If we impose the initial conditions $y(0) = 0.2$, $y'(0) = 0.15$, we see that $c_0 = 0.2$ and $c_1 = 0.15$, in exact agreement with (9.56).

While this method allows us to find as many terms in our solution as we desire, the calculations may become very tedious after a while. Our work would be greatly simplified if we could find a pattern for the values of c_n. To do this we again substitute this series expression in (9.65) into the original differential equation (9.66), but this time we use the summation notation. This gives

$$\sum_{m=2}^{\infty} c_m m(m - 1)x^{m-2} + \sum_{m=0}^{\infty} c_m x^m = 0.$$

In order to combine terms in this equation, we need the same power on x in the two series, as well as the same limits of summation. To accomplish this, we change the summation index in the first series by increasing it by 2 (letting $m - 2 = n$) while simply replacing m by n in the second series. These operations give

$$\sum_{n=0}^{\infty} c_{n+2}(n + 2)(n + 1)x^n + \sum_{n=0}^{\infty} c_n x^n = \sum_{n=0}^{\infty} \left[(n + 2)(n + 1)c_{n+2} + c_n \right] x^n = 0.$$

Because the differential equation is to be satisfied for all values of x, we may equate all of the coefficients of x^n to zero, giving

$$(n + 2)(n + 1)c_{n+2} + c_n = 0, \qquad n = 0, 1, 2, \cdots. \tag{9.67}$$

Because $(n + 2)(n + 1) \neq 0$ for $n = 0, 1, 2, \cdots$, we may rearrange (9.67) as

$$c_{n+2} = -\frac{c_n}{(n + 2)(n + 1)}, \qquad n = 0, 1, 2, \cdots. \tag{9.68}$$

If we now write down this equation for the first few values of n we see that

$$c_2 = -\frac{1}{2 \cdot 1}c_0, \quad c_3 = -\frac{1}{3 \cdot 2}c_1, \quad c_4 = -\frac{1}{4 \cdot 3}c_2, \quad c_5 = -\frac{1}{5 \cdot 4}c_3.$$

Looking at these equations we observe that coefficients with even subscripts are related to coefficients with even subscripts, and coefficients with odd subscripts are related to coefficients with odd subscripts. If we look at (9.68) we discover that it expresses a relationship between coefficients whose subscripts differ by 2. Equation (9.68) is an example of a RECURRENCE RELATIONSHIP, also called a RECURRENCE RELATION.

Even though (9.68) gives us a recipe for finding one coefficient in terms of the coefficient whose subscript is 2 less, it would be more convenient to have an explicit formula for c_n. Computing the first few coefficients with even subscripts — $n = 0, 2,$ and 4 — gives

$$c_2 = -\frac{1}{2}c_0, \quad c_4 = -\frac{1}{4 \cdot 3}c_2 = (-1)^2 \frac{1}{4 \cdot 3 \cdot 2}c_0, \quad c_6 = -\frac{1}{6 \cdot 5}c_4 = (-1)^3 \frac{1}{6 \cdot 5 \cdot 4 \cdot 3 \cdot 2}c_0,$$

where we observe that all these coefficients with even subscripts are written as multiples of c_0 and there are no requirements on c_0. This means that c_0 is an arbitrary constant. To obtain a general formula for coefficients with even subscripts, we return to the recurrence relation (9.68) and let $n + 2 = 2m$, that is, $n = 2m - 2$, obtaining

$$c_{2m} = (-1) \frac{1}{(2m)(2m - 1)} c_{2m-2}.$$

In this equation we now replace the term on the right-hand side by what we obtain by letting $n = 2m - 4$ in (9.68), namely

$$c_{2m-2} = (-1) \frac{1}{(2m - 2)(2m - 3)} c_{2m-4}.$$

This gives

$$c_{2m} = (-1)^2 \frac{1}{2m(2m - 1)(2m - 2)(2m - 3)} c_{2m-4}.$$

Continuing in this manner gives us

$$
\begin{aligned}
c_{2m} &= (-1) \frac{1}{(2m)(2m - 1)} c_{2m-2} \\
&= (-1)^2 \frac{1}{2m(2m - 1)(2m - 2)(2m - 3)} c_{2m-4} \\
&= \cdots \\
&= (-1)^m \frac{1}{2m(2m - 1)(2m - 2)(2m - 3) \cdots 3 \cdot 2 \cdot 1} c_0,
\end{aligned}
$$

or[5]

$$c_{2m} = (-1)^m \frac{1}{(2m)!} c_0, \qquad m = 0, 1, 2, \cdots. \tag{9.69}$$

This is an explicit formula for all the even coefficients c_{2m} in terms of the arbitrary constant c_0.

    The recurrence relation for the odd subscripts is obtained from (9.68) by setting $n + 2 = 2m + 1$, that is, $n = 2m - 1$, as

$$c_{2m+1} = -\frac{1}{(2m + 1) 2m} c_{2m-1}, \qquad m = 1, 2, 3, \cdots,$$

which leads to

$$c_3 = -\frac{1}{3 \cdot 2} c_1, \quad c_5 = -\frac{1}{5 \cdot 4} c_3 = (-1)^2 \frac{1}{5 \cdot 4 \cdot 3 \cdot 2} c_1,$$

$$c_7 = -\frac{1}{7 \cdot 6} c_5 = (-1)^3 \frac{1}{7 \cdot 6 \cdot 5 \cdot 4 \cdot 3 \cdot 2} c_1.$$

Here we see that all of the coefficients with odd subscripts are written in terms of c_1. Since there are no conditions on c_1 it is also an arbitrary constant.

    For the general case for odd subscripts we have

$$c_{2m+1} = (-1) \frac{1}{(2m + 1) 2m} c_{2m-1}$$

---

[5]It is wise to substitute what we think is the solution of a recurrence relation back into the recurrence relation to confirm that it is a solution.

$$= (-1)^2 \frac{1}{(2m+1)\,2m\,(2m-1)\,(2m-2)} c_{2m-3}$$
$$= \cdots$$
$$= (-1)^m \frac{1}{(2m+1)\,2m\,(2m-1)\,(2m-2)\cdots 4 \cdot 3 \cdot 2} c_1,$$

or

$$c_{2m+1} = (-1)^m \frac{1}{(2m+1)!} c_1, \qquad m = 0, 1, 2, \cdots. \tag{9.70}$$

Thus, we can write our solution of (9.66) in the form

$$y(x) = \sum_{n=0}^{\infty} c_n x^n = \sum_{m=0}^{\infty} c_{2m} x^{2m} + \sum_{m=0}^{\infty} c_{2m+1} x^{2m+1},$$

or, using (9.69) and (9.70),

$$y(x) = c_0 \sum_{m=0}^{\infty} (-1)^m \frac{x^{2m}}{(2m)!} + c_1 \sum_{m=0}^{\infty} (-1)^m \frac{x^{2m+1}}{(2m+1)!}, \tag{9.71}$$

where c_0 and c_1 are arbitrary constants. This is the solution of (9.66) in terms of power series. Can we express these power series in terms of familiar functions? If we write out the first few terms of the coefficient of c_0 we find

$$\sum_{m=0}^{\infty} (-1)^m \frac{x^{2m}}{(2m)!} = 1 - \frac{x^2}{2!} + \frac{x^4}{4!} - \frac{x^6}{6!} + \cdots$$

which we recognize as the Taylor series expansion for $\cos x$. Treating the coefficient of c_1 in the same way we recognize

$$\sum_{m=0}^{\infty} (-1)^m \frac{x^{2m+1}}{(2m+1)!} = x - \frac{x^3}{3!} + \frac{x^5}{5!} - \frac{x^7}{7!} + \cdots$$

as the Taylor series expansion for $\sin x$. Both these series converge[6] for all values of x so the solution (9.71) can be written

$$y(x) = c_0 \cos x + c_1 \sin x,$$

valid for all x.

If we use the initial conditions $y(0) = 0.2$, $y'(0) = 0.15$ in (9.71), we find $c_0 = y(0) = 0.2$ and $c_1 = y'(0) = 0.15$, which agrees with (9.53). $\square$

Definition 9.3: **A RECURRENCE RELATION among** $c_0, c_1, c_2, \cdots$, **is an equation of the type**

$$c_{n+m} = f(c_n, c_{n+1}, \cdots c_{n+m-1}),$$

that is, c_{n+m} **is determined by its previous** m **terms.**

Comments about recurrence relations

- Recurrence relations are also called finite difference equations.

---

[6]Power series that converge do so absolutely within the interval of convergence.

- The positive integer m is the ORDER of the recurrence relation. The recurrence relation (9.68), namely,

$$c_{n+2} = -\frac{c_n}{(n+2)(n+1)}, \qquad n = 0, 1, 2, \cdots,$$

  is a second order recurrence relation.

- There is no general way to solve recurrence relations.

In this example we were fortunate in two respects. First, we were able to solve the recurrence relation to find an explicit formula for c_n as a function of n. Second, we were able to recognize the series we obtained as ones that converged to familiar functions. This meant that we did not have to worry about convergence. But how do we know that in general our power series solutions will converge? For linear differential equations with polynomial coefficients, the following theorem answers that question.

Theorem 9.6:   *Consider*

$$a_2(x)\frac{d^2y}{dx^2} + a_1(x)\frac{dy}{dx} + a_0(x)y = 0, \tag{9.72}$$

where $a_2(x)$, $a_1(x)$, and $a_0(x)$ are the polynomials in x remaining after all factors in common to all three terms have been removed. If $a_2(x_0) \neq 0$, the general solution of (9.72) may be obtained as the power series about x_0,

$$y(x) = \sum_{n=0}^{\infty} c_n(x - x_0)^n = C_1 y_1(x) + C_2 y_2(x),$$

where $y_1(x)$ and $y_2(x)$ are linearly independent functions and C_1 and C_2 are arbitrary constants. This series converges at least for $|x - x_0| < R$, where R is the distance between x_0 and the zero of $a_2(x)$ that is closest to x_0. In considering the zeros of $a_2(x)$, we must include both real and complex zeros.[7]

Comments about Theorem 9.6

- There are usually many values of x_0 that could be used in this theorem. If initial conditions are given for a particular value of x, then that is the point to use for x_0.

- In our last example, namely,

$$y'' + y = 0,$$

  we are assured that the two series we obtained converge for all values of x because $a_2(x) = 1$, $a_1(x) = 0$, and $a_0(x) = 1$ are all polynomials and because $a_2(x)$ is never zero. The distance between $x_0 = 0$ and any roots of $a_2(x)$ is infinite, so $R = \infty$.

- This theorem can be extended to nonhomogeneous linear differential equations with polynomial coefficients

$$a_2(x)\frac{d^2y}{dx^2} + a_1(x)\frac{dy}{dx} + a_0(x)y = h(x),$$

---

[7]The distance between x_0 and a complex zero $z_1 = x_1 + iy_1$ is $\sqrt{(x_1 - x_0)^2 + y_1^2}$.

if the forcing term $h(x)$ has a convergent Taylor series about $x = x_0$. The conclusion regarding convergence still holds, and the form of the solution will be $C_1 y_1(x) + C_2 y_2(x) + f(x)$, where $f(x)$ is a power series.

- If the hypotheses of the theorem are not satisfied, the convergence of each series must be examined separately.

How to Find a Power Series Solution of a Linear Differential Equation

Purpose: To find a power series solution of

$$a_2(x)\frac{d^2 y}{dx^2} + a_1(x)\frac{dy}{dx} + a_0(x)y = 0, \tag{9.73}$$

about 0, where $a_2(x)$, $a_1(x)$, and $a_0(x)$ are polynomials in x.

Technique:

1. Factor out any common factors in the three polynomials $a_2(x)$, $a_1(x)$, and $a_0(x)$ and assume a power series solution of the form $y(x) = \sum_{n=0}^{\infty} c_n x^n$.

2. Substitute this series into (9.73), and compute the appropriate derivatives.

3. Combine terms, which may mean changing indices of summation.

4. Have the lower limits of all the series start with the same integer, placing all the extra terms together.

5. Set the coefficient of each power of x equal to zero, obtaining the recurrence relation.

6. Find as many terms in the series solution as desired, or, if possible, find a general expression for c_n.

7. Check convergence using Theorem 9.6 or the ratio test.[8]

Comments about power series solutions of a linear differential equation

- The form of the solution will be $y(x) = C_1 y_1(x) + C_2 y_2(x)$, where $y_1(x)$ and $y_2(x)$ are linearly independent functions and C_1 and C_2 are arbitrary constants.

- We may not be able to find a general expression for c_n.

- For the original expansion of $y(x) = \sum_{n=0}^{\infty} c_n x^n$ with initial conditions, $y(0) = y_0$, $y'(0) = y_0^*$, we have $c_0 = y_0$, and $c_1 = y_0^*$

- If initial values are given for $x_0 \neq 0$, the series expansion to use will be $y(x) = \sum_{n=0}^{\infty} c_n (x - x_0)^n$. There are two different ways to proceed in this case.

  Follow the preceding technique, using $y(x) = \sum_{n=0}^{\infty} c_n (x - x_0)^n$ in place of $y(x) = \sum_{n=0}^{\infty} c_n x^n$. In this case it is useful to express the coefficients $a_2(x)$, $a_1(x)$, and $a_0(x)$ as polynomials in $x - x_0$. [For example, write $x^2 = (x - x_0)^2 + 2x_0(x - x_0) + x_0^2$.]

  Change the independent variable from x to X, where $X = x - x_0$. In this way the point $x = x_0$ becomes the point $X = 0$ and $y(x) = \sum_{n=0}^{\infty} c_n (x - x_0)^n$ becomes $y(X) = \sum_{n=0}^{\infty} c_n X^n$, which is a power series about the origin of X. Now use the preceding technique.

---

[8]See the appendix.

- This method will work on nonhomogeneous linear differential equations

$$a_2(x)\frac{d^2y}{dx^2} + a_1(x)\frac{dy}{dx} + a_0(x)y = h(x),$$

  if $h(x)$ is first expanded in a power series. See Exercise 10 on page 481.
- This method may also work with nonlinear differential equations; the fact that we will have nonlinear algebraic equations for the coefficients may present an additional complication. See Exercises 16 and 17 on page 482.

We demonstrate this procedure on the last example of the previous section.

EXAMPLE 9.16

Find the power series solution of

$$y'' + 2xy' + 2y = 0. \tag{9.74}$$

We are assured by Theorem 9.6 that the series we obtain will converge for all values of x because $a_2(x) = 1$, $a_1(x) = 2x$, and $a_0(x) = 2$ are all polynomials and $a_2(x)$ is never zero. The distance between $x_0 = 0$ and any roots of $a_2(x)$ is infinite, so $R = \infty$.

Substituting the series $y(x) = \sum_{n=0}^{\infty} c_n x^n$ into the differential equation (9.74), we obtain

$$\sum_{n=2}^{\infty} c_n n(n-1)x^{n-2} + 2x\sum_{n=1}^{\infty} c_n n x^{n-1} + 2\sum_{n=0}^{\infty} c_n x^n = 0.$$

In order to combine terms in this equation, we need the same power on x in the three series, as well as the same limits of summation. To accomplish this, we combine the x multiplying the second series with the x^{n-1} inside the summation sign and change the summation index in the first series by raising it by 2. Also note that lowering the limit on the middle series to 0 does not change anything because of the multiplicative factor of n. These operations give

$$\sum_{n=0}^{\infty} c_{n+2}(n+2)(n+1)x^n + \sum_{n=0}^{\infty} 2c_n n x^n + \sum_{n=0}^{\infty} 2c_n x^n$$

$$= \sum_{n=0}^{\infty}\left[(n+2)(n+1)c_{n+2} + 2(n+1)c_n\right]x^n = 0.$$

Because the differential equation is to be satisfied for all values of x, we may equate all of the coefficients to zero, giving the second order recurrence relation

$$(n+2)(n+1)c_{n+2} + 2(n+1)c_n = 0, \qquad n = 0, 1, 2, \cdots. \tag{9.75}$$

Rearranging (9.75) as

$$c_{n+2} = -\frac{2c_n}{n+2}, \qquad n = 0, 1, 2, \cdots \tag{9.76}$$

shows that we have no requirements on c_1 and c_0 and that all the other coefficients are given by (9.76) in terms of c_1 and c_0. We note that all the coefficients with even subscripts are given in terms of c_0, whereas all the coefficients with odd subscripts are given in terms of c_1. We also note that c_0 and c_1 are arbitrary constants, which agrees with the number of arbitrary constants we had earlier for solutions of second order linear differential equations.

Computing the first few coefficients with even subscripts gives

$$c_2 = -c_0, \qquad c_4 = -\frac{1}{2}c_2 = (-1)^2\frac{1}{2}c_0, \qquad c_6 = -\frac{1}{3}c_4 = (-1)^3\frac{1}{3\cdot 2\cdot 1}c_0.$$

For arbitrary even integer $2m$, we have

$$\begin{aligned}
c_{2m} &= (-1)\frac{2}{2m}c_{2m-2} \\
&= (-1)^2\frac{1}{m(m-1)}c_{2m-4} \\
&= (-1)^m\frac{1}{m(m-1)(m-2)\cdots 3\cdot 2\cdot 1}c_0 \\
&= (-1)^m\frac{1}{m!}c_0, \qquad m = 0, 1, 2, \cdots.
\end{aligned}$$

The recurrence relation for the odd subscripts is obtained from (9.76) as

$$c_{2m+1} = -\frac{2}{2m+1}c_{2m-1}, \qquad m = 1, 2, 3, \cdots,$$

which leads to $c_3 = -(2/3)c_1$, $c_5 = -(2/5)c_3 = (-1)^2 2^2 c_1/(5\cdot 3)$, and $c_7 = -(2/7)c_5 = (-1)^3 2^3 c_1/(7\cdot 5\cdot 3)$. For the general case we have

$$\begin{aligned}
c_{2m+1} &= (-1)\frac{2}{2m+1}c_{2m-1} \\
&= (-1)^2\frac{2^2}{(2m+1)(2m-1)}c_{2m-3} \\
&= (-1)^m\frac{2^m}{(2m+1)(2m-1)(2m-3)\cdots 7\cdot 5\cdot 3}c_1 \\
&= (-1)^m\frac{2^m}{1\cdot 3\cdot 5\cdots(2m+1)}c_1, \qquad m = 0, 1, 2, \cdots.
\end{aligned}$$

Thus, we have the solution of (9.74) in the form

$$y(x) = \sum_{m=0}^{\infty} c_{2m}x^{2m} + \sum_{m=0}^{\infty} c_{2m+1}x^{2m+1},$$

which can be written

$$y(x) = c_0\sum_{m=0}^{\infty}(-1)^m\frac{x^{2m}}{m!} + c_1\sum_{m=0}^{\infty}(-1)^m\frac{2^m}{1\cdot 3\cdot 5\cdots(2m+1)}x^{2m+1}.$$

We recognize that the first series converges for all values of x to

$$c_0\sum_{m=0}^{\infty}(-1)^m\frac{(x^2)^m}{m!} = c_0\sum_{m=0}^{\infty}\frac{1}{m!}\left(-x^2\right)^m = c_0 e^{-x^2}$$

but that the second series does not lead to a familiar function (see Exercises 5 and 6 on page 480).

Thus, the general solution of (9.74), valid for all x, is given by

$$y(x) = c_0\sum_{m=0}^{\infty}(-1)^m\frac{x^{2m}}{m!} + c_1\sum_{m=0}^{\infty}(-1)^m\frac{2^m}{1\cdot 3\cdot 5\cdots(2m+1)}x^{2m+1},$$

or

$$y(x) = c_0 e^{-x^2} + c_1 \sum_{m=0}^{\infty} (-1)^m \frac{2^m}{1 \cdot 3 \cdot 5 \cdots (2m+1)} x^{2m+1}. \tag{9.77}$$

If we use the initial conditions $y(0) = y_0$, $y'(0) = y_0^*$ in (9.77), we have $c_0 = y(0) = y_0$ and $c_1 = y'(0) = y_0^*$. If we write out the first few terms of each of the series in (9.77), we find (9.64).

We know what the solution $c_0 e^{-x^2}$ behaves like (see Figure 9.13), but what about the coefficient of c_1? We know from Section 9.1 that this solution cannot oscillate. Furthermore, if we impose the initial conditions $y(0) = 0$, $y'(0) = 1$, then (9.77) reduces to the solution

$$y(x) = \sum_{m=0}^{\infty} (-1)^m \frac{2^m}{1 \cdot 3 \cdot 5 \cdots (2m+1)} x^{2m+1}. \tag{9.78}$$

Figure 9.14 shows the power series polynomials of degrees 1 through 10 for (9.78). Notice how we can begin to see the solution emerging as the common points between the ninth and tenth degree polynomials. However, in spite of appearances, we know that this solution cannot cross the positive x-axis. If it did it would then have two zeros ($x = 0$ being the other one), and so e^{-x^2} would have to vanish between these two zeros, which it does not. It is possible to confirm these observations in other ways (see Exercise 5, page 480).   □

EXAMPLE 9.17: Airy's Equation

Let us return to Airy's equation (9.3), that is,

$$\frac{d^2 y}{dx^2} - xy = 0. \tag{9.79}$$

We already know from Section 9.1 that every nontrivial solution of (9.79) oscillates for $x < 0$ and has at most one zero for $x > 0$. Now we wish to find a power series solution for this equation. Theorem 9.6 guarantees that the series we obtain will converge for all values of x. Substituting the series $y(x) = \sum_{n=0}^{\infty} c_n x^n$ into the differential equation (9.79), we find $c_2 = 0$ and the third order recurrence relation

$$c_{n+2} = \frac{1}{(n+2)(n+1)} c_{n-1}, \qquad n = 1, 2, 3, \cdots.$$

After several calculations we find

$$c_{3m} = (3m-2) \cdot (3m-5) \cdots 7 \cdot 4 \cdot 1 \cdot c_0 / (3m)!, \qquad m = 1, 2, 3, \cdots,$$
$$c_{3m+1} = (3m-1) \cdot (3m-4) \cdots 8 \cdot 5 \cdot 2 \cdot c_1 / (3m+1)!, \qquad m = 0, 1, 2, \cdots,$$
$$c_{3m+2} = 0, \qquad m = 0, 1, 2, \cdots.$$

The general solution of (9.79), valid for all x, is therefore

$$y(x) = c_0 \sum_{m=1}^{\infty} \frac{(3m-2) \cdot (3m-5) \cdots 7 \cdot 4 \cdot 1}{(3m)!} x^{3m} +$$

$$+ c_1 \sum_{m=0}^{\infty} \frac{(3m-1) \cdot (3m-4) \cdots 8 \cdot 5 \cdot 2}{(3m+1)!} x^{3m+1}. \tag{9.80}$$

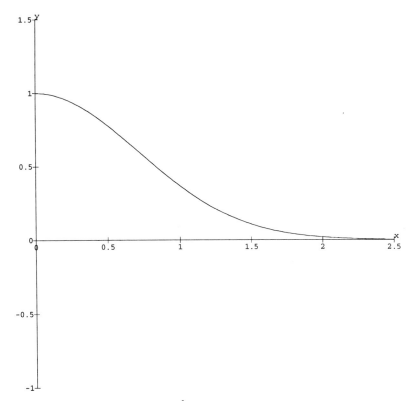

FIGURE 9.13 The solution e^{-x^2} of $y'' + 2xy' + 2y = 0$, $y(0) = 1$, $y'(0) = 0$

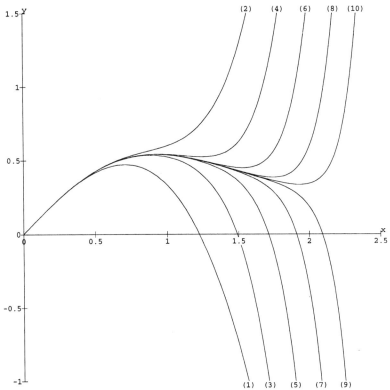

FIGURE 9.14 Power series polynomials of degrees 1 through 10 for $y'' + 2xy' + 2y = 0$, $y(0) = 0$, $y'(0) = 1$

Figure 9.15 shows the power series polynomials of degrees 6 through 27 for the coefficient of c_0 in (9.80). Notice that it corresponds to the initial condition $y(0) = 1$, $y'(0) = 0$, and that we can begin to see the solution emerging as the common points between the 24th and 27th degree polynomials. We also see that for $x > 0$ there are no roots, whereas for $x < 0$, the solution begins to show oscillatory behavior. These observations are consistent with the previous results on oscillations and zeros.

Figure 9.16 shows the power series polynomials of degrees 7 through 28 for the coefficient of c_1 in (9.80). Notice that it corresponds to the initial condition $y(0) = 0$, $y'(0) = 1$, and that we can begin to see the solution emerging as the common points between the 25th and 28th degree polynomials. We also see that for $x > 0$ there are no roots, whereas for $x < 0$, the solution oscillates. These observations are consistent with the previous results on oscillations and zeros.   □

EXAMPLE 9.18

Solve

$$\frac{d^2y}{dx^2} - 2(x - 1)\frac{dy}{dx} + 2y = 0 \qquad (9.81)$$

by using a power series solution about $x = 1$.

The coefficients of y'', y', and y are polynomials, so we know that there is a solution of (9.81) of the form

$$y(x) = \sum_{n=0}^{\infty} c_n (x - 1)^n.$$

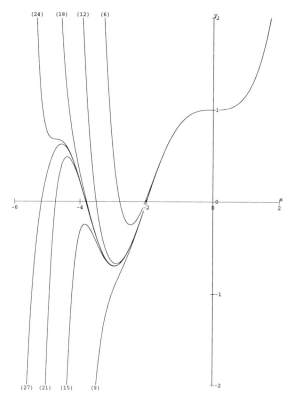

FIGURE 9.15 Power series polynomials of degrees 6 through 27 for Airy's equation with $c_1 = 0$

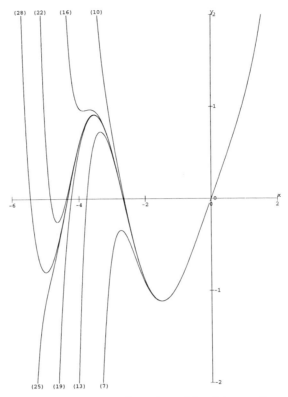

(28) (22) (16) (10)

(25) (19) (13) (7)

FIGURE 9.16 Power series polynomials of degrees 7 through 28 for Airy's equation with $c_0 = 0$

There are two different ways to find the coefficients in this expansion.

We could substitute $y(x) = \sum_{n=0}^{\infty} c_n (x - 1)^n$ into (9.81), collect like powers of $x - 1$ together, equate their coefficients to zero, and proceed as usual (see Exercise 7 on page 481).

A second way is to make the change of variable

$$X = x - 1,$$

so that the point $x = 1$ becomes the point $X = 0$, and the differential equation (9.81) becomes the differential equation

$$\frac{d^2 y}{d X^2} - 2X \frac{dy}{dX} + 2y = 0. \tag{9.82}$$

Thus, the original problem of finding a series solution of (9.81) about $x = 1$ is equivalent to finding a power series solution of (9.82) about the point $X = 0$, that is,

$$y(X) = \sum_{n=0}^{\infty} c_n X^n.$$

In Exercise 8 in this section you will find that the recurrence relation in this case becomes

$$c_{n+2} = \frac{2 (n - 1)}{(n + 2) (n + 1)} c_n, \qquad n = 0, 1, 2, \cdots,$$

which leads to the general solution

$$y(X) = c_0 \left(1 - X^2 - \frac{1}{6}X^4 - \frac{1}{30}X^6 + \cdots \right) + c_1 X.$$

In terms of the original variable x, the solution of (9.81) is

$$y(x) = c_0 \left[1 - (x-1)^2 - \frac{1}{6}(x-1)^4 - \frac{1}{30}(x-1)^6 + \cdots \right] + c_1 (x-1).$$

$\square$

Theorem 9.6 deals with the second order linear differential equation

$$a_2(x)\frac{d^2y}{dx^2} + a_1(x)\frac{dy}{dx} + a_0(x)y = 0,$$

where $a_2(x)$, $a_1(x)$, and $a_0(x)$ are polynomials in x with $a_0(x_0) \neq 0$. But what happens if this is not the case? If we divide by the coefficient of the second derivative, we obtain

$$\frac{d^2y}{dx^2} + p(x)\frac{dy}{dx} + q(x)y = 0, \tag{9.83}$$

which we use as our standard form. It should be clear that in (9.83) we have $p(x) = a_1(x)/a_2(x)$ and $q(x) = a_0(x)/a_2(x)$.

The success of the power series method is ensured by the following theorem:

Theorem 9.7:   *If both $p(x)$ and $q(x)$ are analytic[9] at $x = x_0$, then the general solution of*

$$\frac{d^2y}{dx^2} + p(x)\frac{dy}{dx} + q(x)y = 0$$

may be obtained in the form of a power series about $x = x_0$ with a radius of convergence at least as large as the minimum of the radii of convergence of the series expansions for $p(x)$ and $q(x)$ about $x = x_0$.

Comments about Theorem 9.7

- If the hypothesis of Theorem 9.7 is satisfied, we may confidently proceed with the calculation of a power series solution knowing the interval of convergence even before we start. We could either use the power series method or implicitly differentiate the differential equation as we did in the previous section.

- If both $p(x)$ and $q(x)$ are analytic at the point $x = x_0$, then the point $x = x_0$ is called an ORDINARY POINT of (9.83). If $x = x_0$ is not an ordinary point of (9.83), the point $x = x_0$ is a SINGULAR POINT of the differential equation, and the function with the singular point is said to be SINGULAR and have a SINGULARITY AT $x = x_0$.

- This theorem can be extended to nonhomogeneous differential equations if the radius of convergence of the forcing term is taken into account.

---

[9]A function is analytic at a point if it has a Taylor series that converges in some interval about the point.

EXAMPLE 9.19: Legendre's Differential Equation

Exercise 13 is concerned with LEGENDRE'S DIFFERENTIAL EQUATION, which has the form

$$(1 - x^2)y'' - 2xy' + \lambda y = 0,$$

where λ is a specified constant.

Legendre's differential equation has singular points at 1 and -1, with all other points being ordinary points. Thus, according to either Theorem 9.6 or Theorem 9.7, a power series for the solution about $x = 0$ will converge at least for $-1 < x < 1$. $\square$

EXAMPLE 9.20

Consider the power series solution of the differential equation

$$(4 + x^2)y'' + \tan \frac{x}{4} y' + xy = 0,$$

about (a) $x = 1$, and (b) $x = 4$.

Writing the differential equation in the form

$$y'' + \frac{\tan(x/4)}{4 + x^2} y' + \frac{x}{4 + x^2} y = 0$$

shows that

$$p(x) = \frac{\tan(x/4)}{4 + x^2} \text{ and } q(x) = \frac{x}{4 + x^2}.$$

Here the singular points are at $2i$ and $-2i$, and the Taylor series for $\tan x/4$ converges for $-2\pi < x < 2\pi$. Thus, by Theorem 9.7, a power series solution about $x = 1$ will converge at least for $|x - 1| < \sqrt{1^2 + 2^2} = \sqrt{5}$. A power series solution about $x = 4$ will converge at least for $|x - 4| < 2\pi - 4$. $\square$

EXAMPLE 9.21

Find a power series up to degree 5 for the solution of

$$xy'' + \sin x \, y' + 2xy = 0 \qquad (9.84)$$

about $x = 0$, and check its convergence.

To check the hypothesis of Theorem 9.7, we divide (9.84) by x to put it in the standard form of

$$y'' + \frac{\sin x}{x} y' + 2y = 0. \qquad (9.85)$$

The coefficients

$$p(x) = \frac{\sin x}{x} = \frac{1}{x} \sum_{m=0}^{\infty} (-1)^m \frac{x^{2m+1}}{(2m + 1)!} = \sum_{m=0}^{\infty} (-1)^m \frac{x^{2m}}{(2m + 1)!}$$

and $q(x) = 2$ are analytic at $x = 0$, so the hypothesis of Theorem 9.7 is satisfied with $x_0 = 0$. Because the radii of convergence of $p(x)$ and $q(x)$ are infinite, we are assured that the power series solution of (9.84) converges for all values of x.

If we substitute the series $y(x) = \sum_{n=0}^{\infty} c_n x^n$ into (9.85), we obtain

$$\sum_{n=2}^{\infty} n(n-1) c_n x^{n-2} + \left(\sum_{m=0}^{\infty} (-1)^m \frac{x^{2m}}{(2m+1)!} \right) \sum_{n=1}^{\infty} n c_n x^{n-1} + \sum_{n=0}^{\infty} 2 c_n x^n = 0.$$

If we want to find the power series up to degree 5, we truncate all these series to obtain

$$2c_2 + 6c_3 x + 12c_4 x^2 + 20c_5 x^3 + 30c_6 x^4 +$$
$$+ \left[1 - (1/6) x^2 + (1/120) x^4 \right] (c_1 + 2c_2 x + 3c_3 x^2 + 4c_4 x^3 + 5c_5 x^4) +$$
$$+ 2c_0 + 2c_1 x + 2c_2 x^2 + 2c_3 x^3 + 2c_4 x^4 + 2c_5 x^5 = 0.$$

Equating to zero the coefficients of powers of x^n for $n = 0, 1, 2,$ and 3 gives the following set of homogeneous algebraic equations:

$$2c_2 + c_1 + 2c_0 = 0,$$
$$6c_3 + 2c_2 + 2c_1 = 0,$$
$$12c_4 + 3c_3 - (1/6) c_1 + 2c_2 = 0,$$
$$20c_5 + 4c_4 - (1/3) c_2 + 2c_3 = 0.$$

Note that this system of four equations with six unknowns is a dependent system. As expected, we may choose c_0 and c_1 arbitrarily, and the solution of this system of equations becomes

$$c_2 = -(2c_0 + c_1)/2,$$
$$c_3 = (2c_0 - c_1)/6,$$
$$c_4 = (3c_0 + 5c_1)/36,$$
$$c_5 = -(24c_0 + 7c_1)/360.$$

This means that the solution in the form of a power series, including terms of degree 5 or less, is given by

$$y(x) = c_0 \left[1 - x^2 + \frac{1}{3} x^3 + \frac{1}{12} x^4 - \frac{1}{15} x^5 + \cdots \right] +$$

$$+ c_1 \left[x - \frac{1}{2} x^2 - \frac{1}{6} x^3 + \frac{5}{36} x^4 - \frac{7}{360} x^5 + \cdots \right].$$

(9.86)

Figure 9.17 shows the power series polynomials of degrees 2 through 5 for the coefficient of c_0 in (9.86). Notice that it corresponds to the initial condition $y(0) = 1$, $y'(0) = 0$, and that we can begin to see the solution emerging as the common points between the fourth and fifth degree polynomials. Figure 9.18 shows the power series polynomials of degrees 2 through 5 for the coefficient of c_1 in (9.86). It corresponds to the initial condition $y(0) = 0$, $y'(0) = 1$. Again we can begin to see the solution emerging as the common points between the fourth and fifth degree polynomials. However, from this analysis we cannot tell whether these solutions oscillate. In fact they do oscillate (see Exercise 9 on page 481).　□

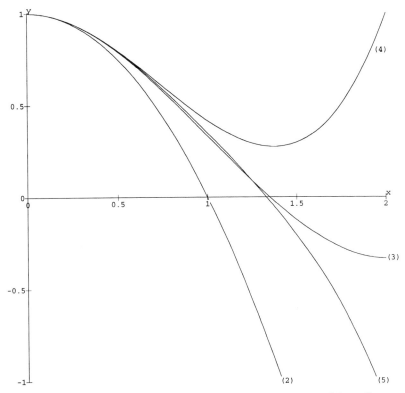

FIGURE 9.17 Power series polynomials of degrees 2 through 5 for $xy'' + \sin x\, y' + 2xy = 0$ subject to $y(0) = 1$, $y'(0) = 0$

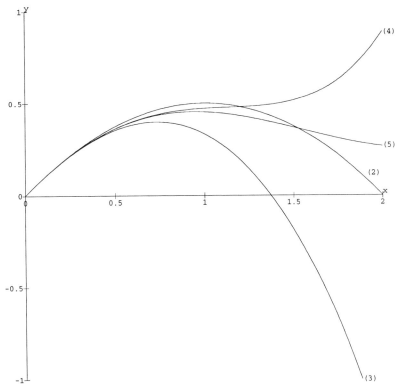

FIGURE 9.18 Power series polynomials of degrees 2 through 5 for $xy'' + \sin x\, y' + 2xy = 0$ subject to $y(0) = 0$, $y'(0) = 1$

EXERCISES

1. Make a change in the index of summation to show the validity of the following equations.

   (a) $\sum_{k=0}^{\infty} c_k x^{k+2} = \sum_{m=2}^{\infty} c_{m-2} x^m$

   (b) $\sum_{k=4}^{\infty} c_{k-1} x^{k-2} = \sum_{m=2}^{\infty} c_{m+1} x^m$

   (c) $\sum_{m=n}^{\infty} c_m x^m = \sum_{k=0}^{\infty} c_{k+n} x^{k+n}$ (That is, if we reduce the lower limit of the summation by n, we increase the index to the right of the summation sign by n.)

   (d) $\sum_{m=n}^{\infty} c_m x^{m-p} = \sum_{k=n-p}^{\infty} c_{k+p} x^k$ (That is, if we increase the exponent by p, we increase the corresponding subscript by p and reduce the lower limit of summation by p.)

2. Show that

   (a) $\sum_{k=2}^{\infty} c_k + A \sum_{k=0}^{\infty} a_k = A(a_0 + a_1) + \sum_{k=2}^{\infty} (c_k + Aa_k)$.

   (b) $\sum_{k=0}^{\infty} c_k = \sum_{k=0}^{n} c_k + \sum_{k=n+1}^{\infty} c_k$ for any positive integer n.

3. Use the power series method to solve

$$y'' - xy' + 3y = 0, \qquad y(0) = 2, \qquad y'(0) = 0.$$

4. Use the power series method to solve

$$y'' + 2x^2 y' + xy = 0, \qquad y(0) = 0, \qquad y'(0) = 3.$$

5. Make the change of dependent variable $y(x) = z(x)e^{-x^2}$ in (9.74), that is,

$$y'' + 2xy' + 2y = 0,$$

   to obtain a second solution in a different form from that given in (9.77), namely,

$$y(x) = e^{-x^2} \int_0^x e^{t^2} \, dt.$$

   (a) Show that this solution cannot vanish for $x > 0$ in agreement with the observation made following (9.78).

   (b) Show that this solution has an extreme value at the point $(x_m, 1/(2x_m))$ where x_m satisfies $y(x_m) = 1/(2x_m)$. Show that at this extreme value we have $y''(x_m) = -2y(x_m)$. Explain why this implies that there is a maximum at $(x_m, 1/(2x_m))$. Can this solution have a minimum? Can it have a second maximum? Find a numerical value for x_m, and compare your answer with Figure 9.14.

   (c) Show that this solution has a point of inflection at the point $\left(x_i, x_i/\left(2x_i^2 - 1\right)\right)$ where x_i satisfies $y(x_i) = x_i/\left(2x_i^2 - 1\right)$. Can this solution have a second point of inflection? Find a numerical value for x_i, and compare your answer with Figure 9.14.

   (d) Use local linearity or L'Hôpital's rule to show that this solution has the property that $y(x) \to 0$ as $x \to \infty$.

   (e) Use local linearity or L'Hôpital's rule to show that this solution has the property that $2xy(x) \to 1$ as $x \to \infty$, so that $y(x) \to 1/(2x)$ as $x \to \infty$.

   (f) Use the information from parts (a) through (e) to sketch this solution for $x > 0$.

   (g) Show that the solution satisfies $y(x) = -y(-x)$. Use this information to sketch this solution for all x.

6. Show that the substitution $u(x) = y' + 2xy$ converts the differential equation (9.74), that is,

$$y'' + 2xy' + 2y = 0,$$

   into $u' = 0$. Use this fact to find the general solution of (9.74). Explain how this idea can be used to solve differential equations of the form

$$y'' + p(x)y' + p'(x)y = 0.$$

7. Use the power series method about $x = 1$ to find the general solution of

$$\frac{d^2 y}{dx^2} - 2(x - 1)\frac{dy}{dx} + 2y = 0.$$

8. Use the power series method about $x = 0$ to find the general solution of

$$\frac{d^2 y}{dx^2} - 2x\frac{dy}{dx} + 2y = 0.$$

9. Show that the differential equation (9.85), namely,

$$y'' + \frac{\sin x}{x}y' + 2y = 0,$$

has, from (9.13), a $Q(x)$ given by

$$Q(x) = q(x) - \frac{1}{2}p'(x) - \frac{1}{4}p^2(x),$$

which can be written in the form

$$Q(x) = \frac{3}{2} + \frac{1}{4x^2}\left[(x - \cos x)^2 + x^2 + 2\sin x - 1\right].$$

Show that $x^2 + 2\sin x - 1 \geq x^2 - 3$, so that if $x \geq \sqrt{3}$, then $Q(x) \geq 3/2$. Explain how this guarantees that nontrivial solutions of (9.85) must oscillate for $x \geq \sqrt{3}$.

10. Find the power series solution, up to terms of degree 4, of the following initial value problems, and give a value of R such that the series solution converges for $|x - x_0| < R$.
    (a) $y'' + x^2 y' + \sin x\, y = 3$, $\quad y(0) = 0$, $\quad y'(0) = 3$
    (b) $y'' + 3xy' + \cos x\, y = x^2$, $\quad y(0) = 1$, $\quad y'(0) = 0$
    (c) $y'' - 2xy' + e^x y = e^x$, $\quad y(0) = 0$, $\quad y'(0) = -2$
    (d) $y'' + \sin x\, y' + xy = 0$, $\quad y(0) = 3$, $\quad y'(0) = 0$
    (e) $y'' - xy' + \ln x\, y = 0$, $\quad y(1) = 0$, $\quad y'(1) = 2$
    (f) $y'' - xy' + y/(1 - 2x) = 0$, $\quad y(0) = 1$, $\quad y'(0) = 0$

11. Find the general solution of the following differential equations as a power series about the given point, and give the radius of convergence.
    (a) $y'' + x^2 y = 0$, $\quad x_0 = 0$
    (b) $y'' - 3xy' + y = 0$, $\quad x_0 = 0$
    (c) $(1 - x^2)y'' + 2xy' + 5y = 0$, $\quad x_0 = 0$

12. Find the first five nonzero terms in a power series solution of the following initial value problems. Give the radius of convergence.
    (a) $xy'' + x^2 y' + y = 0$, $\quad y(2) = 0$, $\quad y'(2) = 1$
    (b) $x^2 y'' + y' + 2y = 0$, $\quad y(1) = 0$, $\quad y'(1) = 1$
    (c) $(x^2 + 1)y'' + xy' + 3y = 0$, $\quad y(2) = 0$, $\quad y'(2) = 3$

13. Legendre polynomials are solutions of Legendre's differential equation

$$(1 - x^2)y'' - 2xy' + \lambda y = 0$$

for special values of λ.
    (a) Show that Legendre's differential equation will have polynomial solutions whenever λ is the product of two consecutive nonnegative integers, that is $\lambda = n(n + 1)$.
    (b) Find the first 5 Legendre polynomials [denoted by $P_0(x)$, $P_1(x)$, $P_2(x)$, $P_3(x)$, and $P_4(x)$] by choosing the arbitrary constants in the solutions for $n = 0, 1, 2, 3$, and 4, so that $P_n(1) = 1$.

14. List the singular points for the following differential equations

(a) $(1 - x)y'' + \dfrac{3x}{x + 2}y' + \dfrac{(1 - x)^2}{x + 3}y = 0$

(b) $(x^2 + x)y'' + \dfrac{x^3}{x - 1}y' + \dfrac{x^4 + 3x}{x + 2}y = 0$

(c) $\dfrac{1}{x}y'' + \dfrac{3(x - 4)}{x + 6}y' + \dfrac{x^2(x - 2)}{x - 1}y = 0$

(d) $(x^2 + 3x + 2)y'' + \dfrac{x + 2}{x - 1}y' + \dfrac{x - 2}{x}y = 0$

(e) $(x^2 + 9)y'' + \dfrac{x - 2}{x + 7}y' + \dfrac{x^2 + 3}{2}y = 0$

(f) $e^x y'' + \dfrac{3x - 4}{x + 4}y' + \dfrac{x}{x - 4}y = 0$

(g) $\sin x\, y'' + \dfrac{x^3}{x - 1}y' + \dfrac{x^4 + 3x}{x + 2}y = 0$

15. Use Theorems 9.6 and 9.7 to determine the interval of convergence for each of the differential equations in Exercise 14 for the following values of x_0.

(a) $x_0 = 1$

(b) $x_0 = 4$

(c) $x_0 = -1$

(d) $x_0 = 2$

(e) $x_0 = 0$

(f) $x_0 = 0$

(g) $x_0 = 3$

16. Find the first six terms in a power series solution of the nonlinear differential equation

$$y'' + \left(y'\right)^2 = e^x y, \qquad y(0) = 0, \qquad y'(0) = 1.$$

17. Find the first six terms in a power series solution of the nonlinear differential equation

$$y' = 2x^2 - y^2, \qquad y(0) = 1.$$

Now make the change of variable $y = u'/u$, and solve $u'' = 2x^2 u$, $u'(0) = u(0)$. Comment on this form of your answer.

What Have We Learned?

MAIN IDEAS

- Two functions $f(x)$ and $g(x)$ are linearly independent on an interval if the only way $c_1 f(x) + c_2 g(x) = 0$ for all values of x in this interval is for $c_1 = c_2 = 0$. See Theorem 9.4 on page 446 for the Wronskian test for independence.

  The following comments refer to the linear differential equation

  $$a_2(x)\frac{d^2 y}{dx^2} + a_1(x)\frac{dy}{dx} + a_0(x)y = 0. \tag{9.87}$$

- The general solution of (9.87) is a linear combination of two of its linearly independent solutions.

- There are ways of determining whether solutions of (9.87) oscillate without constructing the solution. See Theorem 9.2 on page 432 and Theorem 9.3 on page 436.

- If we know one solution of (9.87), we may always obtain a second solution that is linearly independent of the first solution. See *How to Reduce the Order* on page 442.

- If the coefficients in (9.87) have the form of a constant times x raised to the power of the subscript [that is, $a_2(x) = b_2 x^2$, $a_1(x) = b_1 x$, and $a_0(x) = b_0$] we have a Cauchy-Euler equation, which may always be solved. See *How to Solve a Cauchy-Euler Differential Equation* on page 452.

- See page 476 for the definition of ordinary and singular points.

- If $x = x_0$ is an ordinary point of (9.87), then the solution of (9.87) subject to the initial conditions $y(x_0) = y_0$ and $y'(x_0) = y_0^*$ may be obtained as a power series about $x = x_0$. The answer to Exercise 6 of Section 9.5, page 463, explains how to find a Taylor series solution. Also see *How to Find a Power Series Solution of a Linear Differential Equation* on page 469.

CHAPTER 10

Generalized Power Series Solutions

Where Are We Going?

In this chapter we consider second order linear differential equations of the form

$$a_2(x)\frac{d^2y}{dx^2} + a_1(x)\frac{dy}{dx} + a_0(x)y = 0,$$

where a_1/a_2 or a_0/a_2 is not analytic at some point $x = x_0$. With appropriate restrictions on the behavior of a_1/a_2 and a_0/a_2 at x_0 we can modify our techniques of Chapter 9 to find general solutions of the form $y(x) = (x - x_0)^s \sum_{n=0}^{\infty} c_n (x - x_0)^n$, where the exponent s and the coefficients c_n are to be determined. We give theorems regarding the convergence of these series solutions, so the region of convergence is known before we find the specific coefficients.

10.1   Regular Singular Points

In Section 9.6 we used a power series method to find series solutions of differential equations about ordinary points. If we use that technique to find a power series solution about the point $x = 0$ for the linear differential equation

$$x^2 y'' - 2y = 0, \tag{10.1}$$

we find that

$$x^2(2c_2 + 6c_3 x + 12c_4 x^2 + \cdots) - 2(c_0 + c_1 x + c_2 x^2 + c_3 x^3 + c_4 x^4 + \cdots) = 0.$$

Combining like terms gives

$$-2c_0 - 2c_1 x + 4c_3 x^3 + 10c_4 x^4 + \cdots = 0,$$

which requires that $c_0 = c_1 = c_3 = c_4 = \cdots = 0$. This gives us no condition on c_2, so $c_2 x^2$ is a solution, where c_2 is an arbitrary constant. However, we do not get a second solution by this method.

If we try this same method on a similar differential equation,

$$4x^2 y'' - 3y = 0, \tag{10.2}$$

we find that all of the coefficients in $\sum_{n=0}^{\infty} c_n x^n$ must equal zero. (Do this!)

What is happening? If we look at Theorem 9.6 on page 468, we notice that the hypothesis that $a_2(0) \neq 0$ is violated in both these examples, so that theorem does not apply here. Also if we look at Theorem 9.7 on page 476, we notice that the hypothesis that $q(x)$ be analytic at $x = 0$ is violated in both these examples, because $q(x) = -2/x^2$ and $q(x) = -3/(4x^2)$. So the point $x = 0$ is not an ordinary point, but a singular point.

So what can we try if $x = 0$ is not an ordinary point? We note that both (10.1) and (10.2) are Cauchy–Euler differential equations with general solutions $C_1 x^2 + C_2 x^{-1}$ and $C_1 x^{3/2} + C_2 x^{-1/2}$, respectively. Because the exponents in a power series may only be nonnegative integers, only one of the four functions in these general solutions is a power series expansion about $x = 0$; but all these solutions have the form of x to a power. Maybe we should try a solution of the form of a power series multiplied by x to a power — that is, $y(x) = x^s \sum_{n=0}^{\infty} c_n x^n$, where s is a constant.

EXAMPLE 10.1

We now try this form of a series — namely,

$$y(x) = x^s \sum_{n=0}^{\infty} c_n x^n, \tag{10.3}$$

in the first differential equation in this section, (10.1). In this series, s, as well as the coefficients c_n, $n = 0, 1, 2, \cdots$, are to be found in order that (10.3) satisfies the differential equation (10.1).[1] To simplify subsequent derivatives, we move the x^s term inside the series as

$$y(x) = \sum_{n=0}^{\infty} c_n x^{n+s} = c_0 x^s + c_1 x^{s+1} + c_2 x^{s+2} + c_3 x^{s+3} + \cdots. \tag{10.4}$$

Because the series must start somewhere, we always assume that c_0 is the coefficient of the smallest surviving power of x, and so $c_0 \neq 0$. We substitute (10.4) into (10.1) to obtain

$$x^2 \left[c_0 s(s-1)x^{s-2} + c_1(s+1)sx^{s-1} + c_2(s+2)(s+1)x^s + \right. $$
$$ + c_3(s+3)(s+2)x^{s+1} + \cdots \right] + $$
$$ - 2 \left[c_0 x^s + c_1 x^{s+1} + c_2 x^{s+2} + c_3 x^{s+3} + \cdots \right] = 0.$$

If we combine like powers of x, this reduces to

$$[s(s-1) - 2]c_0 x^s + [(s+1)s - 2]c_1 x^{s+1} + [(s+2)(s+1) - 2]c_2 x^{s+2} + $$
$$ + [(s+3)(s+2) - 2]c_3 x^{s+3} + \cdots = 0.$$

---

[1] Here s need not be an integer.

Equating coefficients of each power of x to zero, we find

$$[s(s-1)-2]c_0 = 0,$$
$$[(s+1)s-2]c_1 = 0,$$
$$[(s+2)(s+1)-2]c_2 = 0,$$
$$[(s+3)(s+2)-2]c_3 = 0.$$

$$(10.5)$$

Because $c_0 \neq 0$, the first equation gives

$$s(s-1)-2 = 0,$$

or

$$(s-2)(s+1) = 0, \qquad (10.6)$$

so the possible choices for s are 2 and -1. We now show how these choices lead to our previous general solution.

1. With $s = 2$ in (10.5), we find $c_1 = c_2 = c_3 = \cdots = 0$, so in this case the series (10.3) reduces to $c_0 x^2$, where c_0 is arbitrary. Thus, $C_1 y_1(x)$, where $y_1(x) = x^2$, is one solution of (10.1).

2. With $s = -1$ in (10.5), we also find $c_1 = c_2 = c_3 = \cdots = 0$, so in this case the series (10.3) reduces to $c_0 x^{-1}$, where c_0 is arbitrary. Thus, $C_2 y_2(x)$, where $y_2(x) = x^{-1}$, is another solution of (10.1).

Now $y_1(x)$ and $y_2(x)$ are linearly independent, so the general solution of (10.1) is

$$y(x) = C_1 y_1(x) + C_2 y_2(x) = C_1 x^2 + C_2 \frac{1}{x},$$

in agreement with our previous observation. $\square$

The quadratic equation that determines the possible values for s — in this case (10.6) — is given a special name.

Definition 10.1: **If $x = x_0$ is a singular point of the differential equation**

$$a_2(x)y'' + a_1(x)y' + a_0(x)y = 0$$

and we substitute $y(x) = (x - x_0)^s \sum_{n=0}^{\infty} c_n (x - x_0)^n$ into the differential equation, collect like powers of $x - x_0$, and then set the coefficient of the lowest power of $x - x_0$ equal to zero, the equation so obtained is called THE INDICIAL EQUATION. It is a quadratic equation in s.

EXAMPLE 10.2

We apply the technique used in Example 10.1 to the differential equation given in (10.2),

$$4x^2 y'' - 3y = 0, \qquad (10.7)$$

which has a singular point at $x = 0$.

We assume a solution of the form

$$y(x) = x^s \sum_{n=0}^{\infty} c_n x^n = \sum_{n=0}^{\infty} c_n x^{n+s} = c_0 x^s + c_1 x^{s+1} + c_2 x^{s+2} + c_3 x^{s+3} + \cdots, \quad \textbf{(10.8)}$$

where $c_0 \neq 0$, and substitute (10.8) into (10.7) to obtain

$$4x^2 \left[c_0 s(s-1)x^{s-2} + c_1(s+1)sx^{s-1} + c_2(s+2)(s+1)x^s + \right.$$
$$\left. + c_3(s+3)(s+2)x^{s+1} + \cdots \right] +$$
$$- 3 \left[c_0 x^s + c_1 x^{s+1} + c_2 x^{s+2} + c_3 x^{s+3} + \cdots \right] = 0.$$

If we combine like powers of x, this reduces to

$$[4s(s-1) - 3]c_0 x^s + [4(s+1)s - 3]c_1 x^{s+1} + [4(s+2)(s+1) - 3]c_2 x^{s+2} +$$
$$+ [4(s+3)(s+2) - 3]c_3 x^{s+3} + \cdots = 0.$$

The lowest power of x is x^s, so the indicial equation is

$$[4s(s-1) - 3]c_0 = 0.$$

Because $c_0 \neq 0$, we have the quadratic equation

$$4s^2 - 4s - 3 = 0,$$

with roots $3/2$ and $-1/2$. Equating the coefficients of the higher powers of x to zero, we find

$$[4(s+1)s - 3]c_1 = 0,$$
$$[4(s+2)(s+1) - 3]c_2 = 0, \quad \textbf{(10.9)}$$
$$[4(s+3)(s+2) - 3]c_3 = 0.$$

From the indicial equation the possible choices for s are $3/2$ and $-1/2$. We now use each of these choices in turn.

1. With $s = 3/2$ in (10.9), we find $c_1 = c_2 = c_3 = \cdots = 0$, so in this case the series (10.8) reduces to $c_0 x^{3/2}$, where c_0 is arbitrary. Thus, $C_1 y_1(x)$, where $y_1(x) = x^{3/2}$, is one solution of (10.7).

2. With $s = -1/2$ in (10.9), we also find $c_1 = c_2 = c_3 = \cdots = 0$, so in this case the series (10.8) reduces to $c_0 x^{-1/2}$, where c_0 is arbitrary. Thus, $C_2 y_2(x)$, where $y_2(x) = x^{-1/2}$, is another solution of (10.7).

Now $y_1(x)$ and $y_2(x)$ are linearly independent, so the general solution of (10.7) is

$$y(x) = C_1 y_1(x) + C_2 y_2(x) = C_1 x^{3/2} + C_2 x^{-1/2},$$

again in agreement with our previous observation. □

EXAMPLE 10.3

If we use the preceding technique on the linear differential equation

$$x^4 y'' + 2x^3 y - y = 0, \quad \textbf{(10.10)}$$

which has a singular point at $x = 0$, we find that

$$x^4 \left[c_0 s(s-1)x^{s-2} + c_1(s+1)sx^{s-1} + c_2(s+2)(s+1)x^s + \cdots \right] +$$
$$+ 2x^3 \left[c_0 sx^{s-1} + c_1(s+1)x^s + c_2(s+2)x^{s+1} + \cdots \right] +$$
$$- \left[c_0 x^s + c_1 x^{s+1} + c_2 x^{s+2} + c_3 x^{s+3} + \cdots \right] = 0.$$

Combining like terms gives

$$-c_0 x^s - c_1 x^{s+1} + [c_0 s(s-1) + 2c_0 s + c_2]x^{s+2} + \cdots = 0.$$

The coefficient of the lowest power of x does not depend on s, and the only way this coefficient can equal zero is for $c_0 = 0$. However, $c_0 \neq 0$, so there is no solution of (10.10) of the form $y(x) = x^s \sum_{n=0}^{\infty} c_n x^n$. In fact the solution of (10.10) is

$$y(x) = C_1 e^{1/x} + C_2 e^{-1/x},$$

(see Exercise 5 on page 494) which is of the form $x^s \sum_{n=0}^{\infty} c_n x^{-n}$. Figure 10.1 shows the graphs of $e^{1/x}$ and $e^{-1/x}$.   □

Why did this technique work for the first two differential equations with singular points at $x = 0$, but not for the third in (10.10)? To answer that question, we need to subdivide singular points into two types, regular and irregular.

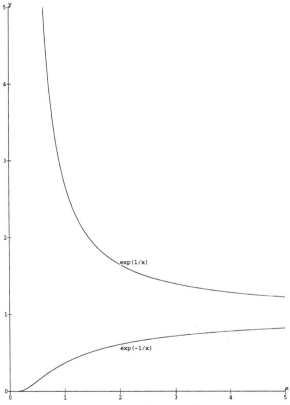

FIGURE 10.1   The functions $e^{1/x}$ and $e^{-1/x}$

Definition 10.2:   **Consider a differential equation in the form**

$$\frac{d^2y}{dx^2} + p(x)\frac{dy}{dx} + q(x)y = 0, \tag{10.11}$$

which has $x = x_0$ as a singular point. If the functions

$$(x - x_0)p(x) \qquad \text{and} \qquad (x - x_0)^2 q(x) \tag{10.12}$$

are both analytic at x_0, then x_0 is called a REGULAR SINGULAR POINT of (10.11). If at least one of the products in (10.12) is not analytic at x_0, then x_0 is called an IRREGULAR SINGULAR POINT of (10.11).

Notice that in the first two differential equations, $x = 0$ is a regular singular point, whereas $x = 0$ is an irregular singular point of (10.10).

EXAMPLE 10.4

In Example 9.19 on page 477 we noted that Legendre's differential equation,

$$(1 - x^2)y'' - 2xy' + \lambda y = 0, \tag{10.13}$$

has singular points at 1 and -1. Let us check to see if these singular points are regular.

We first put (10.13) in standard form as

$$y'' - \frac{2x}{1 - x^2}y' + \frac{\lambda}{1 - x^2}y = 0.$$

Because $(x - 1)p(x) = 2x/(1 + x)$ and $(x - 1)^2 q(x) = -\lambda(x - 1)/(x + 1)$ are both analytic at $x = 1$, then $x = 1$ is a regular singular point. A similar situation holds true for $x = -1$. [Don't forget in that case we would consider $(x + 1)p(x)$ and $(x + 1)^2 q(x)$.] $\square$

EXAMPLE 10.5

As an example with coefficients that are not rational functions of x, we consider

$$\sin x \, y'' + 3(x + 1)y' - \frac{2}{x^2(x - 4)^2}y = 0.$$

We want to find and classify the singular points of this differential equation.

In terms of the differential equation in standard form, we have $p(x) = 3(x + 1)/\sin x$ and $q(x) = -2/[x^2(x - 4)^2 \sin x]$. The singular points occur at $x = 0$, $x = 4$, and places where $\sin x = 0$ — namely, $x = k\pi$, $k = 0, 1, 2, \cdots$.

Because $x^2 q(x) = -2/[(x - 4)^2 \sin x]$ is not analytic at $x = 0$, $x = 0$ is an irregular singular point. Because $(x - 4)p(x) = 3(x - 4)(x + 1)/\sin x$ and $(x - 4)^2 q(x) = -2/(x^2 \sin x)$ are both analytic at $x = 4$, $x = 4$ is a regular singular point.

To determine the nature of the remaining singularities at $x = k\pi$, $k = 1, 2, 3, \cdots$, we examine $(x - k\pi)p(x) = (x - k\pi)3(x + 1)/\sin x$ and $(x - k\pi)^2 q(x) = (x - k\pi)^2(-2)/[x^2(x - 4)^2 \sin x]$. We know that the quotient of analytic functions is analytic at all points, except possibly where the denominator is zero, so we need only check the behavior of both of these expressions near $x = k\pi$, $k = 1, 2, 3, \cdots$. By using either the local linearity property of the sine function near $x = k\pi$ or L'Hôpital's rule, we see that the limits as $x \to k\pi$ of both $(x - k\pi)p(x)$ and $(x - k\pi)^2 q(x)$ exist, being $3(k\pi + 1)/(-1)^k$, and 0 respectively. Thus, we have regular singular points at $x = k\pi$, $k = 1, 2, 3, \cdots$. $\square$

Knowing the nature of the singular point of a differential equation is important because of the following theorem.

Theorem 10.1: *Consider the second order linear differential equation in the standard form*

$$\frac{d^2 y}{dx^2} + p(x)\frac{dy}{dx} + q(x)y = 0, \tag{10.14}$$

where $x = x_0$ is a regular singular point. Let R denote the smallest radius of convergence of the functions $(x - x_0)p(x)$ and $(x - x_0)^2 q(x)$, and let s denote the larger root of the indicial equation. Then (10.14) has a solution of the form $y(x) = (x - x_0)^s \sum_{n=0}^{\infty} c_n (x - x_0)^n$, which converges for $0 < |x - x_0| < R$.

Comments about Theorem 10.1

- This theorem assumes that the roots of the indicial equation are real numbers, and in this book we restrict our attention to this assumption. However, these ideas can be extended to the case in which the indicial equation has complex roots.

- The indicial equation is a quadratic that will have two solutions, s_1 and s_2, where $s_1 \geq s_2$. This theorem says that if we use $s = s_1$ we are sure of one solution. Sometimes the other choice for s — that is, s_2 — generates a second linearly independent solution.

- Whether or not the choice s_2 for s produces a second linearly independent solution, we can always use reduction of order to generate a second solution from the first.

- In general the interval of convergence $0 < |x - x_0| < R$ does not contain $x = x_0$, so there are usually two disconnected intervals of convergence, $x_0 < x < x_0 + R$ and $x_0 - R < x < x_0$. If $x_0 = 0$, then these intervals are $0 < x < R$ and $-R < x < 0$. In all examples we will concentrate on the interval $x_0 < x < x_0 + R$, which in the case of $x_0 = 0$ reduces to $0 < x < R$.

- For cases where the roots of the indicial equation are not integers or ratios of odd integers, we need to replace $(x - x_0)^s$ in our solution with $|x - x_0|^s$. (We would also use this for complex roots.)

Because the behavior of the series solution near a regular singular point depends so heavily on the root of the indicial equation, we seek ways of determining these roots without having to go through the details of trying a series solution. We list the main result here as a theorem; another technique is given as Exercise 4 on page 494.

Theorem 10.2: *If $x = x_0$ is a regular singular point of the second order linear differential equation*

$$a_2(x)y'' + a_1(x)y' + a_0(x)y = 0, \tag{10.15}$$

then the indicial equation is obtained by equating to zero the lowest power of $x - x_0$ that occurs when $(x - x_0)^s$ is substituted into (10.15).

Although we won't prove Theorem 10.2, we note that because x_0 is a regular singular point, $a_1(x)/a_2(x)$ has at most a factor of $1/(x - x_0)$ and $a_0(x)/a_2(x)$ has at most a factor of $1/(x - x_0)^2$. Thus, the indicial equation associated with a regular singular point will be a quadratic equation in s.

EXAMPLE 10.6

Find the roots of the indicial equation associated with a series solution about the regular singular point of

$$2xy'' + 6y' - 9xy = 0,$$

and find the radius of convergence of this series solution.

We can see that $x = 0$ is the only singular point for this differential equation. Because $xp(x) = 3$ and $x^2 q(x) = -9x^2/2$ are both analytic at $x = 0$, we have a regular singular point there. To find the indicial equation, we substitute x^s into the differential equation to obtain

$$2xs(s - 1)x^{s-2} + 6sx^{s-1} - 9xx^s = 2s(s - 1)x^{s-1} + 6sx^{s-1} - 9x^{s+1}.$$

Here the lowest power of x is x^{s-1}, so the indicial equation is given by

$$2s(s - 1) + 6s = 2s^2 + 4s = 0,$$

with roots of 0 and $-1/2$. From Theorem 10.1 we have that the accompanying series solution for the larger root will converge for all values of x and the factor in front of our series will be $1 (= x^0)$. This theorem tells us nothing about the series solution associated with the smaller root, $s = -1/2$. The way to find out if we have a solution for $s = -1/2$ is to try to compute the coefficients. This we do in the next section.   □

EXAMPLE 10.7

Find the roots of the indicial equation associated with a series solution about the regular singular point $x = 0$ of

$$3x^2 y'' + xy' - \frac{2}{2 - x} y = 0, \tag{10.16}$$

and find the radius of convergence of this series solution.

The singular point $x = 0$ is a regular singular point because both $xp(x) = 1/3$ and $x^2 q(x) = -x^2/[3(2 - x)]$ are analytic at $x = 0$. (Are there any other singular points?) To find the indicial equation, we substitute x^s into (10.16) and use the expansion

$$\frac{1}{1 - a} = 1 + a + a^2 + \cdots$$

to find that

$$\frac{2}{2 - x} = \frac{2}{2\left(1 - \frac{x}{2}\right)} = 1 + \frac{x}{2} + \left(\frac{x}{2}\right)^2 + \cdots.$$

This gives

$$3x^2 s(s - 1)x^{s-2} + xsx^{s-1} - \left(1 + \frac{x}{2} + \left(\frac{x}{2}\right)^2 + \cdots\right)x^s,$$

so the indicial equation is

$$3s(s - 1) + s - 1 = 3s^2 - 2s - 1 = 0.$$

The roots of this indicial equation are 1 and $-1/3$, and from Theorem 10.1, the resulting series solution corresponding to $s = 1$ converges at least for the interval $-2 < x < 2$. (Where does this interval come from?) □

EXAMPLE 10.8

Find the roots of the indicial equation associated with a series solution about the regular singular point of

$$(x - 3) y'' + 2xy' + 5y = 0, \tag{10.17}$$

and find the radius of convergence of this series solution.

The only singular point in this differential equation is $x = 3$. We see that it is a regular singular point because both $(x - 3) p(x) = 2x$ and $(x - 3)^2 q(x) = 5 (x - 3)$ are analytic at $x = 3$. To find the indicial equation we substitute $(x - 3)^s$ into (10.17) and obtain

$$(x - 3) s (s - 1) (x - 3)^{s-2} + (2x) s (x - 3)^{s-1} + 5 (x - 3)^s . \tag{10.18}$$

To find the coefficient of the lowest power of $(x - 3)$ we must express each term in (10.18) as a polynomial in $(x - 3)$. Thus, we write $x = (x - 3) + 3$, and the middle term in (10.18) in the form

$$(2x) s (x - 3)^{s-1} = 2 [(x - 3) + 3] s (x - 3)^{s-1} = 2s (x - 3)^s + 6s (x - 3)^{s-1} .$$

This means that (10.18) may be rewritten in the form

$$s (s - 1) (x - 3)^{s-1} + 2s (x - 3)^s + 6s (x - 3)^{s-1} + 5 (x - 3)^s ,$$

so the indicial equation is

$$s (s - 1) + 6s = s^2 + 5s = 0.$$

Here the roots of the indicial equation are 0 and -5. From Theorem 10.1 we have that the series associated with $s = 0$ will converge for all values of x. □

In the next two sections we will use this technique to solve differential equations.

EXERCISES

1. List the singular points for the following differential equations.

   (a) $(1 - x)y'' + \dfrac{3x}{x + 2}y' + \dfrac{(1 - x)^2}{x + 3}y = 0$

   (b) $(x^2 + x)y'' + \dfrac{x^3}{x - 1}y' + \dfrac{x^4 + 3x}{x + 2}y = 0$

   (c) $\dfrac{1}{x}y'' + \dfrac{3(x - 4)}{x + 6}y' + \dfrac{x^2 (x - 2)}{x - 1}y = 0$

   (d) $(x^2 + 3x + 2)y'' + \dfrac{x + 2}{x - 1}y' + \dfrac{x - 2}{x}y = 0$

   (e) $(x^2 + 9)y'' + \dfrac{x - 2}{x + 7}y' + \dfrac{x^2 + 3}{2}y = 0$

   (f) $e^x y'' + \dfrac{3x - 4}{x + 4}y' + \dfrac{x}{x - 4}y = 0$

(g) $\sin x\, y'' + \dfrac{x^3}{x-1} y' + \dfrac{x^4 + 3x}{x+2} y = 0$

2. Determine the regular and irregular singular points for the following differential equations.

(a) $x^2 \left(x^2 - 1\right)^2 y'' + \dfrac{x\,(x+1)}{x-4} y' + \dfrac{3\,(x-1)}{x^2 - 16} y = 0$

(b) $x \left(x^2 - 3x - 10\right) y'' + \dfrac{x+4}{x-2} y' + 16 y = 0$

(c) $x \sin x\, y'' + \dfrac{3\,(x-1)}{x+1} y' + \cos x\, y = 0$

(d) $\sin x\, y'' + \dfrac{x \cos x}{x+1} y' - \dfrac{x^2}{x-2} y = 0$

(e) $x^2 \left(x^2 - 16\right) y'' + \dfrac{(x-2)}{x+2} y' + 32 y = 0$

(f) $e^x y'' + 3xy' + \dfrac{1}{1 - e^x} y = 0$

(g) $x\,(x-1)\, y'' + \dfrac{x+1}{(x-4)^2} y' + \dfrac{1}{x^2} y = 0$

(h) $\left(x^2 + x - 6\right) y'' + \dfrac{14}{1 - x^2} y' + \dfrac{12}{1 + x^2} y = 0$

3. Find the indicial equation and the minimum interval of convergence of series solutions about the regular singular points of the following differential equations.

(a) $4xy'' + 2\,(1+x)\, y' + y = 0$

(b) $3xy'' + y' - y = 0$

(c) $4xy'' + 2y' + y = 0$

(d) $2x^2 y'' + \left(2x^2 + x\right) y' + 2xy = 0$

(e) $2x^2 y'' - xy' + (x - 5)\, y = 0$

(f) $2\,(x-1)\, y'' - y' + e^x y = 0$

(g) $2\,(x+1)^2\, y'' - 3\left(x^2 + 3x + 2\right) y' + (3x + 5)\, y = 0$

(h) $xy'' + (x-1)y' - [2/(1-2x)]y = 0$

(i) $3(x-1)y'' + y' - y = 0$

(j) $(x+1)y'' - (x+4)y' + 2y = 0$

(k) $(x-1)2y'' + (3x^2 - 4x + 1)y' - 2y = 0$

(l) $9x^2 y'' + 9x^2 y' + 2y = 0.$

4. An alternative method of finding the indicial equation applies to the case in which we express our second order linear differential equation with a regular singular point at $x = x_0$ as

$$\left(x - x_0\right)^2 b_2(x)y'' + \left(x - x_0\right) b_1(x)y' + b_0(x)y = 0,$$

where $b_2(x_0) = \alpha \ne 0$. In this case the indicial equation is

$$\alpha s(s - 1) + \beta s + \gamma = 0,$$

where $\beta = b_1(x_0)$ and $\gamma = b_0(x_0)$. Use this result to find the indicial equation for the differential equations in Exercise 3.

5. The purpose of this exercise is to find the general solution of (10.10) — namely,

$$x^4 \dfrac{d^2 y}{dx^2} + 2x^3 y - y = 0. \tag{10.19}$$

(a) Show that the change of independent variable from x to t, where $x = 1/t$, converts the differential equation (10.19) into

$$\frac{d^2y}{dt^2} - y = 0.$$

(b) Solve the differential equation in part (a), finding $y = y(t)$.

(c) Substitute $t = 1/x$ into the solution $y = y(t)$ of part (b) to show that the solution of (10.19) is

$$y(x) = C_1 e^{1/x} + C_2 e^{-1/x}.$$

10.2 The Method of Frobenius, Part 1

In this section we develop solutions of second order linear differential equations by using a series expression of the form

$$y(x) = (x - x_0)^s \sum_{n=0}^{\infty} c_n (x - x_0)^n, \tag{10.20}$$

where x_0 is a regular singular point of the differential equation. In the previous section we discovered how to obtain the indicial equation for the evaluation of the index s, and here we carry on and determine the coefficients. The method of determining solutions of the form (10.20) is called the METHOD OF FROBENIUS.

EXAMPLE 10.9

We now find the general solution of

$$4xy'' + 2y' + y = 0$$

as a series about the origin using the method of Frobenius. It is easily seen that $x = 0$ is a regular singular point, and using Theorem 10.1 we see that the series solution we obtain will converge for all values of x. If we substitute the series in (10.20), with $x_0 = 0$, into this differential equation, we obtain

$$\sum_{n=0}^{\infty} 4(n+s)(n+s-1)c_n x^{n+s-1} + \sum_{n=0}^{\infty} 2(n+s)c_n x^{n+s-1} + \sum_{n=0}^{\infty} c_n x^{n+s} = 0.$$

We now raise the index of summation in the first two terms by 1 (that is, we let $n - 1 = m$) and get

$$4s(s-1)c_0 x^{s-1} + 2sc_0 x^{s-1} + \sum_{m=0}^{\infty}[4(m+1+s)(m+s)c_{m+1} + \tag{10.21}$$

$$+ 2(m+1+s)c_{m+1}]x^{m+s} \sum_{n=0}^{\infty} c_n x^{n+s} = 0.$$

The indicial equation is obtained by setting the coefficient of the lowest power of x to 0, giving

$$[4s(s-1) + 2s]c_0 = 0.$$

Because $c_0 \neq 0$, we have

$$4s^2 - 2s = 2s(2s - 1) = 0,$$

so this gives $1/2$ and 0 as our two values of s.

The recurrence relation is obtained by setting the coefficients of the remaining powers of x in (10.21) to 0. This gives the first order recurrence relation

$$[4(m + 1 + s)(m + s) + 2(m + 1 + s)]c_{m+1} + c_m = 0, \qquad m = 0, 1, 2, \cdots. \quad \textbf{(10.22)}$$

If we use the larger value of s, $s = 1/2$, in this recurrence relation, we obtain

$$(2m + 3)(2m + 2)c_{m+1} + c_m = 0, \qquad m = 0, 1, 2, \cdots.$$

We have no condition on c_0, so it may be chosen arbitrarily, and the rest of the coefficients are determined in terms of c_0 as

$$
\begin{aligned}
c_{m+1} &= -1/[(2m + 3)(2m + 2)]\, c_m \\
&= -1/[(2m + 3)(2m + 2)]\{-1/[(2m + 1)(2m)]c_{m-1}\} \\
&= (-1)^2/[(2m + 3)(2m + 2)(2m + 1)(2m)]\, c_{m-1} \\
&= (-1)^2/[(2m + 3)(2m + 2)(2m + 1)(2m)]\{-1/[(2m - 1)(2m - 2)]c_{m-2}\} \\
&= (-1)^3/[(2m + 3)(2m + 2)(2m + 1)(2m)(2m - 1)(2m - 2)]\, c_{m-2} \\
&= (-1)^{m+1}/[(2m + 3)(2m + 2)(2m + 1)(2m)...(3)(2)]\, c_0 \\
&= (-1)^{m+1}/[(2m + 3)!]\, c_0, \qquad m = 0, 1, 2, \cdots.
\end{aligned}
$$

This gives the series corresponding to $s = 1/2$ as

$$c_0 + \sum_{m=0}^{\infty} \frac{(-1)^{m+1}}{(2m + 3)!} x^{m+1} c_0 = c_0 \sum_{m=0}^{\infty} \frac{(-1)^m}{(2m + 1)!} x^m,$$

and so a solution of (10.20) may be written as

$$y_1(x) = \sqrt{x} \sum_{m=0}^{\infty} \frac{(-1)^m}{(2m + 1)!} x^m = \sum_{m=0}^{\infty} \frac{(-1)^m}{(2m + 1)!} x^{m+1/2},$$

where we have restricted ourselves to the case $x > 0$. This series can be written in the form

$$y_1(x) = \sum_{m=0}^{\infty} \frac{(-1)^m}{(2m + 1)!} \left(x^{1/2}\right)^{2m+1},$$

which we recognize converges to

$$y_1(x) = \sin \sqrt{x}.$$

Because we have found one solution, we can now use reduction of order techniques to obtain a second solution. Recall from Section 9.2 that when we know one solution, say $y_1(x)$, of

$$a_2(x)\frac{d^2 y}{dx^2} + a_1(x)\frac{dy}{dx} + a_0(x)y = 0,$$

then the second one is

$$y(x) = y_1 \int \left[\frac{1}{y_1^2} \exp\left(-\int \frac{a_1}{a_2} dx \right) \right] dx.$$

In our example, $a_1(x)/a_2(x) = 1/(2x)$, so $\int [a_1(x)/a_2(x)] dx = (\ln x)/2$. This means that

$$y_2(x) = y_1 \int \frac{1}{y_1^2} \exp\left(-\frac{1}{2} \ln x \right) dx$$

$$= y_1(x) \int \frac{1}{\sqrt{x} y_1^2} dx,$$

or

$$y_2(x) = \sin \sqrt{x} \int \frac{1}{\sqrt{x} \sin^2 \sqrt{x}} dx.$$

By using the substitution $u = \sqrt{x}$, this integral can easily be evaluated, giving rise to

$$y_2(x) = -2 \cos \sqrt{x}.$$

Thus, the general solution may be written

$$y(x) = C_1 \sin \sqrt{x} + C_2 \cos \sqrt{x}.$$

Although this is the solution we seek, let's see what happens if we use the other solution of the indicial equation ($s = 0$) in the recurrence relation of (10.22). Doing so gives

$$[4(m+1)(m) + 2(m+1)]c_{m+1} + c_m = (2m+2)(2m+1)c_{m+1} + c_m = 0,$$
$$m = 0, 1, 2, \cdots.$$

Again, c_0 is an arbitrary constant, and the rest of the coefficients, after some calculation, are given by

$$c_{m+1} = \frac{(-1)^{m+1}}{(2m+2)!} c_0, \qquad m = 0, 1, 2, \cdots.$$

Thus, we have the series solution corresponding to $s = 0$ as

$$c_0 + \sum_{m=0}^{\infty} \frac{(-1)^{m+1}}{(2m+2)!} x^{m+1} c_0 = c_0 \sum_{m=0}^{\infty} \frac{(-1)^m}{(2m)!} x^m,$$

which we immediately recognize as the series expansion for the function $c_0 \cos \sqrt{x}$.

Figure 10.2 shows the graphs of $\sin \sqrt{x}$ and $\cos \sqrt{x}$. Notice how the roots are interlaced. $\square$

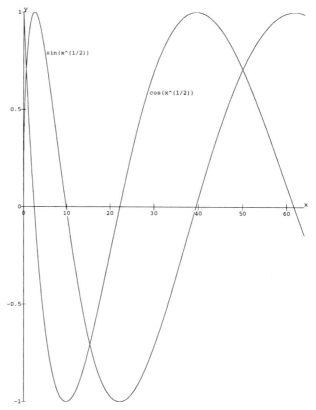

FIGURE 10.2   The functions $\sin x^{1/2}$ and $\cos x^{1/2}$

Comments about using both roots of the indicial equation

- This example, and the previous ones, suggest that we can always obtain the general solution of a differential equation by using both roots of the indicial equation. Unfortunately, that is not always true even if the roots are distinct, because sometimes the smaller root does not yield a series solution different from the first. For example,

$$2xy'' + 6y' - 9xy = 0$$

has two distinct roots of the indicial equation (0 and -2), and each gives the same solution (see Exercise 3 on page 502).

EXAMPLE 10.10: Bessel's Differential Equation of Order 1/2

Bessel's differential equation occurs in many applications, such as vibrations of circular membranes and radiation from right circular cylinders, and is given by

$$x^2 y'' + xy' + (x^2 - \mu^2)y = 0, \tag{10.23}$$

where μ is a constant. On physical grounds we are usually interested only in $x > 0$.

Solutions of this differential equation are called Bessel functions of order μ. It is straightforward to verify that $x = 0$ is a regular singular point of (10.23). In this example, we solve (10.23) for the special case where $\mu^2 = 1/4$ — namely,

$$x^2 y'' + xy' + (x^2 - \frac{1}{4})y = 0. \tag{10.24}$$

Because $x = 0$ is a regular singular point of (10.24), we substitute the proper form of a series solution, $y(x) = x^s \sum_{n=0}^{n=\infty} c_n x^n$, into (10.24) and obtain

$$\sum_{n=0}^{\infty}(n+s)(n+s-1)c_n x^{n+s} + \sum_{n=0}^{\infty}(n+s)c_n x^{n+s} + \sum_{n=0}^{\infty} c_n x^{n+s+2} - \frac{1}{4}\sum_{n=0}^{\infty} c_n x^{n+s} = 0.$$

(10.25)

The indicial equation is found by setting the coefficient of the lowest power of x — namely, x^s — to zero and is given by

$$s(s-1) + s - \frac{1}{4} = s^2 - \frac{1}{4} = 0.$$

This gives $1/2$ and $-1/2$ as roots of the indicial equation.

In (10.25) we see that three of the series have the same power of x, so we lower the index of summation in the remaining series to obtain

$$\left[s^2 - \frac{1}{4}\right]c_0 x^s + \left[(s+1)^2 - \frac{1}{4}\right]c_1 x^{s+1} + \sum_{n=2}^{\infty}\left[(n+s)^2 - \frac{1}{4}\right]c_n x^{n+s} + \sum_{n=2}^{\infty} c_{n-2} x^{n+s} = 0.$$

If we now consider the smaller root of the indicial equation — namely, $s = -1/2$ — we see that the term multiplying c_0 is zero. (Explain why this must be.) The term multiplying c_1 is also 0, giving both c_0 and c_1 as arbitrary constants. The remaining coefficients are obtained from the second order recurrence relation

$$c_n = \frac{-1}{(n-1/2)^2 - 1/4}c_{n-2} = \frac{-1}{n(n-1)}c_{n-2}, \qquad n = 2, 3, 4, \cdots.$$

Using this recurrence relation to obtain a general form for the coefficients gives

$$\begin{aligned}
c_n &= -1/[n(n-1)]c_{n-2}\\
&= -1/[n(n-1)]\{-1/[(n-2)(n-3)]c_{n-4}\}\\
&= (-1)^2/[n(n-1)(n-2)(n-3)]\,c_{n-4}\\
&= (-1)^2/[n(n-1)(n-2)(n-3)]\{-1/[(n-4)(n-5)]c_{n-6}\}\\
&= (-1)^3/[n(n-1)(n-2)(n-3)(n-4)(n-5)]\,c_{n-6}\\
&= (-1)^3/[n(n-1)(n-2)(n-3)(n-4)(n-5)]\{-1/[(n-6)(n-7)]c_{n-8}\}\\
&= (-1)^4/[n(n-1)(n-2)(n-3)(n-4)(n-5)(n-6)(n-7)]\,c_{n-8}.
\end{aligned}$$

If n is an even integer, all the coefficients will be found in terms of c_0, whereas if n is an odd integer, they will be in terms of c_1. Thus, we have

$$c_{2m} = \frac{(-1)^m}{(2m)!}c_0, \text{ and } c_{2m+1} = \frac{(-1)^m}{(2m+1)!}c_1,$$

and the general solution of (10.24) may be written as

$$y(x) = x^{-1/2}[c_0 + c_0 \sum_{m=1}^{\infty}\frac{(-1)^m}{(2m)!}x^{2m} + c_1 x + c_1 \sum_{m=1}^{\infty}\frac{(-1)^m}{(2m+1)!}x^{2m+1}].$$

A more compact way of writing these series is

$$y(x) = c_0 x^{-1/2}\sum_{m=0}^{\infty}\frac{(-1)^m}{(2m)!}x^{2m} + c_1 x^{-1/2}\sum_{m=0}^{\infty}\frac{(-1)^m}{(2m+1)!}x^{2m+1}. \qquad (10.26)$$

We note that the two series in (10.26) seem familiar. In fact they are the Taylor expansions for the cosine and sine functions, so we may write (10.26) as

$$y(x) = c_0 \frac{1}{\sqrt{x}} \cos x + c_1 \frac{1}{\sqrt{x}} \sin x.$$

Because c_0 and c_1 are undetermined, and because $x^{-1/2} \cos x$ and $x^{-1/2} \sin x$ are linearly independent, this is the general solution. This form of our solution is in agreement with the conclusion of Exercise 3 of Section 9.1 that all solutions of (10.24) oscillate.

Solutions of Bessel's differential equation with parameter $\mu = 1/2 + m$, where m is an integer, and specific values for the constants c_0 and c_1 are called SPHERICAL BESSEL FUNCTIONS. Choosing $c_0 = \sqrt{2/\pi}$ and $c_1 = 0$, and then $c_0 = 0$ and $c_1 = \sqrt{2/\pi}$, gives the spherical Bessel functions

$$J_{-1/2}(x) = \sqrt{\frac{2}{\pi x}} \cos x \text{ and } J_{1/2}(x) = \sqrt{\frac{2}{\pi x}} \sin x.$$

These functions are plotted in Figure 10.3. Notice how the roots are interlaced as predicted by Theorem 9.2, see Exercise 3 on page 437.   □

Comments about using one root of the indicial equation

- It may have come as a surprise that we obtained the general solution of our differential equation by only considering one root of the indicial equation. It turns out that in cases where the two roots of the indicial equation differ by an integer, choosing the smaller root of the indicial equation sometimes gives the general solution. For example, we consider the case where the two roots of the indicial equation are -1 and 1. Taking the smaller root our solution would have the form

$$c_0 x^{-1} + c_1 + c_2 x + c_3 x^2 + c_4 x^3 + \cdots.$$

  If, from the recurrence relation associated with $s = -1$, we find that c_1 is determined in terms of c_0, and c_3, c_4, and so on, are determined in terms of c_2, then c_0 and c_2 are the arbitrary constants, and we have found the general solution. The series which is the coefficient of c_2 would have been the same series we would have obtained if we had used the larger root of the indicial equation $s = 1$. Unfortunately, this situation does not always occur.

EXAMPLE 10.11

Find the first few terms in the series solution about 0 of the differential equation from Example 10.7,

$$3x^2 y'' + x y' - \frac{2}{2 - x} y = 0. \tag{10.27}$$

In Example 10.7 we discovered that $x = 0$ is a regular singular point and that a series solution will converge at least for $-2 < x < 2$, except possibly at $x = 0$. We also determined roots of the indicial equation as 1 and $-1/3$.

Let us now find the recurrence relation by substituting the series expression $x^s \sum_{n=0}^{\infty} c_n x^n$ into (10.27) and using the expansion

$$\frac{2}{2 - x} = \frac{2}{2(1 - x/2)} = 1 + \frac{x}{2} - \left(\frac{x}{2}\right)^2 + \cdots.$$

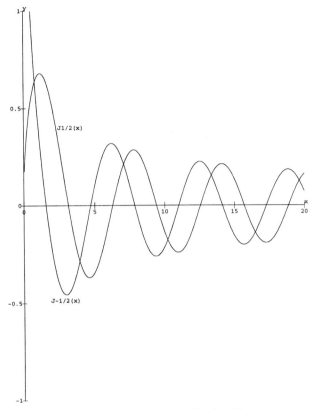

FIGURE 10.3 The Bessel functions of order 1/2

This gives

$$\sum_{n=0}^{\infty} 3(n+s)(n+s-1)c_n x^{n+s} +$$

$$+ \sum_{n=0}^{\infty}(n+s)c_n x^{n+s} - \left[1 + \frac{x}{2} - \left(\frac{x}{2}\right)^2 + \cdots\right]\sum_{n=0}^{\infty} c_n x^{n+s} = 0,$$

or

$$\sum_{n=0}^{\infty}(n+s)(3n+3s-2)c_n x^{n+s} - \left[1 + \frac{x}{2} - \left(\frac{x}{2}\right)^2 + \cdots\right]\sum_{n=0}^{\infty} c_n x^{n+s} = 0.$$

We see that the coefficient of the lowest power of x, $s(3s-2)-1$, equals zero for $s = -1/3$ and $s = 1$. (Explain why.) Equating the coefficients of the next few powers of x to zero gives

$$(1+s)(3+3s-2)c_1 - (c_1 + \frac{1}{2}c_0) = 0,$$

$$(2+s)(6+3s-2)c_2 - \left(c_2 + \frac{1}{2}c_1 - \frac{1}{4}c_0\right) = 0,$$

$$(3+s)(9+3s-2)c_3 - \left(c_3 + \frac{1}{2}c_2 - \frac{1}{4}c_1 + \frac{1}{8}c_0\right) = 0.$$

For both values of s, there will no condition specified for c_0, so it is an arbitrary constant, and all the other constants will be obtained in terms of c_0. For the larger root, $s = 1$, we have

$$c_1 = \frac{1}{14}c_0, \quad c_2 = \frac{1}{5 \cdot 14}c_0, \quad c_3 = \frac{1}{280 \cdot 39}c_0,$$

so the first four terms in this part of our solution are

$$x\left(1 + \frac{1}{14}x + \frac{1}{70}x^2 + \frac{1}{10920}x^3\right)c_0.$$

Similar calculations for the smaller root, $s = -1/3$, lead to a solution of the form

$$x^{-1/3}\left(1 - \frac{1}{2}x - \frac{1}{8}x^2 + \frac{1}{80}x^3\right)c_0.$$

These two series are not proportional, and so they must be linearly independent. Thus, the general solution of (10.27) has the form

$$y(x) = C_1 x\left(1 + \frac{1}{14}x + \frac{1}{70}x^2 + \frac{1}{10920}x^3 + \cdots\right) +$$

$$+ C_2 x^{-1/3}\left(1 - \frac{1}{2}x - \frac{1}{8}x^2 + \frac{1}{80}x^3 + \cdots\right). \qquad \square$$

EXERCISES

1. Use the method of Frobenius to find general solutions to the following differential equations. For series for which the formula for the nth term is not apparent, find terms in the series expansion for the solution up to x^4.

   (a) $4xy'' + 2(1 + x)y' + y = 0$

   (b) $xy'' - (3 + x)y' + 2y = 0$

   (c) $3xy'' + y' - y = 0$

   (d) $xy'' + (x^3 - 1)y' + x^2 y = 0$

   (e) $4xy'' + 2y' + y = 0$

   (f) $4x^2 y'' - 4xy' + (3 - 4x^2)y = 0$

   (g) $2x^2 y'' + (2x^2 + x)y' + 2xy = 0$

   (h) $2x^2 y'' - xy' + (x - 5)y = 0$

   (i) $4x^2 y'' + (4x - 2x^2)y' - (25 + 3x)y = 0$

2. Find the general solution of the following differential equations using the method of Frobenius with expansions about the given point.

   (a) $2(x - 1)y'' - y' + e^x y = 0, \qquad x_0 = 1$

   (b) $2(x - 1)^2 y'' + (5x - 5)y' + xy = 0, \qquad x_0 = 1$

   (c) $2(x + 1)^2 y'' - 3(x^2 + 3x + 2)y' + (3x + 5)y = 0, \qquad x_0 = -1$

   (d) $x(x + 1)^2 y'' - (x^2 + 3x + 2)y' + 9y = 0, \qquad x_0 = -1$

   (e) $(x - 1)^2 y'' - (x - 1)(x^2 - 2x)y' + (2x - x^2)y = 0, \qquad x_0 = 1$

3. Concerning the differential equation

   $$2xy'' + 6y' - 9xy = 0,$$

   (a) Show that $x = 0$ is a regular singular point.

(b) Find the indicial equation associated with a series solution using the method of Frobenius.

(c) Find the series solution associated with the larger root of the indicial equation.

(d) Show that this method for the smaller root of the indicial equation gives the same solution as in part (c).

(e) Show that any solution of this differential equation has at most one zero for $x > 0$.

4. Develop a *How to Use the Method of Frobenius* page that encompasses the use of Theorem 10.1 by adding statements under Purpose, Technique, and Comments that summarize what you discovered in this section.

10.3   The Method of Frobenius, Part 2

In most of the examples and exercises in Section 10.2, using the two roots of the indicial equation gave series solutions in the form of two linearly independent functions. However, this is not always the case. We now consider such an example.

EXAMPLE 10.12

Find the general solution of

$$x(x-1)\frac{d^2y}{dx^2} + (5x-2)\frac{dy}{dx} + 4y = 0 \qquad (10.28)$$

as a series about the origin using the method of Frobenius.

If we divide this expression by $x(x-1)$ we obtain

$$\frac{d^2y}{dx^2} + \frac{5x-2}{x(x-1)}\frac{dy}{dx} + \frac{4}{x(x-1)}y = 0.$$

Here we see that both $xp(x) = x(5x-2)/[x(x-1)] = (5x-2)/(x-1)$ and $x^2q(x) = x^2 4/[x(x-1)] = 4x/(x-1)$ are analytic at $x = 0$, giving the origin as a regular singular point. Because the radius of convergence for both analytic functions is $R = 1$, we expect our solution to be valid for $0 < x < 1$.

To find the indicial equation, we substitute $y = x^s$ into the left-hand side of (10.28) and obtain

$$(x^2 - x)s(s-1)x^{s-2} + (5x-2)sx^{s-1} + 4x^s$$
$$= s(s-1)x^s - s(s-1)x^{s-1} + 5sx^s - 2sx^{s-1} + 4x^s.$$

From this equation we see that the indicial equation (obtained by setting the coefficient of the lowest power of s to 0) is

$$-s(s-1) - 2s = -s^2 - s = -s(s+1) = 0.$$

Thus, we have $s = 0$ and $s = -1$ as the roots, and the method of Frobenius guarantees us one solution if we use the larger root $s = 0$.

We substitute the series

$$y(x) = x^0 \sum_{n=0}^{\infty} c_n x^n = \sum_{n=0}^{\infty} c_n x^n$$

into (10.28) and find that

$$\left(x^2 - x\right) \sum_{n=0}^{\infty} n(n-1)c_n x^{n-2} + (5x - 2) \sum_{n=0}^{\infty} nc_n x^{n-1} + 4 \sum_{n=0}^{\infty} c_n x^n = 0,$$

or

$$\sum_{n=0}^{\infty} n(n-1)c_n x^n - \sum_{n=0}^{\infty} n(n-1)c_n x^{n-1} +$$

$$+ \sum_{n=0}^{\infty} 5nc_n x^n - \sum_{n=0}^{\infty} 2nc_n x^{n-1} + \sum_{n=0}^{\infty} 4c_n x^n = 0.$$

If we now combine series with like exponents, we obtain

$$\sum_{n=0}^{\infty} [n(n-1) + 5n + 4]c_n x^n - \sum_{n=0}^{\infty} [n(n-1) + 2n]c_n x^{n-1}$$

$$= \sum_{n=0}^{\infty} (n+2)^2 c_n x^n - \sum_{n=0}^{\infty} n(n+1) c_n x^{n-1} = 0.$$

We now change the dummy index in the second series and obtain

$$\sum_{n=0}^{\infty} (n+2)^2 c_n x^n - \sum_{n=0}^{\infty} (n+1)(n+2) c_{n+1} x^n$$

$$= \sum_{n=0}^{\infty} (n+2) [(n+2)c_n - (n+1) c_{n+1}]x^n = 0.$$

This gives the first order recurrence relation as

$$(n+1)c_{n+1} - (n+2) c_n = 0, \qquad n = 0, 1, 2, \cdots$$

or

$$c_{n+1} = \frac{n+2}{n+1}c_n, \qquad n = 0, 1, 2, \cdots.$$

From this we see that

$$c_{n+1} = \frac{n+2}{n+1}c_n = \frac{n+2}{n+1}\frac{n+1}{n}c_{n-1} = \frac{n+2}{n+1}\frac{n+1}{n}\frac{n}{n-1}c_{n-2} = \cdots = (n+2)c_0,$$

and our solution is

$$y_1(x) = c_0 \left(1 + 2x + 3x^2 + 4x^3 + \cdots\right).$$

It is possible to express this series in terms of familiar functions, as follows. Because

$$\frac{1}{1-x} = 1 + x + x^2 + x^3 + x^4 + \cdots,$$

by differentiating we obtain

$$\frac{1}{(1-x)^2} = 1 + 2x + 3x^2 + 4x^3 + \cdots,$$

so our first solution is simply

$$y_1(x) = \frac{1}{(1-x)^2}.$$

Exercise 15 on page 515 shows that the smaller root of the indicial equation, $s = -1$, does not give a second solution, but just gives us the same $y_1(x)$ again. However, because we have obtained one solution, we now use reduction of order techniques to obtain a second solution. Recall from Section 9.2 that when we know one solution, say $y_1(x)$, of

$$a_2(x)\frac{d^2y}{dx^2} + a_1(x)\frac{dy}{dx} + a_0(x)y = 0,$$

then the second one is

$$y_2(x) = y_1 \int \left[\frac{1}{y_1^2} \exp\left(-\int \frac{a_1}{a_2} dx \right) \right] dx.$$

In our example, $a_1(x)/a_2(x) = (5x - 2)/[x(x-1)]$, so

$$\int \frac{a_1}{a_2} dx = \int \frac{5x - 2}{x(x-1)} dx$$

$$= \int \frac{2}{x} dx - \int \frac{3}{1-x} dx$$

$$= 2\ln x + 3\ln(1-x), \qquad \text{if } 0 < x < 1,$$

$$= \ln\left[x^2 (1-x)^3 \right].$$

This means that

$$y_2(x) = y_1(x) \int \frac{1}{y_1^2} \exp\left(-\ln\left[x^2(1-x)^3 \right] \right) dx$$

$$= \frac{1}{(1-x)^2} \int \frac{(1-x)^4}{x^2(1-x)^3} dx$$

$$= \frac{1}{(1-x)^2} \int \frac{1-x}{x^2} dx,$$

or

$$y_2(x) = \frac{1}{(1-x)^2} \left(-\frac{1}{x} - \ln x \right). \qquad (10.29)$$

Thus, the general solution of (10.28) is

$$y(x) = C_1 y_1(x) + C_2 y_2(x),$$

so that

$$y(x) = C_1 \frac{1}{(1-x)^2} - C_2 \left[\frac{1}{(1-x)^2} \ln x + \frac{1}{x} \frac{1}{(1-x)^2} \right]. \qquad (10.30)$$

The functions $y_1(x)$ and $-y_2(x)$ are plotted in Figure 10.4.

We used reduction of order because we could not obtain a second series solution by using the second root. However, because usually both of these solutions will be series, we

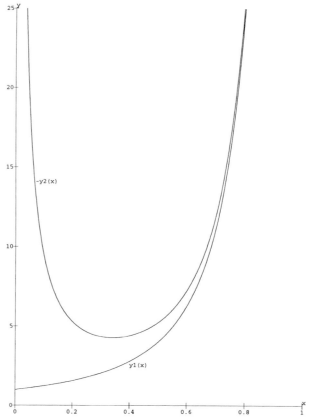

FIGURE 10.4   Solutions of $x(x-1)y'' + (5x-2)y' + 4y = 0$

expand $y_2(x)$ so we can see its form as an infinite series. From our previous calculations we have

$$\frac{1}{(1-x)^2} = 1 + 2x + 3x^2 + 4x^3 + \cdots,$$

so we can write our solution as

$$y(x) = C_1 y_1(x) - C_2 \left[y_1(x) \ln x + x^{s_2} \left(1 + 2x + 3x^2 + 4x^3 + \cdots \right) \right], \qquad \textbf{(10.31)}$$

where s_2 is the smaller root of the indicial equation and $y_1(x)$ is the solution corresponding to the larger root s_1.   ☐

A similar situation, in which we obtain only one solution from $y(x) = x^s \sum_{n=0}^{\infty} c_n x^n$, arises if the indicial equation has equal roots. For example (see Exercise 16 on page 515) the differential equation

$$x^2 \frac{d^2 y}{dx^2} + (x^2 - 3x)\frac{dy}{dx} + (-2x + 4)\, y = 0$$

has $s_1 = s_2 = 2$ as the repeated root of the indicial equation associated with the regular singular point $x_0 = 0$. Using $s = 2$, we find one solution as $y_1(x) = x^2$. The second solution can be obtained by reduction of order, and we have

$$y_2(x) = x^2 \ln x + x^2 \sum_{n=1}^{\infty} \frac{(-1)^n}{n!n} x^n.$$

This general solution can be written in a form similar to (10.31) — namely,

$$y(x) = C_1 y_1(x) + C_2 \left[y_1(x) \ln x + x^s \sum_{n=1}^{\infty} \frac{(-1)^n}{n!n} x^n \right].$$

This procedure works to find a second solution as long as we can find one solution of the differential equation. The last two examples gave simple forms of the solution because the simple forms of $y_1(x)$, $1/(x-1)^2$ and x^2, made it very easy to divide by $[y_1(x)]^2$. In general, when we use the method of Frobenius our first solution will be in the form of an infinite series that we cannot express as a familiar function. Although we can divide by an infinite series to find the second solution by reduction of order, it is not the simplest of operations. What we can do is rely on the following theorem, which gives an alternative to using reduction of order to obtain the second solution.

Theorem 10.3:   *Consider the second order differential equation*

$$a_2(x) \frac{d^2 y}{dx^2} + a_1(x) \frac{dy}{dx} + a_0(x)y = 0, \tag{10.32}$$

with $x = 0$ as a regular singular point. Let s_1 and s_2 be solutions of the indicial equation with $s_1 \geq s_2$. Then there are two linearly independent solutions of (10.32) — namely, $y_1(x)$ and $y_2(x)$ — whose series converge for at least $0 < x < R$, where R is the smaller radius of convergence[2] for $x a_1(x)/a_2(x)$ and $x^2 a_0(x)/a_2(x)$. For all cases

$$y_1(x) = x^{s_1} \sum_{n=0}^{\infty} c_n x^n,$$

while the form for $y_2(x)$ depends on the relationship between the roots of the indicial equation, as follows.

(a) If $s_1 - s_2 \neq$ integer, then

$$y_2(x) = x^{s_2} \sum_{n=0}^{\infty} c_n x^n. \tag{10.33}$$

(b) If $s_1 = s_2$, then

$$y_2(x) = y_1(x) \ln x + x^{s_1} \sum_{n=1}^{\infty} c_n x^n. \tag{10.34}$$

---

[2] Similar results hold for $-R < x < 0$.

(c) If $s_1 - s_2 = $ integer, then

$$y_2(x) = Cy_1(x) \ln x + x^{s_2}(1 + \sum_{n=1}^{\infty} c_n x^n), \qquad \textbf{(10.35)}$$

where, depending on the differential equation, the constant C may or may not be zero.

Comments about Theorem 10.3

- If the regular singular point is at x_0, use the translation $X = x - x_0$ to move the singularity to the origin of the X-axis.

- Note that for the Bessel functions in Example 10.10, the indicial equation had roots $1/2$ and $-1/2$, which differ by an integer. Thus, the second solution, corresponding to $s = -1/2$, has the form of (10.35) with $C = 0$.

- Note that the structure of the second solution in (10.30) has the form of (10.35), with $C = 1$.

- It is worthwhile to return to all the examples in Sections 10.1, 10.2, and 10.3 and see exactly to which case each example belongs.

How to Use the Method of Frobenius

Purpose: To find a series solution of

$$a^2(x)\frac{d^2y}{dx^2} + a_1(x)\frac{dy}{dx} + a_0(x)y = 0 \qquad \textbf{(10.36)}$$

of the form $x^s \sum_{n=0}^{\infty} c_n x^n$, where $x = 0$ is a regular singular point of (10.36).

Technique:

1. Verify that 0 is a regular singular point of (10.36), and use Theorem 10.1 to determine the minimum radius of convergence of the series solution we are seeking.

2. Determine the indicial equation, and solve for its two roots, s_1 and s_2, where $s_1 \geq s_2$. The indicial equation may be determined by using Theorem 10.2, or Exercise 4, Section 10.1, or by equating the lowest power of x to zero in the result of step 3 below.

3. Determine the recurrence relation. This is obtained by substituting the series expression $x^{s_1} \sum_{n=0}^{\infty} c_n x^n$ into (10.36) and equating the coefficient of each power of x to zero.

4. Use s_1 in the recurrence relation to find a general expression for c_n, or as many terms as desired, giving $y_1(x)$.

5. To find the second solution corresponding to s_2 there are three cases.

   (a) If $s_1 - s_2 \neq$ integer, then assume a solution of the form (10.33) and return to steps 3 and 4, using (10.33) in place of $x^{s_1} \sum_{n=0}^{\infty} c_n x^n$ and s_2 in place of s_1.

   (b) If $s_1 = s_2$, then assume a solution of the form (10.34) and return to steps 3 and 4, using (10.34) in place of $x^{s_1} \sum_{n=0}^{\infty} c_n x^n$.

   (c) If $s_1 - s_2 =$ integer, then assume a solution of the form (10.35) and return to steps 3 and 4, using (10.35) in place of $x^{s_1} \sum_{n=0}^{\infty} c_n x^n$. Depending on the differential equation, the constant C may or may not be zero.

In Example 10.12 we found a second solution to a differential equation using reduction of order. In our next example, we use the preceding results to find the second solution.

EXAMPLE 10.13

Find the general solution of

$$x\frac{d^2y}{dx^2} + \frac{dy}{dx} + 4xy = 0 \tag{10.37}$$

as a series about the origin using the method of Frobenius.

If we divide this expression by x we obtain

$$\frac{d^2y}{dx^2} + \frac{1}{x}\frac{dy}{dx} + 4y = 0.$$

Here we see that both $xp(x) = 1$ and $x^2q(x) = 4x^2$ are analytic at $x = 0$, giving the origin as a regular singular point.

To find the indicial equation, we substitute $y = x^s$ into the left-hand side of (10.37) and obtain

$$xs(s-1)x^{s-2} + sx^{s-1} + (4x)x^s = s(s-1)x^{s-1} + sx^{s-1} + 4x^{s+1}.$$

From here we see that the indicial equation (obtained by setting the coefficient of the lowest power of x to 0) is

$$s(s-1) + s = s^2 = 0.$$

Thus, we have $s = 0$ as a double root, and we proceed to find the associated solution.

We substitute the series

$$y = \sum_{n=0}^{\infty} c_n x^n$$

into (10.37) and find that

$$x\sum_{n=0}^{\infty} n(n-1)c_n x^{n-2} + \sum_{n=0}^{\infty} nc_n x^{n-1} + 4x\sum_{n=0}^{\infty} c_n x^n$$

$$= \sum_{n=0}^{\infty} n(n-1)c_n x^{n-1} + \sum_{n=0}^{\infty} nc_n x^{n-1} + \sum_{n=0}^{\infty} 4c_n x^{n+1} = 0.$$

If we now combine series with like exponents, we obtain

$$\sum_{n=1}^{\infty} n^2 c_n x^{n-1} + \sum_{n=0}^{\infty} 4c_n x^{n+1} = 0.$$

We now separate out the first term in the first series and lower by 2 the dummy index of summation in the second series above to obtain

$$c_1 x^0 + \sum_{n=2}^{\infty} n^2 c_n x^{n-1} + \sum_{n=2}^{\infty} 4c_{n-2} x^{n-1} = 0.$$

Because this equation must be an identity x, we see that we must have $c_1 = 0$, and the recurrence relation becomes

$$n^2 c_n + 4c_{n-2} = 0, \qquad n = 2, 3, 4, \cdots.$$

We now rearrange the recurrence relation as

$$c_n = -\frac{4}{n^2} c_{n-2}, \qquad n = 2, 3, 4, \cdots.$$

Because $c_1 = 0$, all the odd coefficients equal 0. The even ones are obtained from

$$c_{2m} = \frac{-4}{(2m)^2} c_{2m-2}, \qquad m = 1, 2, 3, \cdots.$$

Thus, we have

$$c_{2m} = \frac{(-4)^m}{(2m)^2(2m-2)^2(2m-4)^2 \cdots 6^2 \cdot 4^2 \cdot 2^2} c_0, \qquad m = 1, 2, 3, \cdots.$$

This expression may be written in a more compact form by noting that each term in the denominator is an even integer, so we can factor out 2 from each term, leaving us with the result that

$$c_{2m} = \frac{(-4)^m}{(2^2)^m \, m^2(m-1)^2(m-2)^2 \cdots 3^2 \cdot 2^2 \cdot 1^2} c_0 = \frac{(-1)^m}{(m!)^2} c_0, \qquad m = 1, 2, 3, \cdots.$$

This gives one solution of our original differential equation as

$$y_1(x) = 1 + \sum_{m=1}^{\infty} \frac{(-1)^m}{(m!)^2} x^{2m} = \sum_{m=0}^{\infty} \frac{(-1)^m}{(m!)^2} x^{2m}, \tag{10.38}$$

where we have set $c_0 = 1$.

According to Theorem 10.3, our second solution will have the form

$$y_2(x) = y_1(x) \ln x + \sum_{n=1}^{\infty} c_n x^n, \tag{10.39}$$

where $y_1(x)$ is given by (10.38) [see (10.34)]. Because the exact form of $y_1(x)$ is not needed for the first steps in this procedure, we will not use its series expansion until necessary. We first differentiate (10.39) to obtain

$$y_2'(x) = y_1'(x) \ln x + \frac{1}{x} y_1(x) + \sum_{n=1}^{\infty} n c_n x^{n-1},$$

$$y_2''(x) = y_1''(x) \ln x + \frac{1}{x} 2 y_1'(x) - \frac{1}{x^2} y_1(x) + \sum_{n=1}^{\infty} n(n-1) c_n x^{n-1}.$$

Substituting these values into (10.37) and rearranging terms gives

$$\left(x y_1'' + y_1' + 4x y_1 \right) \ln x + 2 y_1' + \sum_{n=1}^{\infty} n(n-1) c_n x^{n-1} + \sum_{n=1}^{\infty} n c_n x^{n-1} + \sum_{n=1}^{\infty} 4 c_n x^{n+1} = 0.$$

Because $y_1(x)$ is a solution of (10.37), the coefficient of $\ln x$ is 0, and the rest of the expression is

$$2y_1' + \sum_{n=1}^{\infty} n^2 c_n x^{n-1} + \sum_{n=1}^{\infty} 4c_n x^{n+1} = 2y_1' + c_1 + 4c_2 x + \sum_{n=3}^{\infty} \left(n^2 c_n + 4c_{n-2}\right) x^{n-1} = 0,$$

where we combined the last two series as before.

We now differentiate our series expression for y_1 from (10.38) and use the result in our last equation above to obtain

$$\sum_{m=1}^{\infty} (-1)^m \frac{4m}{(m!)^2} x^{2m-1} + c_1 + 4c_2 x + \sum_{n=3}^{\infty} \left(n^2 c_n + 4c_{n-2}\right) x^{n-1} = 0,$$

or

$$-4x + \sum_{m=2}^{\infty} (-1)^m \frac{4m}{(m!)^2} x^{2m-1} + c_1 + 4c_2 x + \sum_{n=3}^{\infty} \left(n^2 c_n + 4c_{n-2}\right) x^{n-1} = 0.$$

Equating coefficients of like powers of x to zero gives us

$$c_1 = 0, \qquad -4 + 4c_2 = 0, \qquad n^2 c_n + 4c_{n-2} = 0, \quad n = 3, 5, 7, \cdots,$$

and

$$(-1)^m \frac{4m}{(m!)^2} + (2m)^2 c_{2m} + 4c_{2m-2} = 0, \quad m = 2, 3, 4, \cdots.$$

Thus, all the odd coefficients are equal to zero, and the first few even coefficients are given by

$$c_2 = 1,$$
$$c_4 = \frac{1}{16}\left(-4c_2 - \frac{8}{(2!)^2}\right) = -\frac{3}{8},$$
$$c_6 = \frac{1}{36}\left(-4c_4 + \frac{12}{(3!)^2}\right) = \frac{11}{216},$$
$$c_8 = \frac{1}{64}\left(-4c_6 - \frac{16}{(4!)^2}\right) = -\frac{25}{6912}.$$

This gives our second solution the form

$$y_2(x) = y_1(x) \ln x + x^2 - \frac{3}{8}x^4 + \frac{11}{216}x^6 - \frac{25}{6912}x^8 + \cdots. \tag{10.40}$$

The general solution of (10.37) is therefore given by

$$y(x) = C_1 y_1(x) + C_2 y_2(x),$$

where C_1 and C_2 are arbitrary constants, and $y_1(x)$ and $y_2(x)$ are given by (10.38) and (10.40).

Because $x a_1(x)/a_2(x) = 1$ and $x^2 a_0 x/a_2(x) = 4x^2$ are both polynomials, both series in our solution converge for $0 < x < \infty$. Using Theorems 9.2 and 9.3 with $Q(x) = 4 - 1/2(-1/x)^2 - 1/4(1/x)^2 = 4 + 1/(4x^2) > 0$, we know our solution oscillates for both positive and negative values of x, something not possible to tell from the series solution. The first few terms in our solution for $y_1(x)$ are shown in Figure 10.5. Notice how the oscillatory behavior is beginning to emerge. $\qquad\square$

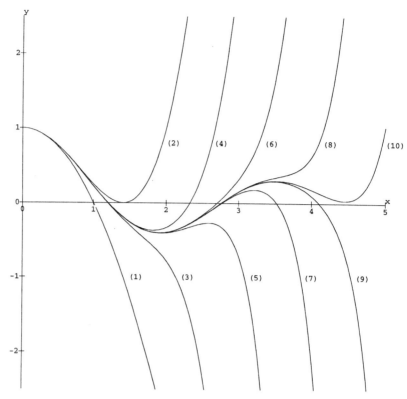

FIGURE 10.5 Approximations to $y_1(x)$, a solution of $xy'' + y' + 4xy = 0$

We conclude this section with an example in which the roots of the indicial equation differ by a positive integer and in which the C in (10.35) of Theorem 10.3 is not zero.

EXAMPLE 10.14

To find the general solution of

$$x\frac{d^2y}{dx^2} + (x-1)\frac{dy}{dx} - 2y = 0 \tag{10.41}$$

as a series about $x = 0$, we first note that $x = 0$ is a regular singular point. Also, because $xp(x) = x - 1$ and $x^2 q(x) = -2x$ are both polynomials, the series solutions we obtain will converge for all values of x.

We now substitute x^s into (10.41) and obtain

$$xs(s-1)x^{s-2} + (x-1)sx^{s-1} - 2x^s = s(s-1)x^{s-1} + sx^s - sx^{s-1} - 2x^s.$$

Thus, we have the indicial equation (the coefficient of the lowest power of x) given as

$$s^2 - s - s = s(s-2) = 0,$$

with roots of 0 and 2. If we let $s = 2$, and substitute the series $\sum_{n=0}^{\infty} c_n x^{n+2}$ into (10.41) we obtain

$$x\sum_{n=0}^{\infty}(n+2)(n+1)c_n x^n + x\sum_{n=0}^{\infty}(n+2)c_n x^{n+1} - \sum_{n=0}^{\infty}(n+2)c_n x^{n+1} - 2\sum_{n=0}^{\infty}c_n x^{n+2} = 0,$$

or

$$\sum_{n=0}^{\infty}(n+2)(n+1)c_n x^{n+1} + \sum_{n=0}^{\infty}(n+2)c_n x^{n+2} - \sum_{n=0}^{\infty}(n+2)c_n x^{n+1} - \sum_{n=0}^{\infty}2c_n x^{n+2} = 0.$$

If we now combine the terms with like powers of x, we obtain

$$\sum_{n=0}^{\infty}(n+2)nc_n x^{n+1} + \sum_{n=0}^{\infty}nc_n x^{n+2} = 0.$$

To combine the two series, we lower the index of summation on the second series to obtain

$$\sum_{n=0}^{\infty}(n+2)nc_n x^{n+1} + \sum_{n=1}^{\infty}(n-1)c_{n-1} x^{n+1} = 0$$

so we may combine the two series into one as

$$\sum_{n=1}^{\infty}\left[(n+2)nc_n + (n-1)c_{n-1}\right] x^{n+1} = 0.$$

This gives the first order recurrence relation as

$$c_n = -\frac{n-1}{(n+2)n}c_{n-1}, \qquad n = 1, 2, 3, \cdots.$$

Now c_0 is arbitrary, and from the recurrence relation we have that $c_1 = 0$ and all the remaining coefficients are also zero. If we choose $c_0 = 1$, we obtain our first solution of (10.41) as

$$y_1(x) = x^2.$$

The proper form for our second solution is found in (10.35), namely,

$$y_2(x) = Cx^2 \ln x + x^0 \left(1 + \sum_{n=1}^{\infty}c_n x^n\right).$$

Differentiation of this expression gives

$$y_2'(x) = C(2x \ln x + x) + \sum_{n=1}^{\infty}nc_n x^{n-1},$$

$$y_2''(x) = C(2\ln x + 3) + \sum_{n=2}^{\infty}n(n-1)c_n x^{n-2},$$

and substitution of these results into (10.41) gives

$$x\left[C(2\ln x + 3) + \sum_{n=2}^{\infty}n(n-1)c_n x^{n-2}\right] +$$

$$+ (x-1)\left[C(2x \ln x + x) + \sum_{n=1}^{\infty}nc_n x^{n-1}\right] +$$

$$- 2\left[Cx^2 \ln x + 1 + \sum_{n=1}^{\infty}c_n x^n\right] = 0.$$

Combining a few terms gives

$$-2 + 2Cx + Cx^2 + \sum_{n=2}^{\infty} n(n-1)c_n x^{n-1} + \sum_{n=1}^{\infty} nc_n x^n - \sum_{n=1}^{\infty} nc_n x^{n-1} - 2\sum_{n=1}^{\infty} c_n x^n = 0$$

or

$$-2 + 2Cx + Cx^2 + \sum_{n=1}^{\infty} n(n-2)c_n x^{n-1} + \sum_{n=1}^{\infty} (n-2)c_n x^n = 0,$$

which can be written

$$-2 + 2Cx + Cx^2 + \sum_{n=1}^{\infty} n(n-2)c_n x^{n-1} + \sum_{n=2}^{\infty} (n-3)c_{n-1} x^{n-1} = 0.$$

We now set the coefficients of like powers of x to zero to obtain

$$-2 - c_1 = 0,$$
$$2C - c_1 = 0,$$
$$C + 3c_3 + 0c_2 = 0,$$
$$n(n-2)c_n = -(n-3)c_{n-1}, \qquad n = 4, 5, 6, \cdots.$$

This gives the coefficients as $c_1 = -2$, $C = -1$, c_2 is arbitrary, $c_3 = 1/3$,

$$c_4 = -\frac{1}{4 \cdot 2} c_3 = -\frac{1}{4 \cdot 3 \cdot 2},$$
$$c_5 = -\frac{2}{5 \cdot 3} c_4 = \frac{(-2)(-1)}{5 \cdot 4 \cdot 3 \cdot 2 \cdot 3},$$
$$c_6 = -\frac{3}{6 \cdot 4} c_5 = \frac{(-3)(-2)(-1)}{6 \cdot 5 \cdot 4 \cdot 3 \cdot 2 \cdot 4 \cdot 3},$$
$$c_n = (-1)^{n-3} \frac{2(n-3)!}{n!(n-2)!} = \frac{(-1)^{n-3}2}{n!(n-2)}, \qquad n = 4, 5, 6, \cdots.$$

Our second solution may now be expressed as

$$y_2(x) = -x^2 \ln x + 1 - 2x + c_2 x^2 + \frac{1}{3}x^3 + 2\sum_{n=4}^{\infty} \frac{(-1)^{n-3}}{n!(n-2)} x^n,$$

where we notice that c_2 is arbitrary. Because $y_1(x) = x^2$, the term $c_2 x^2$ can be combined with the arbitrary constant multiplying $y_1(x)$ in our general solution, which has the form $C_1 x^2 + C_2 y_2(x)$, where C_1 and C_2 are arbitrary constants. Another way would be to set $c_2 = 0$ in the expression for $y_2(x)$ above.

Because one of our solutions is x^2, we know from Theorems 9.2 and 9.3 on pages 432 and 436 that $y_2(x)$ will not oscillate.   □

EXERCISES

Find the general solution to the following differential equations for $x > 0$ using the method of Frobenius. For series for which the formula for the n th term is not apparent, find terms in the series up to x^4. Also find R such that the series converge for $0 < x < R$.

1. $xy'' + y' - xy = 0$

2. $xy'' + y' - 4xy = 0$

3. $(x^2 - x)y'' + 3y' - 2y = 0$

4. $x^2 y'' + 2xy' + xy = 0$

5. $(x^2 + x^4)y'' + (9x + 5x^3)y' + (16 + 4x^2)y = 0$

6. $xy'' + (x - 1)y' - y = 0$

7. $x^2 y'' + (x^2 - x)y' - (x - 1)y = 0$

8. $x^2 y'' + (x^2 - x)y' + y = 0$

9. $xy'' + xy' + y = 0$

10. $(x - x^2)y'' - 3xy' - y = 0$

11. Show that the answer to Exercise 10 may also be written

$$y(x) = c_0 \frac{x}{(1 - x)^2} + c_1 \left[\frac{x}{(1 - x)^2} \ln x + \frac{1}{1 - x} \right].$$

12. Show that one solution of Exercise 8 is $y(x) = xe^{-x}$, and use the reduction of order technique to find the second solution.

13. Show that using $s = 0$ in a Frobenius series solution of (10.41) gives the same solution as $s = 2$.

14. Use the method of Frobenius to obtain the second solution of Example 10.12, and compare your answer with (10.30).

15. Show that trying a Frobenius type solution with $s = -1$ in (10.28),

$$x(x - 1)\frac{d^2 y}{dx^2} + (5x - 2)\frac{dy}{dx} + 4y = 0,$$

yields the same solution as that for $s = 0$.

16. Show that the indicial equation obtained by substituting $y(x) = x^s \sum_{n=0}^{\infty} c_n x^n$ in

$$x^2 \frac{d^2 y}{dx^2} + (x^2 - 3x)\frac{dy}{dx} + (-2x + 4)\,y = 0$$

gives rise to a double root of $s = 2$ and a solution of $y_1(x) = x^2$. Use reduction of order to find the second solution in the form

$$y_2(x) = x^2 \int \frac{1}{x} e^{-x} \, dx.$$

By expanding e^{-x} in a Taylor series about the origin and integrating term by term, show that

$$y_2(x) = x^2 \ln x + x^2 \sum_{n=1}^{\infty} \frac{(-1)^n}{n!n} x^n.$$

17. Use Theorem 10.3 to find a second solution to the differential equation from Exercise 3 of Section 10.2, namely,

$$2xy'' + 6y' - 9xy = 0.$$

What Have We Learned?

MAIN IDEAS

The following comments pertain to the linear differential equation

$$a_2(x)\frac{d^2y}{dx^2} + a_1(x)\frac{dy}{dx} + a_0(x)y = 0. \tag{10.42}$$

- A regular singular point is a point $x = x_0$ where at least one of a_1/a_2 and a_0/a_2 is not analytic, but both $(x - x_0)a_1/a_2$ and $(x - x_0)^2 a_0/a_2$ are analytic. An irregular singular point is a singular point that is not regular. See page 490.

- The indicial equation needed to find a solution of (10.42) using the method of Frobenius may be found in three different ways. See pages 487 and 491 and Exercise 4 on page 494.

- If x_0 is a regular singular point of (10.42), the method of Frobenius always gives one solution of the form

$$y(x) = (x - x_0)^{s_1} \sum_{n=0}^{\infty} c_n (x - x_0)^n ,$$

where s_1 is the larger root of the indicial equation. See Theorem 10.1 on page 491.

- If x_0 is a regular singular point of (10.42), the second solution may have the form

$$y_2(x) = y_1(x) \ln (x - x_0) + (x - x_0)^{s_1} \sum_{n=1}^{\infty} c_n (x - x_0)^n \tag{10.43}$$

or

$$y_2(x) = Cy_1(x) \ln (x - x_0) + (x - x_0)^{s_2} \left[1 + \sum_{n=1}^{\infty} c_n^n (x - x_0)^n \right], \tag{10.44}$$

where (10.43) works for repeated roots of the indicial equation and (10.44) for all other cases. It is possible that in some cases the constant C in (10.44) will be zero. See Theorem 10.3 on page 507.

- The radius of convergence of the series obtained by using the method of Frobenius is at least as large as the radii of convergence of the two functions $(x - x_0)a_1/a_2$ and $(x - x_0)^2 a_0/a_2$. See Theorems 10.1 and 10.3, which have $x = 0$ as a regular singular point.

CHAPTER 11

Nonhomogeneous Second Order Linear Differential Equations

Where Are We Going?

In the last four chapters we have been primarily concerned with solutions of homogeneous second order differential equations. Because many important applications of differential equations deal with nonhomogeneous differential equations, we consider that situation in this chapter.

In Section 6.3 we discovered that the general solution of a nonhomogeneous first order linear differential equation consists of the sum of the general solution of an associated homogeneous differential equation and a particular solution. Here we will discover that this result carries over to second and higher order linear differential equations.

For the second order case we will develop three methods for finding particular solutions: the method of undetermined coefficients, reduction of order, and variation of parameters. These methods also carry over to finding particular solutions for orders of differential equations higher than two. We will also show that our earlier methods of solving linear differential equations with constant coefficients and Cauchy-Euler equations also carry over to higher order linear differential equations.

11.1　The General Solution

Many applications of differential equations involve solving the nonhomogeneous linear differential equation

$$a_2(t)\frac{d^2y}{dt^2} + a_1(t)\frac{dy}{dt} + a_0(t)y = f(t), \tag{11.1}$$

where $a_2(t)$, $a_1(t)$, $a_0(t)$, and $f(t)$ are given functions of t. In order to develop the general method for solving (11.1), we will look at a simple example.

<u>**EXAMPLE 11.1**</u>

Solve

$$\frac{d^2y}{dt^2} + y = t. \tag{11.2}$$

If we stare at (11.2) long enough we may see that $y(t) = t$ is a particular solution. Can we exploit the fact that we have found a particular solution to find the most general $y(t)$ that satisfies (11.2)? In the past we have used two techniques for finding the general solution once we have some information about a specific solution. Both involved using the known solution to change the dependent variable from $y(t)$ to $z(t)$, either by multiplying $z(t)$ by the known solution or by adding $z(t)$ to the known solution. In this case this would mean trying either the change of variable

$$y(t) = tz(t) \tag{11.3}$$

or the change of variable

$$y(t) = t + z(t). \tag{11.4}$$

The first of these suggestions, (11.3), does not simplify the problem. (Try it.) However, if we substitute (11.4) into (11.2) we find

$$\frac{d^2z}{dt^2} + t + z = t,$$

or

$$\frac{d^2z}{dt^2} + z = 0.$$

We recognize this as a homogeneous differential equation with constant coefficients (see page 344) with solution

$$z(t) = C_1 \cos t + C_2 \sin t. \tag{11.5}$$

Substituting (11.5) into (11.4) gives the general solution of (11.2), namely,

$$y(t) = t + C_1 \cos t + C_2 \sin t. \tag{11.6}$$

Let's look at the structure of this solution. It consists of two parts: t, which is a particular solution of the nonhomogeneous equation, and $C_1 \cos t + C_2 \sin t$, which is the general solution of the associated homogeneous equation. This homogeneous equation is obtained from the nonhomogeneous equation by replacing the forcing term with 0, that is

$$\frac{d^2y}{dt^2} + y = 0.$$

In fact this is the same structure that we encountered in the case of first order linear differential equations. In Section 6.3 we discovered that the general solution of first order nonhomogeneous equations consisted of the sum of the general solution of an associated homogeneous differential equation and a particular solution.   □

We want to derive a similar result for the general equation (11.1). Imagine we have found a particular solution $y = y_p(t)$ of (11.1), so that

$$a_2(t)\frac{d^2y_p}{dt^2} + a_1(t)\frac{dy_p}{dt} + a_0(t)y_p = f(t). \qquad (11.7)$$

Now the general solution $y = y(t)$ that we seek must satisfy

$$a_2(t)\frac{d^2y}{dt^2} + a_1(t)\frac{dy}{dt} + a_0(t)y = f(t). \qquad (11.8)$$

If we subtract (11.7) from (11.8) we find

$$a_2(t)\left(\frac{d^2y}{dt^2} - \frac{d^2y_p}{dt^2}\right) + a_1(t)\left(\frac{dy}{dt} - \frac{dy_p}{dt}\right) + a_0(t)\left(y - y_p\right) = 0,$$

or

$$a_2(t)\frac{d^2\left(y - y_p\right)}{dt^2} + a_1(t)\frac{d\left(y - y_p\right)}{dt} + a_0(t)\left(y - y_p\right) = 0.$$

Thus, the quantity

$$y_h(t) = y - y_p \qquad (11.9)$$

satisfies the associated homogeneous equation

$$a_2(t)\frac{d^2y_h}{dt^2} + a_1(t)\frac{dy_h}{dt} + a_0(t)y_h = 0.$$

If we rewrite (11.9) in the form

$$y(t) = y_p(t) + y_h(t)$$

we see that the general solution of (11.8), $y(t)$, is again the sum of $y_p(t)$, a particular solution of the nonhomogeneous equation, and $y_h(t)$, the general solution of the associated homogeneous differential equation.

How to Solve Nonhomogeneous Linear Differential Equations

Purpose: To find the general solution $y = y(t)$ of

$$a_2(t)\frac{d^2y}{dt^2} + a_1(t)\frac{dy}{dt} + a_0(t)y = f(t). \qquad (11.10)$$

Technique:

1. Find the general solution $y = y_h(t)$ of the associated homogenous equation (see *How to Solve Second Order Linear Differential Equations with Constant Coefficients* on page 344 and *How to Solve a Cauchy-Euler Differential Equation* on page 452)

$$a_2(t)\frac{d^2y}{dt^2} + a_1(t)\frac{dy}{dt} + a_0(t)y = 0. \qquad (11.11)$$

2. Find any solution $y = y_p(t)$ of the nonhomogeneous equation (11.10).

3. The general solution of (11.10) is

$$y(t) = y_h(t) + y_p(t). \tag{11.12}$$

Comments about nonhomogeneous linear differential equations

- Chapters 7, 9, and 10 were devoted to developing techniques for solving homogeneous equations, so finding $y_h(t)$ will depend on applying those methods successfully.

- Any solution $y = y_p(t)$ of (11.10) is called a particular solution of (11.10). The differential equation (11.10) has many particular solutions. All we need is one.

- It does not matter how we find a particular solution. We might find it by guesswork, luck, or insight or perhaps by using a systematic approach.

- At present we have no general method for finding a particular solution. The next two sections will be devoted to this topic.

EXERCISES

1. In each of the following check that the given $y_p(t)$ is a particular solution of the differential equation, and then find the general solution of the differential equation.

   (a) $y'' - y = t^2$, $y_p(t) = -2 - t^2$

   (b) $y'' + 3y' + 2y = 20e^{3t}$, $y_p(t) = e^{3t}$

   (c) $y'' + 3y' + 2y = e^{-t}$, $y_p(t) = te^{-t}$

   (d) $y'' + y' - 2y = 4\sin 2t$, $y_p(t) = -(\cos 2t + 3\sin 2t)/5$

   (e) $y'' - 4y' + 4y = 12te^{2t}$, $y_p(t) = 2t^3 e^{2t}$

   (f) $y'' + 2y' + 5y = 4e^{-t}\cos 2t$, $y_p(t) = te^{-t}\sin 2t$

   (g) $t^2 y'' + 2ty' - 2y = 2$, $y_p(t) = -1$. (Hint: Recall the Cauchy-Euler equations in Section 9.4.)

2. Solve (11.2), $y'' + y = t$, by assuming a power series solution $y(t) = \sum_{n=0}^{\infty} c_n t^n$. (See *How to Find a Power Series Solution of a Linear Differential Equation* on page 469.) Compare your answer with (11.6).

3. Show that the two functions t and $t + 2$ are both particular solutions of $y'' + y' = 1$.

   (a) Find the general solution of $y'' + y' = 1$ using t as the particular solution.

   (b) Find the general solution of $y'' + y' = 1$ using $t + 2$ as the particular solution.

   (c) Reconcile the solutions obtained in parts (a) and (b).

4. This problem deals with the differential equation $y'' = t + 1$.

   (a) Solve the differential equation $y'' = t + 1$ by integrating twice with respect to t.

   (b) Find constants a, b, c, and d, so that $at^3 + bt^2 + ct + d$ is a particular solution of the differential equation $y'' = t + 1$. Use this information to find the general solution of $y'' = t + 1$.

   (c) Reconcile the solutions obtained in parts (a) and (b).

11.2 Method of Undetermined Coefficients

In this section we discover methods of determining a particular solution for second order linear differential equations with constant coefficients when the forcing function contains terms involving polynomials, exponentials, the sine or cosine function, or products of these three types of functions.

Consider an electric circuit in which wires connect a resistor, capacitor, inductor, and switch in series with a voltage generator, as shown is Figure 11.1. We denote the current in the circuit as I (units of amperes) and the charge on the capacitor as y (units of coulombs). The voltage drops across each of these three elements are given by $L dI/dt$, RI, and y/C, respectively, where, from Section 8.2, the inductance is given by L (units of henries), the resistance by R (units of ohms), and the capacitance by C (units of farads). If we let the applied voltage be $E(t)$, we may use one of Kirchhoff's law to equate this applied voltage to the sum of the three voltage drops across the three circuit elements. This gives

$$L\frac{d^2y}{dt^2} + R\frac{dy}{dt} + \frac{1}{C}y = E(t). \tag{11.13}$$

Notice that we have also made use of the fact that the current in an electric circuit is the rate of change of charge ($I = dy/dt$).

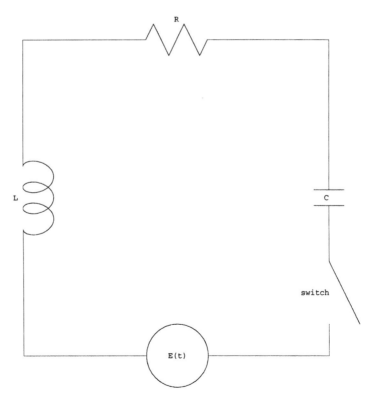

FIGURE 11.1 Series RLC circuit

EXAMPLE 11.2

To simplify our future calculations, we use specific values of L, R, C, and $E(t)$, which give (11.13) the form

$$y'' + 5y' + 6y = 24e^{-t}. \tag{11.14}$$

The forcing function in (11.14) models the case in which the voltage applied to the circuit decreases with time in an exponential manner from 24 to 0.

In Section 6.3 we discovered that the general solution of a first order linear differential equation consisted of two parts — namely, a general solution of the associated homogeneous differential equation plus a particular solution. In Section 11.1 we discovered that such is also the case here.

We know some techniques for finding the general solution of homogeneous second order linear differential equations from Chapters 7, 9, and 10, so in this section we concentrate on finding particular solutions. Recall that in Chapter 6, Tables 6.2 and 6.3 gave appropriate forms for particular solutions for the case of first order linear equations with constant coefficients. These tables applied when the forcing function in the differential equation took the form of exponential, polynomial, the sine or cosine function, or products and sums of these three types of functions. These tables also apply here and so are reproduced as Table 11.1. If any part of the suggested trial solution in Table 11.1 satisfies the associated homogeneous differential equation, then we use t times the usual function indicated on the right-hand side of Table 11.1.

TABLE 11.1   **Particular Solutions of** $a_2 y'' + a_1 y' + a_0 y = f(t)$

$f(t)$	Trial Solution
$a_0 + a_1 t + a_2 t^2 + a_3 t^3 + \cdots + a_n t^n$	$A_0 + A_1 t + A_2 t^2 + A_3 t^3 + \cdots + A_n t^n$
ae^{ct}	Ae^{ct}
$a \sin wt + b \cos wt$	$A \sin wt + B \cos wt$
$e^{ct}(a \sin wt + b \cos wt)$	$e^{ct}(A \sin wt + B \cos wt)$
$e^{ct} \sum_{k=0}^{k=n} a_k t^k$	$e^{ct} \sum_{k=0}^{k=n} A_k t^k$
$\sum_{k=0}^{k=n} a_k t^k (a \sin wt + b \cos wt)$	$\sum_{k=0}^{k=n} A_k t^k \sin wt + \sum_{k=0}^{k=n} B_k t^k \cos wt$
$e^{ct} \sum_{k=0}^{k=n} a_k t^k (a \sin wt + b \cos wt)$	$e^{ct}(\sum_{k=0}^{k=n} A_k t^k \sin wt + \sum_{k=0}^{k=n} B_k t^k \cos wt)$

Following the technique described in the previous section, we first find the solution of the associated homogeneous differential equation given by

$$y'' + 5y' + 6y = 0. \tag{11.15}$$

The characteristic equation for (11.15) is

$$r^2 + 5r + 6 = 0,$$

which factors as $(r + 3)(r + 2)$. This gives e^{-3t} and e^{-2t} as solutions of (11.15), and so

$$y_h(t) = C_1 e^{-3t} + C_2 e^{-2t}. \tag{11.16}$$

From Table 11.1 we have the proper form of a particular solution of (11.14) as $y_p(t) = Ae^{-t}$, with A to be determined so as to satisfy the differential equation (11.14). We substitute this expression into (11.14) and obtain

$$Ae^{-t} - 5Ae^{-t} + 6Ae^{-t} = 24e^{-t}.$$

Dividing by the common factor of e^{-t} and solving for A gives $A = 12$ and our particular solution as $y_p(t) = 12e^{-t}$. Thus, the general solution of our original problem may be written as

$$y(t) = y_h(t) + y_p(t) = C_1 e^{-3t} + C_2 e^{-2t} + 12e^{-t}. \qquad \textbf{(11.17)}$$

If we had an initial value problem, we would evaluate the arbitrary constants in this expression (making sure we included the particular solution). Thus, if we closed the switch when the charge on the capacitor was 4 coulombs, we would have $y(0) = 4$. Because we have two arbitrary constants, we need a second condition. This is obtained from the fact that at the moment the switch is closed there is no current: Thus, $I(0) = y'(0) = 0$. Applying these initial conditions to our general solution in (11.17) and its derivative gives

$$C_1 e^0 + C_2 e^0 + 12e^0 = 4,$$
$$-3C_1 e^0 - 2C_2 e^0 - 12e^0 = 0.$$

Solving this set of algebraic equations gives

$$C_1 = 4, \qquad C_2 = -12,$$

and our solution to this initial value problem is

$$y(t) = 4e^{-3t} - 12e^{-2t} + 12e^{-t}.$$

The graph of this solution is shown in Figure 11.2, in which the decay of the charge to 0 is very apparent. $\qquad\qquad\qquad\qquad\qquad\qquad\qquad\qquad\qquad\qquad\qquad\qquad$ □

EXAMPLE 11.3

Suppose in our last example we had the applied voltage decaying more rapidly — namely, as $24e^{-2t}$. The appropriate differential equation then becomes

$$y'' + 5y' + 6y = 24e^{-2t}, \qquad \textbf{(11.18)}$$

subject to $y(0) = 4$ and $y'(0) = 0$.

Our usual trial solution for this situation would be Ae^{-2t}. However, the solution of the associated homogeneous differential equation, (11.16), contains Ae^{-2t} when $C_1 = 0$ and $C_2 = A$. So we multiply this trial solution by t to obtain

$$y_p(t) = Ate^{-2t}.$$

Substituting this expression into (11.18) gives

$$A(-4e^{-2t} + 4te^{-2t}) + 5A(e^{-2t} - 2te^{-2t}) + 6Ate^{-2t} = 24e^{-2t}.$$

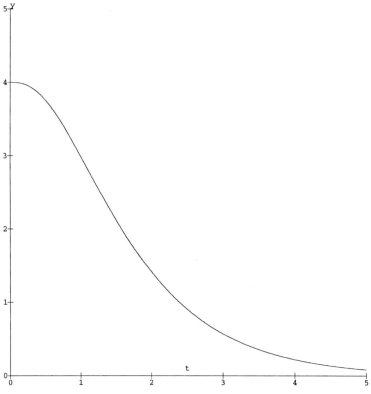

FIGURE 11.2 Graph of $4e^{-3t} - 12e^{-2t} + 12e^{-t}$

We solve this equation for A, finding $A = 24$. This allows us to write the general solution of (11.18) as

$$y(t) = C_1 e^{-3t} + C_2 e^{-2t} + 24te^{-2t}.$$

From the initial conditions, $y(0) = 4$, $y'(0) = 0$, we find $C_1 = 16$ and $C_2 = -12$, so in this case the solution is

$$y(t) = 16e^{-3t} - 12e^{-2t} + 24te^{-2t}.$$

The graph of this solution is shown in Figure 11.3, in which the decay of the charge to 0 is again apparent and more rapid than in the previous example. $\qquad\square$

EXAMPLE 11.4

For our final example, we consider the case in which the applied voltage in the preceding circuit has both a component that decreases exponentially and one that is sinusoidal, giving the differential equation

$$y'' + 5y' + 6y = 24e^{-3t} + 52 \sin 2t. \tag{11.19}$$

In Section 6.2 we discovered the principle of linear superposition for first order linear differential equations which allowed us to find particular solutions for each term in the forcing function and then add the two solutions. This is also the case for second order linear differential equations (see Exercise 3 on page 529) so using Table 11.1, we choose particular solutions as $A_1 te^{-3t}$ and $A \sin 2t + B \cos 2t$. We could substitute each of these two expressions into (11.19) separately, or substitute them in as a sum, as we do now.

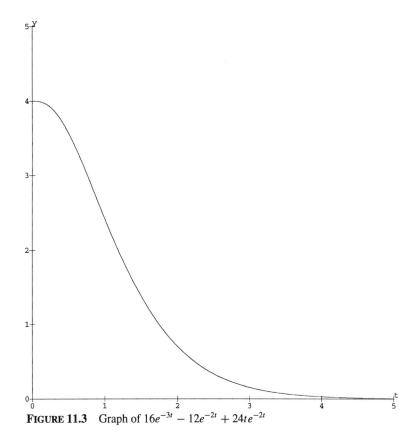

FIGURE 11.3 Graph of $16e^{-3t} - 12e^{-2t} + 24te^{-2t}$

[Be sure to note that because we considered each of the terms on the right-hand side of (11.19) separately, only that part of the particular solution which satisfies the associated homogeneous differential equation is multiplied by t.]

Setting

$$y_p(t) = A_1 te^{-3t} + A \sin 2t + B \cos 2t$$

and differentiating gives

$$\frac{dy_p}{dt} = A_1 (-3t + 1) e^{-3t} + 2A \cos 2t - 2B \sin 2t,$$
$$\frac{d^2 y_p}{dt^2} = A_1 (9t - 6) e^{-3t} - 4A \sin 2t - 4B \cos 2t,$$

whereas substituting these three expressions into (11.19) gives

$$-A_1 e^{-3t} + (2A - 10B) \sin 2t + (10A + 2B) \cos 2t = 24e^{-3t} + 52 \sin 2t.$$

Because we want this to be an identity (that is, be true for all values of t), we may equate coefficients of like terms to obtain the set of algebraic equations

$$-A_1 = 24,$$
$$2A - 10B = 52,$$
$$10A + 2B = 0,$$

with solution $A = 1$, $B = -5$, $A_1 = -24$. We have the same associated homogeneous differential equation as before, so we may write the general solution of (11.19) as

$$y(t) = C_1 e^{-3t} + C_2 e^{-2t} - 24te^{-3t} + \sin 2t - 5\cos 2t. \qquad (11.20)$$

If we use the same initial conditions as Example 11.2, we have $y(0) = 4$, $y'(0) = 0$. Using these initial conditions for the solution in (11.20) gives $C_1 = 40$ and $C_2 = -31$, and the solution of (11.19) becomes

$$y(t) = -40e^{-3t} + 49e^{-2t} - 24te^{-3t} + \sin 2t - 5\cos 2t. \qquad (11.21)$$

From the form of (11.21) we see that the first three terms contain exponential functions with negative arguments. Thus, as time increases they will approach 0 (and be the transient solution) whereas the two periodic terms persist and are the steady state solution. The graphs of $y(t)$ and the steady state solution $\sin 2t - 5\cos 2t$ are shown in Figure 11.4. Notice that after about one complete cycle, the motion appears periodic and indistinguishable from the steady state solution. $\qquad \square$

Although using Table 11.1 will always give the proper form for the particular solution to linear differential equations with constant coefficients, there are two other commonly used methods for determining this form. One of these methods is discussed now; the other is given in Exercise 14 on page 530.

If we look at the differential equation given in (11.14),

$$y'' + 5y' + 6y = 24e^{-t},$$

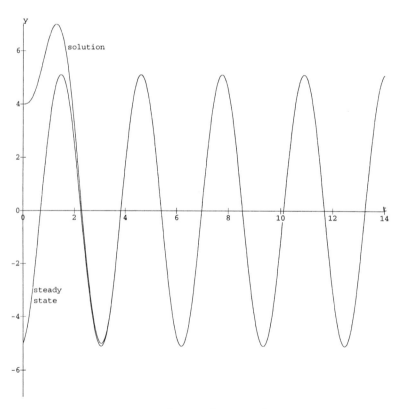

FIGURE 11.4   Graphs of $-40e^{-3t} + 49e^{-2t} - 24te^{-3t} + \sin 2t - 5\cos 2t$ and the steady state solution $\sin 2t - 5\cos 2t$

and think about a particular solution, we could ask ourselves the question, "What type of function could we find such that if we took a linear combination of that function and its first and second derivatives we would come up with $24e^{-t}$?" Because the answer is a constant times e^{-t}, we would try the same function as before. If we now focus our attention on (11.19),

$$y'' + 5y' + 6y = 24e^{-3t} + 52\sin 2t,$$

we can ask, "What type of function could we start with such that a linear combination of itself and its first and second derivatives would end up as $52\sin 2t$?" We would probably answer "a linear combination of $\sin 2t$ and $\cos 2t$" because we could obtain a term like $\sin 2t$ by differentiating $\cos 2t$ once, or $\sin 2t$ twice.

Now consider $e^{at}\sin bt$, and ask what types of terms can we differentiate to obtain this term. It should be clear that the result of differentiating either $e^{at}\sin bt$ or $e^{at}\cos bt$ will contain $e^{at}\sin bt$ as one of its components. Note that in all these three cases, we could obtain the proper terms to include in the trial solution by simply including terms obtained by differentiating the term in question.

Definition 11.1: **The DIFFERENTIAL FAMILY associated with a function is the set of linearly independent functions obtained from the function and its derivatives.**

Comments about differential families

- The differential family associated with e^{-t} is $\{e^{-t}\}$.

- The differential family of $\sin 2t$ is $\{\sin 2t, \cos 2t\}$.

- The differential family of $e^{at}\sin bt$ is $\{e^{at}\sin bt, e^{at}\cos bt\}$.

- The proper form of the particular solution for a specific forcing function is a linear combination of the elements in the differential family associated with that function.

- This rule works for forcing functions of the form $t^n e^{at}\sin bt$ or $t^n e^{at}\cos bt$, where a and b are real constants and n is a nonnegative integer.

Compare the proper trial solutions in Table 11.1 with the set of differential families in Table 11.2.

TABLE 11.2 Family of Functions

Function	Differential Family
te^{at}	$\{e^{at}, te^{at}\}$
$t^2 e^{at}$	$\{e^{at}, te^{at}, t^2 e^{at}\}$
t^3	$\{1, t, t^2, t^3\}$
$t^2 \cos bt$	$\{\cos bt, \sin bt, t\cos bt, t\sin bt, t^2\cos bt, t^2\sin bt\}$

Just as in the use of Table 11.1, if one of the elements in the trial solution is a solution of the associated homogeneous differential equation, we must first multiply the proposed trial solution by t (or t^2, if t times the trial solution still has a term that satisfies this associated homogeneous differential equation).

To illustrate this last point, if we were considering the differential equation

$$y'' + 4y' + 4y = 3e^{-2t},$$

we would normally try a particular solution of the from Ae^{-2t}. However, the general solution of the associated homogeneous differential equation contains both e^{-2t} and te^{-2t}. Thus, the proper trial solution would have the form

$$y_p(t) = At^2 e^{-2t}.$$

We conclude this section by noting the proper form for the particular solution for three nonhomogeneous differential equations. You should provide the reasons for the correctness of each choice.

The proper form for the particular solution of

$$y'' + 16y = 16t^2 + e^{4t} + 16 \sin 4t$$

is

$$y_p(t) = A + Bt + Ct^2 + De^{4t} + Et \sin 4t + Ft \cos 4t.$$

The proper form for the particular solution of

$$y'' + 4y' - 5y = 4te^t + \sin 5t - 3te^{2t}$$

is

$$y_p(t) = Ate^t + Bt^2 e^t + C \sin 5t + D \cos 5t + Ee^{2t} + Fte^{2t}.$$

The proper form for the particular solution of

$$y'' - 6y' + 9y = 3e^{3t} - 2e^{-3t} + 5$$

is

$$y_p(t) = At^2 e^{3t} + Be^{-3t} + C.$$

EXERCISES

1. If $y_1(t)$ and $y_2(t)$ are solutions of

$$a_2(t)y'' + a_1(t)y' + a_0(t)y = q_1(t)$$

and

$$a_2(t)y'' + a_1(t)y' + a_0(t)y = q_2(t),$$

respectively, then their sum, $y_1(t) + y_2(t)$, satisfies

$$a_2(t)y'' + a_1(t)y' + a_0(t)y = q_1(t) + q_2(t).$$

2. Show that if $y_1(t)$ satisfies the differential equation

$$a_2(t)y'' + a_1(t)y' + a_0(t)y = q(t)$$

then $\alpha y_1(t)$ satisfies

$$a_2(t)y'' + a_1(t)y' + a_0(t)y = \alpha q(t),$$

where α is a constant.

3. Combine the results of Exercises 1 and 2 to show that if $y_1(t)$ and $y_2(t)$ are solutions of

$$a_2(t)y'' + a_1(t)y' + a_0(t)y = q_1(t)$$

and

$$a_2(t)y'' + a_1(t)y' + a_0(t)y = q_2(t),$$

respectively, then their linear combination, $\alpha y_1(t) + \beta y_2(t)$, satisfies

$$a_2(t)y'' + a_1(t)y' + a_0(t)y = \alpha q_1(t) + \beta q_2(t),$$

where α and β are specified constants.

4. Find the general solution of
   (a) $$y'' + y' - 6y = 28e^{4t}.$$

   (b) $$y'' + y' - 6y = 8e^{2t}.$$

5. Use the results from Exercises 3 and 4(a) and (b) to find the general solution of

$$y'' + y' - 6y = 280e^{4t} - 4e^{2t}.$$

6. Create a *How to Find a Particular Solution for Second Order Linear Differential Equations with Constant Coefficients*. Add statements under the headings Purpose, Technique, and Comments that summarize what you discovered in this section.

7. Find the general solution of the following differential equations.
   (a) $y'' + y' - 6y = t$
   (b) $y'' + y' - 6y = e^{3t}$
   (c) $y'' + y' - 6y = 2t + \pi e^{3t}$
   (d) $y'' + 3y' - 4y = t^2 + 1$
   (e) $y'' + y' = 6e^{-t}$
   (f) $y'' + 4y = \cos 3t$
   (g) $y'' + 4y = 10\sin 3t$
   (h) $y'' + 4y = 6\cos 3t + 20\sin 3t$
   (i) $y'' + 4y = \sin 2t$
   (j) $y'' + 4y' + 4y = \cos 2t$
   (k) $4y'' - 12y' + 9y = 24te^{3t/2}$

8. Solve the following initial value problems.
   (a) $y'' + y' - 12y = 8e^{3t}$, $\quad y(0) = 0$, $\quad y'(0) = 1$
   (b) $y'' + 6y' + 9y = e^{3t}$, $\quad y(0) = 0$, $\quad y'(0) = 6$
   (c) $y'' - 5y' + 6y = 12te^{-t} - 7e^{-t}$, $\quad y(0) = y'(0) = 0$
   (d) $y'' + 4y = 8\sin 2t + 8\cos 2t$, $\quad y(\pi) = y'(\pi) = 2\pi$

9. Show that if $\omega \neq \alpha$, a particular solution of

$$y'' + \omega^2 y = A\cos\alpha t + B\sin\alpha t$$

is

$$y_p(t) = \frac{A}{\omega^2 - \alpha^2} \cos \alpha t + \frac{B}{\omega^2 - \alpha^2} \sin \alpha t.$$

10. Show that the general solution of

$$y'' + \omega^2 y = A \cos \omega t$$

is

$$y(t) = C_1 \cos \omega t + C_2 \sin \omega t + \frac{A}{2\omega} t \sin \omega t.$$

11. Find $y(t)$ for the electrical circuit described by (11.13) having $L = 1$, $R = 2$, $C = 1/10$, $E(t) = 20$, $y(0) = 2$, and $y'(0) = 6$. Identify the transient and steady state parts of your solution. What is the maximum value of the charge on the capacitor?

12. Find $y(t)$ for the electrical circuit described by (11.13) having $L = 16$, $R = 8$, $C = 1/10$, $E(t) = \sin \gamma t$, $y(0) = 0$, and $y'(0) = 0$, for given γ.

13. What value of γ will maximize the amplitude of the steady-state response in Exercise 12? (In practice, a circuit is "tuned" by varying C until the steady-state amplitude is maximized for a fixed value of γ.)

14. Consider the differential equation in Exercise 4(a), written in operator form, as

$$(D^2 + D - 6)y = 28e^{4t}. \tag{11.22}$$

(a) Operate on both sides of this equation with $(D - 4)$ to obtain the homogeneous differential equation

$$(D - 4)(D^2 + D - 6)y = 0,$$

and show that the characteristic equation for this differential equation is

$$(r - 4)(r + 3)(r - 2) = 0.$$

Note that the roots of this equation include the two roots of the characteristic equation of the homogeneous differential equation associated with (11.22) plus the additional root 4. The particular solution of (11.22) is then a constant times e^{4t}. Because applying the operator $(D - 4)$ to (11.22) resulted in a homogenous differential equation, this method is sometimes called the annihilator method. (The operator $D - 4$ annihilated $28e^{4t}$.)

(b) Find the operator to act on the differential equations in Exercises 4(b) and 5 so the result of the operation will be a homogeneous differential equation. Find the characteristic equation of these new differential equations, and identify the proper form for the particular solutions. Verify that they are the same as you would choose using Table 11.1.

15. Find the particular solution of

$$y'' + 5y' + 6y = 24e^{-2t}.$$

(a) Use the method of Exercise 14.

(b) Use the method of Exercise 16.

16. Another method of obtaining a particular solution is based on a product substitution and the reduction of order technique from Section 9.2. This is now outlined for

$$y'' - 5y' + 4y = e^{5t}.$$

(a) Show that e^{4t} and e^t satisfy the associated homogeneous differential equation.

(b) Make a change of dependent variable as $y = e^t z(t)$, and obtain the resulting second order differential equation for $z(t)$.

(c) Let $w(t) = z'(t)$ to reduce the differential equation to one that is first order.

(d) Solve this differential equation for $w(t)$, and then integrate the result to obtain $z(t)$. We may set the two arbitrary constants of integration equal to 0, because we need only one particular solution.

(e) Multiply our result in part (d) by e^t to obtain a particular solution.

Thus, the procedure is as follows:

- Make a change of variables $y = y_1(t)z(t)$, where $y_1(t)$ is a nontrivial solution of the associated homogeneous differential equation. This results in a differential equation that does not explicitly contain the new dependent variable $z(t)$, but only its derivatives.

- Let $w(t) = z'(t)$, and solve the resulting first order linear differential equation for $w(t)$.

- Integrate this result for $w(t)$ to obtain $z(t)$, and multiply by $y_1(t)$ to obtain a particular solution. (We may set both the integration constants to zero.)

11.3 Reduction of Order and Variation of Parameters

The method of undetermined coefficients of the previous section is guaranteed to work only when the forcing function for linear differential equations with constant coefficients has the forms listed in Table 11.1. However, the method given in Exercise 16 in Section 11.2 requires only that the differential equation be linear, not that it have constant coefficients. Also, the forcing function is not limited to the forms of Table 11.1 for this method to work.

We motivate this method by a result from Section 5.2. There we discovered that the general solution of (5.26), rewritten here with t as the independent variable,

$$\frac{dy}{dt} + p(t)y = q(t), \tag{11.23}$$

was

$$y(t) = u(t) \int \frac{q(t)}{u(t)} \, dt + Cu(t), \tag{11.24}$$

where $u(t)$ was a solution of the associated homogeneous differential equation and C was an arbitrary constant. A crucial observation here (for what we develop in this section) is that the first term on the right-hand side of (11.24) (the one that does not involve the arbitrary constant) is a particular solution of (11.23) and has the form of a function times a solution of the associated homogeneous differential equation. Thus, for second order differential equations, we will also assume a particular solution in the form of an unknown function times a solution of the associated homogeneous differential equation. We illustrate this technique with two examples, one that has a forcing function not covered by Table 11.1 and another that contains a differential equation that does not have constant coefficients. For the general case, see Exercise 4 on page 540.

EXAMPLE 11.5

Consider the differential equation

$$y'' + y = \cot t, \tag{11.25}$$

where we first find the characteristic equation of the associated homogeneous differential equation as

$$r^2 + 1 = 0.$$

The roots of this characteristic equation are i and $-i$, so the corresponding solutions of the associated homogeneous differential equation are $\cos t$ and $\sin t$. We now let one of these functions (say $\sin t$) be included in the definition of a new dependent variable, $z(t)$, as

$$y(t) = \sin t \, z(t). \tag{11.26}$$

Taking derivatives of (11.26) and substituting the results into (11.25) gives

$$\sin t \, z''(t) + 2 \cos t \, z'(t) = \cot t.$$

Substituting $w(t) = z'(t)$ and dividing by $\sin t$ gives

$$w'(t) + 2 \cot t \, w(t) = \frac{\cot t}{\sin t}, \tag{11.27}$$

which is a linear differential equation in standard form. The integrating factor is

$$\exp\left(\int 2 \cot t \, dt\right) = \exp\left(2 \int \frac{\cos t}{\sin t} \, dt\right) = \exp\left(2 \ln |\sin t|\right) = \sin^2 t,$$

so we multiply (11.27) by this factor and collect terms on the left-hand side of the result to obtain

$$\frac{d}{dt}\left[\sin^2 t \, w(t)\right] = \sin t \cot t = \cos t.$$

Integration gives

$$\sin^2 t \, w(t) = \sin t,$$

where, because we need only one particular solution, we set the arbitrary constant to zero. This gives

$$w(t) = \csc t,$$

and because $z'(t) = w(t)$, we use a table of integrals to integrate once more, to obtain

$$z(t) = \int \csc t \, dt = \ln |\csc t - \cot t|.$$

Again we omitted the constant of integration. Thus, our particular solution is $\sin t \ln |\csc t - \cot t|$, and the general solution of (11.25) is

$$y(t) = C_1 \sin t + C_2 \cos t + \sin t \ln |\csc t - \cot t|. \tag{11.28}$$

We are familiar with the behavior of the first two functions in this solution, but we are unfamiliar with the behavior of the particular solution $\sin t \ln |\csc t - \cot t|$. To analyze its behavior we notice that neither $\csc t$ nor $\cot t$ are defined at $t = n\pi, n = 0, \pm 1, \pm 2, \cdots$. This is not surprising because the forcing function in the differential equation (11.25) — namely, $\cot t$ — is not defined at these values of t either. Thus, from the start we know that whatever the behavior of the solutions of (11.25), its regions of validity would be $n\pi < t < (n + 1)\pi$ for $n = 0, \pm 1, \pm 2, \cdots$. The appropriate interval is determined by the value t used in the initial conditions.

Figure 11.5 gives the graph of $\sin t \ln |\csc t - \cot t|$, where the function appears to be continuous everywhere, even though it is not defined at $t = n\pi, n = 0, \pm 1, \pm 2, \cdots$, and so cannot be continuous there. (What does this suggest about $\lim_{t \to n\pi} \sin t \ln |\csc t - \cot t|$? Can you prove it?) The extension in the Comments about Theorem 9.7 on page 476 shows that for an initial condition in the interval $0 < t < \pi$, our solution will converge in the same interval, which is in agreement with the graph. $\square$

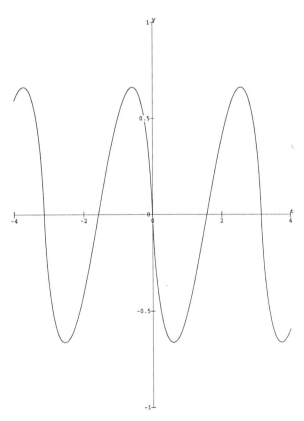

FIGURE 11.5   Graph of $\sin t \ln |\csc t - \cot t|$

How to Find a Particular Solution Using Reduction of Order

Purpose: To find a particular solution of the second order linear differential equation

$$a_2(t)y'' + a_1(t)y' + a_0(t)y = f(t) \tag{11.29}$$

using one solution of the associated homogeneous differential equation.

Technique:

1. Solve the associated homogeneous differential equation, namely,

$$a_2(t)y'' + a_1(t)y' + a_0(t)y = 0,$$

and obtain two linearly independent solutions, say $y_1(t)$ and $y_2(t)$.

2. Assume a particular solution of the form

$$y_p(t) = y_1(t)z(t), \tag{11.30}$$

and substitute (11.30) into (11.29). This will always lead to a first order differential equation in $w(t) = z'(t)$.

3. Solve the resulting first order equation for $w(t)$, and then integrate the result to find $z(t)$.

4. The final form of the particular solution is obtained by substituting the expression for $z(t)$ into (11.30).

EXAMPLE 11.6

Consider the nonhomogeneous Cauchy–Euler differential equation of the form

$$t^2 y'' - ty' + y = t^2 \ln t. \tag{11.31}$$

The solutions of the associated homogeneous differential equation are t and $t \ln t$. Thus, we choose $y_1(t) = t$ and define a new dependent variable, $z(t)$, by

$$y(t) = tz(t).$$

Differentiating this expression and using the results in (11.31) gives

$$t^3 z''(t) + 2t^2 z'(t) - t^2 z'(t) = t^2 \ln t.$$

Letting $w(t) = z'(t)$ and putting the result in the standard form of a first order linear differential equation gives

$$w' + \frac{1}{t}w = \frac{\ln t}{t}. \tag{11.32}$$

Finding the integrating factor as t and then multiplying (11.32) by this integrating factor gives

$$\frac{d}{dt}[tw(t)] = \ln t.$$

Integration gives

$$tw(t) = t \ln t - t + C_1,$$

so

$$w(t) = \ln t - 1 + \frac{1}{t}C_1$$

and one more integration gives

$$z(t) = \int w(t)\,dt = (t\ln t - t) - t + C_1 \ln t + C_2.$$

This gives a particular solution as $t^2 \ln t - 2t^2$, and the general solution as

$$y(t) = C_1 t \ln t + C_2 t + t^2 \ln t - 2t^2.$$

Notice that this technique gave not only a particular solution but also the general solution. This will happen if we include the arbitrary constants at the two steps in the method that involve integration (see Exercises 7 and 8 on page 541). $\square$

Although this method is more general that the one given in Section 11.2, it may have drawbacks. Any one of the three successive integrations required by this method may not be expressible in terms of elementary functions, in which case the particular solution will be left as a sequence of integrals (see Exercise 10 on page 541). Thus, we proceed to develop another technique that does not have this limitation.

In both the preceding examples we had information that we did not exploit in trying to find a particular solution of

$$a_2(t)y'' + a_1(t)y' + a_0(t)y = f(t); \tag{11.33}$$

namely, we used only one of the two independent solutions $y_1(t)$ and $y_2(t)$ of the associated homogeneous differential equation

$$a_2(t)y'' + a_1(t)y' + a_0(t)y = 0.$$

How might we use both? Rather than consider either $y_1(t)z(t)$ or $y_2(t)z(t)$ and try to select $z(t)$ in an appropriate way, we could consider

$$y_p(t) = y_1(t)z_1(t) + y_2(t)z_2(t) \tag{11.34}$$

and seek functions $z_1(t)$ and $z_2(t)$ such that $y_p(t)$ satisfies (11.33). However, we should note that we are seeking only one function, $y_p(t)$, whereas we have two functions, $z_1(t)$ and $z_2(t)$, to determine. This means that we will have some flexibility in selecting $z_1(t)$ and $z_2(t)$. One choice could be to set $z_2(t) = 0$ and then determine $z_1(t)$, which is the reduction of order method we have just dealt with. However, there is an alternative way to exploit this flexibility.

Because we will substitute the expression in (11.34) into the differential equation (11.33), we first calculate derivatives of this y_p. Doing so gives

$$\frac{dy_p}{dt} = y_1'(t)z_1(t) + y_1(t)z_1'(t) + y_2'(t)z_2(t) + y_2(t)z_2'(t),$$

which we can write as

$$\frac{dy_p}{dt} = y_1'(t)z_1(t) + y_2'(t)z_1(t) + \left[y_1(t)z_1'(t) + y_2(t)z_2'(t)\right]. \tag{11.35}$$

From this we now compute

$$\frac{d^2 y_p}{dt^2} = y_1''(t)z_1(t) + y_1'(t)z_1'(t) + y_2''(t)z_2(t) + y_2'(t)z_2'(t) + \left[y_1(t)z_1'(t) + y_2(t)z_2'(t)\right]'. \tag{11.36}$$

If we now substitute the expressions in (11.34), (11.35), and (11.36) into the original differential equation (11.33) we obtain

$$a_2(t) \left\{ y_1'(t)z_1'(t) + y_2'(t)z_2'(t) + \left[y_1(t)z_1'(t) + y_2(t)z_2'(t) \right]' \right\} + \\ + a_1(t) \left[y_1(t)z_1'(t) + y_2(t)z_2'(t) \right] = f(t). \tag{11.37}$$

This is a single equation from which we expect to obtain the two functions $z_1(t)$ and $z_2(t)$. We can exploit the flexibility mentioned earlier if we require that the coefficient of $a_1(t)$ be zero, that is,

$$y_1(t)z_1'(t) + y_2(t)z_2'(t) = 0,$$

in which case (11.37) reduces to

$$a_2(t) \left[y_1'(t)z_1'(t) + y_2'(t)z_2'(t) \right] = f(t).$$

Thus, we have two equations in the two unknowns $z_1'(t)$ and $z_2'(t)$, which we write as the system

$$\begin{aligned} y_1(t)z_1'(t) + y_2(t)z_2'(t) &= 0, \\ y_1'(t)z_1'(t) + y_2'(t)z_2'(t) &= f(t)/a_2(t). \end{aligned} \tag{11.38}$$

The system in (11.38) has a unique solution for $z_1'(t)$ and $z_2'(t)$ because the determinant of the coefficients,

$$\begin{vmatrix} y_1(t) & y_2(t) \\ y_1'(t) & y_2'(t) \end{vmatrix} = W \left[y_1(t), y_2(t) \right],$$

is the Wronskian of $y_1(t)$ and $y_2(t)$. This Wronskian is never zero because they are linearly independent. Thus, we solve for $z_1'(t)$ and $z_2'(t)$ as

$$z_1'(t) = -\frac{f(t)y_2(t)}{a_2(t)W \left[y_1(t), y_2(t) \right]},$$

$$z_2'(t) = \frac{f(t)y_1(t)}{a_2(t)W \left[y_1(t), y_2(t) \right]}.$$

To obtain the final form of the particular solution, we integrate these two equations and substitute the result into the original form of y_p from (11.34). This gives

$$y_p(t) = -y_1(t) \int \frac{f(t)y_2(t)}{a_2(t)W \left[y_1(t), y_2(t) \right]} \, dt + y_2(t) \int \frac{f(t)y_1(t)}{a_2(t)W \left[y_1(t), y_2(t) \right]} \, dt. \tag{11.39}$$

Most people do not find it useful to memorize the form of the particular solution in (11.39), so we will now work an example using a series of three steps.

EXAMPLE 11.7

Find a particular solution of

$$y'' - 3y' + 2y = -\frac{e^{2t}}{e^t + 1}. \tag{11.40}$$

Our first step is to solve the associated homogeneous differential equation by factoring its characteristic equation

$$r^2 - 3r + 2 = (r - 2)(r - 1) = 0.$$

This gives two linearly independent solutions of the associated homogeneous differential equation as e^{2t} and e^t.

The second step is to write the form for a particular solution as

$$y_p(t) = e^{2t} z_1(t) + e^t z_2(t), \tag{11.41}$$

along with the two equations that $z_1'(t)$ and $z_2'(t)$ must satisfy, namely,

$$e^{2t} z_1' + e^t z_2' = 0,$$

$$2e^{2t} z_1' + e^t z_2' = -e^{2t} / \left(e^t + 1\right). \tag{11.42}$$

The last step is to solve this system of equations for $z_1'(t)$ and $z_2'(t)$, integrate, and substitute the result back into (11.41). We may multiply the top equation in (11.42) by 2 and subtract the bottom equation from the result to obtain

$$e^t z_2' = \frac{e^{2t}}{e^t + 1},$$

or

$$z_2' = \frac{e^t}{e^t + 1}. \tag{11.43}$$

Using this expression in the top equation of (11.42), we obtain

$$e^{2t} z_1' = -\frac{e^{2t}}{e^t + 1},$$

or

$$z_1' = -\frac{1}{e^t + 1} = -\frac{e^{-t}}{1 + e^{-t}}. \tag{11.44}$$

Integrating the expressions in (11.43) and (11.44) gives $z_2 = \ln(e^t + 1)$ and $z_1(t) = \ln(1 + e^{-t})$, and our particular solution is obtained from (11.41) as

$$y_p(t) = e^{2t} \ln(1 + e^{-t}) + e^t \ln(e^t + 1).$$

The general solution of (11.40) is then given by

$$y(t) = C_1 e^{2t} + C_2 e^t + e^{2t} \ln(1 + e^{-t}) + e^t \ln(e^t + 1).$$

Notice that because the functions $z_1(t)$ and $z_2(t)$ multiply $y_1(t)$ and $y_2(t)$, the inclusion of the constants of integration when finding $z_1(t)$ and $z_2(t)$ will not change the form of the general solution. Including them will give us the general solution directly.   □

How to Find a Particular Solution Using Variation of Parameters

Purpose: To find a particular solution of the second order linear differential equation

$$a_2(t)y'' + a_1(t)y' + a_0(t)y = f(t) \tag{11.45}$$

using two linearly independent solutions of the associated homogeneous differential equation.

Technique:

1. Solve the associated homogeneous differential equation, namely,

$$a_2(t)y'' + a_1(t)y' + a_0(t)y = 0,$$

   and obtain two linearly independent solutions, say $y_1(t)$ and $y_2(t)$.

2. Assume a particular solution of the form

$$y_p(t) = y_1(t)z_1(t) + y_2(t)z_2(t), \tag{11.46}$$

   and write down the system of equations satisfied by $z_1'(t)$ and $z_2'(t)$,

$$y_1(t)z_1'(t) + y_2(t)z_2'(t) = 0,$$

$$y_1'(t)z_1'(t) + y_2'(t)z_2'(t) = f(t)/a_2(t). \tag{11.47}$$

3. Solve (11.47) for $z_1'(t)$ and $z_2'(t)$, and integrate to obtain $z_1(t)$ and $z_2(t)$. (We may set the arbitrary constants to zero.) The final form of the particular solution is obtained by substituting these values of $z_1(t)$ and $z_2(t)$ into (11.46).

Comments about variation of parameters

- Including the arbitrary constants when integrating the expressions for $z_1'(t)$ and $z_2'(t)$ will result in the expression for y_p also being the general solution of (11.45).

- If we are solving an initial value problem and cannot evaluate the integrals for $z_1(t)$ and $z_2(t)$ in terms of elementary function, it is often convenient to write the integrals with the independent variable as the upper limit and the value of t giving the initial condition as the lower limit. See Exercise 9 on page 541.

- A common mistake in writing down (11.47) is to forget to divide $f(t)$ by $a_2(t)$ on the right-hand side of the bottom equation.

EXAMPLE 11.8

To find a particular solution of

$$t^2 y'' + t y' - y = \frac{2}{t+1}$$

for $t > 0$, our first step is to solve the associated homogeneous differential equation

$$t^2 y'' + t y' - y = 0.$$

This is a Cauchy–Euler equation, so we let $t = e^x$, $ty' = dy/dx$, and $t^2 y'' = d^2 y/dx^2 - dy/dx$ and obtain the differential equation

$$\frac{d^2 y}{dx^2} - y = 0,$$

with solution

$$y(x) = C_1 e^x + C_2 e^{-x}.$$

In terms of t the solution of the associated homogeneous differential equation is thus

$$y(t) = C_1 t + C_2 t^{-1}.$$

In the second step, we let

$$y_p(t) = t z_1(t) + t^{-1} z_2(t), \tag{11.48}$$

where z_1' and z_2' satisfy

$$t z_1' + t^{-1} z_2' = 0,$$

$$z_1' - t^{-2} z_2' = 2 / \left[t^2 (t + 1) \right]. \tag{11.49}$$

Finally, we solve for z_1' by multiplying the top equation in (11.49) by t^{-1} and adding the result to the bottom equation, as

$$z_1' = \frac{1}{t^2 (t + 1)}.$$

This gives z_2' as

$$z_2' = -t^2 z_1' = \frac{-1}{t + 1},$$

which may be integrated immediately to yield

$$z_2 = -\ln(t + 1).$$

To find an antiderivative for z_1, we first rewrite the form for z_1', using partial fractions, as

$$z_1' = -\frac{1}{t} + \frac{1}{t^2} + \frac{1}{t + 1},$$

so

$$z_1 = -\ln t - t^{-1} + \ln(t + 1).$$

Thus, from (11.48), our particular solution is

$$y_p(t) = t \ln(1 + 1/t) - 1 - t^{-1} \ln(t + 1). \qquad \square$$

EXERCISES

1. Show that the particular solution of (11.23),

$$\frac{dy}{dt} + p(t)y = q(t),$$

given by

$$y(t) = u(t) \int \frac{q(t)}{u(t)} \, dt,$$

may be obtained by making the substitution $y = u(t)z(t)$, where $u(t)$ is a solution of the associated homogeneous differential equation

$$\frac{du}{dt} + p(t)u = 0.$$

2. Find the general solution of the following differential equations.
- (a) $y'' + 2y' + y = 2t^{-2}e^{-t}$, $t \neq 0$
- (b) $y'' + 6y' + 9y = t^{-1}e^{-3t}$, $t \neq 0$
- (c) $y'' - y' = \sec^2 t - \tan t$, $-\pi/2 < t < \pi/2$
- (d) $y'' + y = \sec t$, $-\pi/2 < t < \pi/2$
- (e) $y'' + y = \tan t$, $-\pi/2 < t < \pi/2$. (Hint: Use $\sin^2 t + \cos^2 t = 1$.)

3. Find the general solution of the following differential equations. (You might find it useful to review *How to Solve a Cauchy-Euler Differential Equation* on page 452.)
- (a) $t^2 y'' + 7ty' + 5y = 10 - 4t^{-1}$, $t > 0$
- (b) $t^2 y'' - ty' + y = t \ln t$, $t > 0$
- (c) $t^2 y'' - 2y = \ln t$, $t > 0$
- (d) $t^2 y'' + ty' + y = t^3$, $t > 0$

4. Consider the nonhomogeneous second order linear differential equation

$$a_2(t)y'' + a_1(t)y' + a_0(t)y = f(t) \tag{11.50}$$

with $u(t)$ as a solution of the corresponding homogeneous differential equation. Show that the substitution $y = u(t)z(t)$ gives a particular solution of (11.50) as

$$y(t) = \int \left[\frac{1}{u(t)\mu(t)} \int \frac{u(t)\mu(t)f(t)}{a_2(t)} \, dt \right] dt,$$

where[1]

$$\mu(t) = \exp\left(\int \frac{a_1(t)}{a_2(t)} \, dt \right).$$

5. Find a particular solution of the differential equation (11.31) as follows.
- (a) Use Table 11.1 of Section 11.2.
- (b) Use the method of Exercise 4.

6. Use the method of Exercise 4 to find particular solutions for Exercises 2 and 3.

---

[1] Remember $\exp(a) = e^a$.

7. Consider the equation

$$t^3 y'' + t y' - y = e^{1/t}.$$

(a) Show that $y = t$ is a solution of the associated homogeneous differential equation.

(b) Show that the differential equation satisfied by $u(t)$ with the change of variable $y = u(t)t$ is

$$t^4 u'' + (2t^3 + t^2)u' = e^{1/t}.$$

(c) Solve this equation and obtain the general solution of the original differential equation as

$$y(t) = C_1 t + C_2 t e^{1/t} + (1 - t)e^{1/t}.$$

8. Consider the equation

$$(t^2 + t)y'' + (2 - t^2)y' - (2 + t)y = (t + 1)^2.$$

(a) Show that e^t is a solution of the associated homogeneous differential equation.

(b) Find a second solution of this homogeneous equation by making a change of variable $y = u(t)e^t$.

(c) Use the method of variation of parameters to find a particular solution.

(d) Write down the general solution of the original differential equation.

9. Show that using the method of variation of parameters gives the solution of the initial value problem

$$y'' - y = 1/t, \qquad y(1) = A, \qquad y'(1) = B,$$

as

$$y(t) = \frac{1}{2}\left[(A + B)e^{t-1} + (A - B)e^{1-t} + e^t \int_1^t \frac{e^{-s}}{s}\,ds - \int_1^t \frac{e^s}{s}\,ds\right].$$

10. Use the method of Exercise 4 to show that the solution of the initial value problem of Exercise 9 may be written in the form

$$y = e^t \int_1^t e^{-2x}\left(\int_1^x \frac{e^s}{s}\,ds\right)dx - \frac{B}{2}e^{2-t} + (\frac{A}{e} + \frac{B}{2})e^t.$$

11.4   Applications

We will now apply what we have learned in this chapter to various situations.

EXAMPLE 11.9: Tuning Forks

Our first example concerns a tuning fork, a device often shaped like the letter U with a stem. When struck against a solid object a tuning fork emits a sound with a specific frequency.

    We consider an example in which we have two ideal tuning forks, whose natural frequencies differ by a small amount — say, $5/\pi$ and $4.5/\pi$. Suppose that one fork is struck so it emits a signal of the form $19\cos 9t$. If both vibrating forks are then held firmly against a table, one model governing the vibration of the second fork is the differential equation

$$y'' + 100y = 19\cos 9t. \tag{11.51}$$

The initial conditions for this second tuning fork are

$$y(0) = y'(0) = 0, \tag{11.52}$$

signifying no motion in this second tuning fork at the moment the two forks are held against the table.

The characteristic equation of the associated homogeneous differential equation,

$$y'' + 100y = 0,$$

is

$$r^2 + 100 = 0,$$

with solutions of $\pm 10i$. Thus, the solution of the associated homogeneous equation is

$$y_h(t) = C_1 \sin 10t + C_2 \cos 10t.$$

We obtain the proper form for a particular solution of our original differential equation (11.51) using Table 11.1. Thus, we substitute

$$y_p(t) = A \sin 9t + B \cos 9t.$$

into (11.51) to obtain

$$(-81 + 100)A \sin 9t + (-81 + 100)B \cos 9t = 19 \cos 9t.$$

Solving for A and B gives $A = 0$ and $B = 1$, and the general solution is

$$y(t) = C_1 \sin 10t + C_2 \cos 10t + \cos 9t.$$

If we now evaluate the two arbitrary constants by imposing the initial conditions in (11.52), we find that $C_1 = 0$, and $C_2 = -1$, giving our solution as

$$y(t) = \cos 9t - \cos 10t. \tag{11.53}$$

The graph of this explicit solution is shown in Figure 11.6, where the motion appears to have a rapid oscillation within a slower oscillation. A more convenient form to analyze this behavior may be obtained if we use the trigonometric identity

$$\cos \alpha - \cos \beta = 2 \sin \left(\frac{\beta + \alpha}{2} \right) \sin \left(\frac{\beta - \alpha}{2} \right) \tag{11.54}$$

to transform our explicit solution in (11.53) to the form

$$y(t) = 2 \sin \frac{19t}{2} \sin \frac{t}{2}.$$

This form of the solution suggests that the motion is composed of a slow oscillation ($\sin t/2$) with a rapidly time-varying amplitude ($2 \sin 19t/2$). This time-varying amplitude causes the sound emitted by the tuning fork to change intensity with time. $\square$

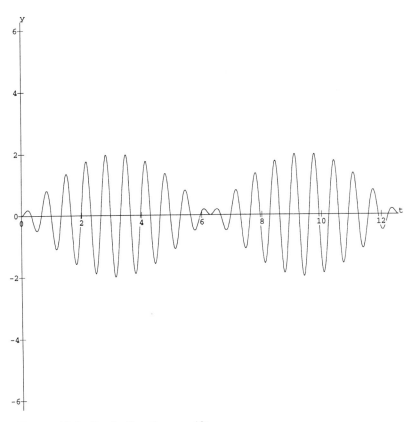

FIGURE 11.6 Graph of $\cos 9t - \cos 10t$

EXAMPLE 11.10: Vibration of a Linear Spring Under the Influence of a Periodic

Forcing Function

In Example 8.5 on page 384 we considered the oscillatory motion of a linear spring. Here we consider the situation in which that spring is subject to a periodic external force of the form $q \cos \gamma t$, where q and γ are constants. This gives the differential equation

$$mx'' + kx = q \cos \gamma t,$$

where m, k, q, and γ are constants. In this example we take $m = 1$, $k = 100$, and $q = 19$ and leave γ as an unspecified parameter. This gives

$$x'' + 100x = 19 \cos \gamma t, \qquad \qquad \textbf{(11.55)}$$

and our task is to consider the effect of changes in γ on the motion of the spring mass system. [Notice the similarity of (11.55) with the equation governing the motion of the tuning fork.]

We know from our previous example that $\sin 10t$ and $\cos 10t$ are two linearly independent solutions of the associated homogeneous differential equation. Table 11.1 gives the proper form for the particular solution of (11.55) as

$$x_p(t) = A \sin \gamma t + B \cos \gamma t,$$

if $\gamma \neq 10$.

Substituting this expression into (11.55) and solving the resulting equation for A and B gives $A = 0$ and $B = 19/(-\gamma^2 + 100)$. (Recall that $\sin \gamma t$ and $\cos \gamma t$ are linearly independent functions over any interval.) This gives our particular solution as $x_p(t) = 19/(100 - \gamma^2) \cos \gamma t$ and the general solution of (11.55) as

$$x(t) = C_1 \sin 10t + C_2 \cos 10t + \frac{19}{100 - \gamma^2} \cos \gamma t. \tag{11.56}$$

Suppose that at $t = 0$ the mass on this spring was at rest in its equilibrium position and subject to the forcing function. The initial conditions appropriate for this situation are $x(0) = 0$, $x'(0) = 0$. Imposing these initial conditions on our solution in (11.56) gives $C_1 = 0$, $C_2 = -19/(100 - \gamma^2)$, so our solution becomes

$$x(t) = \frac{19}{100 - \gamma^2} \left(\cos \gamma t - \cos 10t \right). \tag{11.57}$$

Figures 11.7, 11.8, 11.9, and 11.10 show the graphs of this solution for $\gamma = 9.5$, 9.7, 9.9, and 9.99, respectively. Figure 11.11 shows all four cases (using a condensed vertical scale). It is clear that as γ approaches 10, something is happening to the graphs. To understand this effect we use the trigonometric identity given in (11.54) to change the form of our answer in (11.57) to

$$x(t) = \frac{38}{100 - \gamma^2} \sin \left[(10 + \gamma)t/2 \right] \sin \left[(10 - \gamma)t/2 \right]. \tag{11.58}$$

As in the previous example, we can consider the motion as sinusoidal, $\sin \left[(10 - \gamma)t/2 \right]$, with a time varying amplitude of $38 \sin \left[(10 + \gamma)t/2 \right] / \left(100 - \gamma^2 \right)$. Notice that the closer that γ is to 10, the larger is this amplitude, and the longer the period of $\sin \left[(10 - \gamma)t/2 \right]$. In fact this amplitude is unbounded as γ approaches 10, although setting $\gamma = 10$ in (11.58) gives an indeterminate form of $0/0$. Of course, for this application of the spring-mass system, the displacements cannot get very large, and the spring would break at some point.

The situation in which the forcing frequency equals the natural frequency of a system (in this case when $\gamma = 10$) is called RESONANCE. Of course, the particular solution $x_p(t) = 19/(100 - \gamma^2) \cos \gamma t$ is not valid for $\gamma = 10$, so we return to the differential equation

$$x'' + 100x = 19 \cos 10t \tag{11.59}$$

and now try a particular solution of the form

$$x_p(t) = At \sin 10t + Bt \cos 10t.$$

The t is present in this particular solution because the normal trial solution contains two terms, $\sin 10t$ and $\cos 10t$, which are solutions of the associated homogeneous differential equation.

Substituting this expression into (11.59) gives

$$20A \cos 10t - 20B \sin 10t = 19 \cos 10t,$$

so we need $B = 0$ and $A = 19/20$. This gives our particular solution as $19/20t \sin 10t$ and the general solution of (11.59) as

$$x(t) = C_1 \sin 10t + C_2 \cos 10t + \frac{19}{20}t \sin 10t. \tag{11.60}$$

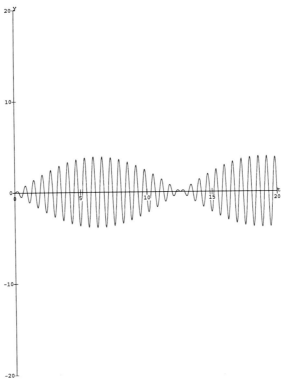

FIGURE 11.7 Graph of $\left[38/(100 - \gamma^2)\right]\sin\left[(10+\gamma)t/2\right]$ $\sin\left[(10-\gamma)t/2\right]$ with $\gamma = 9.5$

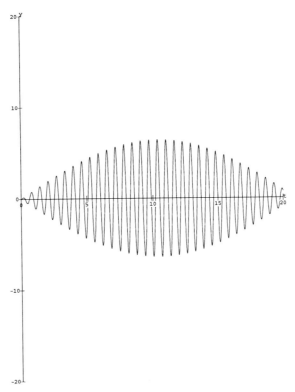

FIGURE 11.8 Graph of $\left[38/(100 - \gamma^2)\right]\sin\left[(10+\gamma)t/2\right]$ $\sin\left[(10-\gamma)t/2\right]$ with $\gamma = 9.7$

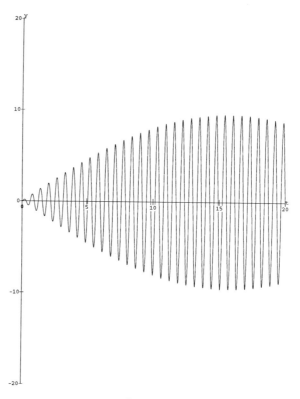

FIGURE 11.9 Graph of $\left[38/(100-\gamma^2)\right]\sin\left[(10+\gamma)t/2\right]$ $\sin\left[(10-\gamma)t/2\right]$ with $\gamma = 9.9$

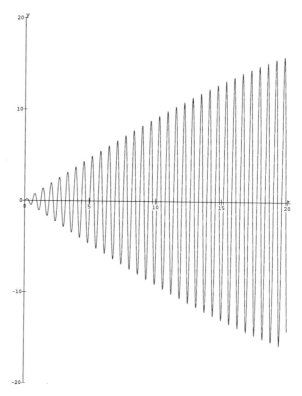

FIGURE 11.10 Graph of $\left[38/(100-\gamma^2)\right]$ $\sin\left[(10+\gamma)t/2\right]\sin\left[(10-\gamma)t/2\right]$ with $\gamma = 9.99$

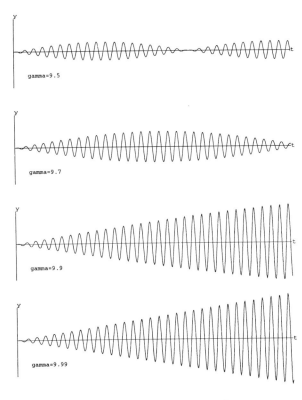

FIGURE 11.11 Graph of $\left[38/(100 - \gamma^2)\right]$ $\sin\left[(10 + \gamma)t/2\right]\sin\left[(10 - \gamma)t/2\right]$ with $\gamma = 9.5$, 9.7, 9.9, and 9.99

Imposing our previous initial conditions on our solution in (11.60) gives $C_1 = C_2 = 0$, so the final form of our solution is

$$x(t) = \frac{19}{20}t \sin 10t.$$

Here the solution will oscillate between t and $-t$ and will be unbounded as $t \to \infty$. Its graph is shown in Figure 11.12. Notice the strong similarity between this figure and Figure 11.10.   □

EXAMPLE 11.11: Forced Vibration of a Damped Spring-Mass System

In Example 8.6 on page 389 we discussed the damped motion of a spring-mass system. Here we consider that same system, which is also subject to an oscillatory force, resulting in the differential equation

$$m\frac{d^2x}{dt^2} + b\frac{dx}{dt} + kx = q\cos\gamma t,$$

where m, b, k, q, and γ are constants. For illustrative purposes, we use the values of $b/m = 2$, $k/m = 17$, and $q/m = Q$, giving

$$\frac{d^2x}{dt^2} + 2\frac{dx}{dt} + 17x = Q\cos\gamma t. \tag{11.61}$$

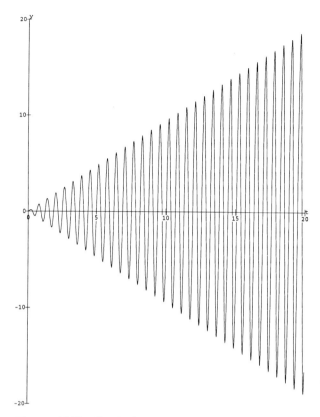

FIGURE 11.12   Graph of $(19t/20)\sin 10t$

We know that our solution will consist of two parts. The first part is the solution of the homogeneous differential equation associated with (11.61). This leads to consideration of the characteristic equation

$$r^2 + 2r + 17 = 0,$$

with solution $r = -1 \pm 4i$. This solution gives our two linearly independent solutions of the associated homogeneous differential equation as $e^{-t}\sin 4t$ and $e^{-t}\cos 4t$, so the first part of our solution is

$$x_h(t) = C_1 e^{-t}\sin 4t + C_2 e^{-t}\cos 4t.$$

The second part of our solution of (11.61) is a particular solution, which from Table 11.1 will have the form

$$x_p(t) = A\sin\gamma t + B\cos\gamma t.$$

Substituting this form for x_p into (11.61) gives the equation

$$-\gamma^2\,(A\sin\gamma t + B\cos\gamma t) + 2\gamma\,(A\cos\gamma t - B\sin\gamma t) + 17\,(A\sin\gamma t + B\cos\gamma t)$$
$$= Q\cos\gamma t.$$

Because $\sin\gamma t$ and $\cos\gamma t$ are linearly independent functions, we may equate coefficients of like terms in this equation to obtain

$$-\gamma^2 A - 2\gamma B + 17A = 0,$$

$$-\gamma^2 B + 2\gamma A + 17B = Q.$$

From the first equation above we have that $B = (\gamma^2 - 17)A/(-2\gamma)$, while using this fact in the second equation yields

$$[2 + (17 - \gamma^2)^2/(2\gamma)]A = Q,$$

giving A and B as

$$A = -2\gamma Q/[4\gamma^2 + (17 - \gamma^2)^2],$$

$$B = (\gamma^2 - 17)Q/[4\gamma^2 + (17 - \gamma^2)^2/(2\gamma)].$$

(11.62)

Thus, our particular solution is given by

$$x_p(t) = A \sin \gamma t + B \cos \gamma t,$$

and our general solution of (11.61) is

$$x(t) = C_1 e^{-t} \sin 4t + C_2 e^{-t} \cos 4t + A \sin \gamma t + B \cos \gamma t, \qquad (11.63)$$

where C_1 and C_2 are arbitrary constants. A and B are given in (11.62). The values of C_1 and C_2 are specified once some initial conditions are given, but regardless of their values, the first two terms in (11.63) become very small as t increases. Because this is so, these terms constitute the transient solution, and the remaining terms that do not decay give the steady state solution. We now change the form of this steady state solution by using some results from trigonometry. We write the steady state solution $x_{ss}(t)$ as

$$x_{ss}(t) = A \sin \gamma t + B \cos \gamma t = \sqrt{A^2 + B^2}\left(\frac{A}{\sqrt{A^2 + B^2}} \sin \gamma t + \frac{B}{\sqrt{A^2 + B^2}} \cos \gamma t\right),$$

(11.64)

where we multiplied and divided by $\sqrt{A^2 + B^2}$. Now the coefficients of $\sin \gamma t$ and $\cos \gamma t$ in (11.64) are such that the sum of their squares add to 1, so we define a phase angle ϕ by

$$\cos \phi = \frac{A}{\sqrt{A^2 + B^2}}, \qquad \sin \phi = \frac{B}{\sqrt{A^2 + B^2}}.$$

This allows us to write our steady state solution as

$$x_{ss}(t) = \sqrt{A^2 + B^2} \sin (\gamma t + \phi).$$

Notice that the amplitude of our steady state solution depends on the value of the forcing frequency and, using (11.62), is given by

$$\sqrt{A^2 + B^2} = \frac{Q}{\sqrt{4\gamma^2 + (17 - \gamma^2)^2}}. \qquad (11.65)$$

The graph of this amplitude as a function of the forcing frequency γ is given in Figure 11.13 and shows that the amplitude has a maximum value. To find this maximum value we take the derivative of the amplitude with respect to γ and set the result to zero. Performing this calculation gives the maximum value of the steady state amplitude as $\gamma = \sqrt{15}$ (see Exercise 9 on page 555).

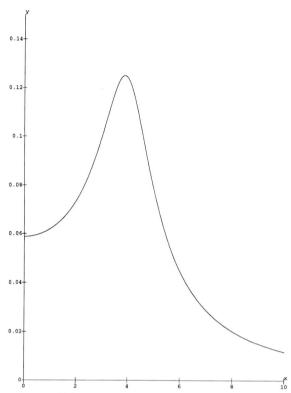

FIGURE 11.13 Amplitude of the steady state solution versus the forcing frequency

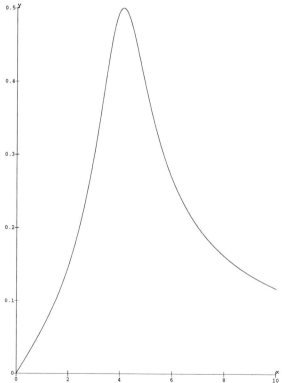

FIGURE 11.14 Velocity of the steady state solution versus the forcing frequency

An interesting property of this steady state solution is the fact that the value of the forcing frequency that maximizes the amplitude of the steady state velocity ($\gamma \sqrt{A^2 + B^2}$) is $\gamma = \sqrt{17}$ (see Figure 11.14). But this is simply the frequency for the same mass and spring constant but with no damping. This means that the frequency that maximizes the steady state velocity of a forced, damped spring-mass system is independent of the damping coefficient. (See Exercise 9 on page 555.) □

EXAMPLE 11.12: Spring-Mass System with Coulomb Friction

If we place a spring-mass system on a horizontal surface, the major damping force is not necessarily proportional to the velocity. One model of this situation is to also have a damping force take on a constant value that is always in opposition to the direction of motion. The differential equation for this model is given by

$$m\frac{d^2x}{dt^2} + b\frac{dx}{dt} + kx = -Qm \, signum\left(\frac{dx}{dt}\right), \tag{11.66}$$

where the value of Q depends on the roughness of the contact between the mass and the surface and where

$$signum(x) = \begin{cases} 1 & \text{if } x > 0, \\ 0 & \text{if } x = 0, \\ -1 & \text{if } x < 0. \end{cases}$$

Thus, the signum function in (11.66) has the value of 1 if $dx/dt > 0$ and -1 if $dx/dt < 0$. Because this differential equation has all constant coefficients, we may transform it to a system of first order differential equations and do an analysis in the phase plane. To do this we define $y = dx/dt$ and obtain the system

$$\frac{dx}{dt} = y, \qquad \frac{dy}{dt} = -\frac{b}{m}y - \frac{k}{m}x - Q \, signum(y), \tag{11.67}$$

so its equation in the phase plane is

$$\frac{dy}{dx} = \frac{-by - kx - Qm \, signum(y)}{my}. \tag{11.68}$$

The most obvious implication of (11.68) is that for positive values of x and y, the slope field will be entirely negative, going towards vertical tangents for $y = 0$. This is clear from the direction field as seen in Figure 11.15 (with $b = 0.1$, $k = 2$, $m = 1$, and $Q = 0.1$). From (11.67) we have that in this first quadrant x is an increasing function and y is a decreasing function. We can also observe, both from (11.67) and the slope field, that the mass is at rest (zero velocity) when the displacement x has its extreme values, but the maximum velocity appears not to occur when $x = 0$. Figure 11.16 shows a trajectory in the phase plane for the initial conditions $x(0) = 0$ and $x'(0) = y(0) = 5$, that is, when the motion is started with an initial positive velocity. □

EXAMPLE 11.13: Sliding Mass with Coulomb Friction

Suppose in the last example we take the value of the spring constant k to be zero, where our initial conditions are still $x(0) = 0$, $x'(0) = 5$. This models the situation in which a mass is given an initial velocity and there is no restoring force. Thus, the mass will move until the damping force brings it to a stop. Here we are interested only in the first quadrant, where both x and y are positive. A typical direction field is shown in Figure 11.17 (with $b = 0.1$, $m = 1$, and $Q = 1.1$), where it is clear that motion is from left to right and is decreasing and

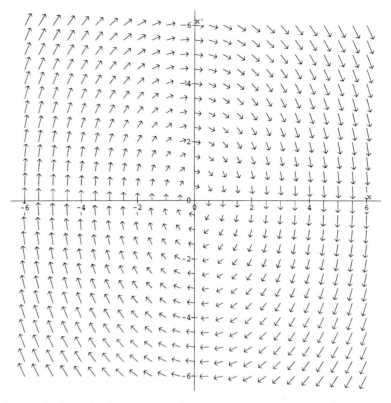

FIGURE 11.15   Direction field for spring-mass system with Coulomb friction

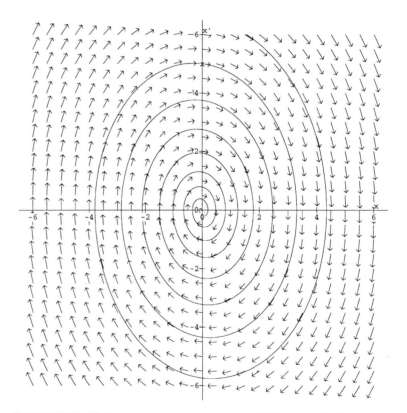

FIGURE 11.16   Trajectory for spring-mass system with Coulomb friction

concave down. For this situation the velocity is never negative, so the differential equation for the trajectories, from (11.68), is simply

$$\frac{dy}{dx} = \frac{-by - Qm}{my}.$$

This is a separable differential equation, which, when solved, gives us our trajectories in the phase plane as

$$y - \frac{Qm}{b} \ln\left(y + \frac{Qm}{b}\right) = -\frac{b}{m}x + C.$$

Initial conditions give $C = 5 - (Qm/b)\ln(5 + Qm/b)$, so our solution may be expressed as

$$y - \frac{Qm}{b} \ln\left(\frac{y + Qm/b}{5 + Qm/b}\right) = 5 - \frac{b}{m}x. \qquad \textbf{(11.69)}$$

From this we can find how far the mass will move before stopping (when $y = 0$), which is

$$x = 5\frac{m}{b} + Q \ln\left(\frac{Qm/b}{5 + Qm/b}\right),$$

or $x \approx 8.78$, using $b = 0.1$, $m = 1$, and $Q = 1.1$.

The graph of the trajectory (11.69) for parameter values of $Q = 1.1$ and $b/m = 0.1$ is also shown in Figure 11.17. From this we can confirm how far the mass will move before

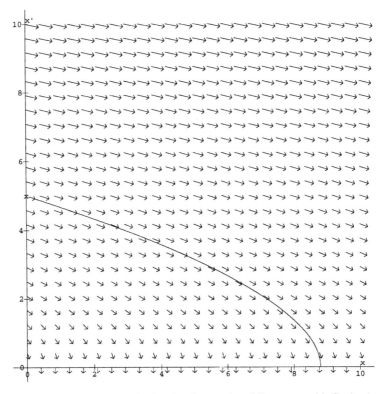

FIGURE 11.17 Direction field and trajectory for sliding mass with Coulomb friction

stopping, but we cannot discover from this graph the time this takes. We need to return to our original second order equation (11.66), which for this situation reduces to

$$m\frac{d^2x}{dt^2} + b\frac{dx}{dt} = -Qm. \tag{11.70}$$

The general solution of this equation is

$$x(t) = C_1 + C_2 e^{-bt/m} - \frac{Q}{b}t.$$

Thus, $x'(t) = -b/m C_2 e^{-bt/m} - Q/b$, so to satisfy our initial conditions, we need $C_1 + C_2 = 0$ and $-b/m C_2 - Q/b = 5$. This gives the distance that the mass moves as a function of time as

$$x(t) = \frac{m}{b}\left(5 + \frac{Q}{b}\right)\left(1 - e^{-bt/m}\right) - \frac{Q}{b}t$$

and the velocity as

$$x'(t) = \left(5 + \frac{Q}{b}\right)e^{-bt/m} - \frac{Q}{b}.$$

This last equation allows us to determine the time it takes for the mass to come to a stop as $t = (m/b)\ln(5b/Q + 1)$, which, for $b = 0.1$, $m = 1$, and $Q = 1.1$, is $t \approx 3.73$. $\quad\square$

EXERCISES

1. Someone has said that for differential equations of the form

$$x'' + \omega^2 x = \cos \gamma t$$

we can find a particular solution from the form

$$x_p(t) = B \cos \gamma t$$

instead of from the two expressions given by Table 11.1. Explain why this could be true.

2. Someone has said that for differential equations of the form

$$x'' + \omega^2 x = \sin \gamma t$$

we can find a particular solution from the form

$$x_p(t) = A \sin \gamma t$$

instead of from the two expressions given by Table 11.1. Explain why this could be true.

3. Consider the solution given by (11.57) for $\gamma \neq 10$, and take the limit as $\gamma \to 10$. (Either use Taylor series or L'Hôpital's rule.) Compare your answer with that given by (11.60).

4. Consider the spring-mass governed by

$$\frac{d^2x}{dt^2} + 16x = 15\sin t, \qquad x(0) = 0, \qquad x'(0) = \sqrt{3}.$$

(a) Find x as a function of time.

(b) What is the period of the solution in part (a)?

5. Consider the spring-mass governed by

$$\frac{d^2x}{dt^2} + 2\frac{dx}{dt} + 17x = 2\sin\gamma t, \qquad x(0) = 0, \qquad x'(0) = 0.$$

(a) Find x as a function of time.

(b) Identify the steady state and transient parts of the solution in part (a).

6. Consider the initial value problem

$$\frac{d^2x}{dt^2} + 2\alpha\frac{dx}{dt} + \left(\omega^2 + \alpha^2\right)x = \sin\gamma t, \qquad x(0) = 0, \qquad x'(0) = 0.$$

(a) Find x as a function of time.

(b) Identify the steady state and transient parts of the solution in part (a).

(c) Determine the amplitude of the steady state portion of the solution, and, considering α and ω as fixed, find the value of γ that maximizes this amplitude. (Hint: Use calculus.)

(d) Find the value of γ in Exercise 5 that maximizes the amplitude of the steady state solution. (When the frequency of the periodic forcing function is chosen to maximize the amplitude of the steady state response, this damped system is often said to be in resonance.)

7. Find the value of γ so that the system governed by

$$\frac{d^2x}{dt^2} + 6\frac{dx}{dt} + 22x = \cos\gamma t$$

is in resonance.

8. According to Robert Ehrlich[2] "You can find the resonant frequency of a hand-held spring from which a hanging weight is suspended, by gently shaking the top of the spring at different frequencies and seeing how the amplitude varies."

(a) Explain how this experiment will find the resonant frequency.

(b) Find the resonant frequency of a Slinky spring, by performing this experiment.

9. Show that the maximum value of the amplitude of the steady state displacement in Example 11.11 [$\sqrt{A^2 + B^2}$ in (11.65)] is obtained when $\gamma = \sqrt{15}$. Show that the maximum value of the amplitude of the steady state velocity in Example 11.11 ($\gamma\sqrt{A^2 + B^2}$) is obtained when $\gamma = \sqrt{17}$.

10. Consider the general damped spring-mass system with a sinusoidal forcing function

$$m\frac{d^2x}{dt^2} + b\frac{dx}{dt} + kx = q\cos\gamma t.$$

(a) Show that the general solution has the form

$$x(t) = C_1 e^{\alpha t}\sin\beta t + C_2 e^{\alpha t}\cos\beta t + A\sin\gamma t + B\cos\gamma t$$

where C_1 and C_2 are arbitrary constants, $\alpha + i\beta = -b/(2m) + i\sqrt{4mk - b^2})/(2m)$, $A = b\gamma q/[(\gamma b)^2 + (k - \gamma^2 m)^2]$, and $B = (k - \gamma^2 m)q/[(\gamma b)^2 + (k - \gamma^2 m)^2]$.

(b) Show that the maximum value of the amplitude of the steady state solution in part (a) is given when the frequency of the forcing function is given by $\gamma = \sqrt{k/m - b^2/(2m^2)}$.

---

[2]"Turning the World Inside Out" by R. Ehrlich, Princeton University Press, 1990, page 93.

(c) Show that the maximum value of the amplitude of the velocity of the steady state solution in part (a) is given by $\gamma = \sqrt{k/m}$.

11. Solve (11.66) for initial conditions $x(0) = 0$, $x'(0) = 10$, $b/m = 2$, $k/m = 10$, and $Q = 2$ for values of t between 0 and when it changes direction the second time. Hint: Initially the sign of dx/dt is positive, so solve (11.66) for x and determine the value of t for which $dx/dt = 0$. Then, for that value of t, say t_s, use $x(t_s)$ as the initial displacement and $x'(t_s) = 0$ as new initial conditions for (11.66) with dx/dt now negative for times slightly greater than t_s.

12. Suppose that a mass is governed by the differential equation

$$m\frac{d^2x}{dt^2} + b\left(\frac{dx}{dt}\right)^{\gamma} = 0,$$

that is, the damping force is proportional to the velocity to a power. If the mass is subject to the initial conditions $x(0) = 0$ and $x'(0) = 10$, will the mass come to rest in a finite time? If so, how far will it travel and what will be its travel time? If it does not come to rest, explain why. You may want to consider four different cases for γ: $0 < \gamma < 1$, $\gamma = 1$, $1 < \gamma < 2$, and $\gamma > 2$.

13. Suppose you have a block of wood on a table of length L, and you push it from one end with an initial velocity of v_0. If the motion follows (11.70), where L, m, b, and Q are all specified, for what initial velocities will the block not reach the end of the table?

14. Consider a slender metal rod, embedded in a solid base, with a mass at the free end. This rod will remain straight when the mass is very small. However, for larger masses, this equilibrium may be unstable, where a small displacement from equilibrium will cause the rod-mass combination to move to a new equilibrium point. If y denotes the displacement from equilibrium and x the distance along the strut, then the shape of the strut $y = y(x)$ is governed (for small values of y) by the differential equation

$$EI\frac{d^2y}{dx^2} = mg(y_0 - y) + (L - x)f,$$

where EI is the flexural rigidity, m is the mass at the end of the rod, g is a gravitational constant, y_0 is the displacement of the rod, L is the length the rod, and f is the horizontal force on the mass. One end of the strut is like a built-in beam, so appropriate boundary conditions are $y(0) = y'(0) = 0$.

(a) Find the solution of this initial value problem.

(b) For a displacement of y_0, compute the resultant force f on the mass.

(c) The critical mass for this strut is defined as the mass for which the force in part (b) is zero. Find the value of this critical mass.

(d) Note that for this critical mass there are three equilibrium positions, two of them stable and one of them unstable. Explain why this is so.

11.5   Higher Order Homogeneous Differential Equations

In this section we examine homogeneous linear differential equations in which the order is greater than two. In Example 8.14, we solved the fourth order differential equation

$$\frac{d^2}{dt^2}\left(EI\frac{d^2y}{dt^2}\right) + P\frac{d^2y}{dt^2} = 0, \tag{11.71}$$

which governs the buckling of a long shaft. Here y is the deflection from equilibrium, t the distance along the shaft, EI is the flexural rigidity, and P is the applied force. Our solution there had the form

$$y(t) = C_1 \cos \lambda t + C_2 \sin \lambda t + C_3 t + C_4, \tag{11.72}$$

where $\lambda = \sqrt{P/(EI)}$, which we obtained by a substitution that reduced the differential equation to one of second order. For second order linear differential equations with constant coefficients, we used the characteristic equation to find solutions of the form e^{rt}. If we try such a solution for (11.71) we obtain

$$EIr^4 + Pr^2 = r^2(EIr^2 + P) = 0.$$

Because this equation has a double root[3], $r = 0$, and complex roots, $r = \pm i \sqrt{P/(EI)}$, we have the solution in (11.72) above.

Notice that we have polynomials and sinusoidal functions in our solution given by (11.72). In earlier chapters we noticed that all solutions of second order linear differential equations with constant coefficients have the form of exponential functions, sine or cosine functions, products of these two types of functions, or (for double roots of the characteristic equation) these products multiplied by t. (See *How to Solve Second Order Linear Differential Equations with Constant Coefficients* on page 344.) Because the sine and cosine functions, using Euler's formula $e^{it} = \cos t + i \sin t$, and constants, $e^0 = 1$, may be written in terms of exponential functions, we want to see if this is true for solutions of higher order linear differential equations with constant coefficients. Thus, we consider the nth order differential equation

$$a_n \frac{d^n y}{dt^n} + a_{n-1} \frac{d^{n-1} y}{dt^{n-1}} + \cdots + a_2 \frac{d^2 y}{dt^2} + a_1 \frac{dy}{dt} + a_0 y = 0, \tag{11.73}$$

where the coefficients $a_n, a_{n-1}, \cdots, a_2, a_1$, and a_0 are constants. In summation notation this becomes

$$\sum_{m=0}^{n} a_m \frac{d^m y}{dt^m} = 0,$$

where the zeroth derivative is just the function itself. Because $de^{rt}/dt = re^{rt}$ for real or complex values of r, it follows that $d^m e^{rt}/dt^m = r^m e^{rt}$ for any positive integer m. Thus, we seek solutions of (11.73) in the form $y = e^{rt}$. Substituting this exponential function into (11.73), using the preceding result, and factoring out the common factor of e^{rt} yields

$$\left(a_n r^n + a_{n-1} r^{n-1} + \cdots + a_2 r^2 + a_1 r + a_0\right) e^{rt} = 0.$$

The exponential function is never zero, so we have

$$a_n r^n + a_{n-1} r^{n-1} + \cdots + a_2 r^2 + a_1 r + a_0 = 0.$$

This nth order algebraic equation is called the characteristic equation associated with (11.73) and will have n roots. Associated with each root, designated by r_i, is the solution

---

[3] A root of multiplicty 2.

$e^{r_i t}$. To form our solution of the differential equation we take linear combinations of these solutions. [Notice that if we write the differential equation in operator form,

$$\left(a_n D^n + a_{n-1} D^{n-1} + \cdots + a_2 D^2 + a_1 D + a_0\right) y = 0,$$

(D represents d/dt), then the characteristic equation may be simply obtained by replacing the derivative operator D with the parameter r.]

We now illustrate this technique with an example.

EXAMPLE 11.14

Consider the fourth order linear differential equation

$$\left(D^4 - 8D^2 + 16\right) y = 0,$$

where the characteristic equation resulting from solutions of the form e^{rt} is

$$r^4 - 8r^2 + 16 = 0.$$

This equation may be factored as

$$(r^2 - 4)(r^2 + 4) = (r - 2)(r + 2)(r - 2i)(r + 2i) = 0,$$

so our solution is given by

$$y(t) = C_1 e^{2t} + C_2 e^{-2t} + C_3 \cos 2t + C_4 \sin 2t. \tag{11.74}$$

If we had the initial conditions $y(0) = 4$, $y'(0) = -2$, $y''(0) = -8$, $y'''(0) = -8$, we would take derivatives of our solution in (11.74) before substituting in the value of $t = 0$. This gives the four algebraic equations

$$\begin{aligned} C_1 + C_2 + C_3 &= 4, \\ 2C_1 - 2C_2 + 2C_4 &= -2, \\ 4C_1 + 4C_2 - 4C_3 &= -8, \\ 8C_1 - 8C_2 - 8C_4 &= -8, \end{aligned}$$

in four unknowns. If we multiply the first equation by -4 and add the result to the third equation, we obtain $C_3 = 3$, whereas if we multiply the second equation by -4 and add the result to the fourth equation, we have $C_4 = 0$. Using these values in the first two equations gives $C_1 = 0$ and $C_2 = 1$, and the solution of our initial value problem is

$$y(t) = e^{-2t} + 3 \cos 2t.$$

The graphs of this solution and $3 \cos 2t$ are shown in Figure 11.18, where we see the effect of the exponential term diminishes quickly with time. We would say that e^{-2t} was the transient solution and $3 \cos 2t$ was the steady state solution, using the same terminology we used for first and second order linear differential equations. □

This technique works for linear differential equations with constant coefficients of any order.

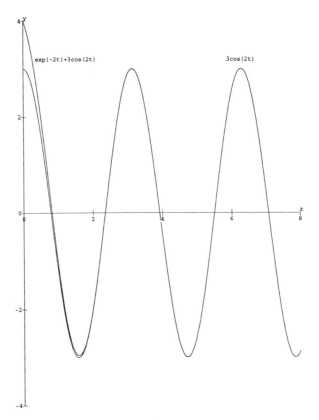

FIGURE 11.18   Graphs of $e^{-2t} + 3 \cos 2t$ and $3 \cos 2t$

How to Solve Higher Order Linear Differential Equations with Constant Coefficients

Purpose: To find the general solution of the linear differential equation

$$\sum_{m=0}^{n} a_m \frac{d^m y}{dt^m} = a_n \frac{d^n y}{dt^n} + a_{n-1} \frac{d^{n-1} y}{dt^{n-1}} + \cdots + a_2 \frac{d^2 y}{dt^2} + a_1 \frac{dy}{dt} + a_0 y = 0,$$

where the coefficients $a_n, a_{n-1}, \cdots, a_2, a_1, a_0$ are constants.

Technique:

1. Write the characteristic equation as

$$\sum_{m=0}^{n} a_m r^m = a_n r^n + a_{n-1} r^{n-1} + \cdots + a_2 r^2 + a_1 r + a_0 = 0.$$

2. Solve the characteristic equation, calling the solutions $r_1, r_2, r_3, \cdots, r_n$, and identify

   (a) All simple roots[4] (real or complex).

   (b) All repeated roots (real or complex).

3. Write the general solution. Here we have two cases:

---

[4] A simple root is a root with multiplicity one.

(a) The roots of the characteristic equation are all simple. In this case, to each root we assign the function

$$y_m(t) = e^{r_m t}, \qquad m = 1, 2, 3, \cdots, n.$$

The general solution is given by

$$y(t) = \sum_{m=1}^{n} C_m e^{r_m t} = C_1 e^{r_1 t} + C_2 e^{r_2 t} + \cdots + C_n e^{r_n t}. \tag{11.75}$$

(b) The characteristic equation has one or more roots that are not simple. We take the sum of arbitrary constants multiplied by the exponential functions associated with each simple root to form part of our solution, as in (11.75). To each nonsimple root r_m of multiplicity k, we assign the function

$$y_m(t) = \left(C_0^* + C_1^* t + \cdots + C_{k-1}^* t^{k-1} \right) e^{r_m t}$$

and add this to the sum associated with the simple roots. Doing this for all the repeated roots gives us the general solution.

Comments about solving higher order linear differential equations with constant coefficients

- Notice that for complex roots, it is usually advantageous to use a linear combination of $e^{\alpha t} \cos \beta t$ and $e^{\alpha t} \sin \beta t$ instead of a linear combination of $e^{\alpha + i\beta}$ and $e^{\alpha - i\beta}$.

- Notice that arbitrary constants are already included in the expression for $y_m(t)$ resulting from multiple roots [in step 3(b)] and that we will not need to multiply by an additional arbitrary constant when forming the solution, as in (11.75).

- This procedure produces n linearly independent functions (see Exercises 3 through 9 on page 565).

- If we choose the arbitrary constants in our general solution so it takes on specified initial values for the function and its first $n - 1$ derivatives [that is, $y(0)$, $y'(0)$, $y''(0)$, $\cdots$, $y^{(n-1)}(0)$ are specified], the resulting solution is unique.

- Higher order Cauchy–Euler differential equations have the form

$$\sum_{m=0}^{n} b_m x^m \frac{d^m y}{dx^m} = b_n x^n \frac{d^n y}{dx^n} + b_{n-1} x^{n-1} \frac{d^{n-1} y}{dx^{n-1}} + \cdots + b_2 x^2 \frac{d^2 y}{dx^2} + b_1 x \frac{dy}{dx} + b_0 y = 0,$$

where the b_m, $m = 0, 1, 2, \cdots, n$ are all constants. In Chapter 9 we used the transformation $x = e^t$ to transform second order Cauchy-Euler differential equations into ones with constant coefficients.[5] In that case we replace $x d/dx$ with d/dt and $x^2 d^2/dx^2$ with $d^2/dt^2 - d/dt$. If we let D represent the operator d/dt, we replace $x d/dx$ with D and $x^2 d^2/dx^2$ with $D^2 - D = D(D - 1)$. For higher order derivatives, these replacements become

$$x^3 \tfrac{d^3}{dx^3} = D(D - 1)(D - 2),$$
$$x^4 \tfrac{d^4}{dx^4} = D(D - 1)(D - 2)(D - 3), \cdots,$$
$$x^n \tfrac{d^n}{dx^n} = D(D - 1)(D - 2)(D - 3) \cdots [D - (n - 1)].$$

---

[5] This assumes that $x > 0$. If $x < 0$ we use $x = -e^t$.

Thus, if we make these substitutions in Cauchy-Euler differential equations, we may obtain their solution by the technique previously outlined (see Example 11.17).

EXAMPLE 11.15

Solve the differential equation

$$\left(D^4 + 8D^3 + 16\right) y = 0, \tag{11.76}$$

subject to the initial conditions

$$y(0) = 0, \qquad y'(0) = 5, \qquad y''(0) = 0, \qquad y'''(0) = -28.$$

Using our previous technique, we note that the characteristic equation associated with (11.76) is

$$r^4 + 8r^2 + 16 = (r^2 + 4)^2 = 0,$$

so both $2i$ and $-2i$ are double roots.

This gives our general solution of (11.76) as

$$y(t) = (C_1 + C_2 t) \cos 2t + (C_3 + C_4 t) \sin 2t.$$

Satisfying the initial conditions gives the following system of four algebraic equations

$$y(0) = C_1 = 0,$$
$$y'(0) = C_2 + 2C_3 = 5,$$
$$y''(0) = -4C_1 + 4C_4 = 0,$$
$$y'''(0) = -12C_2 - 8C_3 = -28.$$

These equations are readily solved, yielding $C_1 = 0$, $C_2 = 1$, $C_3 = 2$, and $C_4 = 0$. Thus, the solution to this initial value problem is

$$y(t) = t \cos 2t + 2 \sin 2t. \qquad \square$$

EXAMPLE 11.16

Find the general solution of the following differential equation, which for convenience we give in factored operator form as

$$\left[\left(D^2 - 9\right)\left(D^2 - 9\right)\left(D^2 - 4\right)\right] y = 0.$$

The characteristic equation for this sixth order equation is

$$\left(r^2 - 9\right)\left(r^2 - 9\right)\left(r^2 - 4\right) = 0,$$

which has roots at $r = 2, -2, 3$, and -3. The roots at 3 and -3 are both double roots, so our general solution is given by

$$y(t) = C_1 e^{2t} + C_2 e^{-2t} + (C_3 + C_4 t)e^{3t} + (C_5 + C_6 t)e^{-3t}. \qquad \square$$

EXAMPLE 11.17

Solve the differential equation

$$x^3 \frac{d^3 y}{dx^3} - 6x^2 \frac{d^2 y}{dx^2} + 18x \frac{dy}{dx} - 24y = 0, \tag{11.77}$$

subject to the initial conditions $y(1) = 0$, $y'(1) = -4$, $y''(1) = -18$.

Because the initial conditions for this third order Cauchy-Euler differential equation are given at $x = 1$, we are interested in values of $x > 0$. Thus, we change our independent variable by $x = e^t$, giving $t = \ln x$. Our rules for replacing derivatives with respect to x in the equation above to ones with respect to t, where we use D to represent d/dt, are $x^3 d^3 / dx^3 \rightarrow D(D-1)(D-2)$, $x^2 d^2 / dx^2 \rightarrow D(D-1)$, and $xd/dx \rightarrow D$. These substitutions result in the differential equation with constant coefficients,

$$[D(D-1)(D-2) - 6D(D-1) + 18D - 24]y = 0,$$

with an associated characteristic equation

$$r(r-1)(r-2) - 6r(r-1) + 18r - 24 = 0.$$

The first two terms have a common factor of $r - 1$, but because the last term does not have this factor, it seems that we must expand this cubic polynomial. Doing so gives

$$r^3 - 9r^2 + 26r - 24 = 0,$$

where a root is not obvious. Here we have two alternatives, one graphical and the other analytical.

In the graphical approach we examine the graph of this cubic polynomial, Figure 11.19. Here it appears that this polynomial has roots at 2, 3, and 4. If so, then the product $(r-2)(r-3)(r-4)$ when expanded, will be proportional to the cubic polynomial. Expansion of these three products shows that

$$(r-2)(r-3)(r-4) = r^3 - 9r^2 + 26r - 24.$$

To follow an algebraic approach that yields rational roots, we recall that our possible integer roots of polynomials of this type (where the coefficient of the highest term is 1) must be factors of the constant term. This means our choices are ± 1, ± 2, ± 3, ± 4, ± 6, ± 8, ± 12, and ± 24, and it turns out that 2 is a root. If we factor out that root, we obtain

$$r^3 - 9r^2 + 26r - 24 = (r-2)(r^2 - 7r + 12) = (r-2)(r-3)(r-4) = 0,$$

giving 3 and 4 as the other two roots.

Thus, the general solution of (11.77) is $C_1 e^{2t} + C_2 e^{3t} + C_3 e^{4t}$, or in terms of our original independent variable,

$$y(x) = C_1 x^2 + C_2 x^3 + C_3 x^4.$$

To satisfy our initial conditions, we substitute $x = 1$ into the above solution and its derivatives to obtain

$$\begin{aligned} y(1) &= C_1 + C_2 + C_3 = 0, \\ y'(1) &= 2C_1 + 3C_2 + 4C_3 = -4, \\ y''(1) &= 2C_1 + 6C_2 + 12C_3 = -18. \end{aligned}$$

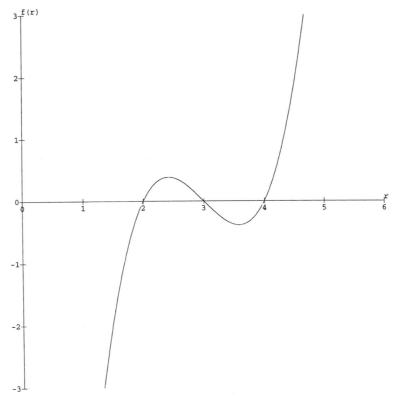

FIGURE 11.19   The cubic $r^3 - 9r^2 + 26r - 24$

If we subtract the last two equations we obtain $-3C_2 - 8C_3 = 14$, whereas if we multiply the top equation by -2 and add the result to the second equation we obtain $C_2 + 2C_3 = -4$. Solving these two equations gives $C_2 = -2$ and $C_3 = -1$. If we use these two values in the top equation we obtain $C_1 = 3$, giving the solution of our initial value problem as

$$y(x) = 3x^2 - 2x^3 - x^4. \qquad \square$$

EXAMPLE 11.18

Find the general solution of

$$2y''' + 3y'' + 4y' + 6y = 0.$$

Here the characteristic equation is

$$2r^3 + 3r^2 + 4r + 6 = 0,$$

where the factors are not apparent. However, the graph of this cubic polynomial (see Figures 11.20 and 11.21) shows that it has a real root between -1 and -2, near $-3/2$. If we substitute $r = -3/2$ into $2r^3 + 3r^2 + 4r + 6$ we find that it vanishes, which means that $(r + 3/2)$ must be a factor of $2r^3 + 3r^2 + 4r + 6$, that is,

$$2r^3 + 3r^2 + 4r + 6 = (r + 3/2)(ar^2 + br + c).$$

An elementary calculation shows that

$$2r^3 + 3r^2 + 4r + 6 = (r + 3/2)(2r^2 + 4) = 2(r + 3/2)(r^2 + 2) = 0.$$

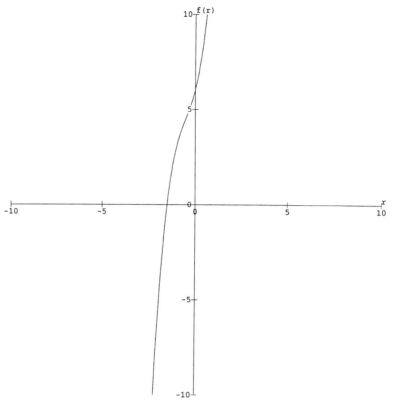

FIGURE 11.20   The cubic $2r^3 + 3r^2 + 4r + 6$

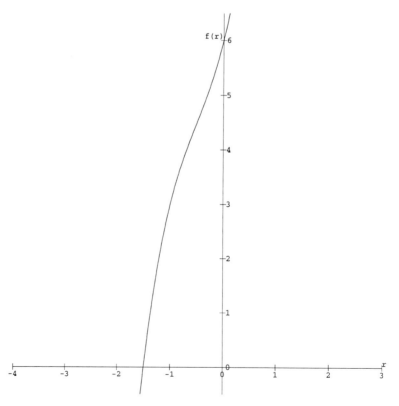

FIGURE 11.21   The cubic $2r^3 + 3r^2 + 4r + 6$

(To use an algebraic approach, we need to recall that the possible rational factors of polynomials are the ratios of the factors of the constant term, 6, to the factors of the coefficient of the leading term, 2. Thus, the possible roots are ± 1, ± 2, ± 3, ± 6, $\pm 1/2$, and $\pm 3/2$. A few trials yields $-3/2$ as a real root, and factorization yields the preceding result.) Thus, our general solution is

$$y(t) = C_1 e^{-3t/2} + C_2 \cos \sqrt{2}t + C_3 \sin \sqrt{2}t.$$

$\square$

EXERCISES

1. Find the general solution of the following differential equations.

   (a) $(D^3 + D)y = 0$

   (b) $(D^3 + 1)y = 0$

   (c) $(D^4 + 4D^2 + 4)y = 0$

   (d) $(D^4 + 2D^2 - 15)y = 0$

   (e) $(D^4 + 2D^2 - 8)y = 0$

   (f) $(D^3 - 7D^2 + 19D - 13)y = 0$

   (g) $y''' - 2y'' - y' + 2y = 0$

   (h) $y'''' + 8y'' - 9y = 0$

   (i) $(D^3 + D^2 + 3D - 5)y = 0$

   (j) $(D^4 - 5D^2 + 4)y = 0$

   (k) $x^3 y''' + 3xy' + y = 0$

   (l) $x^3 y''' + 2x^2 y'' - xy' + y = 0$

2. Solve the following initial value problems.

   (a) $(D^4 + 3D^3 + 2D^2)y = 0$, $y(0) = y'(0) = y''(0) = 0$, $y'''(0) = 8$

   (b) $(D^4 + 6D^2 + 9)y = 0$, $y(0) = y'(0) = y''(0) = 0$, $y'''(0) = 6$

   (c) $(D^3 + 6D^2 + 5D - 12)y = 0$, $y(0) = 0$, $y'(0) = 4$, $y''(0) = -8$

   (d) $(D^4 - 16)y = 0$, $y(0) = y'(0) = y''(0) = 0$, $y'''(0) = 8$ [Hint: Before evaluating the arbitrary constants, write the solution of the differential equation in terms of hyperbolic and trigonometric functions instead of exponentials.]

3. Our earlier ideas of linear independence and dependence for two functions extend to a set of n functions as follows: The set of functions $\{f_1(t),\ f_2(t), \cdots, f_n(t)\}$ is said to be linearly independent for all t in some common domain (say, $a < t < b$) of these functions if the only way a linear combination of these n functions may equal zero for all t in this domain,

$$\sum_{m=1}^{n} b_m f_m(t) = b_1 f_1(t) + b_2 f_2(t) + \cdots + b_n f_n(t) = 0, \tag{11.78}$$

is for all the constants in (11.78) to equal zero, that is,

$$b_1 = b_2 = b_3 = \cdots = b_n = 0.$$

If there is a way to have (11.78) satisfied with at least one of the $b_m \neq 0$, the set of functions is said to be linearly dependent for t in that domain. We define a Wronskian for a set of n functions by

$$W[f_1, f_2, \cdots, f_n] = \begin{vmatrix} f_1 & f_2 & \cdots & f_n \\ f_1' & f_2' & \cdots & f_n' \\ \vdots & \vdots & & \vdots \\ f_1^{(n-1)} & f_2^{(n-1)} & & f_n^{(n-1)} \end{vmatrix}.$$

Now use the Wronskian to show linear independence of a set of functions $\{f_1(t), f_2(t), f_3(t)\}$, by starting start with the linear combination

$$b_1 f_1(t) + b_2 f_2(t) + b_3 f_3(t) = 0 \tag{11.79}$$

and see if you can find nonzero values of the constants that make (11.79) an identity. By differentiating this identity twice, you obtain three equations in the three unknowns b_1, b_2, and b_3 and use this to prove that if the Wronskian $W[f_1, f_2, f_3] \neq 0$ at some point in the common domain of the three functions, then the functions are linearly dependent there.

4. Prove that if the Wronskian $W[f_1, f_2, f_3, f_4] \neq 0$ at some point in the common domain of the four functions $\{f_1(t), f_2(t), f_3(t), f_4(t)\}$, then the functions are linearly dependent there.

5. Are the following sets of functions linearly independent or dependent? (Compute the Wronskian first.)

    (a) $3, \sin t, \cos t$

    (b) $3, \sin^2 t, \cos^2 t$

    (c) $1, 1+t, t+3$

    (d) $1-t, 1+t, t^2$

    (e) $e^t, te^t, t^2 e^t$

    (f) $e^{rt}, te^{rt}, t^2 e^{rt}$

    (g) e^t, e^{-t}, e^{2t}

6. The determinant

$$\begin{vmatrix} 1 & 1 & 1 & \cdots & 1 \\ r_1 & r_2 & r_3 & \cdots & r_n \\ r_1^2 & r_2^2 & r_3^2 & \cdots & r_n^2 \\ \vdots & \vdots & \vdots & & \vdots \\ r_1^{n-1} & r_2^{n-1} & r_3^{n-1} & \cdots & r_n^{n-1} \end{vmatrix}$$

is called VANDERMONDE'S DETERMINANT.

    (a) Show that

$$\begin{vmatrix} 1 & 1 \\ r_1 & r_2 \end{vmatrix} = r_2 - r_1.$$

    (b) Show that

$$\begin{vmatrix} 1 & 1 & 1 \\ r_1 & r_2 & r_3 \\ r_1^2 & r_2^2 & r_3^2 \end{vmatrix} = (r_2 - r_1) \left[(r_3 - r_1)(r_3 - r_2) \right].$$

(c) Show that

$$\begin{vmatrix} 1 & 1 & 1 & 1 \\ r_1 & r_2 & r_3 & r_4 \\ r_1^2 & r_2^2 & r_3^2 & r_4^2 \\ r_1^3 & r_2^3 & r_3^3 & r_4^3 \end{vmatrix} = (r_2 - r_1)\left[(r_3 - r_1)(r_3 - r_2)\right]\left[(r_4 - r_1)(r_4 - r_2)(r_4 - r_3)\right].$$

(d) Use induction to prove that for any positive integer n, this determinant equals

$$(r_2 - r_1)\left[(r_3 - r_1)(r_3 - r_2)\right]\left[(r_4 - r_1)(r_4 - r_2)(r_4 - r_3)\right] \times$$
$$\times \left[\cdots\right]\left[(r_n - r_1)(r_n - r_2)\cdots(r_n - r_{n-1})\right].$$

7. Use the results of Exercise 6 to prove that the following sets of functions are linearly independent.
   (a) e^t, e^{2t}, e^{4t}
   (b) $e^t, e^{2t}, e^{3t}, e^{4t}, e^{5t}$
   (c) $e^t, e^{2t}, e^{3t}, \cdots, e^{nt}$

8. The result we use to show that a set of solutions of a differential equation is linearly independent or dependent is as follows: Let $\{f_i(t),\ i = 1, 2, 3, \cdots n\}$ be a set of solutions of the differential equation

$$\left[\sum_{j=0}^{n} a_j(t)D^j\right]y = 0, \qquad a < x < b. \tag{11.80}$$

Then this set of solutions is linearly independent if and only if the Wronskian

$$W[f_1, f_2, \cdots, f_n] \neq 0$$

for some t in $a < t < b$. Prove this result for $n = 3$ using the following steps.
   (a) Show that

$$\frac{d}{dt} W[f_1, f_2, f_3] = \begin{vmatrix} f_1 & f_2 & f_3 \\ f_1' & f_2' & f_3' \\ f_1''' & f_2''' & f_3''' \end{vmatrix}.$$

   (Recall that the derivative of a determinant is the sum of the determinants obtained by differentiating each row in turn.)
   (b) Use the fact that the f_i, $i = 1, 2, 3$, satisfy (11.80) with $n = 3$ to show that

$$\frac{d}{dt} W[f_1, f_2, f_3] = -\frac{a_2(t)}{a_3(t)} W[f_1, f_2, f_3].$$

   (c) Integrate the differential equation in part (b), and derive Abel's formula,

$$W[f_1, f_2, f_3] = C \exp\left[-\int \frac{a_2(t)}{a_3(t)}\, dt\right].$$

   (d) Use the expression for $W[f_1, f_2, f_3]$ in part (c) to complete the proof.
   (e) Show that if the integral in part (c) is written as a definite integral from t_0 to t, the constant C becomes the Wronskian evaluated at t_0.

9. Determine if the following solutions of a linear differential equation form a linearly independent set.
   (a) $1, \ln x, (\ln x)^2$, for $x > 0$

(b) $e^t, e^{2t}, e^t - e^{2t}$

(c) $1, t, t^2, t^3, \cdots, t^n$

(d) $\sin t, \cos t, t \sin t, t \cos t$

(e) $e^{r_1 t}, e^{r_2 t}, e^{r_3 t}, \cdots, e^{r_n t}$, where $r_1, r_2, r_3, \cdots, r_n$ are all distinct real numbers

11.6   Higher Order Nonhomogeneous Differential Equations

In the previous section we considered higher order homogeneous linear differential equations and found that our techniques from earlier chapters for solving second order linear differential equations applied there as well. Now we consider the case in which these higher order equations are nonhomogeneous, and we will discover that our previous techniques from earlier sections of this chapter need very little modification.

EXAMPLE 11.19: Response of a Beam to a Distributed Oscillatory Force

We now consider the situation in which a beam of length L extends horizontally from its support and is subject to an oscillatory force distributed along its length (see Figure 11.22). The resulting oscillatory motion of the beam is governed by a partial differential equation, with the spatial deflection y obeying the fourth order ordinary differential equation

$$\frac{d^2}{dt^2}\left(EI\frac{d^2 y}{dt^2}\right) - m\omega^2 y = f(t). \tag{11.81}$$

Here t is the distance from the support (so $0 < t < L$), m is the mass per unit length of the beam, ω is the frequency of the oscillatory force, EI is the flexural rigidity, and $f(t)$ gives the spatial distribution of the force.

If we divide (11.81) by EI and define k by $k^4 = m\omega^2/(EI)$, we may rewrite the differential equation as

$$\frac{d^4 y}{dt^4} - k^4 y = \frac{1}{EI}f(t). \tag{11.82}$$

From Section 11.1 we know that general solutions of second order linear differential equations have the form

$$y(t) = y_h(t) + y_p(t),$$

where $y_h(t)$ is the general solution of the associated homogeneous differential equation and $y_p(t)$ is a particular solution. In the last section we developed ways of determining $y_h(t)$, so we now need a method of finding a particular solution. To illustrate this entire procedure, we consider the preceding situation in which the spatial distribution on the beam is such that $f(t)/(EI) = Q \sin \pi t/L$.

Our first task in solving (11.82) is to find the general solution of the associated homogenous differential equation, which has as its characteristic equation

$$r^4 - k^4 = (r - k)(r + k)(r - ki)(r + ki) = 0.$$

Because this equation has four simple roots, we may write the solution of the associated differential equation as

$$y_h(t) = C_1 \cosh kt + C_2 \sinh kt + C_3 \cos kt + C_4 \sin kt,$$

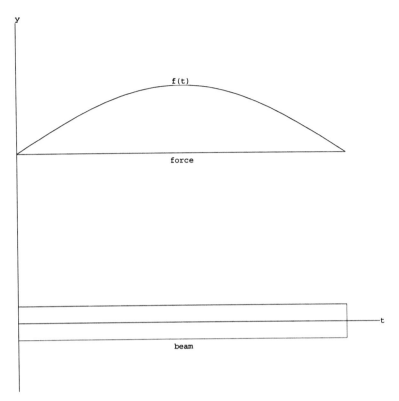

FIGURE 11.22 Horizontal beam and distributed force

where we have used the hyperbolic functions $\cosh kt = (e^{kt} + e^{-kt})/2$ and $\sinh kt = (e^{kt} - e^{-kt})/2$ instead of our usual exponentials.

We now need to determine a particular solution. Recall that in Section 11.2 we discussed the family of functions associated with forcing functions consisting of exponential functions, polynomials, sine and cosine functions, and products of these functions. There we discovered that a linear combination of members of this family of functions was the proper trial solution for y_p for nonhomogeneous linear differential equations that had constant coefficients. The same reasons used for developing that family for second order equations carries over to the higher order equations as well. Because we will be using Table 11.1 in the following development, the table is repeated here for our convenience as Table 11.3.

TABLE 11.3 Trial Solution for Special Forcing Functions

$f(t)$	Trial Solution
$a_0 + a_1 t + a_2 t^2 + a_3 t^3 + \cdots + a_n t^n$	$A_0 + A_1 t + A_2 t^2 + A_3 t^3 + \cdots + A_n t^n$
ae^{ct}	Ae^{ct}
$a \sin wt + b \cos wt$	$A \sin wt + B \cos wt$
$e^{ct}(a \sin wt + b \cos wt)$	$e^{ct}(A \sin wt + B \cos wt)$
$e^{ct} \sum_{k=0}^{k=n} a_k t^k$	$e^{ct} \sum_{k=0}^{k=n} A_k t^k$
$\sum_{k=0}^{k=n} a_k t^k (a \sin wt + b \cos wt)$	$\sum_{k=0}^{k=n} A_k t^k \sin wt + \sum_{k=0}^{k=n} B_k t^k \cos wt$
$e^{ct} \sum_{k=0}^{k=n} a_k t^k (a \sin wt + b \cos wt)$	$e^{ct}(\sum_{k=0}^{k=n} A_k t^k \sin wt + \sum_{k=0}^{k=n} B_k t^k \cos wt)$

The forcing function in (11.82) is $Q \sin \pi t/L$, so Table 11.3 gives the proper form for a particular solution (for $k \neq \pi/L$) as

$$y_p(t) = A \cos \pi t/L + B \sin \pi t/L.$$

Substituting this expression into (11.82) and equating coefficients of like terms gives $A = 0$ and $B = Q/[(\pi/L)^4 - k^4]$. This means that the general solution of our original differential equation is given by

$$y(t) = C_1 \cosh kt + C_2 \sinh kt + C_3 \cos kt + C_4 \sin kt + \frac{Q}{(\pi/L)^4 - k^4} \sin \pi t/L,$$

where C_1, C_2, C_3, and C_4 are arbitrary constants. $\qquad\square$

The fact that we could use Table 11.3 to solve this fourth order differential equation was no accident. The method we used to find particular solutions for forcing functions of the form appearing on the left-hand side of Table 11.3 is valid for linear differential equations with constant coefficients regardless of the order. We outline this method.

How to Find a Particular Solution for Higher Order Linear Differential Equations with Constant Coefficients

Purpose: To find a particular solution of the linear differential equation

$$\sum_{m=0}^{n} a_m \frac{d^m y}{dt^m} = a_n \frac{d^n y}{dt^n} + a_{n-1} \frac{d^{n-1} y}{dt^{n-1}} + \cdots + a_2 \frac{d^2 y}{dt^2} + a_1 \frac{dy}{dt} + a_0 y = f(t),$$

where the coefficients a_n, a_{n-1}, $\cdots$, a_2, a_1, a_0 are constants and $f(t)$ is a polynomial, an exponential function, a sine or cosine function, or a product and sum of any of these three types of functions.

Technique:

1. Solve the associated homogeneous differential equation and obtain n linearly independent solutions, labeled $y_1(t)$, $y_2(t)$, $y_3(t), \ldots, y_n(t)$. See *How to Solve Higher Order Linear Differential Equations with Constant Coefficients* on page 344.

2. Consider one term in $f(t)$ and compare it with the functions listed on the left-hand side of Table 11.3. For this term, write down the proper particular solution as given on the right-hand side of the table.

3. Compare the form of this particular solution with the n linearly independent solutions listed in step 1. If this particular solution contains any of $y_1(t)$, $y_2(t)$, $y_3(t), \ldots, y_n(t)$, multiply this particular solution by the lowest power of t that makes every term in the result different from every one of these linearly independent solutions.

4. If there is more than one term in $f(t)$, repeat steps 2 and 3 for each term.

5. Examine all the expressions for $y_p(t)$ resulting from the preceding four steps and eliminate all duplicates. The proper form for $y_p(t)$ will be the result of this operation.

EXAMPLE 11.20

To illustrate this procedure we seek the general solution of the fourth order differential equation

$$\frac{d^4y}{dt^4} - 5\frac{d^2y}{dt^2} + 4y = 80e^{3t} - 36e^{2t}. \tag{11.83}$$

If we follow the procedure above, we first find the general solution of the associated homogeneous differential equation

$$\frac{d^4y}{dt^4} - 5\frac{d^2y}{dt^2} + 4y = 0. \tag{11.84}$$

Now the characteristic equation may be factored as

$$r^4 - 5r^2 + 4 = (r^2 - 4)(r^2 - 1) = (r - 2)(r + 2)(r - 1)(r + 1) = 0,$$

so the general solution of (11.84) is

$$y_h(t) = C_1 e^{2t} + C_2 e^{-2t} + C_3 e^t + C_4 e^{-t}, \tag{11.85}$$

where C_1, C_2, C_3, and C_4 are arbitrary constants.

To find the proper form to try for a particular solution, we first consider the forcing function $80e^{3t}$ and note from Table 11.3 that the proper trial solution is Ae^{3t}. Because this function does not appear as one of the linearly independent solutions of the homogeneous differential equation as given in (11.85), we keep this as part of our trial solution and consider the term $-36e^{2t}$. The proper trial solution for this term is Be^{2t}, but because this term occurs in (11.85), we multiply the term by t and note that Bte^{2t} does not occur in (11.85). Thus, we have our proper trial solution as

$$y_p(t) = Ae^{3t} + Bte^{2t}.$$

Substituting this expression into (11.83) results in

$$40Ae^{3t} + 36Be^{2t} = 80e^{3t} - 36e^{2t},$$

from which we see that we must choose $A = 2$ and $B = -1$. Thus, the general solution of (11.83) is

$$y(t) = C_1 e^{2t} + C_2 e^{-2t} + C_3 e^t + C_4 e^{-t} + 2e^{3t} - te^{2t}. \tag{11.86}$$

There are two terms in (11.86) which could be classified as transient terms, with the other four terms being unbounded as $t \to \infty$. Which of these four terms dominates as $t \to \infty$? □

The use of Table 11.3 is obviously limited to the types of functions listed on the left-hand side of the table. For other forcing functions, or for differential equations that do not have constant coefficients, we will use a modified version of the variation of parameters technique that we discussed in Section 11.3. Because of the great similarity in the procedure for using variation of parameters for second order and higher order differential equations, we will simply write down the procedure and then give examples.

How to Find a Particular Solution for Higher Order Differential Equations Using Variation of Parameters

Purpose: To find a particular solution of the linear differential equation,

$$\sum_{m=0}^{n} a_m \frac{d^m y}{dt^m} = a_n \frac{d^n y}{dt^n} + a_{n-1} \frac{d^{n-1} y}{dt^{n-1}} + \cdots + a_2 \frac{d^2 y}{dt^2} + a_1 \frac{dy}{dt} + a_0 y = f(t), \quad \textbf{(11.87)}$$

where the coefficients $a_n, a_{n-1}, \cdots, a_2, a_1,$ and a_0 may be functions of t.

Technique:

1. Solve the associated homogeneous differential equation and obtain n linearly independent solutions, labeled $y_1(t), y_2(t), y_3(t), ..., y_n(t)$.

2. Assume a particular solution of the form

$$y_p(t) = \sum_{m=1}^{n} y_m(t) z_m(t) = y_1(t) z_1(t) + y_2(t) z_2(t) + \cdots + y_n(t) z_n(t), \quad \textbf{(11.88)}$$

and write down the system of equations satisfied by $z'_m(t)$, $m = 1, 2, 3, \cdots, n$,

$$y_1(t) z'_1(t) + y_2(t) z'_2(t) + \cdots + y_n(t) z'_n(t) = 0,$$
$$y'_1(t) z'_1(t) + y'_2(t) z'_2(t) + \cdots + y'_n(t) z'_n(t) = 0,$$
$$y''_1(t) z'_1(t) + y''_2(t) z'_2(t) + \cdots + y''_n(t) z'_n(t) = 0, \cdots,$$
$$y_1^{(n-1)}(t) z'_1(t) + y_2^{(n-1)}(t) z'_2(t) + \cdots + y_n^{(n-1)}(t) z'_n(t) = f(t)/a_n(t).$$

3. Solve this system for $z'_m(t)$, $m = 1, 2, 3, \cdots, n$, and integrate to obtain $z_m(t)$, $m = 1, 2, 3, \cdots, n$, where we may set the arbitrary constants equal to zero. The final form of the particular solution is obtained by substituting back into (11.88) these values of $z_m(t)$, $m = 1, 2, 3, \cdots, n$.

Comments about using variation of parameters for higher order differential equations

- If we include the arbitrary constants when integrating the expressions for the derivatives $z'_m(t)$, $m = 1, 2, 3, \cdots, n$, the expression for y_p will also be the general solution of (11.87).

- If we are solving an initial value problem and cannot evaluate the integrals giving $z_m(t)$, $m = 1, 2, 3, \cdots, n$, in terms of elementary functions, it is often convenient to write the integrals with the independent variable as the upper limit and the value of t giving the initial condition as the lower limit.

- A common mistake in writing down the system of equations in step 2 is to forget to divide by $a_n(t)$ on the right-hand side of the bottom equation.

EXAMPLE 11.21

Find the general solution of the differential equation

$$y''' + y' = \sec t. \quad \textbf{(11.89)}$$

We follow the preceding steps by first determining the general solution of the associated differential equation by considering the characteristic equation

$$r^3 + r = r(r^2 + 1) = r(r + i)(r - i) = 0.$$

Thus, the general solution of this homogeneous differential equation is

$$y_h(t) = C_1 + C_2 \sin t + C_3 \cos t, \qquad \textbf{(11.90)}$$

where C_1, C_2, and C_3 are arbitrary constants.

Because the forcing function on the right-hand side of (11.89) does not appear on the left side of Table 11.3, we use the method of variation of parameters. Thus, we write our particular solution as

$$y_p(t) = z_1 + \sin t \, z_2 + \cos t \, z_3,$$

where z_1, z_2, and z_3 are functions of t to be determined. The differential equations satisfied by the derivatives of the three functions are

$$
\begin{aligned}
z_1' + \sin t \, z_2' + \cos t \, z_3' &= 0, \\
\cos t \, z_2' - \sin t \, z_3' &= 0, \\
- \sin t \, z_2' - \cos t \, z_3' &= \sec t.
\end{aligned}
$$

To solve this system of equations for z_1', z_2', and z_3', we multiply the last equation by $\cos t$ and add the result to what we obtain by multiplying the middle equation by $\sin t$. This gives $z_3' = -1$, so from the middle equation, we discover that $z_2' = - \sin t / \cos t$, and from the top equation we have that $z_1' = \sin^2 t / \cos t + \cos t = 1 / \cos t$. With help from a table of integrals, these three functions may integrated to obtain

$$z_1(t) = \ln | \sec t + \tan t |, \qquad z_2(t) = \ln | \cos t |, \qquad z_3(t) = -t.$$

Thus, we have our particular solution as

$$z_p(t) = \ln | \sec t + \tan t | + \sin t \ln | \cos t | - t \cos t \qquad \textbf{(11.91)}$$

and the general solution of (11.89) as the sum of the functions given in (11.90) and (11.91).

$\square$

EXAMPLE 11.22

Find the general solution of the nonhomogeneous Cauchy-Euler differential equation

$$x^3 \frac{d^3 y}{dx^3} - x^2 \frac{d^2 y}{dx^2} + 2x \frac{dy}{dx} - 2y = x^2, \qquad x > 0. \qquad \textbf{(11.92)}$$

We start by finding the general solution of the associated homogeneous differential equation, which, with the change of independent variable $x = e^t$ becomes

$$[D(D-1)(D-2) - D(D-1) + 2D - 2]y = 0.$$

Because this is a linear differential equation with constant coefficients, we first find our characteristic equation as

$$r(r-1)(r-2) - r(r-1) + 2(r-1) = 0.$$

We notice that each term in this equation has a common factor of $r - 1$, so we rewrite it as

$$(r-1)(r^2 - 2r - r + 2) = (r-1)(r^2 - 3r + 2) = (r-1)(r-1)(r-2) = 0.$$

Thus, $r = 1$ is a double root, and $r = 2$ is a single root. This means we may write our solution of the transformed differential equation as

$$y(t) = C_1 e^{2t} + (C_2 + C_3 t)e^t,$$

which, in terms of our original independent variable, becomes

$$y_h(x) = C_1 x^2 + (C_2 + C_3 \ln x)x.$$

To find a particular solution with the method of variation of parameters, we let

$$y_p(x) = y_1(x)z_1(x) + y_2(x)z_2(x) + y_3(x)z_3(x).$$

Using the three linearly independent functions from our homogeneous solution above gives

$$y_p(x) = x^2 z_1(x) + x z_2(x) + x \ln x \, z_3(x).$$

Thus, our three equations for determining these three functions are

$$\begin{aligned}
x^2 z_1'(x) + x z_2'(x) + x \ln x \, z_3'(x) &= 0, \\
2x z_1'(x) + z_2'(x) + (1 + \ln x)\, z_3'(x) &= 0, \\
2 z_1'(x) + \frac{1}{x} z_3'(x) &= \frac{1}{x}.
\end{aligned}$$
(11.93)

Although there are several ways to solve this system, the way we chose takes advantage of the fact that the $z_2'(x)$ term is missing in the bottom equation of (11.93). Thus, we eliminate $z_2'(x)$ from the other two equations by multiplying the top equation by $-1/x$ and adding the result to the middle equation. This gives

$$x z_1'(x) + z_3'(x) = 0.$$
(11.94)

We now multiply this equation by $-1/x$ and add the result to the bottom equation in (11.93) to obtain

$$z_1'(x) = \frac{1}{x}.$$

We now use (11.94) to find that

$$z_3'(x) = -1.$$

To find $z_2'(x)$, we return to the top equation in (11.93) and obtain

$$z_2'(x) = -\ln x \, z_3'(x) - x z_1'(x) = \ln x - 1,$$

so by integration we have $z_2(x) = x \ln x - 2x$. Integrating our other two expressions gives $z_1(x) = \ln x$ and $z_3'(x) = -x$. We now can write a particular solution as

$$\begin{aligned}
y_p(x) &= x^2 z_1(x) + x z_2(x) + x \ln x \, z_3(x) \\
&= x^2 \ln x + x(x \ln x - 2x) - (x \ln x)x \\
&= x^2 \ln x - 2x^2.
\end{aligned}$$

Thus, the general solution of (11.92) is

$$y(x) = C_1 x^2 + (C_2 + C_3 \ln x)x + x^2 \ln x - 2x^2.$$

Notice that part of the particular solution $(-2x^2)$ is a constant times part of the general solution of the associated homogeneous equation. It may therefore be incorporated into the term $C_1 x^2$, yielding

$$y(x) = C_1 x^2 + (C_2 + C_3 \ln x)x + x^2 \ln x.$$

$\square$

EXERCISES

1. Find the general solution of the following differential equations.

(a) $(D^3 + D)y = \sin 2t$

(b) $(D^3 + D)y = t$

(c) $(D^3 + D)y = \cos t$

(d) $(D^4 + 4D^2 + 4)y = 6 - t - e^t$

(e) $(D^4 + 4D^2 + 4)y = 4e^{-2t}$

(f) $(D^4 + 3D^3 + 2D^2)y = t + \sin t$

(g) $(D^4 + 3D^3 + 2D^2)y = e^{-t} + e^t$

(h) $(D^4 + 6D^2 + 9)y = \cos t$

(i) $(D^4 + 6D^2 + 9)y = \cos t + e^{-3t}$

(j) $(D^3 + 6D^2 + 5D - 12)y = e^t + e^{4t}$

What Have We Learned?

How to Solve Nonhomogeneous Linear Differential Equations

Purpose: To find the general solution of the nonhomogeneous linear differential equation,

$$\sum_{m=0}^{n} a_m \frac{d^m y}{dt^m} = a_n \frac{d^n y}{dt^n} + a_{n-1} \frac{d^{n-1} y}{dt^{n-1}} + \cdots + a_2 \frac{d^2 y}{dt^2} + a_1 \frac{dy}{dt} + a_0 y = f(t), \quad \textbf{(11.95)}$$

where the coefficients $a_n, a_{n-1}, \cdots, a_2, a_1$, and a_0 may be functions of t.

Technique:

1. Find the general solution of the associated homogeneous differential equation. See *How to Solve Linear Differential Equations with Constant Coefficients* on page 344 or *How to Solve Higher Order Linear Differential Equations with Constant Coefficients* on page 559, which also deals with Cauchy-Euler Equations in the attached comments.

2. Find a particular solution of (11.95). Here we have a choice.

(a) See Section 11.2, *Method of Undetermined Coefficients*, or *How to Find a Particular Solution for Higher Order Linear Differential Equations with Constant Coefficients* on page 570.

(b) See *How to Find a Particular Solution Using Variation of Parameters* on page 538, or *How to Find a Particular Solution Using Variation of Parameters for Higher Order Differential Equations* on page 572.

3. Add the solutions found in steps 1 and 2 to obtain the general solution of (11.95).

Comments about how to solve nonhomogeneous linear differential equations

- This three-step procedure will work for other types of linear differential equations not covered in this section. The necessary ingredient for success is to find n linearly independent solutions of the associated homogeneous differential equation and then use variation of parameters.

CHAPTER 12

Autonomous Systems

Where Are We Going?

In this chapter we return to the analysis of systems of two linear autonomous differential equations that we began in Chapter 7. For these autonomous systems, in which the coefficients in the differential equations are all constants, we obtain explicit solutions for all the possible cases and use these solutions to analyze their behavior in the phase plane. In order to characterize this behavior we introduce the terms *node, center, focus,* and *saddle point.*

We continue with our graphical analysis by discovering how nullclines, which are isoclines for horizontal and vertical tangents in the phase plane, can aid us in determining the stability of equilibrium points. We start with linear systems, where we always have explicit solutions, and continue with nonlinear systems, where we seldom can find explicit solutions. We apply our nullcline analysis to two nonlinear population models and confirm our results with numerical solutions.

12.1 Solutions of Linear Autonomous Systems

In Chapter 7 we had several examples of systems of two first order linear autonomous[1] differential equations for which we found explicit solutions by changing the system to a single second order differential equation. Here we consider the general situation in which the system of autonomous equations has the form

$$\begin{cases} dx/dt = ax + by, \\ dy/dt = cx + dy, \end{cases} \tag{12.1}$$

where a, b, c, and d are constants. This system has $x(t) = 0$ and $y(t) = 0$ as a solution, called the trivial solution of (12.1). The technique developed in this section is for finding nontrivial solutions of (12.1).

---

[1]Remember that a system of differential equations $x' = P(x, y, t)$, $y' = Q(x, y, t)$ is autonomous if P and Q do not contain t explicitly.

If we consider the case in which $b = c = 0$, the two equations in (12.1) uncouple, and each may be solved separately to obtain a solution as

$$\begin{cases} x(t) = C_1 e^{at}, \\ y(t) = C_2 e^{dt}. \end{cases} \tag{12.2}$$

If the two equations in (12.1) are to be coupled, then at least one of b or c must be different from zero, so we first consider $b \neq 0$. We can differentiate the top equation in (12.1) and substitute into the result the expression for dy/dt from the bottom equation in (12.1) to obtain

$$\begin{aligned} x'' &= ax' + by' \\ &= ax' + b\,(cx + dy) \\ &= ax' + bcx + d\,(by). \end{aligned}$$

We can eliminate the dependence on y in this equation by substituting by from the top equation in (12.1), giving

$$x'' = ax' + bcx + d\left(x' - ax\right),$$

or

$$x'' - (a+d)x' + (ad - bc)x = 0. \tag{12.3}$$

Had we differentiated the bottom equation in (12.1) and substituted for dx/dt from the top equation, the resulting equation for y would have the same coefficients, namely,

$$y'' - (a+d)y' + (ad - bc)y = 0. \tag{12.4}$$

(See Exercise 2 on page 583.)

In Section 7.2 we discovered that using a solution of (12.3) in the form of an exponential e^{rt} led to the characteristic equation

$$r^2 - (a+d)r + ad - bc = 0. \tag{12.5}$$

This quadratic equation has three types of solutions: two real, distinct roots; one real, repeated root; or two complex roots, one the complex conjugate of the other. Thus, the differential equation in (12.3) has three possible forms for its general solution, depending on the relative values of a, b, c, and d and on the sign of the discriminant[2] of the quadratic equation (12.5) — namely,

$$(a+d)^2 - 4(ad - bc) = (a - d)^2 + 4bc.$$

We now cover these three possibilities in turn.

1. **Two real, distinct roots**. If $(a - d)^2 + 4bc > 0$, and we label the roots r_1 and r_2, then these roots are given by

$$r_1 = \frac{1}{2}\left[a + d - \sqrt{(a - d)^2 + 4bc}\right],$$

$$r_2 = \frac{1}{2}\left[a + d + \sqrt{(a - d)^2 + 4bc}\right],$$

---

[2]The discriminant of the quadratic equation $ax^2 + bx + c = 0$ is $b^2 - 4ac$.

and the general solution for $x(t)$ may be written as

$$x(t) = C_1 e^{r_1 t} + C_2 e^{r_2 t}.$$

If we now substitute this expression into the top equation of (12.1) we obtain

$$
\begin{aligned}
by &= x' - ax \\
&= r_1 C_1 e^{r_1 t} + r_2 C_2 e^{r_2 t} - a\left(C_1 e^{r_1 t} + C_2 e^{r_2 t}\right) \\
&= C_1 (r_1 - a) e^{r_1 t} + C_2 (r_2 - a) e^{r_2 t}.
\end{aligned}
$$

This means we can write the general solution of (12.1) in this case as

$$
\begin{cases}
x(t) = C_1 e^{r_1 t} + C_2 e^{r_2 t}, \\
y(t) = C_1 (r_1 - a) e^{r_1 t}/b + C_2 (r_2 - a) e^{r_2 t}/b.
\end{cases}
\tag{12.6}
$$

2. **One real, repeated root.** If $(a - d)^2 + 4bc = 0$, then $r = (a + d)/2$, and the general solution for $x(t)$ can be written as

$$x(t) = C_1 e^{rt} + C_2 t e^{rt}.$$

Substituting this expression into $by = x' - ax$ gives

$$by = C_1 r e^{rt} + (1 + rt) C_2 e^{rt} - a\left(C_1 e^{rt} + C_2 t e^{rt}\right),$$

so this means that the general solution of (12.1) may be written as

$$
\begin{cases}
x(t) = C_1 e^{rt} + C_2 t e^{rt}, \\
y(t) = \left[C_1 (r - a) + C_2\right] e^{rt}/b + C_2 (r - a) t e^{rt}/b.
\end{cases}
\tag{12.7}
$$

3. **Two complex roots.** If $(a - d)^2 + 4bc < 0$, then we can express the roots as

$$
\begin{aligned}
r_1 &= \frac{1}{2}\left[a + d - i\sqrt{-(a - d)^2 - 4bc}\right] = \alpha - i\beta, \\
r_2 &= \frac{1}{2}\left[a + d + i\sqrt{-(a - d)^2 - 4bc}\right] = \alpha + i\beta.
\end{aligned}
$$

Here we have the solution for $x(t)$ given as

$$x(t) = e^{\alpha t}\left(C_1 \sin \beta t + C_2 \cos \beta t\right)$$

and by given as

$$
\begin{aligned}
by &= x' - ax \\
&= e^{\alpha t}\left(\beta C_1 \cos \beta t - \beta C_2 \sin \beta t\right) + \alpha e^{\alpha t}\left(C_1 \sin \beta t + C_2 \cos \beta t\right) + \\
&\quad -a e^{\alpha t}\left(C_1 \sin \beta t + C_2 \cos \beta t\right) \\
&= e^{\alpha t}\left\{\left[(\alpha - a) C_1 - \beta C_2\right] \sin \beta t + \left[\beta C_1 + (\alpha - a) C_2\right] \cos \beta t\right\}.
\end{aligned}
$$

This allows us to write the general solution as

$$
\begin{cases}
x(t) = e^{\alpha t}\left(C_1 \sin \beta t + C_2 \cos \beta t\right), \\
y(t) = e^{\alpha t}\left(\left\{\left[(\alpha - a) C_1 - \beta C_2\right]/b\right\} \sin \beta t + \left\{\left[\beta C_1 + (\alpha - a) C_2\right]/b\right\} \cos \beta t\right).
\end{cases}
\tag{12.8}
$$

The above assumed that $b \neq 0$. If $c \neq 0$, then a similar analysis (see Exercise 2 on page 583) gives the solutions for the three cases as follows.

1. Two real, distinct roots

$$\begin{cases} x(t) = C_1(r_1 - d)e^{r_1 t}/c + C_2(r_2 - d)e^{r_2 t}/c, \\ y(t) = C_1 e^{r_1 t} + C_2 e^{r_2 t}. \end{cases} \qquad \textbf{(12.9)}$$

2. One real, repeated root

$$\begin{cases} x(t) = \left[C_1(r - d) + C_2 \right] e^{rt}/c + C_2(r - d)te^{rt}/c, \\ y(t) = C_1 e^{rt} + C_2 te^{rt}. \end{cases} \qquad \textbf{(12.10)}$$

3. Two complex roots

$$\begin{cases} x(t) = e^{\alpha t} \left(\left\{ \left[(\alpha - d) \, C_1 - \beta C_2 \right]/c \right\} \sin \beta t + \left\{ \left[\beta C_1 + (\alpha - d) \, C_2 \right]/c \right\} \cos \beta t \right), \\ y(t) = e^{\alpha t} \left(C_1 \sin \beta t + C_2 \cos \beta t \right). \end{cases}$$
$$\textbf{(12.11)}$$

Even though we have formulas for the general solution of (12.1) for all possible situations where r satisfies (12.5) — namely, (12.2), (12.6), (12.7), (12.8), (12.9), (12.10), and (12.11) — we do not recommend that you commit any of these to memory. Instead, we use the fact that we know the form of the solution of the system of equations for the three possible cases. This allows us to write down the solution for one of the variables, say $x(t)$ if $b \neq 0$, and then use the first equation in (12.1) to obtain $y(t)$. To take advantage of this knowledge, we need a simple way to obtain the characteristic equation

$$r^2 - (a + d)r + ad - bc = 0$$

without having to commit it to memory. If we rewrite this equation in the form

$$(a - r)(d - r) - bc = 0,$$

we see that it can be expressed in terms of the 2 by 2 determinant

$$\begin{vmatrix} a - r & b \\ c & d - r \end{vmatrix} = 0.$$

We can construct this determinant directly from (12.1) by adding the term $-r$ to the coefficients on the diagonal. This gives us a simple way to obtain the characteristic equation.

> ### *How to to Find the General Solution of* $dx/dt = ax + by$, $dy/dt = cx + dy$
>
> **Purpose:** To solve the autonomous system
>
> $$\begin{cases} dx/dt = ax + by, \\ dy/dt = cx + dy, \end{cases} \qquad \textbf{(12.12)}$$
>
> where a, b, c, and d are constants, for $x = x(t)$, $y = y(t)$.
>
> **Technique:**
>
> **1.** If $b = c = 0$, the equations are not coupled, and the general solution is
>
> $$\begin{cases} x(t) = C_1 e^{at}, \\ y(t) = C_2 e^{dt}. \end{cases}$$

In this case there is nothing else to do.

2. Solve the characteristic equation

$$\begin{vmatrix} a-r & b \\ c & d-r \end{vmatrix} = 0,$$

for r, obtaining the roots r_1 and r_2.

3. If $b \neq 0$, we

   (a) write down the solution for $x(t)$ based on r_1 and r_2 as follows.

      i. If $r_1 \neq r_2$, where r_1 and r_2 are both real, then the general solution for $x(t)$ is

$$x(t) = C_1 e^{r_1 t} + C_2 e^{r_2 t}.$$

      ii. If $r_1 = r_2 = r$, then the general solution for $x(t)$ is

$$x(t) = C_1 e^{rt} + C_2 t e^{rt}.$$

      iii. If r_1 and r_2 are complex, where $r_1 = \alpha - i\beta$ and $r_2 = \alpha + i\beta$, then the general solution for $x(t)$ is

$$x(t) = e^{\alpha t} \left(C_1 \sin \beta t + C_2 \cos \beta t \right).$$

   (b) Calculate the general solution for $y(t)$ from the general solution for $x(t)$ and its derivative dx/dt by rearranging the top equation in (12.12) as

$$y = (dx/dt - ax)/b.$$

4. If $b = 0$ and $c \neq 0$, we

   (a) write down the solution for $y(t)$ based on r_1 and r_2 as follows.

      i. If $r_1 \neq r_2$, where r_1 and r_2 are both real, then the general solution for $y(t)$ is

$$y(t) = C_1 e^{r_1 t} + C_2 e^{r_2 t}.$$

      ii. If $r_1 = r_2 = r$, then the general solution for $y(t)$ is

$$y(t) = C_1 e^{rt} + C_2 t e^{rt}.$$

      iii. If r_1 and r_2 are complex, where $r_1 = \alpha - i\beta$ and $r_2 = \alpha + i\beta$, then the general solution for $y(t)$ is

$$y(t) = e^{\alpha t} \left(C_1 \sin \beta t + C_2 \cos \beta t \right).$$

   (b) Calculate the general solution for $x(t)$ from the general solution for $y(t)$ and its derivative dy/dt by rearranging the bottom equation in (12.12) as

$$x = (dy/dt - dy)/c.$$

Comments about the general solution of $dx/dt = ax + by$, $dy/dt = cx + dy$

- It is possible to extend the preceding analysis to systems of equations of the form

$$\begin{cases} dx/dt = ax + by + e, \\ dy/dt = cx + dy + f, \end{cases}$$

where e and f are constants for the situation where $ad - bc \neq 0$. The major difference between this case and the case when $e = f = 0$ is the fact that $x(t) = 0$ and $y(t) = 0$ is no longer a solution. In this case we have the solution $x(t) = x_0$, $y(t) = y_0$, where x_0 and y_0 are constants that satisfy the two equations $ax_0 + by_0 + e = 0$ and $cx_0 + dy_0 + f = 0$. The change of variables from (x, y) to (u, v) where $u = x - x_0$ and $v = y - y_0$ will convert the preceding system in terms of x and y into an equivalent linear system in terms of u and v.

EXAMPLE 12.1

Solve

$$\begin{cases} dx/dt = -x + y, \\ dy/dt = -kx - ky, \end{cases} \tag{12.13}$$

where (a) $k = 6$, (b) $k = 1$, (c) $k = 3 + 2\sqrt{2}$.

In all three cases the characteristic equation is

$$\begin{vmatrix} -1 - r & 1 \\ -k & -k - r \end{vmatrix} = 0,$$

or

$$(-1 - r)(-k - r) + k = r^2 + (1 + k)r + 2k = 0,$$

with roots

$$r = \frac{1}{2}\left[-(1 + k) \pm \sqrt{(1 + k)^2 - 8k} \right].$$

(a) With $k = 6$ the solutions of the characteristic equation are $r_1 = -4$ and $r_2 = -3$. Thus,

$$x(t) = C_1 e^{-4t} + C_2 e^{-3t}.$$

From this and the first equation in (12.13) we have

$$\begin{aligned} y(t) &= dx/dt + x \\ &= -4C_1 e^{-4t} - 3C_2 e^{-3t} + C_1 e^{-4t} + C_2 e^{-3t} \\ &= -3C_1 e^{-4t} - 2C_2 e^{-3t}. \end{aligned}$$

(b) With $k = 1$ the solutions of the characteristic equation are $r_1 = -1 - i$ and $r_2 = -1 + i$. Thus,

$$x(t) = e^{-t}\left(C_1 \sin t + C_2 \cos t \right).$$

From this and the first equation in (12.13) we have

$$\begin{aligned} y(t) &= dx/dt + x \\ &= e^{-t}\left(-C_1 \sin t - C_2 \cos t + C_1 \cos t - C_2 \sin t \right) + e^{-t}\left(C_1 \sin t + C_2 \cos t \right), \end{aligned}$$

or

$$y(t) = e^{-t}\left(C_1 \cos t - C_2 \sin t \right).$$

(c) With $k = 3 + 2\sqrt{2}$ the solutions of the characteristic equation are $r = r_1 = r_2 = -(1 + k)/2 = -2 - \sqrt{2}$. Thus,

$$x(t) = C_1 e^{rt} + C_2 t e^{rt}.$$

From this and the first equation in (12.13) we have

$$y(t) = dx/dt + x = rC_1 e^{rt} + rC_2 t e^{rt} + C_2 e^{rt} + C_1 e^{rt} + C_2 t e^{rt},$$

or

$$y(t) = \left[(r + 1) C_1 + C_2\right] e^{rt} + (r + 1) C_2 t e^{rt},$$

where $r = -2 - \sqrt{2}$.   □

EXERCISES

1. Solve the following systems of differential equations.

(a)
$$\begin{cases} dx/dt = 2x + 5y \\ dy/dt = x + 6y \end{cases}$$

(b)
$$\begin{cases} dx/dt = 2x - y \\ dy/dt = -6x + y \end{cases}$$

(c)
$$\begin{cases} dx/dt = 5x + 6y \\ dy/dt = x + 4y \end{cases}$$

(d)
$$\begin{cases} dx/dt = 5x - 3y \\ dy/dt = -x - y \end{cases}$$

(e)
$$\begin{cases} dx/dt = 2x + 9y \\ dy/dt = -x - 4y \end{cases}$$

(f)
$$\begin{cases} dx/dt = 2x + y \\ dy/dt = x - 4y \end{cases}$$

(g)
$$\begin{cases} dx/dt = 4x - 2y \\ dy/dt = 5x - 2y \end{cases}$$

(h)
$$\begin{cases} dx/dt = 2x + y \\ dy/dt = -4x + 2y \end{cases}$$

(i)
$$\begin{cases} dx/dt = 3x + 5y \\ dy/dt = -5x + 3y \end{cases}$$

(j)
$$\begin{cases} dx/dt = x + y \\ dy/dt = x + 3y \end{cases}$$

2. Differentiate the second equation of

$$\begin{cases} dx/dt = ax + by \\ dy/dt = cx + dy \end{cases}$$

and substitute for $dx/dt = ax + by$ from the first and $cx = dy/dt - dy$ from the second (assuming $c \neq 0$), to obtain (12.4), namely,

$$y'' - (a + d)y' + (ad - bc)y = 0.$$

Solve

$$y'' - (a+d)y' + (ad - bc)y = 0$$

for $c \neq 0$ and obtain (12.9), (12.10), and (12.11).

12.2   Stability of Linear Autonomous Systems

In the previous section we found the general solution of the linear autonomous system of differential equations

$$\begin{cases} dx/dt = ax + by, \\ dy/dt = cx + dy, \end{cases} \tag{12.14}$$

where a, b, c, and d are constants. These solutions fall into three general categories according to the roots r_1 and r_2 of the characteristic equation

$$r^2 - (a+d)r + ad - bc = 0.$$

The relationship between a, b, c, and d and r_1 and r_2, is

$$\begin{aligned} a+d &= r_1 + r_2, \\ ad - bc &= r_1 r_2. \end{aligned} \tag{12.15}$$

(Verify this.)

The three general categories, and typical solutions, are as follows.

1. Two real, distinct roots

$$\begin{cases} x(t) = C_1 e^{r_1 t} + C_2 e^{r_2 t} \\ y(t) = K_1 e^{r_1 t} + K_2 e^{r_2 t} \end{cases} \tag{12.16}$$

2. One real, repeated root

$$\begin{cases} x(t) = C_1 e^{rt} + C_2 t e^{rt} \\ y(t) = K_1 e^{rt} + K_2 t e^{rt} \end{cases} \tag{12.17}$$

3. Two complex roots

$$\begin{cases} x(t) = e^{\alpha t} \left(C_1 \sin \beta t + C_2 \cos \beta t \right) \\ y(t) = e^{\alpha t} \left(K_1 \sin \beta t + K_2 \cos \beta t \right) \end{cases} \tag{12.18}$$

In these solutions C_1, C_2, K_1, and K_2 are constants related to one another in various ways, so that only two of these four constants are independent. In this section we want to discuss the stability and long-term behavior of these solutions by looking at trajectories in the phase plane. It is important to remember that if $r < 0$, then $e^{rt} \to 0$ and $te^{rt} \to 0$ as $t \to \infty$, while if $r > 0$, then $e^{rt} \to \infty$ and $te^{rt} \to \infty$ as $t \to \infty$. Finally, if $r = 0$, then $e^{rt} = 1$. The reason we are concerned with the limit $t \to \infty$ (and ignore the limit $t \to -\infty$) is that in applications the independent variable is often time. We usually use mathematical models to predict the future, not to discover the past.

1. In the case of **two real, distinct roots** we assume that $r_1 < r_2$. If we look at (12.16) we see that the long-term behavior of $x(t)$ and $y(t)$ will depend on the signs of r_1 and

r_2. There are five possibilities: (a) $r_1 < r_2 < 0$, (b) $r_1 < r_2 = 0$, (c) $r_1 < 0, r_2 > 0$, (d) $r_1 = 0, r_2 > 0$, (e) $0 < r_1 < r_2$.

2. In the case of **one real, repeated root** we have $r = r_1 = r_2$. If we look at (12.17) we see that the long-term behavior of $x(t)$ and $y(t)$ will depend on the sign of r. There are three possibilities: (a) $r < 0$, (b) $r = 0$, (c) $r > 0$.

3. In the case of **two complex roots** we have $r_1 = \alpha - i\beta$, $r_2 = \alpha + i\beta$. If we look at (12.18) we see that the long-term behavior of $x(t)$ and $y(t)$ will depend on the sign of α. There are three possibilities: (a) $\alpha < 0$, (b) $\alpha = 0$, (c) $\alpha > 0$.

From our previous work in Chapter 7, we know that equilibrium points are points in the phase plane corresponding to constant solutions

$$x(t) = x_0, \qquad y(t) = y_0, \tag{12.19}$$

of (12.14). If we substitute (12.19) into (12.14) we find

$$\begin{aligned} ax_0 + by_0 &= 0, \\ cx_0 + dy_0 &= 0. \end{aligned} \tag{12.20}$$

The solution of this equation falls into two cases, depending on the value of $ad - bc$.

1. If $ad - bc \neq 0$, then (12.20) has the unique solution

$$x_0 = 0, \qquad y_0 = 0,$$

and so, in this case, there is exactly one equilibrium point, $(0, 0)$. From (12.15) we see that $ad - bc \neq 0$ implies that neither r_1 nor r_2 can be zero.

2. If $ad - bc = 0$, then the two equations in (12.20) are linearly dependent, that is,

$$cx_0 + dy_0 = k(ax_0 + by_0),$$

and so the solution of (12.20) is

$$ax_0 + by_0 = 0. \tag{12.21}$$

In (12.21) either x_0 or y_0 may be chosen arbitrarily and the other must be such that the equation is satisfied. Equation (12.21) represents a straight line, which might be vertical ($b = 0$), horizontal ($a = 0$), or slanted ($a \neq 0$ and $b \neq 0$) with slope $m = -a/b$. Thus, if $ad - bc = 0$, we have an infinite number of equilibrium points, none of which are isolated.[3] From (12.15) we see that $ad - bc = 0$ implies that at least one of r_1 and r_2 must be zero.

Rather than consider (12.14) in full generality, we will look at typical cases. We will select examples where the origin $(0, 0)$ is the only equilibrium point, so $r_1 \neq 0$ and $r_2 \neq 0$. Thus, we shall look at examples corresponding to cases 1(a), 1(c), 1(e), 2(a), 2(c), 3(a), 3(b), and 3(c). The other cases are dealt with in Exercise 3 on page 598.

---

[3] An equilibrium point (x_0, y_0) is ISOLATED if we can find a circle with center (x_0, y_0) inside which there is no other equilibrium point. The radius of the circle can be as small as we please.

EXAMPLE 12.2

First, we want an example of case 1(a), where $r_1 < r_2 < 0$, (say, $r_1 = -2, r_2 = -1$).[4] From (12.15) we see that to have these roots, a, b, c, and d must satisfy $a + d = -3$, with $bc = 0$. These will be satisfied if we consider $a = -2, d = -1$, and $b = c = 0$, that is,

$$\begin{cases} dx/dt = -2x, \\ dy/dt = -y. \end{cases} \tag{12.22}$$

Here the equilibrium point is $(0, 0)$.

The explicit solution of the system of differential equations in (12.22) is

$$x(t) = C_1 e^{-2t}, \qquad y(t) = C_2 e^{-t}. \tag{12.23}$$

The trajectories of (12.22) in the phase plane will satisfy

$$\frac{dy}{dx} = \frac{-y}{-2x} = \frac{y}{2x}.$$

This is a first order separable differential equation which may be solved [5] to yield

$$y^2 = Cx,$$

where C is an arbitrary constant. [Notice that we could have also obtained this equation by eliminating t in (12.23).] If $C \neq 0$, the trajectories are parabolas, opening to the right if $C > 0$ and to the left if $C < 0$. If $C = 0$, the trajectory is $y = 0$. From (12.23) we see that as $t \to \infty$, both x and y approach the equilibrium point $(0, 0)$. Figure 12.1 shows the phase plane and some trajectories of (12.22), including $x = 0$. Because all trajectories approach the equilibrium point $(0, 0)$ as $t \to \infty$, the equilibrium point is stable.

Notice that in the vicinity of the equilibrium point all trajectories appear to be asymptotic to the line $x = 0$, except the trajectory $y = 0$. We can verify this from (12.23) by observing that if $C_2 \neq 0$, then

$$\frac{x(t)}{y(t)} = \frac{C_1 e^{-2t}}{C_2 e^{-t}} = \frac{C_1}{C_2} e^{-t} \to 0, \text{ as } t \to \infty.$$

If $C_2 = 0$ then $y(t) = 0$, and this is the exceptional trajectory. This equilibrium point is called a NODE. $\qquad\qquad\square$

Figure 12.2 shows another possible example of case 1(a) — namely, $a = -1, d = -2$, and $b = c = 0$ — so that

$$\begin{cases} dx/dt = -x, \\ dy/dt = -2y. \end{cases} \tag{12.24}$$

(See Exercise 2 on page 598.)

---

[4]Of course there are many other possibilities.

[5]When you do this, don't mislay the equilibrium solution $y(x) = 0$.

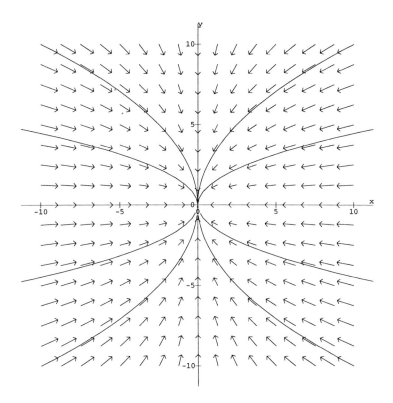

FIGURE 12.1 Trajectories of $dx/dt = -2x$ and $dy/dt = -y$, where $r_1 < r_2 < 0$ and the equilibrium point is a stable node

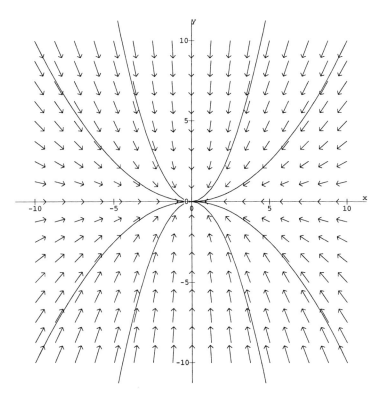

FIGURE 12.2 Trajectories of $dx/dt = -x$ and $dy/dt = -2y$, where $r_1 < r_2 < 0$ and the equilibrium point is a stable node

EXAMPLE 12.3

To construct an example corresponding to case 1(c), we want $r_1 < 0$ and $r_2 > 0$, so we consider $r_1 = -1$, $r_2 = 1$. From (12.15) we see that selecting $a = -1$ and $d = 1$, with $b = c = 0$, gives such an example — that is,

$$\begin{cases} dx/dt = -x, \\ dy/dt = y. \end{cases} \tag{12.25}$$

Here the equilibrium point is again $(0, 0)$.

The explicit solution of the system of differential equations in (12.25) is

$$x(t) = C_1 e^{-t}, \qquad y(t) = C_2 e^{t}. \tag{12.26}$$

The trajectories of (12.25) in the phase plane will satisfy

$$\frac{dy}{dx} = \frac{y}{-x},$$

which is integrated to yield

$$y = \frac{C}{x}, \tag{12.27}$$

where C is an arbitrary constant. [Notice that we could have also obtained (12.27) by eliminating t in (12.26).] If $C \neq 0$, the trajectories given by (12.27) are hyperbolas. If $C = 0$, the trajectory is $y = 0$. From (12.26) we see that if $t \to \infty$, then $x \to 0$. At the same time, either $y \to \infty$ (if $C_2 \neq 0$) or $y \to 0$ (if $C_2 = 0$). Figure 12.3 shows the phase plane of (12.25) and some trajectories. Because the trajectory $y = 0$ approaches the equilibrium point $(0, 0)$ as $t \to \infty$, but all other trajectories move away from the equilibrium point, the equilibrium point is neither stable nor unstable. It is called a SADDLE POINT.   □

Figure 12.4 shows another possible example of case 1(c) — namely, $a = d = 0$ and $b = c = 1$, so that

$$\begin{cases} dx/dt = y, \\ dy/dt = x. \end{cases}$$

We see that if we rotate Figure 12.14 counterclockwise through 45° we have Figure 12.3, so the equilibrium point is again a saddle point.

EXAMPLE 12.4

To construct an example corresponding to case 1(e), we want $0 < r_1 < r_2$, say $r_1 = 1$ and $r_2 = 2$. From (12.15) we see that selecting $a = 2$, $d = 1$, and $b = c = 0$ gives such an example, namely,

$$\begin{cases} dx/dt = 2x, \\ dy/dt = y. \end{cases} \tag{12.28}$$

Here the equilibrium point is again $(0, 0)$. If, in (12.28), we replace t with $-t$ (called "running the differential equation backwards in time"), we see that (12.28) becomes (12.22). Thus, we can construct the phase plane for (12.28) from the phase plane for (12.22) by reversing the arrows, in which case the equilibrium point $(0, 0)$ is an unstable node. This is shown in Figure 12.5.   □

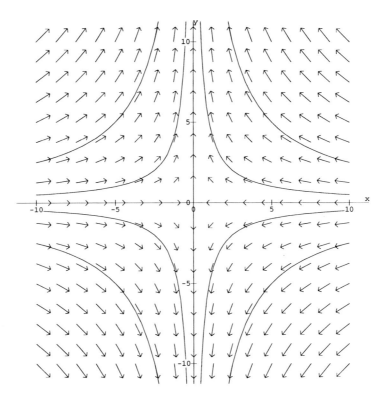

FIGURE 12.3 Trajectories of $dx/dt = -x$ and $dy/dt = y$, where $r_1 < 0$, $r_2 > 0$ and the equilibrium point is a saddle point

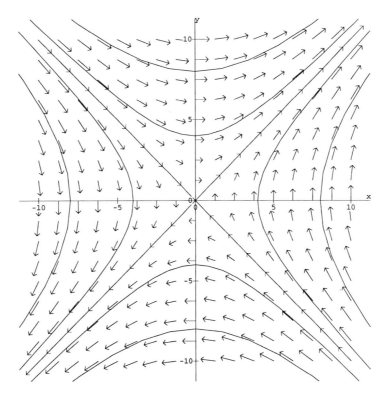

FIGURE 12.4 Trajectories of $dx/dt = y$ and $dy/dt = x$, where $r_1 < 0, r_2 > 0$ and the equilibrium point is a saddle point

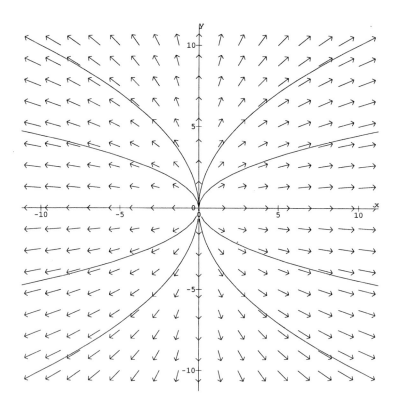

FIGURE 12.5 Trajectories of $dx/dt = 2x$ and $dy/dt = y$, where $0 < r_1 < r_2$ and the equilibrium point is an unstable node

EXAMPLE 12.5

To construct an example corresponding to case 2(a), we want $r_1 = r_2 < 0$, so we choose $r_1 = r_2 = -1$. From (12.15) we see that $a = -1$ and $d = -1$, with $b = c = 0$, gives such an example — that is,

$$\begin{cases} dx/dt = -x, \\ dy/dt = -y. \end{cases} \tag{12.29}$$

Here the equilibrium point is again $(0, 0)$.

The solution of (12.29) is

$$x(t) = C_1 e^{-t}, \qquad y(t) = C_2 e^{-t}. \tag{12.30}$$

The trajectories of (12.29) in the phase plane will satisfy

$$\frac{dy}{dx} = \frac{y}{x},$$

which may be integrated to yield

$$y = Cx, \tag{12.31}$$

where C is an arbitrary constant. The trajectories, (12.31) or $x = 0$, are straight lines through the origin. From (12.30) we see that if $t \to \infty$, then $(x, y) \to (0, 0)$. Figure 12.6 shows the phase plane and some trajectories of (12.29). Because all trajectories approach the equilibrium point $(0, 0)$ as $t \to \infty$, the equilibrium point is a stable node. $\square$

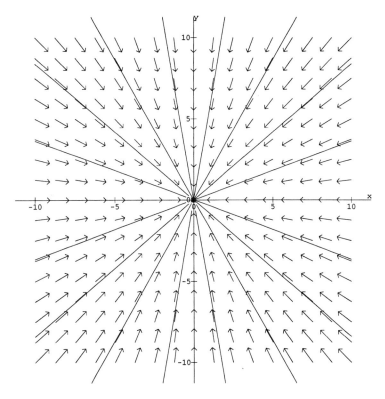

FIGURE 12.6 Trajectories of $dx/dt = -x$ and $dy/dt = -y$, where $r_1 = r_2 < 0$, and the equilibrium point is a stable node

Figure 12.7 shows another possible example of case 2(a), namely, $a = d = -1$, $b = 1$, and $c = 0$, so that

$$\begin{cases} dx/dt = -x + y, \\ dy/dt = \quad -y, \end{cases} \tag{12.32}$$

and the equilibrium point is again a stable node. Notice that in this case, all trajectories appear to be asymptotic to the line $y = 0$. This can be confirmed from the explicit solution of (12.32), namely,

$$x(t) = C_1 e^{-t} + C_2 t e^{-t}, \qquad y(t) = C_2 e^{-t},$$

because

$$\frac{y(t)}{x(t)} = \frac{C_2 e^{-t}}{C_1 e^{-t} + C_2 t e^{-t}} = \frac{C_2}{C_1 + C_2 t} \to 0 \text{ as } t \to \infty,$$

for all choices of C_1, C_2.

Case 2(c), with $r > 0$, can be handled in a similar way to case 2(a), by "reversing the time." The equilibrium point is an unstable node.

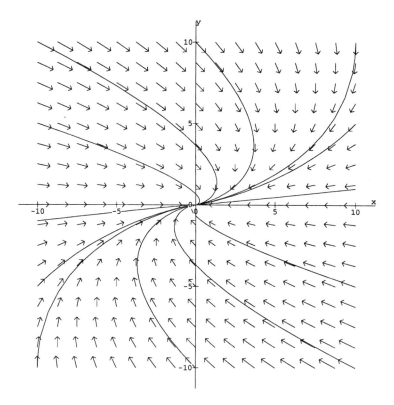

FIGURE 12.7   Trajectories of $dx/dt = -x + y$ and $dy/dt = -y$, where $r_1 = r_2 < 0$ and the equilibrium point is a stable node

EXAMPLE 12.6

To construct an example corresponding to case 3(a), we want $a + d = 2\alpha < 0$ and $ad - bc = \alpha^2 + \beta^2$. These will be satisfied if we consider $a = \alpha < 0$, $d = \alpha$, $b = -\beta$, and $c = \beta$, namely,

$$\begin{cases} dx/dt = \alpha x - \beta y, \\ dy/dt = \beta x + \alpha y. \end{cases} \tag{12.33}$$

Here the equilibrium point is again $(0, 0)$.

The solution of (12.33) is

$$x(t) = e^{\alpha t}\left(C_1 \cos \beta t - C_2 \sin \beta t\right), \qquad y(t) = e^{\alpha t}\left(C_2 \cos \beta t + C_1 \sin \beta t\right), \tag{12.34}$$

which can be rewritten in the form (see Exercise 4 on page 598)

$$x(t) = \sqrt{C_1^2 + C_2^2}\, e^{\alpha t} \cos\left(\beta t + \phi\right), \qquad y(t) = \sqrt{C_1^2 + C_2^2}\, e^{\alpha t} \sin\left(\beta t + \phi\right). \tag{12.35}$$

As $t \to \infty$ we see that $x(t) \to 0$ and $y(t) \to 0$, because $\alpha < 0$. Thus, the equilibrium point is stable. The trajectories of (12.33) in the phase plane will satisfy

$$\frac{dy}{dx} = \frac{\beta x + \alpha y}{\alpha x - \beta y}.$$

Although this can be solved, the final solution is not very illuminating. However, (12.35) represents the parametric form of these trajectories, from which we can obtain a great deal of information. From (12.35) we have

$$x^2 + y^2 = \left(C_1^2 + C_2^2 \right) e^{2\alpha t}, \tag{12.36}$$

so as $t \to \infty$ the trajectories follow "circles" whose radii decrease with time, because $\alpha < 0$. In fact these trajectories are spirals in the xy plane (see Exercise 4 on page 598). Figure 12.8 shows the phase plane and some trajectories of (12.33) for the case $\alpha = -0.1$ and $\beta = 1$. This equilibrium point is called a FOCUS. □

A similar analysis applies to case 3(b), with $\alpha = 0$, where we find the trajectories are circles. Here the equilibrium point is stable (see Figure 12.9) and called a CENTER.

Case 3(c), with $\alpha > 0$, can be handled similar to case 3(a) (by "reversing the time") so the trajectories spiral out. This equilibrium point is unstable (see Figure 12.10) and is again called a focus.

We summarize these results in the following way.

Theorem 12.1: *Consider the system of differential equations*

$$\begin{cases} dx/dt = ax + by \\ dy/dt = cx + dy \end{cases}$$

with characteristic equation

$$r^2 - (a+d)r + ad - bc = 0.$$

1. *If the roots of the characteristic equation are real and distinct, with $r_1 < r_2$, then*
   (a) If $r_1 < r_2 < 0$, the origin is a stable node.
   (b) If $0 < r_1 < r_2$, the origin is an unstable node.
   (c) If $r_1 < 0, r_2 > 0$, then the origin is a saddle point.

2. *If $r_1 = r_2$, then*
   (a) If $r_1 = r_2 < 0$, the origin is a stable node.
   (b) If $0 < r_1 = r_2$, the origin is an unstable node.

3. *If $r_{1,2} = \alpha \pm i\beta$, then*
   (a) If $\alpha < 0$, the origin is a stable focus.
   (b) If $\alpha > 0$, the origin is an unstable focus.
   (c) If $\alpha = 0$, the origin is a center and is neutrally stable.

EXAMPLE 12.7

Identify and classify the equilibrium point of the system

$$\begin{cases} dx/dt = -x + y, \\ dy/dt = -x - y. \end{cases} \tag{12.37}$$

The characteristic equation corresponding to (12.37) is

$$r^2 + 2r + 2 = 0,$$

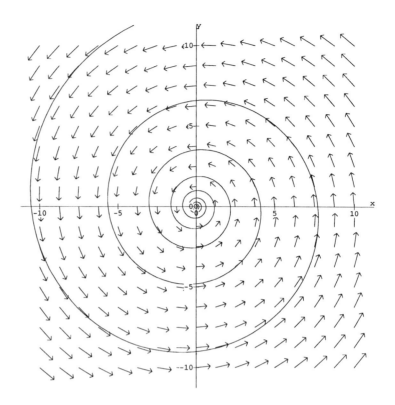

FIGURE 12.8   Trajectories of $dy/dx = (\beta x + \alpha y)/(\alpha x - \beta y)$, where r_1 and r_2 are complex with negative real parts and the equilibrium point is a stable focus

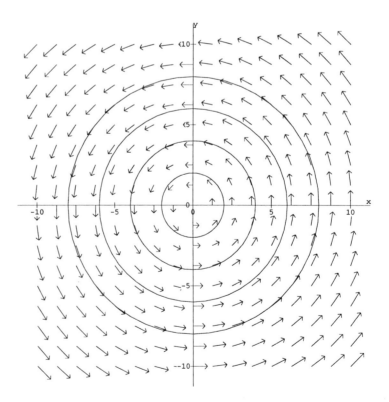

FIGURE 12.9   Trajectories of $dy/dx = -x/y$, where r_1 and r_2 are complex with real parts zero and the equilibrium point is a (stable) center

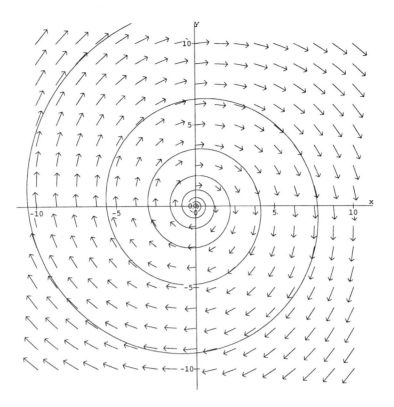

FIGURE 12.10   Trajectories of $dy/dx = (\beta x + \alpha y)/(\alpha x - \beta y)$, where r_1 and r_2 are complex with positive real parts and the equilibrium point is an unstable focus

with roots $r_{1,2} = -1 \pm i$. These are complex roots with negative real part, so the equilibrium point $(0, 0)$ is a stable focus. The phase plane and some of its trajectories are shown in Figure 12.11.   □

EXAMPLE 12.8

Identify and classify the equilibrium point of the system

$$\begin{cases} dx/dt = -x + y, \\ dy/dt = -6x - 6y. \end{cases} \tag{12.38}$$

The characteristic equation corresponding to (12.38) is

$$r^2 + 7r + 12 = 0,$$

with roots $r_1 = -3$, $r_2 = -4$. These are negative distinct real roots, so the equilibrium point $(0, 0)$ is a stable node. The phase plane and some of its trajectories are shown in Figure 12.12.

Notice that in Figure 12.12 the trajectories appear to be asymptotic to a line with negative slope. We can determine the equation of this line from the solution of (12.38), namely,

$$x(t) = C_1 e^{-4t} + C_2 e^{-3t},$$

$$y(t) = -3C_1 e^{-4t} - 2C_2 e^{-3t},$$

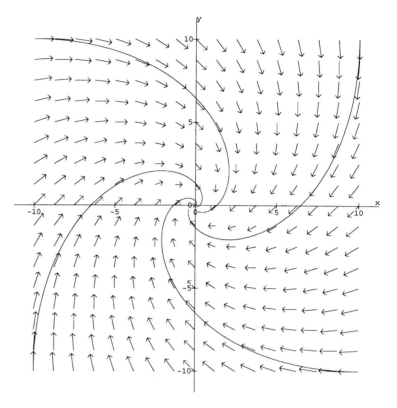

FIGURE 12.11   Trajectories of $dx/dt = -x + y$ and $dy/dt = -x - y$

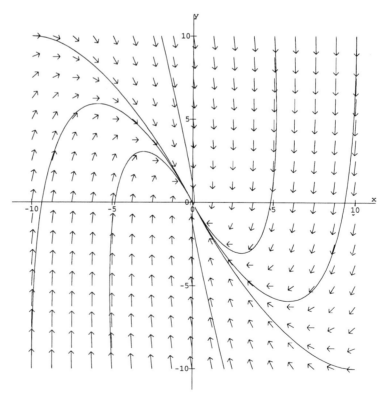

FIGURE 12.12   Trajectories of $dx/dt = -x + y$ and $dy/dt = -6x - 6y$

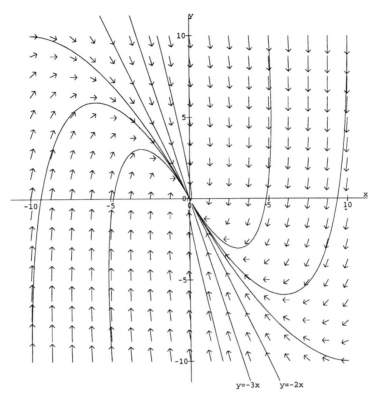

FIGURE 12.13 Trajectories of $dx/dt = -x + y$ and $dy/dt = -6x - 6y$, and the lines $y = -2x$ and $y = -3x$

by computing

$$\frac{y(t)}{x(t)} = \frac{-3C_1 e^{-4t} - 2C_2 e^{-3t}}{C_1 e^{-4t} + C_2 e^{-3t}} = \frac{-3C_1 e^{-t} - 2C_2}{C_1 e^{-t} + C_2} \to -2 \text{ as } t \to \infty,$$

if $C_2 \neq 0$. Thus, all trajectories corresponding to $C_2 \neq 0$ are asymptotic to the line $y = -2x$. (The line $y = -2x$ is also a trajectory. Why?) But what happens if $C_2 = 0$? In this case

$$\frac{y(t)}{x(t)} = \frac{-3C_1 e^{-4t}}{C_1 e^{-4t}} = -3,$$

so this is the line $y = -3x$. (The line $y = -3x$ is also a trajectory. Why?) Notice that it is difficult, if not impossible, to spot the trajectories $y = -2x$ and $y = -3x$ in Figure 12.12. Figure 12.13 shows the trajectories $y = -2x$ and $y = -3x$ superimposed on Figure 12.12. $\qquad\square$

EXERCISES

1. For each of the following differential equations, identify whether the equilibrium points are stable or unstable and whether each is a node, saddle point, center, or focus.

(a)
$$\begin{cases} dx/dt = 2x + 5y \\ dy/dt = x + 6y \end{cases}$$

(b)
$$\begin{cases} dx/dt = 2x - y \\ dy/dt = -6x + y \end{cases}$$

(c) $$\begin{cases} dx/dt = 5x + 6y \\ dy/dt = x + 4y \end{cases}$$

(d) $$\begin{cases} dx/dt = 5x - 3y \\ dy/dt = -x - y \end{cases}$$

(e) $$\begin{cases} dx/dt = 2x + 9y \\ dy/dt = -x - 4y \end{cases}$$

(f) $$\begin{cases} dx/dt = 2x + y \\ dy/dt = x - 4y \end{cases}$$

(g) $$\begin{cases} dx/dt = 4x - 2y \\ dy/dt = 5x - 2y \end{cases}$$

(h) $$\begin{cases} dx/dt = 2x + y \\ dy/dt = -4x + 2y \end{cases}$$

(i) $$\begin{cases} dx/dt = 3x + 5y \\ dy/dt = -5x + 3y \end{cases}$$

(j) $$\begin{cases} dx/dt = x + y \\ dy/dt = x + 3y \end{cases}$$

2. Show that the origin is a stable node of (12.24):

   (a) By finding the explicit solution.

   (b) By finding the trajectories in the phase plane.

3. Show that the phase plane of the system of differential equations

   $$\begin{cases} dx/dt = ax + by \\ dy/dt = cx + dy \end{cases}$$

   where $ad - bc = 0$, falls into one of the following three categories.

   (a) Every point is an equilibrium point.

   (b) There is a line of equilibrium points, with every trajectory being a straight line approaching an equilibrium point.

   (c) There is a line of equilibrium points, with every trajectory being a straight line parallel to the line of equilibrium points.

4. Use the trigonometric identities for $\cos(\beta t + \phi)$ and $\sin(\beta t + \phi)$ to show that (12.34) and (12.35) are true. Then from (12.35) show that

   $$t = \frac{1}{\beta} \left[\arctan(y/x) - \phi \right],$$

   so that (12.36) can be written

   $$x^2 + y^2 = c^2 e^{(2\alpha/\beta)\arctan(y/x)},$$

   where $c^2 = \left(C_1^2 + C_2^2 \right) e^{-2\alpha\phi/\beta}$ is chosen appropriately. Now change to polar coordinates r and θ, where $r = \sqrt{x^2 + y^2}$ and $\theta = \arctan(y/x)$, to obtain the spiral equation

   $$r = c e^{\alpha\theta/\beta}.$$

12.3   Straight Line Trajectories of Linear Autonomous Systems

If we look carefully at Figures 12.1 through 12.12, we see that in some cases there appear to be straight line trajectories through the origin of the phase plane (Figures 12.1 through 12.7, and 12.12), whereas in others there are none (Figures 12.8 through 12.11). In this section we want to find under what circumstances straight line trajectories exist.[6]

To start, we look at Figure 12.1 and notice that the vertical line $x = 0$ is a trajectory. A natural question is under what circumstances is $x(t) = 0$ a solution of

$$\begin{cases} dx/dt = ax + by, \\ dy/dt = cx + dy. \end{cases}$$

If we substitute $x(t) = 0$ into this we are forced to the condition that $b = 0$. Conversely, if $b = 0$, then $dx/dt = x$ contains $x(t) = 0$ as one of its solutions. Thus, $b = 0$ is equivalent to the existence of the straight line trajectory $x = 0$ in the phase plane. We also notice that if $b = 0$, then the characteristic equation reduces to

$$r^2 - (a + d)r + ad = (r - a)(r - d).$$

Thus, the vertical phase plane trajectory $x = 0$ can occur only if $r_1 = a$ and $r_2 = d$ are real. Of course, there may be other straight line trajectories for $b = 0$. Confirm these observations by looking at Figures 12.1 through 12.12.

Now we look at nonvertical straight line trajectories through the origin in the phase plane. We denote these trajectories by $y = mx$ and try to determine m. If we substitute $y = mx$ into the equation for the phase plane trajectories

$$\frac{dy}{dx} = \frac{cx + dy}{ax + by},$$

we find

$$m = \frac{c + dm}{a + bm},$$

or

$$bm^2 + (a - d)m - c = 0.$$

If $b \neq 0$ then we can solve this quadratic equation to give

$$m = \frac{1}{2b}\left[-(a - d) \pm \sqrt{\Delta}\right],$$

where

$$\Delta = (a - d)^2 + 4bc.$$

Thus, there will be two values for the slope m if $\Delta > 0$, one value if $\Delta = 0$, and no values if $\Delta < 0$. But wait a minute, we have seen the quantity $\Delta = (a - d)^2 + 4bc$ before. It is

---

[6]Actually each straight line consists of two trajectories that meet at the equilibrium point. If the equilibrium point is stable, then these trajectories take an infinite time to reach the equilibrium point. If unstable they take an infinite time to reach infinity.

exactly the discriminant of the characteristic equation, where we have real and distinct roots if $\Delta > 0$, repeated roots if $\Delta = 0$, and complex roots if $\Delta < 0$. (Confirm these observations by looking at Figures 12.1 through 12.12.) Thus, we have the following result.

- If $b \neq 0$ then
  — when we have two real distinct roots of the characteristic equation, there are two distinct straight line trajectories through the origin.
  — when we have a repeated root of the characteristic equation, there is one straight line trajectory through the origin.
  — when we have no real roots of the characteristic equation, there are no straight line trajectories through the origin.

Now let's look at the situation when $b = 0$, remembering that we already know that there is always a vertical straight line trajectory through the origin in this case. To decide if there are any other straight line trajectories through the origin, we note that because $b = 0$, the quadratic for m reduces to

$$(a - d)m - c = 0.$$

If $a \neq d$ (that is, $r_1 \neq r_2$) then there is another straight line trajectory with slope $m = c/(a - d)$. If $a = d$ (that is, $r_1 = r_2$) and $c \neq 0$, there are no more straight line trajectories. If $a = d$ (that is, $r_1 = r_2$) and $c = 0$, then there is no restriction on m, and so there are an infinite number of straight line trajectories. Confirm these observations by looking at Figures 12.1 through 12.12. We combine these results in the following statements.

- If $b = 0$ then
  — when we have two real distinct roots of the characteristic equation, there are two distinct straight line trajectories through the origin.
  — when we have a repeated root of the characteristic equation, there is one straight line trajectory through the origin if $c \neq 0$ and an infinite number of straight line trajectories through the origin if $c = 0$.

We thus have the following theorem.

Theorem 12.2: *If the characteristic equation corresponding to*

$$\frac{dy}{dx} = \frac{cx + dy}{ax + by}$$

has two real distinct roots, then there are two distinct straight line trajectories through the origin in the phase plane. If the characteristic equation has no real roots, then there are no straight line trajectories through the origin in the phase plane. If the characteristic equation has a repeated root, then either there are an infinite number of straight line trajectories through the origin if $b = c = 0$, or there is one straight line trajectory through the origin.

EXAMPLE 12.9

Find the straight line trajectories through the origin of the phase plane for

$$\begin{cases} dx/dt = -x + y, \\ dy/dt = -kx - ky, \end{cases} \tag{12.39}$$

where (a) $k = 6$, (b) $k = 3 + 2\sqrt{2}$, (c) $k = 1$.

The characteristic equation corresponding to this system is

$$r^2 + (1 + k)r + 2k = 0,$$

with two real distinct roots if $k = 6$, a repeated real root if $k = 3 + 2\sqrt{2}$, and two complex roots if $k = 1$. Because $b \neq 0$ we expect two straight line trajectories in case (a), one in case (b), and none in case (c).

Substituting $y = mx$ into the equation for the trajectories in the phase plane,

$$\frac{dy}{dx} = \frac{-kx - ky}{-x + y},$$

gives

$$m = \frac{-k - km}{-1 + m},$$

or

$$m^2 + (k - 1)m + k = 0.$$

The discriminant of this quadratic equation in m is $\Delta = (k - 1)^2 - 4k = k^2 - 6k + 1$, with solution

$$m = \frac{1}{2}\left(1 - k \pm \sqrt{\Delta}\right).$$

In case (a), where $k = 6$, we see that $\Delta = 1$, so $m = -3$ and $m = -2$, in agreement with Figure 12.13.

In case (b), where $k = 3 + 2\sqrt{2}$, we see that $\Delta = 0$, so $m = \left(1 - 3 - 2\sqrt{2}\right)/2 = -1 - \sqrt{2}$. Figure 12.14 shows this.

In case (c), where $k = 1$, we see that $\Delta = -4$, so there are no real solutions for m. This agrees with Figure 12.11, which has no straight line solutions.

In fact if we start with $k = 6$ (Figure 12.13) and gradually decrease the value of k, the two straight line trajectories gradually converge to one line when $k = 3 + 2\sqrt{2} \approx 5.828$ (Figure 12.14). If we decrease k further, the line immediately vanishes (Figure 12.11). $\square$

In the previous example we found the straight line trajectories through the origin of the phase plane. However, we have already found the explicit solutions of the system of differential equations (12.39) in Example 12.1 on page 582.

In case (a), where $k = 6$, we found

$$\begin{cases} x(t) = C_1 e^{-4t} + C_2 e^{-3t}, \\ y(t) = -3C_1 e^{-4t} - 2C_2 e^{-3t}. \end{cases}$$

We can see that one of the straight line trajectories through the origin of the phase plane, $y = -3x$, corresponds to $C_2 = 0$. The other straight line trajectory $y = -2x$ corresponds to $C_1 = 0$. If we rewrite the explicit solution in vector notation, that is,

$$\begin{bmatrix} x(t) \\ y(t) \end{bmatrix} = C_1 \begin{bmatrix} 1 \\ -3 \end{bmatrix} e^{-4t} + C_2 \begin{bmatrix} 1 \\ -2 \end{bmatrix} e^{-3t}, \tag{12.40}$$

we see the straight line trajectories in the phase plane are generated by the two vectors

$$\begin{bmatrix} 1 \\ -3 \end{bmatrix} \text{ and } \begin{bmatrix} 1 \\ -2 \end{bmatrix}.$$

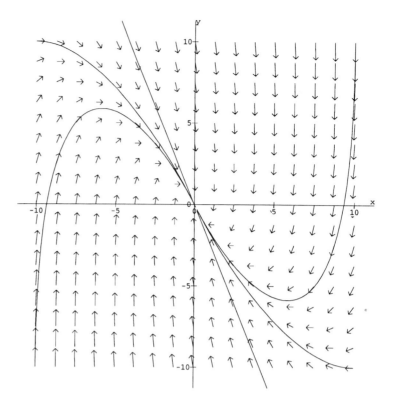

FIGURE 12.14 Trajectories of $dx/dt = -x + y$ and $dy/dt = -kx - ky$, where $k = 3 + 2(2)^{1/2}$ and the line $y = (-1 - 2^{1/2})x$

In case (b), where $k = 3 + 2\sqrt{2}$, we found the explicit solution of the system of differential equations (12.39) as

$$\begin{cases} x(t) = C_1 e^{rt} + C_2 t e^{rt}, \\ y(t) = \left[(r+1)C_1 + C_2 \right] e^{rt} + (r+1)C_2 t e^{rt}, \end{cases}$$

where $r = -2 - \sqrt{2}$. We see that the straight line trajectory through the origin of the phase plane $y = \left(-1 - \sqrt{2} \right)x$ — that is, $y = (r+1)x$ — corresponds to $C_2 = 0$, which is apparent if we rewrite the explicit solution in vector notation as

$$\begin{bmatrix} x(t) \\ y(t) \end{bmatrix} = \begin{bmatrix} C_1 \\ (r+1)C_1 + C_2 \end{bmatrix} e^{rt} + C_2 \begin{bmatrix} 1 \\ r+1 \end{bmatrix} t e^{rt}.$$

In case (c), where $k = 1$, we found explicit solutions of the system of differential equations (12.39) as

$$\begin{cases} x(t) = e^{-t} \left(C_1 \sin t + C_2 \cos t \right), \\ y(t) = e^{-t} \left(C_1 \cos t - C_2 \sin t \right), \end{cases}$$

which can be rewritten in the form

$$\begin{bmatrix} x(t) \\ y(t) \end{bmatrix} = C_1 \begin{bmatrix} \sin t \\ \cos t \end{bmatrix} e^{-t} + C_2 \begin{bmatrix} \cos t \\ -\sin t \end{bmatrix} e^{-t}.$$

Consequently there is an advantage in writing the explicit solutions of systems of differential equations in vector form.

EXERCISES

1. For each of the following differential equations, identify all straight line trajectories through the origin of the phase plane.

(a)
$$\begin{cases} dx/dt = 2x + 5y \\ dy/dt = x + 6y \end{cases}$$

(b)
$$\begin{cases} dx/dt = 2x - y \\ dy/dt = -6x + y \end{cases}$$

(c)
$$\begin{cases} dx/dt = 5x + 6y \\ dy/dt = x + 4y \end{cases}$$

(d)
$$\begin{cases} dx/dt = 5x - 3y \\ dy/dt = -x - y \end{cases}$$

(e)
$$\begin{cases} dx/dt = 2x + 9y \\ dy/dt = -x - 4y \end{cases}$$

(f)
$$\begin{cases} dx/dt = 2x + y \\ dy/dt = x - 4y \end{cases}$$

(g)
$$\begin{cases} dx/dt = 4x - 2y \\ dy/dt = 5x - 2y \end{cases}$$

(h)
$$\begin{cases} dx/dt = 2x + y \\ dy/dt = -4x + 2y \end{cases}$$

(i)
$$\begin{cases} dx/dt = 3x + 5y \\ dy/dt = -5x + 3y \end{cases}$$

(j)
$$\begin{cases} dx/dt = x + y \\ dy/dt = x + 3y \end{cases}$$

2. For the following systems of differential equations (from Section 7.1), identify all straight line trajectories through the origin of the phase plane. (The quantities a, b, c, k, and m are all positive constants.)

(a)
$$\begin{cases} dx/dt = -ay \\ dy/dt = -ax \end{cases}$$

(b)
$$\begin{cases} dx/dt = v \\ dv/dt = -(k/m)x - (b/m)v \end{cases}$$

(c)
$$\begin{cases} dx/dt = ay \\ dy/dt = -bx \end{cases}$$

(d)
$$\begin{cases} dx/dt = ay - cx \\ dy/dt = -bx \end{cases}$$

12.4   Nullcline Analysis of Linear Autonomous Systems

We can frequently sketch the qualitative behavior of the trajectories of the system of equations

$$\begin{cases} dx/dt = ax + by \\ dy/dt = cx + dy \end{cases}$$

without actually solving them. We now show how this can be done using geometrical arguments.

EXAMPLE 12.10

We start with a system of differential equations that we discussed on page 588, namely,

$$\begin{cases} dx/dt = y \\ dy/dt = x \end{cases} \tag{12.41}$$

and its phase plane counterpart

$$\frac{dy}{dx} = \frac{x}{y}.$$

In the phase plane, we construct the special isoclines corresponding to horizontal and vertical tangent lines. These correspond to places where $dy/dt = 0$ and $dx/dt = 0$ and are called NULLCLINES. They intersect at the equilibrium point. The nullclines for (12.41) are the straight lines $x = 0$ (horizontal tangents) and $y = 0$ (vertical tangents). We sketch these in the phase plane and add short horizontal line segments along the nullcline with horizontal tangents and short vertical line segments along the nullcline with vertical tangents. See Figure 12.15. The point where they intersect, $(0, 0)$, is the equilibrium point.

We now want to decide the direction the trajectories will be traveling when they cross a nullcline, and then indicate this by adding arrowheads to the short horizontal and vertical line segments. Let's consider the horizontal isoclines, which occur along $x = 0$. We want to know how x will change with t. From $dx/dt = y$ we see that x increases for $y > 0$ and decreases for $y < 0$. Thus, the arrows along the nullcline $x = 0$ should point from left to right for $y > 0$ and from right to left for $y < 0$. In the same way the arrows along the nullcline $y = 0$ will point up for $x > 0$ and down for $x < 0$. See Figure 12.16.

Thus, any trajectory that enters the first quadrant cannot escape from it. Furthermore, trajectories in this quadrant will move away from the equilibrium point $(0, 0)$. Similar remarks apply to the third quadrant. In the second and fourth quadrants trajectories are being pulled toward the origin. This is indicative of a saddle point. Figure 12.4 shows the trajectories of (12.41). □

Definition 12.1: **The NULLCLINES of the system of autonomous differential equations**

$$\begin{cases} dx/dt = P(x,y) \\ dy/dt = Q(x,y) \end{cases}$$

are the curves along which the trajectories have horizontal or vertical tangent lines.

How to Perform a Nullcline Analysis on a Linear Autonomous System

Purpose: To use nullclines to obtain the qualitative behavior of the solutions of

$$\begin{cases} dx/dt = ax + by \\ dy/dt = cx + dy. \end{cases}$$

Technique:

1. Sketch the line $cx + dy = 0$ in the phase plane. This nullcline is where the trajectories will have horizontal tangents. This is indicated by adding short horizontal lines along the nullcline which is called the horizontal nullcline.

2. Sketch the line $ax + by = 0$ in the phase plane. This nullcline is where the trajectories will have vertical tangents. This is indicated by adding short vertical lines along the nullcline which is called the vertical nullcline.

3. Add arrowheads to the short horizontal line segments of the horizontal nullcline $cx + dy = 0$ as follows. Use the equation $x' = ax + by$ to determine where $x' > 0$ along the horizontal nullcline. Since x is increasing with time on this part of the nullcline the arrows should point from left to right. Now determine where $x' < 0$ along the horizontal nullcline. Since x is decreasing with time, on this part of the nullcline the arrows should point from right to left.

4. Add arrowheads to the short vertical line segments of the vertical nullcline $ax + by = 0$ as follows. Use the equation $y' = cx + dy$ to determine where $y' > 0$ along the vertical nullcline. Since y is increasing with time on this part of the nullcline the arrows should point up. Now determine where $y' < 0$ along the horizontal nullcline. Since y is decreasing with time on this part of the nullcline the arrows should point down.

5. The phase plane is now divided into four regions. If a region only has arrows pointing into it, then trajectories are trapped in that region. If a region only has arrows pointing out of it, then trajectories must leave that region.

Comments about nullcline analysis

- The point where the nullclines intersect, $(0, 0)$, is the equilibrium point.

- This analysis depends on the existence of two nullclines through the equilibrium point $(0, 0)$. The lines $ax + by = 0$ and $cx + dy = 0$ degenerate into a single line if and only if $ad - bc = 0$ — that is, if and only if at least one of the real roots, r_1 or r_2, of the characteristic equation is zero. This is the same condition that guarantees the origin is not an isolated equilibrium point.

- We must be careful when using a nullcline analysis, as can be seen by returning to

$$\begin{cases} dx/dt = -x + y, \\ dy/dt = -kx - ky, \end{cases}$$

where $k > 0$ is a constant. The nullclines are $y = x$ (vertical arrows) and $y = -x$ (horizontal arrows) for all $k > 0$. Furthermore, the arrows point in a clockwise direction. It is tempting to think that this implies that the origin is either a focus or a center because of the suggested spiralling nature. Unfortunately this is not true; for $k \geq 3 + 2\sqrt{2}$ the origin is a node. Figures 12.17 and 12.18 show the situations for $k = 1$ and $k = 6$, respectively. We notice that although the nullclines in both figures are identical the trajectories are very different. Figure 12.18 is characteristic of a node, not a focus or center.

EXAMPLE 12.11

Use a nullcline analysis to obtain the qualitative behavior of the solutions of

$$\begin{cases} dx/dt = -x + y, \\ dy/dt = -2y. \end{cases} \tag{12.42}$$

The nullclines of (12.42) are the straight lines $y = x$ (vertical tangents) and $y = 0$ (horizontal tangents). The equilibrium point is at $(0, 0)$, the intersection of these two lines. The arrows along the line $y = x$ point down for $y > 0$ (because $dy/dt = -2y < 0$ for $y > 0$) and point up for $y < 0$. The arrows on the line $y = 0$ point to the left for $x > 0$ (because $dx/dt = -x + y = -x < 0$ on the positive x-axis) and point to the right for $x < 0$. This is shown in Figure 12.19.

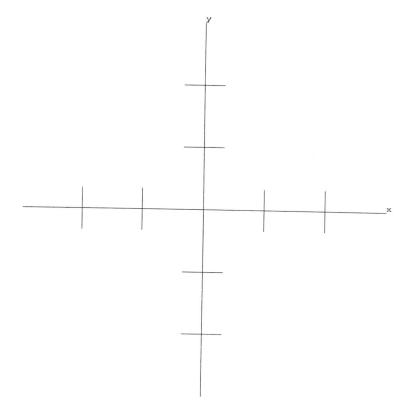

FIGURE 12.15   Nullclines of $dy/dx = x/y$ — step 1

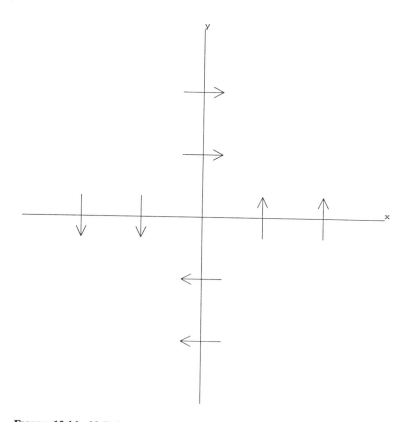

FIGURE 12.16   Nullclines of $dy/dx = x/y$ — step 2

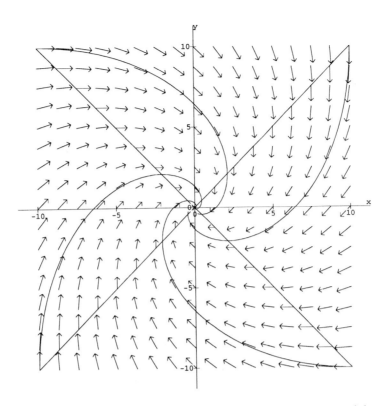

FIGURE 12.17 Trajectories of $dx/dt = -x + y$, $dy/dt = -x - y$, and the lines $y = x$ and $y = -x$

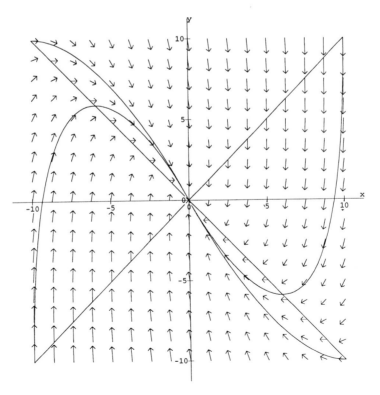

FIGURE 12.18 Trajectories of $dx/dt = -x + y$, $dy/dt = -6x - 6y$, and the lines $y = x$ and $y = -x$

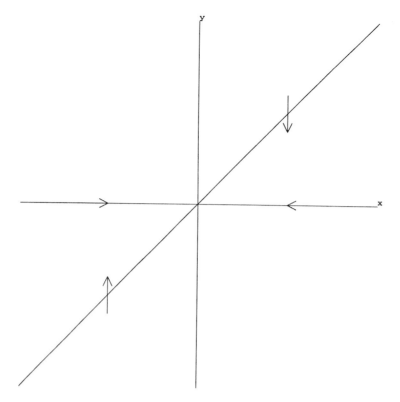

FIGURE 12.19   Nullclines of $dx/dt = -x + y$, $dy/dt = -2y$

Notice that all trajectories that enter the region between the positive x-axis and the line $y = x$ are trapped and have to approach $(0, 0)$ without spiralling. The same is true for the region between the negative x-axis and the line $y = x$. Trajectories in the other two regions are also drawn toward $(0, 0)$. The only way to reach the equilibrium point is asymptotic to the x-axis. Thus, the equilibrium point is a stable node. This can be checked by computing the characteristic equation and finding that its roots are $r_1 = -2$ and $r_2 = -1$, which are real, distinct, and negative. Figure 12.20 shows the phase plane, some trajectories of (12.42), and the lines $y = x$ and $y = 0$.   □

EXERCISES

1. For each of the following differential equations, use a nullcline analysis to identify whether the equilibrium points are stable or unstable and whether each is a node, saddle point, center, or focus. If the analysis fails, explain why.

(a)
$$\begin{cases} dx/dt = 2x + 5y \\ dy/dt = x + 6y \end{cases}$$

(b)
$$\begin{cases} dx/dt = 2x - y \\ dy/dt = -6x + y \end{cases}$$

(c)
$$\begin{cases} dx/dt = 5x + 6y \\ dy/dt = x + 4y \end{cases}$$

(d)
$$\begin{cases} dx/dt = 5x - 3y \\ dy/dt = -x - y \end{cases}$$

(e)
$$\begin{cases} dx/dt = 2x + 9y \\ dy/dt = -x - 4y \end{cases}$$

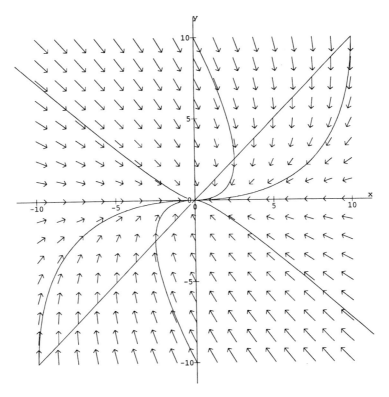

FIGURE 12.20 Trajectories of $dx/dt = -x + y$, $dy/dt = -2y$, and the lines $y = x$ and $y = 0$

(f) $\begin{cases} dx/dt = 2x + y \\ dy/dt = x - 4y \end{cases}$

(g) $\begin{cases} dx/dt = 4x - 2y \\ dy/dt = 5x - 2y \end{cases}$

(h) $\begin{cases} dx/dt = 2x + y \\ dy/dt = -4x + 2y \end{cases}$

(i) $\begin{cases} dx/dt = 3x + 5y \\ dy/dt = -5x + 3y \end{cases}$

(j) $\begin{cases} dx/dt = x + y \\ dy/dt = x + 3y \end{cases}$

2. For the following systems of differential equations (from Section 7.1), use a nullcline analysis to identify whether the equilibrium points are stable or unstable and whether each is a node, saddle point, center, or focus. If the analysis fails, explain why. (The quantities a, b, c, k, and m are all positive constants.)

(a) $\begin{cases} dx/dt = -ay \\ dy/dt = -ax \end{cases}$

(b) $\begin{cases} dx/dt = v \\ dv/dt = -(k/m)x - (b/m)v \end{cases}$

(c) $\begin{cases} dx/dt = ay \\ dy/dt = -bx \end{cases}$

(d) $\begin{aligned} dx/dt &= ay - cx \\ dy/dt &= -bx \end{aligned}$

12.5   Nonlinear Autonomous Systems

We now turn to nonlinear systems of autonomous equations of the form

$$\begin{cases} dx/dt = P(x, y), \\ dy/dt = Q(x, y), \end{cases} \tag{12.43}$$

where the functions $P(x, y)$ and $Q(x, y)$ contain no explicit dependence on t. We will try to use the information from the previous sections on linear autonomous equations to analyze (12.43). We will first look at a simple example to see the type of problems we encounter.

EXAMPLE 12.12: A Simple Two-Population Model

In earlier chapters we discussed the situation in which a population, $x(t)$, grows with time according to the differential equation

$$\frac{1}{x}\frac{dx}{dt} = k,$$

where k is a positive constant. This is exponential growth. We now look at the situation where we have two populations, $x(t)$ and $y(t)$, both growing in this way with the same constant k, but competing for the same finite resources. We use a very simple model where we assume that for the first population x, the rate of change of x per unit population, k, is reduced by an amount proportional to the second population y, so that

$$\frac{1}{x}\frac{dx}{dt} = k - \alpha y,$$

where α is a positive constant. A similar assumption is made for the rate of change of y. Thus, we have

$$\begin{cases} dx/dt = x\,(k - \alpha y), \\ dy/dt = y\,(k - \beta x), \end{cases} \tag{12.44}$$

where α and β are positive constants. Notice that the coefficients of α and β are both xy, and so the system of equations (12.44) is nonlinear. In order to avoid unnecessary algebra, we will consider the case where $k = 5$ and $\alpha = \beta = 1$, so that (12.44) becomes

$$\begin{cases} dx/dt = x\,(5 - y), \\ dy/dt = y\,(5 - x). \end{cases} \tag{12.45}$$

The first thing we might try is to find a differential equation involving only one of the variables x or y. We can do this by the same technique as we used in earlier sections. We differentiate the top equation in (12.45) and into the result substitute the expression for dy/dt from the bottom equation in (12.45) — and y from the top equation in (12.45) — to obtain

$$\begin{aligned} x'' &= x'\,(5 - y) - xy' \\ &= x'\,(x'/x) - xy\,(5 - x) \\ &= (x')^2/x - x\,(5 - x'/x)\,(5 - x) \\ &= (x')^2/x + (5 - x)x' - 5x\,(5 - x) \end{aligned}$$

or

$$x'' - \frac{1}{x}(x')^2 - (5 - x)x' + 5x\,(5 - x) = 0,$$

which is a second order nonlinear differential equation. We have no techniques for solving this equation, so the elimination method does not help. This is usually what happens if we try this method on the general equation (12.43).

Even though we are unable to solve (12.45) for $x(t)$, we may get some useful information by looking at the phase plane. The first thing to do is to find the equilibrium points, which in the case of (12.45) will occur when

$$x(5 - y) = 0$$

and

$$y(5 - x) = 0,$$

that is, when $x = 0$ or $y = 5$ and $y = 0$ or $x = 5$. This gives two equilibrium points — namely, $(0, 0)$ and $(5, 5)$.

If we use a nullcline analysis on (12.45), we find that there are four lines to consider: $x = 0$ and $y = 5$, which have vertical arrows in the phase plane, and $y = 0$ and $x = 5$, which have horizontal arrows. The directions of the arrows are determined from (12.45) and are shown in Figure 12.21.

Notice that the equilibrium points occur only where a nullcline with vertical arrows intersects one with horizontal arrows — namely, at $(0, 0)$ and $(5, 5)$. The points $(5, 0)$ and $(0, 5)$ are not equilibrium points. (Why?)

In order to facilitate the discussion, we have labeled the regions I, II, III, and IV. Notice that if a trajectory enters region I it cannot escape and that it appears to satisfy $x \to 0$ and $y \to \infty$ as $t \to \infty$. Similarly, trajectories entering region III cannot escape and appear to satisfy $x \to \infty$ and $y \to 0$ as $t \to \infty$. Trajectories in region II are either drawn to the equilibrium point $(5, 5)$ or cross a nullcline into regions I or III. Once they enter

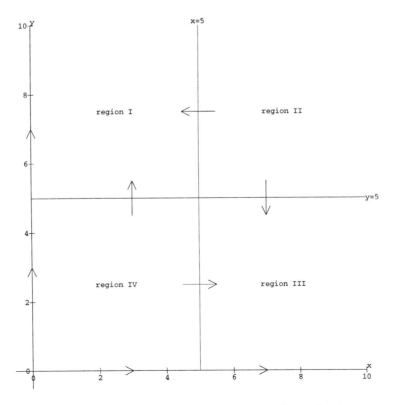

FIGURE 12.21 Nullclines of $dx/dt = x(5 - y)$ and $dy/dt = y(5 - x)$

those regions, we know what happens to them. Trajectories in region IV move away from the equilibrium point $(0, 0)$ and are either drawn to the equilibrium point $(5, 5)$ or cross a nullcline into regions I or III. Thus, the equilibrium point $(0, 0)$ behaves like an unstable node, and the equilibrium point $(5, 5)$ behaves like a saddle point.

We can convince ourselves of this if we concentrate on the areas close to the equilibrium points. For example, when we are near $(0, 0)$, terms like xy can be ignored when compared with x and y. Thus, the system of equations (12.45) can be approximated by the linear system

$$\begin{cases} dx/dt = 5x, \\ dy/dt = 5y, \end{cases}$$

near the equilibrium point $(0, 0)$. This system also has $(0, 0)$ as an equilibrium point, which is an unstable node because the solutions of the characteristic equation are $r_1 = r_2 = 5$. The direction field and some of its trajectories for this linear system are shown in Figure 12.22. By comparison, we show the direction field for (12.45) and some of its trajectories computed numerically in Figure 12.23. Notice how the trajectories for the nonlinear case are distortions of the trajectories in the linear case. These distortions are due to the nonlinear terms in xy.

We now turn to the equilibrium point $(5, 5)$. We are interested in x and y near $x = 5$ and $y = 5$ (that is, we are interested in $u = x - 5$ and $v = y - 5$, where u and v are small). Changing from the variables x and y to the new variables u and v is equivalent to a translation which makes the point $(5, 5)$ in the xy coordinate system the origin of the uv coordinate system. If we make the substitution

$$x = u + 5$$
$$y = v + 5$$

in (12.45), we find

$$\begin{cases} du/dt = -(u + 5)\, v, \\ dv/dt = -(v + 5)\, u. \end{cases} \tag{12.46}$$

For small values of u and v we can neglect terms in uv in comparison with u and v, in which case (12.46) reduces to the linear system

$$\begin{cases} du/dt = -5v, \\ dv/dt = -5u. \end{cases}$$

Here $u = 0$, $v = 0$, is an equilibrium point, which, using the analysis from earlier sections, is a saddle point (the solutions of the characteristic equation are $r_1 = -5$ and $r_2 = 5$). Thus, in the original xy coordinate system, the point $x = 5$, $y = 5$, is a saddle point for points near $(5, 5)$. The direction field and some of its trajectories for this linear system are shown in Figure 12.24. By comparison, we show the direction field for (12.46) and some of its trajectories computed numerically in Figure 12.25. Notice how the trajectories for the nonlinear case are slight distortions of the trajectories in the linear case. These distortions are due to the nonlinear terms in uv. In order to draw attention to these distortions Figure 12.26 is a repeat of Figure 12.25 with the addition of the straight line $v = -u$ from Figure 12.24.

We can try to find the trajectories by solving

$$\frac{dy}{dx} = \frac{y\,(5 - x)}{x\,(5 - y)}, \tag{12.47}$$

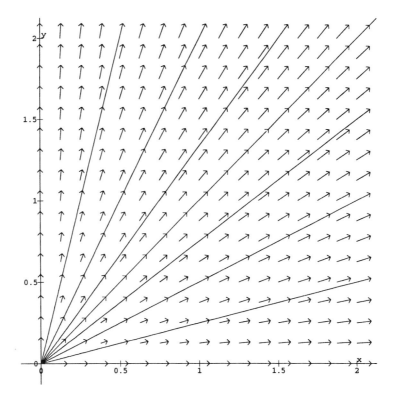

FIGURE 12.22 Trajectories of $dx/dt = 5x$ and $dy/dt = 5y$ near $(0, 0)$

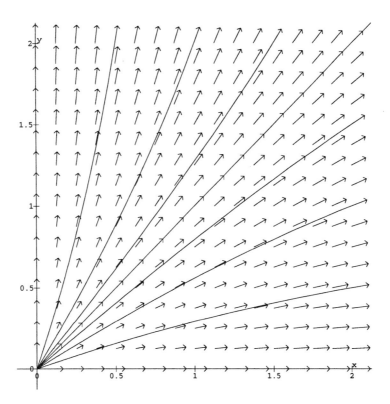

FIGURE 12.23 Numerically computed trajectories of $dx/dt = x(5 - y)$ and $dy/dt = y(5 - x)$ near $(0, 0)$

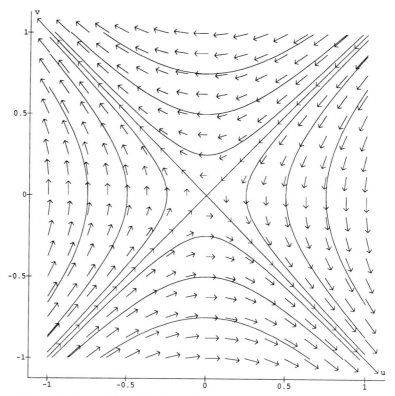

FIGURE 12.24   Trajectories of $du/dt = -5v$ and $dv/dt = -5u$ near $(0, 0)$

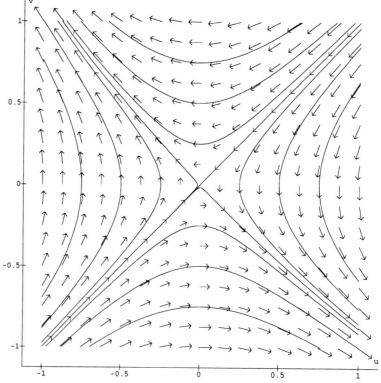

FIGURE 12.25   Numerically computed trajectories of $du/dt = -(u + 5)v$ and $dv/dt = -(v + 5)u$ near $(0, 0)$

which is a separable differential equation with solutions

$$5 \ln y - y = 5 \ln x - x + C. \tag{12.48}$$

This cannot be solved for $y = y(x)$ explicitly but could be sketched using the technique we have developed for drawing solutions of separable differential equations.

However, we can also use technology to evaluate these trajectories numerically, and these trajectories are shown in Figure 12.27. Notice that there appear to be straight line trajectories approaching $(5, 5)$ from both the left and the right. If we substitute $y = mx$ into (12.47) we find $m = 1$, so $y = x$ is a straight line trajectory. This is one of the solutions of (12.48) with $C = 0$. Notice that this particular trajectory divides the xy plane into two regions. An initial condition that starts a trajectory above this trajectory has one type of behavior $(x \to 0, y \to \infty)$, whereas trajectories that start below this trajectory have another type $(x \to \infty, y \to 0)$. This particular trajectory, which separates the behavior of other trajectories, is called a SEPARATRIX.

So, what is the ultimate fate of these populations? If initially the x population exceeds the y population, then the y population becomes extinct, and vice versa. This type of behavior is sometimes called the LAW OF COMPETITIVE EXCLUSION. If it happens that initially $x = y$, then they will tend to coexist and tend to the equilibrium point $(5, 5)$ as $t \to \infty$. However, notice that if by some random event $x \neq y$, then either $x \to \infty$ or $y \to \infty$. (Is this a realistic model?) □

In the preceding example we assumed that in the absence of competition for the same finite resources the populations would grow according to an exponential growth law. In the next two examples, we investigate what happens when the populations grow according to a logistic growth law but have to compete for the same finite resources according to

$$\begin{cases} dx/dt = x \left(k - \gamma x - \alpha y \right), \\ dy/dt = y \left(k - \delta y - \beta x \right), \end{cases} \tag{12.49}$$

where α, β, γ, and δ are positive constants. This system of equations (12.49) is nonlinear. In order to avoid unnecessary algebra, we will consider the case where $k = 5$, $\alpha = \beta = 1$, and $\delta = \gamma$, so that (12.49) becomes

$$\begin{cases} dx/dt = x \left(5 - \gamma x - y \right), \\ dy/dt = y \left(5 - \gamma y - x \right), \end{cases} \tag{12.50}$$

where $\gamma > 0$. This means that in the absence of competition the carrying capacity for both x and y is $5/\gamma$. (Why?)

Our first step is to find the equilibrium points, which in the case of (12.50) occur when

$$x \left(5 - \gamma x - y \right) = 0$$

and

$$y \left(5 - \gamma y - x \right) = 0,$$

that is, when $x = 0$ or $y = 5 - \gamma x$ and $y = 0$ or $y = (5 - x)/\gamma$. If $\gamma \neq 1$ we have four equilibrium points — namely, $(0, 0)$, $(0, 5/\gamma)$, $(5/\gamma, 0)$, and $(5/(\gamma + 1), 5/(\gamma + 1))$. If $\gamma = 1$ we have an infinite number of equilibrium points, namely, $(0, 0)$ and all points on the line $y = 5 - x$. In the two following examples we will consider $\gamma < 1$ and $\gamma > 1$.

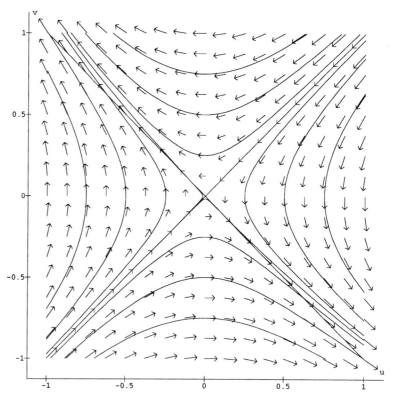

FIGURE 12.26   Numerically computed trajectories of $du/dt = -(u + 5)v$ and $dv/dt = -(v + 5)u$ near $(0, 0)$ and the straight line $v = -u$

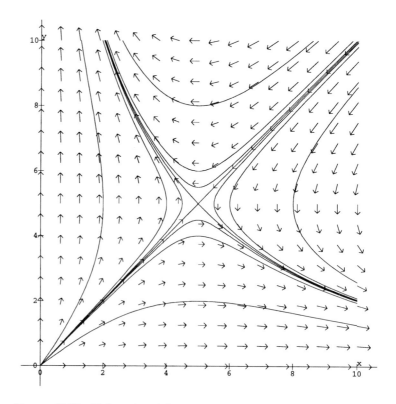

FIGURE 12.27   Trajectories of $dx/dt = x(5 - y)$ and $dy/dt = y(5 - x)$

EXAMPLE 12.13: Another Two Population Model (Part 1)

We illustrate the case $\gamma < 1$ by considering $\gamma = 5/8$, so that (12.50) becomes

$$\begin{cases} dx/dt = x\,(5 - 5x/8 - y)\,, \\ dy/dt = y\,(5 - 5y/8 - x)\,, \end{cases} \qquad (12.51)$$

with equilibrium points at $(0, 0)$, $(0, 8)$, $(8, 0)$, and $(40/13, 40/13)$. If we use a nullcline analysis on (12.51), we find that there are four lines to consider, $x = 0$ and $y = 5 - 5x/8$ (which have vertical arrows in the phase plane) and $y = 0$ and $y = 8(5 - x)/5$ (which have horizontal arrows). The directions of the arrows are determined from (12.51) and are shown in Figure 12.28.

Again we notice that the equilibrium points occur only where a nullcline with vertical arrows intersects one with horizontal arrows — namely, at $(0, 0)$, $(0, 8)$, $(8, 0)$, and $(40/13, 40/13)$. The points $(5, 0)$ and $(0, 5)$ are not equilibrium points. In order to facilitate the discussion, we have labeled the regions I, II, III, and IV. We notice that if a trajectory enters region I it cannot escape and appears to be attracted to the equilibrium point $(0, 8)$ as $t \to \infty$. Similarly, trajectories entering region III cannot escape and appear to be attracted to the equilibrium point $(8, 0)$ as $t \to \infty$. Trajectories in region II are either drawn to one of the equilibrium points $(0, 8)$, $(40/13, 40/13)$, or $(8, 0)$ or cross a nullcline into regions I or III. Once they enter those regions, we know what happens to them. Trajectories in region IV move away from the equilibrium point $(0, 0)$ and are either drawn to one of the equilibrium points $(0, 8)$, $(40/13, 40/13)$, or $(8, 0)$ or cross a nullcline into regions I or III. Thus, the equilibrium point $(0, 0)$ behaves like an unstable node, the equilibrium points $(0, 8)$ and $(8, 0)$ behave like stable nodes, and the equilibrium point $(40/13, 40/13)$ behaves like a saddle point.

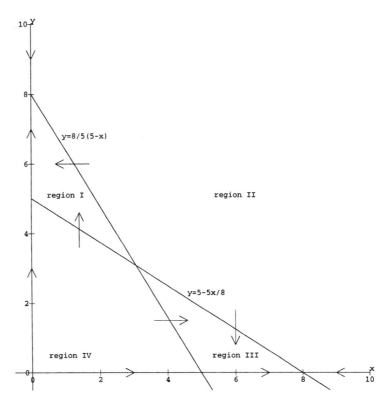

FIGURE 12.28 Nullclines of $dx/dt = x(5 - 5x/8 - y)$ and $dy/dt = y(5 - 5y/8 - x)$

As in the previous example, we can convince ourselves of this if we concentrate on the areas close to the equilibrium points. For example, when we are near $(0, 0)$, terms such as x^2 and xy can be ignored when compared with x and y. Thus, the system of equations (12.51) can be approximated by the linear system

$$\begin{cases} dx/dt = 5x \\ dy/dt = 5y \end{cases}$$

near the equilibrium point $(0, 0)$. We looked at this system in the previous example, and we found that $(0, 0)$ was an unstable node. The direction field and some of its trajectories for this linear system are shown in Figure 12.22.

We now turn to the equilibrium point $(40/13, 40/13)$. We are interested in x and y near $x = 40/13$ and $y = 40/13$; that is, we are interested in $u = x - 40/13$ and $v = y - 40/13$, where u and v are small. Changing from the variables x and y to the new variables u and v is equivalent to a translation, which makes the point $(40/13, 40/13)$ in the xy coordinate system the origin of the uv coordinate system. If we make the substitution

$$\begin{aligned} x &= u + 40/13 \\ y &= v + 40/13 \end{aligned}$$

in (12.51), we find

$$\begin{cases} du/dt = -(5u/8 + v)(40/13 + u), \\ dv/dt = -(5v/8 + u)(40/13 + v). \end{cases} \tag{12.52}$$

For small values of u and v we can neglect terms in u^2, v^2, and uv in comparison with u and v, in which case (12.52) reduces to the linear equation

$$\begin{cases} du/dt = -(25u + 40v)/13, \\ dv/dt = -(25v + 40u)/13. \end{cases}$$

Here $u = 0$, $v = 0$, is an equilibrium point, which, using the analysis from earlier sections, is a saddle point (the solutions of the characteristic equation are $r_1 = -5$ and $r_2 = 15/13$). Thus, in the original xy coordinate system, the point $x = 40/13$, $y = 40/13$, behaves like a saddle point for points near $(40/13, 40/13)$.

Near the equilibrium point $(0, 8)$, we are interested in $u = x$ and $v = y - 8$, where u and v are small. Changing from the variables x and y to the new variables u and v is equivalent to a translation which makes the point $(0, 8)$ in the xy coordinate system the origin of the uv coordinate system. If we make the substitution

$$\begin{aligned} x &= u \\ y &= v + 8 \end{aligned}$$

in (12.51), we find

$$\begin{cases} du/dt = u(-3 - 5u/8 - v), \\ dv/dt = -(v + 8)(u + 5v/8). \end{cases} \tag{12.53}$$

For small values of u and v we can neglect terms in u^2, v^2, and uv, in which case (12.53) reduces to the linear equation

$$\begin{cases} du/dt = -3u, \\ dv/dt = -(8u + 5v). \end{cases}$$

Here $u = 0$, $v = 0$, is an equilibrium point, which, using the analysis from earlier sections, is a stable node (the solutions of the characteristic equation are $r_1 = -5$ and $r_2 = -3$). Thus, in the original xy coordinate system, the point $x = 0$, $y = 8$, is a stable node for points near $(0, 8)$. By symmetry, a similar analysis applies to the equilibrium point $(8, 0)$.

We can also use technology to evaluate these trajectories numerically, and these trajectories are shown in Figure 12.29. We notice that there appear to be straight line trajectories approaching $(40/13, 40/13)$ from both the left and the right. If we substitute $y = mx$ into (12.51) we find $m = 1$, so $y = x$ is a straight line trajectory. This particular trajectory is a separatrix.

So, what is the ultimate fate of these populations? If initially the x population exceeds the y population, then the y population becomes extinct, and $x \to 8$ as $t \to \infty$. If initially the y population exceeds the x population, then the x population becomes extinct, and $y \to 8$ as $t \to \infty$. $\qquad\square$

EXAMPLE 12.14: Another Two Population Model (Part 2)

We now illustrate the case $\gamma > 1$ by considering $\gamma = 5/4$, so that (12.50) becomes

$$\begin{cases} dx/dt = x\left(5 - 5x/4 - y\right), \\ dy/dt = y\left(5 - 5y/4 - x\right), \end{cases} \qquad (12.54)$$

with equilibrium points at $(0,0)$, $(0,4)$, $(4,0)$, and $(20/9, 20/9)$. If we use a nullcline analysis on (12.54), we find that there are four lines to consider, $x = 0$ and $y = 5 - 5x/4$ (which have vertical arrows in the phase plane) and $y = 0$ and $y = 4(5 - x)/5$ (which have horizontal arrows). The directions of the arrows are determined from (12.54) and are shown in Figure 12.30.

Again we notice that the equilibrium points occur only where a nullcline with vertical arrows intersects one with horizontal arrows — namely, at $(0,0)$, $(0,4)$, $(4,0)$ and $(20/9, 20/9)$. The points $(5, 0)$ and $(0, 5)$ are not equilibrium points. In order to facilitate the discussion, we have labeled the regions I, II, III, and IV. Notice that if a trajectory enters region I it cannot escape and appears to be attracted to the equilibrium point $(20/9, 20/9)$ as $t \to \infty$. Similarly, trajectories entering region III cannot escape and appear to be attracted to the equilibrium point $(20/9, 20/9)$ as $t \to \infty$. Trajectories in region II are either drawn directly to the equilibrium point $(20/9, 20/9)$ or cross a nullcline into regions I or III and are again drawn to the equilibrium point $(20/9, 20/9)$. Trajectories in region IV move away from the equilibrium points $(0,0)$, $(0,4)$, and $(4,0)$ and either are drawn to the equilibrium point $(20/9, 20/9)$ directly or cross a nullcline into regions I or III and are again drawn to the equilibrium point $(20/9, 20/9)$. Thus, the equilibrium points $(0,0)$, $(0,4)$, and $(4,0)$ behave like unstable nodes, and the equilibrium point $(20/9, 20/9)$ behaves like a stable node.

As in the last example, we can convince ourselves of this if we concentrate on the areas close to the equilibrium points. When we are near $(0,0)$, terms such as x^2 and xy can be ignored when compared with x and y. Thus, the system of equations (12.51) can be approximated by the linear system

$$\begin{cases} dx/dt = 5x \\ dy/dt = 5y \end{cases}$$

near the equilibrium point $(0,0)$. We found this system in the last example. It has $(0,0)$ as an unstable node, and the direction field and some of its trajectories for this linear system are shown in Figure 12.22.

We now turn to the area near equilibrium point $(20/9, 20/9)$. We are interested in x and y near $x = 20/9$ and $y = 20/9$; that is, we are interested in $u = x - 20/9$ and $v = y - 20/9$, where u and v are small. Changing from the variables x and y to the new

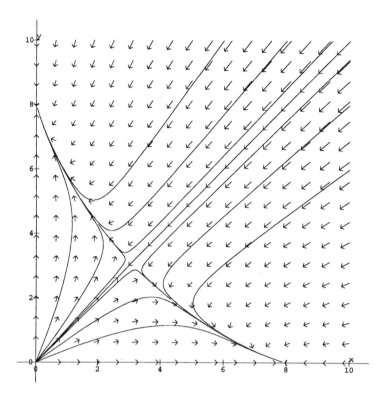

FIGURE 12.29 Numerically computed trajectories of $dx/dt = x(5 - 5x/8 - y)$ and $dy/dt = y(5 - 5y/8 - x)$

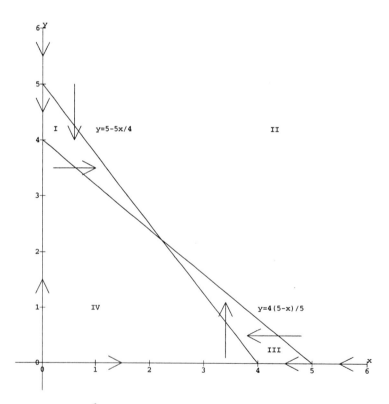

FIGURE 12.30 Nullclines of $dx/dt = x(5 - 5x/4 - y)$ and $dy/dt = y(5 - 5y/4 - x)$

variables u and v is equivalent to a translation, which makes the point $(20/9, 20/9)$ in the xy coordinate system the origin of the uv coordinate system. If we make the substitution

$$x = u + 20/9$$
$$y = v + 20/9$$

in (12.51), we find

$$\begin{cases} du/dt = -(u + 20/9)(5u/4 + v), \\ dv/dt = -(v + 20/9)(5v/4 + u). \end{cases} \tag{12.55}$$

For small values of u and v we can neglect terms in u^2, v^2, and uv in comparison with u and v, in which case (12.55) reduces to the linear equation

$$du/dt = -(25u + 20v)/9,$$
$$dv/dt = -(25v + 20u)/9.$$

Here $u = 0$, $v = 0$, is an equilibrium point, which, using the analysis from earlier sections, is a stable node (the solutions of the characteristic equation are $r_1 = -5$ and $r_2 = -5/9$). Thus, in the original xy coordinate system, the point $x = 20/9$, $y = 20/9$, behaves like a stable node for points near $(20/9, 20/9)$.

Near the equilibrium point $(0, 4)$, we are interested in $u = x$ and $v = y - 4$, where u and v are small. Changing from the variables x and y to the new variables u and v is equivalent to a translation, which makes the point $(0, 4)$ in the xy coordinate system the origin of the uv coordinate system. If we make the substitution

$$x = u,$$
$$y = v + 4,$$

in (12.51), we find

$$\begin{cases} du/dt = u(1 - 5u/4 - v), \\ dv/dt = -(v + 4)(u + 5v/4). \end{cases} \tag{12.56}$$

For small values of u and v we can neglect terms in u^2, v^2, and uv in comparison with u and v, in which case (12.56) reduces to the linear equation

$$\begin{cases} du/dt = u, \\ dv/dt = -(4u + 5v). \end{cases}$$

Here $u = 0$, $v = 0$, is an equilibrium point, which, using the analysis from earlier sections, is a stable node (the solutions of the characteristic equation are $r_1 = -5$ and $r_2 = 1$). Thus, in the original xy coordinate system, the point $x = 0$, $y = 4$, is a saddle point for points near $(20/9, 20/9)$. By symmetry, a similar analysis applies to the equilibrium point $(4, 0)$.

We can also use technology to evaluate these trajectories numerically, and these trajectories are shown in Figure 12.31. In spite of appearances, the approximately circular arc with radius 4 consists of parts of 14 distinct trajectories all of which approach $(20/9, 20/9)$ as $t \to \infty$.

So, what is the ultimate fate of these populations? All populations tend to coexist and tend to the equilibrium point $(20/9, 20/9)$ as $t \to \infty$. ☐

Words of Caution:

- We must be careful when using the preceding analysis on nonlinear equations in the case of centers or spirals. The following three systems of differential equations all have the

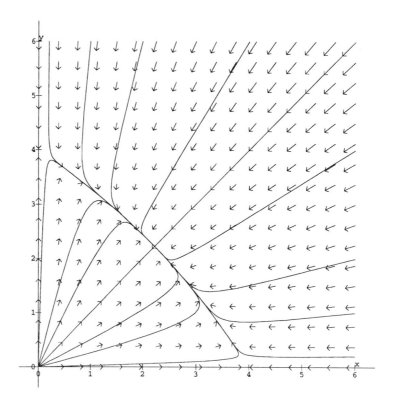

FIGURE 12.31   Numerically computed trajectories of $dx/dt = x(5 - 5x/4 - y)$ and $dy/dt = y(5 - 5y/4 - x)$

origin as their only equilibrium point, and all reduce to the first equation when linearized about this equilibrium point.

$$dx/dt = -y \qquad dy/dt = x,$$
$$dx/dt = -y + x^3 \qquad dy/dt = x + y^3,$$
$$dx/dt = -y - x^3 \qquad dy/dt = x - y^3.$$

However, the actual trajectories in the phase plane of the top equation are circles. Those of the middle equation are outward spirals (see Figure 12.32), and those of the bottom equation are inward spirals (see Figure 12.33).

EXERCISES

1. Use a nullcline analysis to sketch the phase plane trajectories of the following systems of nonlinear equations. Confirm your observations by linearizing these equations about each of the equilibrium points. If possible, check with numerical solutions for the trajectories in the phase plane.

   (a)
   $$\begin{cases} dx/dt = x + y \\ dy/dt = x^2 + y \end{cases}$$

   (b)
   $$\begin{cases} dx/dt = x(y + 1) \\ dy/dt = y(x + 1) \end{cases}$$

   (c)
   $$\begin{cases} dx/dt = -y + x^2 - 1 \\ dy/dt = -y - x^2 + 1 \end{cases}$$

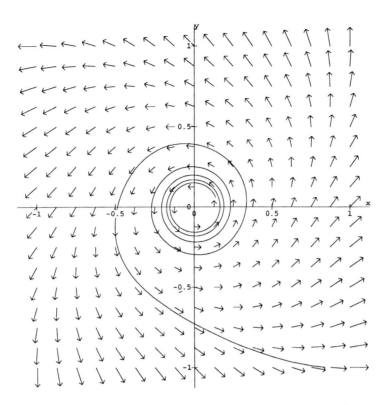

FIGURE 12.32   Numerically computed trajectory of $dx/dt = -y + x^3$ and $dy/dt = x + y^3$

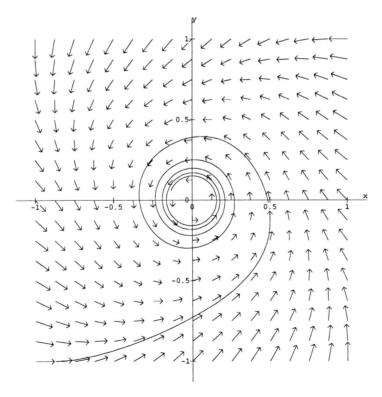

FIGURE 12.33   Numerically computed trajectory of $dx/dt = -y - x^3$ and $dy/dt = x - y^3$

What Have We Learned?

- We may always find explicit solutions for second order autonomous systems of linear differential systems of the form,

$$\begin{cases} dx/dt = ax + by \\ dy/dt = cx + dy, \end{cases}$$

where a, b, c, and d are constants. These solutions are given in terms of roots of the characteristic equation, obtained by expanding the determinant

$$\begin{vmatrix} a-r & b \\ c & d-r \end{vmatrix} = r^2 - (a+d)r + ad - bc = 0.$$

See *How to Find the General Solution of $dx/dt = ax + by$, $dy/dt = cx + dy$* on page 580.

- The stability of equilibrium solutions of linear autonomous systems may be determined from the explicit solution or from the behavior of the roots of the characteristic equation (see Theorem 12.1 on page 593). Terms used to describe the various stability possibilities include *node*, *center*, *focus*, and *saddle point*. Examples are given in Section 12.2 of stable and unstable nodes as well as stable and unstable foci. Saddle points are always unstable, and centers are always neutrally stable. Straight line trajectories in the phase plane for linear systems occur when the roots of the characteristic equation are real (see Theorem 12.2 on page 600).

- It is possible to extend the preceding analysis to systems of equations of the form

$$\begin{cases} dx/dt = ax + by + e, \\ dy/dt = cx + dy + f, \end{cases}$$

where e and f are constants and $ad - bc \neq 0$. The major difference between this case and the case when $e = f = 0$ is that the equilibrium point is no longer $x(t) = 0$, $y(t) = 0$. The equilibrium point will now be $x(t) = x_0$, $y(t) = y_0$, where x_0 and y_0 are constants that satisfy the two equations $ax_0 + by_0 + e = 0$ and $cx_0 + dy_0 + f = 0$. These are two straight lines in the phase plane whose point of intersection will be (x_0, y_0). We can translate coordinates so that this equilibrium point is at the origin if we change variables from (x, y) to (u, v), where $u = x - x_0$ and $v = y - y_0$. This will convert the preceding system in terms of x and y, into an equivalent linear system in terms of u and v.

How to Perform a Nullcline Analysis on an Autonomous System

Purpose: To determine the stability of equilibrium points of

$$\begin{cases} dx/dt = P(x, y) \\ dy/dt = Q(x, y) \end{cases}$$

by using nullclines.

Technique:

1. Determine the equilibrium points by simultaneously solving $P(x, y) = 0$ and $Q(x, y) = 0$ for x, y.

2. Determine the curves (often lines) along which the trajectories in the phase plane have horizontal or vertical tangents.

3. Graph these curves (lines) in the phase plane, and note that they divide the plane into several regions. Add arrows to indicate the direction of the trajectories as they cross these nullclines.

4. Confirm that the equilibrium points occur where the nullclines with horizontal arrows intersect those with vertical arrows.

5. Analyze the stability by considering the direction of trajectories crossing the null-clines. If two nullclines have arrows pointing into a region, then trajectories are trapped in that region. Keep alert for the presence of a separatrix.

CHAPTER 13

Systems of Linear Differential Equations

Where Are We Going?

In Chapter 12 we discussed graphical techniques for analyzing solutions of systems of two linear differential equations with constant coefficients. We also developed explicit solutions for all possible cases. We start this chapter by giving an alternative method for finding these explicit solutions, one that involves simple properties of 2 by 2 matrices. This alternative method introduces eigenvalues and eigenvectors and the notion of a fundamental matrix, which is crucial for a simple algorithm used to solve systems of nonhomogeneous differential equations. The methods for solving systems of two linear differential equations with constant coefficients are then extended to systems of higher order equations.

13.1 Matrix Formulation of Solutions for Autonomous Systems

Chapter 7 contained several examples in which we found explicit solutions of systems of two first order linear differential equations by changing them to single second order differential equations. In Chapter 12 we used this method to obtain the solution of

$$\begin{aligned} dx/dt &= ax + by, \\ dy/dt &= cx + dy, \end{aligned} \tag{13.1}$$

for all possible choices of the constants a, b, c, and d. In this section we will develop an alternative method for solving (13.1) using matrices.

On page 600 we solved (13.1) for the specific choices $a = -1$, $b = 1$, and $c = d = -6$, namely,

$$\begin{aligned} dx/dt &= -x + y, \\ dy/dt &= -6x - 6y. \end{aligned} \tag{13.2}$$

We found that the solution, (12.40), could be written in vector form as

$$\begin{bmatrix} x(t) \\ y(t) \end{bmatrix} = C_1 \begin{bmatrix} 1 \\ -3 \end{bmatrix} e^{-4t} + C_2 \begin{bmatrix} 1 \\ -2 \end{bmatrix} e^{-3t}. \tag{13.3}$$

The fact that we may write the solution in terms of vectors suggests that we write the system of differential equations in terms of vectors. In fact, we can rewrite (13.2) in matrix form[1] as

$$\begin{bmatrix} dx/dt \\ dy/dt \end{bmatrix} = \begin{bmatrix} -1 & 1 \\ -6 & -6 \end{bmatrix} \begin{bmatrix} x \\ y \end{bmatrix} \tag{13.4}$$

or equivalently as

$$\frac{d\mathbf{X}}{dt} = \mathbf{M}\mathbf{X},$$

where $\mathbf{M}$ is the coefficient matrix given by

$$\mathbf{M} = \begin{bmatrix} -1 & 1 \\ -6 & -6 \end{bmatrix}$$

and $\mathbf{X}$ is the vector

$$\mathbf{X} = \begin{bmatrix} x \\ y \end{bmatrix}.$$

Because both terms in the solution (13.3) have the form of a vector of constants times an exponential function, we now explore what happens if we start with that form of the solution. So we seek a solution of (13.2) in the form of

$$\mathbf{X}(t) = \begin{bmatrix} x(t) \\ y(t) \end{bmatrix} = \begin{bmatrix} A \\ B \end{bmatrix} e^{rt},$$

where A, B, and r are constants.

If we substitute this expression into both sides of (13.4), realizing that

$$\frac{d}{dt} \begin{bmatrix} A \\ B \end{bmatrix} e^{rt} = \begin{bmatrix} rA \\ rB \end{bmatrix} e^{rt}$$

(see Exercise 1 on page 641), we obtain

$$\begin{bmatrix} rA \\ rB \end{bmatrix} e^{rt} = \begin{bmatrix} -1 & 1 \\ -6 & -6 \end{bmatrix} \begin{bmatrix} A \\ B \end{bmatrix} e^{rt}.$$

Because the common factor e^{rt} is never zero, we may divide by it and rearrange the preceding equation to obtain

$$\begin{bmatrix} -1 & 1 \\ -6 & -6 \end{bmatrix} \begin{bmatrix} A \\ B \end{bmatrix} - \begin{bmatrix} rA \\ rB \end{bmatrix} = \begin{bmatrix} -1-r & 1 \\ -6 & -6-r \end{bmatrix} \begin{bmatrix} A \\ B \end{bmatrix} = \begin{bmatrix} 0 \\ 0 \end{bmatrix}. \tag{13.5}$$

---

[1]A summary of the properties of matrices is in the appendix.

This system of algebraic equations will have a nontrivial solution only if the determinant of the coefficients is zero, so

$$\begin{vmatrix} -1-r & 1 \\ -6 & -6-r \end{vmatrix} = 0.$$

This gives the quadratic equation

$$(-1-r)(-6-r)+6 = r^2 + 7r + 12 = 0$$

or

$$(r+3)(r+4) = 0. \tag{13.6}$$

This is the characteristic equation we discovered when we solved (13.2) in Example 12.8 on page 595.

The values $r = -3$ and $r = -4$ are the only values of r that give nontrivial solutions of the system of equations in (13.5). If we substitute $r = -4$ in (13.5) we obtain

$$\begin{bmatrix} 3 & 1 \\ -6 & -2 \end{bmatrix} \begin{bmatrix} A \\ B \end{bmatrix} = \begin{bmatrix} 0 \\ 0 \end{bmatrix},$$

so our only condition is that $B = -3A$. This gives the vector

$$\begin{bmatrix} A \\ -3A \end{bmatrix},$$

where A is an arbitrary constant, and the solution

$$\begin{bmatrix} A \\ -3A \end{bmatrix} e^{-4t} = A \begin{bmatrix} 1 \\ -3 \end{bmatrix} e^{-4t} = A \begin{bmatrix} e^{-4t} \\ -3e^{-4t} \end{bmatrix}.$$

If we use $r = -3$ in (13.5) we obtain

$$\begin{bmatrix} 2 & 1 \\ -6 & -3 \end{bmatrix} \begin{bmatrix} A \\ B \end{bmatrix} = \begin{bmatrix} 0 \\ 0 \end{bmatrix},$$

so our only condition is that $B = -2A$. This gives the vector

$$\begin{bmatrix} A \\ -2A \end{bmatrix},$$

where A is an arbitrary constant, and the solution

$$\begin{bmatrix} A \\ -2A \end{bmatrix} e^{-3t} = A \begin{bmatrix} 1 \\ -2 \end{bmatrix} e^{-3t} = A \begin{bmatrix} e^{-3t} \\ -2e^{-3t} \end{bmatrix}.$$

In these two solutions we have two different arbitrary constants, each denoted by the symbol A. If we designate these arbitrary constants C_1 and C_2, we can express our explicit solution of (13.4) as

$$\begin{bmatrix} x(t) \\ y(t) \end{bmatrix} = C_1 \begin{bmatrix} 1 \\ -3 \end{bmatrix} e^{-4t} + C_2 \begin{bmatrix} 1 \\ -2 \end{bmatrix} e^{-3t},$$

which is (13.3).

Using the fact that

$$C_1 \begin{bmatrix} A_1 \\ A_2 \end{bmatrix} + C_2 \begin{bmatrix} B_1 \\ B_2 \end{bmatrix} = \begin{bmatrix} A_1 & B_1 \\ A_2 & B_2 \end{bmatrix} \begin{bmatrix} C_1 \\ C_2 \end{bmatrix},$$

we may also write this solution in matrix form as

$$\begin{bmatrix} x(t) \\ y(t) \end{bmatrix} = \begin{bmatrix} e^{-4t} & e^{-3t} \\ -3e^{-4t} & -2e^{-3t} \end{bmatrix} \begin{bmatrix} C_1 \\ C_2 \end{bmatrix}$$

or

$$\mathbf{X}(t) = \mathbf{UC},$$

where

$$\mathbf{X}(t) = \begin{bmatrix} x(t) \\ y(t) \end{bmatrix}, \mathbf{U} = \begin{bmatrix} e^{-4t} & e^{-3t} \\ -3e^{-4t} & -2e^{-3t} \end{bmatrix}, \text{ and } \mathbf{C} = \begin{bmatrix} C_1 \\ C_2 \end{bmatrix}.$$

Comments about $\mathbf{X}(t) = \mathbf{UC}$

- The matrix $\mathbf{U}$ is called a FUNDAMENTAL MATRIX of our original system of differential equations (13.4). The columns of $\mathbf{U}$ contain our two vector solutions.

- The quadratic equation (13.6) that results from seeking solutions in the form of a constant vector times e^{rt} is called the CHARACTERISTIC EQUATION associated with the system of differential equations (13.1). It is identical with the characteristic equation we discovered in Sections 7.2 and 12.1.

- The two solutions of the characteristic equation, in this case $r = -3$ and $r = -4$, are called EIGENVALUES. The two vectors of constants associated with these eigenvalues, in this case

$$A \begin{bmatrix} 1 \\ -3 \end{bmatrix} \text{ and } A \begin{bmatrix} 1 \\ -2 \end{bmatrix},$$

  are called EIGENVECTORS. The eigenvalues and eigenvectors characterize the behavior of the solution of this system of equations and derive their name from the German word "eigen" which means characteristic.[2] Note the similarity with the use of the words eigenvalue and eigenfunctions in Section 8.4.

Figure 13.1 gives the direction field for the phase plane associated with (13.2), along with trajectories for different initial conditions. Notice that several of the trajectories seem to approach the origin along a straight line. The slope of this straight line appears to be about -2. This is the same direction as the eigenvector associated with the eigenvalue of -3, namely,

$$A \begin{bmatrix} 1 \\ -2 \end{bmatrix}.$$

We also plot the direction of the eigenvector associated with the second eigenvalue of -4, namely,

$$A \begin{bmatrix} 1 \\ -3 \end{bmatrix},$$

---

[2] In fact, eigenvalues and eigenvectors are sometimes called characteristic values and characteristic vectors.

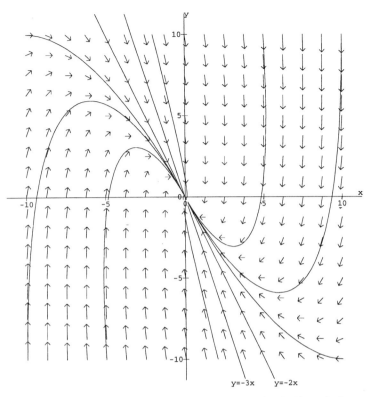

FIGURE 13.1   Direction field and several trajectories for $dy/dx = (-6x - 6y)/(-x + y)$

which gives a straight line through the origin with slope -3. We note that this line is parallel with the direction field in its vicinity. Could these straight lines also be trajectories in the phase plane?

To investigate this further, we examine the differential equation for the trajectories in the phase plane, given by

$$\frac{dy}{dx} = \frac{-6x - 6y}{-x + y}.$$

Straight line trajectories in the phase plane, $y = mx$, are possible if m satisfies

$$m = \frac{-6 - 6m}{-1 + m}$$

or

$$m^2 + 5m + 6 = 0.$$

This gives the directions of straight line trajectories as $m = -3$ and $m = -2$, precisely the same slopes as determined by the eigenvectors we have just found. Thus, we have discovered that the *eigenvectors give the directions of the straight line trajectories in the phase plane*. This is true for all systems of differential equations given by (13.1) that have real eigenvalues. See Exercise 7 on page 643.

EXAMPLE 13.1: Complex Eigenvalues

The details for finding eigenvalues and eigenvectors of systems of differential equations in which the eigenvalues are real and distinct are straightforward, and they follow the technique used in the preceding example. We now consider a system of differential equations in which the eigenvalues are complex numbers, namely,

$$\frac{d}{dt}\begin{bmatrix} x \\ y \end{bmatrix} = \begin{bmatrix} 2 & 1 \\ -1 & 2 \end{bmatrix}\begin{bmatrix} x \\ y \end{bmatrix}. \tag{13.7}$$

In a manner similar to the previous example we seek a solution of the form

$$\mathbf{X}(t) = \begin{bmatrix} x(t) \\ y(t) \end{bmatrix} = \begin{bmatrix} A \\ B \end{bmatrix} e^{rt}$$

and determine the characteristic equation from the determinant of the 2 by 2 matrix in the equation

$$\begin{bmatrix} 2-r & 1 \\ -1 & 2-r \end{bmatrix}\begin{bmatrix} A \\ B \end{bmatrix} = \begin{bmatrix} 0 \\ 0 \end{bmatrix} \tag{13.8}$$

as

$$(2-r)(2-r) + 1 = r^2 - 4r + 5 = 0.$$

This gives our eigenvalues as the complex numbers $r = 2 \pm i$.

If we substitute $r = 2 + i$ into (13.8) we obtain

$$\begin{bmatrix} -i & 1 \\ -1 & -i \end{bmatrix}\begin{bmatrix} A \\ B \end{bmatrix} = \begin{bmatrix} 0 \\ 0 \end{bmatrix},$$

so A and B must be related by $A = -iB$. This gives an eigenvector as

$$B\begin{bmatrix} -i \\ 1 \end{bmatrix}$$

and a solution of (13.7) as

$$B\begin{bmatrix} -i \\ 1 \end{bmatrix} e^{(2+i)t} = B\begin{bmatrix} -i \\ 1 \end{bmatrix} e^{2t}e^{it} = B\begin{bmatrix} -i \\ 1 \end{bmatrix} e^{2t} (\cos t + i \sin t).$$

If we decompose this vector into its real and imaginary parts we obtain

$$B\begin{bmatrix} -i \\ 1 \end{bmatrix} e^{(2+i)t} = B\begin{bmatrix} e^{2t} \sin t \\ e^{2t} \cos t \end{bmatrix} + iB\begin{bmatrix} -e^{2t} \cos t \\ e^{2t} \sin t \end{bmatrix} \tag{13.9}$$

as our solution. However, because our original differential equation (13.7) has only real coefficients and our solution contains both a real part and an imaginary part, each of these parts, namely,

$$\begin{bmatrix} e^{2t} \sin t \\ e^{2t} \cos t \end{bmatrix} \text{ and } \begin{bmatrix} -e^{2t} \cos t \\ e^{2t} \sin t \end{bmatrix},$$

must separately satisfy (13.7). (See Exercise 2 on page 641.) This means we may form our fundamental matrix by having its two columns be the vectors given in (13.9) — that is, the fundamental matrix is

$$\mathbf{U} = \begin{bmatrix} e^{2t}\sin t & -e^{2t}\cos t \\ e^{2t}\cos t & e^{2t}\sin t \end{bmatrix}.$$

Thus, our solution is

$$\mathbf{X}(t) = \begin{bmatrix} x(t) \\ y(t) \end{bmatrix} = \begin{bmatrix} e^{2t}\sin t & -e^{2t}\cos t \\ e^{2t}\cos t & e^{2t}\sin t \end{bmatrix} \begin{bmatrix} C_1 \\ C_2 \end{bmatrix}.$$

In this example we do not seek straight line trajectories in the phase plane because they do not exist for systems of differential equations with complex eigenvalues. (See Exercise 7 on page 643.) The differential equation for trajectories in the phase plane associated with (13.7) is

$$\frac{dy}{dx} = \frac{-x + 2y}{2x + y}.$$

The direction field for this phase plane is shown in Figure 13.2, where we see that there are no straight line trajectories. The direction of the arrows also indicates that the origin is an unstable focus because all solutions spiral away from the origin. □

Our first two examples in this chapter had distinct eigenvalues, the first real and the second complex. Now we examine the remaining possibility in which the characteristic equation has a repeated root. Thus, the eigenvalue is real, but repeated.

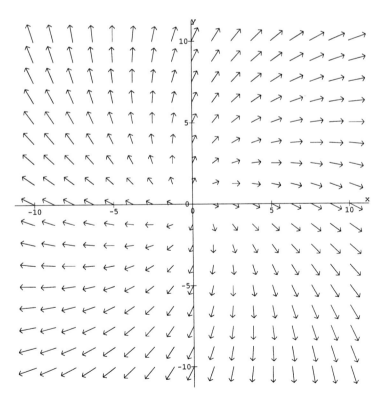

FIGURE 13.2   Direction field for $dy/dx = (-x + 2y)/(2x + y)$

EXAMPLE 13.2: Real, Repeated Eigenvalue

Consider the system of differential equations

$$\frac{d}{dt}\begin{bmatrix} x \\ y \end{bmatrix} = \begin{bmatrix} 1 & 1 \\ -1 & 3 \end{bmatrix}\begin{bmatrix} x \\ y \end{bmatrix}, \tag{13.10}$$

where we seek a solution of the form

$$\mathbf{X}(t) = \begin{bmatrix} x(t) \\ y(t) \end{bmatrix} = \begin{bmatrix} A \\ B \end{bmatrix}e^{rt}.$$

Substituting this expression into (13.10) gives

$$\begin{bmatrix} rA \\ rB \end{bmatrix}e^{rt} = \begin{bmatrix} 1 & 1 \\ -1 & 3 \end{bmatrix}\begin{bmatrix} A \\ B \end{bmatrix}e^{rt}$$

and the associated set of algebraic equations

$$\begin{bmatrix} 1-r & 1 \\ -1 & 3-r \end{bmatrix}\begin{bmatrix} A \\ B \end{bmatrix} = \begin{bmatrix} 0 \\ 0 \end{bmatrix}. \tag{13.11}$$

From the determinant of the preceding 2 by 2 matrix we obtain the characteristic equation as

$$(1-r)(3-r) + 1 = r^2 - 4r + 4 = (r-2)^2 = 0,$$

giving the repeated eigenvalue of $r = 2$.

To find the eigenvector associated with the eigenvalue $r = 2$, we substitute $r = 2$ into (13.11) and obtain

$$\begin{bmatrix} -1 & 1 \\ -1 & 1 \end{bmatrix}\begin{bmatrix} A \\ B \end{bmatrix} = \begin{bmatrix} 0 \\ 0 \end{bmatrix}.$$

This requires that $A = B$, so we have the eigenvector

$$\begin{bmatrix} B \\ B \end{bmatrix}.$$

This eigenvector tells us that the straight line $y = x$ is a trajectory in the phase plane. (Why?) This trajectory is apparent in Figure 13.3, which gives the direction field for the associated phase plane. (Is there another straight line trajectory? Should there be?)

This eigenvector also tells us that we have found one solution of (13.10) as

$$\begin{bmatrix} B \\ B \end{bmatrix}e^{2t},$$

but we need a second solution.

In Chapter 11 we considered repeated roots of the characteristic equation associated with $ay'' + by' + cy = 0$. We discovered that for that case the general solution took the form $C_1 e^{rt} + C_2 t e^{rt}$. However, if we look at a previous example of a system of equations that had a repeated root of the characteristic equation (Example 12.9, page 600), we see that the solution took the form

$$\mathbf{X}(t) = \begin{bmatrix} x(t) \\ y(t) \end{bmatrix} = \begin{bmatrix} C_1 \\ (r+1)C_1 + C_2 \end{bmatrix}e^{rt} + C_2\begin{bmatrix} 1 \\ r+1 \end{bmatrix}te^{rt}.$$

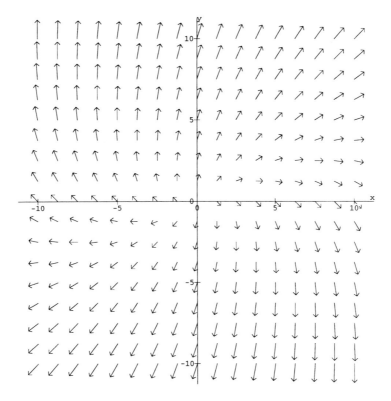

FIGURE 13.3 Direction field for $dy/dx = (-x + 3y)/(2x + y)$

Notice that C_2 not only multiplies the second term containing te^{rt} but also is a part of the first term. Thus, we try a combination of two terms as

$$\mathbf{X}(t) = \begin{bmatrix} x(t) \\ y(t) \end{bmatrix} = \begin{bmatrix} A_1 \\ B_1 \end{bmatrix} e^{2t} + \begin{bmatrix} A_2 \\ B_2 \end{bmatrix} te^{2t}. \tag{13.12}$$

Substituting this expression into (13.10) gives

$$\begin{bmatrix} 2A_1 \\ 2B_1 \end{bmatrix} e^{2t} + \begin{bmatrix} (2t+1)A_2 \\ (2t+1)B_2 \end{bmatrix} e^{2t} = \begin{bmatrix} A_1 + B_1 + t(A_2 + B_2) \\ -A_1 + 3B_1 + t(-A_2 + 3B_2) \end{bmatrix} e^{2t}.$$

Because we want this equation to be valid for all values of t we equate coefficients of e^{2t} and te^{2t} and obtain the system of algebraic equations

$$\begin{aligned} 2A_1 + A_2 &= A_1 + B_1, \\ 2B_1 + B_2 &= -A_1 + 3B_1, \\ 2A_2 &= A_2 + B_2, \\ 2B_2 &= -A_2 + 3B_2. \end{aligned}$$

This system of equations has the solution

$$\begin{aligned} A_1 &= B_1 - B_2, \\ A_2 &= B_2, \end{aligned}$$

where B_1 and B_2 may be chosen arbitrarily, so we denote them by C_1 and C_2. (See Exercise 3 on page 641.) This means that the solution (13.12) may be written in the form

$$\mathbf{X}(t) = \begin{bmatrix} x(t) \\ y(t) \end{bmatrix} = \begin{bmatrix} C_1 - C_2 \\ C_1 \end{bmatrix} e^{2t} + \begin{bmatrix} C_2 \\ C_2 \end{bmatrix} te^{2t},$$

or

$$\mathbf{X}(t) = \begin{bmatrix} x(t) \\ y(t) \end{bmatrix} = \begin{bmatrix} e^{2t} & (t-1)e^{2t} \\ e^{2t} & te^{2t} \end{bmatrix} \begin{bmatrix} C_1 \\ C_2 \end{bmatrix}.$$

Here a fundamental matrix is

$$\mathbf{U} = \begin{bmatrix} e^{2t} & (t-1)e^{2t} \\ e^{2t} & te^{2t} \end{bmatrix}.$$

Notice that the form of our trial solution (13.12) for this differential equation did not make use of the fact that we already knew one solution. Finding the eigenvector in this case is useful because it gives a straight line trajectory in the phase plane, but it is not required to find the explicit solution. □

We now summarize the results of this section.

How to Solve $d\mathbf{X}/dt = \mathbf{MX}$

Purpose: To find the general solution of the system of differential equations

$$\frac{d\mathbf{X}}{dt} = \mathbf{MX}$$

where

$$\mathbf{M} = \begin{bmatrix} a & b \\ c & d \end{bmatrix} \text{ and } \mathbf{X} = \begin{bmatrix} x \\ y \end{bmatrix}.$$

Technique:

1. Substitute

$$\mathbf{X}(t) = \begin{bmatrix} A \\ B \end{bmatrix} e^{rt}$$

into the differential equation and rearrange the result to obtain

$$\begin{bmatrix} a-r & b \\ c & d-r \end{bmatrix} \begin{bmatrix} A \\ B \end{bmatrix} = \begin{bmatrix} 0 \\ 0 \end{bmatrix}. \tag{13.13}$$

2. Solve the characteristic equation

$$\begin{vmatrix} a-r & b \\ c & d-r \end{vmatrix} = 0$$

for r, obtaining the roots (eigenvalues) r_1 and r_2.

3. Find a fundamental matrix for this system of equations. The process in this step depends on the nature of the eigenvalues: whether they are distinct — either real or complex — or repeated.

   (a) For real, distinct eigenvalues r_1 and r_2, substitute one of the eigenvalues — say, r_1 — into (13.13) and solve the resulting algebraic equations for the associated eigenvector. Repeat this process for the other eigenvalue r_2. The columns in a fundamental matrix are formed by these eigenvectors times e^{rt}, for the appropriate value of r.

   (b) For complex eigenvalues r_1 and r_2 substitute one of them — say, $r_1 = \alpha + i\beta$ — into (13.13) and solve the resulting algebraic equations. This gives a complex eigenvector, which when multiplied by $e^{(\alpha+i\beta)t}$ will result in a complex valued solution. Form a fundamental matrix by having its columns as the real and imaginary parts of this solution.

   (c) For a repeated eigenvalue r substitute

$$\mathbf{X}(t) = \begin{bmatrix} x(t) \\ y(t) \end{bmatrix} = \begin{bmatrix} A_1 \\ B_1 \end{bmatrix} e^{rt} + \begin{bmatrix} A_2 \\ B_2 \end{bmatrix} t e^{rt} \qquad \text{(13.14)}$$

   into (13.13) and solve the resulting algebraic equations for A_1, B_1, A_2, and B_2. The result of this operation will contain two arbitrary constants. The columns of a fundamental matrix in this case are composed of the vector functions that multiply the two arbitrary constants.

4. For each of these cases the general solution of the system of differential equations is

$$\mathbf{X}(t) = \mathbf{UC},$$

   where $\mathbf{U}$ is a fundamental matrix and $\mathbf{C}$ is a vector of arbitrary constants

$$\mathbf{C} = \begin{bmatrix} C_1 \\ C_2 \end{bmatrix}.$$

5. For initial value problems — say, $x(0) = x_0$, $y(0) = y_0$ — evaluate the arbitrary constants by solving the system

$$\mathbf{U}(0)\mathbf{C} = \begin{bmatrix} x_0 \\ y_0 \end{bmatrix}$$

   for C_1 and C_2.

Comments about solving $\mathbf{X}' = \mathbf{MX}$

- The arbitrary constants that occur in the eigenvectors may be set equal to any convenient nonzero value. This is because in performing the matrix multiplication $\mathbf{UC}$, the columns of $\mathbf{U}$ — containing the eigenvectors — are multiplied by the arbitrary constants contained in $\mathbf{C}$.

- The order in which the vectors are selected to construct the matrix $\mathbf{U}$ is not important.

- The eigenvectors for distinct roots give the direction of the straight line trajectories in the phase plane.

- There are no straight line trajectories in the phase plane for systems of equations with complex eigenvalues.

- The vector multiplying te^{rt} in (13.14) will be the eigenvector associated with the repeated eigenvalue r. It will give the slope of the only straight line trajectory in the phase plane determined by

$$\frac{dy}{dx} = \frac{cx + dy}{ax + by}.$$

- The reason we use the expression *a* fundamental matrix rather than *the* fundamental matrix is because fundamental matrices are not unique. They may contain arbitrary constants that can be set equal to any convenient nonzero value.

- Eigenvalues and eigenvectors arise in matrix theory in a manner that is completely independent of differential equations. If we are given a matrix **M**, the system of algebraic equations in (13.13) may be obtained without reference to differential equations. In matrix notation it is simply finding values of r and **X** for which

$$\mathbf{MX} = r\mathbf{X}$$

or

$$(\mathbf{M} - r\mathbf{I})\,\mathbf{X} = 0,$$

where **I** is the identity matrix

$$\mathbf{I} = \begin{bmatrix} 1 & 0 \\ 0 & 1 \end{bmatrix}.$$

In this setting the characteristic equation is called the characteristic polynomial.

We conclude this section with an example to illustrate the use of this technique. In Chapters 4 and 6 we determined the concentration of salt in a single container in which the rates of input and output, together with the initial concentration, were known. Now we consider a similar situation but for two containers, connected as shown in Figure 13.4.

EXAMPLE 13.3

Pure water is entering container A at a rate of 3 gallons per minute while the well-stirred mixture is leaving container B at a rate of 3 gallons per minute. There are 100 gallons in each container, and the well-stirred mixture flows from container A to container B at a rate of 4 gallons per minute and leaks back from container B to container A at a rate of 1 gallon per minute. We want to predict the amount of salt in each container at any time.

We let $x(t)$ be the amount of salt in container A at time t, and $y(t)$ be the amount of salt in container B at time t. Both x and y have units of pounds, and t has units of minutes. We may derive the differential equations governing this system by using a continuity equation. For container A we have

$$\begin{aligned} dx/dt &= \text{input} - \text{output} \\ &= (y/100)(1) - (x/100)(4), \end{aligned}$$

or

$$\frac{dx}{dt} = \frac{y}{100} - \frac{x}{25}.$$

Balancing input and output for container B gives

$$\frac{dy}{dt} = \frac{x}{100}4 - \frac{y}{100}1 - \frac{y}{100}3 = \frac{1}{25}x - \frac{1}{25}y.$$

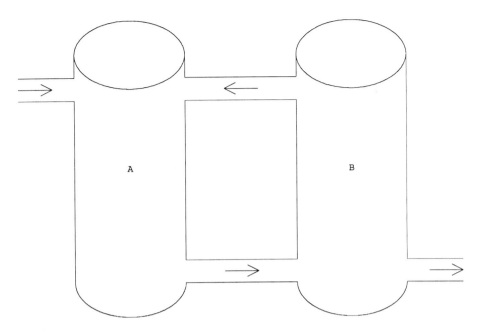

FIGURE 13.4 Two connected containers

To find an explicit solution for this system of equations, we recast them in matrix form as

$$\frac{d\mathbf{X}}{dt} = \mathbf{MX}, \tag{13.15}$$

where

$$\mathbf{X} = \begin{bmatrix} x \\ y \end{bmatrix} \text{ and } \mathbf{M} = \begin{bmatrix} -1/25 & 1/100 \\ 1/25 & -1/25 \end{bmatrix}.$$

If we substitute

$$\mathbf{X}(t) = \begin{bmatrix} A \\ B \end{bmatrix} e^{rt}$$

into (13.15) we obtain

$$\begin{bmatrix} -1/25 - r & 1/100 \\ 1/25 & -1/25 - r \end{bmatrix} \begin{bmatrix} A \\ B \end{bmatrix} = \begin{bmatrix} 0 \\ 0 \end{bmatrix}. \tag{13.16}$$

This gives us the characteristic equation

$$\begin{vmatrix} -1/25 - r & 1/100 \\ 1/25 & -1/25 - r \end{vmatrix} = \left(-\frac{1}{25} - r \right) \left(-\frac{1}{25} - r \right) - \frac{1}{2500} = 0$$

and the eigenvalues as

$$r = -\frac{1}{25} \pm \frac{1}{50},$$

or $-1/50$ and $-3/50$. These eigenvalues are real and distinct, so we seek eigenvectors for each of these eigenvalues in turn.

If we substitute $r = -1/50$ into (13.16) we obtain

$$\begin{bmatrix} -1/50 & 1/100 \\ 1/25 & -1/50 \end{bmatrix} \begin{bmatrix} A \\ B \end{bmatrix} = \begin{bmatrix} 0 \\ 0 \end{bmatrix},$$

so $B = 2A$. Thus, an eigenvector associated with $r = -1/50$ is

$$\begin{bmatrix} A_1 \\ 2A_1 \end{bmatrix},$$

where $A_1 = A$ is an arbitrary constant. A similar calculation for $r = -3/50$ gives the eigenvector

$$\begin{bmatrix} A_2 \\ -2A_2 \end{bmatrix},$$

where $A_2 = A$ is another arbitrary constant.

Thus, a fundamental matrix is

$$\mathbf{U} = \begin{bmatrix} A_1 e^{-t/50} & A_2 e^{-3t/50} \\ 2A_1 e^{-t/50} & -2A_2 e^{-3t/50} \end{bmatrix}$$

or, by choosing $A_1 = A_2 = 1$,

$$\mathbf{U} = \begin{bmatrix} e^{-t/50} & e^{-3t/50} \\ 2e^{-t/50} & -2e^{-3t/50} \end{bmatrix}.$$

The general solution of (13.15) is then

$$\mathbf{X}(t) = \begin{bmatrix} x(t) \\ y(t) \end{bmatrix} = \mathbf{UC} = \begin{bmatrix} e^{-t/50} & e^{-3t/50} \\ 2e^{-t/50} & -2e^{-3t/50} \end{bmatrix} \begin{bmatrix} C_1 \\ C_2 \end{bmatrix}. \tag{13.17}$$

Now consider the situation in which initially there are 16 pounds of salt in container A and 4 pounds of salt in container B. How long will it be until the two containers hold an equal amount of salt?

To answer this question, we evaluate the arbitrary constants in (13.17) by solving

$$\begin{bmatrix} x(0) \\ y(0) \end{bmatrix} = \begin{bmatrix} 16 \\ 4 \end{bmatrix} = \begin{bmatrix} 1 & 1 \\ 2 & -2 \end{bmatrix} \begin{bmatrix} C_1 \\ C_2 \end{bmatrix}.$$

This gives us the two algebraic equations

$$\begin{aligned} 16 &= C_1 + C_2, \\ 4 &= 2C_1 - 2C_2. \end{aligned}$$

If we divide the bottom equation by 2 and add it to the top equation, we obtain $2C_1 = 18$, so $C_1 = 9$. Substituting this value into either of the original equations gives $C_2 = 7$. Thus, the solution to our initial value problem is

$$\mathbf{X}(t) = \begin{bmatrix} x(t) \\ y(t) \end{bmatrix} = \begin{bmatrix} e^{-t/50} & e^{-3t/50} \\ 2e^{-t/50} & -2e^{-3t/50} \end{bmatrix} \begin{bmatrix} 9 \\ 7 \end{bmatrix} = \begin{bmatrix} 9e^{-t/50} + 7e^{-3t/50} \\ 18e^{-t/50} - 14e^{-3t/50} \end{bmatrix}.$$

To determine the time t when the amount of salt in the two tanks is equal, we set $x(t) = y(t)$ and solve for t. This gives

$$9e^{-t/50} + 7e^{-3t/50} = 18e^{-t/50} - 14e^{-3t/50},$$

which, when rearranged, becomes

$$21e^{-3t/50} = 9e^{-t/50}.$$

This yields

$$e^{2t/50} = 7/3,$$

so the required time is

$$t = 25\ln(7/3) \approx 21.18 \text{ minutes.}$$

At that time there will be $12\sqrt{3/7} \approx 7.86$ pounds of salt in each container. (How do we know this?)

Figure 13.5 shows the graphs of $x(t) = 9e^{-t/50} + 7e^{-3t/50}$ and $y(t) = 18e^{-t/50} - 14e^{-3t/50}$. The time when these two curves cross and the value at which they cross are consistent with the preceding analysis. ☐

EXERCISES

1. The derivative of a vector function

$$\mathbf{V} = \begin{bmatrix} x(t) \\ y(t) \end{bmatrix}$$

is defined as

$$\mathbf{V}' = \begin{bmatrix} x'(t) \\ y'(t) \end{bmatrix}.$$

(a) If

$$\mathbf{V} = \begin{bmatrix} A \\ B \end{bmatrix} f(t),$$

where A and B are constants, show that

$$\mathbf{V}' = \begin{bmatrix} A \\ B \end{bmatrix} f'(t).$$

(b) Show that the usual rules for differentiation apply to vector functions, namely,

$$\left(\mathbf{V}_1 + \mathbf{V}_2\right)' = \mathbf{V}_1' + \mathbf{V}_2',$$

$$(z(t)\mathbf{V})' = z'(t)\mathbf{V} + z(t)\mathbf{V}'.$$

2. Show that if $\mathbf{X} = \mathbf{V} + i\mathbf{W}$ is a solution of $\mathbf{X}' = \mathbf{MX}$, where $\mathbf{M}$ is a real 2 by 2 matrix and $\mathbf{V}$ and $\mathbf{W}$ are real vectors, then $\mathbf{V}$ and $\mathbf{W}$ are also solutions.

3. Consider the situation where the eigenvalue of

$$\frac{d\mathbf{X}}{dt} = \mathbf{MX}, \qquad \mathbf{M} = \begin{bmatrix} a & b \\ c & d \end{bmatrix}, \qquad \mathbf{X} = \begin{bmatrix} x \\ y \end{bmatrix} \tag{13.18}$$

is repeated, so $r = (a + d)/2$.

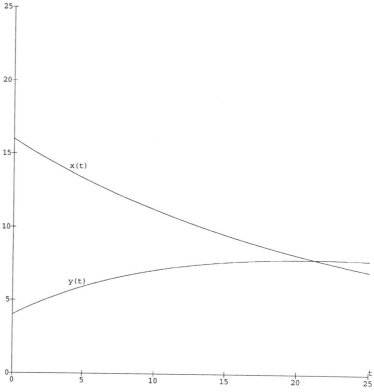

FIGURE 13.5 The functions $x(t) = 9e^{-t/50} + 7e^{-3t/50}$ and $y(t) = 18e^{-t/50} - 14e^{-3t/50}$

(a) Find the algebraic equation satisfied by the components of the associated eigenvector

$$\mathbf{Z}_2 = \begin{bmatrix} A_2 \\ B_2 \end{bmatrix}.$$

(b) Show that trying a solution of (13.18) in the form

$$\begin{bmatrix} A_1 \\ B_1 \end{bmatrix} e^{rt} + \mathbf{Z}_2 t e^{rt}$$

results in the same equations you found for A_2 and B_2 in part (a) and that A_1 and B_1 satisfy

$$\begin{bmatrix} a-r & b \\ c & d-r \end{bmatrix} \begin{bmatrix} A_1 \\ B_1 \end{bmatrix} = \begin{bmatrix} A_2 \\ B_2 \end{bmatrix}.$$

4. Show that the system of equations

$$\frac{d\mathbf{X}}{dt} = \begin{bmatrix} a & 0 \\ 0 & a \end{bmatrix} \mathbf{X}$$

has repeated eigenvalues. Show that the eigenvectors for this case are

$$\begin{bmatrix} C_1 \\ 0 \end{bmatrix} \text{ and } \begin{bmatrix} 0 \\ C_2 \end{bmatrix}.$$

Thus, for this situation a fundamental matrix may have the form

$$\mathbf{U} = \begin{bmatrix} e^{at} & 0 \\ 0 & e^{at} \end{bmatrix}.$$

5. Solve the following systems of differential equations by finding a fundamental matrix. For real eigenvalues find straight line trajectories in the phase plane and verify their agreement with the eigenvectors you found. For complex eigenvalues show that there are no straight line trajectories in the phase plane. Verify your answers by considering the direction field in the phase plane.

(a)
$$\frac{d}{dt}\begin{bmatrix} x \\ y \end{bmatrix} = \begin{bmatrix} 2 & 5 \\ 1 & 6 \end{bmatrix}\begin{bmatrix} x \\ y \end{bmatrix}$$

(b)
$$\frac{d}{dt}\begin{bmatrix} x \\ y \end{bmatrix} = \begin{bmatrix} 2 & -1 \\ -6 & 1 \end{bmatrix}\begin{bmatrix} x \\ y \end{bmatrix}$$

(c)
$$\frac{d}{dt}\begin{bmatrix} x \\ y \end{bmatrix} = \begin{bmatrix} 5 & 6 \\ 1 & 4 \end{bmatrix}\begin{bmatrix} x \\ y \end{bmatrix}$$

(d)
$$\frac{d}{dt}\begin{bmatrix} x \\ y \end{bmatrix} = \begin{bmatrix} 5 & -3 \\ -1 & -1 \end{bmatrix}\begin{bmatrix} x \\ y \end{bmatrix}$$

(e)
$$\frac{d}{dt}\begin{bmatrix} x \\ y \end{bmatrix} = \begin{bmatrix} 2 & 9 \\ -1 & -4 \end{bmatrix}\begin{bmatrix} x \\ y \end{bmatrix}$$

(f)
$$\frac{d}{dt}\begin{bmatrix} x \\ y \end{bmatrix} = \begin{bmatrix} 2 & 1 \\ 1 & -4 \end{bmatrix}\begin{bmatrix} x \\ y \end{bmatrix}$$

(g)
$$\frac{d}{dt}\begin{bmatrix} x \\ y \end{bmatrix} = \begin{bmatrix} 4 & -2 \\ 5 & -2 \end{bmatrix}\begin{bmatrix} x \\ y \end{bmatrix}$$

(h)
$$\frac{d}{dt}\begin{bmatrix} x \\ y \end{bmatrix} = \begin{bmatrix} 2 & 1 \\ -4 & 2 \end{bmatrix}\begin{bmatrix} x \\ y \end{bmatrix}$$

(i)
$$\frac{d}{dt}\begin{bmatrix} x \\ y \end{bmatrix} = \begin{bmatrix} 3 & 5 \\ -5 & 3 \end{bmatrix}\begin{bmatrix} x \\ y \end{bmatrix}$$

(j)
$$\frac{d}{dt}\begin{bmatrix} x \\ y \end{bmatrix} = \begin{bmatrix} 1 & 1 \\ 1 & 3 \end{bmatrix}\begin{bmatrix} x \\ y \end{bmatrix}$$

6. In Example 13.1 we found that the solution of our system of equations was given by $X(t) = UC$, where

$$U = \begin{bmatrix} e^{2t}\sin t & -e^{2t}\cos t \\ e^{2t}\cos t & e^{2t}\sin t \end{bmatrix}. \tag{13.19}$$

(a) Evaluate the constant vector C so this solution satisfies the initial conditions $x(0) = x_0$, $y(0) = y_0$.

(b) Now choose a different fundamental matrix — perhaps just interchange the columns of U in (13.19) — and evaluate the constant vector C in this case. Show that your resulting solution is the same as the one you obtained in part (a).

7. Consider the system of equations

$$\frac{dX}{dt} = MX, \qquad M = \begin{bmatrix} a & b \\ c & d \end{bmatrix}, \qquad X = \begin{bmatrix} x \\ y \end{bmatrix}. \tag{13.20}$$

(a) Find the equation for the eigenvalues for solutions of the form

$$X(t) = \begin{bmatrix} A \\ B \end{bmatrix} e^{rt}.$$

(b) For the case of a real, repeated eigenvalue, $(a + d)^2 = 4bc$, find the associated eigenvector.

(c) Find the equation for straight line trajectories in the phase plane associated with (13.20), and show that they are consistent with the direction given by the eigenvector in part (b).

(d) Repeat parts (b) and (c) for real, distinct eigenvalues.

(e) Show that there are no straight line trajectories in the phase plane associated with (13.20) when the eigenvalues are complex numbers.

8. Consider the two containers of Figure 13.4, each of which holds 100 gallons of liquid.

   (a) Initially, container A has 10 pounds of salt and container B has none. How long after all the valves are opened will the number of pounds of salt in the two containers be equal?

   (b) Determine the amount of salt in each container after 25 minutes.

   (c) Determine the amount of salt in each container as time approaches infinity. Could you determine this result by knowing the eigenvalues? Explain fully.

9. Three mystery direction fields of phase planes associated with the system of differential equations $x' = ax + by$, $y' = cx + dy$, are given in Figures 13.6, 13.7, and 13.8. By looking for straight line trajectories, decide which of these direction fields has the following properties. Explain the reasons behind each choice, and, wherever possible, estimate the components of the eigenvalues.

   (a) Real, distinct eigenvalues

   (b) A real, repeated eigenvalue

   (c) Complex eigenvalues

13.2   Nonhomogeneous Systems of First Order Linear Differential Equations

We now develop methods for solving nonhomogeneous systems of differential equations. They are based on the methods developed for solving second order linear differential equations in Chapter 11. This section will parallel Section 11.2, so we will consider forcing functions that are constants, exponentials, and sine and cosine functions.

We start with a slight modification of our last example in Section 13.1 — Example 13.3 — in which, instead of adding pure water to one of the containers, we add a salt solution.

EXAMPLE 13.4

A solution with concentration of 4 pounds of salt per gallon is entering container A at a rate of 3 gallons per minute, and the well-stirred mixture is leaving container B at a rate of 3 gallons per minute. There are 100 gallons in each container, and the well-stirred mixture flows from container A to container B at a rate of 4 gallons per minute and leaks back from container B to container A at a rate of 1 gallon per minute. We want to predict the amount of salt in each container at any time.

We let $x(t)$ be the amount of salt in container A at time t, and $y(t)$ be the amount of salt in container B at time t. Both x and y have units of pounds, and t has units of minutes. The appropriate differential equations governing this system are obtained from continuity considerations. For container A we have

$$dx/dt = \text{input} - \text{output}$$
$$= (4)(3) + (y/100)(1) - (x/100)(4),$$

or

$$\frac{dx}{dt} = 12 + \frac{y}{100} - \frac{x}{25},$$

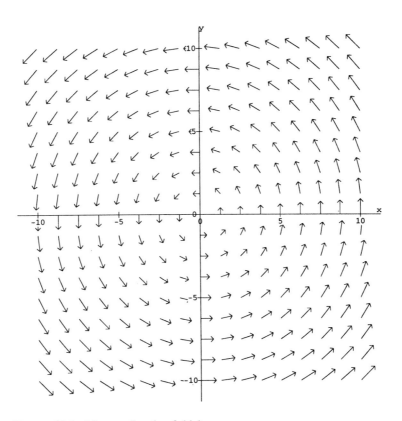

FIGURE 13.6 Mystery direction field A

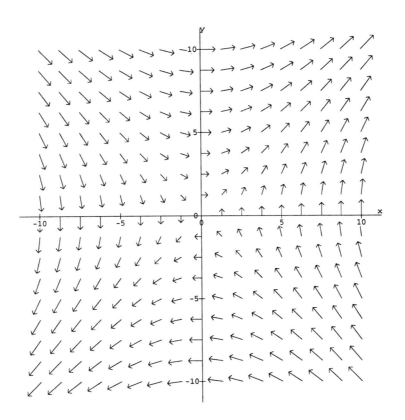

FIGURE 13.7 Mystery direction field B

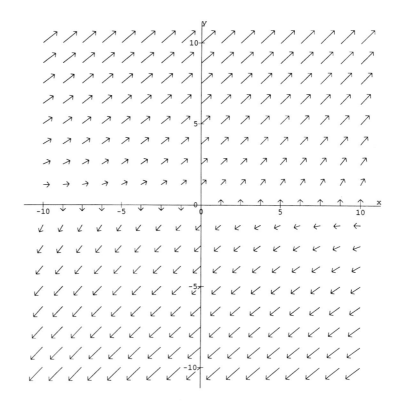

FIGURE 13.8   Mystery direction field C

whereas for container B we have

$$\frac{dy}{dt} = \frac{x}{100}4 - \frac{y}{100}1 - \frac{y}{100}3 = \frac{1}{25}x - \frac{1}{25}y.$$

To find an explicit solution of this system of equations, we recast them in matrix form as

$$\frac{d\mathbf{X}}{dt} = \mathbf{M}\mathbf{X} + \mathbf{V}, \tag{13.21}$$

where

$$\mathbf{X} = \begin{bmatrix} x \\ y \end{bmatrix}, \qquad \mathbf{M} = \begin{bmatrix} -1/25 & 1/100 \\ 1/25 & -1/25 \end{bmatrix}, \qquad \mathbf{V} = \begin{bmatrix} 12 \\ 0 \end{bmatrix}.$$

In Chapter 11 we found general solutions of nonhomogeneous linear differential equations as the sum of the general solution of the associated homogeneous differential equation and a particular solution. This is the situation here as well (see Exercise 2, page 655), so we first need to solve the homogeneous system

$$\frac{d\mathbf{X_h}}{dt} = \mathbf{M}\mathbf{X_h}. \tag{13.22}$$

In Example 13.3 on page 638 we developed the general solution of (13.22) as

$$\mathbf{X_h}(t) = \mathbf{U}\mathbf{C},$$

where a fundamental matrix $\mathbf{U}$ is given by

$$\mathbf{U} = \begin{bmatrix} e^{-t/50} & e^{-3t/50} \\ 2e^{-t/50} & -2e^{-3t/50} \end{bmatrix}$$

and $\mathbf{C}$ is a vector of arbitrary constants.

Now we need to find a particular solution of the original equation (13.21). Because $\mathbf{V}$ is a constant vector, we try a particular solution in the form of a constant vector. Thus, we try

$$\mathbf{X_p}(t) = \begin{bmatrix} K_1 \\ K_2 \end{bmatrix}.$$

Substituting this trial solution into (13.21) gives the set of algebraic equations

$$\begin{bmatrix} -1/25 & 1/100 \\ 1/25 & -1/25 \end{bmatrix} \begin{bmatrix} K_1 \\ K_2 \end{bmatrix} = \begin{bmatrix} -12 \\ 0 \end{bmatrix}$$

with solution

$$K_1 = K_2 = 400,$$

so

$$\mathbf{X_p}(t) = \begin{bmatrix} 400 \\ 400 \end{bmatrix}.$$

This gives the general solution of our original system of differential equations as

$$\mathbf{X}(t) = \begin{bmatrix} x(t) \\ y(t) \end{bmatrix} = \begin{bmatrix} e^{-t/50} & e^{-3t/50} \\ 2e^{-t/50} & -2e^{-3t/50} \end{bmatrix} \begin{bmatrix} C_1 \\ C_2 \end{bmatrix} + \begin{bmatrix} 400 \\ 400 \end{bmatrix}, \qquad (13.23)$$

or in matrix form as

$$\mathbf{X}(t) = \mathbf{UC} + \mathbf{X_p}(t),$$

where $\mathbf{U}$ and $\mathbf{X_p}(t)$ are defined above and

$$\mathbf{C} = \begin{bmatrix} C_1 \\ C_2 \end{bmatrix}.$$

Had this process started with no salt in either container, the initial conditions at $t = 0$ would be $x(0) = y(0) = 0$. If we substitute these values into (13.23) we obtain

$$\begin{bmatrix} 0 \\ 0 \end{bmatrix} = \begin{bmatrix} 1 & 1 \\ 2 & -2 \end{bmatrix} \begin{bmatrix} C_1 \\ C_2 \end{bmatrix} + \begin{bmatrix} 400 \\ 400 \end{bmatrix}$$

with solution $C_1 = -300$ and $C_2 = -100$. In this case the amount of salt in containers A and B at time t would be

$$\mathbf{X}(t) = \begin{bmatrix} x(t) \\ y(t) \end{bmatrix} = \begin{bmatrix} e^{-t/50} & e^{-3t/50} \\ 2e^{-t/50} & -2e^{-3t/50} \end{bmatrix} \begin{bmatrix} -300 \\ -100 \end{bmatrix} + \begin{bmatrix} 400 \\ 400 \end{bmatrix},$$

or

$$\mathbf{X}(t) = \begin{bmatrix} x(t) \\ y(t) \end{bmatrix} = \begin{bmatrix} 400 - 300e^{-t/50} - 100e^{-3t/50} \\ 400 - 600e^{-t/50} + 200e^{-3t/50} \end{bmatrix}.$$

Figure 13.9 shows the graphs of $x(t) = 400 - 300e^{-t/50} - 100e^{-3t/50}$ and $y(t) = 400 - 600e^{-t/50} + 200e^{-3t/50}$.

On the other hand, had this process started with no salt in container A but 300 pounds in container B, the initial conditions would be $x(0) = 0$ and $y(0) = 300$. If we substitute these values into (13.23) we obtain

$$\begin{bmatrix} 0 \\ 300 \end{bmatrix} = \begin{bmatrix} 1 & 1 \\ 2 & -2 \end{bmatrix} \begin{bmatrix} C_1 \\ C_2 \end{bmatrix} + \begin{bmatrix} 400 \\ 400 \end{bmatrix},$$

with solution $C_1 = -225$ and $C_2 = -175$. In this case the amount of salt in containers A and B at time t would be

$$\mathbf{X}(t) = \begin{bmatrix} x(t) \\ y(t) \end{bmatrix} = \begin{bmatrix} e^{-t/50} & e^{-3t/50} \\ 2e^{-t/50} & -2e^{-3t/50} \end{bmatrix} \begin{bmatrix} -225 \\ -175 \end{bmatrix} + \begin{bmatrix} 400 \\ 400 \end{bmatrix},$$

or

$$\mathbf{X}(t) = \begin{bmatrix} x(t) \\ y(t) \end{bmatrix} = \begin{bmatrix} 400 - 225e^{-t/50} - 175e^{-3t/50} \\ 400 - 450e^{-t/50} + 350e^{-3t/50} \end{bmatrix}.$$

Figure 13.10 shows the graphs of $x(t) = 400 - 225e^{-t/50} - 175e^{-3t/50}$ and $y(t) = 400 - 450e^{-t/50} + 350e^{-3t/50}$.

How much of the information contained in these explicit solutions could we have obtained from considerations in the phase plane? Looking at our original differential equations,

$$\frac{dx}{dt} = 12 + \frac{1}{100}y - \frac{1}{25}x,$$

$$\frac{dy}{dt} = \frac{1}{25}x - \frac{1}{25}y,$$

we see that equilibrium points exist at values of x and y that satisfy

$$0 = 12 + \frac{1}{100}y - \frac{1}{25}x,$$

$$0 = \frac{1}{25}x - \frac{1}{25}y.$$

This gives $x = 400$, $y = 400$ as the only equilibrium point. The differential equation in the phase plane is

$$\frac{dy}{dx} = \frac{x/25 - y/25}{12 + y/100 - x/25} = \frac{4(x - y)}{1200 + y - 4x}, \tag{13.24}$$

so the nullclines are given by $y = x$ (horizontal tangents) and by $y = 4x - 1200$ (vertical tangents). These nullclines are plotted in Figure 13.11 along with the direction field for the phase plane from (13.24). The direction of the arrows on the direction field suggests that the point $x = 400$, $y = 400$ is a stable equilibrium. The behavior of trajectories in the

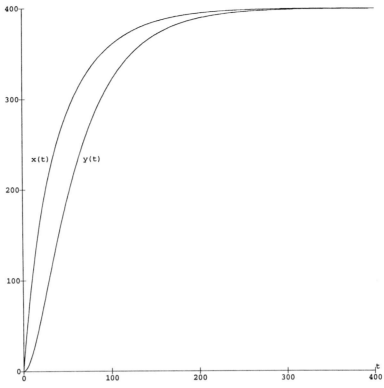

FIGURE 13.9 The functions $x(t) = 400 - 300e^{-t/50} - 100e^{-3t/50}$ and $y(t) = 400 - 600e^{-t/50} + 200e^{-3t/50}$

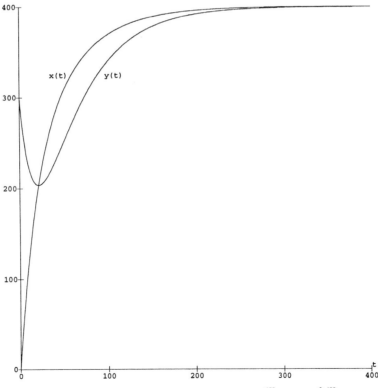

FIGURE 13.10 The functions $x(t) = 400 - 225e^{-t/50} - 175e^{-3t/50}$ and $y(t) = 400 - 450e^{-t/50} + 350e^{-3t/50}$

phase plane for various initial conditions can be determined in a manner we now outline for the preceding two sets of initial conditions.

If we start with both containers free of the substance, we start at the origin ($x = y = 0$). This point is on the isocline for horizontal tangents, and because in the region $x - y > 0$ and $1200 + y - 4x > 0$ both x and y must be increasing, we must move to the right and up. However, because we cannot cross the isocline for vertical tangents (both x and y must increase in this region), we are forced to proceed to the equilibrium point at the intersection of these two isoclines.

On the other hand, if we had initial conditions $x = 0$, $y = 300$, we would start in a region where y must decrease and x must increase. This behavior would continue until we cross the isocline for horizontal tangents, where our previous reasoning takes over and we would end up at the equilibrium solution.

In fact if we look at the form of the general solution in (13.23) we see that the values of C_1 and C_2 are determined by the initial conditions. However, the presence of $e^{-t/50}$ and $e^{-3t/50}$ means that in the limit as $t \to \infty$, x and y approach their equilibrium value regardless of the initial condition. Thus, the equilibrium point is asymptotically stable. Figure 13.12 shows numerical solutions for the trajectories in the phase plane for many initial conditions. Note how they all approach the equilibrium point. The trajectory corresponding initial conditions of our solutions curves in Figures 13.9 and 13.10 are among those shown.   □

Notice that had we made a change of variables from x and y to X and Y so that the equilibrium point was transformed from $(400, 400)$ to $(0, 0)$, we could have used the results of the previous section. (See Exercise 1, page 655.)

Our next example has an exponential forcing function.

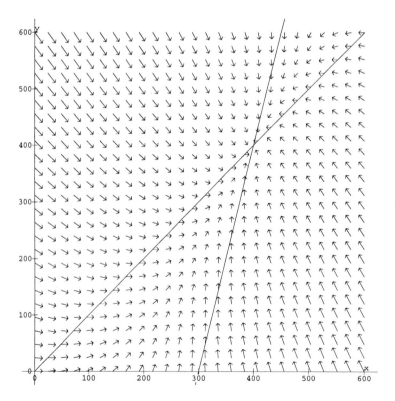

FIGURE 13.11   Direction field and nullclines for $dy/dx = 4(x - y)/(1200 + y - 4x)$

EXAMPLE 13.5

Consider the task of finding the solution of the system of nonhomogeneous first order differential equations given by

$$\frac{d}{dt}\begin{bmatrix} x \\ y \end{bmatrix} = \begin{bmatrix} 1 & -2 \\ 1 & 4 \end{bmatrix}\begin{bmatrix} x \\ y \end{bmatrix} + \begin{bmatrix} 4e^{-t} \\ -4e^{-t} \end{bmatrix}. \tag{13.25}$$

Written in matrix form, this system of equations becomes

$$\frac{d\mathbf{X}}{dt} = \mathbf{MX} + \mathbf{V},$$

where

$$\mathbf{M} = \begin{bmatrix} 1 & -2 \\ 1 & 4 \end{bmatrix}$$

and **V** is the vector given by

$$\mathbf{V} = \begin{bmatrix} 4e^{-t} \\ -4e^{-t} \end{bmatrix} = \begin{bmatrix} 4 \\ -4 \end{bmatrix}e^{-t}.$$

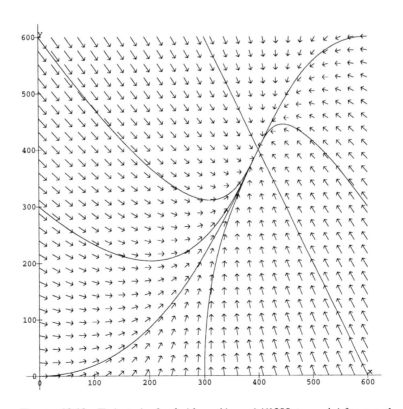

FIGURE 13.12 Trajectories for $dy/dx = 4(x - y)/(1200 + y - 4x)$ for several inital conditions

Now the general solution of (13.25) is the sum of the general solution of the associated homogeneous differential equation and a particular solution of (13.25). Thus, we first consider the homogeneous system of differential equations

$$\frac{d\mathbf{X_h}}{dt} = \mathbf{MX_h} \tag{13.26}$$

and find that a solution of the form

$$\mathbf{X_h}(t) = \begin{bmatrix} A \\ B \end{bmatrix} e^{rt}$$

yields

$$\begin{bmatrix} rA \\ rB \end{bmatrix} e^{rt} = \begin{bmatrix} 1 & -2 \\ 1 & 4 \end{bmatrix} \begin{bmatrix} A \\ B \end{bmatrix} e^{rt}.$$

Because the exponential term is never zero, we may divide both sides by it and obtain the set of algebraic equations

$$\begin{bmatrix} 1-r & -2 \\ 1 & 4-r \end{bmatrix} \begin{bmatrix} A \\ B \end{bmatrix} = \begin{bmatrix} 0 \\ 0 \end{bmatrix}. \tag{13.27}$$

This gives the characteristic equation as

$$(1-r)(4-r) + 2 = r^2 - 5r + 6 = 0$$

with solutions (eigenvalues) of 2 and 3.

Using $r = 2$ in (13.27) gives

$$\begin{bmatrix} -1 & -2 \\ 1 & 2 \end{bmatrix} \begin{bmatrix} A \\ B \end{bmatrix} = \begin{bmatrix} 0 \\ 0 \end{bmatrix},$$

so $A = -2B$, and the associated eigenvector is

$$\begin{bmatrix} -2 \\ 1 \end{bmatrix} B.$$

Using $r = 3$ in (13.27) gives

$$\begin{bmatrix} -2 & -2 \\ 1 & 1 \end{bmatrix} \begin{bmatrix} A \\ B \end{bmatrix} = \begin{bmatrix} 0 \\ 0 \end{bmatrix},$$

so $A = -B$, and the associated eigenvector is

$$\begin{bmatrix} -1 \\ 1 \end{bmatrix} B.$$

Thus, the solution of the associated homogeneous differential equation is

$$\mathbf{X_h}(t) = \mathbf{UC},$$

where $\mathbf{C}$ is a vector of arbitrary constants and $\mathbf{U}$ is a fundamental matrix given by

$$\mathbf{U} = \begin{bmatrix} -2e^{2t} & -e^{3t} \\ e^{2t} & e^{3t} \end{bmatrix}. \tag{13.28}$$

Because the right-hand side of (13.25) contains a constant vector times e^{-t}, and the derivative of a constant times e^{-t} is a constant times e^{-t}, we try a particular solution of the form

$$\mathbf{X}_p(t) = \begin{bmatrix} K_1 \\ K_2 \end{bmatrix} e^{-t}.$$

(This is also what we did for second order nonhomogeneous equations in Chapter 11.)

Substituting this expression into (13.25) gives

$$\begin{bmatrix} -K_1 \\ -K_2 \end{bmatrix} e^{-t} = \begin{bmatrix} 1 & -2 \\ 1 & 4 \end{bmatrix} \begin{bmatrix} K_1 \\ K_2 \end{bmatrix} e^{-t} + \begin{bmatrix} 4e^{-t} \\ -4e^{-t} \end{bmatrix}.$$

Because the expression e^{-t} is common to all terms and is never zero, we may divide through by it and rearrange the resulting system of algebraic equations as

$$\begin{bmatrix} -2 & 2 \\ -1 & -5 \end{bmatrix} \begin{bmatrix} K_1 \\ K_2 \end{bmatrix} = \begin{bmatrix} 4 \\ -4 \end{bmatrix}. \tag{13.29}$$

Because the solution of (13.29) is $K_1 = -1$, $K_2 = 1$, we may write the general solution of (13.25) as

$$\mathbf{X}(t) = \mathbf{U}\mathbf{C} + \mathbf{X}_p(t),$$

where $\mathbf{U}$ is given by (13.28) and

$$\mathbf{X}_p(t) = \begin{bmatrix} -e^{-t} \\ e^{-t} \end{bmatrix}. \tag{13.30}$$

$\square$

The method of finding a particular solution we used in these two examples works for many other situations. In fact the particular solution for a system of nonhomogeneous differential equations

$$\frac{d\mathbf{X}}{dt} = \mathbf{M}\mathbf{X} + \mathbf{V},$$

with $\mathbf{V}$ being a constant vector times $e^{\omega t}$, has the form of

$$\mathbf{X}_p(t) = \begin{bmatrix} K_1 \\ K_2 \end{bmatrix} e^{\omega t}.$$

This is the proper form provided that ω is not an eigenvalue of the matrix $\mathbf{M}$. (Explain what would happen if we used such an $\mathbf{X}_p(t)$ with ω an eigenvalue of $\mathbf{M}$.)

This last result concerning the proper form of a particular solution is also of use for cases where the vector $\mathbf{V}$ in

$$\frac{d\mathbf{X}}{dt} = \mathbf{M}\mathbf{X} + \mathbf{V}$$

contains a sine function or a cosine function. This use is not obvious at first, because a single derivative of a sine function gives a cosine function, and viceversa. However, the sine and cosine functions are related to the exponential function through Euler's identity

$$e^{i\omega t} = \cos \omega t + i \sin \omega t,$$

and the derivative of $e^{i\omega t}$ is $i\omega e^{i\omega t}$.

The key result may be stated in the following manner:

- If **X** is a solution of

$$\frac{d\mathbf{X}}{dt} = \mathbf{M}\mathbf{X} + \mathbf{B}e^{i\omega t}, \tag{13.31}$$

where **M** and **B** contain only real numbers, then the real and imaginary parts of **X** — namely, $Re[\mathbf{X}]$ and $Im[\mathbf{X}]$ — satisfy the differential equations

$$\frac{d}{dt}Re[\mathbf{X}] = \mathbf{M}Re[\mathbf{X}] + \mathbf{B}\cos\omega t, \tag{13.32}$$

$$\frac{d}{dt}Im[\mathbf{X}] = \mathbf{M}Im[\mathbf{X}] + \mathbf{B}\sin\omega t. \tag{13.33}$$

In other words, if **X** satisfies (13.31), then the real part of **X** satisfies (13.32) and the imaginary part of **X** satisfies (13.33).

An example should make clear how to use this idea.

EXAMPLE 13.6

To illustrate the use of the complex exponential function to find the general solution of a nonhomogeneous system of differential equations containing a sine or cosine function we consider

$$\frac{d\mathbf{X}}{dt} = \mathbf{M}\mathbf{X} + \begin{bmatrix} 2\cos t \\ 0 \end{bmatrix}, \text{ where } \mathbf{M} = \begin{bmatrix} -2 & 4 \\ 1 & 1 \end{bmatrix}. \tag{13.34}$$

We can write a three-step procedure as follows:

1. Express a companion differential equation as

$$\frac{d\mathbf{X}}{dt} = \mathbf{M}\mathbf{X} + \begin{bmatrix} 2 \\ 0 \end{bmatrix}e^{it}. \tag{13.35}$$

2. Solve the companion differential equation. The eigenvalues of **M** are determined by solving the quadratic equation obtained from the determinant of the matrix

$$\begin{bmatrix} -2-r & 4 \\ 1 & 1-r \end{bmatrix}.$$

This gives

$$(-2-r)(1-r) - 4 = r^2 + r - 6 = (r+3)(r-2) = 0,$$

so the eigenvalues are 2 and -3. Because the coefficient of t in the exponential forcing function is not an eigenvalue, we may try a particular solution of (13.35) of the form

$$\begin{bmatrix} K_1 \\ K_2 \end{bmatrix}e^{it}.$$

Substituting this expression into (13.35) yields

$$\begin{bmatrix} iK_1 \\ iK_2 \end{bmatrix} e^{it} = \begin{bmatrix} -2 & 4 \\ 1 & 1 \end{bmatrix} \begin{bmatrix} K_1 \\ K_2 \end{bmatrix} e^{it} + \begin{bmatrix} 2 \\ 0 \end{bmatrix} e^{it}.$$

Because the common factor e^{it} is not zero, we may divide by it and rearrange the result to obtain the system of algebraic equations

$$\begin{bmatrix} i+2 & -4 \\ -1 & i-1 \end{bmatrix} \begin{bmatrix} K_1 \\ K_2 \end{bmatrix} = \begin{bmatrix} 2 \\ 0 \end{bmatrix}. \tag{13.36}$$

The solution of (13.36) is

$$\begin{bmatrix} K_1 \\ K_2 \end{bmatrix} = \begin{bmatrix} (8-6i)/25 \\ (-7-i)/25 \end{bmatrix},$$

giving

$$\begin{bmatrix} K_1 \\ K_2 \end{bmatrix} e^{it} = \begin{bmatrix} (8-6i)/25 \\ (-7-i)/25 \end{bmatrix} (\cos t + i \sin t). \tag{13.37}$$

3. Take the real and imaginary parts of this particular solution as appropriate. In this case we need the real part of (13.37), which gives

$$\mathbf{X}_p(t) = \begin{bmatrix} 8/25 \cos t + 6/25 \sin t \\ -7/25 \cos t + 1/25 \sin t \end{bmatrix}$$

as the appropriate particular solution of (13.34). $\square$

EXERCISES

1. Consider the initial value problem

$$\frac{dx}{dt} = 12 + \frac{y}{100} - \frac{x}{25},$$

$$\frac{dy}{dt} = \frac{1}{25}x - \frac{1}{25}y,$$

where $x(0) = y(0) = 0$.

(a) Show that there is an equilibrium point at $(400, 400)$.

(b) Make the change of variables from x and y to X and Y given by $x = X + a$, $y = Y + b$, and choose the constants a and b so the equilibrium point is transformed from $x = y = 400$ to $X = Y = 0$. Find the general solution of this system of equations in X and Y using the results of Section 13.1. Choose the arbitrary constants in your solution so the initial conditions are satisfied. Transform back to the original variables and compare your answer with that found in Example 13.3.

2. Show that if $\mathbf{X}_p(t)$ is a particular solution of

$$\frac{d\mathbf{X}}{dt} = \mathbf{MX} + \mathbf{V} \tag{13.38}$$

and $\mathbf{X}_1(t)$ and $\mathbf{X}_2(t)$ are solutions of the associated homogeneous differential equation, then $C_1\mathbf{X}_1(t) + C_2\mathbf{X}_2(t) + \mathbf{X}_p(t)$, where C_1 and C_2 are arbitrary constants, is a solution of (13.38). [Note that a fundamental matrix has $\mathbf{X}_1(t)$ and $\mathbf{X}_2(t)$ as its column vectors, and your solution is $\mathbf{UC} + \mathbf{X}_p(t)$.]

3. Consider

$$\frac{d\mathbf{X}}{dt} = \mathbf{MX} + \mathbf{V},$$

where $\mathbf{V} = \mathbf{Z}e^{rt}$, with $\mathbf{Z}$ a constant vector. If r is an eigenvector of $\mathbf{M}$, show that there are no particular solutions of the form

$$\mathbf{X}_p(t) = \begin{bmatrix} K_1 \\ K_2 \end{bmatrix} e^{rt}.$$

4. Show that if $\mathbf{X}_{p_1}$ and $\mathbf{X}_{p_2}$ are particular solutions of

$$\frac{d\mathbf{X}}{dt} = \mathbf{MX} + \mathbf{V}_1$$

and

$$\frac{d\mathbf{X}}{dt} = \mathbf{MX} + \mathbf{V}_2,$$

respectively, and $\mathbf{X}_1$ and $\mathbf{X}_2$ are solutions of the associated homogeneous differential equation, then $C_1\mathbf{X}_1 + C_2\mathbf{X}_2 + \mathbf{X}_{p_1} + \mathbf{X}_{p_2}$, where C_1 and C_2 are arbitrary constants, is a solution of

$$\frac{d\mathbf{X}}{dt} = \mathbf{MX} + \mathbf{V}_1 + \mathbf{V}_2.$$

This contains the Principle of Linear Superposition for systems of linear differential equations.

5. Show that if $\mathbf{X}_1$ and $\mathbf{X}_2$ are solutions of

$$\frac{d\mathbf{X}}{dt} = \mathbf{MX},$$

then $C_1\mathbf{X}_1 + C_2\mathbf{X}_2 + A\mathbf{X}_1$ is not a solution of

$$\frac{d\mathbf{X}}{dt} = \mathbf{MX} + \mathbf{X}_1$$

for any choice of the constant A.

6. Create a *How to Find Particular Solutions Using Complex Exponential Functions* by adding statements under Purpose, Technique, and Comments which generalizes the ideas given in Example 13.6.

7. Find a particular solution of

$$\frac{d\mathbf{X}}{dt} = \mathbf{MX} + \mathbf{V}$$

for the following $\mathbf{M}$ and $\mathbf{V}$.

(a)
$$\mathbf{M} = \begin{bmatrix} -2 & 4 \\ 1 & 1 \end{bmatrix} \qquad \mathbf{V} = \begin{bmatrix} 7 \\ 1 \end{bmatrix}$$

(b)
$$\mathbf{M} = \begin{bmatrix} -2 & 4 \\ 1 & 1 \end{bmatrix} \qquad \mathbf{V} = \begin{bmatrix} 7e^{3t} \\ -2e^{3t} \end{bmatrix}$$

(c)
$$\mathbf{M} = \begin{bmatrix} -2 & 4 \\ 1 & 1 \end{bmatrix} \qquad \mathbf{V} = \begin{bmatrix} 0 \\ -\sin 2t \end{bmatrix}$$

(d)
$$\mathbf{M} = \begin{bmatrix} -2 & 4 \\ 1 & 1 \end{bmatrix} \qquad \mathbf{V} = \begin{bmatrix} e^{-t} \\ -2e^{-2t} \end{bmatrix}$$

(e)
$$\mathbf{M} = \begin{bmatrix} -2 & 4 \\ 1 & 1 \end{bmatrix} \qquad \mathbf{V} = 2\begin{bmatrix} 7e^{3t} \\ -2e^{3t} \end{bmatrix} + \pi\begin{bmatrix} 0 \\ -\sin 2t \end{bmatrix}$$

(f)
$$\mathbf{M} = \begin{bmatrix} 2 & 1 \\ -1 & 2 \end{bmatrix} \qquad \mathbf{V} = \begin{bmatrix} e^{\pi t} \\ 3e^{\pi t} \end{bmatrix}$$

(g)
$$\mathbf{M} = \begin{bmatrix} 2 & 1 \\ -1 & 2 \end{bmatrix} \qquad \mathbf{V} = \begin{bmatrix} 3e^t \\ -e^t \end{bmatrix}$$

(h)
$$\mathbf{M} = \begin{bmatrix} 2 & 1 \\ -1 & 2 \end{bmatrix} \qquad \mathbf{V} = \begin{bmatrix} \cos t \\ 0 \end{bmatrix}$$

(i)
$$\mathbf{M} = \begin{bmatrix} 2 & 1 \\ -1 & 2 \end{bmatrix} \qquad \mathbf{V} = \begin{bmatrix} 3e^t \\ -e^t \end{bmatrix} + 17\begin{bmatrix} e^{\pi t} \\ 3e^{\pi t} \end{bmatrix}$$

8. Consider the situation in which the concentration of salt in the solution entering container A in Example 13.4 is given by $4 - 2e^{-t}$. If all the other specifications of that example remain the same, solve the following initial value problem.

$$\frac{d\mathbf{X}}{dt} = \mathbf{M}\mathbf{X} + \mathbf{V},$$

where

$$\mathbf{M} = \begin{bmatrix} -1/25 & 1/100 \\ 1/25 & -1/25 \end{bmatrix} \qquad \mathbf{V} = \begin{bmatrix} 12 \\ 0 \end{bmatrix} - \begin{bmatrix} 6e^{-t} \\ 0 \end{bmatrix}$$

subject to the initial condition

$$\mathbf{X}(0) = \begin{bmatrix} 0 \\ 0 \end{bmatrix}.$$

9. If the forcing function in Exercise 8 is changed to

$$\mathbf{V} = \begin{bmatrix} 12 \\ 0 \end{bmatrix} - \begin{bmatrix} \sin t \\ 0 \end{bmatrix},$$

find the solution of this initial value problem.

13.3 Variation of Parameters

The method we used in Section 13.2 for finding particular solutions of

$$\frac{d\mathbf{X}}{dt} = \mathbf{M}\mathbf{X} + \mathbf{V} \tag{13.39}$$

will not work if the constant r in $\mathbf{V} = \mathbf{Z}e^{rt}$, $\mathbf{Z}$ a constant vector, is an eigenvalue of the matrix $\mathbf{M}$, or with some other forms for $\mathbf{V}$. (See Exercise 5 on page 656.) However, as with the nonhomogeneous second order differential equations in Chapter 11, we have the method of variation of parameters that we can use.

In this case we start with the general solution of the associated homogeneous differential system, which we write as $\mathbf{U}(t)\mathbf{C}$, where $\mathbf{U}$ is a fundamental matrix. Instead of having $\mathbf{C}$ as a constant vector, we allow it to be a function of time. Thus, we have

$$\mathbf{X_p}(t) = \mathbf{U}(t)\mathbf{C}(t),$$

which we substitute into (13.39) to obtain

$$\frac{d}{dt}[\mathbf{U}(t)\mathbf{C}(t)] = \mathbf{M}\mathbf{U}(t)\mathbf{C}(t) + \mathbf{V}(t),$$

or

$$\mathbf{U}(t)\mathbf{C}'(t) + \mathbf{U}'(t)\mathbf{C}(t) = \mathbf{M}\mathbf{U}(t)\mathbf{C}(t) + \mathbf{V}(t). \tag{13.40}$$

(Exercise 3 on page 660 gives the product rule for a matrix times a vector.)

Also, because a fundamental matrix $\mathbf{U}(t)$ of the matrix $\mathbf{M}$ satisfies the differential equation

$$\mathbf{U}'(t) = \mathbf{M}\mathbf{U}(t)$$

(see Exercise 5, page 600), we may multiply both sides of this matrix equation on the right by $\mathbf{C}(t)$ to obtain

$$\mathbf{U}'(t)\mathbf{C}(t) = \mathbf{M}\mathbf{U(t)}\mathbf{C}(t).$$

Using this fact, (13.40) reduces to

$$\mathbf{U}(t)\mathbf{C}'(t) = \mathbf{V}(t). \tag{13.41}$$

Because a fundamental matrix always has an inverse (see Exercise 6 on page 661), we may multiply both sides of the expression in (13.41) by $\mathbf{U}^{-1}(t)$ to obtain

$$\mathbf{U}^{-1}(t)\mathbf{U}(t)\mathbf{C}'(t) = \mathbf{U}^{-1}(t)\mathbf{V}(t).$$

Because $\mathbf{U}^{-1}(t)\mathbf{U}(t) = \mathbf{I}$, the identity matrix, this equation simplifies to

$$\mathbf{C}'(t) = \mathbf{U}^{-1}(t)\mathbf{V}(t).$$

Thus, if we integrate the vector $\mathbf{C}'(t)$ we obtain

$$\mathbf{C}(t) = \int \mathbf{U}^{-1}(t)\mathbf{V}(t)\,dt$$

and find

$$\mathbf{X_p}(t) = \mathbf{U}(t)\mathbf{C}(t) = \mathbf{U}(t)\int \mathbf{U}^{-1}(t)\mathbf{V}(t)\,dt.$$

We include no constant of integration, because we need only one particular solution. Sometimes it is convenient to write limits on the integration in the above equation from 0 to t, so the value of this integral is zero at $t = 0$. If we do this, our particular solution is

$$\mathbf{X_p}(t) = \mathbf{U}(t)\mathbf{C}(t) = \mathbf{U}(t)\int_0^t \mathbf{U}^{-1}(s)\mathbf{V}(s)\,ds.$$

EXAMPLE 13.7

Let us use this technique to find a particular solution of the nonhomogeneous problem in Example 13.5, page 651. The pertinent system of differential equations is

$$\frac{d}{dt}\begin{bmatrix} x \\ y \end{bmatrix} = \begin{bmatrix} 1 & -2 \\ 1 & 4 \end{bmatrix}\begin{bmatrix} x \\ y \end{bmatrix} + \begin{bmatrix} 4e^{-t} \\ -4e^{-t} \end{bmatrix}$$

or

$$\frac{d\mathbf{X}}{dt} = \mathbf{M}\mathbf{X} + \mathbf{V},$$

where

$$\mathbf{V} = \begin{bmatrix} 4e^{-t} \\ -4e^{-t} \end{bmatrix}.$$

In the previous section we discovered that a fundamental matrix for the associated homogeneous equation was

$$\mathbf{U}(t) = \begin{bmatrix} -2e^{2t} & -e^{3t} \\ e^{2t} & e^{3t} \end{bmatrix}, \tag{13.42}$$

which has an inverse of

$$\mathbf{U}^{-1}(t) = -e^{-5t}\begin{bmatrix} e^{3t} & e^{3t} \\ -e^{2t} & -2e^{2t} \end{bmatrix}.$$

Thus, the particular solution will be $\mathbf{U}(t)\mathbf{C}(t)$, where $\mathbf{U}(t)$ is given by (13.42) and where

$$\mathbf{C}(t) = \int \mathbf{U}^{-1}(t)\mathbf{V}(t)\,dt = \int -e^{-5t}\begin{bmatrix} e^{3t} & e^{3t} \\ -e^{2t} & -2e^{2t} \end{bmatrix}\begin{bmatrix} 4e^{-t} \\ -4e^{-t} \end{bmatrix}\,dt,$$

which we integrate to obtain

$$\mathbf{C}(t) = \int -e^{-5t}\begin{bmatrix} 0 \\ 4e^{t} \end{bmatrix}\,dt = \int \begin{bmatrix} 0 \\ -4e^{-4t} \end{bmatrix}\,dt = \begin{bmatrix} 0 \\ e^{-4t} \end{bmatrix}.$$

Performing the indicated multiplication gives our particular solution as

$$\mathbf{X_p}(t) = \mathbf{U}(t)\mathbf{C}(t) = \begin{bmatrix} -2e^{2t} & -e^{3t} \\ e^{2t} & e^{3t} \end{bmatrix}\begin{bmatrix} 0 \\ e^{-4t} \end{bmatrix} = \begin{bmatrix} -e^{-t} \\ e^{-t} \end{bmatrix},$$

in agreement with (13.30). $\qquad\square$

EXERCISES

1. Show that if $ad - bc \neq 0$, the inverse of the matrix

$$\begin{bmatrix} a & b \\ c & d \end{bmatrix}$$

is

$$\frac{1}{ad - bc}\begin{bmatrix} d & -b \\ -c & a \end{bmatrix}.$$

2. Find a particular solution of

$$\frac{d\mathbf{X}}{dt} = \mathbf{MX} + \mathbf{V}$$

for the following **M** and **V**.

(a)
$$\mathbf{M} = \begin{bmatrix} -2 & 4 \\ 1 & 1 \end{bmatrix} \qquad \mathbf{V} = \begin{bmatrix} 7e^{-3t} \\ -2e^{-3t} \end{bmatrix}$$

(b)
$$\mathbf{M} = \begin{bmatrix} -2 & 4 \\ 1 & 1 \end{bmatrix} \qquad \mathbf{V} = \begin{bmatrix} e^{2t} \\ -2e^{2t} \end{bmatrix}$$

(c)
$$\mathbf{M} = \begin{bmatrix} 2 & 1 \\ -3 & -2 \end{bmatrix} \qquad \mathbf{V} = \begin{bmatrix} 3e^{t} \\ -e^{t} \end{bmatrix}$$

(d)
$$\mathbf{M} = \begin{bmatrix} 2 & 1 \\ -3 & -2 \end{bmatrix} \qquad \mathbf{V} = \begin{bmatrix} e^{-t} \\ 2e^{-t} \end{bmatrix}$$

(e)
$$\mathbf{M} = \begin{bmatrix} 2 & 1 \\ -3 & -2 \end{bmatrix} \qquad \mathbf{V} = \begin{bmatrix} (2t + 1)e^{t^2} \\ 0 \end{bmatrix}$$

3. The derivative of a matrix

$$\mathbf{N}(t) = \begin{bmatrix} f(t) & g(t) \\ h(t) & k(t) \end{bmatrix}$$

is defined as

$$\mathbf{N}'(t) = \begin{bmatrix} f'(t) & g'(t) \\ h'(t) & k'(t) \end{bmatrix}.$$

Show that

(a)
$$[C\mathbf{N}(t)]' = C\mathbf{N}'(t)$$

(b)
$$[\mathbf{N}(t) + \mathbf{M}(t)]' = \mathbf{N}'(t) + \mathbf{M}'(t)$$

(c)
$$[z(t)\mathbf{N}(t)]' = z'(t)\mathbf{N}(t) + z(t)\mathbf{N}'(t)$$

(d)
$$[\mathbf{N}(t)\mathbf{X}(t)]' = \mathbf{N}'(t)\mathbf{X}(t) + \mathbf{N}(t)\mathbf{X}'(t)$$

4. Given

$$\mathbf{A}(t) = \begin{bmatrix} t & e^{t} \\ 3 & 3t^2 \end{bmatrix}, \qquad \mathbf{B}(t) = \begin{bmatrix} e^{3t} & e^{2t} \\ e^{-t} & e^{4t} \end{bmatrix}.$$

(a) Compute $t^2\mathbf{A}(t)$.

(b) Compute $\mathbf{A}'(t)$.

(c) Verify that $[t^2\mathbf{A}(t)]' = 2t\mathbf{A}(t) + t^2\mathbf{A}'(t)$.

(d) Compute $e^{-2t}\mathbf{B}(t)$.

(e) Compute $\mathbf{B}'(t)$.

(f) Verify that $[e^{-2t}\mathbf{B}(t)]' = -2te^{-2t}\mathbf{B}(t) + e^{-2t}\mathbf{B}'(t)$.

5. If **U** is a fundamental matrix of the homogeneous system of differential equations

$$\frac{d\mathbf{X}}{dt} = \mathbf{MX}, \tag{13.43}$$

show that **U** satisfies the differential equation

$$\frac{d\mathbf{U}}{dt} = \mathbf{MU}.$$

[Recall that a fundamental matrix of (13.43) has columns composed of vectors that are solutions of (13.43).]

6. Prove that every fundamental matrix has an inverse.

7. Consider the situation in which the concentration of salt in the solution entering container A in Example 13.4 is given by $4 - 2e^{-t/50}$. If all the other specifications of that example remain the same, solve the following initial value problem.

$$\frac{d\mathbf{X}}{dt} = \mathbf{MX} + \mathbf{V},$$

where

$$\mathbf{M} = \begin{bmatrix} -1/25 & 1/100 \\ 1/25 & -1/25 \end{bmatrix} \qquad \mathbf{V} = \begin{bmatrix} 12 \\ 0 \end{bmatrix} - \begin{bmatrix} 6e^{-t/50} \\ 0 \end{bmatrix}$$

subject to the initial condition

$$\mathbf{X}(0) = \begin{bmatrix} 0 \\ 0 \end{bmatrix}.$$

13.4   Applications

We now use our previous methods of solution in three applications. The first application deals with the diffusion of a substance across a membrane. The second examines the movement of an injected substance in a body, and the third is an application to electrical circuits.

EXAMPLE 13.8: Diffusion Across a Membrane

Consider two compartments of volumes V_1 and V_2 that are separated by a membrane allowing substances to diffuse from one compartment to the other (see Figure 13.13). In constructing a system of differential equations to model this situation, we let x and y be the concentrations of a chemical in the compartment with volumes V_1 and V_2, respectively. Because diffusion takes place at a rate that is proportional to the difference in concentrations across the membrane, we have the system of differential equations

$$\begin{aligned} dx/dt &= a(y - x), \\ dy/dt &= b(x - y), \end{aligned} \tag{13.44}$$

where the constants a and b are given by μ/V_1 and μ/V_2, respectively. Here μ is the positive constant that gives the permeability of the membrane. We now write (13.44) in matrix form as

$$\frac{d}{dt} \begin{bmatrix} x \\ y \end{bmatrix} = \begin{bmatrix} -a & a \\ b & -b \end{bmatrix} \begin{bmatrix} x \\ y \end{bmatrix} \tag{13.45}$$

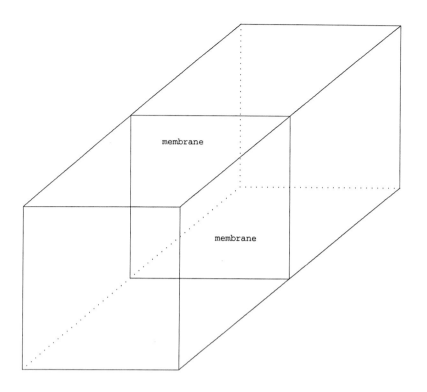

FIGURE 13.13   Two compartments separated by a membrane

and seek solutions in the form

$$\mathbf{X}(t) = \begin{bmatrix} x(t) \\ y(t) \end{bmatrix} = \begin{bmatrix} A \\ B \end{bmatrix} e^{rt}. \tag{13.46}$$

Substituting (13.46) into (13.45) results in

$$\begin{bmatrix} rA \\ rB \end{bmatrix} e^{rt} = \begin{bmatrix} -a & a \\ b & -b \end{bmatrix} \begin{bmatrix} A \\ B \end{bmatrix} e^{rt}$$

or, after dividing by the nonzero factor e^{rt} and rearranging,

$$\begin{bmatrix} -a - r & a \\ b & -b - r \end{bmatrix} \begin{bmatrix} A \\ B \end{bmatrix} = \begin{bmatrix} 0 \\ 0 \end{bmatrix}. \tag{13.47}$$

This system of algebraic equations will have a nontrivial solution only if the determinant of the coefficients is not zero. This gives the characteristic equation

$$(-a - r)(-b - r) - ab = 0,$$

or

$$r^2 + (a + b)r = 0.$$

The two solutions of this equation give the eigenvalues as $r = 0$ and $r = -a - b$. The eigenvector associated with $r = 0$ is found by putting $r = 0$ in (13.47), namely,

$$\begin{bmatrix} -a & a \\ b & -b \end{bmatrix} \begin{bmatrix} A \\ B \end{bmatrix} = \begin{bmatrix} 0 \\ 0 \end{bmatrix},$$

and solving to find

$$\begin{bmatrix} A \\ A \end{bmatrix}.$$

Substituting $r = -a - b$ into (13.47) yields

$$\begin{bmatrix} b & a \\ b & a \end{bmatrix} \begin{bmatrix} A \\ B \end{bmatrix} = \begin{bmatrix} 0 \\ 0 \end{bmatrix}.$$

This gives an eigenvector as

$$\begin{bmatrix} -aA/b \\ A \end{bmatrix}.$$

Thus, the solution of our original system of equations (13.44) is

$$\mathbf{X}(t) = \begin{bmatrix} x(t) \\ y(t) \end{bmatrix} = \begin{bmatrix} 1 & -(a/b)e^{-(a+b)t} \\ 1 & e^{-(a+b)t} \end{bmatrix} \begin{bmatrix} C_1 \\ C_2 \end{bmatrix}. \tag{13.48}$$

The two arbitrary constants in the preceding solution will now be chosen to satisfy the initial conditions $x(0) = x_0$, $y(0) = y_0$. If we set $t = 0$ in (13.48) and use these initial conditions, we obtain the system of algebraic equations

$$\begin{bmatrix} x_0 \\ y_0 \end{bmatrix} = \begin{bmatrix} 1 & -a/b \\ 1 & 1 \end{bmatrix} \begin{bmatrix} C_1 \\ C_2 \end{bmatrix}$$

with solution

$$\begin{bmatrix} C_1 \\ C_2 \end{bmatrix} = \begin{bmatrix} (bx_0 + ay_0)/(a+b) \\ b(y_0 - x_0)/(a+b) \end{bmatrix}.$$

This gives the final form for our answer to the initial value problem as

$$\mathbf{X}(t) = \begin{bmatrix} x(t) \\ y(t) \end{bmatrix} = \begin{bmatrix} 1 & -(a/b)e^{-(a+b)t} \\ 1 & e^{-(a+b)t} \end{bmatrix} \begin{bmatrix} (bx_0 + ay_0)/(a+b) \\ b(y_0 - x_0)/(a+b) \end{bmatrix}. \tag{13.49}$$

From (13.49) we see that for these the initial concentrations, as $t \to \infty$, the concentrations in the two compartments approach the same value of $(bx_0 + ay_0)/(a+b)$. However, this is not an equilibrium point because its value depends on the initial conditions. (Note that $x = y = 0$ is an unstable equilibrium point.) Figure 13.14 shows the graphs of $x(t)$ and $y(t)$ from (13.49) for $a = 0.1$, $b = 0.2$, $x_0 = 1$, and $y_0 = 10$. Figure 13.15 graphs the same functions for $a = 0.1$, $b = 0.2$, $x_0 = 10$, and $y_0 = 1$. Notice the way each function approaches its limiting value.

Let us now see how much information we could have obtained about the solution by analyzing the original system of differential equations. Because a and b are positive, from

$$\begin{aligned} dx/dt &= a(y - x), \\ dy/dt &= b(x - y), \end{aligned} \tag{13.50}$$

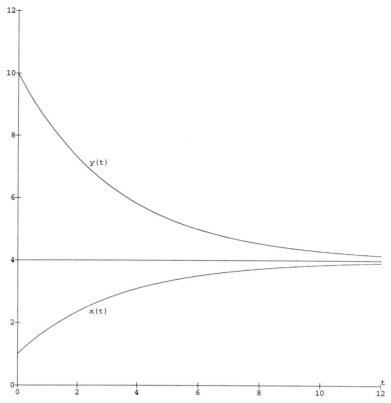

FIGURE 13.14   The functions $x(t)$ and $y(t)$ and the limiting value of 4

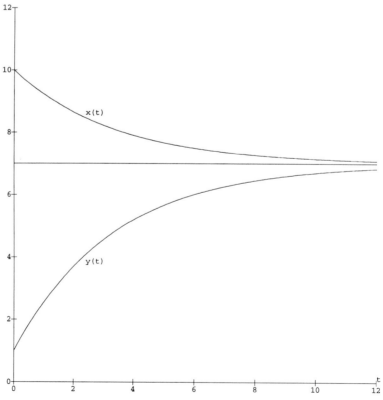

FIGURE 13.15   The functions $x(t)$ and $y(t)$ and the limiting value of 7

we see that x is an increasing function of t when $y > x$ and a decreasing function of t when $y < x$. If x is increasing, y is decreasing. The differential equation in the phase plane is simply

$$\frac{dy}{dx} = -\frac{b}{a},$$

which we may integrate to obtain

$$y(x) = -\frac{b}{a}x + C. \tag{13.51}$$

The arbitrary constant C is obtained by using the initial conditions $x(0) = x_0$, $y(0) = y_0$, as

$$C = y_0 + \frac{b}{a}x_0. \tag{13.52}$$

From (13.50) we see that there will be no change in x and y if they equal each other. Setting $x = y$ in (13.51), solving for y, and using the value of C from (13.52), we obtain the limiting value of

$$y = \frac{bx_0 + ay_0}{a + b},$$

as we discovered earlier. $\qquad\square$

EXAMPLE 13.9: Movement of Inulin

Compartmental models are used to describe the dynamics of many processes in the body. The compartments in models that simulate a process are not chosen randomly, but take into account anatomical locations in the body and the function of various organs. In each compartment a differential equation is derived using continuity arguments — equating the net change of a substance to the difference between what is entering the compartment and what is leaving the compartment. A schematic drawing of a two-compartment model for such a system is shown in Figure 13.16.

One use of a compartmental model is to examine the movement of an injected substance, such as an antibiotic, or a tracer, such as inulin, within the body (often called

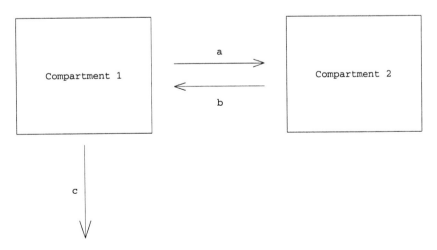

FIGURE 13.16 A two-compartment model

tracer kinetics). Experiments have shown that their rate of elimination may be modeled by a system of linear differential equations. Usually the purpose of such a model is to determine body fluid sizes and rates of clearance by the kidneys.

Here we consider the movement of inulin, which is administered intravenously into the bloodstream of a rabbit, where the two compartments are the blood plasma (compartment 1) and the extracellular fluid (compartment 2). We let $x(t)$ be the concentration of inulin in the plasma and $y(t)$ be the concentration of inulin in the extracellular fluid and use the differential equations

$$dx/dt = -ax + by - cx,$$
$$dy/dt = ax - by.$$

In this model the inulin is transported back and forth between the plasma and the extracellular fluid at different rates, a and b, with the term $-cx$ representing the excretion of the inulin via the urinary tract. Initial conditions appropriate to giving an intravenous injection are $x(0) = x_0$, $y(0) = 0$.

We now find the solution of this initial value problem for the experimentally determined values of the rate constants $a = b = 11$ and $c = 12$. This gives the system of differential equations as

$$\mathbf{X'} = \mathbf{MX}, \text{ where } \mathbf{M} = \begin{bmatrix} -23 & 11 \\ 11 & -11 \end{bmatrix}. \tag{13.53}$$

To find the general solution of (13.53), we use

$$\mathbf{X}(t) = \begin{bmatrix} A \\ B \end{bmatrix} e^{rt}$$

and find that

$$\begin{bmatrix} rA \\ rB \end{bmatrix} e^{rt} = \begin{bmatrix} -23 & 11 \\ 11 & -11 \end{bmatrix} \begin{bmatrix} A \\ B \end{bmatrix} e^{rt}.$$

Because e^{rt} is never zero, we divide by it and rearrange to obtain

$$\begin{bmatrix} -23 - r & 11 \\ 11 & -11 - r \end{bmatrix} \begin{bmatrix} A \\ B \end{bmatrix} = \begin{bmatrix} 0 \\ 0 \end{bmatrix}. \tag{13.54}$$

This gives our characteristic equation as

$$r^2 - 34r + 132 = 0.$$

Using the resulting eigenvalues of -29.5 and -4.5 in (13.54), in turn, gives eigenvectors of

$$\begin{bmatrix} -11 \\ 6.5 \end{bmatrix} \text{ and } \begin{bmatrix} 6.5 \\ 11 \end{bmatrix},$$

respectively. Thus, our general solution of (13.53) is $\mathbf{X}(t) = \mathbf{UC}$, where

$$\mathbf{U} = \begin{bmatrix} -11e^{-29.5t} & 6.5e^{-4.5t} \\ 6.5e^{-29.5t} & 11e^{-4.5t} \end{bmatrix}$$

and $\mathbf{C}$ is a vector of arbitrary constants.

If we now use our initial conditions to evaluate these arbitrary constants, we have that

$$\begin{bmatrix} x_0 \\ 0 \end{bmatrix} = \begin{bmatrix} -11 & 6.5 \\ 6.5 & 11 \end{bmatrix} \begin{bmatrix} C_1 \\ C_2 \end{bmatrix}.$$

Solving these equations gives $C_1 = -0.068x_0$, $C_2 = 0.040x_0$, and the solution of our initial value problem is

$$\mathbf{X}(t) = \begin{bmatrix} x(t) \\ y(t) \end{bmatrix} = \begin{bmatrix} \left(0.74e^{-29.5t} + 0.26e^{-4.5t} \right) x_0 \\ \left(-0.44e^{-29.5t} + 0.44e^{-4.5t} \right) x_0 \end{bmatrix}.$$

Figure 13.17 shows the concentrations in the two compartments as a function of time for $x_0 = 1$. The curve that starts at the origin and increases to a maximum corresponds to compartment 2 (the y value). The fact that both functions decay to zero for any initial condition means that the origin is a stable equilibrium. (Verify this from the fundamental matrix.) □

Our next example considers a nonhomogeneous system of differential equations describing an electric circuit.

EXAMPLE 13.10: A Parallel RLC Circuit

In Chapter 8 we examined the discharge of a capacitor in a parallel RLC circuit. Here we consider the RLC circuit as shown in Figure 13.18 , which contains a voltage source, $E(t)$.

If we consider the voltage drops around the outer loop of the circuit, we obtain

$$V_L + V_R = E(t). \tag{13.55}$$

(Recall that the algebraic sum of the voltages is zero for elements in series.) In the loop containing the resistor and capacitor we have that the voltage drop across each element must be the same, so

$$V_R = V_C. \tag{13.56}$$

We now have two equations for our three unknowns. If we consider the voltage drops around the loop consisting of the inductor and the capacitor, we obtain $V_L + V_C = E(t)$, which is also the result of combining the first two equations.

To find a third, independent, equation, we use Kirchhoff's current law for the junction in which wires from the three circuit elements meet — at the top of Figure 13.18. Kirchhoff's current law states that the current entering a junction must equal the current leaving the junction. (Thus, we see that this is a type of continuity equation.) If we consider the current to be positive for flow in a clockwise direction, we have that

$$I_L = I_R + I_C. \tag{13.57}$$

These three equations may be combined to obtain a system of differential equations in two unknowns as follows. Recalling that for inductors and resistors, voltage drops are given by $L\,dI/dt$ and RI, we rewrite (13.55) as

$$L\frac{dI_L}{dt} + RI_R = E(t). \tag{13.58}$$

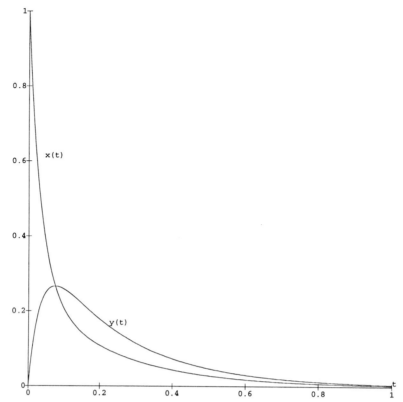

FIGURE 13.17   Concentrations of inulin for a two-compartment model

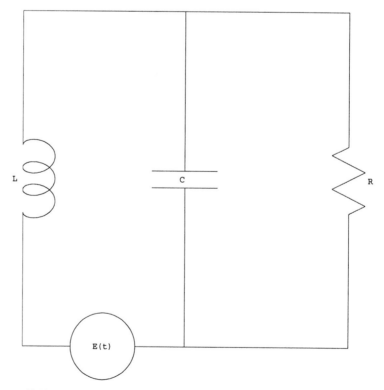

FIGURE 13.18   Parallel RLC circuit with a voltage source

Because the voltage drop across a capacitor is $(1/C) \int I(t)\, dt$, we differentiate (13.56) to obtain $R\, dI_R/dt = (1/C)I_C$. Then we substitute the value of I_C from (13.57) into this result to obtain

$$R\frac{dI_R}{dt} = \frac{1}{C}(I_L - I_R). \tag{13.59}$$

To use our usual notation we let $x = I_L$, $y = I_R$, which gives the system of differential equations from (13.58) and (13.59) as

$$\begin{aligned} dx/dt &= -(R/L)y + (1/L)E(t), \\ dy/dt &= 1/(RC)\,(x - y). \end{aligned}$$

If we consider the situation in which $L = 1$, $R = 5$, $C = 1/20$, and $E(t) = 20\cos 2t$, this gives the system of differential equations

$$\begin{aligned} dx/dt &= -5y + 20\cos 2t, \\ dy/dt &= 4x - 4y. \end{aligned}$$

In matrix form this becomes $\mathbf{X}' = \mathbf{MX} + \mathbf{V}$, where

$$\mathbf{M} = \begin{bmatrix} 0 & -5 \\ 4 & -4 \end{bmatrix} \text{ and } \mathbf{V} = \begin{bmatrix} 20\cos 2t \\ 0 \end{bmatrix}.$$

In Section 13.2 we discovered that the general solution of such nonhomogeneous differential equations is given by $\mathbf{X}(t) = \mathbf{UC} + \mathbf{X}_p(t)$, where $\mathbf{U}$ is a fundamental matrix for the associated homogeneous system $\mathbf{X}' = \mathbf{MX}$, $\mathbf{C}$ is a vector of arbitrary constants, and $\mathbf{X}_p(t)$ is a particular solution.

Thus, our first step is to find a solution of this associated homogeneous system of differential equations, $\mathbf{X}' = \mathbf{MX}$. We recall that when $\mathbf{M}$ has constant coefficients, we always have solutions of such a system in the form

$$\mathbf{X}(t) = \begin{bmatrix} A \\ B \end{bmatrix} e^{rt},$$

where r, A, and B are determined from

$$\begin{bmatrix} -r & -5 \\ 4 & -4-r \end{bmatrix} \begin{bmatrix} A \\ B \end{bmatrix} = \begin{bmatrix} 0 \\ 0 \end{bmatrix}. \tag{13.60}$$

This gives rise to the characteristic equation $r^2 + 4r + 20 = 0$ and eigenvalues $r = -2 \pm 4i$. Substituting $r = -2 + 4i$ into the top equation in (13.60), $-rA - 5B = 0$, allows us to choose the components of our eigenvector as $A = 5$, $B = 2 - 4i$. Because our solution had the form of this eigenvector times e^{rt}, it may also be written as

$$\begin{bmatrix} 5 \\ 2-4i \end{bmatrix} e^{-2t+4it} = \begin{bmatrix} 5e^{-2t}\cos 4t \\ 2e^{-2t}\cos 4t + 4e^{-2t}\sin 4t \end{bmatrix} + i\begin{bmatrix} 5e^{-2t}\sin 4t \\ 2e^{-2t}\sin 4t - 4e^{-2t}\cos 2t \end{bmatrix}.$$

Our original system of differential equations had real coefficients, so both the real and imaginary parts of this vector must be solutions. Thus, we may form a fundamental matrix as

$$\mathbf{U} = \begin{bmatrix} 5e^{-2t}\cos 4t & 5e^{-2t}\sin 4t \\ 2e^{-2t}\cos 4t + 4e^{-2t}\sin 4t & 2e^{-2t}\sin 4t - 4e^{-2t}\cos 2t \end{bmatrix}. \tag{13.61}$$

In Section 13.2 we discovered that in finding particular solutions for trigonometric forcing functions it was convenient to replace such functions with related complex valued exponential functions. (See also Example 13.6, page 654.) Using this idea in this example means we replace $20 \cos 2t$ with $20e^{2it}$ and seek a particular solution of

$$\frac{d\mathbf{X}}{dt} = \mathbf{M}\mathbf{X} + \begin{bmatrix} 20e^{2it} \\ 0 \end{bmatrix}. \qquad (13.62)$$

The particular solution of our original differential equation will be the real part of the particular solution of (13.62).

Because $2i$ is not an eigenvalue of $\mathbf{X}' = \mathbf{M}\mathbf{X}$, we try a particular solution of (13.62) in the form

$$\begin{bmatrix} K_1 \\ K_2 \end{bmatrix} e^{2it}.$$

Substituting this expression into (13.62) gives

$$\begin{bmatrix} 2i K_1 \\ 2i K_2 \end{bmatrix} e^{2it} = \begin{bmatrix} 0 & -5 \\ 4 & -4 \end{bmatrix} \begin{bmatrix} K_1 \\ K_2 \end{bmatrix} e^{2it} + \begin{bmatrix} 20 \\ 0 \end{bmatrix} e^{2it}.$$

Because the common factor e^{2it} is never zero, we may divide by it to obtain the algebraic system of equations

$$2i K_1 = -5K_2 + 20, \qquad (2i + 4)K_2 = 4K_1.$$

The solution of these two equations is $K_1 = 5$, $K_2 = 4 - 2i$, so our particular solution for the forcing function $20e^{2it}$ is

$$\begin{bmatrix} 5 \\ 4 - 2i \end{bmatrix} e^{2it} = \begin{bmatrix} 5 \cos 2t \\ 4 \cos 2t + 2 \sin 2t \end{bmatrix} + i \begin{bmatrix} 5 \sin 2t \\ 4 \sin 2t - 2 \cos 2t \end{bmatrix}.$$

We now have the particular solution of our original system as the real part of this vector, namely,

$$\mathbf{X_p}(t) = \begin{bmatrix} 5 \cos 2t \\ 4 \cos 2t + 2 \sin 2t \end{bmatrix}. \qquad (13.63)$$

This gives our general solution as $\mathbf{X}(t) = \mathbf{U}\mathbf{C} + \mathbf{X_p}(t)$, where $\mathbf{U}$ is the fundamental matrix from (13.61), $\mathbf{C}$ is a vector of arbitrary constants, and $\mathbf{X_p}(t)$ is the particular solution from (13.63).

Note that at this stage, we have also found a particular solution for the differential equation

$$\mathbf{X}' = \begin{bmatrix} 0 & -5 \\ 4 & -4 \end{bmatrix} \mathbf{X} + \begin{bmatrix} 20 \sin 2t \\ 0 \end{bmatrix}$$

as

$$\mathbf{X_p}(t) = \begin{bmatrix} 5 \sin 2t \\ 4 \sin 2t - 2 \cos 2t \end{bmatrix}.$$

If we examine the behavior of this circuit subject to the initial conditions $x(0) = y(0) = 0$ — that is, there is no current in any part of the circuit at time $t = 0$ — we have $\mathbf{0} = \mathbf{U}(0)\mathbf{C} + \mathbf{X_p}(0)$, or

$$\begin{bmatrix} 0 \\ 0 \end{bmatrix} = \begin{bmatrix} 5 & 0 \\ 2 & -4 \end{bmatrix} \begin{bmatrix} C_1 \\ C_2 \end{bmatrix} + \begin{bmatrix} 5 \\ 4 \end{bmatrix}.$$

This gives $C_1 = -1$, $C_2 = 1/2$, and the solution of this initial value problem is

$$\mathbf{X}(t) = \begin{bmatrix} 5e^{-2t}\cos 4t & 5e^{-2t}\sin 4t \\ 2e^{-2t}\cos 4t + 4e^{-2t}\sin 4t & 2e^{-2t}\sin 4t - 4e^{-2t}\cos 4t \end{bmatrix} \begin{bmatrix} -1 \\ 1/2 \end{bmatrix} +$$

$$+ \begin{bmatrix} 5\cos 2t \\ 4\cos 2t + 2\sin 2t \end{bmatrix},$$

or

$$\mathbf{X}(t) = \begin{bmatrix} 5e^{-2t}(-\cos 4t + 1/2\sin 4t) + 5\cos 2t \\ e^{-2t}(-4\cos 4t - 3\sin 4t) + 4\cos 2t + 2\sin 2t \end{bmatrix}. \tag{13.64}$$

We see that the terms involving e^{-2t} are the transient part of the solution, and that the terms containing $\cos 2t$ and $\sin 2t$ are the steady state part. Figure 13.19 shows the solution given in (13.64). The upper right-hand box shows $y(t)$, the lower right-hand box $x(t)$, and the upper left-hand box the phase plane. Notice that the amplitude of the steady state oscillations for $x(t)$ is 5, whereas that for $y(t)$ is $\sqrt{4^2 + 2^2} = \sqrt{20}$. Notice also that periodic steady state oscillations give rise to closed trajectories in the phase plane. $\square$

EXERCISES

1. Find the solution of the initial value problem for the circuit of Figure 13.18 for the following values of the parameters: $L = 1/5$, $C = 1/4$, $R = 1/3$, and $E(t) = 4\sin 2t$. The system of differential equations is therefore $\mathbf{X}' = \mathbf{MX} + \mathbf{V}$, where

$$\mathbf{M} = \begin{bmatrix} 0 & -5/3 \\ 12 & -12 \end{bmatrix}, \quad \mathbf{V} = \begin{bmatrix} 20\sin 2t \\ 0 \end{bmatrix},$$

and the initial condition is $\mathbf{X}(0) = \mathbf{0}$.

2. Find the solution of the initial value problem for the circuit of Figure 13.18 for the following values of the parameters: $L = 1$, $C = 1/20$, $R = 5$, and $E(t) = 4(1 - e^{-t})$. The system of differential equations is therefore $\mathbf{X}' = \mathbf{MX} + \mathbf{V}$, where

$$\mathbf{M} = \begin{bmatrix} 0 & -5 \\ 4 & -4 \end{bmatrix}, \quad \mathbf{V} = \begin{bmatrix} 4(1 - e^{-t}) \\ 0 \end{bmatrix},$$

and the initial condition is $\mathbf{X}(0) = \mathbf{0}$. Find and classify all equilibrium points.

3. The system of differential equations

$$\begin{aligned} dx/dt &= -ax + by - cx + f(t), \\ dy/dt &= ax - by, \end{aligned}$$

models the situation in Example 13.9, page 665, where the inulin is administered intravenously with a time dependence described by $f(t)$. If $f(t) = x_0 e^{-t}$, solve the system of differential equations. Graph your explicit solution for $x(t)$ and $y(t)$ for $0 \leq t \leq 1$, and compare your results with Figure 13.17 on page 668. Comment on any similarities or differences.

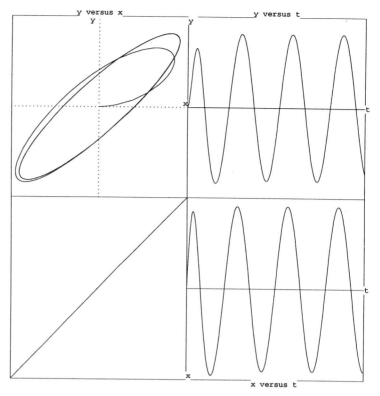

FIGURE 13.19   Currents in a parallel RLC circuit and the associated phase plane

4. Let x denote the number of fish in a lake and y the number of people fishing at the lake. If the rate of stocking the lake is denoted by $S(t)$, the differential equations describing x and y are

$$dx/dt = ax - by + S(t),$$
$$dy/dt = cx - dy. \qquad\qquad (13.65)$$

Here a is the difference between the natural birth and death rates of the fish and b is the catch rate. The constants c and d are both positive because the rate of change of people fishing should increase with the number of fish present and decrease with the number of people fishing.

(a) If $a = c = 2, b = 8, d = 6$, and the stocking rate is approximated by $S(t) = S_0 [1 - \cos(t/10)]$, where S_0 is a constant, solve (13.65) with initial conditions $x(0) = A$, $y(0) = 0$.

(b) Note that the solution for part (a) has terms that decrease rapidly with time (the transient solutions) as well as terms that do not (the steady state solution). Using the steady state solution of this model, determine what values of S_0 will guarantee that the number of people fishing will always be less than 300?

(c) Solve (13.65) if $a = 4, c = 5, b = d = 2, S(t) = S_0 (1 - \cos t), x(0) = A, y(0) = 0$.

5. Patients with severe kidney disease sometimes use a dialyzer, which removes wastes from their blood. A simple model for a dialyzer has the patient's blood flowing along a membrane at a fixed rate while a purifying liquid is simultaneously flowing in the opposite direction on the other side of this membrane at a different rate. This purifying liquid has an affinity for the impurities in the blood, and the rate of change of the impurities across the membrane is proportional to the difference in concentrations. If we let x represent the concentration of impurity in the blood and

y represent the concentration of impurity in the dialyzer liquid at a distance along the membrane, given by t, we have the following system[3] of differential equations:

$$dx/dt = a(y - x)/v,$$
$$dy/dt = a(x - y)/V.$$

Here a is a positive constant, and v and V are the volume flow rates of the blood and dialyzer liquid, respectively. The constant a measures the effectiveness of the dialyzer liquid: Larger a values correspond to a more rapid removal of impurities.

(a) Observe that the direction field in the phase plane is independent of a but depends strongly on the ratio V/v. Comment on the effect on the direction field of changing this ratio.

(b) The usual initial condition is that at the start of the membrane, where $t = 0$, the blood is saturated with impurities. At the other end of the membrane the dialyzer liquid is pure. This gives $x(0) = x_m$, and $y(L) = 0$, where L is the length of the membrane. Find the equation of the trajectory in the phase plane for these initial conditions. Which direction along this trajectory will the solution take as time increases from 0?

(c) Find the explicit solution of this system of differential equations, subject to the initial condition given in part (b).

(d) The efficiency of the dialyzer — called the "clearance" — is given by the ratio $v[x(0) - x(L)]/x(0)$. Use the results of parts (c) and (d) to find the optimum value of V (assuming L, v, and a are all fixed).

13.5 Higher Order Systems

Thus far in this chapter we have considered solution methods for systems of only two linear differential equations in two unknown functions. There are many applications that require more differential equations and more unknown functions, so we consider such situations in this section. We start with an example that leads to a system of four differential equations in four unknown functions and discover that our techniques for solving second order systems still work. The only additional requirement is to find the determinant of a matrix that has four rows and four columns.

EXAMPLE 13.11

A simple physical example that leads to a system of equations that contains more than two first order differential equations with more than two unknown functions is that of two masses suspended on two springs, as shown on the left-hand side of Figure 13.20.

We treat the two masses m_1 and m_2 as point masses and assume that the two springs obey Hooke's law with respective spring constants k_1 and k_2. If there is no motion, the system is said to be in equilibrium. We consider vertical oscillations of this system and let x and y be the displacements of the upper and lower masses, respectively, from equilibrium. Positive values of x and y are in the downward direction. If we displace each mass from its equilibrium position, as shown on the right-hand side of Figure 13.20, then the net force acting on m_1 is $-k_1 x + k_2(y - x)$ and the net force acting on m_2 is $-k_2(y - x)$. Using these values in Newton's second law of motion gives

$$m_1 x'' = -k_1 x + k_2(y - x),$$
$$m_2 y'' = -k_2(y - x).$$

---

[3]Based on "Mathematical Model of a Kidney Machine", by D.M. Burley, Mathematical Spectrum 8, 1975/76, pages 69–75.

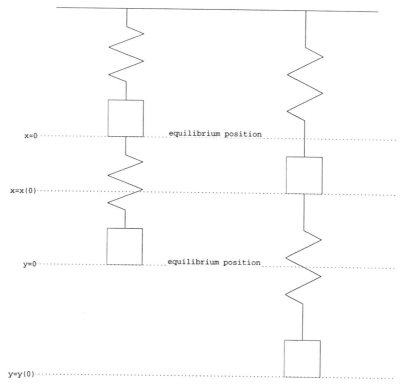

FIGURE 13.20   Two springs and two masses — in equilibrium and displaced

We now solve this system of equations subject to the initial conditions $x(0) = 0$, $x'(0) = 0$, $y(0) = y_0$, and $y'(0) = 0$. These initial conditions model holding the top mass at its equilibrium position, the bottom mass a distance y_0 below its equilibrium position, and then releasing the two masses from rest. One way of solving this system is to convert it to a system of four first order differential equations by defining two new variables u and v, by $x' = u$ and $y' = v$. This gives us the fourth order system

$$
\begin{aligned}
x' &= u, \\
y' &= v, \\
m_1 u' &= -k_1 x + k_2(y - x), \\
m_2 v' &= -k_2(y - x),
\end{aligned}
$$

which may be put in the matrix form

$$\mathbf{X'} = \mathbf{MX}, \tag{13.66}$$

where

$$
\mathbf{M} =
\begin{bmatrix}
0 & 0 & 1 & 0 \\
0 & 0 & 0 & 1 \\
-(k_1 + k_2)/m_1 & k_2/m_1 & 0 & 0 \\
k_2/m_2 & -k_2/m_2 & 0 & 0
\end{bmatrix}.
$$

Because (13.66) has the form of the systems of differential equations we solved in the previous sections — the only difference is the size of the vectors and matrix in (13.66) — we seek a solution in the form of a vector of constants times an exponential function of t.

We simplify the resulting algebra if we take the following specific values of the two masses and spring constants as $m_1 = m_2 = 1$, $k_1 = 5$, and $k_2 = 6$. This gives our matrix **M** as

$$\mathbf{M} = \begin{bmatrix} 0 & 0 & 1 & 0 \\ 0 & 0 & 0 & 1 \\ -11 & 6 & 0 & 0 \\ 6 & -6 & 0 & 0 \end{bmatrix}.$$

Now we substitute the vector

$$\mathbf{X} = \begin{bmatrix} A \\ B \\ C \\ D \end{bmatrix} e^{rt}$$

into (13.66) and obtain

$$\begin{bmatrix} rA \\ rB \\ rC \\ rD \end{bmatrix} = \begin{bmatrix} 0 & 0 & 1 & 0 \\ 0 & 0 & 0 & 1 \\ -11 & 6 & 0 & 0 \\ 6 & -6 & 0 & 0 \end{bmatrix} \begin{bmatrix} A \\ B \\ C \\ D \end{bmatrix},$$

or

$$\begin{bmatrix} -r & 0 & 1 & 0 \\ 0 & -r & 0 & 1 \\ -11 & 6 & -r & 0 \\ 6 & -6 & 0 & -r \end{bmatrix} \begin{bmatrix} A \\ B \\ C \\ D \end{bmatrix} = \begin{bmatrix} 0 \\ 0 \\ 0 \\ 0 \end{bmatrix}. \tag{13.67}$$

This system of algebraic equations will have a nontrivial solution only when the determinant of the coefficient matrix is zero. This gives

$$r^4 + 17r^2 + 30 = (r^2 + 15)(r^2 + 2) = 0,$$

which is called the characteristic equation associated with the matrix **M**. The roots of this equation are called eigenvalues and are given by $r = \pm i\sqrt{2}$ and $\pm i\sqrt{15}$.

If we substitute the eigenvalue $r = i\sqrt{2}$ into (13.67), we obtain the dependent system of algebraic equations

$$\begin{bmatrix} -i\sqrt{2} & 0 & 1 & 0 \\ 0 & -i\sqrt{2} & 0 & 1 \\ -11 & 6 & -i\sqrt{2} & 0 \\ 6 & -6 & 0 & -i\sqrt{2} \end{bmatrix} \begin{bmatrix} A \\ B \\ C \\ D \end{bmatrix} = \begin{bmatrix} 0 \\ 0 \\ 0 \\ 0 \end{bmatrix}.$$

This means that an eigenvector may be chosen with $A = 1$, $B = 3/2$, $C = i\sqrt{2}$, and $D = i3\sqrt{2}/2$. Thus, our solution is

$$\mathbf{X} = \begin{bmatrix} 1 \\ 3/2 \\ i\sqrt{2} \\ i3\sqrt{2}/2 \end{bmatrix} e^{i\sqrt{2}t} = \left(\begin{bmatrix} 1 \\ 3/2 \\ 0 \\ 0 \end{bmatrix} + i \begin{bmatrix} 0 \\ 0 \\ \sqrt{2} \\ 3\sqrt{2}/2 \end{bmatrix} \right) \left(\cos\sqrt{2}t + i\sin\sqrt{2}t \right),$$

or

$$\mathbf{X} = \begin{bmatrix} \cos \sqrt{2}t \\ (3/2)\cos \sqrt{2}t \\ -\sqrt{2}\sin \sqrt{2}t \\ -(3\sqrt{2}/2)\sin \sqrt{2}t \end{bmatrix} + i \begin{bmatrix} \sin \sqrt{2}t \\ (3/2)\sin \sqrt{2}t \\ \sqrt{2}\cos \sqrt{2}t \\ \left(3\sqrt{2}/2\right)\cos \sqrt{2}t \end{bmatrix}. \qquad (13.68)$$

Because the original differential equation $\mathbf{X}' = \mathbf{M}\mathbf{X}$ contained all real numbers, both the real and imaginary parts of our solution in (13.68) must each be a solution of (13.66).

Next we consider the eigenvalue $r = i\sqrt{15}$, which may be substituted into (13.67) to yield

$$\begin{bmatrix} -i\sqrt{15} & 0 & 1 & 0 \\ 0 & -i\sqrt{15} & 0 & 1 \\ -11 & 6 & -i\sqrt{15} & 0 \\ 6 & -6 & 0 & -i\sqrt{15} \end{bmatrix} \begin{bmatrix} A \\ B \\ C \\ D \end{bmatrix} = \begin{bmatrix} 0 \\ 0 \\ 0 \\ 0 \end{bmatrix}.$$

This dependent system of algebraic equations has a solution consisting of $A = 1$, $B = -2/3$, $C = i\sqrt{15}$, and $D = -i2\sqrt{15}/3$. Thus, our solution is

$$\mathbf{X} = \begin{bmatrix} 1 \\ -2/3 \\ i\sqrt{15} \\ -i2\sqrt{15}/3 \end{bmatrix} e^{i\sqrt{15}t} = \left(\begin{bmatrix} 1 \\ -2/3 \\ 0 \\ 0 \end{bmatrix} + i \begin{bmatrix} 0 \\ 0 \\ \sqrt{15} \\ -2\sqrt{15}/3 \end{bmatrix} \right) \left(\cos \sqrt{15}t + i \sin \sqrt{15}t \right),$$

or

$$\mathbf{X} = \begin{bmatrix} \cos \sqrt{15}t \\ -(2/3)\cos \sqrt{15}t \\ -\sqrt{15}\sin \sqrt{15}t \\ (2\sqrt{15}/3)\sin \sqrt{15}t \end{bmatrix} + i \begin{bmatrix} \sin \sqrt{15}t \\ -(2/3)\sin \sqrt{15}t \\ \sqrt{15}\cos \sqrt{15}t \\ -\left(2\sqrt{15}/3\right)\cos \sqrt{15}t \end{bmatrix}.$$

Again, both the real and imaginary components of this solution are solutions of our original system of differential equations. This allows us to form a fundamental matrix and give the general solution of (13.66) as $\mathbf{X}(t) = \mathbf{U}\mathbf{C}$, where

$$\mathbf{U} = \begin{bmatrix} \cos \sqrt{2}t & \sin \sqrt{2}t & \cos \sqrt{15}t & \sin \sqrt{15}t \\ (3/2)\cos \sqrt{2}t & (3/2)\sin \sqrt{2}t & -(2/3)\cos \sqrt{15}t & -(2/3)\sin \sqrt{15}t \\ -\sqrt{2}\sin \sqrt{2}t & \sqrt{2}\cos \sqrt{2}t & -\sqrt{15}\sin \sqrt{15}t & \sqrt{15}\cos \sqrt{15}t \\ -(3\sqrt{2}/2)\sin \sqrt{2}t & (3\sqrt{2}/2)\cos \sqrt{2}t & (2\sqrt{15}/3)\sin \sqrt{15}t & -(2\sqrt{15}/3)\cos \sqrt{15}t \end{bmatrix}$$

and $\mathbf{C}$ is a vector containing four arbitrary constants. Notice that the top two rows in $\mathbf{U}\mathbf{C}$ contain the solution for x and y and the bottom two rows give their derivatives.

If we now choose the constant in $\mathbf{C}$ using our initial conditions $x(0) = x'(0) = y'(0) = 0$, $y(0) = y_0$, we obtain $C_1 = -C_3 = (6/13) y_0$, $C_2 = C_4 = 0$. Thus, the displacement of the two springs is given by

$$\begin{aligned} x(t) &= (6/13)\left(\cos \sqrt{2}t - \cos \sqrt{15}t\right) y_0, \\ y(t) &= \left[(9/13)\cos \sqrt{2}t + (4/13)\cos \sqrt{15}t\right] y_0. \end{aligned} \qquad (13.69)$$

The functions $x(t)$ and $y(t)$ in (13.69) are graphed in Figure 13.21 for $y_0 = 8$. $\qquad \square$

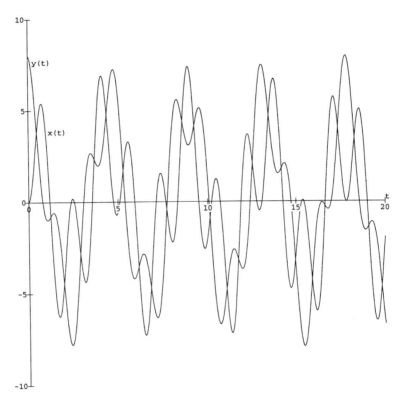

FIGURE 13.21   The functions $x(t) = (6/13)\left(\cos 2^{1/2}t - \cos 15^{1/2}t\right) y_0$ and $y(t) = \left[(9/13)\cos 2^{1/2}t + (4/13)\cos 15^{1/2}t\right] y_0$

We now give a general procedure for finding the solution of a system of linear differential equations with constant coefficients.

How to Solve $d\mathbf{X}/dt = \mathbf{MX}$

Purpose: To find the general solution of the system of differential equations

$$\frac{d\mathbf{X}}{dt} = \mathbf{MX}, \tag{13.70}$$

where $\mathbf{M}$ is a square matrix of constants and $\mathbf{X}$ is a vector of unknown functions. Both $\mathbf{M}$ and $\mathbf{X}$ have the same number of rows, denoted by n.

Technique:

1. Substitute the vector $\mathbf{X} = \mathbf{A}e^{rt}$, where $\mathbf{A}$ is a vector containing n unknown constants, into (13.70) and rearrange the result to obtain the system of algebraic equations

$$(\mathbf{M} - r\mathbf{I})\mathbf{A} = \mathbf{0}. \tag{13.71}$$

2. Solve for the eigenvalues. These are the solutions of the characteristic equation obtained by setting the determinant of the coefficient matrix in (13.71) to 0 — that is, $\det(\mathbf{M} - r\mathbf{I}) = 0$. There will be n eigenvalues if we include repeated roots of the characteristic equation.

3. Find a fundamental matrix for this system of differential equations. The process in this step depends on the nature of the eigenvalues: whether they are distinct or repeated, real or complex.

(a) For the case where we have n real, distinct eigenvalues $r_1, r_2, r_3, \cdots, r_n$, substitute each eigenvalue, in turn, into (13.71) and find the associated eigenvector. The columns in a fundamental matrix are formed by these eigenvectors times e^{rt}, for the appropriate value of r.

(b) For the case in which we have n complex, distinct eigenvalues, we need use only one half of these eigenvalues. (This is true because in this situation complex roots occur in pairs, one being the complex conjugate of the other.) Substitute one of each of the pairs of complex eigenvalues, in turn, into (13.71) and find the associated eigenvector. These eigenvectors will contain complex numbers and, when multiplied by the complex exponential function e^{rt}, will give a complex valued vector function whose real and imaginary parts each satisfy the original differential equation. Two columns of a fundamental matrix are formed by the components of these vector functions. Repeat this procedure for all pairs of complex eigenvalues.

(c) For the case in which all of the eigenvalues are distinct, either real or complex, follow the procedure in part (a) for the real eigenvalues and part (b) for the complex eigenvalues. A fundamental matrix may be formed as columns of the solutions found by each procedure.

(d) The case for repeated eigenvalues utilizes linear combinations of powers of t times e^{rt} times vectors of constants. (This case will not be extensively dealt with in this book. See Exercise 6 on page 682 for an example with repeated real eigenvalues.)

Comments about solving $X' = MX$

- One way to check for mistakes in forming a fundamental matrix is to compute the determinant of the matrix. If this determinant is zero, you have made a mistake, because all fundamental matrices have nonzero determinants. Another way is to see if it satisfies $U' = MU$. It should!

EXAMPLE 13.12

To illustrate the use of this procedure, we consider the problem of finding the general solution of $dX/dt = MX$, where

$$M = \begin{bmatrix} 2 & 0 & 9 \\ 0 & 3 & 0 \\ 1 & 0 & 2 \end{bmatrix} \text{ and } X = \begin{bmatrix} u \\ v \\ w \end{bmatrix}.$$

We start by substituting the vector $X = Ae^{rt}$, where

$$A = \begin{bmatrix} A \\ B \\ C \end{bmatrix}$$

and A, B, C, and r are unknown constants, into the differential equation. Rearranging the result, we obtain the system of algebraic equations

$$(M - rI)A = \begin{bmatrix} 2-r & 0 & 9 \\ 0 & 3-r & 0 \\ 1 & 0 & 2-r \end{bmatrix} \begin{bmatrix} A \\ B \\ C \end{bmatrix} = \begin{bmatrix} 0 \\ 0 \\ 0 \end{bmatrix}. \tag{13.72}$$

From the determinant of this coefficient matrix we obtain the characteristic equation

$$(2 - r)(3 - r)(2 - r) - 9(3 - r) = (3 - r)[(2 - r)(2 - r) - 9] = 0.$$

Expanding the term in brackets on the right in this last equation gives the eigenvalues as $r = 3, -1$, and 5.

Substituting the eigenvalue $r = 3$ into (13.72) gives

$$\begin{bmatrix} -1 & 0 & 9 \\ 0 & 0 & 0 \\ 1 & 0 & -1 \end{bmatrix} \begin{bmatrix} A \\ B \\ C \end{bmatrix} = \begin{bmatrix} 0 \\ 0 \\ 0 \end{bmatrix},$$

so an eigenvector may be chosen as

$$\begin{bmatrix} 0 \\ 1 \\ 0 \end{bmatrix}.$$

Similar calculations for $r = -1$ and $r = 5$ yield the eigenvectors

$$\begin{bmatrix} 3 \\ 0 \\ -1 \end{bmatrix} \text{ and } \begin{bmatrix} 3 \\ 0 \\ 1 \end{bmatrix},$$

respectively.

Thus, we have a fundamental matrix

$$\mathbf{U} = \begin{bmatrix} 0 & 3e^{-t} & 3e^{5t} \\ 3e^{3t} & 0 & 0 \\ 0 & -e^{-t} & e^{5t} \end{bmatrix}, \tag{13.73}$$

and the general solution is given by $\mathbf{X}(t) = \mathbf{UC}$, where $\mathbf{U}$ is our fundamental matrix and $\mathbf{C}$ is a vector of arbitrary constants. □

We now consider the task of finding particular solutions of nonhomogeneous systems of linear differential equations. The methods used for higher order systems are identical in form to those used in Section 13.2 for second order systems. To illustrate these methods we first consider the nonhomogeneous system

$$\frac{d\mathbf{X}}{dt} = \mathbf{MX} + \mathbf{V}, \tag{13.74}$$

where $\mathbf{M}$ is an n by n matrix of constants and $\mathbf{V}$ has the form of a vector of constants times $e^{\omega t}$. If ω is not an eigenvalue of the matrix $\mathbf{M}$, then a proper form for a particular solution is that of a vector containing n unknown constants times $e^{\omega t}$ — namely, $\mathbf{K}e^{\omega t}$. Substitution of such a vector into (13.74) will result in a system of n algebraic equations in n unknowns. The solution of this algebraic system gives the components of the vector $\mathbf{K}$ in the particular solution.

EXAMPLE 13.13

As an example of this method, we find a particular solution of $\mathbf{X}' = \mathbf{MX} + \mathbf{V}$, where

$$\mathbf{M} = \begin{bmatrix} 2 & 0 & 9 \\ 0 & 3 & 0 \\ 1 & 0 & 2 \end{bmatrix} \text{ and } \mathbf{V} = \begin{bmatrix} 4 \\ 12 \\ 4 \end{bmatrix} e^{-3t}. \tag{13.75}$$

If we substitute the trial particular solution

$$\mathbf{X_p}(t) = \begin{bmatrix} K_1 \\ K_2 \\ K_3 \end{bmatrix} e^{-3t}$$

into the differential equation and rearrange the resulting expression, we obtain

$$\begin{bmatrix} -3K_1 \\ -3K_2 \\ -3K_3 \end{bmatrix} = \begin{bmatrix} 2 & 0 & 9 \\ 0 & 3 & 0 \\ 1 & 0 & 2 \end{bmatrix} \begin{bmatrix} K_1 \\ K_2 \\ K_3 \end{bmatrix} + \begin{bmatrix} 4 \\ 12 \\ 4 \end{bmatrix}.$$

The solution of this system of algebraic equations is $K_1 = 1$, $K_2 = -2$, and $K_3 = -1$, so our particular solution is

$$\mathbf{X_p}(t) = \begin{bmatrix} 1 \\ -2 \\ -1 \end{bmatrix} e^{-3t}. \tag{13.76}$$

However, if $\mathbf{V}$ is not of the form of a vector of constants times $e^{\omega t}$, or if it is but ω is an eigenvalue of $\mathbf{M}$, this method of finding a particular solution will not work. In this case we need to use the method of variation of parameters — it always works.

To use the method of variation of parameters in the current case, we would write our particular solution as

$$\mathbf{X_p}(t) = \mathbf{U}(t) \int \mathbf{U}^{-1}(t) \mathbf{V}(t) \, dt,$$

where $\mathbf{U}$ is given by (13.73) and $\mathbf{V}$ is given by (13.75). We compute the inverse of $\mathbf{U}$ as

$$\mathbf{U}^{-1} = \frac{1}{6} \begin{bmatrix} 0 & 2e^{-3t} & 0 \\ e^t & 0 & -3e^t \\ e^{-5t} & 0 & 3e^{-5t} \end{bmatrix},$$

so the integral in the expression for $\mathbf{X_p}(t)$ is given by

$$\int \frac{1}{6} \begin{bmatrix} 0 & 2e^{-3t} & 0 \\ e^t & 0 & -3e^t \\ e^{-5t} & 0 & 3e^{-5t} \end{bmatrix} \begin{bmatrix} 4 \\ 12 \\ 4 \end{bmatrix} e^{-3t} \, dt = \int \frac{1}{6} \begin{bmatrix} 24e^{-6t} \\ -8e^{-2t} \\ 16e^{-8t} \end{bmatrix} dt = \frac{1}{6} \begin{bmatrix} -4e^{-6t} \\ 4e^{-2t} \\ -2e^{-8t} \end{bmatrix}.$$

To obtain $\mathbf{X_p}(t)$ we multiply this vector by $\mathbf{U}$ to obtain

$$\mathbf{X_p}(t) = \begin{bmatrix} 0 & 3e^{-t} & 3e^{5t} \\ 3e^{3t} & 0 & 0 \\ 0 & -e^{-t} & e^{5t} \end{bmatrix} \begin{bmatrix} -(2/3)\, e^{-6t} \\ (2/3)\, e^{-2t} \\ -(1/3)\, e^{-8t} \end{bmatrix}.$$

Performing the indicated multiplications yields the same particular solution as we found in (13.76). $\qquad\square$

EXERCISES

1. Find the general solution of the following problems in terms of a fundamental matrix.

   (a) $\mathbf{X}' = \mathbf{MX}$, with $\mathbf{M} = \begin{bmatrix} 1 & 2 & 1 \\ 0 & -1 & 0 \\ 1 & 2 & 1 \end{bmatrix}$

   (b) $\mathbf{X}' = \mathbf{MX}$, with $\mathbf{M} = \begin{bmatrix} 12 & -3 & -3 \\ -3 & 9 & 0 \\ -3 & 0 & 9 \end{bmatrix}$

   (c) $\mathbf{X}' = \mathbf{MX}$, with $\mathbf{M} = \begin{bmatrix} 3 & -1 & -1 \\ -1 & 3 & -1 \\ -1 & -1 & 3 \end{bmatrix}$

   (d) $\mathbf{X}' = \mathbf{MX}$, with $\mathbf{M} = \begin{bmatrix} 2 & 0 & 9 \\ 0 & 3 & 0 \\ 1 & 0 & 2 \end{bmatrix}$

   (e) $\mathbf{X}' = \mathbf{MX}$, with $\mathbf{M} = \begin{bmatrix} 2 & -2 & 0 \\ 1 & -2 & -1 \\ -2 & 1 & -2 \end{bmatrix}$

   (f) $\mathbf{X}' = \mathbf{MX}$, with $\mathbf{M} = \begin{bmatrix} 2 & 0 & 5 \\ 0 & 1 & 2 \\ -4 & 5 & 0 \end{bmatrix}$

   (g) $\mathbf{X}' = \mathbf{MX}$, with $\mathbf{M} = \begin{bmatrix} 0 & 3 & 3 \\ 2 & -1 & -3 \\ 0 & -1 & -1 \end{bmatrix}$

   (h) $\mathbf{X}' = \mathbf{MX}$, with $\mathbf{M} = \begin{bmatrix} -1 & 0 & 0 \\ 2 & -1 & 0 \\ 3 & 5 & -1 \end{bmatrix}$

2. Solve the following initial value problems.

   (a) Exercise 1(c) with $\mathbf{X}(0) = \begin{bmatrix} 1 \\ 0 \\ 0 \end{bmatrix}$

   (b) Exercise 1(d) with $\mathbf{X}(0) = \begin{bmatrix} 1 \\ 0 \\ 1 \end{bmatrix}$

   (c) Exercise 1(f) with $\mathbf{X}(0) = \begin{bmatrix} 0 \\ 1 \\ 1 \end{bmatrix}$

   (d) Exercise 1(g) with $\mathbf{X}(0) = \begin{bmatrix} 3 \\ 0 \\ 7 \end{bmatrix}$

3. For the RLC parallel circuit of Figure 13.18, consider the situation in which the voltage source is shifted from being below the inductor to being below the resistor.

   (a) Show that the resulting circuit is described by

   $$\begin{aligned} L\,dI_L/dt + RI_R &= V(t), \\ R\,dI_R/dt + (1/C)I_C &= V'(t), \\ I_R &= I_C + I_L. \end{aligned}$$

   (b) Solve the system of equations in part (a) if $R = 50$, $C = 0.02$, $L = 0.004$, $V(t) = 100$, and $I_R(0) = I_L(0) = 0$. Show that the system has an equilibrium point, and determine its stability.

   (c) Solve the system of equations if R, L, C, and the initial conditions are as in part (b) and $V(t) = 100 \sin 100t$.

4. If the springs and masses in Figure 13.20 are given by $m_1 = m_2 = 2, k_1 = 6, k_2 = 4$, find solutions of (13.66) that satisfy the initial conditions $x_1(0) = 0$, $x_1'(0) = 4$, $x_2(0) = 0$, $x_2'(0) = -7$.

5. Show that for positive values of m_1, m_2, k_1, and k_2, the solutions of (13.66) will always be oscillatory.

6. Consider the system of differential equations $d\mathbf{X}/dt = \mathbf{MX}$, where

$$\mathbf{M} = \begin{bmatrix} 1 & 1 & 0 & 0 \\ 0 & 1 & 2 & 0 \\ 0 & 0 & 1 & 0 \\ 0 & -2 & 0 & 1 \end{bmatrix} \text{ and } \mathbf{X} = \begin{bmatrix} x \\ y \\ u \\ v \end{bmatrix}.$$

(a) By substituting an appropriate expression for $\mathbf{X}$ as a vector of constants times e^{rt}, find the eigenvalues of the resulting characteristic equation. (They will be repeated.)

(b) Discover the two linearly independent eigenvectors associated with this repeated eigenvalue.

(c) Find the two remaining columns of a fundamental matrix by trying a solution of the form $\mathbf{X}(t) = \left(t^2 \mathbf{A}_2 + t \mathbf{A}_1 + \mathbf{A}_0\right) e^{rt}$, where r is the repeated eigenvalue from part (a). $\mathbf{A}_2$, $\mathbf{A}_1$, and $\mathbf{A}_0$ are vectors, each containing four constants that are to be determined so $\mathbf{X}(t)$ is a solution.

(d) Combine the answers from parts (b) and (c) to find a fundamental matrix, and give the general solution of the original system of differential equations.

7. Find the general solution of $\mathbf{X}' = \mathbf{MX} + \mathbf{V}$, where

(a) $\mathbf{M} = \begin{bmatrix} 1 & 2 & 1 \\ 0 & -1 & 0 \\ 1 & 2 & 1 \end{bmatrix}, \quad \mathbf{V} = \begin{bmatrix} 0 \\ 1 \\ e^t \end{bmatrix}$

(b) $\mathbf{M} = \begin{bmatrix} 12 & -3 & -3 \\ -3 & 9 & 0 \\ -3 & 0 & 9 \end{bmatrix}, \quad \mathbf{V} = \begin{bmatrix} e^{-t} \\ e^{-t} \\ e^{-t} \end{bmatrix}$

(c) $\mathbf{M} = \begin{bmatrix} 3 & -1 & -1 \\ -1 & 3 & -1 \\ -1 & -1 & 3 \end{bmatrix}, \quad \mathbf{V} = \begin{bmatrix} e^t \\ 0 \\ 0 \end{bmatrix}$

(d) $\mathbf{M} = \begin{bmatrix} 2 & 0 & 9 \\ 0 & 3 & 0 \\ 1 & 0 & 2 \end{bmatrix}, \quad \mathbf{V} = \begin{bmatrix} \sin 2t \\ 0 \\ 0 \end{bmatrix}$

(e) $\mathbf{M} = \begin{bmatrix} 2 & 0 & 9 \\ 0 & 3 & 0 \\ 1 & 0 & 2 \end{bmatrix}, \quad \mathbf{V} = \begin{bmatrix} \cos 2t \\ 0 \\ 0 \end{bmatrix}$

(f) $\mathbf{M} = \begin{bmatrix} 2 & 0 & 9 \\ 0 & 3 & 0 \\ 1 & 0 & 2 \end{bmatrix}, \quad \mathbf{V} = \begin{bmatrix} e^{-t} \\ e^{3t} \\ e^{5t} \end{bmatrix}$

(g) $\mathbf{M} = \begin{bmatrix} 2 & 0 & 5 \\ 0 & 1 & 2 \\ -4 & 5 & 0 \end{bmatrix}, \quad \mathbf{V} = \begin{bmatrix} \cos 2t \\ 0 \\ \sin 2t \end{bmatrix}$

8. Solve the following initial value problems.

(a) Exercise 1(c) with $\mathbf{X}(0) = \begin{bmatrix} 1 \\ 0 \\ 0 \end{bmatrix}$

(b) Exercise 1(d) with $\mathbf{X}(0) = \begin{bmatrix} 1 \\ 0 \\ 1 \end{bmatrix}$

(c) Exercise 1(e) with $\mathbf{X}(0) = \begin{bmatrix} 0 \\ 1 \\ 1 \end{bmatrix}$

(d) Exercise 1(f) with $\mathbf{X}(0) = \begin{bmatrix} 3 \\ 0 \\ 7 \end{bmatrix}$

What Have We Learned?

- In this chapter we discovered ways of solving linear differential equations $\mathbf{X}' = \mathbf{MX} + \mathbf{V}$, where $\mathbf{X}$ is a vector of n independent variables, $\mathbf{M}$ is an n by n matrix of constants, and $\mathbf{V}$ is a vector of n given functions.

- A fundamental matrix , $\mathbf{U}$, is a square matrix whose columns are solution vectors of the homogeneous differential equation $\mathbf{X}' = \mathbf{MX}$. The general solution of such equations is $\mathbf{X}(t) = \mathbf{UC}$, where $\mathbf{C}$ is a vector of arbitrary constants.

- Eigenvalues of the matrix $\mathbf{M}$ are the zeros of the polynomial in r found by expanding the determinant of the matrix obtained by subtracting the parameter r from each of the elements of the diagonal of $\mathbf{M}$.

- We used two methods of finding particular solutions of $\mathbf{X}' = \mathbf{MX} + \mathbf{V}$.

  — If $\mathbf{V}$ has the form of a constant vector times $e^{\omega t}$ and ω is not an eigenvalue of $\mathbf{M}$, then $\mathbf{X}_p(t) = \mathbf{K}e^{\omega t}$, where $\mathbf{K}$ is a constant vector.

  — If $\mathbf{V}$ does not have this form or if ω is an eigenvalue of $\mathbf{M}$, then variation of parameters will give a particular solution.

  The entire chapter is summarized by the following:

How to Solve Nonhomogeneous Systems of Linear Differential Equations with Constant Coefficients

Purpose: To find the general solution of the nonhomogeneous system of linear differential equations of the form

$$\frac{d\mathbf{X}}{dt} = \mathbf{MX} + \mathbf{V}, \tag{13.77}$$

where $\mathbf{M}$ is an n by n matrix of real valued constants, $\mathbf{X}$ is a vector of dependent variables, and $\mathbf{V}$ is a forcing function.

Technique:

1. Find the general solution of the associated homogeneous differential equation. See *How to Solve $d\mathbf{X}/dt = \mathbf{MX}$* on page 636 for $n = 2$ and page 677 for higher order equations.

2. Find a particular solution of (13.77). Here we have a choice.

   (a) If the forcing function has the form of a vector of constants times $e^{\omega t}$, where ω is not an eigenvalue of $\mathbf{M}$, then the proper form for a particular solution is $\mathbf{X}_p(t) = \mathbf{K}e^{\omega t}$, where $\mathbf{K}$ is a vector of unknown constants. Substitute this expression into (13.77) and choose the constants so the system of differential equations is satisfied.

   (b) If the forcing function is not of the form mentioned in part (a), or if in that form ω is an eigenvalue of the matrix $\mathbf{M}$, we use variation of parameters. Here a particular solution is given by

$$\mathbf{X}_p(t) = \mathbf{U}(t) \int \mathbf{U}^{-1}(t)\mathbf{V}(t)\, dt,$$

where **U** is the fundamental matrix found in the general solution of the homogeneous problem in step 1.

3. The general solution is given by

$$\mathbf{X}(t) = \mathbf{U}\mathbf{C} + \mathbf{X}_\mathbf{p}(t),$$

where **C** is a vector of arbitrary constants.

CHAPTER **14**

The Laplace Transform

Where Are We Going?

In this chapter we develop properties of the Laplace transform and show how it is used to solve linear differential equations with constant coefficients. We will see that this transform applies to all orders of such differential equations as well as to systems of linear differential equations. Because the existence of a table of transforms and inverse transforms of many functions greatly eases the use of Laplace transforms, we develop results that help establish such tables.

14.1   Introduction

We introduce Laplace transforms as another way of solving initial value problems for linear differential equations with constant coefficients. This technique relies on converting the task of solving a differential equation to that of solving an algebraic equation. We will begin this section by demonstrating the use of this technique with a simple example.

EXAMPLE 14.1

Consider the initial value problem

$$\frac{dy}{dt} + y = e^{at}, \tag{14.1}$$

where a is a constant, subject to

$$y(0) = 0. \tag{14.2}$$

Using the techniques from Chapter 6, we already know that our solution is

$$y(t) = \frac{1}{a+1}\left(e^{at} - e^{-t}\right) \tag{14.3}$$

if $a \neq -1$ and

$$y(t) = te^{-t} \tag{14.4}$$

if $a = -1$. (Verify that these are the solutions.)

We will now solve this initial value problem in a completely different way, which at first will appear very arbitrary and unmotivated. Many of the initial value problems that arise in applications have time as the independent variable. The initial condition is given at $t = 0$ and the region of interest is $t \geq 0$. Thus, as we solve this problem we focus attention on the interval $0 \leq t < \infty$.

First we multiply (14.1) by e^{-st}, where s is a positive constant, and integrate the result from 0 to ∞, obtaining

$$\int_0^\infty e^{-st} \left(\frac{dy}{dt} + y \right) dt = \int_0^\infty e^{-st} e^{at} \, dt,$$

or

$$\int_0^\infty e^{-st} \frac{dy}{dt} \, dt + \int_0^\infty e^{-st} y \, dt = \int_0^\infty e^{(a-s)t} \, dt. \tag{14.5}$$

We next evaluate the integral on the right-hand side as

$$\int_0^\infty e^{(a-s)t} \, dt = \lim_{b \to \infty} \int_0^b e^{(a-s)t} \, dt = \lim_{b \to \infty} \left[\frac{1}{a-s} e^{(a-s)t} \Big|_0^b \right] = \lim_{b \to \infty} \frac{1}{a-s} e^{(a-s)b} - \frac{1}{a-s},$$

or, if $s > a$,

$$\int_0^\infty e^{(a-s)t} \, dt = \frac{1}{s-a}. \tag{14.6}$$

Substituting (14.6) into (14.5) gives

$$\int_0^\infty e^{-st} \frac{dy}{dt} \, dt + \int_0^\infty e^{-st} y \, dt = \frac{1}{s-a}. \tag{14.7}$$

We now look at the left-hand side of (14.7). If we use integration by parts, that is,

$$\int u \frac{dv}{dt} \, dt = uv - \int v \frac{du}{dt} \, dt$$

on the first integral, by letting $u = e^{-st}$ and $dv = (dy/dt)\,dt$ so $v = y$, we find

$$\int e^{-st} \frac{dy}{dt} \, dt = e^{-st} y - \int y \frac{de^{-st}}{dt} \, dt,$$

or

$$\int e^{-st} \frac{dy}{dt} \, dt = e^{-st} y + s \int e^{-st} y \, dt.$$

Adding the limits of integration gives

$$\int_0^\infty e^{-st} \frac{dy}{dt} \, dt = \lim_{b \to \infty} \left(e^{-st} y \Big|_0^b \right) + s \int_0^\infty e^{-st} y \, dt,$$

or

$$\int_0^\infty e^{-st} \frac{dy}{dt} \, dt = \lim_{b \to \infty} e^{-sb} y(b) - y(0) + s \int_0^\infty e^{-st} y \, dt. \tag{14.8}$$

Using the initial condition $y(0) = 0$, we find

$$\int_0^\infty e^{-st} \frac{dy}{dt}\, dt = \lim_{b \to \infty} e^{-sb} y(b) + s \int_0^\infty e^{-st} y\, dt. \qquad (14.9)$$

If we look at the term $\lim_{b \to \infty} e^{-sb} y(b)$ and realize that $\lim_{b \to \infty} e^{-sb} = 0$, we would expect that $\lim_{b \to \infty} e^{-sb} y(b) = 0$ for many functions. For the time being, let's assume that this is true for our solution, in which case (14.9) becomes

$$\int_0^\infty e^{-st} \frac{dy}{dt}\, dt = s \int_0^\infty e^{-st} y\, dt. \qquad (14.10)$$

We now substitute this equation into (14.7) to find

$$s \int_0^\infty e^{-st} y\, dt + \int_0^\infty e^{-st} y\, dt = \frac{1}{s-a},$$

or

$$(s+1) \int_0^\infty e^{-st} y\, dt = \frac{1}{s-a}.$$

This means that the solution $y = y(t)$ that we seek must satisfy

$$\int_0^\infty e^{-st} y\, dt = \frac{1}{(s+1)(s-a)}. \qquad (14.11)$$

One way to find a solution would be to compute $\int_0^\infty e^{-st} y\, dt$ for many different functions $y(t)$, such as 1, t, $\sin t$, $\cos t$, e^t, and so on, or more generally for a, t^n, $\sin at$, $\cos at$, and e^{at}. Then if we are lucky we might find that for one of these functions we had (14.11). For example, in (14.6) we found that

$$\int_0^\infty e^{-st} e^{at}\, dt = \frac{1}{s-a}, \qquad (14.12)$$

so that if the right-hand side of (14.11) were $1/(s-a)$ our solution would be $y(t) = e^{at}$. In the present case, we can take advantage of this knowledge as follows. For the case $a \neq -1$ we use partial fractions to change the form of the right-hand side of (14.11) to

$$\frac{1}{(s+1)(s-a)} = \frac{1}{a+1} \left(\frac{1}{s-a} - \frac{1}{s+1} \right).$$

Substituting this expression into (14.11) yields

$$\int_0^\infty e^{-st} y\, dt = \frac{1}{a+1} \left(\frac{1}{s-a} - \frac{1}{s+1} \right),$$

or, by using (14.12) twice — once with arbitrary a and once with $a = -1$ — we find

$$\int_0^\infty e^{-st} y\, dt = \frac{1}{a+1} \left(\int_0^\infty e^{-st} e^{at}\, dt - \int_0^\infty e^{-st} e^{-t}\, dt \right),$$

or

$$\int_0^\infty e^{-st} y\, dt = \int_0^\infty e^{-st} \left[\frac{1}{a+1} \left(e^{at} - e^{-t} \right) \right] dt.$$

Thus, the solution of our initial value problem is

$$y(t) = \frac{1}{a+1}\left(e^{at} - e^{-t}\right),$$

which agrees with (14.3).   □

Now let's look at what we have just done, first in broad outline, and then with a more critical eye. We started with (14.1) and, after multiplying by e^{-st} and integrating, obtained (14.5). Notice that there are two terms on the left-hand side of (14.5), one involving y' and the other y. We used integration by parts to obtain (14.8), which relates the term $\int_0^\infty e^{-st} y' \, dt$ to the term $\int_0^\infty e^{-st} y \, dt$. This led to (14.11), which we "solved" for y by luck. That is the broad outline.

To look at what we have done a little more critically, there are three questions that we have avoided discussing.

- Under what circumstances do integrals such as $\int_0^\infty e^{-st} y \, dt$ exist?

- Under what circumstances can we guarantee $\lim_{b \to \infty} e^{-sb} y(b) = 0$?

- Is $\left(e^{at} - e^{-t}\right)/(a+1)$ the only function that satisfies (14.11)? The associated general question is, if

$$\int_0^\infty e^{-st} f(t) \, dt = \int_0^\infty e^{-st} g(t) \, dt,$$

  is $f(t) = g(t)$?

We will return to these points in Section 14.8, but for the time being, in order to concentrate on the broad outline, we are going to make the following **working assumptions** for the functions which we will encounter.[1]

1. $\int_0^\infty e^{-st} y \, dt$ exists.

2. $\lim_{b \to \infty} e^{-sb} y(b) = 0$.

3.
$$\text{If } \int_0^\infty e^{-st} f(t) \, dt = \int_0^\infty e^{-st} g(t) \, dt \text{ then } f(t) = g(t). \tag{14.13}$$

The quantity $\int_0^\infty e^{-st} y(t) \, dt$ appears repeatedly and is given a special name and notation.

Definition 14.1:   **The integral**

$$Y(s) = \mathcal{L}\{y(t)\} = \int_0^\infty e^{-st} y(t) \, dt \tag{14.14}$$

is called the LAPLACE TRANSFORM of $y(t)$.

Comments about Laplace transforms

- Unstated in this definition is a qualification that the integral must exist.

---

[1] We stress that the full mathematical conditions under which these assumptions are justified will be established in Section 14.8.

- Because of the ∞ in the upper limit, the integral is an improper integral and is defined — and evaluated — by

$$\int_0^\infty e^{-st} y(t)\, dt = \lim_{b \to \infty} \int_0^b e^{-st} y(t)\, dt. \tag{14.15}$$

- The two different notations, $Y(s)$ and $\mathcal{L}\{y(t)\}$, may appear confusing at first. The $Y(s)$ notation is the standard function notation, where the only variable that remains after the right-hand side of (14.14) is evaluated will be s. The $\mathcal{L}\{y(t)\}$ notation is used when we want to draw attention to the function $y(t)$ in the integrand. If we change $y(t)$, then the integral will change, which is what the $\mathcal{L}\{y(t)\}$ notation is pointing out.

- The pattern of $Y(s) = \mathcal{L}\{y(t)\}$ is followed for other functions. For example, $F(s) = \mathcal{L}\{f(t)\}$, $G(s) = \mathcal{L}\{g(t)\}$, and so on.

- If we use the assumption that $\lim_{b \to \infty} e^{-sb} y(b) = 0$, then (14.8) can be written as

$$\mathcal{L}\{y'\} = -y(0) + s\mathcal{L}\{y\}. \tag{14.16}$$

This allows us to relate the Laplace transform of y' to the Laplace transform of y.

- We can rewrite the assumption given in (14.13) as

$$\text{if } \mathcal{L}\{f(t)\} = \mathcal{L}\{g(t)\} \text{ then } f(t) = g(t). \tag{14.17}$$

- We can rewrite the equation

$$\int_0^\infty e^{-st} [af(t) + bg(t)]\, dt = a \int_0^\infty e^{-st} f(t)\, dt + b \int_0^\infty e^{-st} g(t)\, dt,$$

which we used to find (14.5) from the equation preceding it, as

$$\mathcal{L}\{af(t) + bg(t)\} = a\mathcal{L}\{f(t)\} + b\mathcal{L}\{g(t)\}. \tag{14.18}$$

A useful special case of this is

$$\mathcal{L}\{af(t)\} = a\mathcal{L}\{f(t)\}.$$

- We can rewrite the result (14.12) in the form

$$\mathcal{L}\{e^{at}\} = \frac{1}{s-a} \tag{14.19}$$

if $s > a$, while

$$\int_0^\infty e^{-st} 0\, dt = 0$$

is

$$\mathcal{L}\{0\} = 0. \tag{14.20}$$

EXAMPLE 14.2

With these changes of notation, we will now repeat the technique used to solve (14.1). We first take the Laplace transform of

$$y' + y = e^{at},$$

obtaining

$$\mathcal{L}\{y' + y\} = \mathcal{L}\{e^{at}\}.$$

Using (14.18) allows us to rewrite this as

$$\mathcal{L}\{y'\} + \mathcal{L}\{y\} = \mathcal{L}\{e^{at}\}.$$

We substitute for $\mathcal{L}\{y'\}$ from (14.16) and $\mathcal{L}\{e^{at}\}$ from (14.19) to find

$$-y(0) + s\mathcal{L}\{y\} + \mathcal{L}\{y\} = \frac{1}{s-a}.$$

Using the initial condition $y(0) = 0$ and solving for $\mathcal{L}\{y\}$ leads to

$$\mathcal{L}\{y\} = \frac{1}{(s+1)(s-a)},$$

which, for the case $a \neq -1$, can be rewritten as

$$\mathcal{L}\{y\} = \frac{1}{a+1}\left(\frac{1}{s-a} - \frac{1}{s+1}\right). \tag{14.21}$$

From (14.18) and (14.19) we have

$$\mathcal{L}\{y\} = \mathcal{L}\left\{\frac{1}{a+1}\left(e^{at} - e^{-t}\right)\right\}, \tag{14.22}$$

which finally gives $y(t) = \left(e^{at} - e^{-t}\right)/(a+1)$ from (14.17). □

If we look at the preceding example for the case when $a = -1$, then the counterpart of (14.21) is

$$\mathcal{L}\{y\} = \frac{1}{(s+1)^2}. \tag{14.23}$$

Here we have a problem: We don't have an equation similar to (14.19) for the function $1/(s+1)^2$.

The crucial step for the success of this analysis in the case $a \neq -1$ is going from (14.21) to (14.22). We were lucky to have the result in (14.19) at hand. The reason we are unsuccessful in the case $a = -1$ is the fact that we were not lucky enough to have at hand a result similar to (14.19) for the function $1/(s+1)^2$.

However, because we know that the solution of this initial value problem for $a = -1$ is te^{-t}, this suggests we compute the integral

$$\int_0^\infty e^{-st} t e^{-t}\, dt = \int_0^\infty t e^{-(s+1)t}\, dt = \lim_{b\to\infty} \int_0^b t e^{-(s+1)t}\, dt,$$

which, using integration by parts, gives

$$\int_0^\infty e^{-st} t e^{-t}\, dt = \lim_{b\to\infty} \left[-\frac{1}{s+1} t e^{-(s+1)t} \Big|_0^b \right] + \int_0^\infty \frac{1}{s+1} e^{-(s+1)t}\, dt.$$

The first term on the right-hand of this equation is 0 for $s+1 > 0$, and, with (14.6), the second term evaluates as $1/(s+1)^2$, so this gives us

$$\mathcal{L}\left\{ t e^{-t} \right\} = \frac{1}{(s+1)^2},$$

as we expected.

In order to be so lucky in the future, we need to construct a table of Laplace transforms $\mathcal{L}\{y\}$ for many different functions $y(t)$. In the next example we compute the Laplace transform of $\sin t$.

EXAMPLE 14.3: The Laplace Transform of $\sin t$

To find the Laplace transform of $\sin t$, we need to calculate

$$\mathcal{L}\{\sin t\} = \int_0^\infty e^{-st} \sin t\, dt.$$

We could integrate this by parts twice to evaluate this integral (see Exercise 1 on page 694). Instead, we will use a different approach that foreshadows the techniques used in the next section — namely, use an existing result to generate new Laplace transforms. In this case we take advantage of (14.16) — that is,

$$\mathcal{L}\left\{ y' \right\} = -y(0) + s\mathcal{L}\{y\}. \tag{14.24}$$

To find $\mathcal{L}\{\sin t\}$ from this equation we could try $y'(t) = \sin t$, in which case we should choose $y(t) = -\cos t$. In this way we find

$$\mathcal{L}\{\sin t\} = -(-1) + s\mathcal{L}\{-\cos t\} = 1 - s\mathcal{L}\{\cos t\}, \tag{14.25}$$

which relates $\mathcal{L}\{\sin t\}$ to $\mathcal{L}\{\cos t\}$ but does not find $\mathcal{L}\{\sin t\}$ explicitly. However, perhaps we can use (14.24) again to find $\mathcal{L}\{\cos t\}$. If we try $y'(t) = \cos t$, so that $y(t) = \sin t$, in (14.24), we now find

$$\mathcal{L}\{\cos t\} = s\mathcal{L}\{\sin t\}. \tag{14.26}$$

Substituting (14.26) into (14.25), we see that

$$\mathcal{L}\{\sin t\} = 1 - s^2 \mathcal{L}\{\sin t\}.$$

We solve this equation for $\mathcal{L}\{\sin t\}$, finally finding the Laplace transform of $\sin t$, namely,

$$\mathcal{L}\{\sin t\} = \frac{1}{s^2+1}.$$

We notice that from (14.26), we can also write down the Laplace transform of $\cos t$ as

$$\mathcal{L}\{\cos t\} = \frac{s}{s^2+1}.$$

$\square$

The Laplace transforms of a few other familiar functions are developed in the exercises, and the results are listed in Table 14.1. In Table 14.2 we also summarize the elementary properties of Laplace transforms that we have found thus far.

TABLE 14.1 Laplace transforms of several functions

$f(t)$	$\mathcal{L}\{f(t)\} = F(s)$
1	$1/s$
t	$1/s^2$
e^{at}	$1/(s-a)$
te^{at}	$1/(s-a)^2$
$\sin t$	$1/(s^2+1)$
$\cos t$	$s/(s^2+1)$
$\sinh t$	$1/(s^2-1)$
$\cosh t$	$s/(s^2-1)$

TABLE 14.2 Simple properties of Laplace transforms

If $\mathcal{L}\{f(t)\} = F(s)$, then

$$\mathcal{L}\{af(t) + bg(t)\} = a\mathcal{L}\{f(t)\} + b\mathcal{L}\{g(t)\}$$
$$\mathcal{L}\{af(t)\} = a\mathcal{L}\{f(t)\}$$
$$\mathcal{L}\{f'(t)\} = s\mathcal{L}\{f(t)\} - f(0)$$

Now that we have a table of Laplace transforms of several functions, let us solve another initial value problem.

EXAMPLE 14.4

Solve the initial value problem

$$y' + 2y = 10\sin t,$$
$$y(0) = 1.$$

Taking the Laplace transform of both sides of this differential equation and using the property given in (14.18), we have

$$\mathcal{L}\{y' + 2y\} = \mathcal{L}\{10\sin t\},$$

or

$$\mathcal{L}\{y'\} + 2\mathcal{L}\{y\} = 10\mathcal{L}\{\sin t\}. \tag{14.27}$$

We now use results from Table 14.1, the fact that $\mathcal{L}\{y'\} = s\mathcal{L}\{y\} - y(0)$, and the initial condition $y(0) = 1$, to write (14.27) as

$$s\mathcal{L}\{y\} - 1 + 2\mathcal{L}\{y\} = \frac{10}{s^2+1}.$$

Solving for $\mathcal{L}\{y\}$ gives

$$\mathcal{L}\{y\} = \frac{1}{s+2} + \frac{10}{(s+2)(s^2+1)}. \tag{14.28}$$

In order to rearrange (14.28) to exploit Tables 14.1 and 14.2, we use partial fractions[2] as follows:

$$\frac{10}{(s+2)(s^2+1)} = \frac{A}{s+2} + \frac{Bs+C}{s^2+1}$$

or

$$\frac{10}{(s+2)(s^2+1)} = \frac{A(s^2+1)+(Bs+C)(s+2)}{(s+2)(s^2+1)},$$

where A, B, and C are constants to be determined. Because the denominators on the two sides of this equation are the same, the numerators must also be equal, giving

$$10 = A(s^2+1)+(Bs+C)(s+2) = (A+B)s^2 + (2B+C)s + (A+2C).$$

This equation is an identity in s, so we may equate the coefficients of like powers of s to obtain

$$10 = A+2C, \qquad 0 = 2B+C, \qquad \text{and} \qquad 0 = A+B.$$

Solving these equations gives $A=2$, $B=-2$, and $C=4$ so we have the equivalent expression

$$\frac{10}{(s+2)(s^2+1)} = \frac{2}{s+2} + \frac{-2s+4}{s^2+1}.$$

Substituting this expression into (14.28), we find

$$\mathcal{L}\{y\} = \frac{3}{s+2} - \frac{2s}{s^2+1} + \frac{4}{s^2+1}.$$

Using Tables 14.1 and 14.2 allows us to give the solution of our initial value problem as

$$y(t) = 3e^{-2t} - 2\cos t + 4\sin t.$$

□

We now formalize the procedure we have just used.

---

[2]There is a discussion of partial fractions in the appendix.

How to Solve First Order Linear Differential Equations Using Laplace Transforms

Purpose: To find $y(t)$ that satisfies the initial value problem

$$a_1 \frac{dy}{dt} + a_0 y = f(t), \qquad y(0) = y_0$$

by using Laplace transforms. Here a_1, a_0, and y_0 are all constants.

Technique:

1. Put the linear differential equation in standard form by dividing by a_1, then multiply both sides of the result by e^{-st} and integrate with respect to t from 0 to ∞.

2. Use results from Tables 14.1 and 14.2 to change all of the integrated expressions to those involving the transformed dependent variable

$$Y(s) = \mathcal{L}\{y(t)\} = \int_0^\infty e^{-st} y(t)\, dt$$

and known functions of s.

3. Solve for the transformed dependent variable, $Y(s)$, and rearrange the result so all the terms have the form of some expression on the right-hand side of Table 14.1.

4. Use Tables 14.1 and 14.2 to write down your solution for $y(t)$.

Comments about using Laplace transforms to solve initial value problems

- The process of multiplying both sides of a differential equation by e^{-st} and integrating on t from 0 to ∞ is called taking the Laplace transform of the equation.

- The broad outline for this technique is

    We transform the original differential equation to its Laplace transform counterpart.

    We use algebraic manipulations to solve for the unknown Laplace transform.

    We then transform from the Laplace transform variables back to the original differential equation variables.

- If no initial condition is given, you may leave your answer in terms of the arbitrary constant $y(0)$.

- Partial fractions are sometimes useful in step 3 of the technique.

- Extensive lists of Laplace transforms are found in many books. [3]

- If the initial condition is not at $t = 0$ but at $t = t_0$, we first make the change of variable $\tau = t - t_0$, and then we use Laplace transforms on the differential equation in τ.

EXERCISES

1. Use integration by parts to evaluate the $\mathcal{L}\{\sin t\}$ and $\mathcal{L}\{\cos t\}$.

2. Complete Table 14.1 by evaluating the Laplace transforms of the following functions.

---

[3] For example, "Handbook of Mathematical Functions" by M. Abramowitz and I. A. Stegun, Dover 1964, page 1020.

   (a) 1

   (b) t

   (c) te^{at}

   (d) $\sinh t = \left(e^t - e^{-t}\right)/2$

   (e) $\cosh t = \left(e^t + e^{-t}\right)/2$

3. Use (14.24) to derive

   (a) $\mathcal{L}\{\sin at\}$.

   (b) $\mathcal{L}\{\cos at\}$.

   (c) $\mathcal{L}\{\sinh at\}$.

   (d) $\mathcal{L}\{\cosh at\}$.

4. Use Tables 14.1 and 14.2 to evaluate the Laplace transforms of the following functions.

   (a) $\sin 2t + \cos 2t$

   (b) $2e^t + 3e^{-t}$

   (c) $4te^{at} - t$

   (d) $e^{-t} + 2\sinh t$ (Are you surprised at this result?)

5. Use Table 14.1, Table 14.2, and (14.17) to find functions whose Laplace transforms are as follows.

   (a) $1/(s-2) - 1/(s+3)$

   (b) $6/[s(s+3)]$

   (c) $1/[(s+2)(s+3)]$

   (d) $5/\left[(s+2)\left(s^2+1\right)\right]$

   (e) $6/\left[(s-2)\left(s^2-1\right)\right]$

6. Solve the following initial value problems.

   (a) $y' + 3y = 5e^{2t}$, $y(0) = 0$

   (b) $y' + 3y = 6 - e^{-2t}$, $y(0) = 1$

   (c) $y' + 2y = 5\sin t$, $y(0) = 0$

   (d) $y' - 2y = 6\sinh t$, $y(0) = 2$

   (e) $y' + 3y = e^{2t}$, $y(2) = 0$

   (f) $y' + 3y = t - e^{-2t}$, $y(-2) = 1$

7. In deriving $\mathcal{L}\left\{e^{at}\right\} = 1/(s-a)$, we used $\lim_{b\to\infty}[e^{(a-s)b}/(a-s)] = 0$ for $s > a$. If a and s are complex numbers it is also true that $\lim_{b\to\infty}[e^{(a-s)b}/(a-s)] = 0$ provided that the real part of $s - a$, namely, $Re[s-a]$, is positive — that is, $Re[s-a] > 0$. Thus, we have that $\mathcal{L}\left\{e^{it}\right\} = 1/(s-i)$ for $Re[s-i] = Re[s] > 0$ and $\mathcal{L}\left\{e^{-it}\right\} = 1/(s+i)$ for $Re[s+i] = Re[s] > 0$. Use these results to derive the formulas for $\mathcal{L}\{\sin t\}$ and $\mathcal{L}\{\cos t\}$.

14.2   Constructing New Laplace Transforms from Old

In the previous section we found Laplace transforms of several functions. We also solved two first order linear differential equations with constant coefficients using these transforms, and our success depended in part on our ability to recast the original differential equation in terms of Laplace transforms of known functions. However, at the moment we have a very modest list of Laplace transforms — Table 14.1. Thus, in this section, we will create a larger list of functions than is given in this table.

We could always use the definition of the Laplace transform of $f(t)$, namely,

$$\mathcal{L}\{f(t)\} = \int_0^\infty e^{-st} f(t)\, dt,$$

to calculate a Laplace transform using integration techniques. However, we will try to avoid this lengthy process by using Tables 14.1 and 14.2 as our starting points to generate Laplace transforms of more functions. This means that the methods we discover will be largely disjointed but will stem from relevant observations and questions. We must remember — as we pointed out in the previous section — that in all cases we will assume that we are dealing only with functions whose Laplace transforms exist.

If we look at the entries under $f(t)$ in Table 14.1, perhaps the most glaring omission is that we have no Laplace transform for t^2, or any higher power of t. We could evaluate this Laplace transform from its definition, namely,

$$\mathcal{L}\{t^2\} = \int_0^\infty e^{-st} t^2\, dt,$$

but this will involve integrating by parts twice — and n times in the case of t^n. Instead we follow our previous suggestion and try to use old results to generate new results.

EXAMPLE 14.5: The Laplace Transform of t^n

We know that

$$\mathcal{L}\{y'(t)\} = s\mathcal{L}\{y(t)\} - y(0),$$

so this could lead to new results if we choose $y(t)$ in different ways.

If we try $y(t) = t^2$ and note that $y(0) = 0$ and $y'(t) = 2t$, we see that

$$\mathcal{L}\{2t\} = s\mathcal{L}\{t^2\}.$$

Thus, solving for $\mathcal{L}\{t^2\}$ and using Tables 14.1 and 14.2 gives

$$\mathcal{L}\{t^2\} = \frac{1}{s}\mathcal{L}\{2t\} = \frac{2}{s}\mathcal{L}\{t\} = \frac{2}{s^3}.$$

Notice that we have found a new Laplace transform from an old one. We will add this to our extended list of Laplace transforms.

If we try the same technique on $y(t) = t^n$ for a positive integer n, we find

$$\mathcal{L}\{nt^{n-1}\} = s\mathcal{L}\{t^n\}$$

or

$$\mathcal{L}\{t^n\} = \frac{n}{s}\mathcal{L}\{t^{n-1}\}. \tag{14.29}$$

This gives $\mathcal{L}\{t^n\}$ in terms of $\mathcal{L}\{t^{n-1}\}$. Thus, for example, if we put $n = 3$ we have

$$\mathcal{L}\{t^3\} = \frac{3}{s}\mathcal{L}\{t^2\} = \frac{3}{s}\frac{2}{s^3} = \frac{3 \cdot 2}{s^4}.$$

Continuing in this way, or by recognizing that (14.29) is a first order recurrence relation — see Section 9.6 — we find

$$\mathcal{L}\left\{t^n\right\} = \frac{n!}{s^{n+1}}. \tag{14.30}$$

$\square$

We can also use the property $\mathcal{L}\left\{y'(t)\right\} = s\mathcal{L}\left\{y(t)\right\} - y(0)$ on the function $y(t) = f'(t)$ to find

$$\mathcal{L}\left\{f''(t)\right\} = s\mathcal{L}\left\{f'(t)\right\} - f'(0).$$

If we substitute for $\mathcal{L}\left\{f'(t)\right\}$, where

$$\mathcal{L}\left\{f'(t)\right\} = s\mathcal{L}\{f(t)\} - f(0),$$

into this expression for $\mathcal{L}\left\{f''(t)\right\}$, we find

$$\mathcal{L}\left\{f''(t)\right\} = s\left[s\mathcal{L}\{f(t)\} - f(0)\right] - f'(0)$$

or

$$\mathcal{L}\left\{f''(t)\right\} = s^2\mathcal{L}\{f(t)\} - \left[sf(0) + f'(0)\right].$$

We can continue this process as often as we like to find the Laplace transform of the nth derivative of $f(t)$ — namely, $f^{(n)}(t)$.

Theorem 14.1:   *If the Laplace transform of $f(t)$ and its first $n - 1$ derivatives exist, then*

$$\mathcal{L}\left\{f^{(n)}(t)\right\} = s^n\mathcal{L}\{f(t)\} - \left[s^{n-1}f(0) + s^{n-2}f'(0) + \cdots + sf^{(n-2)}(0) + f^{(n-1)}(0)\right]. \tag{14.31}$$

Comments about the Laplace transform of $f^{(n)}(t)$

- We have essentially used this theorem, with $n = 2$, in Example 14.3 on page 691 when we calculated the Laplace transform of $\sin t$.

- This property is crucial for solving higher order linear differential equations using Laplace transforms. It will be used repeatedly in Section 14.6.

EXAMPLE 14.6: The Laplace Transform of $t \sin t$

We apply (14.31), with $n = 2$, to the function $f(t) = t \sin t$. We will need $f'(t)$ and $f''(t)$ which are $f'(t) = \sin t + t \cos t$ and $f''(t) = 2\cos t - t \sin t$, respectively. Thus, from (14.31), and because in this case $f(0) = f'(0) = 0$, we have

$$\mathcal{L}\{2\cos t - t\sin t\} = s^2\mathcal{L}\{t\sin t\}.$$

Solving for $\mathcal{L}\{t\sin t\}$ gives

$$\mathcal{L}\{t\sin t\} = \frac{1}{s^2 + 1}\mathcal{L}\{2\cos t\}.$$

From our previous results we know that

$$\mathcal{L}\{2\cos t\} = 2\mathcal{L}\{\cos t\} = \frac{2s}{s^2 + 1}.$$

This leads to

$$\mathcal{L}\{t\sin t\} = \frac{2s}{(s^2 + 1)^2}.$$

To appreciate the power of this technique, you should try to evaluate $\mathcal{L}\{t\sin t)\}$ from its definition — namely,

$$\mathcal{L}\{t\sin t\} = \int_0^\infty e^{-st} t\sin t\, dt.$$

Good luck! $\square$

We now turn to a different question. Our list of Laplace transforms in Table 14.1 includes that of $\sin t$, but does not include that of $\sin at$. Is there an easy way for us to use the Laplace transform of $\sin t$ to find the Laplace transform of $\sin at$?

EXAMPLE 14.7: The Laplace Transform of $\sin at$

To explore this idea we write the expression for $\mathcal{L}\{\sin at\}$, namely

$$\mathcal{L}\{\sin at\} = \int_0^\infty e^{-st}\sin at\, dt. \tag{14.32}$$

Again, we could evaluate this integral directly, but that involves integrating by parts twice. Instead we will try to use our old results. Because we know the Laplace transform of $\sin t$, namely,

$$F(s) = \mathcal{L}\{\sin t\} = \int_0^\infty e^{-st}\sin t\, dt,$$

we make the change of variable $at = z$ in (14.32) to find

$$\mathcal{L}\{\sin at\} = \int_0^\infty e^{-sz/a}\sin z\, dz/a = \frac{1}{a}\int_0^\infty e^{-(s/a)z}\sin z\, dz. \tag{14.33}$$

If we compare the integral in (14.33) with the integral giving the Laplace transform of $\sin t$, we see that

$$\mathcal{L}\{\sin at\} = \frac{1}{a}F\left(\frac{s}{a}\right).$$

Because from Table 14.1 we have

$$F(s) = \mathcal{L}\{\sin t\} = \frac{1}{s^2 + 1},$$

this means that

$$\mathcal{L}\{\sin at\} = \frac{1}{a}F\left(\frac{s}{a}\right) = \frac{1}{a}\frac{1}{(s/a)^2 + 1} = \frac{a}{s^2 + a^2}. \tag{14.34}$$

This shows that we are able to find the Laplace transform of $\sin at$ from the Laplace transform of $\sin t$. □

This same procedure works for any two functions $f(t)$ and $f(at)$ that have Laplace transforms as stated in the following result.

Theorem 14.2: *If the Laplace transform of $f(t)$ exists — that is, $\mathcal{L}\{f(t)\} = F(s)$ — then*

$$\mathcal{L}\{f(at)\} = \frac{1}{a}F\left(\frac{s}{a}\right)$$

if $a > 0$.

Comments about Theorem 14.2

- The primary use of this theorem is to supplement our list of Laplace transforms. For example, because we know that $\mathcal{L}\{\cos t\} = s/(s^2 + 1)$, we can use this theorem to conclude that

$$\mathcal{L}\{\cos at\} = \frac{1}{a}\frac{(s/a)}{(s/a)^2 + 1} = \frac{s}{s^2 + a^2}.$$

- We restricted the value of a to be positive to guarantee the existence of integrals such as $\int_0^\infty e^{-sz/a}\sin z\,dz$. See Exercise 3 on page 702.
- Because the details in the proof of this theorem closely follow the procedure used in Example 14.7, they are left to Exercise 3 on page 702.

We now turn to an unrelated observation. Notice in Table 14.1 that we have $\mathcal{L}\{t\} = 1/s^2$, while $\mathcal{L}\{te^t\} = 1/(s-1)^2$. Because the forms of these two transforms are similar, we might be tempted to ask if there is some general principle lurking here.

EXAMPLE 14.8: The Laplace Transform of te^t

To check this out, we write down the Laplace transform of te^t, namely,

$$\mathcal{L}\{te^t\} = \int_0^\infty e^{-st}te^t\,dt = \int_0^\infty e^{-(s-1)t}t\,dt, \qquad (14.35)$$

and compare it with the Laplace transform of t, namely,

$$\mathcal{L}\{t\} = \int_0^\infty e^{-st}t\,dt = \frac{1}{s^2}. \qquad (14.36)$$

We see that the integrand in (14.35) differs from the one in (14.36) only in that there is a factor of $s - 1$ in the exponent of e rather than just s. This means that we can evaluate (14.35) using (14.36) by replacing s by $s - 1$ throughout, giving

$$\mathcal{L}\{te^t\} = \int_0^\infty e^{-(s-1)t}t\,dt = \frac{1}{(s-1)^2}.$$

□

This result may be generalized to the following theorem, with the proof left to Exercise 4 on page 702.

Theorem 14.3:   *If the Laplace transform of $f(t)$ exits for $s > \alpha$, so that $\mathcal{L}\{f(t)\} = F(s)$ for $s > \alpha$, then*

$$\mathcal{L}\left\{e^{at} f(t)\right\} = F(s - a)$$

for $s > \alpha + a$.

Comments about Theorem 14.3

- The primary use of this theorem is to supplement our list of Laplace transforms. For example, because we know from (14.34) that $\mathcal{L}\{\sin bt\} = b/(s^2 + b^2)$, we can use this theorem to conclude that

$$\mathcal{L}\left\{e^{at} \sin bt\right\} = \frac{b}{(s - a)^2 + b^2}.$$

In a similar way

$$\mathcal{L}\left\{e^{at} \cos bt\right\} = \frac{s - a}{(s - a)^2 + b^2}.$$

- The property of Laplace transforms given in this theorem is often called the SHIFTING PROPERTY. (Explain why this terminology is appropriate.)

We again turn to a new question. A previous property dealt with what happens when we differentiate $y(t)$ with respect to t. We could ask whether there are any useful properties obtained by differentiating the Laplace transform of $f(t)$, namely,

$$F(s) = \mathcal{L}\{f(t)\} = \int_0^\infty e^{-st} f(t)\, dt,$$

with respect to s. If we differentiate both sides with respect to s (assuming it is legal to interchange the order of differentiation and integration) we obtain

$$\frac{dF(s)}{ds} = \frac{d}{ds}\left[\int_0^\infty e^{-st} f(t)\, dt\right] = \int_0^\infty \frac{d}{ds}\left[e^{-st} f(t)\right] dt = \int_0^\infty -t e^{-st} f(t)\, dt,$$

or

$$\frac{dF(s)}{ds} = -\mathcal{L}\{tf(t)\}.$$

Differentiating once more yields

$$\frac{d^2 F(s)}{ds^2} = \int_0^\infty (-t)^2\, e^{-st} f(t)\, dt = (-1)^2 \mathcal{L}\left\{t^2 f(t)\right\}.$$

This process may be repeated as many times as needed, until we find the following result.

Theorem 14.4:   *If $\mathcal{L}\{f(t)\} = F(s)$, then for any integer $n \geq 0$, we have*

$$\mathcal{L}\left\{t^n f(t)\right\} = (-1)^n \frac{d^n F(s)}{ds^n}.$$

Comments about Theorem 14.4

- The primary use of this theorem is to supplement our list of Laplace transforms. For example, because we know from (14.34) that $\mathcal{L}\{\sin at\} = a/(s^2 + a^2)$, we can use this theorem to conclude that

$$\mathcal{L}\{t \sin at\} = -\frac{d}{ds}\left(\frac{a}{s^2 + a^2}\right) = \frac{2as}{\left(s^2 + a^2\right)^2}$$

and

$$\mathcal{L}\left\{t^2 \sin at\right\} = \frac{d^2}{ds^2}\left(\frac{a}{s^2 + a^2}\right) = \frac{2a\left(a^2 - 3s^2\right)}{\left(s^2 + a^2\right)^3}.$$

In the same way we can find

$$\mathcal{L}\{t \cos at\} = -\frac{d}{ds}\left(\frac{s}{s^2 + a^2}\right) = \frac{s^2 - a^2}{\left(s^2 + a^2\right)^2}.$$

Using this theorem to find these Laplace transforms is considerably less daunting than the prospect of evaluating the integrals $\int_0^\infty e^{-st}t \sin at\, dt$, $\int_0^\infty e^{-st}t^2 \sin at\, dt$, and $\int_0^\infty e^{-st}t \cos at\, dt$ by standard integration techniques.

- We can use this theorem as yet another way of calculating $\mathcal{L}\left\{te^{at}\right\}$. [Use it to show that $\mathcal{L}\left\{te^{at}\right\} = 1/(s - a)^2$.]

We collect some of these Laplace transforms and properties together in Tables 14.3 and 14.4.

TABLE 14.3   Laplace transforms of several functions

$f(t)$	$\mathcal{L}\{f(t)\} = F(s)$
a	a/s
t	$1/s^2$
t^n	$n!/s^{n+1}$
e^{at}	$1/(s-a)$
te^{at}	$1/(s-a)^2$
$\sin at$	$a/(s^2 + a^2)$
$\cos at$	$s/(s^2 + a^2)$
$\sinh at$	$a/(s^2 - a^2)$
$\cosh at$	$s/(s^2 - a^2)$
$t \sin at$	$2as/(s^2 + a^2)^2$
$t \cos at$	$(s^2 - a^2)/(s^2 + a^2)^2$

TABLE 14.4   Properties of Laplace transforms

If $\mathcal{L}\{f(t)\} = F(s)$, then
$\mathcal{L}\{af(t) + bg(t)\} = a\mathcal{L}\{f(t)\} + b\mathcal{L}\{g(t)\}$
$\mathcal{L}\{af(t)\} = a\mathcal{L}\{f(t)\}$
$\mathcal{L}\{f'(t)\} = s\mathcal{L}\{f(t)\} - f(0)$
$\mathcal{L}\{f''(t)\} = s^2\mathcal{L}\{f(t)\} - [sf(0) + f'(0)]$
$\mathcal{L}\{f(at)\} = (1/a)F(s/a)$
$\mathcal{L}\{e^{at}f(t)\} = F(s - a)$
$\mathcal{L}\{t^n f(t)\} = (-1)^n F^{(n)}(s)$

EXERCISES

1. Evaluate $\mathcal{L}\{t\cos at\}$ by

    (a) the method of Example 14.6.

    (b) integration by parts.

2. Use Theorem 14.1 to evaluate

    (a) $\mathcal{L}\{t\sinh t\}$.

    (b) $\mathcal{L}\{t\cosh t\}$.

3. Prove Theorem 14.2 by following the procedure used in Example 14.7. If $a < 0$, what complications are introduced in trying to prove this theorem?

4. Prove Theorem 14.3 by using the definition of the Laplace transform.

5. Use Theorem 14.3 to evaluate

    (a) $\mathcal{L}\{e^{at}\sinh at\}$.

    (b) $\mathcal{L}\{e^{at}\cosh at\}$.

    (c) $\mathcal{L}\{t^2 e^{at}\}$.

6. Use Theorem 14.4 to evaluate

    (a) $\mathcal{L}\{t^2\cos at\}$.

    (b) $\mathcal{L}\{t\sinh at\}$.

    (c) $\mathcal{L}\{t\cosh at\}$.

    (d) $\mathcal{L}\{t^2 e^{at}\}$.

7. Another property of Laplace transforms is as follows: If $\mathcal{L}\{f(t)\} = F(s)$ for $s > \alpha$, and $f(t)/t$ is bounded for $t > 0$, then

$$\mathcal{L}\left\{\frac{1}{t}f(t)\right\} = \int_s^\infty F(z)\,dz.$$

    (a) Verify this result by writing $F(z) = \int_0^\infty e^{-zt}f(t)\,dt$ and integrating both sides from $z = s$ to $z = \infty$. Interchange the orders of the resulting integration.

    (b) This theorem is used to calculate Laplace transforms for which straightforward integration techniques may fail. Use this theorem to evaluate $\mathcal{L}\{(\sin t)/t\}$ and $\mathcal{L}\{(\sinh t)/t\}$.

    (c) Use the definition of the Laplace transform to observe the difficulty of evaluating $\mathcal{L}\{(\sin t)/t\}$ by integration techniques.

14.3   The Inverse Laplace Transform and the Convolution Theorem

In Section 14.1 we outlined a procedure for solving first order linear differential equations with constant coefficients using Laplace transforms. Our success at finding explicit solutions of such initial value problems depends on two things: being able to transform the original differential equation into an algebraic equation involving the Laplace transform of our explicit solution, and then being able to return to familiar functions using our tables. In the previous section we developed techniques that allow us to go from the differential equation to the transformed equation. In this section we concentrate on the inverse problem: With the Laplace transform $F(s) = \mathcal{L}\{f(t)\}$, what is $f(t)$?

The key to making this process work is Table 14.3. We should notice the parallel between using this table and using a typical table of integrals. With a table of integrals the integral we want to evaluate may not appear explicitly. We then try to use the properties of integrals to convert the integral we want into integrals that are already in the table. Exactly the same reasoning applies here, so the correct way to view Table 14.3 in order to answer our question — with a Laplace transform $F(s) = \mathcal{L}\{f(t)\}$, what is $f(t)$? — is from right to left. From this viewpoint, there are many functions missing from the right-hand side of Table 14.3. Do you notice anything unusual about the functions that are there? All of them are rational functions — that is, one polynomial in s divided by another. In fact, they are proper rational functions, where the degree of the numerator is always less than that of the denominator.

We now look at some examples in which we are given the Laplace transform of $f(t)$, namely, $F(s) = \mathcal{L}\{f(t)\}$, and want to find $f(t)$. This process is used so often that we give it a special name.

Definition 14.2:   If $\mathcal{L}\{f(t)\} = F(s)$ then the INVERSE LAPLACE TRANSFORM of $F(s)$ is $f(t)$ and is denoted by

$$\mathcal{L}^{-1}\{F(s)\} = f(t).$$

Comments about inverse Laplace transforms

- In Section 14.8 we will discuss the existence and uniqueness of inverse Laplace transforms. Until then, we will assume that any particular inverse Laplace transform exists and is unique.

- The process of solving a differential equation by Laplace transforms is to take the Laplace transform of the differential equation, manipulate the Laplace transform of the unknown solution, and then use the inverse Laplace transform to return to the original variables.

- From $\mathcal{L}\{af(t) + bg(t)\} = a\mathcal{L}\{f(t)\} + b\mathcal{L}\{g(t)\} = aF(s) + bG(s)$ we immediately have

$$\mathcal{L}^{-1}\{aF(s) + bG(s)\} = a\mathcal{L}^{-1}\{F(s)\} + b\mathcal{L}^{-1}\{G(s)\}. \qquad \textbf{(14.37)}$$

- Table 14.3 can be rewritten in terms of inverse Laplace transforms as Table 14.5.

TABLE 14.5 The inverse Laplace transforms of several functions

$F(s)$	$\mathcal{L}^{-1}\{F(s)\} = f(t)$
$1/s$	1
$1/s^2$	t
$n!/s^{n+1}$	t^n
$1/(s-a)$	e^{at}
$1/(s-a)^2$	te^{at}
$a/(s^2+a^2)$	$\sin at$
$s/(s^2+a^2)$	$\cos at$
$a/(s^2-a^2)$	$\sinh at$
$s/(s^2-a^2)$	$\cosh at$
$2as/(s^2+a^2)^2$	$t\sin at$
$(s^2-a^2)/(s^2+a^2)^2$	$t\cos at$

- Theorem 14.2 from the previous section can be useful when we are looking for inverse Laplace transforms. We restate it using the current terminology:

$$\text{If } \mathcal{L}^{-1}\{F(s)\} = f(t)\,, \text{ then } \mathcal{L}^{-1}\{F(s/a)\} = af(at). \qquad \textbf{(14.38)}$$

So this will allow us to compute $\mathcal{L}^{-1}\{F(s/a)\}$ if we know $\mathcal{L}^{-1}\{F(s)\}$.

- Theorem 14.3 from the previous section can also be useful when looking for inverse Laplace transforms. We restate it using the current terminology:

$$\text{If } \mathcal{L}^{-1}\{F(s)\} = f(t)\,, \text{ then } \mathcal{L}^{-1}\{F(s-a)\} = e^{at}f(t). \qquad \textbf{(14.39)}$$

This will allow us to compute $\mathcal{L}^{-1}\{F(s-a)\}$ if we know $\mathcal{L}^{-1}\{F(s)\}$.

- In many cases the quantity $F(s)$ will be a proper rational function. Consequently, in this case we could always rewrite $F(s)$ using partial fractions. The typical components of a partial fraction decomposition are $A/(s-a)^n$, $B/(s^2+a^2)^n$, and $Cs/(s^2+a^2)^n$. Some of these are already in our list of inverse Laplace transforms — Table 14.5. However, we should be cognizant of the fact that blindly applying partial fractions may make the problem more difficult than it needs to be. For example, it would be silly to decompose $1/(s^2-a^2)$ using partial fractions, because Table 14.5 already has its inverse Laplace transform, $\mathcal{L}^{-1}\{1/(s^2-a^2)\} = \sinh at$.

We will look at a few examples in which we compute inverse Laplace transforms.

EXAMPLE 14.9: The Inverse Laplace Transform of $1/[(s-a)(s-b)]$

If

$$F(s) = \mathcal{L}\{f(t)\} = \frac{1}{(s-a)(s-b)},$$

where $a \neq b$, what is $f(t)$? That is, what is $\mathcal{L}^{-1}\{F(s)\}$?

We use partial fractions and write

$$\frac{1}{(s-a)(s-b)} = \frac{1}{a-b}\left(\frac{1}{s-a} - \frac{1}{s-b}\right),$$

which from (14.37) yields

$$\mathcal{L}^{-1}\left\{\frac{1}{(s-a)(s-b)}\right\} = \frac{1}{a-b}\left(\mathcal{L}^{-1}\left\{\frac{1}{s-a}\right\} - \mathcal{L}^{-1}\left\{\frac{1}{s-b}\right\}\right).$$

The inverse Laplace transform of $1/(s-a)$ occurs in Table 14.5, so we have

$$f(t) = \mathcal{L}^{-1}\left\{\frac{1}{(s-a)(s-b)}\right\} = \frac{e^{at} - e^{bt}}{a-b}. \tag{14.40}$$

□

EXAMPLE 14.10: The Inverse Laplace Transform of $s/[(s-a)(s-b)]$

If

$$F(s) = \mathcal{L}\{f(t)\} = \frac{s}{(s-a)(s-b)},$$

where $a \neq b$, what is $f(t) = \mathcal{L}^{-1}\{F(s)\}$?

We could decompose $s/[(s-a)(s-b)]$ using partial fractions, and there is nothing wrong with that. However, we can frequently use known inverse Laplace transforms to calculate as yet unknown ones. In this example we can add and subtract a from the numerator to find

$$\frac{s}{(s-a)(s-b)} = \frac{(s-a)+a}{(s-a)(s-b)} = \frac{1}{s-b} + \frac{a}{(s-a)(s-b)}.$$

Thus, taking the inverse Laplace transform yields

$$\mathcal{L}^{-1}\left\{\frac{s}{(s-a)(s-b)}\right\} = \mathcal{L}^{-1}\left\{\frac{1}{s-b}\right\} + a\mathcal{L}^{-1}\left\{\frac{1}{(s-a)(s-b)}\right\},$$

or

$$\mathcal{L}^{-1}\left\{\frac{s}{(s-a)(s-b)}\right\} = e^{bt} + a\frac{e^{at} - e^{bt}}{a-b} = \frac{ae^{at} - be^{bt}}{a-b}. \tag{14.41}$$

□

EXAMPLE 14.11: The Inverse Laplace Transform of $1/\left(s^2 + 1\right)^2$

If

$$F(s) = \mathcal{L}\{f(t)\} = \frac{1}{\left(s^2 + 1\right)^2},$$

what is $f(t) = \mathcal{L}^{-1}\{F(s)\}$?

There is no point in trying to decompose $1/\left(s^2 + 1\right)^2$ by partial fractions — it is already decomposed. (Why?) Thus, the only method we have is to try to rewrite $1/\left(s^2 + 1\right)^2$ in terms of expressions that appear on the left-hand side of Table 14.5. After much thought and experimenting we find

$$\frac{1}{\left(s^2 + 1\right)^2} = \frac{\left(s^2 + 1\right) - s^2}{\left(s^2 + 1\right)^2} = \frac{1}{s^2 + 1} - \frac{s^2}{\left(s^2 + 1\right)^2}.$$

The first term on the right-hand side of this equation is listed in Table 14.5, so we concentrate on the second. We notice that it is part of $\left(s^2 - 1\right) / \left(s^2 + 1\right)^2$, which is in the table, so we try to use this information by writing

$$\frac{1}{\left(s^2 + 1\right)^2} = \frac{1}{s^2 + 1} - \frac{\left(s^2 - 1\right) + 1}{\left(s^2 + 1\right)^2} = \frac{1}{s^2 + 1} - \frac{s^2 - 1}{\left(s^2 + 1\right)^2} - \frac{1}{\left(s^2 + 1\right)^2}.$$

Ah! If we move the final term on the right-hand side to the left-hand side and divide by 2 we find

$$\frac{1}{\left(s^2 + 1\right)^2} = \frac{1}{2}\left[\frac{1}{s^2 + 1} - \frac{s^2 - 1}{\left(s^2 + 1\right)^2}\right].$$

Thus, we finally have

$$\mathcal{L}^{-1}\left\{\frac{1}{\left(s^2 + 1\right)^2}\right\} = \frac{1}{2}\left[\mathcal{L}^{-1}\left\{\frac{1}{s^2 + 1}\right\} - \mathcal{L}^{-1}\left\{\frac{s^2 - 1}{\left(s^2 + 1\right)^2}\right\}\right]$$

or

$$\mathcal{L}^{-1}\left\{\frac{1}{\left(s^2 + 1\right)^2}\right\} = \frac{1}{2}\left(\sin t - t\cos t\right). \tag{14.42}$$

$\square$

It would be an understatement to say we were lucky to find this inverse Laplace transform. In fact, while looking down Table 14.5 you will have noticed that $\mathcal{L}^{-1}\left\{1/\left(s^2 + 1\right)\right\} = \sin t$, and you may have been tempted to think that $\mathcal{L}^{-1}\left\{1/\left(s^2 + 1\right)^2\right\} = \mathcal{L}^{-1}\left\{1/\left(s^2 + 1\right)\right\} \times \mathcal{L}^{-1}\left\{1/\left(s^2 + 1\right)\right\}$. If this were true, then we would have found $\mathcal{L}^{-1}\left\{1/\left(s^2 + 1\right)^2\right\} = \sin^2 t$, which does not agree with the correct value of $\left(\sin t - t\cos t\right)/2$, which we just obtained. Although this technique is incorrect, it does raise the following question: Is there a formula for $\mathcal{L}^{-1}\left\{F(s)G(s)\right\}$? If there is a formula, then we may be able to use it on this example, without relying on luck.

There is a formula, but it is very unlikely anyone would guess what it is. So, without motivation we state the result.

Theorem 14.5: *If $F(s) = \mathcal{L}\{f(t)\}$ and $G(s) = \mathcal{L}\{g(t)\}$ then*

$$\mathcal{L}^{-1}\left\{F(s)G(s)\right\} = \int_0^t f(z)g(t - z)\, dz.$$

Comments about Theorem 14.5

- This theorem is usually called the CONVOLUTION THEOREM. An outline of its proof is given in Exercise 6.

- The integral $\int_0^t f(z)g(t - z)\, dz$ is called the CONVOLUTION of the two functions $f(t)$ and $g(t)$ and is denoted by $f * g$. Thus,

$$f * g = \int_0^t f(z)g(t - z)\, dz,$$

and the theorem states that

$$\mathcal{L}^{-1}\left\{F(s)G(s)\right\} = f * g.$$

- The convolution $f * g$ is symmetric in the sense that $f * g = g * f$. See Exercise 7 on page 710.

- This theorem may also be stated in the form

$$\mathcal{L}\left\{\int_0^t f(z)g(t-z)\,dz\right\} = \mathcal{L}\{f\}\mathcal{L}\{g\}$$

or

$$\mathcal{L}\{f * g\} = \mathcal{L}\{f\}\mathcal{L}\{g\}.$$

- A useful special case of this theorem is obtained by choosing $g(t) = 1$ so that $G(s) = 1/s$. We then have the following statements:

$$\text{If } F(s) = \mathcal{L}\{f(t)\}, \text{ then } \mathcal{L}^{-1}\left\{\frac{1}{s}F(s)\right\} = \int_0^t f(z)\,dz, \qquad \textbf{(14.43)}$$

or, equivalently,

$$\text{If } \mathcal{L}\{f(t)\} = F(s), \text{ then } \mathcal{L}\left\{\int_0^t f(z)\,dz\right\} = \frac{1}{s}F(s).$$

We will now use this theorem to recalculate $\mathcal{L}^{-1}\left\{1/\left(s^2+1\right)^2\right\}$.

EXAMPLE 14.12: The Inverse Laplace Transform of $1/\left(s^2+1\right)^2$ Revisited

We apply the theorem to $f(t) = g(t) = \sin t$ and use the fact that $\mathcal{L}^{-1}\left\{1/\left(s^2+1\right)\right\} = \sin t$ to obtain

$$\mathcal{L}^{-1}\left\{\frac{1}{\left(s^2+1\right)^2}\right\} = \int_0^t \sin z \sin(t-z)\,dz.$$

Using the identity

$$\sin A \sin B = \frac{1}{2}\left[\cos\left(A-B\right) - \cos\left(A+B\right)\right]$$

gives

$$\mathcal{L}^{-1}\left\{\frac{1}{\left(s^2+1\right)^2}\right\} = \frac{1}{2}\int_0^t \left[\cos\left(2z-t\right) - \cos t\right]\,dz = \frac{1}{2}\int_0^t \cos\left(2z-t\right)\,dz - \frac{1}{2}\cos t\int_0^t dz.$$

(Remember we are integrating with respect to z, so t plays the role of a constant in this process.) The first integral on the right-hand side can be evaluated with the change of variable $u = 2z - t$, so that

$$\int_0^t \cos\left(2z-t\right)\,dz = \frac{1}{2}\int_{-t}^t \cos u\,du = \sin t.$$

The second integral can be evaluated immediately, so we have

$$\mathcal{L}^{-1}\left\{\frac{1}{\left(s^2+1\right)^2}\right\} = \frac{1}{2}\sin t - \frac{1}{2}t\cos t, \qquad (14.44)$$

as anticipated. Notice that we can use (14.38) and (14.44) to show that

$$\mathcal{L}^{-1}\left\{\frac{1}{\left(s^2+a^2\right)^2}\right\} = \frac{1}{2a^3}\left(\sin at - at\cos at\right).$$

(See Exercise 2 on page 709.)   □

Whenever we have the term $1/s$ in a Laplace transform, we should consider using the special case of the Convolution Theorem, namely (14.43). We now give an example of its use.

EXAMPLE 14.13: The Inverse Laplace Transform of $1/[s(s-1)^2]$

We seek $f(t)$ for which

$$\mathcal{L}\{f(t)\} = \frac{1}{s(s-1)^2}.$$

If we write this as

$$f(t) = \mathcal{L}^{-1}\left\{\frac{1}{s(s-1)^2}\right\} = \mathcal{L}^{-1}\left\{\frac{1}{s}\frac{1}{(s-1)^2}\right\},$$

we recognize that this is in the form of (14.43). Because we know that $\mathcal{L}\left\{te^t\right\} = 1/(s-1)^2$, we have

$$f(t) = \mathcal{L}^{-1}\left\{\frac{1}{s}\frac{1}{(s-1)^2}\right\} = \int_0^t ze^z\,dz.$$

Integration by parts gives $\int_0^t ze^z\,dz = te^t - e^t + 1$, so

$$f(t) = te^t - e^t + 1$$

and

$$\mathcal{L}^{-1}\left\{\frac{1}{s(s-1)^2}\right\} = te^t - e^t + 1. \qquad (14.45)$$

We can use (14.38) and (14.45) to show that

$$\mathcal{L}^{-1}\left\{\frac{1}{s(s-a)^2}\right\} = \frac{1}{a^2}\left(ate^{at} - e^{at} + 1\right).$$

(See Exercise 3 on page 709.)   □

Table 14.6 lists all these results.

TABLE 14.6   The inverse Laplace transform list

$F(s)$	$\mathcal{L}^{-1}\{F(s)\} = f(t)$
a/s	a
$1/s^2$	t
$n!/s^{n+1}$	t^n
$1/(s-a)$	e^{at}
$1/(s-a)^2$	te^{at}
$1/[(s-a)(s-b)]$	$(e^{at}-e^{bt})/(a-b) \qquad a \neq b$
$s/[(s-a)(s-b)]$	$(ae^{at}-be^{bt})/(a-b)$
$a/(s^2+a^2)$	$\sin at$
$s/(s^2+a^2)$	$\cos at$
$a/(s^2-a^2)$	$\sinh at$
$s/(s^2-a^2)$	$\cosh at$
$s/[(s^2-a^2)(s-b)]$	$[e^{at}/(a-b)-e^{-at}/(a+b)-2be^{bt}/(a^2-b^2)]/2 \qquad a \neq b$
$b/[(s-a)(s^2+b^2)]$	$(be^{at}-a\sin bt-b\cos bt)/(a^2+b^2)$
$1/(s^2+a^2)^2$	$(\sin at - at\cos at)/(2a^3)$
$1/(s^2-a^2)^2$	$(-\sinh at + at\cosh at)/(2a^3)$
$s^2/(s^2+a^2)^2$	$(\sin at + at\cos at)/(2a)$
$2as/(s^2+a^2)^2$	$t\sin at$
$(s^2-a^2)/(s^2+a^2)^2$	$t\cos at$
$1/[s(s-a)^2]$	$(ate^{at}-e^{at}+1)/a^2$

EXERCISES

1. Find the inverse Laplace transform of the following functions.

   (a) $1/(s^2-3s+2)$

   (b) $1/(s^2+2s+1)$

   (c) $1/(s^3+2s^2-s-2)$

   (d) $1/[s^2(s+1)]$

   (e) $s/[(s-1)(s^2+a^2)]$

   (f) $1/[(s+1)(s^2+1)]$

   (g) $1/[s(s^2+4s+13)]$

   (h) $1/[s(s^2+16)]$

   (i) $a/[(s-1)(s^2+a^2)]$

   (j) $1/[(s-b)(s^2-a^2)] \qquad a \neq b$

   (k) $s/[(s-b)(s^2-a^2)] \qquad a \neq b$

   (l) $s^2/(s^2+a^2)^2$

   (m) $1/(s^2-a^2)^2$

2. From (14.38) and (14.44) show that

$$\mathcal{L}^{-1}\left\{\frac{1}{(s^2+a^2)^2}\right\} = \frac{1}{2a^3}(\sin at - at\cos at).$$

3. From (14.38) and (14.45) show that

$$\mathcal{L}^{-1}\left\{\frac{1}{s(s-a)^2}\right\} = \frac{1}{a^2}(ate^{at} - e^{at} + 1).$$

4. The property $F'(s) = -\mathcal{L}\{tf(t)\}$ may be used to invert Laplace transforms if we write it as $tf(t) = \mathcal{L}^{-1}\{F'(s)\}$. Thus, if we know the inverse Laplace transform of $F'(s)$, we also know the inverse Laplace transform of $F(s)$. Use this property to find

(a) $\mathcal{L}^{-1}\left\{\ln\left|\dfrac{s-a}{s-b}\right|\right\}$

(b) $\mathcal{L}^{-1}\left\{\ln\left|1+\dfrac{1}{s^2}\right|\right\}$

(c) $\mathcal{L}^{-1}\left\{\dfrac{\pi}{2} - \arctan\dfrac{s}{2}\right\}$

5. The Laplace transform may also be used to solve certain types of integral equations. One type, called a VOLTERRA INTEGRAL EQUATION, has the form

$$y(t) = f(t) + \int_0^t y(z)g\,(t-z)\,dz,$$

where f and g are known functions, and $y(t)$ is to be determined. Volterra integral equations may be solved with the aid of the convolution theorem. Solve the following equations for $y(t)$.

(a) $y(t) = 4t - 3 - \int_0^t \sin(t-z)\,y(z)\,dz$

(b) $y(t) = 3\sin t - 2\int_0^t \cos(t-z)\,y(z)\,dz$

(c) $y(t) = \sin t + 4e^{-t} - 2\int_0^t \cos(t-z)\,y(z)\,dz$

(d) $y(t) = t - \int_0^t (t-z)\,y(z)\,dz$

(e) $y(t) = t - \int_0^t e^{t-z}y(z)\,dz$

(f) $y(t) = \cos t + \int_0^t e^{t-z}y(z)\,dz$

6. An outline of the proof of the convolution theorem uses the following steps.

(a) Show that

$$\mathcal{L}\{f(t) * g(t)\} = \iint_R e^{-st} f(z)g(t-z)\,dz\,dt,$$

where R is the region shown in Figure 14.1.

(b) Make the change of variables $u = t - z$, $v = z$, to show that

$$\mathcal{L}\{f(t) * g(t)\} = \int_0^\infty \int_0^\infty e^{-s(u+v)} f(v)g(u)\,du\,dv.$$

(c) Write the integral in part (b) as the product of two integrals, and note that this product contains the Laplace transform of f and g.

7. By making the substitution $u = t - z$, prove that $f * g = g * f$.

14.4   The Unit Step Function

In the first section we showed how the Laplace transform is used to solve first order linear differential equations with constant coefficients. Because we have already covered two methods for solving such equations in Chapter 6, you may wonder why we introduce a third. There are two reasons. The first is that using the Laplace transform process reduces the problem of solving a differential equation to a purely algebraic one. The second is the ease with which the Laplace transform handles certain types of forcing functions, especially ones that suddenly "jump" from one value to another value at a particular time.

An example of this is the situation in which a forcing function in an electric circuit is 0 from $t = 0$ to $t = 2$ and then jumps to 12 at $t = 2$ and remains at 12 thereafter. (This

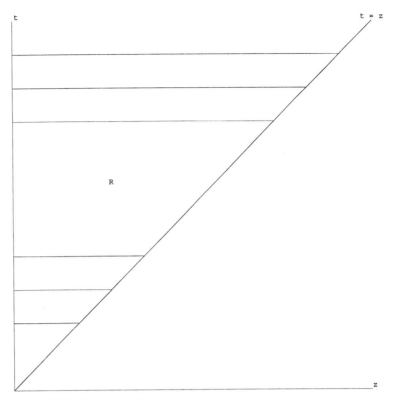

FIGURE 14.1 The region R

would correspond to a switch being turned on at $t = 2$ when a constant voltage of 12 is applied.) If we denote such a function by $E(t)$, then

$$E(t) = \begin{cases} 0 & \text{if } 0 \le t < 2, \\ 12 & \text{if } 2 \le t. \end{cases}$$

The graph of $E(t)$ is shown in Figure 14.2. Sometimes when graphing functions with jumps, it is difficult to see the major features because of the gaps. For this reason we will always fill a gap with a vertical line, as we have done in Figure 14.3. However, these vertical lines are merely aids to the eye. They are not part of the function.

Because functions with jumps, such as shown by $E(t)$, are common in applications, we define a special function with this property.

Definition 14.3: **The UNIT STEP FUNCTION[4] $u(t)$ is given by**

$$u(t) = \begin{cases} 0 & \text{if } t < 0, \\ 1 & \text{if } 0 \le t. \end{cases}$$

Comments about the unit step function

- The graph of the unit step function is shown in Figure 14.4.
- If we are given a function $f(t)$, then the function $f(t)u(t)$ is 0 for $t < 0$, and the original function $f(t)$ for $t \ge 0$. The graph of $tu(t)$ is shown in Figure 14.5.

---

[4]This function is sometimes called the Heaviside function.

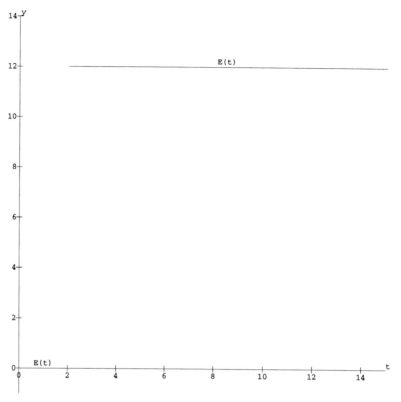

FIGURE 14.2   The graph of $E(t)$

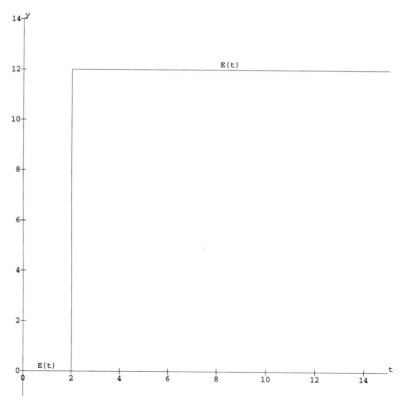

FIGURE 14.3   The graph of $E(t)$ with a vertical line added

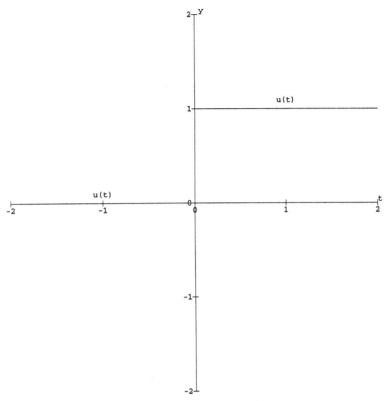

FIGURE 14.4 The graph of the unit step function $u(t)$

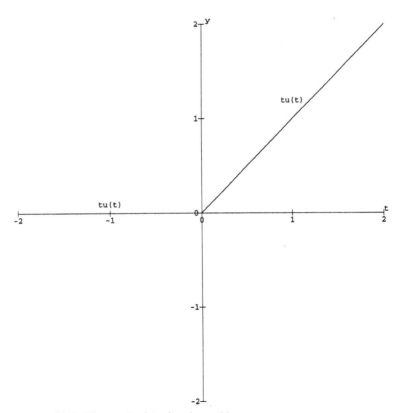

FIGURE 14.5 The graph of the function $tu(t)$

- Although $u(t)$ has its jump at $t = 0$, the function $u(t - a)$ has its jump at $t = a$. (Why?) The graph of $u(t - 1)$ is shown in Figure 14.6.

- If we are given a function $f(t)$, then the function $f(t)u(t - a)$ is 0 for $t < a$ and the original function $f(t)$ for $t \geq a$. Thus, the function $E(t)$ can be written in terms of the unit step function by $E(t) = 12u(t - 2)$.

- If we are given a function $f(t)$ that is 0 for $t < 0$, then the function $f(t - a)u(t - a)$ is the original function $f(t)$ shifted to the right by a units, if $a > 0$. If $a < 0$ the function is shifted to the left. The graph of the $(t - 1)u(t - 1)$ is shown in Figure 14.7.

EXAMPLE 14.14: The Function $u(t - 1) - u(t - 3)$

To sketch the graph of $u(t - 1) - u(t - 3)$, we first recognize that there are two possible jumps, at $t = 1$ and $t = 3$. Thus, we should analyze $u(t - 1) - u(t - 3)$ in the three regions $t < 1$, $1 < t < 3$, and $3 < t$.

If $t < 1$, then both $u(t - 1)$ and $u(t - 3)$ contribute 0, so the function is 0 here. At $t = 1$ the contribution from $u(t - 1)$ is 1. If $1 < t < 3$, then $u(t - 1) = 1$, but $u(t - 3) = 0$, so their difference is 1. If $3 \leq t$, both $u(t - 1)$ and $u(t - 3)$ contribute 1 so their difference is 0. Combining this information, we find

$$u(t - 1) - u(t - 3) = \begin{cases} 0 & \text{if } t < 1, \\ 1 & \text{if } 1 \leq t < 3, \\ 0 & \text{if } 3 \leq t. \end{cases}$$

The graph of the $u(t - 1) - u(t - 3)$ is shown in Figure 14.8. Physically this might correspond to a pulse of magnitude 1 lasting for 2 time units, starting at $t = 1$ and ending at $t = 3$. ◻

This example can be generalized to give

$$f(t)\,[u(t - a) - u(t - b)] = \begin{cases} 0 & \text{if } t < a, \\ f(t) & \text{if } a \leq t < b, \\ 0 & \text{if } b \leq t, \end{cases} \qquad \textbf{(14.46)}$$

EXAMPLE 14.15

Write down a formula for the square wave function $f(t)$ whose graph is shown in Figure 14.9. The graph continues in the same fashion forever.

We see that $f(t) = 1$ if $0 < t < 1$ and that $f(t) = 0$ if $1 < t < 2$. Then the function repeats itself indefinitely — usually called PERIODIC [5] of period 2 . From (14.46) we see that $u(t) - u(t - 1)$ will be 1 if $0 \leq t < 1$ and 0 if $1 \leq t$. In the same way, the second bump can be obtained from $u(t - 2) - u(t - 3)$, so $u(t) - u(t - 1) + u(t - 2) - u(t - 3)$ gives the first two bumps. From this we see that

$$f(t) = u(t) - u(t - 1) + u(t - 2) - u(t - 3) + \cdots + u(t - 2n) - u(t - 2n - 1) + \cdots$$

or

$$f(t) = \sum_{k=0}^{\infty} [u(t - 2k) - u(t - 2k - 1)].$$

---

[5] A function $f(t)$ is periodic of period P if $f(t + P) = f(t)$ for every t, where P is the smallest positive number with this property.

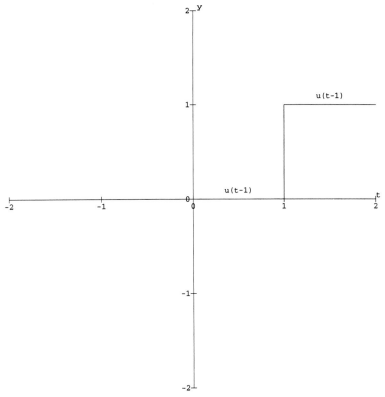

FIGURE 14.6   The graph of the function $u(t-1)$

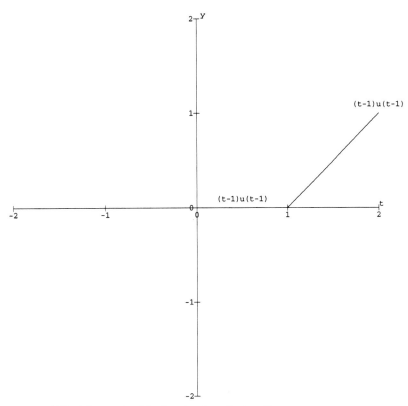

FIGURE 14.7   The graph of the function $(t-1)u(t-1)$

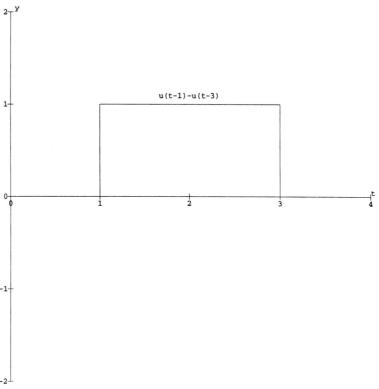

FIGURE 14.8   The graph of the function $u(t-1) - u(t-3)$

Because of the appearance of the unit step function in various applications, we need to compute its Laplace transform.

EXAMPLE 14.16: The Laplace Transform of $u(t)$

The Laplace transform of $u(t)$ is

$$\mathcal{L}\{u(t)\} = \int_0^\infty e^{-st} \, dt = \lim_{b \to \infty} -\frac{1}{s} e^{-st} \Big|_0^b,$$

or

$$\mathcal{L}\{u(t)\} = \frac{1}{s}.$$

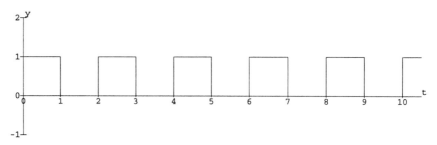

FIGURE 14.9   What function is this?

Thus, we also have

$$\mathcal{L}^{-1}\left\{\frac{1}{s}\right\} = u(t).$$

$\square$

EXAMPLE 14.17: The Laplace Transform of $u(t-a)$

The Laplace transform of $u(t-a)$ is

$$\mathcal{L}\{u(t-a)\} = \int_0^a e^{-st}0\,dt + \int_a^\infty e^{-st}\,dt = \lim_{b\to\infty} -\frac{1}{s}e^{-st}\Big|_a^b,$$

or, for $s > 0$,

$$\mathcal{L}\{u(t-a)\} = \frac{1}{s}e^{-sa}.$$

Thus, we also have

$$\mathcal{L}^{-1}\left\{\frac{1}{s}e^{-sa}\right\} = u(t-a).$$

$\square$

EXAMPLE 14.18: The Laplace Transform of $f(t-a)u(t-a)$

The Laplace transform of $f(t-a)u(t-a)$ is

$$\mathcal{L}\{f(t-a)u(t-a)\} = \int_0^a e^{-st}0\,dt + \int_a^\infty e^{-st}f(t-a)\,dt.$$

We now make the change of variable $t - a = z$ in the second integral to find that

$$\mathcal{L}\{f(t-a)u(t-a)\} = \int_0^\infty e^{-s(z+a)}f(z)\,dz = e^{-sa}\int_0^\infty e^{-sz}f(z)\,dz = e^{-sa}\mathcal{L}\{f(t)\}.$$

Thus, we have that if $F(s) = \mathcal{L}\{f(t)\}$ and $a > 0$, then

$$\mathcal{L}\{f(t-a)u(t-a)\} = e^{-sa}\mathcal{L}\{f(t)\}.$$

We also have

$$\mathcal{L}^{-1}\left\{e^{-sa}F(s)\right\} = f(t-a)u(t-a). \tag{14.47}$$

Although the derivation of (14.47) used $a > 0$, a different change of variables will show that (14.47) is also true for $a < 0$. $\square$

EXAMPLE 14.19

In order to illustrate the use of these results, we consider an electric circuit that has a 12 volt battery connected in series with a resistor and a capacitor. At $t = 0$ the battery is bypassed, allowing the capacitor to begin discharging. Two seconds later the battery is inserted back into the circuit. (See Figure 14.10.) We want to discover how the charge on the capacitor behaves under this situation.

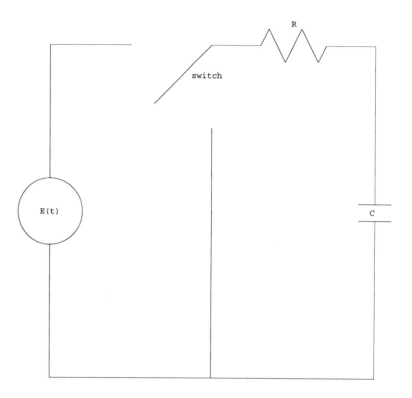

FIGURE 14.10   Electric circuit

Our pertinent initial value problem is

$$R\frac{dq}{dt} + \frac{1}{C}q = \begin{cases} 0 & \text{if } 0 \le t < 2, \\ 12 & \text{if } 2 \le t, \end{cases}$$

with $q(0) = 12C$. In this example we take $R = 10$ and $C = 1/10$, so we want to solve

$$\frac{dq}{dt} + q = \begin{cases} 0 & \text{if } 0 \le t < 2, \\ 1.2 & \text{if } 2 \le t, \end{cases}$$

with $q(0) = 1.2$.

We could solve this using the techniques of Chapter 6. We would proceed in two steps. First solve $q' + q = 0$ subject to $q(0) = 1.2$. This solution will be valid for $0 \le t < 2$. From this we would evaluate $q(2) = \lim_{t \to 2^-} q(t)$. Then we would solve $q' + q = 1.2$ subject to the initial value of $q(2)$. This solution will be valid for $2 \le t$. Although we could solve the problem in this way, it is very clumsy.

Instead we use Laplace transforms. We first express the forcing function in terms of the unit step function. This gives our initial value problem as

$$\frac{dq}{dt} + q = 1.2u(t - 2) \tag{14.48}$$

with $q(0) = 1.2$.

We take the Laplace transform of both sides of (14.48) and use our initial condition and some of the properties of Laplace transforms to obtain

$$\mathcal{L}\{q'\} + \mathcal{L}\{q\} = 1.2\frac{1}{s}e^{-2s}.$$

This simplifies to

$$s\mathcal{L}\{q\} - 1.2 + \mathcal{L}\{q\} = 1.2\frac{1}{s}e^{-2s},$$

so we may solve for $\mathcal{L}\{q\}$ as

$$\mathcal{L}\{q\} = \frac{1.2}{s+1} + \frac{1.2e^{-2s}}{s(s+1)}. \tag{14.49}$$

We know the function that has a Laplace transform of $1.2/(s+1)$ is $1.2e^{-t}$. But what about the second term in (14.49)? What function has a Laplace transform of $1.2e^{-2s}/[s(s+1)]$? In order to use our result in (14.47) we need to know what function has a Laplace transform of $1/[s(s+1)]$. From Table 14.6 we have that

$$\mathcal{L}\left\{\frac{1}{a}\left(e^{at}-1\right)\right\} = \frac{1}{s(s-a)},$$

so

$$\mathcal{L}\left\{-\left(e^{-t}-1\right)\right\} = \frac{1}{s(s+1)}.$$

Using (14.47), with $f(t) = 1 - e^{-t}$, gives our solution to (14.48) as

$$q(t) = 1.2\left\{e^{-t} + \left[1 - e^{-(t-2)}\right]u(t-2)\right\}. \tag{14.50}$$

This could be expressed as

$$q(t) = \begin{cases} 1.2e^{-t} & \text{if } 0 \le t < 2, \\ 1.2\left\{e^{-t} + \left[1 - e^{-(t-2)}\right]\right\} & \text{if } 2 \le t. \end{cases}$$

The graph of (14.50) is shown in Figure 14.11. (Does its behavior agree with what you anticipated?)   □

EXAMPLE 14.20: The Laplace Transform of $\sum_{k=0}^{\infty}[u(t-2k) - u(t-2k-1)]$

If a forcing function such as the one we used in our previous example turns on and off indefinitely, we will need the Laplace transform of the square wave function

$$f(t) = \sum_{k=0}^{\infty}[u(t-2k) - u(t-2k-1)],$$

namely, $\mathcal{L}\{f(t)\} = \mathcal{L}\left\{\sum_{k=0}^{\infty}[u(t-2k) - u(t-2k-1)]\right\}$. Assuming it is legal to extend (14.18) to an infinite number of terms, we find

$$\mathcal{L}\left\{\sum_{k=0}^{\infty}[u(t-2k) - u(t-2k-1)]\right\} = \sum_{k=0}^{\infty}\mathcal{L}\{u(t-2k)\} - \sum_{k=0}^{\infty}\mathcal{L}\{u(t-2k-1)\}.$$

From

$$\mathcal{L}\{u(t-a)\} = \frac{1}{s}e^{-sa}$$

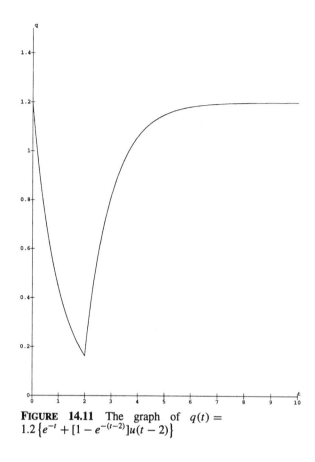

FIGURE 14.11 The graph of $q(t) = 1.2\{e^{-t} + [1 - e^{-(t-2)}]u(t-2)\}$

we have

$$\mathcal{L}\left\{\sum_{k=0}^{\infty}[u(t-2k) - u(t-2k-1)]\right\} = \sum_{k=0}^{\infty}\frac{1}{s}e^{-s2k} - \sum_{k=0}^{\infty}\frac{1}{s}e^{-s(2k+1)}.$$

Using $e^{-s(2k+1)} = e^{-s2k}e^{-s}$ in the last term on the right-hand side allows us to combine both terms in the form

$$\mathcal{L}\left\{\sum_{k=0}^{\infty}[u(t-2k) - u(t-2k-1)]\right\} = \frac{1-e^{-s}}{s}\sum_{k=0}^{\infty}e^{-s2k}. \qquad (14.51)$$

The infinite series $\sum_{k=0}^{\infty}e^{-s2k}$ is a geometric series $\sum_{k=0}^{\infty}r^k$ (where $r = e^{-s2}$ or e^{-2s}) with sum $1/(1-r)$, that is, $1/(1-e^{-2s})$. Finally, we have the Laplace transform of the square wave as

$$\mathcal{L}\left\{\sum_{k=0}^{\infty}[u(t-2k) - u(t-2k-1)]\right\} = \frac{1-e^{-s}}{s}\frac{1}{1-e^{-2s}}.$$

Alternatively, using $1 - e^{-2s} = (1-e^{-s})(1+e^{-s})$, we have

$$\mathcal{L}\left\{\sum_{k=0}^{\infty}[u(t-2k) - u(t-2k-1)]\right\} = \frac{1}{s}\frac{1}{1+e^{-s}}.$$

Sometimes it is useful to write this result in the form

$$\mathcal{L}\left\{\sum_{k=0}^{\infty}[u(t-2k)-u(t-2k-1)]\right\} = \frac{1}{s}\sum_{k=0}^{\infty}(-1)^k e^{-sk}. \qquad \textbf{(14.52)}$$

(How did we do this?) □

Notice that we were able to simplify the form of the Laplace transform of this periodic function considerably. Was this simplification due to the periodicity or to the fact that it consists of parts of horizontal lines? To investigate further we look at the transform of a general function with period P. Thus, we consider $\mathcal{L}\{f(t)\}$ where $f(t+P) = f(t)$ and are able to use the standard results from calculus (see Exercise 7 on page 728) to prove the following result.

Theorem 14.6: *If $f(t)$ is periodic with period P, then*

$$\mathcal{L}\{f(t)\} = \frac{1}{1-e^{-Ps}}\int_0^P e^{-st}f(t)\,dt.$$

EXAMPLE 14.21

Let us now consider the RC circuit in Example 14.19 in which the battery is connected and disconnected every 2 seconds following the pattern started in the first 4 seconds. (See Figure 14.10.) Thus, our initial value problem for the charge $q(t)$ is now

$$\frac{dq}{dt} + q = E(t), \qquad q(0) = 1.2, \qquad \textbf{(14.53)}$$

where

$$E(t) = \begin{cases} 0 & \text{if } 0 \le t < 2, \\ 1.2 & \text{if } 2 \le t < 4, \end{cases} \quad \text{and } E(t+4) = E(t) \text{ for } t \ge 0.$$

In terms of unit step functions we could write $E(t) = 1.2\sum_{k=0}^{\infty}[u(4k+2) - u(4k+4)]$.
We take the Laplace transform of (14.53) to obtain

$$\mathcal{L}\{q'\} + \mathcal{L}\{q\} = \mathcal{L}\{E(t)\}. \qquad \textbf{(14.54)}$$

Rather than compute $\mathcal{L}\{E(t)\}$ as we did in Example 14.20, we use the results of Theorem 14.6 to find

$$\mathcal{L}\{E(t)\} = \frac{1}{1-e^{-4s}}\left(\int_0^2 e^{-st}0\,dt + \int_2^4 e^{-st}1.2\,dt\right),$$

or

$$\mathcal{L}\{E(t)\} = 1.2\frac{1}{s}\frac{e^{-2s}-e^{-4s}}{1-e^{-4s}}.$$

We can simplify this expression once we realize that $e^{-2s} - e^{-4s} = e^{-2s}\left(1-e^{-2s}\right)$ and $1-e^{-4s} = \left(1-e^{-ss}\right)\left(1+e^{-2s}\right)$, obtaining

$$\mathcal{L}\{E(t)\} = 1.2\frac{1}{s}\frac{e^{-2s}}{1+e^{-2s}}.$$

Using some properties of Laplace transforms in Table 14.4, we can rewrite (14.54) as

$$s\mathcal{L}\{q(t)\} - 1.2 + \mathcal{L}\{q(t)\} = 1.2\frac{1}{s}\frac{e^{-2s}}{1+e^{-2s}},$$

or, solving for $\mathcal{L}\{q(t)\}$,

$$\mathcal{L}\{q(t)\} = 1.2\frac{1}{s+1} + 1.2\frac{1}{s(s+1)}\frac{e^{-2s}}{1+e^{-2s}}. \tag{14.55}$$

Looking at the right-hand side of (14.55), we see that we know the inverse Laplace transform of $1/(s+1)$ and $1/[s(s+1)]$. But what do we do with the term involving $e^{-2s}/(1+e^{-2s})$? Our clue comes from Example 14.20, in which we transformed the infinite series $\sum_{k=0}^{\infty} e^{-s2k}$ to $1/(1-e^{-2s})$. Because we have formulas for the inverse Laplace transform of products of $e^{as}F(s)$, using $1/(1-r) = \sum_{k=0}^{\infty} r^k$ allows us to write $e^{-2s}/(1+e^{-2s})$ as a series of exponential functions. Thus, we use $1/(1+e^{-2s}) = \sum_{k=0}^{\infty}(-1)^k e^{-2sk}$ to find that

$$\frac{e^{-2s}}{1+e^{-2s}} = e^{-2s}\sum_{k=0}^{\infty}(-1)^k e^{-2sk} = \sum_{k=1}^{\infty}(-1)^{k+1} e^{-2sk}.$$

Using this result in (14.55) gives

$$\mathcal{L}\{q(t)\} = \frac{1.2}{s+1} + 1.2\frac{1}{s(s+1)}\sum_{k=0}^{\infty}(-1)^{k+1} e^{-2sk}.$$

From Table 14.6 we have $\mathcal{L}^{-1}\{1/(s+1)\} = e^{-t}$ and $\mathcal{L}^{-1}\{1/[s(s+1)]\} = 1 - e^{-t}$. These results and (14.47) give us the solution of our initial value problem:

$$q(t) = 1.2e^{-t} + 1.2\sum_{k=0}^{\infty}(-1)^{k+1}\left[1 - e^{-(t-2k)}\right]u(t-2k). \tag{14.56}$$

If we expand the right-hand side of our solution for the first four time periods, we find

$$q(t) = \begin{cases} 1.2e^{-t} & \text{if } 0 \leq t < 2, \\ 1.2\left\{e^{-t} + \left[1 - e^{-(t-2)}\right]\right\} & \text{if } 2 \leq t < 4, \\ 1.2\left\{e^{-t} + \left[1 - e^{-(t-2)}\right] - \left[1 - e^{-(t-4)}\right]\right\} & \text{if } 4 \leq t < 6, \\ 1.2\left\{e^{-t} + \left[1 - e^{-(t-2)}\right] - \left[1 - e^{-(t-4)}\right] + \left[1 - e^{-(t-6)}\right]\right\} & \text{if } 6 \leq t < 8, \end{cases} \tag{14.57}$$

which agrees with our earlier results in Example 14.19 for the common domain of $0 \leq t < 4$. The graph of the charge given by (14.56) is shown in Figure 14.12 for $0 \leq t < 10$.

The figure suggests that we have periodic behavior with constant amplitude. Do we? To answer this question we need to look at the solution $q(t)$ more carefully. If we simplify (14.57), we find

$$q(t) = \begin{cases} 1.2e^{-t} & \text{if } 0 \leq t < 2, \\ 1.2\left[1 + e^{-t}\left(1 - e^2\right)\right] & \text{if } 2 \leq t < 4, \\ 1.2e^{-t}\left(1 - e^2 + e^4\right) & \text{if } 4 \leq t < 6, \\ 1.2\left[1 + e^{-t}\left(1 - e^2 + e^4 - e^6\right)\right] & \text{if } 6 \leq t < 8. \end{cases}$$

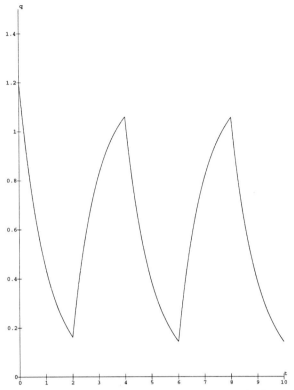

FIGURE 14.12   The charge from a square wave forcing
function

From this we can see a pattern emerge depending on whether t satisfies $4n \leq t < 4n + 2$
or $4n + 2 \leq t < 4n + 4$ for $n = 0, 1, 2, \cdots$. That pattern is

$$q(t) = \begin{cases} 1.2e^{-t}\left(1 - e^2 + e^4 - \cdots + e^{4n}\right) & \text{if } 4n \leq t < 4n + 2, \\ 1.2\left[1 + e^{-t}\left(1 - e^2 + e^4 - \cdots - e^{4n+2}\right)\right] & \text{if } 4n + 2 \leq t < 4n + 4. \end{cases} \quad \textbf{(14.58)}$$

We recognize that $1 - e^2 + e^4 - \cdots + e^{4n}$ can be simplified by means of the geometric
series property

$$1 + z + z^2 + \cdots + z^m = \frac{1 - z^{m+1}}{1 - z}$$

as

$$1 - e^2 + e^4 - \cdots + e^{4n} = \frac{1 - (-e^2)^{2n+1}}{1 - (-e^2)} = \frac{1 + e^{4n+2}}{1 + e^2}.$$

In a similar way we find

$$1 - e^2 + e^4 - \cdots - e^{4n+2} = \frac{1 + e^{4n+4}}{1 + e^2},$$

so we can write (14.58) as

$$q(t) = \begin{cases} 1.2e^{-t}\left(1 + e^{4n+2}\right)/\left(1 + e^2\right) & \text{if } 4n \leq t < 4n + 2, \\ 1.2\left[1 + e^{-t}\left(1 + e^{4n+4}\right)/\left(1 + e^2\right)\right] & \text{if } 4n + 2 \leq t < 4n + 4. \end{cases} \quad \textbf{(14.59)}$$

The successive maxima of $q(t)$ will occur at $t = 4n$, $n = 0, 1, 2, \cdots$. That is, they will occur at

$$q(4n) = 1.2 e^{-4n} \frac{1 + e^{4n+2}}{1 + e^2} = 1.2 \frac{e^{-4n} + e^2}{1 + e^2},$$

which is a decreasing function that tends to $1.2 e^2 / \left(1 + e^2\right)$ as $t \to \infty$. In the same way the minima will occur at $t = 4n + 2$, $n = 0, 1, 2, \cdots$, which is at

$$q(4n + 2) = 1.2 \left[1 + e^{-4n-2} \frac{1 - e^{4n+4}}{1 + e^2}\right] = 1.2 \left[1 + \frac{e^{-4n-2} - e^2}{1 + e^2}\right] = 1.2 \left[\frac{e^{-4n-2} + 1}{1 + e^2}\right].$$

This expression is a decreasing function that tends to $1.2 / \left(1 + e^2\right)$ as $t \to \infty$. Thus, the maxima will lie on the curve $1.2 \left(e^{-t} + e^2\right) / \left(1 + e^2\right)$ and the minima on the curve $1.2 \left(e^{-t} + 1\right) / \left(1 + e^2\right)$. Figure 14.13 shows these two curves superimposed on Figure 14.12. In spite of its appearance, the amplitude of $q(t)$ is not constant.   □

Notice that in Examples 14.19 and 14.21 we had discontinuous forcing functions but continuous solutions, as shown in Figures 14.11 and 14.12. (How can you tell from the figures that the solutions are continuous?) We will return to this comment in Section 14.8, where we will discuss the circumstances under which this situation occurs.

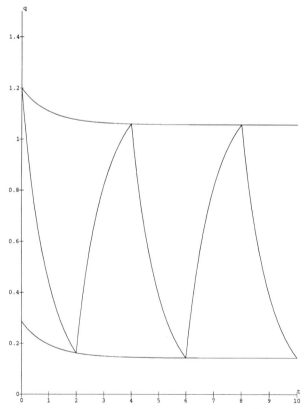

FIGURE 14.13   The functions $1.2(e^{-t} + e^2)/(1 + e^2)$ and $1.2(e^{-t} + 1)/(1 + e^2)$ and the charge from a square wave forcing function

EXERCISES

1. Graph the following functions, and then write them in terms of the unit step function.

   (a) $f(t) = \begin{cases} 1 & \text{if } 0 \le t < 4 \\ 0 & \text{if } 4 \le t \end{cases}$

   (b) $f(t) = \begin{cases} 1 & \text{if } 0 \le t < 2 \\ 0 & \text{if } 2 \le t < 3 \\ 3 & \text{if } 3 \le t \end{cases}$

   (c) $f(t) = \begin{cases} t & \text{if } 0 \le t < 2 \\ 2 & \text{if } 2 \le t < 4 \\ 6 - t & \text{if } 4 \le t < 6 \\ 0 & \text{if } 6 \le t \end{cases}$

   (d) $f(t) = \begin{cases} 1 & \text{if } 0 \le t < 2 \\ -1 & \text{if } 2 \le t < 4 \end{cases}$ where $f(t+4) = f(t)$ for $t \ge 0$

   (e) $f(t) = \begin{cases} 1 & \text{if } 0 \le t < a \\ 0 & \text{if } a \le t < 2a \\ -1 & \text{if } 2a \le t < 3a \end{cases}$ where $f(t+3a) = f(t)$ for $t \ge 0$

2. Evaluate the Laplace transforms of the functions in Exercise 1.

3. The function $g(t) = \sin t$ if $0 \le t < \pi$, where $g(t+\pi) = g(t)$ for $t \ge 0$, is shown in Figure 14.14. This is called the FULL-WAVE RECTIFICATION of $\sin t$.[6] Express this function in terms of the unit step function, and then evaluate its Laplace transform.

4. Find the equations, in terms of the unit step function, for the graphs in the following figures.

   (a) Figure 14.15
   (b) Figure 14.16
   (c) Figure 14.17
   (d) Figure 14.18

5. Evaluate the Laplace transforms of the functions in Exercise 4.

6. Evaluate the inverse Laplace transforms of the following functions.

   (a) $F(s) = e^{-s}/(s+1)^2$
   (b) $F(s) = e^{-2s}/(s+1)$
   (c) $F(s) = e^{-3s}/[s(s+1)]$
   (d) $F(s) = e^{-4s}/(s^2 - 1)$

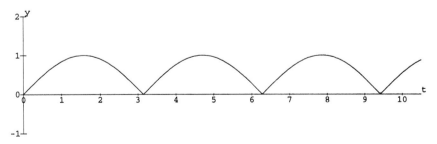

FIGURE 14.14 The full-wave rectification of $\sin t$

---

[6]The full-wave rectification of $f(t)$ is the function $|f(t)|$. The half-wave rectification of $f(t)$ is the function $f(t)$ when $f(t) \ge 0$, and 0 when $f(t) < 0$.

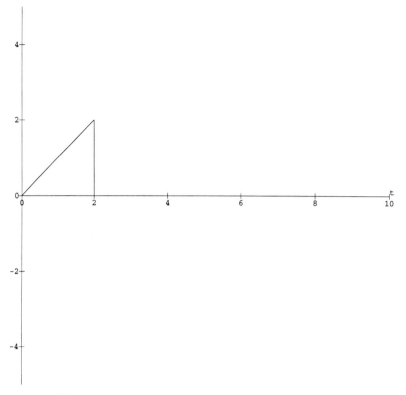

FIGURE 14.15 Mystery function A

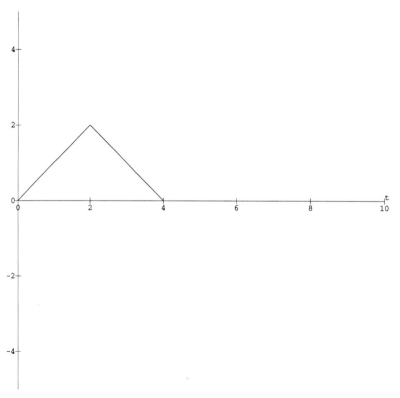

FIGURE 14.16 Mystery function B

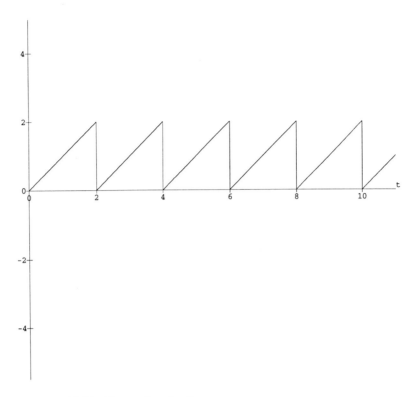

FIGURE 14.17 Mystery function C

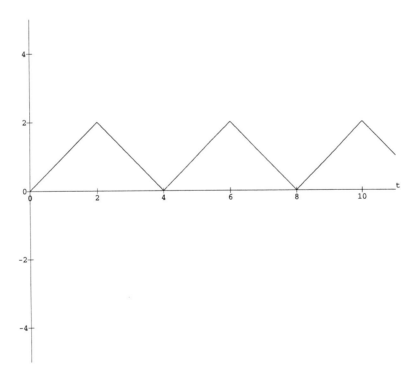

FIGURE 14.18 Mystery function D

(e) $F(s) = \left(e^{-s} - e^{-2s}\right)/(s-4)^2$

(f) $F(s) = \left(e^{-2s} + e^{-3s}\right)/(s^2 - 3s + 2)$

(g) $F(s) = e^{-3s}/(s^2 + 6s + 25)$

7. Prove Theorem 14.6 using the following outline.

(a)   Write the Laplace transform of the function $f(t)$ with period P in the form

$$\mathcal{L}\{f(t)\} = \int_0^\infty e^{-st} f(t)\, dt = \int_0^P e^{-st} f(t)\, dt + \int_P^\infty e^{-st} f(t)\, dt.$$

(b)   Make the change of variable $z = t - P$ in the second integral to find

$$\mathcal{L}\{f(t)\} = \int_0^\infty e^{-st} f(t)\, dt = \int_0^P e^{-st} f(t)\, dt + \int_0^\infty e^{-s(z+P)} f(z+P)\, dz.$$

(c)   Use the periodicity of $f(t)$ in the form $f(z + P) = f(z)$ to factor out an e^{-sP} to remove the dependence on P in the integrand.

(d)   Finish the argument.

14.5   Applications to First Order Linear Differential Equations

Now that we have developed several properties of Laplace transforms and tables containing transforms of familiar functions, we move on to solve linear first order differential equations with constant coefficients. We start by revisiting two applications we covered earlier to show how using Laplace transforms simplifies some of our calculations. We finish with an application from forensic pathology.

EXAMPLE 14.22: Population of Botswana, with and without Emigration

In two different parts of this book we considered the population growth of Botswana, once to obtain a model assuming no emigration, and later where we included hypothetical emigration. Here we combine these two into a single mathematical model by starting with a population in 1975 of 0.755 million people and assuming that there is no significant emigration or immigration until 1990. At that time we suppose that emigration commences in a linear manner with 1.285 million people leaving over a 20-year period (see Example 5.4). With these assumptions our mathematical model requires us to find the population $P(t)$ such that

$$\frac{dP}{dt} - kP = \begin{cases} 0 & \text{if } 0 < t < 15, \\ -a(t-15) & \text{if } 15 \le t, \end{cases}$$

with $P(0) = 0.755, k = 0.0355,$ and $a = 1.60625 \times 10^{-3}$. In order to use prior results from Laplace transforms, we first recast this equation in terms of unit step functions, namely,

$$\frac{dP}{dt} - kP = -a(t-15)u(t-15). \tag{14.60}$$

If we take the Laplace transform of (14.60), we obtain

$$\mathcal{L}\left\{P' - kP\right\} = \mathcal{L}\{-a(t-15)u(t-15)\},$$

or, by using the properties listed in (14.47), Table 14.3, and Table 14.4,

$$s\mathcal{L}\{P(t)\} - 0.755 - k\mathcal{L}\{P(t)\} = -\frac{a}{s^2}e^{-15s}.$$

Solving for $\mathcal{L}\{P(t)\}$ gives

$$\mathcal{L}\{P(t)\} = \frac{0.755}{s-k} - \frac{a}{(s-k)s^2}e^{-15s}. \qquad \textbf{(14.61)}$$

The inverse Laplace transform of the first term on the right-hand side of (14.61) is apparent, but with the term e^{-15s} present in the second term, we need to use (14.47). However, to do that we need the inverse Laplace transform of $1/[(s-k)s^2]$. Although this function does not appear on our largest list of inverse Laplace transforms, Table 14.6, both $1/(s-k)$ and $1/s^2$ do appear. Thus, we can use the convolution theorem with $F(s) = 1/(s-k)$, so $f(t) = e^{kt}$, and $G(s) = 1/s^2$, so $g(t) = t$. This gives us

$$\mathcal{L}^{-1}\left\{\frac{1}{(s-k)s^2}\right\} = \int_0^t e^{kz}(t-z)\,dz.$$

We use integration by parts to find

$$\mathcal{L}^{-1}\left\{\frac{1}{(s-k)s^2}\right\} = \frac{1}{k}e^{kz}(t-z)\Big|_0^t - \int_0^t -\frac{1}{k}e^{kz}\,dz,$$

or

$$\mathcal{L}^{-1}\left\{\frac{1}{(s-k)s^2}\right\} = -\frac{t}{k} + \frac{e^{kt}-1}{k^2}.$$

Now we use (14.47) to write our solution for the population as

$$P(t) = 0.755e^{kt} - a\left\{-\frac{t-15}{k} + \frac{1}{k^2}\left[e^{k(t-15)} - 1\right]\right\}u(t-15). \qquad \textbf{(14.62)}$$

The graph of (14.62) is shown in Figure 14.19 for a 45-year period until 2020. (Does it look like you expected?) □

EXAMPLE 14.23: The Yam in the Oven Revisited

In Example 6.14, we had another situation in which the forcing function in a differential equation had different forms for different time intervals. In that example the oven temperature was described by

$$T_a(t) = \begin{cases} 70 + 66t & \text{if } 0 < t < 5, \\ 400 & \text{if } 5 \leq t, \end{cases}$$

or in terms of our unit step function,

$$T_a(t) = (70 + 66t)\left[1 - u(t-5)\right] + 400u(t-5) = (70 + 66t) + (330 - 66t)u(t-5).$$

We will now solve that initial value problem, namely,

$$\frac{dT}{dt} = k\left[T - T_a(t)\right] = k\left[T - (70 + 66t) - (330 - 66t)u(t-5)\right], \qquad \textbf{(14.63)}$$

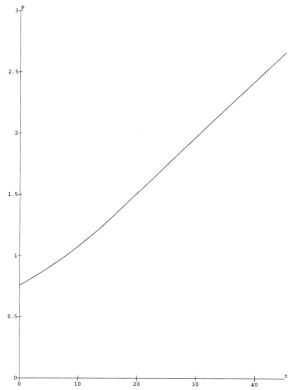

FIGURE 14.19 The population of Botswana for the period 1975 through 2020

with $k = -0.04$ and $T(0) = 70$, using Laplace transforms.

We start by taking the Laplace transform of (14.63) as

$$\mathcal{L}\left\{T'(t)\right\} = \mathcal{L}\left\{k\left[T - (70 + 66t) - (330 - 66t)u(t - 5)\right]\right\}. \qquad \textbf{(14.64)}$$

We have results to evaluate the Laplace transform of all of the terms in (14.64) except the last one. There, in order to use (14.47), we need the term involving $u(t - 5)$ to have the form $f(t - a)u(t - a)$. Thus, we write $t = (t - 5) + 5$ and rewrite $(330 - 66t)u(t - 5)$ as $-66(t - 5)u(t - 5)$. Using this result allows us to write (14.64) as

$$\mathcal{L}\left\{T'(t)\right\} = \mathcal{L}\left\{k\left[T - (70 + 66t) + 66(t - 5)u(t - 5)\right]\right\}. \qquad \textbf{(14.65)}$$

Now we may use the properties in (14.47), Table 14.3, and Table 14.4 to find

$$s\mathcal{L}\left\{T(t)\right\} - 70 = k\left(\mathcal{L}\left\{T(t)\right\} - \frac{70}{s} - \frac{66}{s^2} + \frac{66}{s^2}e^{-5s}\right).$$

Solving for $\mathcal{L}\left\{T(t)\right\}$ gives

$$\mathcal{L}\left\{T(t)\right\} = \frac{70}{s - k} - \frac{70k}{s(s - k)} - \frac{66k}{s^2(s - k)} + \frac{66k}{s^2(s - k)}e^{-5s}. \qquad \textbf{(14.66)}$$

The first two terms on the right-hand side of (14.66) are included in Table 14.6, and we have just evaluated the inverse Laplace transform of the third term in our previous example. Thus, from (14.66) we have

$$T(t) = 70 - 66[f(t) - f(t-5)u(t-5)], \tag{14.67}$$

where

$$f(t) = -t + \frac{1}{k}(e^{kt} - 1).$$

The graph of $T(t)$ from (14.67) is shown in Figure 14.20. □

EXAMPLE 14.24: Estimating the Time of Death of a Murder Victim

Consider a crime scene in Alaska in which the owner of a jewelry store was killed. The body was discovered in the jewelry store early in the morning and at 7:00 A.M. the coroner measured its temperature as 72.5^oF. One hour later the body temperature was 72^oF. During the night the temperature of the store was a constant 70^oF, and we want to find the time the person died.

We assume that the temperature of the body $T(t)$ is governed by Newton's law of cooling,

$$\frac{dT}{dt} = k\left[T(t) - T_a(t)\right], \tag{14.68}$$

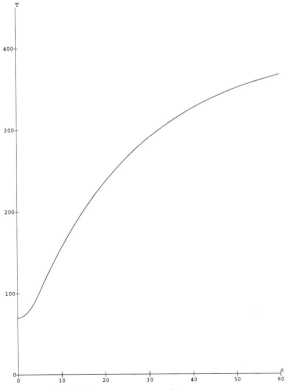

FIGURE 14.20 The temperature of the yam

where k is a negative constant and $T_a(t)$ is the ambient temperature, and that time t is measured in hours from the time of death. We also assume that the temperature of the person at the time of death was $98.6°$F so that $T(0) = 98.6$.

The temperature data taken by the corner give us $T(t_c) = 72.5$ and $T(t_c + 1) = 72$, where t_c is the number of hours after death. We will use these two pieces of information to evaluate k and t_c.

We start by taking the Laplace transform of (14.68), which gives us

$$\mathcal{L}\left\{T'(t)\right\} = \mathcal{L}\left\{k\left[T(t) - T_a(t)\right]\right\}.$$

If we use the properties of Laplace transforms in Table 14.4, we find that

$$s\mathcal{L}\{T(t)\} - T(0) = k\mathcal{L}\{T(t)\} - k\mathcal{L}\left\{T_a(t)\right\},$$

and we may solve for $\mathcal{L}\{T(t)\}$ as

$$\mathcal{L}\{T(t)\} = \frac{T(0)}{s - k} - \frac{k}{s - k}\mathcal{L}\left\{T_a(t)\right\}. \tag{14.69}$$

We now use the convolution theorem to find the inverse Laplace transform of (14.69) as

$$T(t) = T(0)e^{kt} - \int_0^t kT_a(z)e^{k(t-z)}\,dz,$$

or

$$T(t) = T(0)e^{kt} - ke^{kt}\int_0^t T_a(z)e^{-kz}\,dz. \tag{14.70}$$

The ambient temperature was a constant $70°$F after the person's death. Thus, we take $T_a(t) = 70$ and integrate (14.70) to obtain

$$T(t) = 98.6e^{kt} + 70\left(1 - e^{-kt}\right),$$

or

$$T(t) = 28.6e^{kt} + 70. \tag{14.71}$$

From the coroner's measurements we have

$$T(t_c) = 72.5 = 28.6e^{kt_c} + 70 \tag{14.72}$$

and

$$T(t_c + 1) = 72 = 28.6e^{k(t_c+1)} + 70.$$

We want to solve these two equations for t_c and k. If we subtract 70 from both equations and then divide the resulting equations, we obtain $2.5/2 = e^{-k}$, so $k = \ln 0.8 \approx -0.223$. If we substitute this value of k into (14.72), we obtain $e^{-0.223t_c} \approx 2.5/28.6$, or

$t_c \approx 10.92$. Thus, the coroner thinks the person died about 10.92 hours before 7 A.M. — that is, about 8 P.M. the previous evening.[7] Figure 14.21 shows the graph of the temperature from (14.71).   □

EXAMPLE 14.25: Reexamining the Crime Scenario

Consider the previous crime scene through the eyes of the perpetrator, who in fact committed the crime at midnight when the owner of the jewelry store was locking the front door of the store. However, the killer left the victim's body outside the store for some time, where the temperature was a constant 30°F. Later he moved the body inside the store, where it was discovered under the circumstances described in the previous example. How long was the body outside?

The main difference between this example and the previous one is that the ambient temperature $T_a(t)$ changes from 30 to 70 at the time b when the body is moved inside the store. This suggests we let midnight be when $t = 0$ and use the unit step function $u(t)$ to construct $T_a(t)$, which is 30 for $0 \leq t < b$ and 70 for $b \leq t$, that is, $T_a(t) = 30 + 40u(t - b)$. Thus, b represents the number of hours the body was outside. We return to our explicit solution in (14.70) and do the indicated integration for this form of $T_a(t)$ — namely,

$$T(t) = T(0)e^{kt} - ke^{kt} \int_0^t [30 + 40u(t - b)]\, e^{-kz}\, dz.$$

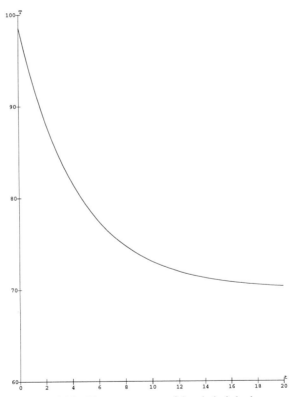

FIGURE 14.21   The temperature of the victim's body

---

[7]Notice that contrary to the situation frequently depicted on television crime shows, we need to know the temperature of the room and need to take the temperature of the body at two different times to estimate the time of death.

For $b \leq t$ this becomes

$$T(t) = T(0)e^{kt} - ke^{kt} \left(\int_0^b 30e^{-kz} \, dz + \int_b^t 70e^{-kz} \, dz \right),$$

or

$$T(t) = T(0)e^{kt} - ke^{kt} \left[-\frac{1}{k} \left(30e^{-kb} - 30 \right) - \frac{1}{k} \left(70e^{-kt} - 70e^{-kb} \right) \right].$$

We can rewrite this as

$$T(t) = e^{kt} \left[T(0) - 40e^{-kb} - 30 \right] + 70, \tag{14.73}$$

where we know $T(0) = 98.6$.

There are two unknowns in (14.73): k and b. But we also have two conditions to be met: $T(7) = 72.5$ and $T(8) = 72$. When we substitute these conditions into (14.73), we find

$$2.5 = e^{7k} \left[68.6 - 40e^{-kb} \right] \tag{14.74}$$

and

$$2 = e^{8k} \left[68.6 - 40e^{-kb} \right].$$

Dividing the lower equation by the upper gives $e^k = 2/2.5$, or $k = \ln 0.8 \approx -0.223$. This is exactly the same value for k as we found in the previous example. This is not surprising because k is the cooling constant for the body. For the same body it should have the same value. To find the value of b, we use this value of k in (14.74),

$$2.5 = e^{7 \ln 0.8} \left[68.6 - 40e^{-(\ln 0.8)b} \right],$$

or, solving for b,

$$b = -\frac{1}{\ln 0.8} \ln \left[\frac{68.6 - 2.5e^{-7 \ln 0.8}}{40} \right] \approx 2.15.$$

Thus, the body was moved at about 2:09 A.M. Figure 14.22 shows the graph of the temperature of the body. □

We have spent some time using the Laplace transform to solve first order linear differential equations with constant coefficients, so now we look at the general initial value problem associated with such differential equations, namely,

$$y' + ky = f(t), \qquad y(0) = y_0, \tag{14.75}$$

where $f(t)$ is a given function and k and y_0 are constants. If we follow the same pattern that we used in the previous examples — take the Laplace transform of (14.75) and solve for $Y(s) = \mathcal{L}\{y(t)\}$ — we get

$$Y(s) = \frac{y_0}{s + k} + \frac{1}{s + k} F(s). \tag{14.76}$$

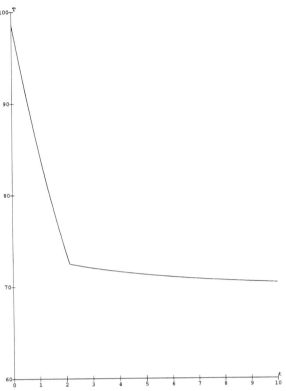

FIGURE 14.22   The temperature of the victim's body from midnight

We then apply the inverse Laplace transform to this equation and then use the convolution theorem to find

$$y(t) = y_0 e^{-kt} + \int_0^t e^{-k(t-z)} f(z) \, dz,$$

or

$$y(t) = y_0 e^{-kt} + e^{-kt} \int_0^t e^{kz} f(z) \, dz. \tag{14.77}$$

This is the general solution of the initial value problem (14.75), which we found, using different methods, in Chapter 5 [see (5.34) where $p(x) = k$]. Notice in (14.77) that the first term on the right-hand side is a solution of the associated homogeneous equation and that the second term is a particular solution of (14.75). Hence, we can associate these two terms with their counterparts in the transformed equation (14.76).

EXERCISES

1. Use (14.43) along with the fact that $\mathcal{L}\left\{(e^{kt} - 1)/k\right\} = 1/[s(s-k)]$ to find $\mathcal{L}^{-1}\left\{1/\left[s^2/(s-k)\right]\right\}$.

2. Use partial fractions to find $\mathcal{L}^{-1}\left\{1/\left[s^2/(s-k)\right]\right\}$.

3. Solve the equation $y' + y = f(t)$ for the following functions $f(t)$. Compare your answers with those you found for Exercise 9 in Section 6.4.

(a) $f(t) = \begin{cases} 2 & \text{if } 0 \le t < 1 \\ 1 & \text{if } t \ge 1 \end{cases}$ where $y(0) = 0$

(b) $f(t) = \begin{cases} 2 & \text{if } 0 \le t < 1 \\ 0 & \text{if } t \ge 1 \end{cases}$ where $y(0) = 0$

(c) $f(t) = \begin{cases} 5 & \text{if } 0 \le t < 10 \\ 1 & \text{if } t \ge 10 \end{cases}$ where $y(0) = 6$

(d) $f(t) = \begin{cases} e^{-t} & \text{if } 0 \le t < 2 \\ e^{-2} & \text{if } t \ge 2 \end{cases}$ where $y(0) = y_0$

4. Solve the following initial value problems.

(a) $y' + 4y = e^{-t} + u(t - 2)e^{-t+2}$,   where $y(0) = 1$

(b) $y' - y = [1 - u(t - 2)]e^t$,   where $y(0) = 4$

(c) $y' - y = [u(t - 1) - u(t - 3)]e^t$,   where $y(0) = 2$

(d) $y' + y = [1 - u(t - \pi)]\sin t$,   where $y(0) = 1$

(e) $y' + 4y = f(t)$,   where $y(0) = a$, $f(t)$ is a given function, and a is a constant

5. The differential equation for the current $I(t)$ in a series circuit with inductance L and resistance R is

$$LI' + RI = E(t), \tag{14.78}$$

where $E(t)$ is the applied voltage. Solve (14.78) subject to the initial condition $I(0) = 0$ for the following situations. (E_0 is a given constant.)

(a) $E(t) = E_0[u(t - 1) - u(t - 2)]$ (a square pulse)

(b) $E(t) = E_0[1 - u(t - \pi)]\sin t$ (a single pulse of a sine wave)

(c) $E(t) = E_0 f(t)$ where $f(t)$ is the full square wave in Example 14.20

(d) $E(t) = E_0 g(t)$ where $g(t)$ is the full-wave rectification of $\sin t$ from Exercise 3 on page 725

6. The differential equation

$$V\frac{dy}{dt} = r_i f(t) - r_o y$$

may be used to describe a compartmental model of the absorption of a drug by a body organ of volume V. The function $y(t)$ is the concentration of the drug in the organ's fluid at time t, r_i and r_o are the respective rates of fluid into and out of the organ, and $f(t)$ is the concentration of the drug entering the organ.

(a) Find $y(t)$ if $y(0) = 0$, $r_i = r_o$, and $f(t) = t[1 - u(t - a)]$.

(b) Find $y(t)$ if $y(0) = 1$, $r_i \ne r_o$, and $f(t) = t[1 - u(t - a)]$.

7. An exercise in Section 6.4 (Exercise 10) dealt with cooking a pumpkin pie. Here we consider the same problem with a different pattern to the oven temperature. A pumpkin pie recipe says to place the ingredients in a preheated oven at $425°$F for 15 minutes, then to turn the thermostat to $350°$F and continue baking for an additional 45 minutes. Assume that the temperature of the oven decreases linearly from $425°$F to $350°$F in 10 minutes after the thermostat is changed and that Newton's law of heating, namely,

$$\frac{dT}{dt} = k[T - T_a(t)],$$

applies to this situation.

(a) Find an expression for the oven temperature, $T_a(t)$, in terms of the unit step function.

(b) If the initial temperature of the uncooked pie is $70°$F, find the temperature of the pie as a function of t.

(c) For $T(0) = 70$, find the temperature of the pie as a function of t if the temperature of the oven changes instantly from $450°F$ to $350°F$ after the first 15 minutes.

(d) Compare the graphs of your two solutions in parts (b) and (c), and comment on their similarities and differences.

14.6 Applications to Higher Order Linear Differential Equations

We now focus our attention on solving second order linear differential equations with constant coefficients using Laplace transforms. Here we find that the only change from the previous section is that we need to use the result for $\mathcal{L}\left\{y''\right\}$ in addition to that for $\mathcal{L}\left\{y'\right\}$. We revisit the application areas covered in Chapter 8 — namely electric circuits, spring-mass systems, and the linear pendulum.

EXAMPLE 14.26: A Spring-Mass System

The differential equation that describes a spring-mass system that is subjected to an external forcing function is given by

$$m\frac{d^2x}{dt^2} + b\frac{dx}{dt} + kx = f(t),$$

where m is the mass, b is a damping coefficient, k is the spring constant, and $f(t)$ is the forcing function.

We consider the situation in which m, b, k, and $f(t)$ are chosen so the differential equation has the form

$$\frac{d^2x}{dt^2} + 4\frac{dx}{dt} + 13x = 9e^{-2t} \tag{14.79}$$

and where initially the mass is pulled 3 units from equilibrium and released from rest. This means that our initial conditions are $x(0) = 3$, $x'(0) = 0$.

We start by taking the Laplace transform of (14.79) to obtain

$$\mathcal{L}\left\{\frac{d^2x}{dt^2} + 4\frac{dx}{dt} + 13x\right\} = \mathcal{L}\left\{9e^{-2t}\right\}. \tag{14.80}$$

Using the properties of Laplace transforms from Tables 14.3 and 14.4 and our initial conditions, we can reduce (14.80) to

$$s^2\mathcal{L}\left\{x\right\} - 3s + 4\left(s\mathcal{L}\left\{x\right\} - 3\right) + 13\mathcal{L}\left\{x\right\} = \frac{9}{s+2}.$$

Solving for $\mathcal{L}\left\{x\right\}$, we find that

$$\mathcal{L}\left\{x\right\} = \frac{3s+12}{s^2+4s+13} + \frac{9}{(s+2)(s^2+4s+13)}. \tag{14.81}$$

Because the quadratic in the denominator may be written as $s^2 + 4s + 13 = (s + 2)^2 + 9$, and because from Table 14.6 we have

$$\mathcal{L}^{-1}\left\{\frac{s+2}{(s+2)^2+9}\right\} = e^{-2t}\cos 3t$$

and

$$\mathcal{L}^{-1}\left\{\frac{3}{(s+2)^2+9}\right\} = e^{-2t}\sin 3t,$$

we rewrite $3s + 12$ as $3(s + 2) + 2 \times 3$. Then we can recognize the inverse Laplace transform of the first term on the right-hand side of (14.81) as $e^{-2t}(3\cos 3t + 2\sin 3t)$. To evaluate the inverse Laplace transform of the remaining term, we could use either partial fractions or the convolution theorem. We use the latter with $G(s) = 1/(s + 2)$, so $g(t) = e^{-2t}$, and $F(s) = 1/(s^2 + 4s + 13)$, so $f(t) = (1/3)e^{-2t}\sin 3t$. This gives us

$$\mathcal{L}^{-1}\left\{\frac{9}{(s+2)(s^2+4s+13)}\right\} = 9\int_0^t \frac{1}{3}e^{-2z}\sin 3z\, e^{-2(t-z)}\, dz,$$

or

$$\mathcal{L}^{-1}\left\{\frac{9}{(s+2)(s^2+4s+13)}\right\} = 3e^{-2t}\int_0^t \sin 3z\, dz = 3e^{-2t}\left(-\frac{1}{3}\cos 3z\Big|_0^t\right).$$

Thus, we have

$$\mathcal{L}^{-1}\left\{\frac{9}{(s+2)(s^2+4s+13)}\right\} = e^{-2t}(-\cos 3t + 1). \tag{14.82}$$

Combining all our inverse transforms gives our solution of the initial value problem as

$$\begin{aligned} x(t) &= e^{-2t}(3\cos 3t + 2\sin 3t) + e^{-2t}(-\cos 3t + 1) \\ &= e^{-2t}(2\cos 3t + 2\sin 3t + 1). \end{aligned} \tag{14.83}$$

$\square$

Notice that had we used our previous method of finding the general solution of the associated homogeneous differential equation and adding to this a particular solution, we would have derived the characteristic equation

$$r^2 + 4r + 13 = 0.$$

This gives us the same roots for r as we found for s when factoring the quadratic expression in the denominator of (14.81). Thus, some of the details in solving initial value problems using Laplace transforms are the same as those we encountered with methods used earlier.

One advantage of using Laplace transforms is the fact that the initial conditions are satisfied automatically. Another advantage is in treating forcing functions that are periodic or given by different formulas over different domains.

EXAMPLE 14.27

Recall that a series RLC circuit is governed by the differential equation

$$L\frac{d^2q}{dt^2} + R\frac{dq}{dt} + \frac{1}{C}q = E(t),$$

where L is the inductance, R the resistance, and C the capacitance. The charge at any time t is given by $q(t)$, and the applied voltage is given by $E(t)$.

We want to solve the initial value problem associated with this circuit where $L = 1$, $R = 6, C = 1/10, E(t) = 2[u(t-1) - u(t-2)], q(0) = 0$, and $q'(0) = 1$. This gives our initial value problem as

$$\frac{d^2q}{dt^2} + 6\frac{dq}{dt} + 10q = 2[u(t-1) - u(t-2)], \qquad q(0) = 1, \qquad q'(0) = 0.$$

The graph of the applied voltage is shown in Figure 14.23. (What action in the electric circuit does $2[u(t-1) - u(t-2)]$ represent?)

If we use these given values in the preceding equation and take the Laplace transform of the result, we obtain

$$\mathcal{L}\left\{\frac{d^2q}{dt^2} + 6\frac{dq}{dt} + 10q\right\} = \mathcal{L}\left\{2[u(t-1) - u(t-2)]\right\}.$$

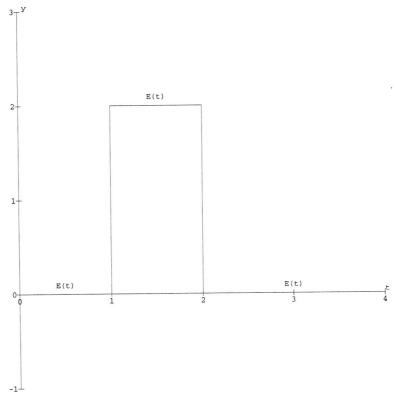

FIGURE 14.23 The applied voltage $E(t)$

Using some properties of Laplace transforms changes this equation to

$$s^2 \mathcal{L}\{q\} - 1 + 6s\mathcal{L}\{q\} + 10\mathcal{L}\{q\} = 2\left(\frac{e^{-s}}{s} - \frac{e^{-2s}}{s}\right).$$

Solving this equation for $\mathcal{L}\{q\}$ gives

$$\mathcal{L}\{q\} = \frac{1}{s^2 + 6s + 10} + \frac{1}{s^2 + 6s + 10}\frac{2}{s}\left(e^{-s} - e^{-2s}\right). \qquad (14.84)$$

We now examine the first term on the right-hand side of (14.84) and express it as $1/(s^2 + 6s + 10) = 1/[(s+3)^2 + 1]$, so

$$\mathcal{L}^{-1}\left\{\frac{1}{s^2 + 6s + 10}\right\} = e^{-3t}\sin t.$$

In the second term on the right-hand side of (14.84) we have $[1/(s^2 + 6s + 10][2/s]$, for which we may use the special case of the convolution theorem to find

$$\mathcal{L}^{-1}\left\{\frac{1}{s^2 + 6s + 10}\frac{2}{s}\right\} = 2\int_0^t e^{-3z}\sin z \, dz.$$

We designate this function by $h(t)$ and use a table of integrals, or integration by parts (twice), to find that

$$h(t) = 2\int_0^t e^{-3z}\sin z \, dz = -0.2e^{-3t}(3\sin t + \cos t) + 0.2.$$

Combining these results gives the solution of our initial value problem as

$$q(t) = e^{-3t}\sin t + h(t-1)u(t-1) - h(t-2)u(t-2). \qquad (14.85)$$

This can be expressed in the form

$$q(t) = \begin{cases} e^{-3t}\sin t & \text{if } t < 1, \\ e^{-3t}\sin t + h(t-1) & \text{if } 1 \leq t < 2, \\ e^{-3t}\sin t + h(t-1) - h(t-2) & \text{if } 2 \leq t. \end{cases}$$

The graph of $q(t)$ for $0 \leq t \leq 4$ is shown in Figure 14.24.   □

EXAMPLE 14.28

Here we consider the solution of the initial value problem

$$\frac{d^2 x}{dt^2} + x = f(t), \qquad x(0) = 1, \qquad x'(0) = 1,$$

where $f(t)$ is the full square wave from Example 14.20. This initial value problem models small undamped oscillations of a

$$\left\{ \begin{array}{l} \text{spring-mass system} \\ \text{LC electrical circuit} \\ \text{linear pendulum} \end{array} \right\}$$

depending on the interpretation of x, $f(t)$, $x(0)$, and $x'(0)$. Regardless of the physical situation, we will use the Laplace transform to solve this problem.

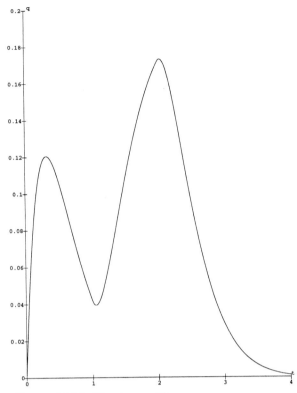

FIGURE 14.24   The charge $q(t)$

Taking the Laplace transform of our differential equation gives

$$\mathcal{L}\left\{\frac{d^2x}{dt^2}+x\right\}=\mathcal{L}\{f(t)\}.$$

Using some properties of Laplace transforms changes this equation to

$$s^2\mathcal{L}\{x\}-s-1+\mathcal{L}\{x\}=\mathcal{L}\{f(t)\}.$$

Solving this equation for $\mathcal{L}\{x\}$ we get

$$\mathcal{L}\{x\}=\frac{s}{s^2+1}+\frac{1}{s^2+1}+\frac{1}{s^2+1}\mathcal{L}\{f(t)\}.\tag{14.86}$$

The inverse Laplace transforms of the first two terms in (14.86) are in Table 14.6, whereas $\mathcal{L}\{f(t)\}$ is given by (14.52) as $(1/s)\sum_{k=0}^{\infty}(-1)^k e^{-sk}$. Because we know that $\mathcal{L}^{-1}\{e^{-as}F(s)\}=f(t-a)u(t-a)$, we use the special case of the convolution theorem to evaluate $\mathcal{L}^{-1}\{1/[s(s^2+1)]\}$ as

$$\mathcal{L}^{-1}\left\{\frac{1}{s\left(s^2+1\right)}\right\}=\int_0^t\sin z\,dz=1-\cos t.$$

Thus, we may write our solution of this initial value problem as

$$x(t)=\cos t+\sin t+\sum_{k=0}^{\infty}(-1)^k\left[1-\cos(t-k)\right]u(t-k).$$

Figure 14.25 shows the effect of the square wave forcing function. Without the forcing function, we have oscillatory motion as given in the bottom curve in this figure.   □

Notice that in Examples 14.27 and 14.28 we had forcing functions that were discontinuous but solutions that were not only continuous but also differentiable, see Figures 14.24 and 14.25. (How can you tell from the figures that the solutions are differentiable?) We will return to this comment in Section 14.8, where we will discuss the circumstances under which this occurs.

We have spent some time using the Laplace transform to solve second order linear differential equations with constant coefficients, so now we look at the general initial value problem associated with such differential equations, namely,

$$ay'' + by' + cy = f(t), \qquad y(0) = y_0, \qquad y'(0) = y_0^*, \qquad \textbf{(14.87)}$$

where $f(t)$ is a given function and a, b, c, y_0, and y_0^* are constants. If we follow the same pattern that we used in the previous examples — take the Laplace transform of (14.87) and solve for $Y(s) = \mathcal{L}\{y(t)\}$ — we get

$$Y(s) = \frac{asy_0 + ay_0^* + by_0}{as^2 + bs + c} + \frac{1}{as^2 + bs + c} F(s). \qquad \textbf{(14.88)}$$

Next we introduce the two functions

$$h(t) = \mathcal{L}^{-1}\left\{\frac{1}{as^2 + bs + c}\right\}$$

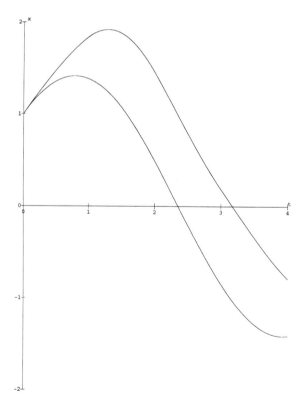

FIGURE 14.25   The effect of the square wave forcing function

and

$$g(t) = \mathcal{L}^{-1} \left\{ \frac{s}{as^2 + bs + c} \right\}.$$

We then apply the inverse Laplace transform to (14.88) and use convolution theorem to find

$$y(t) = ay_0 g(t) + \left(ay_0^* + by_0\right) h(t) + \int_0^t h(t - z) f(z) \, dz. \tag{14.89}$$

This is the general solution of the nonhomogeneous second order linear differential equation with constant coefficients. Notice that the functions $h(t)$ and $g(t)$ are solutions of the associated homogeneous equation and that they correspond to the first term on the right-hand side of (14.88). The second term on the right-hand side of (14.88) corresponds to the particular solution $\int_0^t h(t - z) f(z) \, dz$.

In addition to the function f, the explicit solution (14.89) will depend on the factors of the quadratic expression $as^2 + bs + c$. There are three possibilities.

1. The quadratic expression $as^2 + bs + c$ has two real distinct factors, r_1 and r_2, so that $as^2 + bs + c = a\left(s - r_1\right)\left(s - r_2\right)$. In this case we have

$$h(t) = \mathcal{L}^{-1} \left\{ \frac{1}{as^2 + bs + c} \right\} = \mathcal{L}^{-1} \left\{ \frac{1}{a\left(s - r_1\right)\left(s - r_2\right)} \right\} = \frac{1}{a} \frac{1}{r_1 - r_2} \left(e^{r_1 t} - e^{r_2 t}\right)$$

and

$$g(t) = \mathcal{L}^{-1} \left\{ \frac{s}{as^2 + bs + c} \right\} = \mathcal{L}^{-1} \left\{ \frac{s}{a\left(s - r_1\right)\left(s - r_2\right)} \right\} = \frac{1}{a} \frac{1}{r_1 - r_2} \left(r_1 e^{r_1 t} - r_2 e^{r_2 t}\right).$$

The linear combination of $h(t)$ and $g(t)$ — written as $C_1 e^{r_1 t} + C_2 e^{r_2 t}$ — is the general solution of the associated homogeneous solution. This is consistent with the results we found in Chapter 7 — namely, (7.42).

2. The quadratic expression $as^2 + bs + c$ has one real repeated factor, r, so that $as^2 + bs + c = a\left(s - r\right)^2$. In this case we have

$$h(t) = \mathcal{L}^{-1} \left\{ \frac{1}{as^2 + bs + c} \right\} = \mathcal{L}^{-1} \left\{ \frac{1}{a\left(s - r\right)^2} \right\} = \frac{1}{a} t e^{rt}$$

and

$$g(t) = \mathcal{L}^{-1} \left\{ \frac{s}{as^2 + bs + c} \right\} = \mathcal{L}^{-1} \left\{ \frac{s}{a\left(s - r\right)^2} \right\} = \frac{1}{a} \left(e^{rt} + rt e^{rt}\right).$$

The linear combination of $h(t)$ and $g(t)$ — written as $C_1 e^{rt} + C_2 t e^{rt}$ — is the general solution of the associated homogeneous solution. This is consistent with the results we found in Chapter 7 — namely, (7.47).

3. The quadratic expression $as^2 + bs + c$ has two complex factors, $\alpha \pm i\beta$, so that $as^2 + bs + c = a\left[(s - \alpha)^2 + \beta^2\right]$. In this case we have

$$h(t) = \mathcal{L}^{-1} \left\{ \frac{1}{as^2 + bs + c} \right\} = \mathcal{L}^{-1} \left\{ \frac{1}{a\left[(s - \alpha)^2 + \beta^2\right]} \right\} = \frac{1}{a\beta} e^{\alpha t} \sin \beta t$$

and

$$g(t) = \mathcal{L}^{-1}\left\{\frac{s}{as^2+bs+c}\right\} = \mathcal{L}^{-1}\left\{\frac{s}{a\left[(s-\alpha)^2+\beta^2\right]}\right\} = \frac{1}{a}e^{\alpha t}\left(\frac{\alpha}{\beta}\sin\beta t + \cos\beta t\right).$$

The linear combination of $h(t)$ and $g(t)$ — written as $e^{\alpha t}\left(C_1\cos\beta t + C_2\sin\beta t\right)$ — is the general solution of the associated homogeneous solution. This is consistent with the results we found in Chapter 7 — namely, (7.43).

EXERCISES

1. Create a *How to Solve Second Order Linear Differential Equations Using Laplace Transforms.* Add statements under the headings Purpose, Technique, and Comments that summarize what you discovered in this section.

2. Solve the following initial value problems using Laplace transforms. Then compare your answers with the ones you found in Exercise 1, Section 7.3.
   (a) $y'' + y' - 2y = 0$, $y(0) = 0$, $y'(0) = 3$
   (b) $y'' + 2y' - 10y = 0$, $y(0) = 0$, $y'(0) = 4$
   (c) $y'' + 6y' + 9y = 0$, $y(0) = 2$, $y'(0) = 0$
   (d) $12y'' + y' - y = 0$, $y(0) = 4$, $y'(0) = 0$
   (e) $y'' + 3y' = 0$, $y(0) = 4$, $y'(0) = 3$
   (f) $y'' - 2\pi y' + 2\pi^2 y = 0$, $y(0) = 0$, $y'(0) = -2\pi$
   (g) $y'' + 10y' + 100y = 0$, $y(0) = 15$, $y'(0) = 4$
   (h) $y'' + 10y' + 100y = 0$, $y(1) = 0$, $y'(1) = 0$

3. Solve the following initial value problems using Laplace transforms. Then compare your answers with the ones you found in Exercise 8, Section 11.2.
   (a) $y'' + y' - 12y = 8e^{3t}$, $y(0) = 0$, $y'(0) = 1$
   (b) $y'' + 6y' + 9y = e^{3t}$, $y(0) = 0$, $y'(0) = 6$
   (c) $y'' - 5y' + 6y = 12te^{-t} - 7e^{-t}$, $y(0) = y'(0) = 0$
   (d) $y'' + 4y = 8\sin 2t + 8\cos 2t$, $y(\pi) = y'(\pi) = 2\pi$

4. Solve the differential equation (14.79)

$$\frac{d^2x}{dt^2} + 4\frac{dx}{dt} + 13x = 9e^{-2t}$$

subject to $x(0) = 3$, $x'(0) = 0$, using the methods of Chapter 11. Compare your answer with (14.83).

5. Solve the following initial value problems.
   (a) $x'' + x = e^{-2t}\sin t$, $x(0) = x'(0) = 0$
   (b) $x'' + 2x' + 2x = tu(t-1)$, $x(0) = 1$, $x'(0) = 0$
   (c) $x'' + 4x' + 4x = e^{-3t}$, $x(0) = x'(0) = 0$
   (d) $x'' + x' - 2x = e^{-t}\sin t$, $x(0) = x'(0) = 0$
   (e) $x'' + 3x' + 2x = t^2[1 - u(t-1)]$, $x(0) = x'(0) = 0$
   (f) $x'' + 4x' + 4x = 4\cos 2t$, $x(0) = x'(0) = 1$
   (g) $x'' + 4x' + 4x = f(t)$, $x(0) = 1$, $x'(0) = 0$, where $f(t)$ is a given function
   (h) $x'' + \omega^2 x = f(t)$, $x(0) = 1$, $x'(0) = 0$, where ω is a constant and $f(t)$ is a given function

(i) $x'' + \omega^2 x = f(t)$, $\quad x(0) = 0$, $\quad x'(0) = 1$, where ω is a constant and $f(t)$ is a given function

(j) $x'' + \omega^2 x = u(t - 2) - u(t - 5)$, $\quad x(0) = x'(0) = 1$, where ω is a constant

(k) $x'' + \omega^2 x = [1 - u(t - \pi)] \sin t$, $\quad x(0) = 0$, $\quad x'(0) = 1$, where ω is a constant

(l) $x'' + \omega^2 x = g(t)$, where ω is a constant and $g(t)$ is the full-wave rectification of $\sin t$ from Exercise 3 on page 725, $\quad x(0) = x'(0) = 0$

6. A particular forced vibration of a mass m at the end of a vertical spring, with spring constant k, is described by

$$mx''(t) + kx(t) = f(t),$$

where

$$f(t) = 1 + u(t - 1) - 2u(t - 2).$$

(a) Explain what $f(t)$ represents concerning the motion of the top of the spring.

(b) Solve this differential equation for $m = 1$, $k = 4$, $x(0) = x'(0) = 0$.

(c) Plot your solution for the interval $0 \leq t \leq 3$ as well as the graph of $f(t)$. What do you observe?

7. An initial value problem for a series RLC circuit consists of the differential equation

$$L \frac{d^2 q}{dt^2} + R \frac{dq}{dt} + \frac{1}{C} q = E(t),$$

with initial conditions $q(0) = 1$, $q'(0) = 0$. Solve this initial value problem for the following situations.

(a) $R/L = 4$, $1/(LC) = 20$, $E(t)/L = 10e^{-t/2}$

(b) $R/L = 4$, $1/(LC) = 20$, $E(t)/L = 8 \sin 4t$

(c) $R/L = 5$, $1/(LC) = 6$, $E(t)/L = 2[1 - u(t - 2)]$

(d) $R/L = 5$, $1/(LC) = 6$, $E(t)/L = 2[1 - u(t - 2)] + 4u(t - 4)$

(e) $R/L = 5$, $1/(LC) = 6$, $E(t)/L = 2[1 - u(t - 2)]$, $E(t + 4) = E(t), t \geq 0$

14.7 Applications to Systems of Linear Differential Equations

Initial value problems involving systems of linear differential equations with constant coefficients may also be solved using Laplace transforms. The procedure is similar to that used many times in this chapter. Thus, we use the Laplace transform to transform our initial value problem for a system of differential equations into a system of algebraic equations. After solving this system of algebraic equations for the Laplace transform of our dependent variables, we use tables of inverse Laplace transforms to obtain our explicit solution. The advantages of the Laplace transform here include having the initial conditions satisfied automatically and avoiding the need to find either particular solutions or fundamental matrices. The procedure is illustrated with three examples.

EXAMPLE 14.29

Consider the system of differential equations

$$\begin{aligned} x' &= x - 2y, \\ y' &= 5x - y, \end{aligned} \qquad (14.90)$$

subject to the initial conditions $x(0) = -1$, $y(0) = 2$. If we apply the Laplace transform to both of these differential equations, we obtain

$$\mathcal{L}\left\{x'\right\} = \mathcal{L}\left\{x - 2y\right\},$$
$$\mathcal{L}\left\{y'\right\} = \mathcal{L}\left\{5x - y\right\}.$$

Using some results in Table 14.4 we can transform this system into

$$sX(s) + 1 = X(s) - 2Y(s),$$
$$sY(s) - 2 = 5X(s) - Y(s), \tag{14.91}$$

where $X(s) = \mathcal{L}\left\{x\right\}$ and $Y(s) = \mathcal{L}\left\{y\right\}$. We may solve for $Y(s)$ from the top equation, $Y(s) = (1/2)\left[(1 - s)X(s) + 1\right]$, and substitute the result into the bottom equation to find

$$X(s) = -\frac{s + 5}{s^2 + 9}. \tag{14.92}$$

If we rewrite this expression as

$$X(s) = -\frac{s}{s^2 + 9} - \frac{5}{3}\frac{3}{s^2 + 9},$$

we see that $x(t)$ is

$$x(t) = -\cos 3t - \frac{5}{3}\sin 3t. \tag{14.93}$$

We have two choices for determining $y(t)$. One is to substitute the expression for $X(s)$ from (14.92) into one of the equations in (14.91) to obtain an expression for $Y(s)$, and then find its inverse Laplace transform. However, here it is easier simply to use the explicit solution for $x(t)$ from (14.93) in the first of our original differential equations in (14.90) to obtain

$$y(t) = \frac{1}{2}\left(x - x'\right) = \frac{1}{2}\left(-\cos 3t - \frac{5}{3}\sin 3t - 3\sin 3t + 5\cos 3t\right),$$

or

$$y(t) = 2\cos 3t - \frac{7}{3}\sin 3t.$$

$\square$

EXAMPLE 14.30

We will now consider a situation in which using a fundamental matrix led to a repeated eigenvalue. Thus, we consider the initial value problem

$$\begin{aligned} x' &= -x + y + 25\sin t, \\ y' &= -x - 3y, \end{aligned} \tag{14.94}$$

subject to the initial conditions $x(0) = -1$, $y(0) = -1$.

If we apply the Laplace transform to both of these differential equations, we obtain

$$\mathcal{L}\left\{x'\right\} = \mathcal{L}\left\{-x + y + 25\sin t\right\},$$
$$\mathcal{L}\left\{y'\right\} = \mathcal{L}\left\{-x - 3y\right\}.$$

Using some results in Tables 14.3 and 14.4 transforms this to

$$sX(s) + 1 = -X(s) + Y(s) + 25/(s^2 + 1),$$
$$sY(s) + 1 = \qquad -X(s) - 3Y(s).$$

We may solve for $X(s)$ from the bottom equation, $X(s) = -(3 + s)Y(s) - 1$, and substitute the result into the top equation to find

$$(s + 1)[-(3 + s)Y(s) - 1] + 1 = Y(s) + \frac{25}{s^2 + 1}.$$

We may rewrite this expression as

$$-[(s + 1)(3 + s) + 1]Y(s) = s + \frac{25}{s^2 + 1}$$

or, solving for $Y(s)$,

$$Y(s) = -\frac{s}{(s + 2)^2} - \frac{25}{(s + 2)^2 (s^2 + 1)}.$$

Here the inverse Laplace transform of the first term is $e^{-2t}(2t - 1)$. The second may be calculated in several ways. We will use partial fractions and write

$$\frac{25}{(s + 2)^2 (s^2 + 1)} = \frac{A}{s + 2} + \frac{B}{(s + 2)^2} + \frac{Cs + D}{(s^2 + 1)}$$
$$= \frac{A(s + 2)(s^2 + 1) + B(s^2 + 1) + (Cs + D)(s + 2)^2}{(s + 2)^2 (s^2 + 1)}.$$

Because the denominators are identical, the numerators must also be identical. This gives

$$25 = A(s + 2)(s^2 + 1) + B(s^2 + 1) + (Cs + D)(s + 2)^2,$$

or

$$25 = (A + C)s^3 + (2A + B + 4C + D)s^2 + (A + 4C + 4D)s + (2A + B + 4D).$$

Because this is true for all values of s, we may equate the coefficients of like powers of s to find the four equations

$$A + C = 0,$$
$$2A + B + 4C + D = 0,$$
$$A + 4C + 4D = 0,$$
$$2A + B + 4D = 25.$$

From the first equation we find $A = -C$, which when substituted into the third equation gives $D = -3C/4$. When we substitute these equations for A and D into the third equation, we find $B = 5C + 25$. We now use these equations for A, B, and D in the second equation, obtaining $C = -4$. Thus, the remaining constants are $A = 4$, $B = 5$, and $D = 3$. This gives us

$$\mathcal{L}^{-1}\left\{\frac{25}{(s + 2)^2 (s^2 + 1)}\right\} = 4e^{-2t} + 5te^{-2t} - 4\cos t + 3\sin t,$$

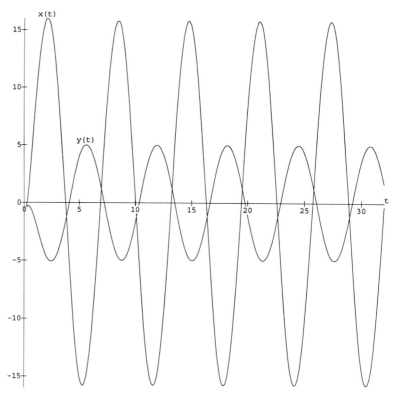

FIGURE 14.26   The two functions $x(t) = e^{-2t}(3t + 8) - 9\cos t + 13\sin t$ and $y(t) = -e^{-2t}(3t + 5) + 4\cos t - 3\sin t$

and the explicit form of $y(t)$ is

$$y(t) = -e^{-2t}(3t + 5) + 4\cos t - 3\sin t.$$

To find $x(t)$ we return to (14.94) and solve for x as $x = -3y - y'$. Using the value of y and its derivative gives

$$x(t) = e^{-2t}(3t + 8) - 9\cos t + 13\sin t.$$

The graphs of $x(t)$ and $y(t)$ appear in Figure 14.26, where $x(t)$ and $y(t)$ both quickly achieve steady state motion. □

EXAMPLE 14.31

As our final example, we will find the solution of the nonhomogeneous system of differential equations

$$\begin{aligned} x' &= -2x + y + u(t - 1), \\ y' &= 4x - 2y + e^{2t}, \end{aligned} \qquad \textbf{(14.95)}$$

subject to the initial conditions $x(0) = x_0$, $y(0) = y_0$. Taking the Laplace transform and using our familiar properties gives

$$\begin{aligned} sX(s) - x_0 &= -2X(s) + Y(s) + e^{-s}/s, \\ sY(s) - y_0 &= 4X(s) - 2Y(s) + 1/(s - 2). \end{aligned} \qquad \textbf{(14.96)}$$

We may solve for $Y(s)$ from the top equation in (14.96) as

$$Y(s) = (s+2)X(s) - x_0 - \frac{1}{s}e^{-s}$$

and substitute this equation into the bottom equation to obtain

$$(s+2)\left[(s+2)X(s) - x_0 - \frac{1}{s}e^{-s}\right] - y_0 = 4X(s) + \frac{1}{s-2}.$$

Solving this equation for $X(s)$ gives

$$X(s) = \frac{1}{s(s+4)}\left[(s+2)x_0 + y_0 + \frac{s+2}{s}e^{-s} + \frac{1}{s-2}\right].$$

Thus, the functions for which we need inverse Laplace transforms are

$$\frac{s+2}{s(s+4)}x_0, \qquad \frac{1}{s(s+4)}y_0, \qquad \frac{s+2}{s^2(s+4)}e^{-s}, \quad \text{and} \quad \frac{1}{s(s+4)(s-2)}.$$

The first two terms in this list may be written as

$$\frac{s+2}{s(s+4)}x_0 + \frac{1}{s(s+4)}y_0 = \frac{1}{s+4}x_0 + \frac{1}{s(s+4)}(2x_0 + y_0)$$

both of which occur in our list of inverse Laplace transforms in Table 14.6. Thus,

$$\mathcal{L}^{-1}\left\{\frac{s+2}{s(s+4)}x_0 + \frac{1}{s(s+4)}y_0\right\} = x_0 e^{-4t} + \frac{1}{4}(2x_0 + y_0)\left(1 - e^{-4t}\right).$$

The last term in our list may be evaluated using a partial fraction decomposition as

$$\frac{1}{s(s+4)(s-2)} = \frac{A}{s} + \frac{B}{s+4} + \frac{C}{s-2} = \frac{A(s+4)(s-2) + Bs(s-2) + Cs(s+4)}{s(s+4)(s-2)}.$$

We then equate the numerators in the first and last terms to obtain

$$1 = A(s+4)(s-2) + Bs(s-2) + Cs(s+4).$$

Because this is to be an identity in s, it must be true for all values of s. Thus, it is true for the particular values of 0, 2, and -4. Using these three values allows us to find $A = -1/8$, $B = 1/24$, and $C = 1/12$, so

$$\mathcal{L}^{-1}\left\{\frac{1}{s(s+4)(s-2)}\right\} = -\frac{1}{8} + \frac{1}{24}e^{-4t} + \frac{1}{12}e^{2t}.$$

The third term in our list has e^{-s} times a rational function of s. We first find the inverse Laplace transform of the rational function and then use a result involving the unit step function. We write

$$\frac{s+2}{s^2(s+4)} = \frac{s}{s^2(s+4)} + \frac{2}{s^2(s+4)} = \frac{1}{s(s+4)} + \frac{2}{s^2(s+4)}.$$

The first of these expressions occurs in Table 14.6, so

$$\mathcal{L}^{-1}\left\{\frac{1}{s(s+4)}\right\} = \frac{1}{4}\left(1 - e^{-4t}\right).$$

We now use the special case of the convolution theorem to write

$$\mathcal{L}^{-1}\left\{\frac{2}{s^2(s+4)}\right\} = \mathcal{L}^{-1}\left\{\frac{2}{s(s+4)}\frac{1}{s}\right\} = \int_0^t \frac{1}{4}\left(1-e^{-4z}\right)\,dz = \frac{1}{8}(4t+e^{-4t}-1)$$

If we define $h(t)$ by $h(t) = \mathcal{L}^{-1}\left\{\frac{s+2}{s^2(s+4)}\right\}$, we have

$$h(t) = \frac{1}{4}\left(1-e^{-4t}\right) + \frac{1}{8}\left(4t+e^{-4t}-1\right) = \frac{1}{8}\left(4t-e^{-4t}+1\right).$$

(See Exercise 2 for an alternative way of computing this inverse Laplace transform.) Finally, we collect our results to find our explicit solution for $x(t)$ as

$$x(t) = x_0 e^{-4t} + \frac{1}{4}\left(2x_0+y_0\right)\left(1-e^{-4t}\right) - \frac{1}{8} + \frac{1}{24}e^{-4t} + \frac{1}{12}e^{2t} + h(t-1)u(t-1).$$

(14.97)

To find $y(t)$ we return to the first equation in (14.95), where we have that $y = x' + 2x - u(t-1)$. Differentiating $x(t)$ in (14.97) and making the appropriate substitutions gives

$$y(t) = \left(\frac{1}{2}y_0 - x_0 - \frac{1}{12}\right)e^{-4t} + x_0 + \frac{1}{2}y_0 - \frac{1}{4} + \frac{1}{3}e^{2t} +$$
$$+ \left[h'(t-1) + 2h(t-1) - 1\right]u(t-1).$$

$\square$

EXERCISES

1. Create a *How to Solve Systems of Linear Differential Equations Using Laplace Transforms*. Add statements under the headings Purpose, Technique, and Comments that summarize what you discovered in this section.

2. Use partial fractions to find

$$\mathcal{L}^{-1}\left\{\frac{2}{s^2(s+4)}\right\}.$$

3. Use Laplace transforms to solve the following initial value problems. Then compare your answers to the ones you found for Exercise 5, Section 7.4.

   (a) $\begin{aligned} x' &= 2x - y, \\ y' &= -x + 2y, \end{aligned}$ $x(0) = 2,$ $y(0) = 0$

   (b) $\begin{aligned} x' &= 2x - 2y, \\ y' &= 3x - 2y, \end{aligned}$ $x(0) = 0,$ $y(0) = 2$

   (c) $\begin{aligned} x' &= 3x - 2y, \\ y' &= 2x - 2y, \end{aligned}$ $x(0) = 2,$ $y(0) = 1$

   (d) $\begin{aligned} x' &= 5x - 2y, \\ y' &= 4x - 2y, \end{aligned}$ $x(0) = 0,$ $y(0) = 6$

4. Use Laplace transforms to solve the following initial value problems.

   (a) $\begin{aligned} x' &= -2x + 4y + 7e^{3t}, \\ y' &= x + y - 2e^{3t}, \end{aligned}$ $x(0) = 1,$ $y(0) = 0$

   (b) $\begin{aligned} x' &= 2x + y + 3e^t, \\ y' &= -3x - 2y + e^{-t}, \end{aligned}$ $x(0) = 1,$ $y(0) = 2$

   (c) $\begin{aligned} x' &= -2x + 4y + e^{2t}, \\ y' &= x + y - e^{2t}, \end{aligned}$ $x(0) = 0,$ $y(0) = 1$

(d) $\begin{aligned} x' &= 2x + y + \cos t, \\ y' &= -3x - 2y, \end{aligned}$ $\quad x(0) = 1, \quad y(0) = 0$

(e) $\begin{aligned} x' &= -2x + 4y + u(t-1) - t(t-3), \\ y' &= x + y, \end{aligned}$ $\quad x(0) = 0, \quad y(0) = 1$

(f) $\begin{aligned} x' &= 2x + y + f(t), \\ y' &= -3x - 2y, \end{aligned}$ $\quad x(0) = 1, \quad y(0) = 0$ where $f(t)$ is the full square wave from
Exercise 14.20.

5. The differential equations that describe the electric circuit in Figure 14.27 are

$$LI_1' + RI_2 = E(t),$$
$$RI_2' + (I_2 - I_1)/C = 0.$$

(a) Solve these equations if $I_1(0) = I_2(0) = 0$, $E(t) = \sin \omega t$, $L = 1$, $R = 5$, $C = 1/20$.

(b) Repeat part (a) if the initial conditions are changed to $I_1(0) = 0$, $I_2(0) = 1$.

6. Another advantage in using the Laplace transform is that the system of equations need not be in the form

$$x' = ax + by + f(t),$$
$$y' = cx + dy + g(t).$$

(a) Show that if $X(s) = \mathcal{L}\{x(t)\}$ and $Y(s) = \mathcal{L}\{y(t)\}$, using the Laplace transform converts the initial value problem

$$\begin{aligned} 2x' + y' - y &= t, \\ x' + y' &= t^2, \end{aligned} \quad x(0) = 1, \quad y(0) = 0$$

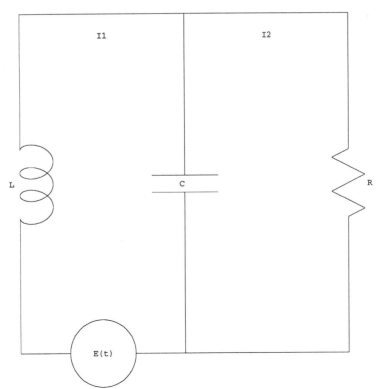

FIGURE 14.27 Electric circuit

into the algebraic system

$$2s\,X(s) + (s-1)\,Y(s) = 2 + 1/s^2,$$
$$s\,X(s) + s\,Y(s) \quad\; = 1 + 2/s^3.$$

(b) Solve the equations in part (a) for $Y(s)$ as

$$Y(s) = \frac{4-s}{s^3\,(s+1)} = \frac{4}{s^3\,(s+1)} - \frac{1}{s^2\,(s+1)},$$

and use partial fractions or the convolution theorem to show that

$$y(t) = 5 - 5t + 2t^2 - 5e^{-t}.$$

(c) Show that $X(s) + Y(s) = 1/s + 2/s^4$, and solve for $x(t)$.

7. Use Laplace transforms to solve the following initial value problems.

(a) $\begin{aligned} x' + 2x + y' &= 16e^{-2t}, \\ 2x' + 3y' + 5y &= 15, \end{aligned}$ $x(0) = 4, \qquad y(0) = -3$

(b) $\begin{aligned} 2x' - 2x + y' &= 1, \\ x' - 3x + y' - 3y &= 2, \end{aligned}$ $x(0) = 0, \qquad y(0) = 0$

(c) $\begin{aligned} x' - y' &= -e^t, \\ 2x' - 2y' - y &= 8, \end{aligned}$ $x(0) = -1, \qquad y(0) = -10$

(d) $\begin{aligned} x' - y' - 6y &= 0, \\ x' - 3x + 2y' &= 0, \end{aligned}$ $x(0) = 2, \qquad y(0) = 3$

(e) $\begin{aligned} x'' + 2x - 4y' &= 0, \\ x' + y'' - 4y &= 0, \end{aligned}$ $x(0) = -4, \qquad y(0) = 1 \qquad x'(0) = 8, \qquad y'(0) = 2$

14.8   When Do Laplace Transforms Exist?

We now turn to the all-important question: What conditions must functions satisfy in order to guarantee the existence of a Laplace transform and the validity of the theorems that we have been using so frequently? In Section 14.1 we made the following working assumptions:

1. $\int_0^\infty e^{-st} y\, dt$ exists.

2. $\lim_{t\to\infty} e^{-st} y(t) = 0.$

3. $\text{If } \int_0^\infty e^{-st} f(t)\, dt = \int_0^\infty e^{-st} g(t)\, dt, \text{ then } f(t) = g(t).$

These led to a formal process that seemed to work very well for the examples we considered.[8] However, not everything is as straight-forward as it looks. We will demonstrate this with an example.

EXAMPLE 14.32: The Inverse Laplace Transform of 1

An obvious omission from Table 14.5 is a function $f(t)$ for which $F(s) = 1$. Let us try to find the function $f(t)$ for which $\mathcal{L}\{f(t)\} = 1$ — that is, find $\mathcal{L}^{-1}\{1\}$.

     If we assume that such a function exists, then, because $F(s) = 1$, we also have $F(s - a) = 1$. From 14.3 — which states that if $\mathcal{L}\{f(t)\} = F(s)$, then $\mathcal{L}\{e^{at} f(t)\} =$

---

[8]Mathematicians use the word formal to mean "we hope it works, and so far everything seems to be OK, but before we trust it we need a proof."

$F(s - a)$ — we then find $\mathcal{L}\{f(t)\} = \mathcal{L}\{e^{at}f(t)\}$. This leads to $f(t) = e^{at}f(t)$, for every $a > 0$, which implies that $f(t) = 0$. Thus, we have $\mathcal{L}\{0\} = 1$. However, we already know that $\mathcal{L}\{0\} = 0$, so $0 = 1$. $\square$

Clearly, something has gone wrong with our formal analysis. The mistake we made in this example was assuming that the Laplace transform exists. In fact, the correct logic to use in the previous example is to start with two possibilities: Either the Laplace transform exists or it does not. If it exists, then the previous example leads to a contradiction. Therefore, the first possibility cannot happen, and so we are led inevitably to the second — the Laplace transform of a function that gives us 1 does not exist.

So we are forced to look at conditions under which Laplace transforms exist. Let's look at the assumption $\lim_{t \to \infty} e^{-st}y(t) = 0$ and think of functions $y(t)$ for which this is true. In fact, most of the functions we work with, such as t^n, $\sin at$, $\cos at$, e^{at}, the unit step function $u(t)$, and so on, have this property. Of these, the function e^{at} grows the most rapidly as $t \to \infty$. However, a little thought shows that there are functions that will violate this condition, such as e^{t^2}. Thus, we are forced to restrict ourselves to functions that don't grow more rapidly than exponentials. These functions are given a special name.

Definition 14.4: **The function $f(t)$ is said to be of EXPONENTIAL ORDER α as $t \to \infty$ if there exists a constant M such that**

$$|f(t)| \le Me^{\alpha t}$$

for all t greater than some T.

Comments about functions of exponential order α

- It is common to omit the phrase "as $t \to \infty$" when referring to a function being "of exponential order α as $t \to \infty$." We usually say the function is "of exponential order α," or just "of exponential order."

- This definition essentially says that a function is of exponential order α if it grows no faster than $e^{\alpha t}$.

- This definition is equivalent to the statement that a function is of exponential order α if $e^{-\alpha t}|f(t)|$ is bounded for $t > T$.

- The functions $\sin at$ and $\cos at$ are of exponential order 0, because $|\sin at| \le 1 = e^{0t}$ and $|\cos at| \le 1 = e^{0t}$ for all t.

- The unit step function $u(t)$ is of exponential order 0, because $|u(t)| \le 1 = e^{0t}$ for all t.

- The function t is of exponential order 1 because $|t| \le e^t = e^{1t}$ for all t. The function t^n is also of exponential order 1.

- The function e^{at} is of exponential order a because $|e^{at}| \le e^{at}$ for all t.

- The function e^{t^2} is not of exponential order because $e^{t^2}/e^{\alpha t} \to \infty$ as $t \to \infty$ for every α.

- When talking about a function being of exponential order, the exact values of α, M, and T are usually not relevant.

We have calculated many Laplace transforms, not all of them of continuous functions. The unit step function $u(t)$ is an example of a discontinuous function that has a Laplace transform. Discontinuities can come in various forms, ranging from mild ones, where the function jumps from one finite value to another, to extreme ones where the function goes to infinity at a finite value of t. It is common to distinguish between these types of discontinuities.

Definition 14.5:   **A function $f(t)$ is said to have a JUMP DISCONTINUITY AT $t = a$ if $\lim_{t \to a^-} f(t)$ and $\lim_{t \to a^+} f(t)$ exist, but $\lim_{t \to a^-} f(t) \neq \lim_{t \to a^+} f(t)$.**

Comments about jump discontinuities

- This definition says that if $f(t)$ goes to a finite value as we approach a from the left of a, and if $f(t)$ goes to a finite value as we approach a from the right of a, but these finite values are not the same, then $f(t)$ has a jump discontinuity at $t = a$.

- The unit step function $u(t)$ has a jump discontinuity at $t = 0$ because $\lim_{t \to 0^-} u(t) = 0$ and $\lim_{t \to 0^+} u(t) = 1$.

- The function $1/t$ has a discontinuity at $t = 0$, but it is not a jump discontinuity.

In order to accommodate the possibility of a function having a reasonable number of jump discontinuities, we introduce another definition.

Definition 14.6:   **A function $f(t)$ is said to be PIECEWISE CONTINUOUS ON THE INTERVAL $a \leq t \leq b$ if this interval can be divided into a finite number of subintervals such that $f(t)$ is continuous in each of these subintervals and has jump discontinuities at the interior endpoints of the intervals.**

Comments about piecewise continuous functions

- The idea behind this definition is as follows: If we can find a finite number of values of t, say, $a_1, a_2, \cdots, a_n$ (where $a < a_1 < a_2 < \cdots < a_n < b$), at which $f(t)$ has jump discontinuities and $f(t)$ is continuous within each of these subintervals, then $f(t)$ is piecewise continuous on $a \leq t \leq b$.

- If a function is continuous on the interval $a \leq t \leq b$, then it is piecewise continuous on the interval $a \leq t \leq b$.

- Every function we have considered in this chapter is piecewise continuous on any interval $a \leq t \leq b$. In particular, the square wave is piecewise continuous on every finite interval.

- If f is piecewise continuous in every finite interval, then $\int_0^t f(z)\, dz$ is continuous in every finite interval.

The main result concerning the existence of the Laplace transform is as follows.

Theorem 14.7:   *Let $f(t)$ be a piecewise continuous function on every finite subinterval for $t \geq 0$ and be of exponential order. Then the Laplace transform of $f(t)$, $\mathcal{L}\{f(t)\} = F(s)$, exists.*

Comments about Theorem 14.7

- All the functions in this chapter for which we calculated the Laplace transform are piecewise continuous on every finite subinterval of $t \geq 0$. (Most of them are continuous.) Also, all these functions are of exponential order. Thus, this theorem guarantees that the Laplace transforms exist and are as calculated.

- It can be shown (see Exercise 8 on page 757), that if the conditions of this theorem are met, then

$$\lim_{s \to \infty} F(s) = 0.$$

Thus, many functions $F(s)$ do not have an inverse Laplace transform. We had already discovered this for the function 1.

- This takes care of the conditions that we need to satisfy for the first two working assumptions.

Assuming we are dealing with functions that satisfy the conditions of Theorem 14.7, we will turn to the third working assumption we made:

$$\text{if } \int_0^\infty e^{-st} f(t)\, dt = \int_0^\infty e^{-st} g(t)\, dt, \text{ then } f(t) = g(t).$$

If we consider the two functions $f(t) = 1$ and

$$g(t) = \begin{cases} 1 & \text{if } t \neq 1, \\ 0 & \text{if } t = 1, \end{cases}$$

they are both piecewise continuous on every finite subinterval for $t \geq 0$, they are both of exponential order, and

$$\int_0^\infty e^{-st} f(t)\, dt = \int_0^\infty e^{-st} g(t)\, dt = \frac{1}{s}.$$

However, $f(t) \neq g(t)$. Thus, the assertion that

$$\text{If } \int_0^\infty e^{-st} f(t)\, dt = \int_0^\infty e^{-st} g(t)\, dt, \text{ then } f(t) = g(t),$$

needs some qualifications. Notice in this example that $f(t) = g(t)$ at all points where $f(t)$ and $g(t)$ are continuous.

Theorem 14.8: *If $\int_0^\infty e^{-st} f(t)\, dt = \int_0^\infty e^{-st} g(t)\, dt$, then $f(t) = g(t)$ at all points where $f(t)$ and $g(t)$ are continuous.*

Comments about Theorem 14.8

- This theorem guarantees the existence of inverse Laplace transforms.

We are now in a position to comment on those situations in which we found that piecewise continuous forcing functions gave rise to continuous solutions. In the first order case we found that the general solution of

$$y' + ky = f(t), \qquad y(0) = y_0.$$

was (14.77) — namely,

$$y(t) = y_0 e^{-kt} + e^{-kt} \int_0^t e^{kz} f(z)\, dz,$$

The only possible discontinuities in this solution could occur in the integral. However, $e^{kz} f(z)$ is piecewise continuous, and because the integral of a piecewise continuous function is continuous, the solution is continuous.

The second order case is a little more complicated. We found that the general solution of

$$ay'' + by' + cy = f(t), \qquad y(0) = y_0, \qquad y'(0) = y_0^*$$

was (14.89) — namely,

$$y(t) = ay_0 g(t) + (ay_0^* + by_0) h(t) + \int_0^t h(t-z) f(z)\, dz,$$

where $h(t)$ is one of $(e^{r_1 t} - e^{r_2 t}) / [a(r_1 - r_2)]$, te^{rt}/a, or $e^{\alpha t} \sin \beta t / (a\beta)$. As in the first order case, the only possible discontinuities in this solution could occur in the integral. However, $h(t-z) f(z)$ is piecewise continuous, and because the integral of a piecewise continuous function is continuous, the solution is continuous.

We now investigate the derivative of the solution. Again, the only possible discontinuities in the derivative of this solution could occur because of the integral. If we concentrate on the particular solution,

$$y_p(t) = \int_0^t h(t-z) f(z)\, dz,$$

and differentiate it,[9] we find

$$y_p'(t) = h(0) f(t) + \int_0^t h'(t-z) f(z)\, dz.$$

However, all three possibilities for $h(t)$ share the common property that $h(0) = 0$, so we have

$$y_p'(t) = \int_0^t h'(t-z) f(z)\, dz.$$

Again, all three possibilities for $h(t)$ share the common property that h'' exists; this requires that h' be continuous. Thus, $h'(t-z) f(z)$ is piecewise continuous, and because the integral of a piecewise continuous function is continuous, the derivative of the solution is continuous. We have just shown that $y_p'(t)$ is also continuous. Thus, $y(t)$ is differentiable everywhere and cannot have any "corners."

EXERCISES

1. Let $f(t)$ and $g(t)$ be of exponential order as $t \to \infty$. Show that their sum, $f(t) + g(t)$, also is of exponential order as $t \to \infty$.

2. Let $f(t)$ and $g(t)$ be piecewise continuous on $a \le t \le b$. Show that their sum, $f(t) + g(t)$, also is piecewise continuous on $a \le t \le b$.

3. Using the results of Exercises 1 and 2, show that $\mathcal{L}\{f(t) + g(t)\} = \mathcal{L}\{f(t)\} + \mathcal{L}\{g(t)\}$.

4. Which of the following functions are piecewise continuous for $0 \le t \le 100$?
(a) $f(t) = t^{-1}$
(b) $f(t) = (t+1)^{-1}$
(c) $f(t) = (t-2)^{-1}$

---

[9]If $y(t) = \int_0^t g(t, z)\, dz$ then

$$\frac{dy}{dt} = g(t, t) + \int_0^t \frac{\partial g(t, z)}{\partial t}\, dz.$$

(d) $f(t) = (\cos t + 1)^{-1}$

(e) $f(t) = (\cos t - 2)^{-1}$

(f) $f(t) = \tan t$

(g) $f(t) = \begin{cases} t^2 & \text{if } 0 \le t \le 1 \\ 2t^{-1} & \text{if } 1 < t \le 100 \end{cases}$

(h) $f(t) = \begin{cases} \tan t & \text{if } 0 \le t \le 1 \\ (2t - 1)^{-1} & \text{if } 1 < t \le 100 \end{cases}$

5. Find the values of α, M, and T from Definition 14.4 for the following functions.

   (a) $f(t) = e^{3t} \cos 2t$

   (b) $f(t) = \begin{cases} 16 & \text{if } 0 \le t \le 10 \\ 100t^{-1} & \text{if } 10 < t \end{cases}$

   (c) $f(t) = e^{7t}$

   (d) $f(t) = 2t$

6. Show that e^{t^2} is not of exponential order.

7. Show that the function $f(t) = 3 \sin \left(e^{t^2} \right)$ is of exponential order but that its derivative, $f'(t)$, is not. List at least two other functions with this same property.

8. Prove that if $f(t)$ is piecewise continuous and of exponential order, then $\lim_{s \to \infty} F(s) = 0$. [Hint: Write

$$\mathcal{L}\{f(t)\} = \int_0^T e^{-st} f(t)\, dt + \int_T^\infty e^{-st} f(t)\, dt,$$

where T is taken from Definition 14.4, and use the facts that $\left| e^{-st} f(t) \right|$ is bounded for $t > T$ and that the absolute value of an integral is less than or equal to the integral of the absolute value of the integrand.]

9. The solution of $ay'' + by' + cy = f(t)$, where $f(t)$ is continuous everywhere except where it has jump discontinuities at $t = 1$ and $t = 3$, is allegedly represented by Figure 14.28. Comment on this claim.

10. For $\mathcal{L}\{f(t)\}$, show that $\mathcal{L}\{tf'(t)\} = -sF'(s) - F(s)$ and $\mathcal{L}\{t^2 f''(t)\} = s^2 F''(s) + 4s F'(s) + 2F(s)$.

11. Solve the initial value problem

$$f''(t) + tf'(t) - 2f(t) = 1, \qquad f(0) = f'(0) = 0, \qquad \textbf{(14.98)}$$

in the following manner.

(a) Show that taking the Laplace transform of (14.98) and using the initial conditions and the results of Exercise 10 gives

$$F'(s) + \left(\frac{3}{s} s - s \right) F(s) = -\frac{1}{s^2}.$$

(b) Find the integrating factor for this first order linear differential equation, and show that it has the solution

$$F(s) = \frac{1}{s^3} + C \frac{1}{s^3} e^{s^2/2}.$$

(c) Evaluate the arbitrary constant in part (b) by using the result of Exercise 8, and find the resulting solution $f(t)$. Verify that your answer is indeed the solution.

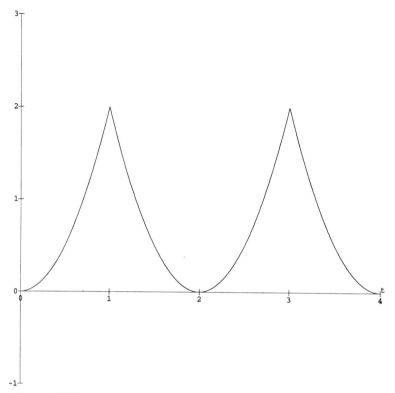

FIGURE 14.28   Mystery solution

12. Solve the initial value problem

$$f''(t) - 2tf'(t) + 2f(t) = 0, \qquad f(0) = 0, \qquad f'(0) = 1.$$

13. For what value of the constant a will the differential equation resulting from taking the Laplace transform of

$$f''(t) - atf'(t) + 3f(t) = 0$$

be of the same form as the original differential equation?

What Have We Learned?

MAIN IDEAS

In this chapter we have seen how Laplace transforms are used in solving a variety of equations.

- We can use Laplace transforms to solve linear differential equations with constant coefficients, of both first and second order. See *How to Solve First Order Linear Differential Equations Using Laplace Transforms* on page 694 and Exercise 1 on page 744.

- Laplace transforms also solve systems of linear differential equations with constant coefficients. See Exercise 1 on page 750.

- Certain types of linear differential equations without constant coefficients may be solved with Laplace transforms. See Exercises 10 through 13 on page 757 and 758.

- Laplace transforms can also solve a special type of integral equation. See Exercise 5 on page 710.

There are two main advantages of using Laplace transforms to solve differential equations:

- They reduce differential equations to algebraic equations.

- They easily handle discontinuous forcing functions, or ones that have a different representation on different parts of the domain.

Other advantages are that they satisfy initial conditions automatically and easily handle periodic boundary conditions.

Appendices

You should remember the following piece of advice. There is so little that is true in mathematics, that anything you make up is likely to be wrong.

There are eight appendices.

1. **Background Material**. The material in this section is critical for success in any ODE course. You will have seen all of this material in previous courses, but, if you are a typical student, you will have forgotten much of it. In fact, you might even think that you haven't seen some of it before.

2. **Partial Fractions**. This material should also be familiar to you. You will use it when integrating and applying Laplace transforms.

3. **Infinite Series, Power Series, and Taylor Series**. This material should also be familiar to you, but most students don't really master series until they use it in differential equations.

4. **Complex Numbers**. Much of this material will be new to most students.

5. **Elementary Matrix Operations**. Most students will have seen special cases of these results.

6. **Least Squares Approximation**. This will be new to most students. It shows how to find the best straight line approximation $y = mx + b$ to a data set consisting of n points.

7. **Numerical Techniques for Solving Differential Equations**. This is new material.

8. **Proof of Comparison Theorems**. This contains the proofs of the theorems stated in Chapter 9.

A.1   Background Material

Solving Quadratic Equations:

The solutions of the quadratic equation

$$ax^2 + bx + c = 0$$

are

$$x = \frac{-b \pm \sqrt{b^2 - 4ac}}{2a}.$$

If the discriminant, that is, $b^2 - 4ac$, is positive, the equation has distinct real roots. If $b^2 - 4ac = 0$, the roots are real, but repeated. If $b^2 - 4ac < 0$, the roots are complex, and are complex conjugates.

Perpendicular Lines:

The two straight lines $y = m_1 x + b_1$ and $y = m_2 x + b_2$ are perpendicular if $m_1 m_2 = -1$.

Properties of Trigonometric Functions:

All angles are measured in radians where $180^o = \pi$ radians. Thus $90^o = \pi/2$ radians, $60^o = \pi/3$ radians, $45^o = \pi/4$ radians, $30^o = \pi/6$ radians. An easy way to remember the values of the trig functions at frequently-used angles is to use the following table:

	0°	30°	45°	60°	90°
$x =$	0	$\pi/6$	$\pi/4$	$\pi/3$	$\pi/2$
$\sin x =$	$\sqrt{0}/2$	$\sqrt{1}/2$	$\sqrt{2}/2$	$\sqrt{3}/2$	$\sqrt{4}/2$
$\cos x =$	$\sqrt{4}/2$	$\sqrt{3}/2$	$\sqrt{2}/2$	$\sqrt{1}/2$	$\sqrt{0}/2$

which simplifies to

	0°	30°	45°	60°	90°
$x =$	0	$\pi/6$	$\pi/4$	$\pi/3$	$\pi/2$
$\sin x =$	0	$1/2$	$1/\sqrt{2}$	$\sqrt{3}/2$	1
$\cos x =$	1	$\sqrt{3}/2$	$1/\sqrt{2}$	$1/2$	0

The graphs of $\sin x$ and $\cos x$ are shown in Figure A.1.

$$\sin(-x) = -\sin x$$
$$\cos(-x) = \cos x$$
$$\sin^2 x + \cos^2 x = 1$$
$$\sin(x + y) = \sin x \cos y + \cos x \sin y$$
$$\cos(x + y) = \cos x \cos y - \sin x \sin y$$

$$\tan x = \frac{\sin x}{\cos x}$$

$$\tan(x + y) = \frac{\tan x + \tan y}{1 - \tan x \tan y}$$

$$\frac{d}{dx} \sin x = \cos x$$

$$\frac{d}{dx} \cos x = -\sin x$$

$$\frac{d}{dx} \tan x = \sec^2 x$$

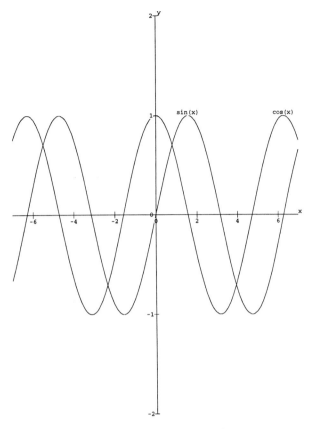

FIGURE A.1 Graphs of the functions $\sin x$ and $\cos x$

$$\int \sin x \, dx = -\cos x + C$$

$$\int \cos dx = \sin x + C$$

$$\int \tan x \, dx = -\ln|\cos x| + C$$

$$\sin x = x - \frac{x^3}{3!} + \frac{x^5}{5!} - \cdots, \text{ for } -\infty < x < \infty$$

$$\cos x = 1 - \frac{x^2}{2!} + \frac{x^4}{4!} - \cdots, \text{ for } -\infty < x < \infty$$

The Inverse Trigonometric Functions:

If $x = \sin\theta$ and $-\pi/2 \le \theta \le \pi/2$ then $\theta = \arcsin x$. The function $\arcsin x$ is sometimes written $\sin^{-1} x$. If $x = \tan\theta$ and $-\pi/2 < \theta < \pi/2$ then $\theta = \arctan x$. The function $\arctan x$ is sometimes written $\tan^{-1} x$.

$$\frac{d}{dx} \arcsin x = 1/\sqrt{1 - x^2}$$

$$\frac{d}{dx} \arctan x = 1/\left(1 + x^2\right)$$

Properties of Exponential Functions:

$$e^0 = 1$$
$$e^{x+y} = e^x e^y$$
$$e^{x-y} = e^x e^{-y}$$
$$e^{-x} = 1/e^x$$
$$e^{ax} = \left(e^a\right)^x$$
$$a^x = e^{x \ln a}, \text{ for } a > 0$$

$$\frac{d}{dx} e^{ax} = a e^{ax}$$
$$\int e^{ax}\, dx = \frac{1}{a} e^{ax} + C$$

$$e^x = 1 + x + \frac{x^2}{2!} + \frac{x^3}{3!} + \cdots, \text{ for } -\infty < x < \infty$$

The graphs of e^x and e^{-x} are shown in Figure A.2.

Properties of Logarithmic Functions:

$\ln x$ is defined only for $x > 0$

$$\ln 1 = 0$$
$$\ln e = 1$$
$$\ln xy = \ln x + \ln y$$
$$\ln (x/y) = \ln x - \ln y$$
$$\ln x^n = n \ln x$$
$$\ln x^{-1} = -\ln x$$
$$e^{\ln x} = x, \text{ if } x > 0$$
$$\ln e^x = x$$

$$\frac{d}{dx} \ln x = \frac{1}{x}$$

$$\int \frac{1}{x}\, dx = \ln |x| + C$$

$$\int \ln x\, dx = x \ln x - x + C$$

There are **NO** general formulas that simplify either $\ln(x + y)$ or $\ln(x - y)$.

WRONG: $\ln(x + y) = \ln x + \ln y$

WRONG: $\ln(x - y) = \ln x - \ln y$

The graphs of e^x and $\ln x$ are shown in Figure A.3.

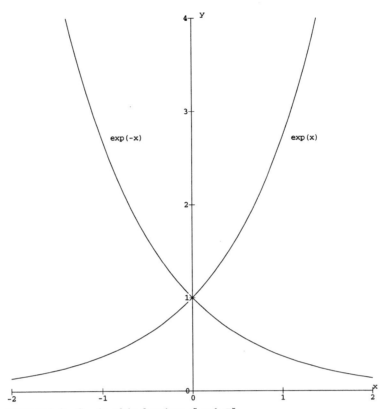

FIGURE A.2 Graphs of the functions e^x and e^{-x}

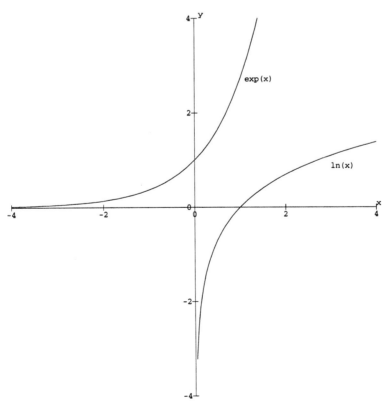

FIGURE A.3 Graphs of the functions e^x and $\ln x$

The Hyperbolic Functions:

$$\sinh x = \frac{1}{2}\left(e^x - e^{-x}\right)$$

$$\cosh x = \frac{1}{2}\left(e^x + e^{-x}\right)$$

$$\tanh x = \frac{\sinh x}{\cosh x} = \frac{e^x - e^{-x}}{e^x + e^{-x}}$$

$$\sinh 0 = 0$$

$$\cosh 0 = 1$$

$$\frac{d}{dx}\sinh x = \cosh x$$

$$\frac{d}{dx}\cosh x = \sinh x$$

Properties of Derivatives:

$$\frac{d}{dx}[cf(x)] = c\frac{d}{dx}f(x)$$

$$\frac{d}{dx}[f(x) \pm g(x)] = \frac{d}{dx}f(x) \pm \frac{d}{dx}g(x)$$

$$\frac{d}{dx}[f(x)g(x)] = \frac{d}{dx}[f(x)]g(x) + f(x)\frac{d}{dx}[g(x)]$$

$$\frac{d}{dx}\left[\frac{f(x)}{g(x)}\right] = \frac{f'(x)g(x) - f(x)g'(x)}{g^2(x)}$$

If $y = f(u(x))$ then $\dfrac{dy}{dx} = \dfrac{df}{du}\dfrac{du}{dx}$.

$$\frac{d}{dx}\int_a^x f(t)\,dt = f(x), \text{ if } f(t) \text{ is continuous at } t = x.$$

If $f(x)$ is differentiable at $x = a$ then $f(x)$ is continuous at $x = a$.

If $f'(x) > 0$ in the interval $a < x < b$ then $f(x)$ is increasing in that interval.

If $f'(x) < 0$ in the interval $a < x < b$ then $f(x)$ is decreasing in that interval.

If $f''(x) > 0$ in the interval $a < x < b$ then $f(x)$ is concave up in that interval.

If $f''(x) < 0$ in the interval $a < x < b$ then $f(x)$ is concave down in that interval.

The function $f(x)$ has a local, or relative, maximum at x_0, if $f(x_0) \geq f(x)$ for all x near x_0.

The function $f(x)$ has a local, or relative, minimum at x_0, if $f(x_0) \leq f(x)$ for all x near x_0.

TABLE A.1 Table of derivatives

$f(x)$	$f'(x)$
c	0
x^n	nx^{n-1}
e^x	e^x
$\sin x$	$\cos x$
$\cos x$	$-\sin x$
$\tan x$	$\sec^2 x = 1/\cos^2 x$
$\cot x$	$-\csc^2 x = -1/\sin^2 x$
$\sec x = 1/\cos x$	$\sec x \tan x = \sin x/\cos^2 x$
$\csc x = 1/\sin x$	$-\csc x \cot x = -\cos x/\sin^2 x$
$\ln x$	$1/x$
$\sinh x = (e^x - e^{-x})/2$	$\cosh x$
$\cosh x = (e^x + e^{-x})/2$	$\sinh x$
$\arcsin x$	$1/\sqrt{1-x^2}$
$\arctan x$	$1/(1+x^2)$

Properties of Integrals:

$$\int cf(x)\,dx = c\int f(x)\,dx$$

$$\int [f(x) \pm g(x)]\,dx = \int f(x)\,dx \pm \int g(x)\,dx$$

$$\int f(x)g'(x)\,dx = f(x)g(x) - \int f'(x)g(x)\,dx$$

$$\int u\,dv = uv - \int v\,du$$

$$\int_a^b f(x)g'(x)\,dx = f(x)g(x)\big|_a^b - \int_a^b f'(x)g(x)\,dx$$

$$\int f(g(x))g'(x)\,dx = \int f(u)\,du \text{ where } u = g(x)$$

There are **NO** general formulas that simplify either $\int [f(x)g(x)]\,dx$ or $\int [f(x)/g(x)]\,dx$.

$$\textbf{WRONG}: \int f(x)g(x)\,dx = \int f(x)\,dx \int g(x)\,dx$$

$$\textbf{WRONG}: \int \frac{f(x)}{g(x)}\,dx = \frac{\int f(x)\,dx}{\int g(x)\,dx}$$

TABLE A.2   **Table of integrals**

$f(x)$	$\int f(x)\,dx$
x^n	$x^{n+1}/(n+1) + C, n \neq -1$
$1/x$	$\ln|x| + C$
e^x	$e^x + C$
$\sin x$	$-\cos x + C$
$\cos x$	$\sin x + C$
$\tan x$	$-\ln|\cos x| + C = \ln|\sec x| + C$
$\cot x$	$\ln|\sin x| + C$
$\sec x = 1/\cos x$	$\ln|\sec x + \tan x| + C$
$\csc x = 1/\sin x$	$\ln|\csc x - \cot x| + C$
$\ln x$	$x \ln x - x + C$
$\sinh x = (e^x - e^{-x})/2$	$\cosh x + C$
$\cosh x = (e^x + e^{-x})/2$	$\sinh x + C$
$1/(1 + x^2)$	$\arctan x + C$
$1/\sqrt{1 - x^2}$	$\arcsin x + C$
$e^{ax} \sin bx$	$e^{ax}(a \sin bx - b \cos bx)/(a^2 + b^2) + C$
$e^{ax} \cos bx$	$e^{ax}(b \sin bx + a \cos bx)/(a^2 + b^2) + C$

A.2   Partial Fractions

We sometimes need to express a rational polynomial, that is, a function of the type

$$R(x) = \frac{P(x)}{Q(x)} \tag{A.1}$$

where $P(x)$ and $Q(x)$ are polynomials, in an alternative form. The standard technique, known as partial fractions, goes as follows. (We should point out that the general explanation is much more involved than doing a particular example.)

1. If the degree of the polynomial $Q(x)$ is less than or equal to the degree of $P(x)$ then divide $Q(x)$ into $P(x)$, obtaining a polynomial plus a term similar to $R(x)$ in (A.1) but where the degree of $Q(x)$ is greater than the degree of the new $P(x)$. From now on we concentrate on this new $R(x)$.

2. Factor $Q(x)$ into linear factors and quadratic factors (that cannot be written as the product of linear factors with real coefficients), so that

$$Q(x) = \left(x - r_1\right)^{n_1} \cdots \left(x - r_p\right)^{n_p} \left(a_1 x^2 + b_1 x + c_1\right)^{m_1} \cdots \left(a_q x^2 + b_q x + c_q\right)^{m_q}$$

where n_1 through n_p and m_1 through m_q are positive integers and $a_1 x^2 + b_1 x + c_1$, and so on, have no real roots. For example, if $Q(x) = x^3 - x$ then $Q(x) = x(x - 1)(x + 1)$, whereas, if $Q(x) = x^3 + x$, then $Q(x) = x(x^2 + 1)$.

3. For each linear factor of $Q(x)$ of degree n, say $(x - r)^n$, write down a contribution to $R(x)$ which is an expansion with n terms, namely,

$$\frac{A_1}{x - r} + \frac{A_2}{(x - r)^2} + \cdots + \frac{A_n}{(x - r)^n},$$